Lecture Notes in Artificial Intelligence 7026

Subseries of Lecture Notes in Computer Science

Luis Antunes   H. Sofia Pinto (Eds.)

Progress in Artificial Intelligence

15th Portuguese Conference
on Artificial Intelligence, EPIA 2011
Lisbon, Portugal, October 10-13, 2011
Proceedings

 Springer

Series Editors

Randy Goebel, University of Alberta, Edmonton, Canada
Jörg Siekmann, University of Saarland, Saarbrücken, Germany
Wolfgang Wahlster, DFKI and University of Saarland, Saarbrücken, Germany

Volume Editors

Luis Antunes
GUESS/LabMAg/Universidade de Lisboa
Faculdade de Ciências, Departamento de Informática
Campo Grande, 749-016 Lisboa, Portugal
E-mail: xarax@di.fc.ul.pt

H. Sofia Pinto
Instituto Superior Técnico, IST
Department of Computer Science and Engineering, INESC-ID
Avenida Rovisco Pais, 1049-001 Lisboa, Portugal
E-mail: sofia@ontol.inesc-id.pt

ISSN 0302-9743   e-ISSN 1611-3349
ISBN 978-3-642-24768-2   ISBN 978-3-642-24769-9 (eBook)
DOI 10.1007/978-3-642-24769-9
Springer Heidelberg Dordrecht London New York

Library of Congress Control Number: 2011938667

CR Subject Classification (1998): I.2, J.4, H.3, H.5.2, I.5, I.4, F.4.1, K.4

LNCS Sublibrary: SL 7 – Artificial Intelligence

Typesetting: Camera-ready by author, data conversion by Scientific Publishing Services, Chennai, India

Printed on acid-free paper

Springer is part of Springer Science+Business Media (www.springer.com)

Preface

The Portuguese Conference on Artificial Intelligence (EPIA) was first established in 1985, and since 1989 it has occurred biannually and is run as an international conference. It aims to promote research in artificial intelligence and has been a lively scientific event for the exchange of ideas for more than 20 years. Research in artificial intelligence is deeply grounded in Portugal and its importance is visible not only by the sheer number of researchers holding a PhD (over 100), but also by the success of numerous important conferences organized in Portugal. EPIA has been organized all over the country, raising considerable success and interest in the several local scientific communities, while never ceasing to attract researchers and practitioners from the whole country and a considerable participation of foreign scientists.

The 15th Portuguese Conference on Artificial Intelligence, EPIA 2011, took place in Lisbon at the Faculdade de Ciências da Universidade de Lisboa (FCUL) and was co-organized by LabMAg/FCUL, INESC-ID, IST, and ISEL.

As in previous recent editions, EPIA 2011 was organized as a set of selected and dedicated tracks. The tracks considered in this edition were:

- Affective Computing
- Ambient Intelligence Environments
- Artificial Intelligence Methodologies for Games
- Artificial Intelligence in Transportation Systems
- Artificial Life and Evolutionary Algorithms
- Computational Logic with Applications
- General Artificial Intelligence
- Intelligent Robotics
- Knowledge Discovery and Business Intelligence
- Multi-Agent Systems: Theory and Applications
- Social Simulation and Modeling
- Text Mining and Applications
- The Doctoral Symposium on Artificial Intelligence

In this edition, a total of 203 contributions were received from 32 countries. All papers were reviewed in a double-blind process by at least three different reviewers. In some cases, contributions were reviewed by up to five reviewers. This book groups together a set of 50 selected papers that were accepted for the 15th Portuguese Conference on Artificial Intelligence, EPIA 2011, covering a wide range of topics and perspectives. This corresponds to a 25% acceptance rate, ensuring the high quality of the event. Geographically, accepted papers came from Portugal, Spain, Brazil, France, Italy, Luxembourg, Mexico, and Poland (ordered by acceptance rate). However, the authors of accepted papers included researchers from institutions from Canada, Denmark, France, Iran, Sweden, and the United Kingdom.

We wish to thank the members of all committees involved in the event, in particular the Advisory Board, the Program Committee and the Organizing Committee. We would also like to thank all the Track Chairs, the reviewers and above all the authors, without whom this event would not be possible. We would also like to acknowledge and thank the use of EasyChair, which greatly eased the management of the submissions, reviews and materials.

August 2011   Luis Antunes
                                                                               H. Sofia Pinto

Organization

The 15th Portuguese Conference on Artificial Inteligence (EPIA 2011) was co-organized by the *Laboratório de Modelação de Agentes (LabMAg)/Faculdade de Ciências da Universidade de Lisboa (FC/UL)*; *Instituto de Engenharia de Sistemas e Computadores (INESC-ID)/Instituto Superior Técnico da Universidade Técnica de Lisboa (IST/UTL)*; *Instituto Superior de Engenharia de Lisboa (ISEL)*.

General and Program Co-chairs

Luis Antunes (GUESS/LabMAg/UL)
H. Sofia Pinto (INESC-ID/IST)

Organization Co-chairs

Rui Prada (INESC-ID/IST)
Paulo Trigo (LabMAg/ISEL)

Advisory Board

Salvador Abreu	Universidade de Évora/CENTRIA
José Júlio Alferes	Universidade Nova de Lisboa
Pedro Barahona	Universidade Nova de Lisboa
Pavel Brazdil	LIAAD/Universidade do Porto
Amilcar Cardoso	Universidade de Coimbra
Helder Coelho	Universidade de Lisboa
Luis Correia	Universidade de Lisboa
Ernesto Costa	Universidade de Coimbra
Gaël Dias	Universidade da Beira Interior
Pedro Rangel Henriques	Universidade do Minho
José Gabriel Lopes	Universidade Nova de Lisboa
Ernesto Morgado	Siscog/Instituto Superior Técnico
José Neves	Universidade do Minho
Eugénio Oliveira	LIACC/Universidade do Porto
Arlindo Oliveira	INESC-ID/Instituto Superior Técnico/Cadence Research Labs.
Ana Paiva	Instituto Superior Técnico
Carlos Ramos	Instituto Superior de Engenharia do Porto
Luis M. Rocha	Indiana University
Luis Seabra Lopes	Universidade de Aveiro
Manuela Veloso	Carnegie Mellon University

Conference Track Chairs

- Affective Computing
  - Goreti Marreiros
  - Andrew Ortony
  - Ana Paiva
- Ambient Intelligence Environments
  - Paulo Novais
  - Ana Almeida
  - Sara Rodríguez González
- Artificial Intelligence Methodologies for Games
  - Luís Paulo Reis
  - Carlos Martinho
  - Pedro Miguel Moreira
  - Pedro Mariano
- Artificial Intelligence in Transportation Systems
  - Rosaldo Rossetti
  - Elisabete Arsénio
  - Jorge Lopes
  - Matteo Vasirani
- Artificial Life and Evolutionary Algorithms
  - Sara Silva
  - Francisco B. Pereira
  - Leonardo Vannesch
- Computational Logic with Applications
  - Paulo Moura
  - Vitor Beires Nogueira
- General Artificial Intelligence
  - H. Sofia Pinto
  - Luis Antunes
- Intelligent Robotics
  - Luís Paulo Reis
  - Luís Correia
  - Nuno Lau
- Knowledge Discovery and Business Intelligence
  - Paulo Cortez
  - Nuno Marques
  - Luís Cavique
  - João Gama
  - Manuel Filipe Santos
- Multi-Agent Systems: Theory and Applications
  - Paulo Urbano
  - Cesar Analide
  - Fernando Lopes
  - Henrique Lopes Cardoso

- Social Simulation and Modeling
  - João Balsa
  - Antônio Carlos da Rocha Costa
  - Armando Geller
- Text Mining and Applications
  - Joaquim Francisco Ferreira Da Silva
  - Vitor Jorge Ramos Rocio
  - Gaël Dias
  - José Gabriel Pereira Lopes
- Doctoral Symposium on Artificial Intelligence
  - Paulo Novais
  - Cesar Analide
  - Pedro Henriques

Program Committee

Affective Computing

César Analide	Universidade do Minho
Antonio Camurri	University of Genoa
Amilcar Cardoso	University of Coimbra
Cristiano Castelfranchi	Institute of Cognitive Sciences and Technologies
Helder Coelho	University of Lisbon
Laurence Devillers	LIMSI-CNRS, Université P11
Anna Esposito	Second University of Naples
Hatice Gunes	Imperial College London
Jennifer Healey	
Ian Horswill	Northwestern University
Eva Hudlicka	
Kostas Karpouzis	National Technical University of Athens
José Machado	Universidade Minho
José Neves	Universidade do Minho
Paulo Novais	Universidade do Minho
Juan Pavón	Universidad Complutense Madrid
Frank Pollick	University of Glasgow
Boon-Kiat Quek	
Carlos Ramos	Instituto Superior de Engenharia do Porto
Ricardo Santos	ESTGF/IPP

Ambient Intelligence Environments

César Analide	Universidade do Minho
Cecilio Angulo	Universitat Politecnica de Catalunya
Juan Augusto	University of Ulster
Javier Bajo	Universidad Pontificia de Salamanca
Carlos Bento	Universidade de Coimbra

Lourdes Borrajo	University of Vigo
Davide Carneiro	Universidade do Minho
Diane Cook	Washington State University
Juan Corchado	University of Salamanca
Ricardo Costa	ESTGF.IPP
José Danado	YDreams, YLabs Research
Eduardo Dias	CITI - FCT/UNL
Antonio Fernández	University of Castilla-La Mancha
Lino Figueiredo	
Diego Gachet	Universidad Europea de Madrid
Junzhong Gu	East China Normal University
Hans Guesgen	Massey University
Javier Jaen	Polytechnic University of Valencia
Rui José	University of Minho
Joyca Lacroix	Philips Research
Kristof Laerhoven	TU Darmstadt
Guillaume Lopez	The University of Tokyo
José Machado	Universidade Minho
Goreti Marreiros	ISEP
Rene Meier	
José Molina	Universidad Carlos III de Madrid
José Neves	Universidade do Minho
Francisco Pereira	Universidade de Coimbra
Davy Preuveneers	K.U. Leuven
Carlos Ramos	Instituto Superior de Engenharia do Porto
Andreas Riener	Johannes Kepler University Linz
Florentino Riverola	University of Vigo
Teresa Romão	DI/FCT/UNL
Ichiro Satoh	National Institute of Informatics
Francisco Silva	Universidade Federal do Maranhão
Dante Tapia	University of Salamanca
Martijn Vastenburg	TU Delft
Yu Zheng	Microsoft Research Aisa

Artificial Intelligence Methodologies for Games

Pedro G. Calero	Universidad Complutense de Madrid
Marc Cavazza	University of Teesside
António Coelho	University of Oporto
Frank Dignum	Utrecht University
Stefan Edelkamp	University of Bremen
Kostas Karpouzis	National Technical University of Athens
Peter Kissmann	TZI Universität Bremen
Nuno Lau	University of Aveiro
Carlos Linares López	Universidad Carlos III
Luiz Moniz	University of Lisbon

Alexander Nareyek   National University of Singapore
João Pedro Neto   University of Lisbon
Jeff Orkin   MIT
Ana Paiva   Instituto Superior Técnico
Filipe Pina   Seed Studios
Guilherme Raimundo   Instituto Superior Técnico
Rui Rodrigues   University of Oporto
Licinio Roque   Universidade de Coimbra
Luis Seabra Lopes   University of Aveiro
A. Augusto de Sousa   University of Oporto
Michael Thielscher   University of New South Wales
Julian Togelius   IT University of Copenhagen
Marco Vala   Instituto Superior Técnico
Giorgios Yannakakis   IT University of Copenhagen
Nelson Zagalo   University of Minho

Artificial Intelligence in Transportation Systems

Ana Almeida   Instituto Superior de Engenharia do Porto
Constantinos Antoniou   National Technical University of Athens
Ramachandran Balakrishna   Caliper Corporation
Federico Barber   DSIC / Technical University of Valencia
Ana Bazzan   Universidade Federal do Rio Grande do Sul
Carlos Bento   Universidade de Coimbra
Vicent Botti   DSIC
Eduardo Camponogara   Federal University of Santa Catarina
António Castro   University of Oporto
Hilmi Berk Celikoglu   Technical University of Istanbul
Hussein Dia   Aecom
Juergen Dunkel   Hannover University for Applied Sciences and
                             Arts
Alberto Fernandez   CETINIA, Rey Juan Carlos University
Adriana Giret   Technical University of Valencia
Franziska Kluegl   Örebro University
Maite Lopez-Sanchez   Universitat de Barcelona
Helen Ma   TU/e
José Manuel Menendez   Universidad Politécnica de Madrid - UPM
Luís Nunes   ISCTE
Cristina Olaverri   University of Oporto
Eugénio Oliveira   University of Oporto LIACC
Luís Osório   ISEL
Sascha Ossowski   Rey Juan Carlos University
Francisco Pereira   Universidade de Coimbra
Francisco Reinaldo   UnilesteMG
Luis Paulo Reis   FEUP
Miguel A. Salido   DSIC / Technical University of Valencia

Majid Sarvi	Monash University
Juergen Sauer	University of Oldenburg
Thomas Strang	DLR German Aerospace Center
Man-Chun Tan	
José Telhada	University of Minho
Harry Timmermans	Eindhoven University of Technology
Giuseppe Vizzari	University of Milano-Bicocca
Fei Wang	Xian Jiaotong University
Danny Weyns	Katholieke Universiteit Leuven

Artificial Life and Evolutionary Algorithms

Wolfgang Banzhaf	Memorial University of Newfoundland
Helio J. C. Barbosa	Laboratório Nacional de Computação Científica
Daniela Besozzi	University of Milan
Christian Blum	Universitat Politècnica de Catalunya
Stefano Cagnoni	University of Parma
Philippe Caillou	LRI, University of Paris Sud 11
Luis Correia	University of Lisbon
Ernesto Costa	University of Coimbra
Carlos Cotta	Universidad de Málaga
Ivanoe De Falco	ICAR - CNR
Kalyanmoy Deb	Indian Institute of Technology Kanpur
Antonio Della Cioppa	University of Salerno
Anikó Ekárt	Aston University
Anna I. Esparcia-Alcázar	S2 Grupo & Universidad Politécnica de Valencia
Carlos M. Fernandes	University of Granada
James A. Foster	University of Idaho
António Gaspar-Cunha	University of Minho
Carlos Gershenson	Universidad Nacional Autónoma de México
Mario Giacobini	University of Turin
Jin-Kao Hao	Université Angers
Inman Harvey	University of Sussex
William B. Langdon	University of Essex
Arnaud Liefooghe	Université Lille 1
Fernando Lobo	University of Algarve
Penousal Machado	University of Coimbra
Ana Madureira	Instituto Superior de Engenharia do Porto
Pedro Mariano	University of Lisbon
Rui Mendes	CCTC - University of Minho
Telmo Menezes	CNRS, Paris
Juan Julián Merelo Guervós	University of Granada
Zbigniew Michalewicz	University of Adelaide
Alberto Moraglio	University of Kent
Shin Morishita	University of Yokohama
Una-May O'Reilly	MIT

Luís Paquete	University of Coimbra
Agostinho Rosa	Technical University of Lisbon
Marc Schoenauer	INRIA
Roberto Serra	University of Modena e Reggio Emilia
Anabela Simões	Polytechnic Institute of Coimbra
Thomas Stuetzle	Université Libre de Bruxelles
Ricardo H. C. Takahashi	Universidade Federal de Minas Gerais
Jorge Tavares	University of Coimbra
Leonardo Trujillo Reyes	Instituto Tecnológico de Tijuana

Computational Logic with Applications

Salvador Abreu	Universidade de Évora and CENTRIA
José Júlio Alferes	Universidade Nova de Lisboa
Roberto Bagnara	University of Parma
Vítor Costa	Universidade do Porto
Bart Demoen	K.U. Leuven
Daniel Diaz	Université de Paris I
Paulo Gomes	CMS - CISUC, University of Coimbra
Gopal Gupta	University of Texas at Dallas
Angelika Kimmig	K.U. Leuven
João Leite	Universidade Nova de Lisboa
Axel Pollers	DERI, National University of Ireland, Galway
Enrico Pontelli	New Mexico State University
Peter Robinson	The University of Queensland
Joachim Schimpf	Monash University
David Warren	University of Stony Brook

General Artificial Intelligence

Salvador Abreu	Universidade de Évora and CENTRIA
José Júlio Alferes	Universidade Nova de Lisboa
Pedro Barahona	Universidade Nova de Lisboa
Pavel Brazdil	LIAAD, University of Oporto
Amilcar Cardoso	University of Coimbra
Helder Coelho	University of Lisbon
Luis Correia	University of Lisbon
Ernesto Costa	University of Coimbra
Gaël Dias	University of Beira Interior
Pedro Rangel Henriques	Universidade do Minho
José Gabriel Lopes	Universidade Nova de Lisboa
Ernesto Morgado	Siscog and IST
José Neves	Universidade do Minho
Eugénio Oliveira	University of Oporto LIACC
Arlindo Oliveira	IST/INESC-ID and Cadence Research Laboratories

Ana Paiva	Instituto Superior Técnico
Carlos Ramos	Instituto Superior de Engenharia do Porto
Luis M. Rocha	Indiana University
Luis Seabra Lopes	University of Aveiro
Manuela Veloso	Carnegie Mellon University

Intelligent Robotics

César Analide	Universidade do Minho
Kai Arras	University of Freiburg
Stephen Balakirsky	NIST
Jacky Baltes	University of Manitoba
Reinaldo Bianchi	Centro Universitario da FEI
Rodrigo Braga	FEUP
Carlos Carreto	Instituto Politécnico da Guarda
Xiaoping Chen	University of Science and Technology of China
Anna Helena Costa	University of Sao Paulo
Augusto Loureiro Da Costa	Universidade Federal da Bahia
Jorge Dias	University of Coimbra
Marco Dorigo	Université Libre de Bruxelles
Paulo Gonçalves	Polytechnic Institute of Castelo Branco
John Hallam	University of Southern Denmark
Kasper Hallenborg	Maersk Institute, University of Southern Denmark
Huosheng Hu	University of Essex
Fumiya Iida	
Luca Iocchi	Sapienza University of Rome
Pedro Lima	Instituto Superior Técnico
António Paulo Moreira	FEUP
António J. R. Neves	IEETA, University of Aveiro
Urbano Nunes	University of Coimbra
Fernando Osório	USP - ICMC - University of Sao Paulo
Enrico Pagello	University of Padua
Armando J. Pinho	University of Aveiro
Mikhail Prokopenko	CSIRO
Isabel Ribeiro	Instituto Superior Técnico / ISR
Martin Riedmiller	
Luis Seabra Lopes	University of Aveiro
Saeed Shiry Ghidary	Amirkabir University of Technology
Armando Sousa	ISR-P, DEEC, FEUP
Guy Theraulaz	CNRS CRCA
Flavio Tonidandel	Centro Universitario da FEI
Paulo Urbano	University of Lisbon
Manuela Veloso	Carnegie Mellon University

Knowledge Discovery and Business Intelligence

Carlos Alzate	K.U. Leuven
Orlando Belo	University of Minho
Albert Bifet	University of Waikato
Agnes Braud	
Rui Camacho	LIACC/FEUP University of Oporto
Logbing Cao	University of Technology Sydney
André Carvalho	Universidade de Sao Paulo
Ning Chen	GECAD, Instituto Superior de Engenharia do Porto
José Costa	Universidade Federal do Rio Grande do Norte
Carlos Ferreira	LIAAD INESC Porto
Peter Geczy	AIST
Paulo Gomes	CMS - CISUC, University of Coimbra
Beatriz Iglesia	University of East Anglia
Elena Ikonomovska	IJS - Department of Knowledge Technologies
Alípio Jorge	FCUP / LIAAD, INESC Porto
Stéphane Lallich	University of Lyon 2
Phillipe Lenca	Telecom Bretagne
Stefan Lessmann	
Hongbo Liu	Dalian Maritime University
Vítor Lobo	CINAV - Escola Naval
José Machado	Universidade do Minho
Patrick Meyer	Institut Télécom Bretagne
Susana Nascimento	Universidade Nova de Lisboa
Fatima Rodrigues	Institute of Engineering of Porto
Joaquim Silva	FCT/UNL
Carlos Soares	University of Oporto
Murate Testik	Hacettepe University
Theodore Trafalis	
Armando Vieira	
Aline Villavicencio	UFRGS and University of Bath

Multi-Agent Systems: Theory and Applications

Huib Aldewereld	University of Utrecht
Reyhan Aydogan	
Javier Bajo	Universidad Pontificia de Salamanca
João Balsa	GUESS/LabMAg/University of Lisbon
Olivier Boissier	ENS Mines Saint-Etienne
Luís Botelho	ISCTE-IUL
Juan Burgillo	University of Vigo
Javier Carbó	UC3M
Amilcar Cardoso	University of Coimbra
Cristiano Castelfranchi	Institute of Cognitive Sciences and Technologies
Yun-Gyung Cheong	Samsung Advanced Institute of Technology
Helder Coelho	University of Lisbon

Juan M. Corchado   University of Salamanca
Luis Correia   University of Lisbon
Yves Demazeau   CNRS - Laboratoire LIG
Frank Dignum   Utrecht University
Virginia Dignum   TU Delft
Amal Elfallah   LIP6 - University of Pierre and Marie Curie
Marc Esteva   IIIA-CSIC
Paulo Ferreira Jr.   Universidade Federal de Pelotas
Michael Fisher   University of Liverpool
Nicoletta Fornara   University of Lugano
Ya'Akov Gal   Ben-Gurion University of the Negev
Alejandro Guerra   Universidad Veracruzana
Jomi Hübner   Federal University of Santa Catarina
Wojtek Jamroga   University of Luxembourg
F. Jordan Srour
Nuno Lau   University of Aveiro
João Leite   Universidade Nova de Lisboa
Christian Lemaître   Universidad Autonoma Metropolitana, UAM
Luis Macedo   University of Coimbra
Pedro Mariano   University of Lisbon
John-Jules Meyer   Utrecht University
Luis Moniz   University of Lisbon
Pavlos Moraitis   Paris Descartes University
Jörg Müller   Clausthal University of Technology
Pablo Noriega   IIIA
Paulo Novais   Universidade do Minho
Eugénio Oliveira   University of Oporto LIACC
Andrea Omicini   Alma Mater Studiorum Università di Bologna
Santiago Ontañón Villar   IIIA-CSIC
António Pereira   LIACC - DEI - FEUP
Alexander Pokahr   University of Hamburg
Luis Paulo Reis   FEUP
Ana Paula Rocha   FEUP
Antônio Carlos da Rocha
  Costa   Universidade Federal do Rio Grande
Jordi Sabater-Mir   IIIA-CSIC
Murat Sensoy   University of Aberdeen
Onn Shehory   IBM Haifa Research Lab
Jaime Sichman   University of Sao Paulo
Paolo Torroni   University of Bologna
Wamberto Vasconcelos   University of Aberdeen
Laurent Vercouter   ISCOD Team, Ecole des Mines de Saint-Étienne
Rosa Vicari   Federal University of Rio Grande do Sul
Neil Yorke-Smith   American University of Beirut and SRI
                             International

Social Simulation and Modeling

Diana Adamatti	Universidade Federal do Rio Grande
Frederic Amblard	Université des Sciences Sociales Toulouse 1
Pedro Andrade	
Luis Antunes	GUESS/LabMAg/University of Lisbon
Pedro Campos	
Amilcar Cardoso	University of Coimbra
Cristiano Castelfranchi	Institute of Cognitive Sciences and Technologies
Helder Coelho	University of Lisbon
Rosaria Conte	Institute of Cognitive Sciences and Technologies
Nuno David	ISCTE
Paul Davidsson	Blekinge Institute of Technology
Graçaliz Dimuro	Universidade Federal do Rio Grande
Bruce Edmonds	Manchester Metropolitan University Business School
Nigel Gilbert	University of Surrey
Laszlo Gulyas	Aitia International, Inc.
Samer Hassan	Universidad Complutense de Madrid
Rainer Hegselmann	Bayreuth University
Wander Jager	University of Groningen
Marco Janssen	Arizona State University
Maciej Latek	George Mason University
Jorge Louçã	ISCTE
Gustavo Lugo	Federal University of Technology-Paraná
Luís Macedo	University of Coimbra
Jean-Pierre Muller	CIRAD
Fernando Neto	University of Pernambuco
Fabio Okuyama	IFRS - Campus Porto Alegre
Juan Pavón	Universidad Complutense Madrid
Juliette Rouchier	CNRS-GREQAM
David Sallach	Argonne National Laboratory and the University of Chicago
Jaime Sichman	University of Sao Paulo
Patricia Tedesco	Center for Informatics / UFPE
Oswaldo Teran	Universidad de Los Andes
Takao Terano	
Klaus Troitzsch	University of Koblenz-Landau
Natalie Van Der Wal	
Harko Verhagen	Stockholm University/KTH

Text Mining and Applications

Helena Ahonen-Myka	
Sophia Ananiadou	University of Manchester
João Balsa	GUESS/LabMAg/University of Lisbon

António Branco	University of Lisbon
Pavel Brazdil	LIAAD, University of Oporto
Luisa Coheur	
Bruno Cremilleux	University of Caen
Walter Daelemans	University of Antwerp
Eric De La Clergerie	INRIA
Antoine Doucet	University of Caen
Tomaz Erjavec	Jožef Stefan Institute
Marcelo Finger	Universidade de Sao Paulo
Pablo Gamallo	University of Santiago de Compostela
João Graça	IST
Brigitte Grau	LIMSI (CNRS)
Gregory Grefenstette	Exalead
Diana Inkpen	University of Ottawa
Mark Lee	University of Birmingham
Nadine Lucas	GREYC CNRS Caen University
Belinda Maia	University of Oporto
Nuno Marques	Universidade Nova de Lisboa
Adeline Nazarenko	LIPN UMR CNRS - Université Paris Nord
Manuel Palomar	University of Alicante
Paulo Quaresma	Universidade de Evora
Irene Rodrigues	Universidade de Evora
Antonio Sanfilippo	Pacific Northwest National Laboratory
Frédérique Segond	Xerox
Isabelle Tellier	LIFO - University of Orléans
Renata Vieira	PUC-RS
Manuel Vilares Ferro	University of Vigo
Aline Villavicencio	UFRGS and University of Bath
Christel Vrain	LIFO - University of Orléans
Pierre Zweigenbaum	LIMSI-CNRS

Doctoral Symposium on Artificial Intelligence

Salvador Abreu	Universidade de Évora and CENTRIA
José Júlio Alferes	Universidade Nova de Lisboa
Victor Alves	Universidade do Minho
Pedro Barahona	Universidade Nova de Lisboa
Amilcar Cardoso	University of Coimbra
Luis Correia	University of Lisbon
Paulo Cortez	University of Minho
Gaël Dias	University of Beira Interior
Eduardo Fermé	Universidade da Madeira
Joaquim Filipe	EST-Setubal/IPS
João Gama	University of Oporto
Paulo Gomes	CMS - CISUC, University of Coimbra
João Leite	Universidade Nova de Lisboa

José Gabriel Lopes	Universidade Nova de Lisboa
José Machado	Universidade do Minho
Sara Madeira	IST
Pedro Mariano	University of Lisbon
Goreti Marreiros	ISEP
Paulo Oliveira	UTAD
Paulo Quaresma	Universidade de Evora
Carlos Ramos	Instituto Superior de Engenharia do Porto
Luis Paulo Reis	FEUP
Ana Paula Rocha	FEUP
Manuel Filipe Santos	University of Minho
Luís Torgo	Universidade do Porto
Paulo Urbano	University of Lisbon

Additional Reviewers

Aiguzhinov, Artur	Lobo, Jorge
Almajano, Pablo	Magalhães, João
Calejo, Miguel	Maheshwari, Nandan
Cazzaniga, Paolo	Melo, Dora
Costa, Francisco	Migliore, Davide
Craveirinha, Rui	Neves Ferreira Da Silva, Ana Paula
Darriba Bilbao, Victor Manuel	Nojoumian, Peyman
Day, Jareth	Oliveira, Marcia
Dodds, Ricardo	Pimentel, Cesar
Eitan, Oran	Pretto, Alberto
Faria, Brigida Monica	Ribadas Pena, Francisco
Ferreira, Ligia	Saldanha, Ricardo
Fišer, Darja	Silveira, Sara
Garcia, Marcos	Slota, Martin
Gebhardt, Birthe	Spanoudakis, Nikolaos
Gonçalo Oliveira, Hugo	Vale, Alberto
Gonçalves, Patricia	Ventura, Rodrigo
Gunura, Keith	Wagner, Markus
Kurz, Marc	Werneck, Nicolau
Leach, Derek	Wichert, Andreas
Ljubešić, Nikola	Wilkens, Rodrigo
Lloret, Elena	Xue, Feng

Table of Contents

Affective Computing

Sentiment Analysis of News Titles: The Role of Entities and a New
Affective Lexicon 1
    Daniel Loureiro, Goreti Marreiros, and José Neves

Ambient Intelligence Environments

Providing Location Everywhere 15
    Ricardo Anacleto, Lino Figueiredo, Paulo Novais, and Ana Almeida

Modeling Context-Awareness in Agents for Ambient Intelligence:
An Aspect-Oriented Approach 29
    Inmaculada Ayala, Mercedes Amor Pinilla, and Lidia Fuentes

Developing Dynamic Conflict Resolution Models Based on the
Interpretation of Personal Conflict Styles 44
    Davide Carneiro, Marco Gomes, Paulo Novais, and José Neves

Organizations of Agents in Information Fusion Environments 59
    *Dante I. Tapia, Fernando de la Prieta, Sara Rodríguez González,
    Javier Bajo, and Juan M. Corchado*

Artificial Intelligence Methodologies for Games

Wasp-Like Agents for Scheduling Production in Real-Time Strategy
Games 71
    Marco Santos and Carlos Martinho

Artificial Intelligence in Transportation Systems

Operational Problems Recovery in Airlines – A Specialized
Methodologies Approach 83
    Bruno Aguiar, José Torres, and António J.M. Castro

Solving Heterogeneous Fleet Multiple Depot Vehicle Scheduling
Problem as an Asymmetric Traveling Salesman Problem 98
    Jorge Alpedrinha Ramos, Luís Paulo Reis, and Dulce Pedrosa

Artificial Life and Evolutionary Algorithms

Evolving Numerical Constants in Grammatical Evolution with the
Ephemeral Constant Method 110
  *Douglas A. Augusto, Helio J.C. Barbosa, André M.S. Barreto, and
  Heder S. Bernardino*

The Evolution of Foraging in an Open-Ended Simulation
Environment 125
  Tiago Baptista and Ernesto Costa

A Method to Reuse Old Populations in Genetic Algorithms 138
  Mauro Castelli, Luca Manzoni, and Leonardo Vanneschi

Towards Artificial Evolution of Complex Behaviors Observed in Insect
Colonies 153
  Miguel Duarte, Anders Lyhne Christensen, and Sancho Oliveira

Network Regularity and the Influence of Asynchronism on the
Evolution of Cooperation 168
  Carlos Grilo and Luís Correia

The Squares Problem and a Neutrality Analysis with ReNCoDe 182
  Rui L. Lopes and Ernesto Costa

Particle Swarm Optimization for Gantry Control: A Teaching
Experiment 196
  *Paulo B. de Moura Oliveira, Eduardo J. Solteiro Pires, and
  José Boaventura Cunha*

Evolving Reaction-Diffusion Systems on GPU 208
  Lidia Yamamoto, Wolfgang Banzhaf, and Pierre Collet

Computational Logic with Applications

Optimal Divide and Query 224
  David Insa and Josep Silva

A Subterm-Based Global Trie for Tabled Evaluation of Logic
Programs 239
  João Raimundo and Ricardo Rocha

General Artificial Intelligence

Intention-Based Decision Making with Evolution Prospection 254
  The Anh Han and Luís Moniz Pereira

Unsupervised Music Genre Classification with a Model-Based
Approach   268
    Luís Barreira, Sofia Cavaco, and Joaquim Ferreira da Silva

Constrained Sequential Pattern Knowledge in Multi-relational
Learning   282
    Carlos Abreu Ferreira, João Gama, and Vítor Santos Costa

Summarizing Frequent Itemsets via Pignistic Transformation   297
    Francisco Guil-Reyes and María-Teresa Daza-Gonzalez

A Simulated Annealing Algorithm for the Problem of Minimal Addition
Chains   311
    Adan Jose-Garcia, Hillel Romero-Monsivais,
    Cindy G. Hernandez-Morales, Arturo Rodriguez-Cristerna,
    Ivan Rivera-Islas, and Jose Torres-Jimenez

Novelty Detection Using Graphical Models for Semantic Room
Classification   326
    André Susano Pinto, Andrzej Pronobis, and Luis Paulo Reis

Intelligent Robotics

A Reinforcement Learning Based Method for Optimizing the Process
of Decision Making in Fire Brigade Agents   340
    Abbas Abdolmaleki, Mostafa Movahedi, Sajjad Salehi,
    Nuno Lau, and Luís Paulo Reis

Humanoid Behaviors: From Simulation to a Real Robot   352
    Edgar Domingues, Nuno Lau, Bruno Pimentel, Nima Shafii,
    Luís Paulo Reis, and António J.R. Neves

Market-Based Dynamic Task Allocation Using Heuristically Accelerated
Reinforcement Learning   365
    José Angelo Gurzoni Jr., Flavio Tonidandel, and
    Reinaldo A.C. Bianchi

Shop Floor Scheduling in a Mobile Robotic Environment   377
    Andry Maykol Pinto, Luís F. Rocha, António Paulo Moreira, and
    Paulo G. Costa

Humanized Robot Dancing: Humanoid Motion Retargeting Based in a
Metrical Representation of Human Dance Styles   392
    Paulo Sousa, João L. Oliveira, Luis Paulo Reis, and Fabien Gouyon

Knowledge Discovery and Business Intelligence

Bankruptcy Trajectory Analysis on French Companies Using
Self-Organizing Map 407
  Ning Chen, Bernardete Ribeiro, and Armando S. Vieira

Network Node Label Acquisition and Tracking 418
  Sarvenaz Choobdar, Fernando Silva, and Pedro Ribeiro

Learning to Rank for Expert Search in Digital Libraries of Academic
Publications 431
  Catarina Moreira, Pável Calado, and Bruno Martins

Thematic Fuzzy Clusters with an Additive Spectral Approach 446
  Susana Nascimento, Rui Felizardo, and Boris Mirkin

Automatically Enriching a Thesaurus with Information from
Dictionaries 462
  Hugo Gonçalo Oliveira and Paulo Gomes

Visualizing the Evolution of Social Networks 476
  Márcia Oliveira and João Gama

Using Data Mining Techniques to Predict Deformability Properties of
Jet Grouting Laboratory Formulations over Time 491
  Joaquim Tinoco, António Gomes Correia, and Paulo Cortez

Multi-Agent Systems: Theory and Applications

Doubtful Deviations and Farsighted Play 506
  Wojciech Jamroga and Matthijs Melissen

Uncertainty and Novelty-Based Selective Attention in the Collaborative
Exploration of Unknown Environments 521
  Luis Macedo, Miguel Tavares, Pedro Gaspar, and Amílcar Cardoso

A Dynamic Agents' Behavior Model for Computational Trust 536
  Joana Urbano, Ana Paula Rocha, and Eugénio Oliveira

The BMC Method for the Existential Part of RTCTLK and Interleaved
Interpreted Systems 551
  *Bożena Woźna-Szcześniak, Agnieszka Zbrzezny, and
  Andrzej Zbrzezny*

Social Simulation and Modeling

Building Spatiotemporal Emotional Maps for Social Systems 566
  Pedro Catré, Luis Cardoso, Luis Macedo, and Amílcar Cardoso

Text Mining and Applications

An Exploratory Study on the Impact of Temporal Features on the
Classification and Clustering of Future-Related Web Documents 581
   Ricardo Campos, Gaël Dias, and Alípio Jorge

Using the Web to Validate Lexico-Semantic Relations 597
   Hernani Pereira Costa, Hugo Gonçalo Oliveira, and Paulo Gomes

A Resource-Based Method for Named Entity Extraction and
Classification 610
   Pablo Gamallo and Marcos Garcia

Measuring Spelling Similarity for Cognate Identification 624
   Luís Gomes and José Gabriel Pereira Lopes

Identifying Automatic Posting Systems in Microblogs................ 634
   Gustavo Laboreiro, Luís Sarmento, and Eugénio Oliveira

Determining the Polarity of Words through a Common Online
Dictionary 649
   António Paulo-Santos, Carlos Ramos, and Nuno C. Marques

A Bootstrapping Approach for Training a NER with Conditional
Random Fields 664
   Jorge Teixeira, Luís Sarmento, and Eugénio Oliveira

Doctoral Symposium on Artificial Intelligence

Domain-Splitting Generalized Nogoods from Restarts................. 679
   Luís Baptista and Francisco Azevedo

A Proposal for Transactions in the Semantic Web 690
   Ana Sofia Gomes and José Júlio Alferes

Author Index 705

Sentiment Analysis of News Titles
The Role of Entities and a New Affective Lexicon

Daniel Loureiro[1], Goreti Marreiros[1], and José Neves[2]

[1] ISEP, GECAD - Knowledge Engineering and Decision Support Group, Portugal
`loureiro97@gmail.com`, `mgt@isep.ipp.pt`
[2] Minho University, Portugal
`jneves@di.uminho.pt`

Abstract. The growth of content on the web has been followed by increasing interest in opinion mining. This field of research relies on accurate recognition of emotion from textual data. There's been much research in sentiment analysis lately, but it always focuses on the same elements. Sentiment analysis traditionally depends on linguistic corpora, or common sense knowledge bases, to provide extra dimensions of information to the text being analyzed. Previous research hasn't yet explored a fully automatic method to evaluate how events associated to certain entities may impact each individual's sentiment perception. This project presents a method to assign valence ratings to entities, using information from their Wikipedia page, and considering user preferences gathered from the user's Facebook profile. Furthermore, a new affective lexicon is compiled entirely from existing corpora, without any intervention from the coders.

1 Introduction

The ability to recognize and classify emotion has been the topic of much research effort over the history of computer science. Pioneers in artificial intelligence, such as Minsky[16] and Picard[22], have described it as one of the hallmarks of their field. The growth of interest in opinion mining for content on the web (e.g., product reviews, news articles, etc.)[20] combined with the permanence of textually based computer interfaces reinforces the need for a computer system capable of classifying emotion accurately[9]. The intricacies of emotion expression stand as a challenge both for the computer systems that try to decode it and the psychological theories that try to explain it.

Much research into sentiment analysis uses psychological models of emotion. The OCC emotion model[18] has found widespread usage, as it provides a cognitive theory for the classification of discrete emotions (e.g. joy, relief) that can be characterized by a computationally tractable set of rules. Sometimes, other models such as Ekman's six "basic" categories of emotions are used instead[4], though that one isn't associated to a cognitive theory and lacks a well-defined rule-based structure. The preferred psychological theory for the classification

L. Antunes and H.S. Pinto (Eds.): EPIA 2011, LNAI 7026, pp. 1–14, 2011.

of emotion is usually accompanied by natural language processing (NLP) techniques for a detailed understanding of the text.

This project uses the OCC emotion model and several techniques for linguistic parsing, word stemming, word sense disambiguation and named entity recognition. This linguistic data is enriched using large databases in an attempt to extract enough information for a substantiated classification. The two main novelties introduced in this paper are the following: (1) the automatic generation of an objective affective lexicon, and (2) a user centric emotion evaluation technique that considers personal preferences towards entities.

The rest of the paper is organized as follows. Section 2 starts by exposing the motivation behind this project. Section 3 briefly explains some similar approaches for sentiment analysis. Sections 5 through 8 explore the implementation, discussing our methods for parsing the news title, word-level valence assignment, recognizing entities and classifying emotion. A description of the results has been detailed in section 9.

2   Background

This project is a follow-up to the work done by Vinagre, on user-centric emotion classification of news headlines[17]. Vinagre proposed a system for gathering news items of a specific emotion based on a user's preferences and personality. It classified the emotion of a news' headline according to the OCC emotion model and assessed a user's personality based on its Big-Five[8] measurement. Theoretically, this provided the system with the ability to predict the user's reaction to the news item. However, the sentiment analysis procedure was handmade for each news' headline. This project is an effort to implement a method for automatic emotion classification that was lacking in the aforementioned work.

3   Related Work

Previous approaches in sentiment analysis have employed many techniques like keyword spotting, lexical affinity, machine learning or knowledge based approaches[9]. Each technique has its strength and weaknesses as described in [9, 20]. Here we'll be looking at some examples most similar to our approach.

3.1   Lexicon-Based Approaches

The first works in sentiment analysis classified text based exclusively on the presence of affective words. This method, called keyword spotting, requires a word list with affective word valences, known as an affective lexicon. Among the most used are the General Inquirer[27] and Ortony's Affective Lexicon[19], that provide sets of words from different emotions categories and can be divided into positive or negative types. The keyword spotting method relies on the occurrence of words recognized by the affective lexicon and can be easily tricked by negation.

For instance, while the sentence "We won the lottery." would be assigned a positive valence, its negation "We didn't win the lottery." would likely have the same valence assigned to it. Moreover, sentences like "We're always late for meetings." would be assigned a neutral valence due to the lack of affective words. Wilson et al.[30] proposed an interesting solution to handle the issue of negation by identifying contextual polarity and polarity shifters.

The development of linguistic corpora with information on the semantics of words brought with it new methods for sentiment analysis, more specifically, content-based approaches.

3.2   Content-Based Approaches

The most used resources in content-based sentiment analysis are WordNet[6] and ConceptNet[10]. WordNet is a lexical database for the English language with over 150,000 nouns, verbs, adjectives and adverbs organized by a number of semantic relations such as synonym sets (known as synsets), hierarchies, and others. ConceptNet is a common sense knowledge base with over a million facts organized as different concepts, interconnected through a set of relations, such as "UsedFor", "IsA", "Desires", and more. Both databases are sufficiently complex to derive an affective lexicon from a group of affective keywords.

The authors of ConceptNet developed a popular approach for sentiment analysis that used the knowledge base to classify sentences into one of Ekman's six "basic" categories of emotions[9]. A set of affective concepts were manually annotated for the emotion categories, and then used in propagation models to infer emotion in other concepts related to them. The accuracy of this approach has not yet been investigated.

A few years later, SenseNet[25] was developed. This approach used WordNet to detect polarity values of words and sentence-level textual data. The system works from a set of assumptions and one of them describes some hand-crafted equations that count the positive and negative senses from the definitions in WordNet. SenseNet also turns to ConceptNet when processing a concept not found on its database. It only produces valenced results, therefore it can only differentiate between positive, negative or neutral emotions.

There's a significant problem with approaches dependent on ConceptNet. The common sense knowledge base has a very uneven distribution of specificity. This increases the noise in affect sensing from related concepts. We estimate that WordNet suffers from the same problem to some extent, but its impact is lessened from a distribution across a small group of categories (hypernyms, hyponyms, meronyms, etc.).

3.3   News Sentiment Analysis

There's already some research in sentiment analysis for news titles specifically. The UPAR7[3] is a content-based approach that classifies emotion using both valence ratings and a set of "booster" categories of words associated to one or more

of Ekman's six basic emotions. The valence ratings are extracted from Word-
Net Affect[29] and SentiWordNet[5], while the booster categories are taken from
a selected group of hypernyms from WordNet. For instance, words that have
"weapon system" as an hypernym, boost anger, fear and sadness emotions. This
is an interesting approach and it performed adequately at the SemEval-2007[28]
affective text task. However, it's unknown if UPAR7's reliance on hypernyms
and selected categories performs as well outside the SemEval task.

The same researchers that developed SenseNet later developed the Emotion Sensitive News Agent (ESNA)[26]. ESNA uses the sentiment ratings from
SenseNet to determine the appropriate classification for news items according
to the OCC model of emotion. It requires users to select news sources from
RSS feeds according to their interests, and rate a list of named entities according to their personal feelings towards them. This approach is very similar to
Vinagre's[17], from which our project stems from. We're only aware of these two
approaches to integrate user preferences and interests into the classification of
emotion with the OCC model. Notice that while Vinagre didn't provide a method
for automatic emotion classification, Shaikh et al. didn't provide a method for
automatic gathering of user preferences.

4   Architecture Overview

Determining the emotional classification of a news title requires parsing the sentence and analyzing its components. A title must be put through a pipeline
where many different linguistic aspects are scrutinized. To get an accurate assessment we need to look at the semantics in work and have knowledge about
the content and its listener. After analyzing these aspects we look at a model
of emotions and find out the appropriate classification. Our approach pursued a
pipelined architecture with the following stages: Linguistic Parsing, Word-level
Valence Assignment, Entity Processing and Emotion Classification, as illustrated
in Figure 1.

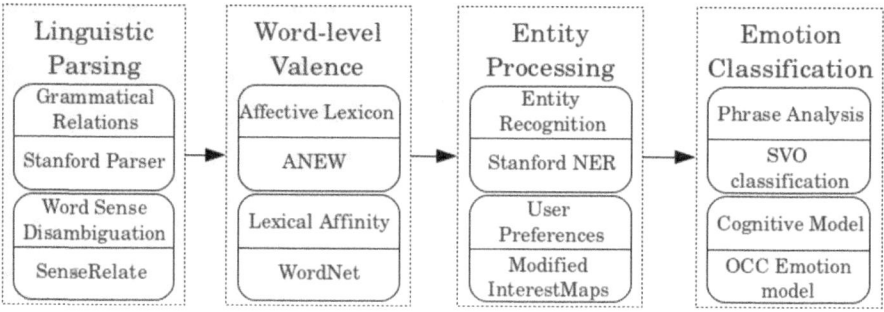

Fig. 1. Four stage pipeline architecture. Each step contains the methods and resources
used.

5   Linguistic Parsing of the News Title

Natural Language Processing (NLP) is used to get a handle on the semantic relations of the phrases in the headline. In this stage we also look at the different senses that a word can assume and try to map it correctly to the appropriate word on the lexicon.

5.1   Grammatical Relations

Grammatical relations are explored using a natural language parser, the Stanford Parser[14], which is typically used to examine the grammatical structure of sentences. The Stanford Parser is a probabilistic parser that produces the most likely analysis based on knowledge gained from hand-parsed sentences. It has some problems with the capitalization of consecutive words, as sometimes found in news titles, that aren't addressed in this paper. We can't use a Part-of-Speech (POS) tagger to decapitalize words because it would interfere with the entity recognition method used later. The solution is to manually fix or avoid improperly capitalized news titles. After fixing that capitalization issue, the parser performs adequately and we use of a specific structure it creates called typed-dependencies. This structure provides a simple representation of the grammatical relations between pairs of words, with over 55 types of relations. The sentence "Michael Phelps won eight gold medals at Beijing." generates the following typed-dependencies:

```
nn('Phelps', 'Michael')      nsubj('won', 'Phelps')
num('medals', 'eight')       amod('medals', 'gold')
dobj('won', 'medals')        prep('won', 'at')
pobj('at', 'Beijing')
```

The dependencies are relations between a governor (also known as a head) and a dependent. The grammatical relations are defined in[13]. This project uses 22 of the most frequent typed-dependencies relations to organize the sentence under a subject-verb-object (SVO) phrase typology with details on prepositions, adjectives, modifiers, possessives, and other relations the parser may find.

The subject phrase is extracted from the governors of the 'nominal subject' (nsubj) and 'passive nominal subject'(nsubjpas) relations. The object phrase is extracted from the governors of the 'direct object' (dobj) and 'object of a preposition' (pobj) relations. Finally, the verb phrase is extracted from the dependents of all those. When the parser can't find the expected relations necessary build the SVO, it still tries to build it from some unused relations, such as prepositional relations. The parser is robust enough to handle most syntactic structures.

The SVO is still compounded with information from other relations. The 'noun compound modifier' (nn) extends nouns, the 'adjectival modifier' (amod) modifies the meaning of nouns, the 'possessive modifier' (poss) associaties nouns to their possessives, among others. These are encoded in the SVO using special characters. The prepositional phrases use square brackets, the possessives use curly brackets and the modifiers use parenthesis.

The sentence parsed before ends up looking like this when wrapped under the SVO structure:

```
Michael Phelps / won [ at Beijing ] / eight medals (gold)
```

5.2   Word Sense Disambiguation

A central issue in NLP is the process of assigning the correct sense to a word, according to the context in which it is found. This is a key obstacle to surpass in text-based sentiment analysis, since assigning the wrong sense to a word with multiple meanings can result in a vastly different classification of the whole sentence. This can be seen in published news titles such as "Drunk Gets Nine Months in Violin Case" or "Complaints About NBA Referees Growing Ugly", taken from[23].

Much research in the field of word sense disambiguation (WSD) uses WordNet's large database to assign senses to words. One such solution is SenseRelate[21], which finds the correct sense for each word by determining the similarity between that word and its neighbors. The result is a set of senses with maximum semantic relatedness. This method doesn't require any training and was shown to perform competitively.

This project gives preference to WordNet driven WSD, because the disambiguated senses can be affectively rated with the same method for word-level valence assignment used for all words in the news title.

6   Word-Level Valence Assignment

Most research on sentiment analysis uses a set of affectively annotated words. This normally involves a word list where each word has a numerical valence rating assigned to it. Sometimes the authors compile a set of affective words and rate them individually[25, 9], or group them into emotion categories. This practice may be suitable when there's some expectation about which words will be analyzed, but a general purpose affective lexicon sensor requires a word list that is as objective as possible.

The results of an early study by Pang et al.[20] on movie reviews suggest that coming up with the right set of keywords might be less trivial than one might initially think.

6.1   Corpus of Choice

The Affective Norms for English Words (ANEW)[2] corpus provides an extensive word list (1034 words) with numerical ratings (1-9) compiled by several experimental focus groups. It considers three dimensions of affect: valence, arousal and dominance. Ratings for each range semantically from 'pleasant' to 'unpleasant', 'excited' to 'calm' and 'controlling' to 'submissive', respectively. A rating below 4 is low, above 6 is high, and neutral in between. Low valences are interpreted as negative and high valences as positive, According to ANEW, the word 'assault' has a low valence (unpleasant), high arousal (excited) and low dominance (submissive) while the word 'comfort' has high valence (pleasant), low arousal (calm) and neutral dominance.

This three dimensional rating allows for finer distinctions between words and combining dimensions results in improved performances. Combining valence and arousal reinforces the difference between valences of similar ratings, for instance, the words 'kindness' and 'excitement' have very similar positive valence ratings but the higher arousal rating for 'excitement' reinforces its positivity. In our work we only use valence and arousal dimensions, as dominance ratings are too scarcely differentiated to have an impact in this implementation.

Furthermore, the ANEW corpus has each rating accompanied by a standard-deviation value. This opens up the possibility to integrate the Big Five model[8] into the affective assessment so that depending on the user's profile, different ratings can be adopted within the standard-deviation. For example, a higher agreeableness profile could use ratings within the upper range of the standard-deviation.

6.2   Beyond the Corpus

Despite ANEW being a fairly large corpus, it was found to be insufficient for reliable sensing. Word stemming enabled broader coverage of the ratings found in the corpus, so that different words with the same word stem will return the same rating in the word list. Mihalcea et. al[15] argued that lemmatization isn't appropriate for words in the ANEW corpus because it contains more than one word for each lemma and its different ratings would be lost. Word stemming shares the same problem. Thus, it was decided to simply include both representations in the lexicon,the word stem and the regular word.

Finally, we included some tokens that are part of English grammar and aren't covered by neither the ANEW corpus nor WordNet. These include pronouns, prepositions, determiners and numerals, totaling an addition of nearly 300 neutrally rated tokens.

The end result is a word list with more than 5000 words rated in two dimensions where, besides the nearly 300 words for grammar neutralization, it was fully derived from the ANEW corpus. Through this process a larger corpus was derived from a scientifically validated one without adding any unnecessary subjectivity.

6.3   Affinity Rating

The affective lexicon can't possibly cover all words in the English dictionary. Regardless of whatever techniques used to expand it, there will always be affective words left unrated. To cover these omissions we turn once again to WordNet, but now to infer valence ratings using the affective lexicon presented before.

There have been many approaches to rate the valences of unknown words from a preexisting lexicon using WordNet. Some assigned the unknown word's rating by averaging the valences of the words in its synonym sets[25]. Others considered more relations, like hypernyms and hyponyms[3]. Our approach averages the synonym sets, and also the affective words contained in its definition. We consider the word's definition because its content is often more affectively coherent than the words on its hierarchies, which quickly get off topic. If we wish to make the system more reactive, we apply a reducing factor to the neutral ratings. The calculations for valence assignment are detailed below.

$$PositiveWeight = AvgPosValence * AvgPosArousal \qquad (1)$$

$$NeutralWeight = AvgNtrValence * AvgNtrArousal * ReducingFactor \quad (2)$$

$$NegativeWeight = AvgNegValence * AvgNegArousal \qquad (3)$$

$$WordValenceLabel = Max(PositiveWeight, NeutralWeight, NegativeWeight) \qquad (4)$$

$$WordValenceIntensity = AvgXValence \text{ where } X \text{ is given in (4)}. \qquad (5)$$

7   Recognizing Entities

An often ignored aspect of sentiment analysis is the role that entities play in the attribution of emotion. The desirability of events is tied to the subjects and objects involved. For instance, the headline "Saddam Hussein arrested in Iraq." evokes a positive emotion only when considering that the entity in case is associated to negative emotions. Other examples like, "Netherlands loses World Cup to Spain" can have opposite emotion classifications depending on the reader's preferences. This section explains how we solve this issue.

7.1   Extracting Entities

This project uses the Stanford Named Entity Recognizer[7] to extract entities from the news titles. It can recognize the names of persons, organizations and locations and label sequences of words accordingly. The "perceived" affective valence of a recognized person or organization entity is determined by a method specifically for sentiment analysis of entities. We determined that locations should always have a neutral valence rating to avoid biased stances in cases of polemic locations.

7.2   Sentiment Analysis of Entities

After extracting the entities found on news titles, they're classified using information on their Wikipedia page. Most entities that people have an opinion about are likely to be relevant enough to have a page on Wikipedia. This includes celebrities, politicians, scientists, writers, organizations, sports teams, and many more. These pages have lots of information, and there's bound to be some overlap between the words on the pages and the words in the affective lexicon. By averaging the valence ratings of these overlapping words, it's possible to assign a valence rating to entities. This process is essentially keyword spotting for wikipages, but here its issues with reliance on surface features are mitigated by Wikipedia's community effort on writing clear and descriptive text. This project uses DBPedia[1] to gather information from Wikipedia pages. DBPedia is an online knowledge base that extracts structured information from Wikipedia and provides its own formal ontology. Using DBPedia we're also not abusing Wikipedia's resources.

The classification of entities according to the user's preferences relies on InterestMaps[11], a semantic network of interests collected from profile pages on social networks. We use a variation on InterestMaps that gathers user information from their Facebook profile through the website's download backup profile feature, and then builds a network of interests using DBPedia's ontology[12]. The user's 'likes' and interests are explored in Wikipedia to provide the system information on other entities that may also be of interest.

The inclusion of InterestMaps into sentiment analysis allows more personalized emotion classifications. Whichever entities are mentioned on the user's profile page can reinforce or reverse the intensity of the associated emotions. For instance, consider a fan of the Rolling Stones (as expressed on his user profile). A news title like "Mick Jagger releases new single." would evoke a positive emotion with high intensity, because his InterestMap would have information on all things related to the band, namely "Mick Jagger". This applies to all content that appears on Facebook and has a page on Wikipedia.

8   Rule-Based Emotion Classification

This project uses the OCC emotion model's rule based structure to determine what emotions are associated to a news title. There are a total of six emotion categories in the model (e.g. well-being, attribution, attraction, etc.) and each is related to a set of variables (e.g. desirability, appealingness, familiarity, etc.). These variables are resolved by analyzing the properties of the subject, verb and object phrases. Afterwards, the model is applied and the sentiment analysis procedure reaches its end.

8.1   Phrase Sentiment Analysis

The linguistic parsing stage built a SVO structure with information on the most relevant linguistic relations of the news title. Among these, we have the subject, verb and object in addition to all modifiers, possessive determiners and prepositional phrases acting on them. Since every word already has a valence rating assigned to it from the word-level valence assignment and entity processing stages, the valence ratings for the phrases are calculated following these rules:

- If the phrase doesn't contain modifiers, possessive determiners or prepositional phrases, then its rating is calculated from the average of all word ratings.
- If there are prepositional phrases present, their ratings are scaled down by a factor of 0.8 and added to the average of the other words.
- If there are modifiers or possessive determiners present, then the phrase inherits their valence ratings. The impact of prepositional phrases is reduced, because it often strays from the main topic of the sentence.

8.2   Cognitive Model of Emotion

After decoding a news title's linguistic structure and affective contents, we look at underlying cognitive structure through the OCC emotion model. This rule-based model uses knowledge and intensity variables to characterize 22 different emotion types. The values for each variable can be assigned using a conditional set of rules. Shaikh et.al presented a method for variable attribution from a linguistic and affective standpoint[24]. This method considered 16 variables from the OCC emotion model and described how to assign their values using linguistic and affective data.

The following code demonstrates how we implement Shaikh's et. al interpretation of the OCC model.

```
if Valenced_reaction == True:
  if Self_reaction == ''Pleased":
    if Self_presumption == ''Desirable":
      if Cognitive_strength == ''Self":
        Joy = True
      if Prospect  == ''Positive":
        if Status == ''Unconfirmed":
          if Object_familiarity == ''High":
            Hope = True
```

9   Evaluation

The results of this project's method for word-level valence assignment were measured using affective words extracted from the General Inquirer as the gold standard. The General Inquirer contains a positive and a negative word list with about 3500 words combined. In order to make results comparable, we labelled our numerical ratings above 6 as positive, below 4 as negative and neutral in between. Table 1 shows the results. The variable recall measures the ratio of recognized words, while accuracy and precision measure the binary classifications of 'positive' and 'negative' valence ratings. As expected, the recall values for the ANEW corpus were very low because of its significantly smaller size than the word lists it's being measured against.

Table 1. Measure of overlap between our method for word-level valence assignment, and the positive and negative word lists from General Inquirer

Lexicon	Accuracy	Precision	Recall
ANEW	90.44%	87.12%	33.9%
ANEW & WordNet	76.20%	69.45%	91.96%

For measuring the results of sentiment analysis for news titles, the trial tests from SemEval 2007 affective text task were used as the gold standard. These trials divide emotions into the regular "joy", "sadness", "fear", "surprise", "anger" and "disgust" categories. Since our method classifies different emotion types, we had to reduce both ours and SemEval's classifications into positive and negative emotion groups. From SemEval, the "joy" category joined the positive group and the rest (except "surprise") joined the negative group. Since the OCC model divides each emotion category into positive and negative types, each type joins the corresponding group. Table 2 shows the results. We noticed that many of the mistakes that occurred were not only involved with the affective aspects, but also the linguistic analysis often failed to produce a valid SVO structure, hence the somewhat low recall.

Table 2. Performance results for sentiment analysis of news titles

Comparison	Accuracy	Precision	Recall
SemEval 2004	57.78%	51.92%	60.52%

Here's the output of an example from SemEval as processed by the system.

```
Headline: ''Nigeria hostage feared dead is freed."
Word-level valence assignment:
'Nigeria' is an entity of type 'Location': Neutral (5.00)
'hostage' is a 'Noun' included in the lexicon: Negative (2.2)
'feared' is a 'Verb' included in the lexicon: Negative (2.25)
'dead' is a 'Adjective' included in the lexicon: Negative (1.94)
'is' is a 'Verb' included in the lexicon: Neutral (5.00)
'freed' is a 'Verb' disambiguated('free.v.06'): Positive (8.14)
     (from 43 tokens)
```

```
Phrase analysis:
   Subject: - Nigeria hostage ( feared dead )  v: negative
   Verb: - freed  v: positive
   Object: -  v:

Entities = ['Nigeria'], no preference
Object familiarity: Low (0)
Emotion_intensity: Neutral (0.018)
Emotion Classified: 'Relief', 'Joy'
```

Notice that the word/phrase/sentence analysis makes it possible to understand which elements contributed to the emotion classification. A purely probablistic classifier couldn't provide this sort of insight into sentiment analysis.

10   Conclusion

In summary, this paper presented an approach for sentiment analysis that provides an alternative affective lexicon and introduces the element of InterestMaps to consider user preferences. While the latter didn't have the opportunity to be tested, the lexicon showed satisfactory results. The emotion classification process reproduced and combined techniques from other works in a unique configuration, showing some encouraging results. Our approach relied on the cognitive structure of sentences and analyzed many linguistic and emotional aspects. The pipeline nature of this process helps to understand what elements associate headlines to particular emotions. In the future, we'll compare our approach to other affective resources and explore some applications.

Acknowledgements. This work was funded by a research integration grant from Fundacao para a Ciencia e Tecnologia (FCT).

References

[1]   Bizer, C., Lehmann, J., Kobilarov, G., Auer, S., Becker, C., Cyganiak, R., Hellmann, S.: DBpedia A Crystallization Point for the Web of Data. Journal of Web Semantics: Science, Services and Agents on the World Wide Web (7), 154–165 (2009)

[2]   Bradley, M.M., Lang, P.J., Cuthbert, B.N.: Affective Norms for English Words in Center for the Study of Emotion and Attention National Institute of Mental Health, University of Florida (1997)

[3]   Chaumartin, F.: UPAR7: A knowledge-based system for headline sentiment tagging. In: SemEval 2007, Prague, ACL, pp. 422–425 (2007)

[4]   Ekman, P.: Facial expression of emotion. American Psychologist 48, 384–392 (1993)

[5]   Esuli, A., Sebastiani, F.: SentiWordnet: A Publicly Available Lexical in Resource for Opinion Mining. In: LREC 2006 (2006)
[6]   Fellbaum, C.: WordNet: An Electronic Lexical Databases. MIT Press, Cambridge (1999)
[7]   Finkel, J., Grenager, T., Manning, C.: Incorporating Non-local Information into Information Extraction Systems by Gibbs Sampling. In: Proceedings of the 43nd Annual Meeting of the Association for Computational Linguistics (ACL 2005), pp. 363–370 (2005)
[8]   Goldberg, R.: The structure of phenotypic personality traits. American Psychologist 48(1), 26–34 (1993)
[9]   Liu, H., Lieberman, H., Selker, T.: A Model of Textual Affect Sensing using Real-World Knowledge. In: Proceedings of the 2003 International Conference on Intelligent User Interfaces, pp. 125–132 (2003)
[10] Liu, H., Singh, P.: ConceptNet: A Practical Commonsense Reasoning Toolkit. BT Technology Journal 22(4), 211–226 (2004)
[11] Liu, H., Maes, P.: Interestmap: Harvesting social network profiles for recommendations in Beyond Personalization - IUI (2005)
[12] Loureiro, D.: Facebook's hidden feature: User Models and InterestMaps. In: 4th Meeting of Young Researchers, UP, IJUP (2011)
[13] Marneffe, M., Manning, C.: Stanford typed dependencies manual (2008)
[14] Marneffe, M., MacCartney, B., Manning, C.D.: Generating Typed Dependency Parses from Phrase Structure Parses in LREC (2006)
[15] Mihalcea, R., Liu, H.: A corpus-based approach to finding happiness. In: The AAAI Spring Symposium on Computational Approaches to Weblogs (2006)
[16] Minsky, M.: The Emotion Machine. Simon & Schuster, New York (2006)
[17] Vinagre, E., Marreiros, G., Ramos, C., Figueiredo, L.: An Emotional and Context-Aware Model for Adapting RSS News to Users and Groups. In: Lopes, L.S., Lau, N., Mariano, P., Rocha, L.M. (eds.) EPIA 2009. LNCS, vol. 5816, pp. 187–198. Springer, Heidelberg (2009)
[18] Ortony, A., Clore, G., Collins, A.: The Cognitive Structure of Emotions. Cambridge University Press, New York (1988)
[19] Ortony, A., Clore, L., Foss, A.: The referential structure of the affective lexicon. Cognitive Science 11, 341–364 (1987)
[20] Pang, B., Lee, L.: Opinion mining and sentiment analysis. Foundations and Trends. Information Retrieval 1(1-2), 1–135 (2008)
[21] Patwardhan, S., Banerjee, S., Pedersen, T.: SenseRelate: TargetWord A generalized framework for word sense disambiguation. In: Proc. of AAAI 2005 (2005)
[22] Picard, R.: Affective Computing. The MIT Press, Massachusetts (1997)
[23] Pinker, S.: The Language Instinct Perennial. HarperCollins (1994)
[24] Shaikh, M.A.M., Prendinger, H., Mitsuru, I.: Rules of Emotions: A Linguistic Interpretation of an Emotion Model for Affect Sensing from Texts. In: Paiva, A.C.R., Prada, R., Picard, R.W. (eds.) ACII 2007. LNCS, vol. 4738, pp. 737–738. Springer, Heidelberg (2007)
[25] Shaikh, M., Ishizuka, M., Prendinger, H.: SenseNet: A Linguistic Tool to Visualize Numerical Valence Based Sentiment of Textual Data. In: Proc. ICON 2007 5th Int'l Conf. on Natural Language (2007)
[26] Shaikh, M., Prendinger, H., Ishizuka, M.: Emotion Sensitive News Agent: An Approach Towards User Centric Emotion Sensing from the News. In: ACM International Conference on Web Intelligence, pp. 614–620 (2007)

[27] Stone, J., Dunphy, D., Smith, M., Ogilvie, D.: The General Inquirer: A Computer Approach to Content Analysis. MIT Press, Cambridge (1966)

[28] Strapparava, C., Mihalcea, R.: SemEval-2007 Task 14: Affective Text. In: Proceedings of the 45th Aunual Meeting of ACL (2007)

[29] Strapparava, C., Valitutti, A.: Wordnet-affect: an affective extension of wordnet. In: Proceedings of the 4th Internation Conference on Language Resources and Evaluation, Lisbon, pp. 1083–1086 (2004)

[30] Wilson, T., Wiebe, J., Hoffmann, P.: Recognizing contextual polarity in phrase-level sentiment analysis. In: Proceedings of the HLT 2005 (2005)

Providing Location Everywhere

Ricardo Anacleto*, Lino Figueiredo, Paulo Novais, and Ana Almeida

GECAD - Knowledge Engineering and Decision Support,
R. Dr. Antonio Bernardino de Almeida, 431,
4200-072 Porto, Portugal
{rmao,lbf,amn}@isep.ipp.pt,
pjon@di.uminho.pt

Abstract. The ability to locate an individual is an essential part of
many applications, specially the mobile ones. Obtaining this location
in an open environment is relatively simple through GPS (Global Positioning System), but indoors or even in dense environments this type of
location system doesn't provide a good accuracy. There are already systems that try to suppress these limitations, but most of them need the
existence of a structured environment to work. Since Inertial Navigation
Systems (INS) try to suppress the need of a structured environment we
propose an INS based on Micro Electrical Mechanical Systems (MEMS)
that is capable of, in real time, compute the position of an individual
everywhere.

Keywords: Location Systems, Dead Reckoning, GPS, MEMS, Machine
Learning, Optimization.

1 Introduction

Location information is an important source of context for ubiquitous computing
systems. Using this information, applications can provide richer, more productive and more rewarding user experiences. Also, location-awareness brings many
possibilities that make mobile devices even more effective and convenient, at
work and in leisure.

The ability of mobile applications to locate an individual can be exploited
in order to provide information to help or to assist in decision-making.Some
examples include electronic systems to help people with visual impairments [10],
support systems for a tourist guide at an exhibition [5] and navigation systems
for armies [7] [19].

In an open environment we use GPS (Global Positioning System) to retrieve
users' location with good accuracy. However, indoors or in a more dense environment (big cities with tall buildings, dense forests, etc) GPS doesn't work or
doesn't provide satisfactory accuracy. Consequently, location-aware applications
sometimes don't have access to the user location.

---

* The authors would like to acknowledge FCT, FEDER, POCTI, POSI, POCI
and POSC for their support to GECAD unit, and the project PSIS
(PTDC/TRA/72152/2006).

L. Antunes and H.S. Pinto (Eds.): EPIA 2011, LNAI 7026, pp. 15–28, 2011.

The aim of this work is to design a system capable of, in real time, determining the position of an individual (preferably in a non-structured environment), where GPS is not capable.

The motivation for this project comes from previous publications, where a recommendation system to support a tourist when he goes on vacations, was presented [1] [2]. Mobile devices currently available on the market already have built-in GPS, which provides the necessary location context to recommend places of interest and to aid the tourist in planning trip visits [17]. However, this assistance, with current technology, can only be used in environments with GPS signal. In closed or dense environments the system can't easily retrieve the user location context. For example, in an art gallery if the system knows the tourist current position inside the building, it can recommend artworks to view and learn even more about the tourist tastes.

To remove this limitation, a system that provides precise indoor location of people becomes necessary. Actually, there are already some proposed systems that retrieve indoor location with good precision. However most of these solutions force the existence of a structured environment. This can be a possible solution when GPS isn't available, but only indoors. In a dense forest we don't have this kind of systems. To retrieve the location on this type of terrain can be very useful for position knowledge of a fireman's team. Also, the implementation of a structured environment using this type of technologies is very expensive and it becomes unfeasible to incorporate this type of systems in all the buildings of the world.

Since location without using a structured environment remains an open research problem, the main goal of the proposed project is to minimize deployment and infrastructure costs and provide location everywhere.

As users of these types of devices (and applications) are on foot, the INS is one of the most appropriate solutions to be used. This system consists of Micro Electrical Mechanical Systems (MEMS) devices, which can be accelerometers, gyroscopes and other types of sensors. MEMS are small in size, which allows easy integration into clothing. Usually, they communicate with a central module using a wireless network (*e.g.*, Bluetooth). These devices obtain individual movements information independently of the building infrastructure. All this sensory set requires the implementation of a sensor fusion. This includes algorithms that can interpret the sensors information and thereby determine the individual position. The collected information, in addition to the motion speed and direction, must also be able to determine the step width and the individual position (sitting, lying or standing).

This type of system uses the PDR (Pedestrian Dead Reckoning) technique [4]. PDR normally is composed by three key technologies: tracking of the sensor's behavior, walking locomotion detection and walking velocity estimation. But unfortunately, large deviations of these sensors can affect performance, as well as the various forms in which a human can move, so this is the project's biggest challenge: to correct the sensor deviations.

A module working only with PDR is not able to ensure that the geographical positions are accurate within a few meters. In fact, although these deviations may be small for every millisecond, the positioning error caused by a sustainable use of the system can exceed one meter in 10 seconds [23].

In order to provide a better contextualization of the existing work on the field, Section 2 gives a brief description of the most relevant INS and indoor location systems. In Section 3 we present our main objectives to implement an INS and how we intend to complete them, presenting the main challenges and the methodology we will use to suppress them. Finally, Section 4 presents some conclusions regarding the presented problem.

2   State of the Art

Indoor localization technologies hold promise for many ambient intelligence applications to facilitate, for example, decision-making. However, many of the existent systems can't work indoors or in too dense environments. This happens because they work based on the location obtained by GPS.

Several localization techniques already exist to give indoor positioning, but some of them doesn't provide good accuracy or are too difficult to implement. From the selected articles we have divided them into two main categories based on their localization techniques: Radio Frequency Waves and Pedestrian Dead Reckoning.

Radio frequency waves are solutions that estimate the location of a mobile target in the environment by measuring one or more properties of an electromagnetic wave radiated by a transmitter and received by a mobile station. These properties typically depend on the distance traveled by the signal and the surrounding environment characteristics. In radio frequency we include technologies such as IEEE 802.11, Infrared, Bluetooth and Radio-Frequency IDentification (RFID) tags.

As examples of these systems we have: RADAR (Wireless LAN) [3], Active Badge (Infrared) [25], a home localization system (Bluetooth) [12], a project for the location of objects and people using RFID tags [22], among others [11]. The main problem with these systems is that they need a structured environment in order to work. This makes them dependent on the particular context and still impractical and expensive.

Pedestrian Dead Reckoning systems use sensors to provide location updates, calculated using information about a previously-estimated location. This position estimation is commonly based on inertial sensors. Since they yield relative positioning information only, an absolute reference is required to specify the displacement reported by an inertial measurement in absolute coordinates.

We have divided the PDR systems into two groups: the pure ones and the hybrid solutions. The first use only inertial sensors to give position (after having a reference point), and the latter use WLAN, Bluetooth and other type of structured indoor location systems to correct the inertial sensors data.

2.1 Dead Reckoning with Structured Environments

The solutions based on inertial sensing are subject to big measurement deviations caused by thermal changes in the sensor circuit. This is because the longer the time indoors, the greater will be the diversion from virtual trajectory to the real trajectory. In other words, there is error accumulation and to make a recalibration of the sensors, some projects use a structured environment.

This structured environment can be constituted by surveillance cameras like Kourogi [15] proposes. In his paper two enhancements for PDR performance are introduced: map matching and dynamic estimation of walking parameters. The users' location and orientation are updated by fusing the measurements from the PDR estimation and the maps. The surveillance cameras are used to measure walking velocity in order to recalibrate the sensor parameters. Also a particle filter [18] is used for probabilistic data fusing (based on a Bayesian filter). Probability distribution of the users' location is predicted from the estimated position, orientation and its uncertainties. The results are satisfactory, but the existence of lots of people in the building will, certainly, confuse the system. Another problem is the cost of this system (over $1500).

Also Wi-Fi signal strength can be used to tackle the traditional drift problems associated with inertial tracking [26]. Woodman proposes a framework constituted by a hip-mounted mobile PC which is used to log data obtained from a foot-mounted Inertial Measurement Unit (IMU). The IMU contains three orthogonal gyroscopes and accelerometers, which report angular velocity and acceleration respectively. The logs are then post processed on a desktop machine (so it isn't a real-time location system). Like Kourogi's system, this uses Bayesian filters to probabilistically estimate the state of a dynamic system based on noisy measurements. The particle filter update consists of three steps: Re-sampling, Propagation and Correction. To reduce the cubic-in-time drift problem of the foot-mounted IMU, they apply a Zero Velocity Update (ZVU) [27], in which the known direction of acceleration due to gravity is used to correct tilt errors which are accumulated during the previous step. Besides Wi-Fi, it also uses a map that acts like a collection of planar floor polygons. The system obtains the Received Signal Strength Indication (RSSI) information by querying the Wi-Fi hardware. Each query returns a list of visible Wi-Fi access points and corresponding RSSI measurements. Each floor polygon corresponds to a surface in the building on which a pedestrians foot may be grounded. Each edge of a floor polygon is either an impassable wall or a connection to the edge of another polygon. Then the system tries to make a correspondence between the Wi-Fi signals, the map and the IMU. A system evaluation was performed in a three floor building, with a total area of $8725m^2$, and the error proved to be $0.73m$ in 95% of the time.

The system proposed by Evennou and Marx [8], it is also based on an IEEE 802.11 wireless network. The absolute positional information obtained from the wireless network was combined with the relative displacements and rotations reported by a gyroscope, a dual-axis accelerometer and a pressure sensor.

The information obtained from all the sensors was combined through Kalman [21] and particle filtering. To calculate the user displacement, an accelerometer was used to count the number of steps taken. Then, a constant estimate of the users' step length was used to calculate the user total displacement within the environment. The authors showed that combining the information from the bank of sensors yielded improved localization when compared to the usage of each sensor separately. The authors conducted an experiment using their multi-sensor localization system. The localization accuracies reported during such experiment ranged from 1.5 to 3.3 meters.

Renaudin [20] proposes a solution based on RFID tags and inertial MEMS. This system was developed with the intuition to be used by firemen's. MEMS and RFID are hybridized in a structure based on an Extended Kalman Filter and a geographical database. While progressing indoors, the fire fighters deploy RFID tags that are used to correct the large errors affecting MEMS performances. The first team of fire fighters attaches a RFID tag each time it passes a door and when moving from one floor to another, at the beginning and at the end of the stairway. Upon installation, the geographical coordinates of the tag are associated with the tag ID. The RFID tag database is a collection of location coordinates of all the building doors and stairs. This information is then built over the evacuation emergency maps. So, the Extended Kalman Filter uses the 3D coordinates of each detected RFID tag to relocate the trajectory.

The IMU is composed by three sensor modules. The module attached on the shank has a gyroscope, measuring the angular rate of the shank in the sagittal plane, and an accelerometer oriented in the vertical plane that allows the gait analysis. The trunk module contains a triad of gyroscopes, magnetometers and accelerometers providing the orientation information. The thigh module contains an accelerometer measuring the frontal acceleration, which permits posture analysis. The algorithms that compute the route use all the information available in a database to extract the optimum walking path.

The proposed pedestrian navigation solution has been tested on the campus of the "Ecole Polytechnique Federale de Lausanne", in Switzerland. As expected, the PDR position error grows with time, whereas this hybrid positioning solution remains under a certain limit close to 5 meters.

2.2   Pure Dead Reckoning

Like we have already seen, there are already systems prepared for indoor positioning. But if we want a system that can provide our location anywhere and doesn't rely in a structured environment, another solution should be used.

One possibility is the use of hybrid solutions based on INS and GPS signal. INS complements the GPS giving location where GPS can't (indoor or in dense environment). The INS uses as start location point the GPS last known coordinate.

As example of an INS system of this type we have NavMote [9] that integrates, MEMS and GPS, in a wireless-device that is in the person's abdominal area. This device consists of an accelerometer and a magnetic compass integrated in a generic wireless controller board, with the radio, a processing unit, and power storage all integrated. This device records user movements and stores them in the system. Sensor data are constantly being stored in a 4MB flash memory which allows an operation time of 1.7 hours with a sampling frequency of 30 Hz. Subsequently, the compressed data is transferred to the main system when the wireless-device comes within range of a sensor network (called NetMote). These sensor networks can be spread along the building. On the main system, data are processed into an estimate of the pedestrian trajectory based on a PDR algorithm. The main system is also responsible for trajectory displaying, map matching and other purposes.

This PDR approach uses the acceleration signal pattern to detect the step occurrences and the magnetic compass to provide continuous azimuth information. Based on a simplified kinematic model of a person's gait, the walked distance is provided by summing up the size of each step over the step count. The system showed an error of only 3%. However, this solution doesn't provide user location in real time. For a 3 minute walk, 45 seconds of download time from the wireless-device to the main system and 25 seconds of system filtration time are typical. So the system is mainly used to track or monitor the user's position with potential security applications rather than to provide high-level interactive travel support.

Walder [24] proposes a system to be used on emergency situations, in large buildings and underground structures. It combines inertial measurements (using 3D accelerometer, gyroscope and 3D magnetometer) of moving persons with building floor plans (CAD drawings) enriched with additional data (*e.g.*, positions of fire extinguishers, room allocation). Because the proposed INS is not accurate enough, the system tries to improve positioning by a permanent interaction between the mobile positioning sensors and floor plans stored in the building information model. The system is composed by four components: position determination, position verification and correction, user interface and communication. The position determination reads raw sensor data and then computes position coordinates. Once computed, a position is verified and corrected, based on the tagged floor plans, if necessary.

The correction of noise and drift is performed by a comparison of real integration with an assumed step length after each step. A normalized step length can be defined as a system parameter, taking into account the user's body size and weight. This step length will be adapted during the movement, depending on the movement pattern, the moving direction (*e.g.*, curve radius) or the recognized environment (*e.g.*, stairs). In the worst case scenario, when the automatic correction doesn't work properly, the user can communicate his current position to the system. Identifying his position by giving information related to his current location (like near a fire extinguisher). For the location of the mobile units to be spread among themselves and to the control and command centers, mechanisms from mobile ad hoc networks and wireless sensor networks are fused to build up a communication network.

The system has two user interfaces: a conventional one to be used on computer terminals equipped with a screen and a pointing device (mouse, touch screen), and one to be used on wearable computers with head mounted display and speech driven control for action forces.

The system was tested in an outdoor and in an indoor scenario. In the first case the position error equals to a 0.77% mean value and 0.10% standard deviation. On the second case the error increases to 2.11% mean value and 0.95% standard deviation. This is a good solution, but only for indoor use in buildings. In a forest fire where GPS isn't available, this system doesn't work, since there isn't any CAD drawing plants, and also indoors it only works if CAD drawing exists, which are unfeasible to deploy in all the existing buildings in the world.

Castaneda [6] proposes a shoe-mounted inertial navigation system for pedestrian tracking using a fuzzy logic procedure for better foot stance phase detection, that is applied to the IMU outputs (gyroscope and accelerometer), and an indirect Kalman filter for drift correction based on the typical zero-updating measurement. In its most basic implementation, the shoe-mounted INS estimates and corrects drift errors via an assisted ZVU Kalman filter.

In order to test the system a real time application was developed. The hardware was composed of a laptop PC wire-linked to the IMU. The software consists of the fuzzy logic step detector and an assisted ZVU Indirect Kalman filter. Three different tests were made, involving various walking scenarios composed by forward walks, turns and stairs. The error for each scenario was 30cm, 15cm, and 27cm respectively. Also, positioning the sensors only on the shoes can bring more accuracy errors mainly on the gyroscope, because we move the feet a lot of times to the left or right and not necessarily all the body moves together with them.

The work proposed by [14] presents a system based on modular sensor units, which can be attached to a person and contains various sensors, such as range sensors, inertial and magnetic sensors, a GPS receiver and a barometer. The measurements are processed using Bayesian Recursive Estimation algorithms and combined with available a priori knowledge, such as, map information or human motion models and constraints. Each sensor module contains, apart from its sensors, a low power microprocessor which collects and processes the sensor readings and sends them to the central module.

The current version of the sensor stack consists of an inertial sensor module with a three axis accelerometer, a three axis gyroscope and a three axis magnetometer; and a ranging module which is also used as a wireless data transmission unit and a GPS module. The system is attached to the person's hip area. The inertial and magnetic sensors are used to estimate the heading of the person and to detect walking steps and the corresponding stride length. These steps are detected using a step detection algorithm, which is based on a set of thresholds applied to the vector length of the three accelerometer readings.

A Zero-Velocity-Update method isn't used since they require the sensors to be mounted on the person's feet. However, although the authors state that this method promises a better accuracy, they also claim that it is more feasible to wear the sensor unit at the waist, the chest or to hold it in the hand. Like other systems, an indoor map is used as an additional input to the localization algorithm.

While the person is moving through the tracking area, all sensor data are locally preprocessed and then transmitted to a PC for further processing and visualization. Like other systems, the localization is not given in real-time to the user. This is a big limitation if we want to implement a system like the one that we described to help tourists. Most of these systems have this limitation since they use very heavy algorithms that need lots of computational power to process all the retrieved information. One solution can be the inclusion of more sensors to have more information about what is happening on the lower limbs of the user, to reduce the computational power for those algorithms.

Because smartphone's are largely benefiting from the ever increasing rate of integration of mobile IT components, Lukianto et al [16] studied a pedestrian indoor navigation system based on a custom INS and a smartphone. The system is also designed to select the optimal set of auxiliary information from available infrastructure such as WLAN and Bluetooth to calibrate the INS, when necessary. The INS provides a continuous estimate of its current position, speed and orientation. These data are sent to the smartphone to track the current position and for visualization. The smartphone, in turn, collects sensory inputs available to its various communication interfaces and to the (A)GPS receiver system. All the additional sensor information is then processed for plausibility and quality, and used to correct the INS. The updated position is then sent back to the INS.

This INS uses an angular rate sensor, an accelerometer, a magnetic field sensor and a static pressure sensor to obtain the step length. The INS can be connected to the smartphone via Bluetooth or by USB cable. In the future, Kalman filter algorithms will be investigated firstly as a means of INS/GPS-integration, and also the concept of particle filtering will be investigated for viability, providing a means of 2D localization if a detailed map of the building is available. A big benefit of this system is that it tries to re-use the equipment that the users already have with them (a smartphone) to complement its AGPS module and provide location where the GPS can't. A limitation, is that it only uses one INU, which is not sufficient to provide a good localization accuracy.

The main challenge in this area is to provide an effective position of a person in an environment where GPS is not available and no structured environment exists. To overcome these challenges INS are commonly used, but they have great drift and provide sometimes inaccurate information. In that sense, the next challenge is to reduce these errors so deviations between a virtual trajectory and the real user path can be avoided.

3   System Overview

The project that we are entitling as "PLASYS - All Over the Place Location System", aims to study and create a system that, combined with GPS, provides location everywhere, including indoor and dense environments.

More specifically, this project is divided into three objectives:

- Analyze how activity recognition can be improved (adding different types of sensors) and finding an optimal sensor position on the body;
- Development of methods to process the data acquired from all the sensors, making a sensor fusion; use of probabilistic algorithms to learn the walking/moving behaviors for real-time sensor data correction;
- Expand the existing tourism mobile application to work with the proposed INS.

In order to solve the existing problem, we need to tackle the challenges present in these objectives. In the next sections we will explain each objective, presenting the main challenges and which methodology we will use to tackle them.

3.1   Sensors

First of all we will explore the type of sensors to use, since it is intended that, in addition to providing accurate data, they must be also comfortable (and imperceptible) enough to allow their integration into the user's clothing. With this approach we don't need any external sensors, avoiding expensive costs on structured environments.

As stated earlier, an Inertial Navigation System can bring several problems, especially because of the drift that the sensors can have. This is the biggest problem/challenge of the whole project. Also, because of people different sizes, the ideal position of the sensors to one person can be different to another. This leads to another big challenge: the discovery of an ideal spot for the sensors to work in diverse types of persons.

Small sensors will be distributed by the lower limbs (legs and hip area) to collect data. These data will be sent to a central module that will handle the calculations to determine, in real time, the position of an individual. The interconnection of the various modules with the core module will be run through a wireless body network. For now, we don't know which technology, Bluetooth (new version, v4) or ZigBee [13], will be used. The final decision will rely in a series of system tests designed to evaluate which is the most reliable technology.

We pretend to distribute the sensors like this: force sensors and accelerometers in the feet; and a gyroscope and a pressure sensor in the abdominal area (see figure 1).

In the abdominal area, there will be a central module that communicates with a PDA. This PDA will show to the user his current location. The force sensors are essential to determine when the user puts his feet on the ground, that combined with the accelerometer (to get the step acceleration) provides a more exact step length. The gyroscope gets the body travel direction. The pressure sensor on the abdominal area is useful to get the user elevation inside the building.

To have a better perception of the comfortableness of the system a survey research will be done. This type of research method is associated with the use of questionnaires and statistical data for analyze user answers. Using questionnaires is useful to provide a better perception of the proposed system usage.

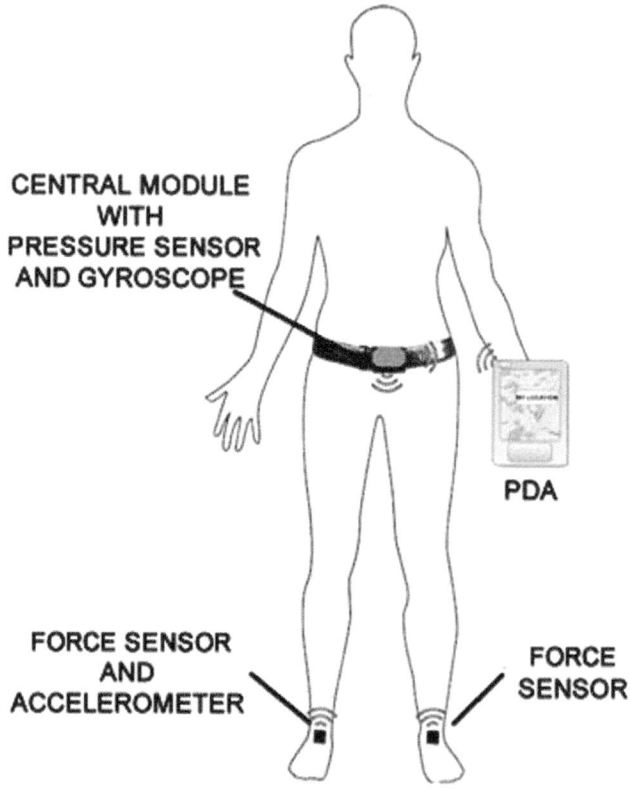

Fig. 1. Disposition of the sensors on the human body

3.2 Sensor Fusion and Probabilistic Algorithms

All the sensory set (pressure sensors and accelerometers in the feet, gyroscope and accelerometer in the abdominal area - see figure 1) requires the implementation of a sensor fusion. This sensor fusion will be implemented using models based on Kalman filters.

In order to reduce errors provided by MEMS, heuristic and probabilistic algorithms have to be implemented. They are necessary to make corrections over the provided errors.

These probabilistic algorithms are used to calibrate the system based on GPS. They will work like this, when the user is in environments where the GPS signal is very good, the system can self-learn how the user makes the steps, and the gaps (errors) that exist in those steps. So when the GPS signal is not available, the collected sensor information can be self-corrected by the accuracy algorithms.

As an applicability example for this system, we have a tourist who is on vacation in a city (without the need to have in every building a structured location system) and will visit a museum. With this system, he can navigate inside the museum and get information about what he is seeing at the moment (*e.g.*, paintings, sculptures) without other tools besides his PDA. The system, in an outdoor environment learns the pattern of user's stride, and the deviation in sensor readings for future accurate indoor positioning calculation. This is called the training phase, the system first requires a set of data to determine the parameters to detect the walking behaviors. From the training data, the unit motion recognizer automatically extracts parameters such as threshold values and the mean and standard deviations of the sensor signals. The system takes as its starting point the last obtained coordinate from the GPS.

In the future, extensive user experiments will be necessary in order to improve the validity and reliability of the research. These experiments will, not only, allow us to precisely assess how accurate the obtained position of the user is, but also, most importantly, to obtain valuable data about the differences in the step patterns originated from a diverse set of users. This knowledge can give more insight on what the best practices to predict the step length and direction are, and also help the implementation of the automatic calibration of these values.

Another problem that can be originated from all the complexity of the project is the delay between the real location and the processed one (that appears to the user on the mobile device). The process is complex: it starts on the sensors that gather the values, which are afterwards sent to the central module. This process has already communication delays. When the central module has the sensors values, it must combine them according to the data timestamps to process the sensor fusion and estimate the walking path, which takes some processing time. Afterwards, it the probabilistic algorithms are executed to correct the estimated walking path. This algorithm can run for a large amount of time. The algorithms should be efficient to process the data in a short period of time, in order for the delay to be minimized. The last step of the process is the presentation the user's current location on his mobile device. Concluding, as can be imagined, this process with so many steps can bring a big delay on estimating the real location of the user.

After the algorithm implementation, an experiment will be performed. The experiment will focus on investigating the variables and the way in which these can affect the experimental work. It will be used to verify the previously formulated hypothesis. Some experimental walking paths will be created to test the variables of the algorithms and the position of the sensors to see if minor changes can affect the system positively or not.

At least three scenarios, represented on figure 2, will be created. The first one will be as simply as possible, only a straight line with 10 meters, to see if the system can correctly measure the distance (Heavy Dash line on figure 2); The second case will involve straight lines combined with changes of direction (for example, in a building a person must travel through two or more offices) Dash Double line on figure 2; The third one (Dotted line on figure 2) will combine the use of outdoor location with the proposed system: the person is outside of a building guided by GPS and when he enters the building, the INS must provide the exact location inside the building.

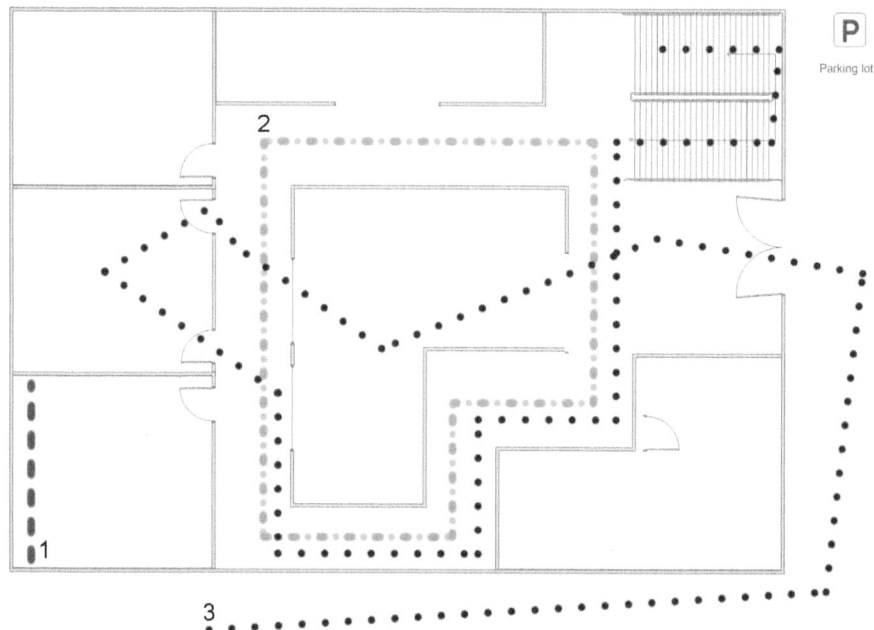

Fig. 2. Experience scenarios - Heavy Dash line (1) represents the first scenario; Dash Double line (2) represents the second scenario; Dotted line (3) represents the third scenario

4   Conclusions

Retrieving location using an inertial navigation system still remains an open research problem, since there isn't still an approach with a good reliability to use in conjunction with GPS, to retrieve location when GPS is not available. This is a complex problem because of the sensors drift and people's different sizes. This is also a problem when selecting the most appropriate spot to put the sensors.

To tackle these challenges, we propose a system based on probabilistic learning algorithms. The person's walking behavior to predict more effectively the walking path. The project objectives were specified along with the research methods that will be used.

Three quantifiable success criteria were identified to give to the project a success degree. The first one is the accuracy of the estimated location that must be between 90% and 95% or, in other words, for each 100 meters traveled the system must have an error of only 5 (to 10) meters. Also, the accumulated error must be kept to the minimum possible. The second criterion, is the delay between the readings and presentation to the user of his current location, that must be less than 2 seconds. For higher delay values, positioning will not be considered real-time. The third criterion focuses in the intrusion that the system may have to the people's day-to-day life. This is difficult to measure and not totally quantifiable. However, it is intended for the sensors to be totally wearable and imperceptible to the user. The system should be as easy to use in a point that the user forgets that he's wearing the sensors.

In order to test this system in a real environment with real people, integration with the PSiS system is envisaged, to support tourists on their visits.

References

1. Anacleto, R., Luz, N., Figueiredo, L.: Personalized sightseeing tours support using mobile devices. IFIP - World Computer Congress 2010, 301–304 (2010)
2. Anacleto, R., Luz, N., Figueiredo, L.: PSiS mobile. In: International Conference on Wireless Networks (ICWN 2010), Las Vegas, USA (2010)
3. Bahl, P., Padmanabhan, V.N.: RADAR: an in-building RF-based user location and tracking system. In: Proceedings of IEEE Nineteenth Annual Joint Conference of the IEEE Computer and Communications Societies, INFOCOM 2000, vol. 2, pp. 775–784 (2000)
4. Beauregard, S., Haas, H.: Pedestrian dead reckoning: A basis for personal positioning. In: Proceedings of the 3rd Workshop on Positioning, Navigation and Communication (WPNC 2006), pp. 27–35 (2006)
5. Bieber, G., Giersich, M.: Personal mobile navigation systems-design considerations and experiences. Computers & Graphics 25(4), 563–570 (2001)
6. Castaneda, N., Lamy-Perbal, S.: An improved shoe-mounted inertial navigation system. In: 2010 International Conference on Indoor Positioning and Indoor Navigation, IPIN, pp. 1–6 (2010)
7. Elwell, J.: Inertial navigation for the urban warrior. In: Proceedings of SPIE, vol. 3709, p. 196 (1999)
8. Evennou, F., Marx, F.: Advanced integration of WiFi and inertial navigation systems for indoor mobile positioning. Eurasip Journal on Applied Signal Processing, 164 (2006)
9. Fang, L., Antsaklis, P.J., Montestruque, L.A., McMickell, M.B., Lemmon, M., Sun, Y., Fang, H., Koutroulis, I., Haenggi, M., Xie, M., et al.: Design of a wireless assisted pedestrian dead reckoning system - the NavMote experience. IEEE Transactions on Instrumentation and Measurement 54(6), 2342–2358 (2005)
10. Hashimoto, H., Magatani, K., Yanashima, K.: The development of the navigation system for visually impaired persons. In: Proceedings of the 23rd Annual International Conference of the IEEE Engineering in Medicine and Biology Society, vol. 2, pp. 1481–1483 (2001)
11. Hightower, J., Borriello, G.: Location systems for ubiquitous computing. IEEE Computer 34(8), 57–66 (2001)
12. Kelly, D., McLoone, S., Dishongh, T.: A bluetooth-based minimum infrastructure home localisation system. In: Wireless Communication Systems, ISWCS 2008, pp. 638–642 (2008)
13. Kinney, P., et al.: Zigbee technology: Wireless control that simply works. In: Communications Design Conference, vol. 2 (2003)
14. Klingbeil, L., Romanovas, M., Schneider, P., Traechtler, M., Manoli, Y.: A modular and mobile system for indoor localization. In: 2010 International Conference on Indoor Positioning and Indoor Navigation, IPIN, pp. 1–10 (2010)

15. Kourogi, M., Ishikawa, T., Kameda, Y., Ishikawa, J., Aoki, K., Kurata, T.: Pedestrian dead reckoning and its applications. In: Proceedings of "Let's Go Out" Workshop in Conjunction with ISMAR, vol. 9 (2009)
16. Lukianto, C., Honniger, C., Sternberg, H.: Pedestrian smartphone-based indoor navigation using ultra portable sensory equipment. In: 2010 International Conference on Indoor Positioning and Indoor Navigation, IPIN, pp. 1–5 (2010)
17. Luz, N., Anacleto, R., Almeida, A.: Tourism mobile and recommendation systems - a state of the art. In: International Conference on e-Learning, e-Business, Enterprise Information Systems, and e-Government, Las Vegas, USA (2010)
18. Maskell, S., Gordon, N.: A tutorial on particle filters for on-line nonlinear/non-Gaussian bayesian tracking. In: Target Tracking: Algorithms and Applications, IEEE, pp. 1–2 (2002)
19. Murray, J.: Wearable computers in battle: recent advances in the land warrior system. In: The Fourth International Symposium on Wearable Computers, pp. 169–170 (2000)
20. Renaudin, V., Yalak, O., Tomé, P., Merminod, B.: Indoor navigation of emergency agents. European Journal of Navigation 5(3), 36–45 (2007)
21. Sorenson, H.W.: Kalman filtering: theory and application. IEEE (1985)
22. Tesoriero, R., Gallud, J., Lozano, M., Penichet, V.: Using active and passive RFID technology to support indoor location-aware systems. IEEE Transactions on Consumer Electronics 54(2), 578 (2008)
23. Thong, Y.K., Woolfson, M.S., Crowe, J.A., Hayes-Gill, B.R., Challis, R.E.: Dependence of inertial measurements of distance on accelerometer noise. Measurement Science and Technology 13, 1163 (2002)
24. Walder, U., Wießflecker, T., Bernoulli, T.: An indoor positioning system for improved action force command and disaster management. In: Proceedings of the 6th International ISCRAM Conference, Gothenburg, Sweden (2009)
25. Want, R., Hopper, A., Falcao, V., Gibbons, J.: The active badge location system. ACM Transactions on Information Systems (TOIS) 10(1), 91–102 (1992)
26. Woodman, O., Harle, R.: Pedestrian localisation for indoor environments. In: Proceedings of the 10th International Conference on Ubiquitous Computing, pp. 114–123 (2008)
27. Yue-yang, B.E.N., Feng, S.U.N., Wei, G.A.O., Ming-hui, C.: Study of zero velocity update for inertial navigation. Journal of System Simulation 17 (2008)

Modeling Context-Awareness in Agents for Ambient Intelligence: An Aspect-Oriented Approach*

Inmaculada Ayala, Mercedes Amor Pinilla, and Lidia Fuentes

E.T.S.I. Informática, Universidad de Málaga
{ayala,pinilla,lff}@lcc.uma.es

Abstract. Ambient Intelligence (AmI) systems are inherently context aware, since they should be able to react to, adapt to and even anticipate user actions or events occurring in the environment in a manner consistent with the current context. Software agents and especially the BDI architecture are considered to be a promising approach to deal with AmI systems development. However current agent models do not offer a proper support for developing AmI systems because they do not offer support to model explicitly the interaction between the agent, context sources and effectors, and the context-awareness features are scattered in the system model. To solve these problems we propose an aspect-oriented agent metamodel for AmI systems, which encourages modularity in the description of context-aware features in AmI systems. This metamodel achieves better results than other metamodels in separation of concerns, size, coupling and cohesion.

Keywords: AmI, agents, metamodel, context-awareness, aspect-modeling.

1 Introduction

Ambient Intelligence (AmI) environments represent a new generation of computing systems equipped with devices with special capabilities that make people aware of the environment and react to it, in a more natural way [19]. AmI systems are inherently *context aware*, since AmI devices should be able to react, adapt and even anticipate user actions or events occurring in the environment in a manner consistent with the current context. Software agents are considered to be a promising approach to deal with AmI systems development [12,17] because of their capacity to naturally adapt themselves to changing circumstances. These features together with social behaviour, autonomy, pro-activity and reactivity facilitate the modelling of context-aware behaviour.

The BDI (from Belief-Intention-Desire) agent model [9] integrates an internal representation of the context and a reasoning engine able to process context changes, and generate and/or execute plans in order to satisfy agent goals according to current context. However, BDI agent metamodels do not offer a proper support for developing other special features and requirements of AmI systems. For example,

---

* This work has been supported by the Spanish Ministry Project RAP TIN2008-01942 and the regional project FamWare P09-TIC-5231.

L. Antunes and H.S. Pinto (Eds.): EPIA 2011, LNAI 7026, pp. 29–43, 2011.

there are particular components present in AmI systems, such as sensors and effectors and relations, such as how the contextual information flows between the components of the system, that cannot be specified or modelled explicitly, being impossible to model, for example, how the capture of the context data is performed. Then, our goal is to enrich software agent models, to naturally express concepts related with AmI systems, focusing on context awareness related concerns.

One of the most widely accepted definitions of context was proposed by Dey and Abowd in [8]. They define context as (sic.) *"any information that can be used to characterize the situation of an entity. An entity is a person, place or object that is considered relevant to the interaction between a user and an application, including the user an application themselves"*. This definition explicitly states the necessity of representing all kinds of contextual data for a given application domain, like AmI systems. If a piece of information can be used to characterise the current state of an entity, then that information must be modelled as part of the context.

Due to the recent popularity of AmI systems, there is an increasing interest in modeling and characterising context awareness [12] and context aware applications [7,8]. ContextUML approach [7] is the most relevant one, the others being proposed mechanisms as special cases or variations of it. So, our purpose is to add concepts expressed mainly in ContextUML to our software agent model for AmI systems, as first order entities. Therefore, we will complement BDI agents with concepts present in other well-known context models, but also with other concerns typical of AmI systems.

At design level, our goal is to represent all these new concerns in a modular way, in order to facilitate their design separately. In current BDI agent models context-related data and functions have to be scattered among diverse elements of the agent. Then, it is not possible to reason how a certain context variable (e.g. location, time) affects agent behaviour. In addition, the separation of AmI-specific concerns improves the design, understanding and evolution of agent models in AmI systems.

Aspect-Oriented Software Development (AOSD[1]) promotes the separation of concerns, by defining a new first-order entity called *aspect*. We will use aspects to model context awareness, which we have identified as a *crosscutting concern*, separate to other base entities of the agent. In particular, we propose an aspect-oriented agent metamodel for AmI systems, which encourages modular high level descriptions of such complex systems. Other approaches have also applied AOSD to improve the internal modularisation of agent models [18], but none of them is focused on context-awareness property or AmI systems. In addition, there are other concerns typical of distributed systems that can benefit from the use of aspects. For example, fault tolerance, which is a very important property in AmI systems, since sensors or network connections normally fail during system execution, and can be modelled in terms of aspects. By applying aspect-orientation the agent modeler has the possibility to modularise any property that normally crosscuts several agent entities using aspect-oriented mechanisms.

Our approach benefits from aspect orientation in that: (1) separating crosscutting concerns in aspects contributes to the modularisation, reconfiguration and adaptability of the proposed metamodel; (2) context-awareness related concerns are explicitly

---

[1] Aspect Oriented Software Development (AOSD), `http://aosd.net`

modeled, making possible their use by other agent concerns (e.g. security, fault-tolerance); (3) it is possible to reason about the global behaviour of the system, considering the combination of concerns.

The content of this paper is organised as follows: In section 2, we provide the background to our work. Section 3 presents our main contribution, the metamodel for context-aware agents. Then we describe a context-aware recommender system as an example (Section 4). Section 5 provides an empirical and quantitative evaluation of our metamodel with other well-known generic agent metamodel and a UML profile for aspect oriented modeling (AOM for short). The paper concludes with a brief summary and comments on future work.

2   Background

2.1   Requirements of Ambient Intelligence System

An AmI system refers to a ubiquitous electronic environment that helps people in their daily tasks, in a proactive, but invisible and non-intrusive manner. The development of AmI systems is a very complex task that poses new requirements, some of which have not been addressed adequately. Specifically, for the modeling phase we can find the following challenges: (1) *embedded software functions and data* for a diversity of small devices with different capacities and resources (i.e. sensors and actuators) have to be modelled explicitly; (2) *environmental awareness*, which means that it is necessary to model the reaction of AmI devices to user actions or events occurring in the environment in a manner consistent with the current context; (3) *personalisation*, which means that it is necessary to model the customisation of AmI systems taking into consideration user preferences, and even anticipate users' wishes without conscious interaction; (4) *adaptiveness* to both user actions and changes produced in the environment, including those not explicitly considered in the system modeling; (5) *mobility*, which means both hardware (e.g. sensors) and software can be moved to a different location.

In addition, the development of an AmI system development implies the consideration of other non functional concerns typical of distributed systems, such as security, fault tolerance and usability [13]. Another important issue for an AmI system is its performance. Despite the fact that modeling is not directly related to this issue, the design of a system can influence the final performance of the implemented system. For example, the execution time of a system can be negatively affected by a design presenting a great number of indirections or a poor modularisation. However, it is also true two systems with the same design can have different performances because the election of the implementation of their internal components.

2.2   Context Modeling in BDI Agents

A BDI agent is composed by three basic elements that may be explicitly represented in its internal architecture: (i) *beliefs*, that is, an internal representation of the world (including itself, other agents and the environment).; (ii) *desires* or goals represent objectives or situations that the agent would like to achieve; and (iii) *intentions* represent desires that the agent has chosen to fulfil by performing a set of running actions (these sets of actions are usually organised in plans).

According to these definitions, the context property is modelled in different ways in BDI approaches, for example AgentSpeak [21] uses a first order logic to model context features, however this method has the disadvantage that sometimes it is difficult to express conditions in complex systems. Therefore, Jack [10] uses relational models for this purpose and Jadex [4] uses objects. Additionally, agent metamodels for BDI agents provide elements to model the context, but usually they are modeled as a set of facts about the environment and do not show dependencies of the context with application specific functionality. This is not the case of PASSI [22], that associates the environment with the resources of an organisation of agents or FAML [3], which uses an element named Facet to model actions over the context, but this metamodel does not link these actions with the context.

2.3 Aspect Oriented Modeling

As stated before, we will follow an aspect-oriented approach to model non functional concerns present in AmI environments. Then, we will apply AOM concepts in the agent metamodel. AOM approaches try to translate aspect oriented programming concepts to an earlier stage of the software development process. An AOM approach comprises of a base model and a set of mechanisms that allow the appropriate encapsulation and composition of crosscutting concerns. These mechanisms usually contain the following elements: (a) *aspects* and *advices,* to encapsulate and represent the functionality of the crosscutting concerns; (b) *joint points* to model points in system execution in which aspects are weaved with basic system functionality; and (c) *pointcuts* to specify how to compose aspects at *joint points.*

As explained before, AmI systems include numerous properties that crosscut system functionality. Moreover, modeling these concerns is not a simple task because there are dependencies and interactions between them. Aspect orientation can be used to simplify the development of AmI systems. Using an AOM approach we can model the basic functionality of our system and encapsulate its crosscutting concern as aspects, later specifying how they are composed with the base model.

2.4 Related Work

In recent years, different approaches have appeared to cope with the design of AmI systems and context-awareness. There are generic approaches, such as PervML[2], CMP [15] or CONON [14]. PervML is a domain specific language for the development of pervasive systems that provides a set of conceptual primitives that allow the description of the system irrespective of the technology. This approach offers tools for supporting the modelling of pervasive systems, code generation or an execution framework and drivers used to integrate devices. Moreover, CONON is OWL encoded context ontology for modelling context in pervasive computing environments, and for supporting logic based context reasoning.

---

[2] PervML: Pervasive System Development. http://www.pros.upv.es/labs/

Other context aware modelling approaches are focused on a specific application domain. For example, ContextUML [7] is a modeling language for the model-driven development of context-aware web services based on the Unified Modeling Language. Additionally, a modified version of this metamodel [16] is used to compose context aware web services with aspects.

There exists numerous metamodels for modelling MAS but none of them is focused on the development of AmI. However, they can be easily used because of the suitability of this paradigm to develop open intelligent distributed systems like AmI environments. Most agent metamodels are focused on a specific facet of MAS and only a few of them can be used to design generic MAS. PIM4Agents [2] is a generic metamodel for MAS that allows modelling BDI and reactive agents, has an IDE and two MDD process to generate code for Jack and JADE agent platforms. FAML is a metamodel for BDI agents that defines 4 viewpoints to design MAS (internal design, external design, internal runtime and external runtime), it has concepts from popular agent methodologies such as Tropos, Ingenias or Adelfe. The later is a methodology for open, dynamic and distributed MAS that also contains a metamodel to design this kind of system. There exists other metamodels that use aspects to enhance the modularity of the agent architecture, like the work presented in [18].

3   An Aspect-Oriented Metamodel for Context-Aware Agents

In order to propose a metamodel for context-aware Multi-Agent Systems (MAS for short), we have studied several agent metamodels, such as PIM4Agents or FAML, methodologies for contextual modeling, e.g. ContextUML, and also agent platforms and architectures, such as Jadex, Malaca [5] or the work presented in [6]. So, the metamodel presented in this paper has concepts from these three domains that will be explained in the proof in this section. The description of the metamodel (*MASDesignDescription*) has been divided in three parts (Fig. 1): (i) agent and MAS metamodel (*MultiAgentSystem*); (ii) *aspect* metamodel (*Aspect*) and (iii) the *pointcut* metamodel (*Pointcut*) that models the composition of the agent behaviour with the aspects. In addition, there are elements to bind aspect with agents (*AspectBinding*).

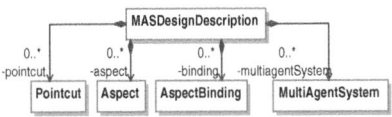

Fig. 1. Main building blocks of our metamodel for MAS design in a UML class diagram

3.1   A Metamodel for Context-Aware Agents

The root concept in the metamodel is *MultiAgentSystem* (Fig. 2), which contains the MAS main blocks: agents, behaviours, capabilities, organisations and roles. The agent model is influenced by the BDI agent model but incorporates elements to represent AmI concepts.

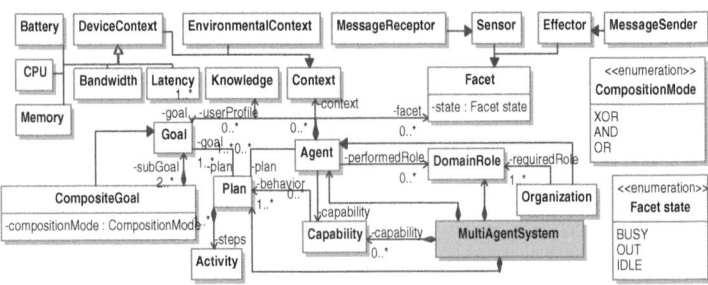

Fig. 2. A partial view of agent metamodel in a UML class diagram

An *Agent* (in the centre of the figure) is composed of a set of *Context* elements, *Goal* elements, *Knowledge* and *Plan* elements. A *Context* element represents information about the environment. There exists different kinds of context that can be modeled: *DeviceContext* and its specialisations refer to device features in which an agent is running (e.g. battery, memory...) and *EnvironmentalContext* to features of the real world (e.g. location, time...). In addition, it is possible to model any kind of context by instantiating *Context* metaclass. *Knowledge* represents information about user preferences allowing the encapsulation of the user profile. The components that sense and manipulate the physical environment are modelled by the *Sensor* and *Effector* elements, which are subclasses of the *Facet* element. In addition, these subclasses have specialisations that enable message sending (*MessageSender*) and reception (*MessageReceptor*).

The internal information of the agent (goal, context, knowledge and plans) is manipulated by means of *Plan* and *Aspect* elements. *Plan* elements are composed by a set of ordered *Activity* elements. There exists different specialisations of the *Activity* elements, but due to space limitations will not be explained, but they offer functionality to handle agent *Knowledge* (*userProfile*), goals (add or remove), plans (execute) and context (update and get). Activities for context model a mechanism named *context-binding*, which is used to model the usage of context information and is present in the ContextUML metamodel.

Moreover, there are special kinds of *Activity* elements that allow the usage of methods from *Facet Effector* elements (facet task), e.g. message sending using the *MessageSender* element or to print information on a screen. *Facet Sensor* elements cannot be used from *Plans* because for our metamodel, they are composed with autonomous behaviour using aspect-oriented weaving mechanisms that will be explained in the next section. *StructuredActivity* is a specialisation of *Activity* that represents control structure such as loops or decisions. Likewise, *Plan* elements can be clustered in *Capability* elements that normally provide functionality to *MultiagentSystem* elements.

As has already been discussed, goals or *desires* in BDI architectures represent information about the objectives to be accomplished [9]. Although goals are not always explicitly represented in BDI agent platforms or metamodels [10], in our approach they are modelled in order to make it possible to think about goals and, from a software engineering point of view, to enable the traceability of goals from higher-level specifications to implementation frameworks. A *Goal* in our metamodel has a name, a set of sub-goals that can be composed with different operations (*CompositionMode*) and a set of *Plan* elements that can be executed to fulfil this goal.

The *Organization* is taken from the PIM4Agents metamodel and has the same characteristics of an *Agent*, i.e. it can perform *DomainRoles* and have *Capabilities* which can be performed by its members. Moreover, *Organization* elements can act in an autonomous manner and provide advantages to the AmI system such as the possibility of ad-hoc organisation (discovery of other devices and the interaction with them) [20] or collaboration to fulfil goals.

With the metamodel presented we have modelled some of the elements to fulfil requirements for the AmI system presented in section 2.1. Requirement (1) is fulfilled by means of *Facet* elements and the base of requirement (2) is set by means of *Context* element. The remainder of the requirements are satisfied using aspects presented in the following sections.

3.2   A Metamodel for Aspect Modelling

In a similar fashion to [1], the aspect features of our metamodel (Fig. 3) has concepts for (a) definition of modelling aspects and their associated elements, such as advices, (b) modelling join points and (c) describing the aspect composition using pointcuts.

We model an aspect as a class with special operations named advices. Advices differ from common operations in that they are never invoked explicitly and they are executed by the aspect-oriented weaver without the knowledge of the base class designer around a join point. In addition, Aspect (in grey in Fig. 3) has associated to it a set of components that provide the functionality required by the aspect. The aspects FunctionalQuality and its specialisations, ContextAwareness, Consistency and Coordination are inherited from Aspect and will be explained later.

A BDI agent contains a set of properties that crosscut several elements of the agent model. Since these properties affect several internal elements of an agent model such as *Context, Goal* or *Plan* elements, they must be properly modularised to ensure the system's maintenance and evolution. These properties are modularised by means of aspects, i.e. separately from other agent concerns. We model as aspects, concerns related to coordination, context awareness, consistency and functional quality attributes (QA).

The first concern we model separately is *Coordination* because it is widely accepted that it is a concern usually intertwined with domain-specific functionality [5] and that it interacts with other concerns, such as *ContextAwareness. Coordination* aspect models agent coordination protocols. An agent has associated with it a set of coordination aspects that represent protocol roles which the agent can carry out in different interactions with the environment, including with other agents. This aspect models role behaviour as a finite state machine (FSM) whose transitions are driven by messages. In order to model the FSM, the aspect keeps a set of message templates and plans that must be executed when a message that follows the template is received (*Transition* elements in Fig. 3). To decide what plan will be executed, it has an advice named *answer*.

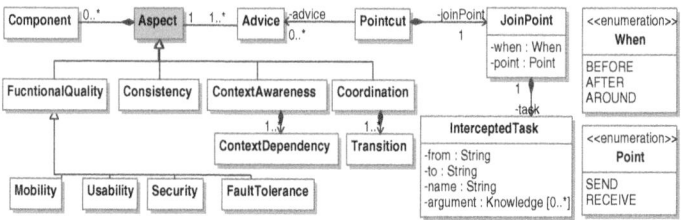

Fig. 3. Partial view of aspect metamodel in UML

Context awareness is a central concern in BDI agents because it depends on agent beliefs (*Context* element in our metamodel) and effects plan execution. The *ContextAwareness* aspect models agent context dependant behaviour. This aspect is composed of the *execute* advice and a set of context dependencies that relates plans with the list of context conditions that should be held in order for the plan to be executed (*ContextDependency* in Fig. 3). This mechanism implements *context-triggering*, which models context triggered actions as in ContextUML, being also very similar to plan preconditions. Note that we achieve the decoupling of the plan from the context condition that triggers its execution.

Consistency has been separated as an aspect because we identified it as a non functional property present in BDI architectures. Specifically we refer to the consistency of the internal agent information (goals and beliefs). In addition, this concern depends on the agent behaviour and application specific functionality. *Consistency* achieves its goals using two advices named *goalRevision* and *contextRevision*. When a plan is executed this aspect removes all goals that have been accomplished using an advice named *goalRevision*. Additionally, when a *Sensor* gets new information about the context, this aspect updates the specific *Context*.

Finally, *FunctionalQuality* models functional quality attributes, typical of distributed systems, which normally crosscut several elements of the agent model. *FunctionalQuality* has many specialisations that may also have dependencies with the context-related concerns of the agent metamodel. For example, the fault tolerance QA may handle failures differently depending on the contextual information. Also, usability is strongly influenced by the user profile modeled as part of the context.

Finally, in our metamodel, we define a non-invasive join point model in the sense that it only allows agent designers to intercept observable behaviour, i.e. actions performed in agent plans (e.g., invoke a web service) or facet actions (e.g., reception of sensor data). In our metamodel a *Joinpoint* is represented by: (1) the intercepted task that has a *name*, a set or arguments, the class name that calls the method and also of that which receives the call (*from* and *to* attributes respectively); (2) *point,* which indicates whether the interception point is either the sending (*SEND*) or the reception (*RECEIVE*) of the method; and (3) *when*, which specifies when an advice is executed related to the join point (*BEFORE, AFTER, AROUND*). Wildcards are available in class and method names: "*" represent any sequence of characters and "..." any sequence of arguments.

3.3 *Pointcut* Modelling

Finally, to complete the aspect-oriented part of our metamodel, we need concepts to construct the pointcuts that specify how to compose the crosscutting concerns modelled as aspects. A pointcut expression is a pattern that matches several join points and associates them with one or more aspect advices. In addition, a pointcut may express some constraints (e.g. the intercepted join point has to be in a specific execution flow) that must be satisfied in order to execute the associated advices. At modeling level, the common practice for specifying pointcuts is to use UML sequence diagram with wildcards [1].

According to our metamodel, a *Pointcut* has associated with it an ordered collection of advices, which will be executed in the order specified on the join points selected by the pointcuts. Using the concepts presented in the metamodel, a set of aspects and pointcuts have been modelled to enable BDI and context-aware behaviour to the agent. In order to add context-awareness to the agent a specific *Pointcut* element is modelled (Fig. 4 left side). This specifies that the *execute* advice from *ContextAwareness* aspect, must be executed after the execution of *updateContext(...)* agent internal action.

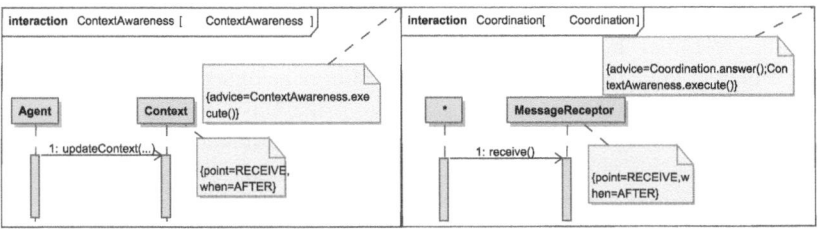

Fig. 4. Pointcuts for coordination and context-awareness in UML sequence diagram

In the pointcut to model message answering (Fig. 4 right side), the *ContextAwareness* aspect is woven with the *Coordination* aspect. When a message is received, i.e. *AFTER* the reception of a message ("*receive(...)*" wildcard combination) from *MessageReceptor Facet*, two advices are sequentially applied: *answer* advice, to decide what plan or plans could be executed and *execute* advice to check preconditions.

Finally, the consistency aspect needs two pointcuts. The pointcut for goal consistency (Fig. 5 right side) specifies that the *goalRevision* advice is applied after an *executePlan()* operation is sent from an unknown source to a plan. The pointcut for context consistency (Fig. 5 left side) specifies that when a *Sensor* gets new information, i.e. sends and *set*(...)*, *contextRevision* advice is applied *AROUND*.

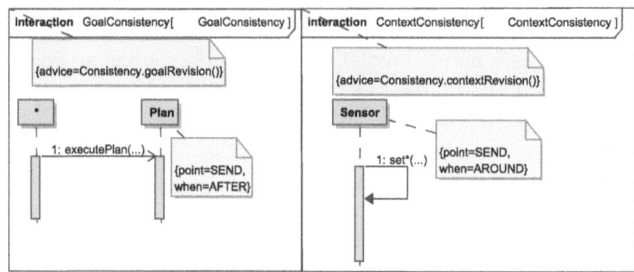

Fig. 5. Pointcuts for consistency in UML sequence diagram

4   Running Example

In order to illustrate the application of our metamodel, let us consider the VANET case study presented in [11]. In this MAS, agents are used to represent and simulate meaningful entities in VANETs. According to this, a MAS for VANETs encompasses two different agents: agents inside a vehicle (*VehicleAgent*), and agents representing services (here, only agents for a gas station (*GasStationAgent*) are considered). Vehicle agents run on board to provide safety and comfort-related services to the vehicle's occupants. These agents are context-aware, since they deal with the reception of events from different information sources (such as the user and internal sensors), and they are able to adapt their behaviour to current context. Gas station agents represent the gas stations in the road network.

Due to space limitations, this paper only focuses on the design of the *VehicleAgent*. This agent presents a context-aware behaviour which decides when and where the vehicle has to refuel. In a specific scenario, when the agent detects that the level of fuel is running low, it decides to refuel. Additional context information, such as vehicle speed, is taken into consideration to make this decision. Vehicle agents interact with service agents in gas stations to receive information about gas stations in the proximity. When refuelling is needed, autonomously, the agent chooses a gas station that meets user preferences (i.e. a specific gas station chain).

VehicleAgent (in dark grey in the centre of the Fig. 6) is composed by a set of internal elements, that model agent knowledge and functionality, and a set of aspects which model interaction between these elements. So, the agent stores information on its goals, user profile, when (petrol tank level) and where (specific gas station chain) the user prefers to refuel, a representation of the relevant contextual information and domain-specific functionality to deal with context.

In our example, the *VehicleAgent* has two goals: *Manage offers* and *Ensure tank level is never lower than a threshold*. Since, our agent offers a context-aware recommendation about refuelling; its context is composed of information about internal sensors of the vehicle (location, petrol tank and speed) and the nearest gas station (*GasStationIsNear*). Moreover, *VehicleAgent* has application specific functionality for dealing with the context and other agents. The former is provided by a set of *Sensor* elements, which represent components that provide context data, i.e. *GPS*, *Velocimeter* and *PetrolSensor*, and the later is provided by *JADE*

MessageReceptor element, which models interface with a Jade agent platform. In addition, the model contains a *Screen* to represent the screen of the device in which the agent is running.

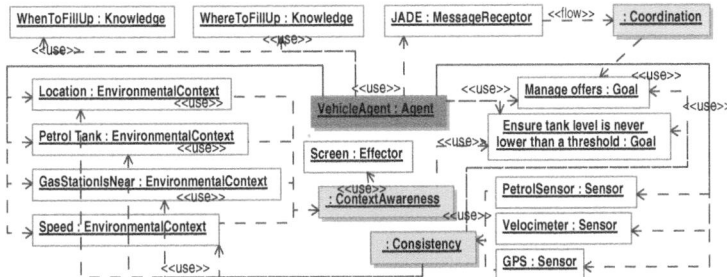

Fig. 6. Implementation diagram in UML of *VehicleAgent*

According to our metamodel we define a set of aspects. *Coordination* aspect intercepts messages and processes information about offers that fulfil the *Manage offers* goal. When something in context changes (location, petrol tank level...), the *ContextAware* aspect executes plans that act over *Screen* and fulfil *Ensure tank level is never lower than a threshold* goal. Finally, the *Consistency* aspect intercepts changes in *Sensor* to update context information and when a *Goal* is accomplished, it removes from the agent, internal information. Finally, *Pointcut* elements are used to specify how to compose the aspects for *VehicleAgent*.

5 Validation

In this section the metamodel presented is evaluated using the VANET case study. To validate the correctness of our approach, we compare our metamodel with a metamodel for MAS modelling, PIM4Agents and with a UML profile for aspect modeling [1]. This comparison is useful to compare the architectural metrics of two pure approaches in isolation (for agents and aspects) and the improvement obtained with their combination. This validation process is done only for the design obtained in the modeling phase. The validation of the implementation of the system is out of the scope of this paper. Additionally, the evaluated system designs (in the three modeling approaches) are provided with a similar level of detail and are complete and correct with the system specification.

Our initial hypothesis was that current agent metamodels do not allow any agent property (such as context awareness) to evolve independently of the agent domain-specific functionality. The reason for this is that such agent properties (crosscutting concerns) are not well modularised since they are tangled and scattered across the different views of the metamodel. Therefore, what we have to show is that our metamodel achieves a better separation of concerns. In order to evaluate our proposal we apply a modified version of the metric suite [6], which is being widely used to compare aspect-oriented approaches as opposite to non-aspect-oriented ones. Some concepts must be changed to get valuable information at the modeling stage.

Separation of concerns (SoC) metrics are applied to calculate the degree to which a single concern or property maps the architectural components. For the modeling phase, we quantify the number of plans (this concept is presented in both agent oriented metamodels), aspects or metamodel specific elements to describe the concern (e.g. in PIM4Agents there exists a specific element to design the coordination protocol) (CDC) and the number of elements (i.e. metaclass instantiation or elements in the UML diagram) used to model this concern (CDE). In the quantitative evaluation, the data collected for them shows favourable results for the AO and agent version for the SoC metrics. Table 1 shows that PIM4Agents (P) requires 5 elements to address coordination and context-awareness concern. Moreover, the aspect UML profile (U) and our metamodel (X) only require one component (aspect) for each concern. For CDE metric, the results for our metamodel are also better than PIM4Agents and the UML profile. In Table 1, we can see the percentage difference of each metric between our metamodel and the others (row labelled as "Percentage difference" above P and A). The difference between our approach and the UML profile is due to this profile is for general purpose and requires more entities to model the same concern as our approach (see [1] for more details).

Table 1. Separation of concerns metrics

Concern	CDC			CDE		
	X	P	U	X	P	U
Coordination	1	5	1	18	41	30
Context-awareness	1	5	1	22	96	19
Average	1	5	1	20	68.5	24.5
Percentage difference		80%	0%		70.8%	20%

The model size physically measures the extension of a MAS's design and consequently, the effort to accomplish this task. Number of lines of code is the traditional measure of size, but it could be biased with different programming styles and does not reflect developing effort when we use graphical tools as UML or EMF. We have overcome this problem measuring the number of drawn elements (NDE) in the diagrams and to weight the results, we apply the used vocabulary size (UVS) metric, which count the number of different elements used to model the case study. Table 2 shows the result of the application of these metrics and we can see that the results are favourable for our metamodel. To design the MAS we need 114 NDE, this is less than the UML profile, which has the lowest UVS and PIM4Agents that also have a lower vocabulary but a similar detail level to model MAS. We can state that these two metamodels (our metamodel and PIM4Agents) have a similar level of detail because PIM4Agents and our metamodel shares concepts (plans, organisations, MAS, ...) and in addition, they provide the necessary information to generate code for agent platforms (Jade, Malaca and Jack for PIM4Agents and Malaca for our metamodel) that does not happen for the UML profile. Additionally, PIM4Agents and the UML profile do not have specialised constructions to model AmI concepts and we need to use more elements than in our approach to model the same concept. Moreover, because we are modelling a MAS in PIM4Agents and in our metamodel, we need to model more issues (e.g. organisations, roles, etc.) than in the UML profile, which is for general purposes. Therefore, we can conclude that to model the system using our metamodel requires less effort than in other metamodels with a similar level of detail.

Table 2. Size metrics

Metamodel	NDE	UVS
Our metamodel	114	96
PIM4Agents	154	79
UML profile for AOM	153	19
Percentage difference	-25.7%	67.5%

Now we will analyse the results of the coupling and cohesion metrics (Table 3). Low coupling is often a sign of a good design, and when combined with high cohesion, supports the general goals of high readability and maintainability of a system. Coupling is evaluated using the Fan-in and Fan-out metrics for our metamodel and PIM4Agents. Due to limitations of space, we discard the UML profile for AOM because it returns very similar results to our metamodel. These metrics count the number of concern specific metamodel elements to implement a specific concern which require services from the assessed element (fan-in metric) and the number of elements from which the assessed element requires services (fan-out metric). Given that we use the same case study, the two agent models contain similar elements or components for representing similar concepts. Hence, the classes considered for measuring the coupling and cohesion are those representing the plans of the agent for vehicle, coordination and organisation. We also considered in the evaluation the *Facet* element since it includes communication functions (with the environment and other agents) addressed by different elements and aspects in our model.

Table 3 shows the average for coupling and cohesion measurement per component in both models (rows labelled "Average"). The metrics demonstrate that there is a coupling reduction in the AO agent model. This happens because in (i) in the AO agent model there does not exist explicit references to aspects (they are bound in the *AspectBinding* element), (ii) domain specific functionality is well defined and can be reused by different aspects, this does not happen in PIM4Agent in which domain specific functionality are black boxes (*InternalTask* element) and (iii) organisation concern is better modularised because it is not related (there does not exist an explicit reference) with coordination concern unlike the case of PIM4Agents. Additionally, the percentage difference is 51.38% and 62.5% in favour of our metamodel.

Cohesion is measured using the Lack of Concern-based Cohesion (LCC). This metric counts the number of different agent properties addressed by each plan (in PIM4Agents), elements and aspects being considered (in our metamodel). The values gathered by this metric demonstrate that the cohesion of the components and aspects in the AO model is much better (1.33 vs. 0.5) than the plans of PIM4Agents model. This superiority is justified by the fact that the behavioural classes (plans) and the agent organisation and coordination elements of the non-AO version need to maintain explicit references with the two concerns considered. In our metamodel the LCC of the functional components is always zero since our design goal was precisely aimed at separating these properties from application specific functionality. So, for more complex agents with several functional elements, this value will remain zero, the minimum expression. Obviously, for each of the aspects considered, the LCC value is one. So the average of the cohesion metrics in our metamodel is 0.6 (less than one), which means that our model elements are very cohesive (percentage difference is 62.4% in favour of our metamodel).

Table 3. Coupling and cohesion metrics for PIM4Agents and Our metamodel

PIM4Agents	Fan in	Fan out	LCC
Organisational elements	1	3	1
Coordination elements	3	1	1.5
Plans	1.4	3.2	1.5
Average	1.8	2.4	1.33
Our Metamodel	**Fan in**	**Fan out**	**LCC**
Organisational elements	1	0	0
Coordination aspect	1	1	1
Context-aware aspect	1	2	1
Facet	0.5	0.6	0
Average	0.875	0.9	0.5
Percentage difference	51.38%	62.5%	62.4%

6 Conclusions and Future Work

In this paper, we have presented an aspect-oriented agent metamodel for AmI systems focusing on the design of context-awareness. On the one hand, the agents are a good way to model AmI system because their elements can be naturally designed as meaningful entities. On the other hand, the aspect orientation improves the modularisation of the crosscutting concerns that are both in agents and AmI systems. The metamodel has been evaluated using a suite of architectural metrics and we can conclude that using our metamodel: (1) the separation of concerns of the designed system is improved; (2) to model an AmI system requires less effort (NDE and UVS metrics) than in a generic agent metamodel; and (3) the design of the AmI system manifests a better coupling and cohesion, which facilitates the evolution and maintenance.

As for future work, we are developing a visual IDE for our metamodel, new examples to check the properties of our metamodel and a transformation process from our metamodel to specific agent platform such as Malaca.

References

1. Fuentes, L., Sánchez, P.: Designing and weaving aspect-oriented executable UML models. Journal of Object Technology 6(7), 109–136 (2007)
2. Hahn, C., Madrigal-Mora, C., Fischer, K.: A platform-independent metamodel for multiagent systems. Autonomous Agents and Multi-Agent Systems 18(2) (2008)

3. Beydoun, G., et al.: FAML: a generic meta-model for MAS development. IEEE Trans. Softw. Eng. 35(6), 841–863 (2009)
4. Pokahr, A., Braubach, L., Lamersdorf, W.: Jadex: A BDI reasoning engine. In: Multi-agent Programming: Languages, Platforms and Applications. Kluwer (2005)
5. Amor, M., Fuentes, L.: Malaca: A component and aspect-oriented agent architecture. Information and Software Technology 51, 1052–1065 (2009)
6. Garcia, A., et al.: Agents in Object-Oriented Software Engineering. Software: Practice and Experience. Elsevier (2004)
7. Sheng, Q.Z., et al.: ContextUML: A UML based modelling language for model-driven development of context-aware web services. In: Proc. of ICMB 2005, pp. 206–212 (2005)
8. Dey, A.K., et al.: Towards a better understanding of context and context-awareness. In: CHI 2000 Workshop on the What, Who, Where, When, and How of Context-Awareness (2000)
9. Rao, A.S., Georgeff, M.P.: BDI agents: from theory to practice. In: Proc. of the First Int. Conf. on Multi-Agent Systems, San Francisco, CA, pp. 312–319 (1995)
10. Howden, N., Ronnquist, R., Hodgson, A., Lucas, A.: JACK Intelligent Agents-Summary of an Agent Infrastructure. In: Proc. 5th ACM Int. Conf. on Autonomous Agents (2001)
11. Amor, M.: Inmaculada Ayala y Lidia Fuentes. A4VANET: context-aware JADE-LEAP agents for VANETS. In: Proc. of 8th PAAMS, pp. 279–284
12. Hong, J., Suh, E., Kim, S.: Context-aware systems: a literature review and classification. Expert Systems With Applications 36(4), 8509–8522 (2009)
13. Bohn, J., et al.: Social, Economic, and Ethical Implications of Ambient Intelligence and Ubiquitous Computing. Journal of Hum Ecol Risk Assess 10(5), 763–786 (2004)
14. Gu,T., et al.: Ontology Based Context Modeling and Reasoning using OWL. In: Proc. of the CNDS 2004, San Diego, CA, USA (January 2004)
15. Simons, C.: CMP: A UML Context Modeling Profile for Mobile Distributed Systems. In: Proceedings of the HICSS 2007, Hawaii, USA, January 3-6 (2007)
16. Prezerakos, et al.: Model-driven composition of context-aware web services using ContextUML and aspects. In: Proceedings of ICWS 2007, pp. 320–329 (2007)
17. Muldoon, C., et al.: Agent Factory Micro Edition: A Framework for Ambient Applications. In: Proceedings of ICCS 2006, Reading, May 28-31 (2006)
18. Silva, C., et al.: Support for Aspectual Modeling to Multiagent System Architecture. In: Proc. of the ICSE Workshop EA 2009, pp. 38–43. IEEE Computer Society, Washington (2009)
19. Cook, D.J., Augusto, J.C., Jakkula, V.R.: Ambient Intelligence: Technologies, applications and opportunities. Pervasive and Mobile Computing (2009)
20. Geihs, K.: Middleware Challenges Ahead. Computer 34(6), 24–31 (2001)
21. Rao, A.S.: AgentSpeak(L): BDI Agents Speak Out in a Logical Computable Language. In: Perram, J., Van de Velde, W. (eds.) MAAMAW 1996. LNCS, vol. 1038, pp. 42–55. Springer, Heidelberg (1996)
22. Cossentino, M.: From Requirements to Code with PASSI Methodology. Agent-Oriented Methodologies. IGI Global, 79–106 (2005)

Developing Dynamic Conflict Resolution Models Based on the Interpretation of Personal Conflict Styles

Davide Carneiro, Marco Gomes, Paulo Novais, and José Neves

Department of Informatics, University of Minho, Braga, Portugal
dcarneiro@di.uminho.pt, pg18373@alunos.uminho.pt,
{pjon,jneves}@di.uminho.pt

Abstract. Conflict resolution is a classic field of Social Science research. However, with conflicts now also emerging in virtual environments, a new field of research has been developing in which Artificial Intelligence and particularly Ambient Intelligence are interesting. As result, the field of Online Dispute Resolution emerged as the use (in part or entirely) of technological tools to solve disputes. In this paper we focus on developing conflict resolution models that are able to adapt strategies in real time according to changes in the personal conflict styles of the parties. To do it we follow a novel approach in which an intelligent environment supports the lifecycle of the conflict resolution model with the provision of important context knowledge. The presented framework is able to react to important changes in the context of interaction, resulting in a conflict resolution approach that is able to perceive the parties and consequently achieve better outcomes.

Keywords: Online Dispute Resolution, Intelligent Environments, Conflict Styles, Profiling.

1 Introduction

The topic of conflict resolution is a quite classic and old one, as old as conflicts themselves. Conflicts are natural and emerge as a consequence of our complex society, in which individuals focus on the maximization of the own gain, sometimes disregarding the other's rights. A conflict can be seen as an opposition of interests or values which, in a certain way, disturbs or blocks an action or a decision making process. Consequently, in order for the action to be carried out, the conflict has to be solved first [1]. The concept of conflict and its resolution has traditionally been addressed by Social Science, although in the last decades Information Science also stepped in. The intersection of these two fields is of great interest as it combines all the established theory about conflict resolution with new methodologies and support tools.

Moreover, we must consider that nowadays most of the conflicts are generated in virtual settings, most of the times supported by an electronic contract. However, very few tools exist to settle conflicts inside their context. As a consequence, conflicting parties have to resort to traditional conflict resolution methods, throwing away significant advantages of the technological environments. The use of technology to develop tools that can support the conflict resolution process, together with the

L. Antunes and H.S. Pinto (Eds.): EPIA 2011, LNAI 7026, pp. 44–58, 2011.

creation of virtual environments for that purpose, is thus of interest. Moreover, Pitt et al. address the issue of the costs of conflicts and the need for alternatives to traditional litigation in court [2]. In particular, the authors argue that litigation is a slow and costly process which may have a special impact on the business of companies and governments. The potential for appeals also adds to the amount of delay and cost.

The work described in this paper is framed in this context. Specifically, after analyzing the current state of the art of conflict resolution platforms, we concluded that most of the processes are static and make no use of context information. In that sense, we are developing a new approach, in line with the concept of Ambient Intelligence. Our aim is to develop conflict resolution methods that make use of context information to adapt strategies in real time, in order to more efficiently achieve satisfactory outcomes. This information may include the conflict style of the parties, the level of escalation, their attitude or even the emotional state. In order to implement a framework able to encompass this kind of information, we are following an approach in line with the concept of Intelligent Environments, in which an intelligent environment supports the conflict resolution platform with context information, as envisioned by [3]. In this work we take into consideration the work of Lewiki et al. [5] and Goldberg et al. [6] on the dynamics and processes of conflict resolution, and the work of Raiffa [7] on decision theory and negotiation analysis.

1.1 Alternative Dispute Resolution

Alternative Dispute Resolution refers to mechanisms that aim to solve disputes without recurring to the traditional judicial process, i.e. litigation in courts. This already traditional approach includes mechanisms such as negotiation, mediation or arbitration. Online Dispute Resolution (ODR) [4], on the other hand, refers to the use of these mechanisms in a technological context, either supported by technology or under a virtual computational environment.

Negotiation [7] is a collaborative and informal process by means of which parties communicate and, without external influence, try to achieve an outcome that can satisfy both. Negotiation is widely used in the most different fields, including legal proceedings, divorces, parental disputes or even hostage situations. From the perspective of Walton and McKersie [8], negotiation can be classified as being distributive or integrative, being integrative negotiation more desirable than distributive. Another collaborative form of conflict resolution is mediation [9]. Here, parties in dispute are guided by a 3rd neutral and independent entity who tries to guide the process to an outcome that may satisfy both disputing parties. In this approach, as in negotiation, parties decide about the outcome instead of it being imposed by the nonaligned one, although using its assistance. The nonaligned is chosen by the parties and has no authority for deciding on the outcome of the dispute but only for guiding and assisting them throughout it. Finally, we can also mention arbitration [10], a method in which the two parties also use the help of a 3rd independent and neutral entity for solving a dispute. However, this entity has no active role on helping the parties throughout the whole process. Instead, the arbitrator simply hears the parties and, based on the facts presented, takes a decision without influencing the parties during their presentations. Traditionally, the outcome of an arbitration process is binding, i.e., there is a final enforceable award that the parties will respect. However, arbitration can also be non-binding.

With the technological evolution new needs appeared in the field of conflict resolution, especially due to the new forms of dispute caused essentially by electronic contracting. New ways to solve disputes are hence appearing, so that the disputant parties neither need to travel nor to meet in courtrooms or in front of arbitrators or mediators. Different forms or methods of alternative dispute resolution for electronic environments have been pointed out by legal doctrine. As a result, we can now speak of Online Dispute Resolution as any method of dispute resolution in which wholly or partially an open or closed network is used as a virtual location to solve a dispute [4].

From a technological point of view, a relevant issue is to determine in what way and to what point traditional mechanisms can be transplanted or adapted to the new telematic environments, taking advantage of all the resources made available by the newest information and communication technologies, namely Artificial Intelligence models and techniques that include but are not limited to Argumentation, Game Theory, Heuristics, Intelligent Agents and Group Decision Systems, as described by Peruginelli and Chiti [11] and Lodder and Thiessen [12]. Moreover, contrary to previous approaches, in Online Dispute Resolution it must be considered not only the disputant parties and the eventual third party but also what Ethan Katsh and Janet Rifkin call the fourth party, i.e., the technological elements involved.

The ultimate goal of AI research in this field is to accomplish a technological threshold, resulting in computational systems that are indeed the 3rd party. In this sweeping approach, there is no major human intervention on the outcome or in guiding the parties to a specific situation. There is, on the other hand, a computational system that performs that major role. This is usually known as an electronic mediator or arbitrator. This is evidently the most challenging approach to follow as computational systems that implement the cognitive capabilities of a Human expert are not easy to accomplish, especially if we include the ability to perceive the emotions and desires of the parties involved.

Depending on the importance of the role that computer systems play on ODR systems, they can be categorized as first or second generation [11]. While in first generation ODR systems technology is a mere tool and has no autonomy, second generation ODR systems are essentially defined by a more autonomous and effective use of technical tools. For the implementation of such services, one can look at fields as diverse as Artificial Intelligence, Mathematics or Philosophy. In the intersection of these fields one can find a range of technologies that will significantly empower the previous generation of ODR tools, namely Artificial Neural Networks, software agents, Case-based Reasoning mechanisms, methods for Knowledge Representation and Reasoning, Argumentation, Learning, and Negotiation. Thus, we move forward from a paradigm in which reactive communication tools are used by parties to share information, to an immersive intelligent environment [3] which proactively supports the lifecycle of the conflict resolution mechanism with important knowledge.

1.2 Important Knowledge

The ideal dispute resolution process is one in which the two parties are better at the end than they were at the beginning. Unfortunately, not all disputes have such conclusion. In order to improve this, we believe that it is of ultimate importance to: (1) provide the parties with important knowledge about the dispute and (2) potentiate the role of the parties throughout all the process. In fact, parties that have poor access to important information generally make bad choices or, at least, they hardly make the best ones.

An important step on the development of conflict resolution mechanisms is thus to identify the knowledge that is meaningful for the parties, according to the legal domain of the dispute. In a first instance, it would be interesting for a party to determine to which extent is it reasonable to engage in a dispute resolution process. That is, are there any significant advantages against litigation? This question can be analyzed from several points of view. On the one hand, alternative dispute resolution processes are generally faster, cheaper, more private and personalized [9]. There is however another important factor: the possible outcome reached through each of the processes. That is, will I reach a better outcome using an alternative dispute resolution process instead of litigation?

It would be really important for each party to know its BATNA - Best Alternative to a Negotiated Agreement, or the possible best outcome "along a particular path if I try to get my interests satisfied in a way that does not require negotiation with the other party" [13]. A party should then understand the notion of a BATNA and what role it should play in ODR. Doing so will, at least, contribute to the acknowledgement that an agreement may be disadvantageous [14]. In fact, the position of the parties may become much more unclear if they are not foreseeing the possible results in case the negotiation / mediation fails. As stated by [6], if you are unaware of what results you could obtain if the negotiations are unsuccessful, you run the risk of entering into an agreement that you would be better off rejecting or rejecting an agreement that you would be better off entering into. That is to say, the parties, by determining their BATNA, would on one side become better protected against agreements that should be rejected and, on the other side, be in a better condition to reach an agreement that better satisfy their interests [15]. But, besides that, a BATNA may play additional interesting features for the parties. For instance, it may be used as a way to put pressure on the other party, especially in dispute resolution procedures allowing the choice of going to court [15].

However, the use of the BATNA alone is not enough to take informed decisions as parties often tend to develop an overly optimistic view on their chances in disputes [15]. This may lead parties to calculate unrealistic BATNAs, which will influence later decisions, leading them reject generous offers from the other parties or to stand stubbornly fixed in some unrealistic positions [15]. It is thus important to also consider the other side of the coin, embodied by the concept of WATNA, or the Worst Alternative to a Negotiated Agreement [13, 16, 17]. A WATNA intends to estimate the worst possible outcome along a litigation path. It can be important in the calculation of the real risks that parties will face in a judicially determined litigation, imagining the worst possible outcome for the party. Considering both these concepts, a party would be aware of the best and worst scenario if the dispute is to be solved in a court.

This helps establish two important boundaries. However, it could also be interesting to consider the whole space between the BATNA and WATNA as a useful element to be taken into account before making or accepting proposals. Indeed, the less space there is between the BATNA and the WATNA, the less dangerous it becomes for the party not to accept the agreement (unless, of course, their BATNA is really disadvantageous). On the other hand, a wider space between the BATNA and the WATNA would usually mean that it can become rather dangerous for the party not to accept the ODR agreement (except in situations when the WATNA is really not inconvenient at all for the party). We can thus argue that the knowledge about the space between the BATNA and the WATNA is also very important. This space is evidently related to the Zone of Possible Agreement proposed by Raiffa [18].

More than that, it would also be interesting for a party to be aware of the region of this space in which an outcome is more likely. That is, if the parties are to solve the dispute through litigation, what is the most likely outcome? In fact, sticking only with the BATNA and WATNA may be unrealistic as these are usually not the most likely outcomes but merely informative boundary values. Thus, an informed party should also consider the MLATNA – Most Likely Alternative to a Negotiated Agreement [17]. Following the same line of thought, we can additionally state that the existence of metrics that measure the probability of each possible outcome could also be extremely useful for a party in an attempt to understand how likely each scenario is [22].

2   Conflict Resolution Styles

In alternative conflict resolution processes in which humans have a preponderant role, specifically in negotiation and mediation, the style of dealing with the conflict of each party will certainly influence the course of action and, consequently, the outcome. On the process of developing conflict resolution mechanisms one should thus regard personal conflict styles as key information. Kenneth Thomas and Ralph Kilmann formalized the way we respond to conflict situations into five different modes in terms of individual's assertiveness and cooperativeness [19]. In this context, assertiveness denotes the extent to which the person attempts to satisfy his/her own interests while cooperativeness denotes the extent to which the person attempts to satisfy the other person's interests. The conflict styles are:

- Competing - This is an uncooperative style by means of which an individual aims at maximizing his/her own gain at the other's expenses. This is a power-oriented style in which an individual will use whatever power seems appropriate to win his/her position (e.g. ability to argue, rank, economic sanctions);
- Accommodating – This style is the opposite of competing, i.e., it is cooperative. When an individual shows an accommodating behavior, he/she neglects his/her own gain to maximize the gain of the other. Under this behavior one founds an element of self-sacrifice. Accommodating includes well-known behaviors such as selfless generosity or charity, obeying another individual's order when we may prefer not to do so or accepting another's point of view;

- Avoiding - The individual that shows an avoiding style of conflict tries to satisfy neither his/her own interests nor those of the other individual. It can be said that he/she is not dealing with the conflict. This style may be evidenced by behaviors such as diplomatically sidestepping an issue, postponing an issue until a better opportunity arises, or simply withdrawing from a threatening situation;
- Collaborating – This cooperative style is the complete opposite of avoiding. When an individual collaborates, he/she attempts to work with the other party to find some solution that fully satisfies the interests of both parties. In this process, the individual explores an issue to discover the underlying desires and fears of the two individuals. An individual that is collaborating might try to explore a disagreement to learn from other's insights;
- Compromising – When an individual has a compromising style of dealing with a conflict, he/she tries to find some expedient, mutually acceptable solution that can partially satisfy both parties. This style is somewhat an intermediate one between competing and accommodating. Generally, compromising can mean splitting the differences between the two positions, exchanging concessions, or seeking a quick middle-ground solution.

Whether it is because of past experiences or because of our temperament, each of us is capable of using all of these conflict-handling styles. Moreover, none of us can be characterized as having one single style of dealing with a conflict. Nevertheless, certain individuals rely on some modes more than others and, therefore tend to use them more often. From the point of view of a mediator and even from the point of view of a conflict resolution platform, it is important to determine the parties' conflict style in an attempt to define how each party will be affected by a given issue. Once the conflict styles are identified, strategies can be implemented that aim at improving the success rate of the conflict resolution process. Namely, we are interesting in developing dynamic processes that adapt strategies based on changes on the conflict styles of the parties. For instance, it is usual for parties to show an avoiding conflict style at the beginning of the process. However, they tend to gradually advance into a more cooperative style. When the conflict resolution platform detects these changes, it may start proposing more "audacious" outcomes since parties will more likely accept them.

3 UMCourt

The work described in this paper is being developed in the context of the TIARAC project – Telematics and Artificial Intelligence in Alternative Conflict Resolution. In that sense, a conflict resolution platform is being developed in which these ideas are being applied: the UMCourt. UMCourt is an agent-based conflict resolution platform that implements two high level functionalities. On the one hand, there is a significant focus on the building and provision of useful knowledge that allows both parties and platform to take better decisions. On the other hand, in UMCourt we are researching novel approaches to negotiation and mediation, in line with the concept of intelligent environments. In that sense, our objective is to develop environments able to provide the conflict resolution platform with important information including the level of stress of the parties, the conflict style or even the emotional state. In this section we focus on the issue of the compilation of important knowledge. The actual architecture of UMCourt will not be depicted here as that has already been done in previous work [20, 21]. The dynamic nature of the conflict resolution will be detailed in the following sections.

3.1 Building Important Knowledge

There is a whole set of knowledge that parties in conflict may use in order to take better and more informed decisions, as addressed before. Under the framework of the TIARAC project we are developing methods for compiling this kind of knowledge. Considering the BATNA and the WATNA, its estimation is usually well defined in the rulings of The Law, in the form of norms. These norms can be implemented in rule-based systems, which efficiently determine, according to the characteristics of the case under evaluation, the legal boundaries of the outcome. As an example of the drawing on of such rules, it is presented below a listing of Def_Rule 396, an abstract description of the procedures that allow the computation of the BATNA and WATNA for the Portuguese Labor Law, as it is stated in Decree of Law (DL) 7/2009 (Portuguese Laws). This simplified rule considers only the case in which a worker ends the contract with a just cause.

```
Def_Rule 396
if RULE_394 then
      WATNA := 3 * (M_SALARY + SENIORITY)
  if TEMPORARY_CONTRACT then
      if WATNA < M_REMAINING *(M_SALARY + SENIORITY)
then
         WATNA := M_REMAINING *(M_SALARY + SENIORITY)
  if WATNA < 15 * (D_SALARY + SENIORITY) then
      WATNA := 15 * (D_SALARY + SENIORITY)
  BATNA := 45 * (D_SALARY + SENIORITY)
  if BATNA < DAMAGE then
      BATNA := +DAMAGE
```

Considering the MLATNA, a slightly different approach is being followed. In fact, in order to determine the most likely outcome, the system needs to analyze past cases in a given context. In that sense, to determine the MLATNA, UMCourt follows a Case-based approach. In a few words, the most similar cases are selected and sorted according to their degree of similarity. The region of the MLATNA will be defined by the most similar cases as, in the legal domain, one may assume that similar cases have similar outcomes, pointing out where an outcome for a case with given characteristics is likely.

As stated before, it is also important for parties to have access to past cases, so that they can analyze them and gain a better understanding about the domain of the problem. In that sense, the framework is able to present the litigant parties with cases that may be relevant, according to their degree of similarity. Besides that, for each selected case, the system also computes the utility of its outcome according to the characteristics of the current case, i.e., the users may acknowledge how much they would gain or lose if the outcome of their cases were the same. Indeed, similar cases may have different outcomes, depending on (in the case of Labor Law) attributes such as worker seniority, wage, and existence or not of extra hours of work not yet paid, among others. In order to be able to compute the utility of the solutions of other cases with respect to the new case, they are structured so that they may be applied to different cases in order to compute its outcomes.

At the end, all this information is presented in a graphical form to the user. Figure 1 depicts a prototype of this interface. Looking at this representation, the user is able to acknowledge the distance between the BATNA and the WATNA (allowing him/her to assess the risk of his/her decisions), to analyze the maximum and minimum utility and similarity values or analyze the similarity versus the utility [22].

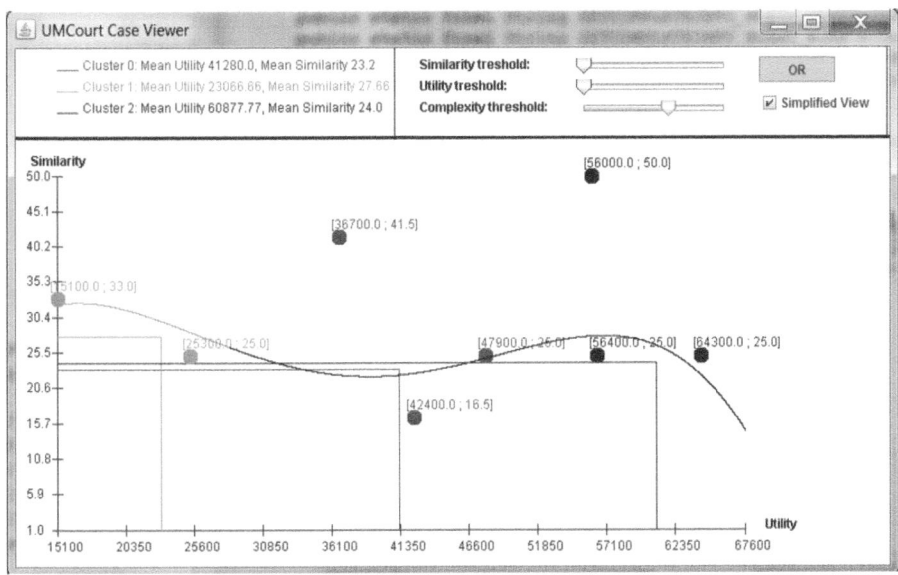

Fig. 1. The prototype of the interface depicting the important knowledge compiled for a party. According to the complexity threshold slider, only a smaller number of cases (colored circles) are presented. The user can click on a case to see its details. The linear regression shows the MLATNA in green. Different colors represent different information clusters.

4 Interpretation of Conflict Styles

It is reasonable to state that in a negotiated process, the most significant factor is the behavior of the parties. In that sense, by knowing in advance how each party behaves, it is possible to draw the best strategy in order to increase the possibility of achieving a successful outcome. The behavior of the parties can essentially be determined in two different ways: by questioning the parties and by analyzing their behavior. The first provides information before the start of the process although it is easy to lie and fake a behavior. Moreover, when we are under stressful situations (as a negotiated process potentially is), we tend to behave differently than we usually do. The second one takes some time to gather enough information to be accurate although it reflects the behavior of the parties in a more reliable way.

In this work, we focus on the interpretation of conflict styles during the negotiated process, by analyzing the behavior of the parties in real time. In that sense, we analyze the actions of the parties in each stage of the negotiation, in which a party may ignore, accept, refuse, exit, reply with a new proposal or reply with a counterproposal. Moreover, we also take into consideration the nature of the solutions proposed (e.g. is a party being too greedy?, is a party being realistic?). The approach proposed consists in analyzing these factors together with the BATNA and the WATNA of each party as well as the ZOPA – the Zone of Potential Agreement, in order to classify the behavior of each party.

Basically, during the negotiation process, parties make successive proposals and counterproposals in order to achieve a mutually agreeable solution. We can thus analyze the proposals of each party in each round according to the actions of the parties and a space defined by the BATNA and WATNA of each party (Figure 2).

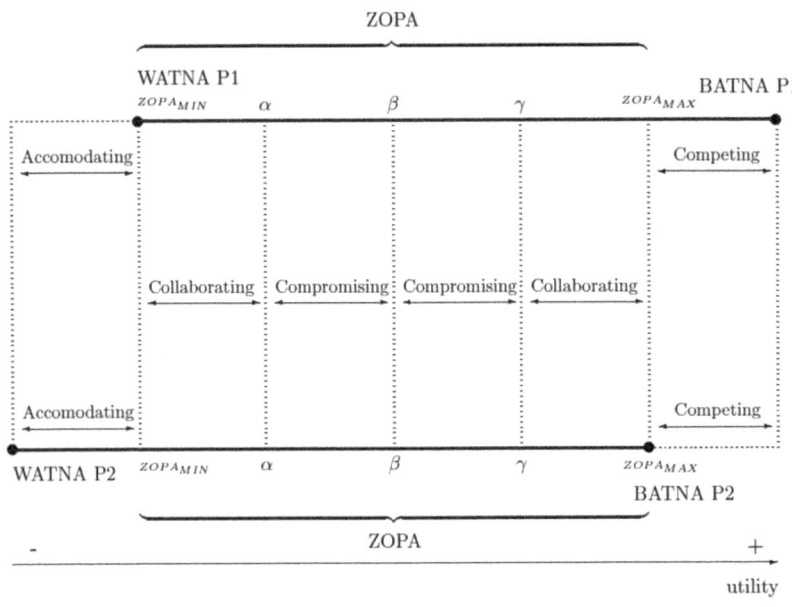

Fig. 2. The space that defines the personal conflict styles in function of the utility of the proposals and the values of the BATNA, BATNA and ZOPA

In each round, each action of a party will contribute to the overall characterization of this conflict style. Thus, the personal conflict style that is computed for each party in each round is a result of all the previous interactions. Two main scenarios are possible: the party ignores the proposal or the party answers to the proposal. If a party, upon receiving a proposal for a solution, simply ignores it, he is not satisfying his interests nor the ones of the other party. In such a scenario, the conflict style evidenced is the *Avoiding* one.

If the party makes a proposal or a counterproposal, he is cooperating on the process. However, the nature of the proposal must be analyzed, namely in terms of its utility for each party. If the utility of the proposal is higher than the BATNA of the other party, he is clearly showing a *Competing* style as he is trying to maximize his own gain, probably in an unrealistically way, completely disregarding the other party. On the other hand, if the utility of the proposal is lower than the WATNA of the other party, he is neglecting his own gain or even maximizing the gain of the other party. In such a scenario, it is reasonable to state that the party is evidencing an *Accommodating* behavior.

When the utility of the proposal falls within the range of the ZOPA, it indicates that the party is being reasonable and try to propose a settlement in which both parties will not win everything but will not lose everything either. In such a scenario, the conflict style is determined according to the distance to the meant point of the ZOPA, defined in (1).

$$\beta = \left(\frac{ZOPA_{MIN} + ZOPA_{MAX}}{2} \right) \tag{1}$$

Two additional points can be defined that will allow to classify the remaining conflict styles. These points, depicted in (2) and (3), allow defining additional intervals to classify the personal conflict styles.

$$\alpha = \left(ZOPA_{MIN} + \frac{\beta - ZOPA_{MIN}}{2} \right) = \left(\frac{ZOPA_{MIN} + \beta}{2} \right) \tag{2}$$

$$\gamma = \left(ZOPA_{MAX} - \frac{ZOPA_{MAX} - \beta}{2} \right) = \left(\frac{ZOPA_{MAX} + \beta}{2} \right) \tag{3}$$

Namely, when the utility of a proposal falls within the range $[\alpha, \gamma]$, it means that the proposing party is negotiating in an intermediary points of the ZOPA. That is, the party is trying to work out compromise that implies a loss from both parties. In such a scenario, it may be said that the party is evidencing a *Compromising* behavior.

On the other hand, if the value of the utility belongs to the range defined by $[ZOPA_{MIN}, \alpha[\cup]\gamma, ZOPA_{MAX}]$, the party is proposing a solution that is closer to the limits of the ZOPA. This may mean that although the party is trying to work out a mutually agreeable solution, he may be trying to explore the weaknesses of the opposing party trying to force him to accept a given solution. Under this scenario, the conflict style of the party may be defined as *Collaborating*.

However, we are aware that we do not make use of a single conflict style at a time. In that sense, we propose a more accurate approach in which a main conflict style is inferred, together with a trend style, meaning that a party shows a given style with a possible tendency towards another one. The following notation is used to denote a *main* conflict style with a trend to a *secondary* one: $Main_{\rightarrow secondary}$.

Let φ be the value of the utility of a proposal. The following personal conflict styles are defined:

$Collaborating_{\rightarrow Accomodating}$ $if\ \varphi \in [ZOPA_{MIN}, \frac{ZOPA_{MIN}+\alpha}{2}[$

$Collaborating_{\rightarrow Compromising}$ $if\ \varphi \in [\frac{ZOPA_{MIN}+\alpha}{2}, \alpha[$

$Compromising_{\rightarrow Collaborating-Accomodating}$ $if\ \varphi \in [\alpha, \beta[$

$Compromising_{\rightarrow Collaborating-Competing}$ $if\ \varphi \in [\beta, \gamma[$

$Collaborating_{\rightarrow Compromising}$ $if\ \varphi \in [\gamma, \frac{ZOPA_{MAX}+\gamma}{2}[$

$Collaborating_{\rightarrow Collaborating-Competing}$ $if\ \varphi \in [\frac{ZOPA_{MAX}+\gamma}{2}, ZOPA_{MAX}]$

By determining the personal conflict style of each party in each round, it is possible to analyze its evolution throughout the conflict resolution process (Figure 3). This will allow the framework to determine the best moments to adapt strategies, as will be seen in the following section.

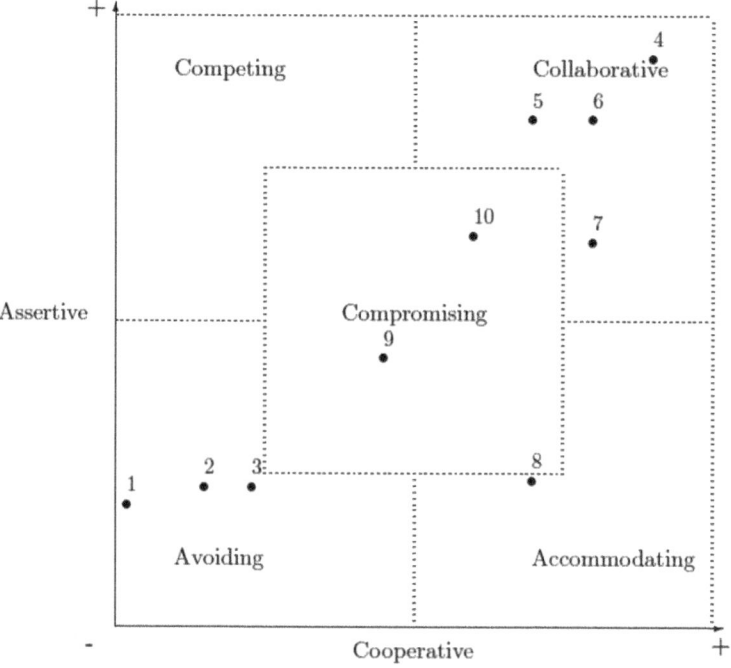

Fig. 3. The evolution of the conflict style of a party in 10 rounds

5   A Dynamic Conflict Resolution Model

In line with the recent trend of Intelligent Environments in which computer systems gradually merge into the environment [3], we aim to develop dynamic conflict resolution methods to be used in the context of a proactive environment. In fact, conflict resolution methods that are run by human experts are generally dynamic as experts have the ability to understand changes in the context of interaction (e.g. a party is getting stressed, a party does not like the current state of affairs) and change the strategy before it is too late (e.g. by making a pause in the process). However, the problem is that under ODR settings, such context information is not available as parties are, generally, "hidden" behind a web interface while they are studying and making proposals for negotiation. This, we believe, is the main problem with current ODR platforms. In that sense, we aim at a new vision on the ODR issue, in line with the concept of Ambient Intelligence. Thus, in this new approach, parties are not simply interacting with web forms. Instead, parties use ODR tools in the context of an intelligent environment that can provide the conflict resolution platform with important context information like the level of stress, the conflict style or even the emotional state.

With this information, the conflict resolution model can dynamically adjust to changes in the context of interaction. As an example, if the conflict resolution platform detects that one of the parties is getting stressed, it may temporarily pause the process or assume a mediator role in which all the communication goes through it and no direct communication between the parties takes place. When the party calms down, the platform may once again allow direct contact between the parties.

However, given the scope of this paper, we are more interested on the role of personal conflict styles in these dynamic models. As said before, each person tends to use more some conflict styles than others, either because of their personality or of past experiences. Moreover, it is common for parties to change the conflict style during the conflict resolution process, according to how it is developing. As an example, it is common for a party to exhibit an avoiding behavior at the outset of the process and then start being more cooperative as confidence on the process grows. Moreover, it is also common for parties to start by being competitive and with high expectations and then, as the process develops and they gain a more realistic view that includes the desires and rights of the others, tend to be more compromising. Evidently the opposite may also happen, i.e., parties that start fully cooperative but that don't like the way that the process is going and start moving towards a more uncooperative style.

The work developed in this context focus on detecting this kind of changes in order to adapt strategies in real time. Basically this adds a new step to the conflict resolution model of UMCourt, making it a dynamic conflict resolution model (Figure 4). This model starts by building all the important knowledge mentioned before, which will be important for parties to develop realistic views about their problem. Then, the platform builds a strategy. In a first iteration, this consists in selecting a group of possible outcomes that will sequentially be suggested to the users. In order to build this first strategy, the platform only takes into consideration the group of similar cases that was selected. Then, the process advances to the actual conflict resolution, either by means of negotiation or mediation.

During this phase, the platform constantly receives information from the environment concerning the personal conflict styles, determined as described above. Whenever the platform detects that a significant change is occurring, an adaptation in the strategy takes place. At this moment, adapting strategies consists in changing the list of outcomes to be proposed to the parties. In order to do it, one very important issue is taken into consideration: the utility of each outcome for each party. The utility of an outcome depicts, as stated before, how good each outcome is for a given party. Thus, the platform looks at the utility of the outcomes of the similar cases and changes the order by which outcomes will be proposed according to the state of the parties. Let us take as example a setting with two parties, in which one party is consistently exhibiting a collaborative behavior while the other is moving from a collaborative to an avoiding one. This may be indicative that the second party is not appreciating the way that the process is going. In that sense, in order to prevent that party from abandoning the process, the system will suggest an outcome whose utility is better for that party than the previous one. This is expected to increase the satisfaction of the party, maintaining him interested in continuing with the process and probably taking him back to a collaborative style.

We are however aware that the information about the personal conflict styles alone is not enough. In that sense, our work now focuses on additional sources of information that the platform can use to adapt strategies. Namely, we are considering the use of information about the keystroke intensity of the parties in order to determine the level of stress as well as some linguistic features. Information about stress is very important, namely to assess the level of escalation of the parties, depicting when the process should be interrupted before emotions running high. We are also considering several ways of determining the emotional state of the parties as this information is of ultimate importance to determine how each issue affects each party. Namely, we are considering non-invasive methods that include image and speech analysis.

With the combination of all this important information we will be able to develop context-aware conflict resolution models that take advantage of technological tools without however losing the richness of face-to-face interaction. This way, we expect to achieve more efficient conflict resolution mechanisms, able to achieve more mutually satisfactory outcomes.

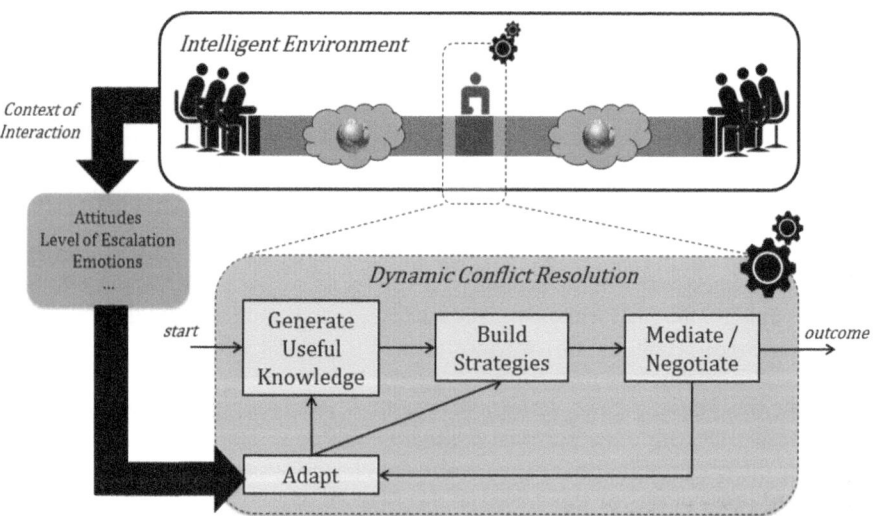

Fig. 4. High level view of the dynamic conflict resolution model presented

6 Conclusions

Current approaches on Online Dispute Resolution are heavily based on technology, as it would be expected. However, this tends to leave aside some important advantages of traditional human-centered approaches. Specifically, the ability of human mediators to deal with context information such as the emotional state of the parties or their personal conflict resolution styles is completely disregarded. This results in conflict resolution platforms that are insensible, unable to perceive the fears and desires of the parties in conflict. In order to reach the so-called second generation ODR, we believe it is mandatory to consider not only all the important context information but also methods that can make use of it in order to more accurately understand the parties and achieve outcomes that are more satisfactory.

In that sense we presented in this paper a methodology for determining the personal conflict styles of the parties, by analyzing their behavior during a negotiated or mediated conflict resolution process. All this is done in a non-intrusive way. Merging this with additional context information such as the levels of stress or even the emotional state, will allow the development of conflict resolution methods that are able to adapt, in real time, to significant changes in the context of interaction. This approach, in line with the vision of Ambient Intelligence, will bring significant advantages for the field of conflict resolution in the sense that it can empower cold and insensitive processes with context-aware abilities usually associated to human experts.

Acknowledgments. The work described in this paper is included in TIARAC - *Telematics and Artificial Intelligence in Alternative Conflict Resolution Project* (PTDC/JUR/71354/2006), which is a research project supported by FCT (Science & Technology Foundation), Portugal. The work of Davide Carneiro is also supported by a doctoral grant by FCT (SFRH/BD/64890/2009).

References

1. Lewin, K.: Resolving social conflicts: Field theory in social science. American Psychological Association (1997) ISBN 1557984158
2. Pitt, J., Ramirez-Cano, D., Kamara, L., Neville, B.: Alternative Dispute Resolution in Virtual Organizations. In: Artikis, A., O'Hare, G.M.P., Stathis, K., Vouros, G.A. (eds.) ESAW 2007. LNCS (LNAI), vol. 4995, pp. 72–89. Springer, Heidelberg (2008)
3. Aarts, E., Grotenhuis, F.: Ambient Intelligence 2.0: Towards Synergetic Prosperity. Journal of Ambient Intelligence and Smart Environments 3, 3–11 (2011)
4. Katsch, E., Rifkin, J.: Online dispute resolution – resolving conflicts in cyberspace. Jossey-Bass Wiley Company, San Francisco (2001)
5. Lewicki, R.J., Barry, B., Saunders, D.M., Minton, J.W.: Negotiation, 4th edn. McGraw-Hill/Irwin, New York (2003)
6. Goldberg, S.B., Sander, F.E., Rogers, N., Cole, S.R.: Dispute Resolution: Negotiation, Mediation and Other Processes. Aspen Publishers, New York (2003)
7. Raiffa, H.: The Art and Science of Negotiation. Harvard University Press (2002)
8. Walton, P.R.E., McKersie, R.B.: A behavioral theory of labor negotiations. McGraw-Hill (1965)

9. Brown, H., Marriott, A.: ADR Principles and Practice. Sweet and Maxwell (1999)
10. Bennett, S.C.: Arbitration: essential concepts. ALM Publishing (2002)
11. Peruginelli, G., Chiti, G.: Artificial Intelligence in Online Dispute Resolution. In: Proceedings of the Workshop on the Law of Electronic Agents – LEA (2002)
12. Lodder, A., Thiessen, E.: The role of artificial intelligence in online dispute resolution. In: Workshop on Online Dispute Resolution at the International Conference on Artificial Intelligence and Law, Edinburgh, UK (2003)
13. Notini, J.: Effective Alternatives Analysis in Mediation: "BATNA/WATNA" Analysis Demystified (2005), http://www.mediate.com/articles/notini1.cfm (last accessed May, 2011)
14. Klaming, L., Van Veenen, J., Leenes, R.: I want the opposite of what you want: summary of a study on the reduction of fixed-pie perceptions in online negotiations. In: Expanding the Horizons of ODR Proceedings of the 5th International Workshop on Online Dispute Resolution (ODR Workshop 2008), Firenze, Italy, pp. 84–94 (2004)
15. De Vries, B.R., Leenes, R., Zeleznikow, J.: Fundamentals of providing negotiation support online: the need for developing BATNAs. In: Proceedings of the Second International ODR Workshop, Tilburg, Wolf Legal Publishers, pp. 59–67 (2005)
16. Fisher, R., Ury, W.: Getting To Yes: Negotiating Agreement Without Giving. Houghton Mifflin, Boston (1981)
17. Steenbergen, W.: Rationalizing Dispute Resolution: From best alternative to the most likely one. In: Proceedings 3rd ODR Workshop, Brussels (2005)
18. Raiffa, H.: The art and science of negotiation: how to resolve conflicts and get the best out of bargaining. The Belknap Press of Harvard University Press, Cambridge (1982)
19. Thomas, K., Kilmann, R.: Conflict and Conflict Management (1974), http://www.kilmann.com/conflict.html (last accessed May, 2011)
20. Carneiro, D., Novais, P., Costa, R., Neves, J.: Developing Intelligent Environments with OSGi and JADE. In: Bramer, M. (ed.) IFIP AI 2010. IFIP Advances in Information and Communication Technology, vol. 331, pp. 174–183. Springer, Heidelberg (2010) ISBN 978-3-642-15285-6
21. Costa, N., Carneiro, D., Novais, P., Barbieri, D., Andrade, F.: An Advice System for Consumer's Law Disputes. In: Filipe, J., Cordeiro, J. (eds.) ICEIS 2010. LNBIP, vol. 73, pp. 237–248. Springer, Heidelberg (2011) ISBN: 978-3-642-19802-1
22. Andrade, F., Novais, P., Carneiro, D., Zeleznikow, J., Neves, J.: Using BATNAs and WATNAs in Online Dispute Resolution. In: Nakakoji, K., Murakami, Y., McCready, E. (eds.) JSAI-isAI 2009. LNCS (LNAI), vol. 6284, pp. 5–18. Springer, Heidelberg (2010)

Organizations of Agents in Information Fusion Environments

Dante I. Tapia[1], Fernando de la Prieta[1], Sara Rodríguez González[1],
Javier Bajo[2], and Juan M. Corchado[1]

[1] University of Salamanca, Salamanca, Spain
{dantetapia,fer,srg,corchado}@usal.es
[2] Pontifical University of Salamanca, Salamanca, Spain
jbajope@upsa.es

Abstract. Information fusion in a context-aware system is understood as a process that assembles assessments of the environment based on its goals. Advantages of intelligent approaches such as Multi-Agent Systems (MAS) and the use of Wireless Sensor Networks (WSN) within the information fusion process are emerging, especially in context-aware scenarios. However, it has become critical to propose improved and efficient ways to handle the enormous quantity of data provided by these approaches. Agents are a suitable option because they can represent autonomous entities by modeling their capabilities, expertise and intentions. In this sense, virtual organizations of agents are an interesting option/possibility because they can provide the necessary capacity to handle open and heterogeneous systems such as those normally found in the information fusion process. This paper presents a new framework that defines a method for creating a virtual organization of software and hardware agents. This approach facilitates the inclusion of context-aware capabilities when developing intelligent and adaptable systems, where functionalities can communicate in a distributed and collaborative way. Several tests have been performed to evaluate this framework and preliminary results and conclusions are presented.

Keywords: Information Fusion, Wireless Sensor Networks, Multi-Agent Systems, Virtual Organizations.

1 Introduction

At present there are small, portable and non-intrusive devices that allow agents to gather context-information in a dynamic and distributed way [1]. However, the integration of such devices is not an easy task. Therefore, it is necessary to develop innovative solutions that integrate different approaches in order to create open, flexible and adaptable systems.

The scientific community within the realm of information fusion remains heir to the traditions and techniques of sensor fusion, which is primarily concerned with the use of sensors to provide information to decision systems. This has led to most models of information fusion processes being directed by data fusion in which the

L. Antunes and H.S. Pinto (Eds.): EPIA 2011, LNAI 7026, pp. 59–70, 2011.
© Springer-Verlag Berlin Heidelberg 2011

sensors and data are the central core. One way to accomplish this process is to apply an intelligent approach such as MAS within the fusion process. Agents are suitable for fusion because they can represent autonomous fusion entities by modeling their capabilities, expertise and intentions [22] [2].

MAS allow the participation of agents within different architectures and even different languages. The development of open MAS is still a recent field in the MAS paradigm, and its development will enable the application of agent technology in new and more complex application domains. However, this makes it impossible to trust agent behavior unless certain controls based on social rules are imposed. To this end, developers have focused on the organizational aspects of agent societies to guide the development process of the system.

This article describes an agent approach to fusion applied to dynamic contexts based on the HERA (*Hardware-Embedded Reactive Agents*) [1] and OVAMAH (*Adaptive Virtual Organizations: Mechanisms, Architectures and Tools*) platforms [14].

In HERA agents are directly embedded on the WSN nodes and their services can be invoked from other nodes (including embedded agents) in the same WSN, or another WSN connected to the former one. The OVAMAH platform allows the framework to incorporate the self-adaptive organizational capabilities of multi-agent systems and create open and heterogeneous systems.

This article is structured as follows: the next section presents the related approaches. Section 3 shows the framework proposal, including the description of the HERA and OVAMAH platforms, the core of the system. Sections 4 and 5 present some results and conclusions obtained.

2 Technological Approaches

Recent trends have shown a number of MAS architectures that utilize data merging to improve their output and efficiency. Such is the case of Castanedo et al. [4], who propose the CS-MAS architecture to incorporate dynamic data fusion through the use of an autonomous agent, locally fused within the architecture. Other models, such as HiLIFE [17], cover all of the phases related to information fusion by specifying how the different computational components can work together in one coherent system.

Despite having all the advantages of MAS, these kinds of systems are monolithic. In an environment in which data heterogeneity is a key feature, it is necessary to use systems with advanced capacities for learning and adaptation. In this regard, an approach within the field of MAS that is gaining more weight in recent times is the consideration of organizational aspects [3], and more concretely those based on virtual organizations (VO).

Currently, there are no virtual organization-based applications oriented to fusion information. However it is possible to find some approaches that try to propose advances in this way. For example, the e-Cat System [10] focuses on the distribution and integration of information. This system is based on enhancing the skills or abilities of members of the organization by defining the different types of skills and relationships that exist between them. This organization aims to ensure the maximum independence between the different partnerships created, and information privacy.

Another example, perhaps more centralized in the fusion of information, is the KRAFT (Knowledge Reuse and Fusion / Transform) architecture [12], which proposes an implementation of agents where organizational aspects are considered to support the processes of heterogeneous knowledge management.

The approach proposed in this article presents an innovative model where MAS and VO are combined to obtain a new architecture specifically oriented to construct information fusion environments.

The following section discusses some of the most important problems of existing approaches that integrate agents into wireless sensor networks, including their suitability for constructing intelligent environments. It also describes the proposed integration of information fusion systems that use the capabilities of multi-agent systems for a particular activity, including reading data from sensors and reacting to them.

3 Integration Framework

3.1 Motivation

An intelligent fusion system has to take contextual information into account. The information may be gathered by sensor networks. The context includes information about the people and their environment. The information may consist of many different parameters such as location, the building status (e.g. temperature), vital signs (e.g. heart rhythm), etc. Each element that forms part of a sensor network is called a node. Each sensor node is habitually formed by a microcontroller, a transceiver for radio or cable transmission, and a sensor or actuator mechanism [11]. Some nodes act as routers, so that they can forward data that must be delivered to other nodes in the network. There are wireless technologies such as Wi-Fi, IEEE 802.15.4/ZigBee and Bluetooth that enable easier deployments than wired sensor networks [16]. WSN nodes must include some type of power manager and certain smart features that increase battery lifetime by having offering? worse throughput or transmission delay through the network [16].

In a centralized architecture, most of the intelligence is located in a central node. That is, the central node is responsible for managing most of the functionalities and knowing the existence of all nodes in a specific WSN. That means that a node belonging to a certain WSN does not know about the existence of another node forming part of a different WSN, even though this WSN is also part of the system. For this reason, it is difficult for the system to dynamically adapt its behavior to changes in the infrastructure.

The combination of agents and WSNs is not easy due to the difficulty in developing, debugging and testing distributed applications for devices with limited resources. The interfaces developed for these distributed applications are either too simple or, in some case, do not even exist, which further complicates their maintenance. Therefore, there are researches [20] [23] [8] [5] [6] [13] that develop methodologies for the systematic development of MAS for WSNs.

The HERA platform tackles some of these issues by enabling an extensive integration of WSNs and optimizing the distribution, management and reutilization of the available resources and functionalities in its networks. As a result of its underlying platform, SYLPH [18], HERA contemplates the possibility of connecting wireless sensor networks based on different radio and link technologies, whereas other approaches do not.

HERA allows the agents embedded into nodes to work in a distributed way and does not depend on the lower stack layers related to the WSN formation (i.e. network layer), or the radio transmission amongst the nodes that form part of the network (i.e. data link and physical layers).

HERA can be executed over multiple wireless devices independently of their microcontroller or the programming language they use. HERA allows the interconnection of several networks from different wireless technologies, such as ZigBee or Bluetooth. Thus, a node designed over a specific technology can be connected to a node from a different technology. This facilitates the inclusion of context-aware capabilities into intelligent fusion systems because developers can dynamically integrate and remove nodes on demand.

On the other hand, if current research on agents is taken into account, it is possible to observe that one of the most prevalent alternatives in distributed architectures are MAS. An agent, in this context, is anything with the ability to perceive its environment through sensors, and to respond through actuators. A MAS is defined as any system composed of multiple autonomous agents incapable of solving a global problem, where there is no global control system, the data is decentralized and the computing is asynchronous [22]. There are several agent frameworks and platforms [20] that provide a wide range of tools for developing distributed MAS.

The development of agents is an essential component in the analysis of data from distributed sensors, and gives those sensors the ability to work together. Furthermore, agents can use reasoning mechanisms and methods in order to learn from past experiences and to adapt their behavior according to the context [22]. Given these capacities, the agents are very appropriate to be applied in information fusion.

The most well-known agent platforms (like Jade) offer basic functionalities to agents, but designers must implement nearly all organizational features, such as the communication constraints imposed by the organization topology. In order to model open and adaptive VO, it becomes necessary to have an infrastructure than can use agent technology in the development process and apply decomposition, abstraction and organization techniques. OVAMAH [3] is the name given to an abstract architecture for large-scale, open multi-agent systems. It is based on a services oriented approach and primarily focuses on the design of VO.

In HERA agents are directly embedded on the WSN nodes and their services can be invoked from other nodes (including embedded agents) in the same WSN or other WSNs connected to the former one. By using OVAMAH, the framework can incorporate the self-adaptive organizational capabilities of multi-agent systems and create open and heterogeneous systems.

3.2 Proposed Information Fusion Framework

The way in which information fusion is held together is the key to this type of system. In general, data can be fused at different levels [9]:

- sensor level fusion, where multiple sensors measuring correlated parameters can be combined;
- feature level fusion, where analysis information resulting from independent analysis methods can be combined;
- decision level fusion, where diagnostic actions can be combined.

These levels generally depend on many factors. In order to provide the most generic and expandable system that can be applied to a wide variety of engine applications with varied instrumentation and data sources, we have chosen to perform the information fusion at the three levels shown in Figure 1.

At the level sensors, HERA makes it possible to work with different WSNs in a way that is transparent to the user. A node in a specific type of WSN (e.g. ZigBee) can directly communicate with a node in another type of WSN (e.g. Bluetooth).

At higher levels (features and decision), it is possible to detect changes in the environment and its consequent action in the system. This consequent action can be managed on the platform as a result of the services and functions that comprise the agents of the organization. For example, if a change within a node (a change of light for instance) is detected at sensor level, the agents at a higher level can decide to send a warning message or perform an action.

This scheme provides for the potential inclusion of a variety of sensors as well as other devices of diagnostic relevant information that might be in the form of maintenance records, monitoring and observations. The framework provides for information synchronization and high-level fusion [21].

The principal objective of the high level fusion shown in Figure 1 is to transform multiple sources of several kinds of sensors and performance information into a monitoring knowledge base. Embedded in this transformation process is a fundamental understanding of node of WSN functions, as well as a systematic methodology for inserting services to support a specific action according to information received by the node.

The framework proposes a new and easier method to develop distributed multi-agent systems, where applications and services can communicate in a distributed way, independent of a specific programming language or operating system. The core of the architecture is a group of deliberative agents acting as controllers and administrators for all applications and services. The functionalities of the agents are not inside their structure, but modeled as services. This approach provides the systems with a higher ability to recover from errors, and a better flexibility to change their behavior at execution time.

The agents in the organization can carry out complex tasks as well as react to changes that occur in the environment. To do this, the agents incorporate an innovative planning model that provides the organization with advances self-adaptive capacities [15].

3.3 The HERA Platform

As indicated in the previous section, this paper aims to describe a new framework for information fusion based on the concept of virtual organizations of agents and multi-agents. This framework uses a sensor platform (HERA) in order to gather the data. Consequently, the description of HERA is general, and describes only the most relevant aspect related to the framework.

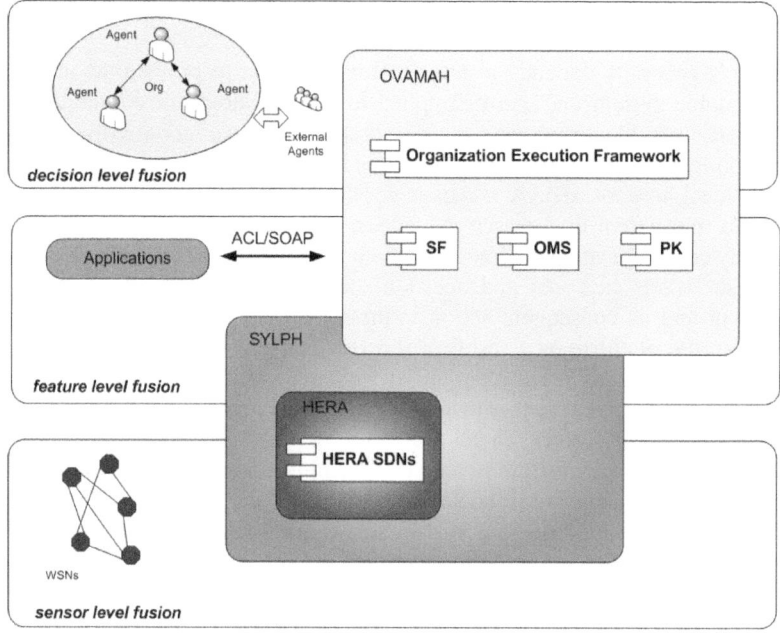

Fig. 1. Proposed framework

HERA facilitates agents, applications and services communication through of using dynamic and self-adaptable heterogeneous WSNs. In HERA, agents are directly embedded in the WSN nodes and their services can be invoked from other nodes in the same network or from other networks connected to the former one. HERA is an evolution of the SYLPH platform [18]. SYLPH follows a SOA model for integrating heterogeneous WSNs in intelligent systems. HERA goes a step beyond SYLPH by embedding agents directly into the wireless nodes and allowing them to be invoked from other nodes either in the same network or in another network connected to the original.

The HERA agent platform adds its own agent layer over SYLPH [18][19]. Thus, HERA takes advantage of one of the primary features of SYLPH: it can be run over any wireless sensor node regardless of its radio technology or the programming language used for its development.

The HERA Agents Layer (or just HERA) can run HERA agents, which are specifically intended to run on devices with reduced resources. To communicate with each other, HERA agents use HERACLES, the agent communication language designed for use with the HERA platform. Each HERA agent is an intelligent piece of code running over a node. As explained below, there must be at least one facilitator agent in every agent platform. This agent is the first one created in the platform and acts as a directory for searching agents. In HERA, this agent is referred to as the HERA-SDN (HERA Spanned Directory Node).

The HERA Communication Language Emphasized to Simplicity (HERACLES) is directly based on the SSDL language. As with SSDL, HERACLES does not use intermediate tags and the order of its elements is fixed to constrain the resource needs of the nodes. This makes its human-readable representation, used by developers for coding, very similar to SSDL. When HERACLES is translated to HERACLES frames, the actual data transmitted amongst nodes, they are encapsulated into simple SSDL frames using "HERA" as their service id field.

Every agent platform needs some kind of facilitator agent that must be created before other agents are instantiated in the platform. Facilitator agents act as agent directories. This way, every time an agent is created, it is registered to one of the existing facilitator agents. This allows other agents to request one of the facilitator agents in order to know where an agent with certain functionalities is located, and how to invoke its (desired?) functionalities. As HERA is intended to run on machines that are not more complex than the sensor nodes themselves, it was necessary to design some hardware facilitator agents that do not need more CPU complexity and memory size than what a regular sensor node has. In HERA, the hardware agents communicate with each other through the HERA Communication Language Emphasized to Simplicity (HERACLES).

Because HERA is implemented over SYLPH through the addition of new layers and protocols (HERA Agents and HERACLES), it can be used over several heterogeneous WSNs in a transparent way.

3.4 OVAMAH

OVAMAH is the chosen platform for the creation of the organization of agents in the proposed framework. The most well-known agent platforms (e.g. JADE) offer basic functionalities to agents, such as AMS (Agent Management System) and DF (Directory Facilitator) services; but designers must implement nearly all of the organizational features by themselves, such as the communication constraints imposed by the organization topology. In order to model open and adaptive virtual organizations, it becomes necessary to have an infrastructure that can use agent technology in the development process and apply decomposition, abstraction and organization techniques, while keeping in mind all of the requirements cited in the previous section. OVAMAH is the name given to an abstract architecture for large-scale, open multi-agent systems. It is based on a service oriented approach and primarily focuses on the design of virtual organizations. The architecture is essentially formed by a set of services that are modularly structured. It uses the FIPA architecture, expanding its capabilities with respect to the design of the organization, while also expanding the services capacity. The architecture has a module with the sole objective of managing organizations that have been introduced into the architecture, and incorporates a new definition of the FIPA Directory Facilitator that is capable of handling services in a much more elaborate way, following service-oriented architecture directives.

OVAMAH consists of three principal components: Service Facilitator (SF), Organization Manager Service (OMS) and Platform Kernel (PK).

The SF primarily provides a place where autonomous entities can register service descriptions as directory entries. The OMS component is primarily responsible for specifying and administrating its structural components (role, units and norms) and its execution components (participating agents and the roles they play, units that are active at each moment). In order to manage these components, OMS handles the following lists: UnitList: maintains the relationship between existing units and the immediately superior units (SuperUnit), objectives and types; RoleList: maintains the relationships between existing roles in each unit, which roles the unit inherits and what their attributes are (accessibility, position); NormList: maintains the relationship between system rules; EntityPlayList: maintains the relationship between the units that register each agent as a member, as well as the role that they play in the unit. Each virtual unit in OVAMAH is defined to represent the "world" for the system in which the agents participate by default. Additionally, the roles are defined in each unit. The roles represent the functionality that is necessary for obtaining the objective of each unit. The PK component directs the basic services on a multi-agent platform and incorporates mechanisms for transporting messages that facilitate interaction among entities.

Open-systems are highly complex and current technology to cover all the described functionalities is lacking. There are some new requirements that still need to be solved. These requirements are imposed mainly by: (i) computation as an inherently social activity; (ii) emergent software models as a service; (iii) a non-monolithic application; (iv) computational components that form virtual organizations, with an autonomous and coordinated behavior; (v) distributed execution environments; (vi) multi-device execution platforms with limited resources and (vii) security and privacy policies for information processing. In order to satisfy all of those requirements, the architecture must provide interaction features between independent (and usually intelligent) entities, that can adapt, coordinate and organize themselves [14] From a global perspective, the architecture offers total integration, enabling agents to transparently offer and request services from other agents or entities and at the same time, and allowing external entities to interact with agents in the architecture by using the services provided. Reorganization and adaptation features in the agent´s behavior are necessary for this platform, for which we have proposed a social planning model [15]. This social planning model offers the possibility of deliberative and social behavior. It is worth mentioning that this is a unique model that incorporates its own reorganization and social adaptation mechanism. The architecture facilitates the development of MAS in an organizational paradigm and the social model adds reorganization and adaptation functions.

4 Results

The case study shows the potential of VO in the design and development of systems for information fusion. In order to evaluate the proposed framework, different tests were performed. In this section we present the results obtained with the aim of valuating two main features: Firstly, the advantages obtained with the integration of the HERA platform as related to the use of resource constrained devices, and secondly the impact and performance of the organizational structure.

An organization is implemented by using the model proposed in [7]. The simulation within the virtual world represents an e-health environment, and the roles that were identified within the case study are: Communicator, SuperUser, Scheduler, Admin, Device Manager, Incident Manager.

In order to evaluate the impact of the development of the MAS using an organizational paradigm, it is necessary to revise the behavior of the MAS in terms of its performance. A prototype was constructed based on OVAMAH, which could be compared to the previous existing system [7]. The MAS shown in this study is not open and the re-organizational abilities are limited, since the roles and norms cannot be dynamically adapted. As can be seen in the Table 1, the system proposed in this study provides several functional, taxonomic, normative, dynamics and adaptation properties. The organizational properties are a key factor in an architecture of this kind, but the capacity for dynamic adaptation in execution time can be considered as a differential characteristic of the architecture.

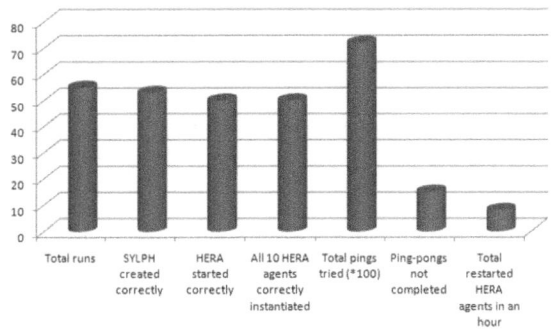

Fig. 2. Results of the HERA performance experiments

Table 1. Comparison of organizational and no organizational systems

Features		No organization system	Organizational system
Functional	BDI Model	Yes	Yes
Taxonomic	Group		Yes
	Topology		Yes
	Roles		Yes
	Interactions	Yes	Yes
Normative	Norms		Yes
Dynamics	Agent Joining	Yes	Yes
	Role Enactment		Yes
	Behaviour control	Yes	Yes
	Org. Joining		Yes
Adaptation	Taxonomic		Yes
	Normative		Yes
	Functional	Yes	Yes

In order to test the HERA platform, a distributed WSN infrastructure with HERA running over it was developed. This experiment consisted of trying to start a platform with HERA over a ZigBee SYLPH network infrastructure. The infrastructure consisted of a ZigBee network with 31 nodes (sensors of actuators).

The nodes were distributed in a short-range simple mesh, with less than 10 meters between any router and the coordinator. Each time the ZigBee network was formed, nodes were powered on different times, so that the mesh topology was different each time. However, there were some constraints: the maximum depth was 5 and the maximum number of neighbors or children for each node was 8.

After the entire network was correctly created, the coordinator and SDN tried to instance a HERA-SDN. HERA-SDN instanced itself and started the HERA platform registering a special SYLPH service called "HERA" on the SDN. Then, 10 of the 30 SYLPH nodes tried to instance one HERA agent, each of them in the HERA platform. Once the HERA-SDN and the 10 HERA agents were successfully instantiated, HERA-SDN started to "ping" each of the ten HERA Agents with a request HERACLES frame including an inform-if command and waiting for a inform frame as a "pong" response. This experiment was run 50 times to measure the success ratio of the platform start and the agent instantiation. However, if the SYLPH network could not be correctly created, or if the HERA platform could not be completely started and created, these runs were also discarded and not taken into account as forming part of the 50 runs. Any HERA agent that crashed was immediately restarted. HERACLES messages were registered to measure when a ping-pong failed and if a HERA agent had to be restarted.

The results (Figure 2) indicate that it is necessary to improve SYLPH creation and the instantiation of HERA Agents. In the first case, a better ARQ (Automatic Repeat Request) mechanism could increase SSP-over-WSN transmissions. In the second case, it is necessary to debug the implementation of the agents and fix errors. In addition, the robustness of the HERA agents should be improved by introducing a mechanism to ping and keep running.

5   Conclusions

In summary, this paper proposes a new perspective for information fusion where intelligent agents can manage the workflow. These intelligent agents collaborate inside a model based on VO. The agents take advantage of their learning and adaptation capabilities in order to provide information fusion models. Moreover, HERA facilitates and speeds up the integration between agents and sensors. A totally distributed approach and the use of heterogeneous WSNs provides a platform that is better capable of recovering from errors, and more flexible to adjust its behavior in execution time.

In conclusion, within the proposed framework, a new infrastructure supporting seamless interactions among hardware and software agents, and capable of recognizing and self-adapting to diverse environments is being designed and developed.

Acknowledgments. This work has been supported by the Spanish JCyL project JCRU / 463AC03.

References

[1] Alonso, R.S., De Paz, J.F., García, Ó., Gil, Ó., González, A.: HERA: A New Platform for Embedding Agents in Heterogeneous Wireless Sensor Networks. In: Corchado, E., Graña Romay, M., Manhaes Savio, A. (eds.) HAIS 2010. LNCS (LNAI), vol. 6077, pp. 111–118. Springer, Heidelberg (2010)

[2] Borrajo, M.L., Corchado, J.M., Corchado, E.S., Pellicer, M.A., Bajo, J.: Multi-Agent Neural Business Control System. Information Sciences (Informatics and Computer Science Intelligent Systems Applications An International Journal) 180(6), 911–927 (2010)

[3] Carrascosa, C., Giret, A., Julian, V., Rebollo, M., Argente, E., Botti, V.: Service Oriented MAS: An open architecture (Short Paper). In: Decker, Sichman, Sierra, Castelfranchi (eds.) Proc. of 8th Int. Conf. on Autonomous Agents and Multiagent Systems (AAMAS 2009), Budapest, Hungary, May, 10–15, pp. 1291–1292 (2009)

[4] Castanedo, F., Patricio, M.A., García, J., Molina, J.M.: Data Fusion to Improve Trajectory Tracking in Cooperative Surveillance Multi-Agent Architecture. Information Fusion. An International Journal. Special Issue on Agent-Based Information Fusion 11, 243–255 (2010), doi:10.1016/j.inffus.2009.09.002

[5] Chen, M., Kwon, T., Yuan, Y., Choi, Y., Leung, V.C.M.: Mobile agent-based directed diffusion in wireless sensor networks. EURASIP J. Appl. Signal Process, 219–219 (2007)

[6] Chen, M., Kwon, T., Yuan, Y., Leung, V.C.: Mobile Agent Based Wireless Sensor Networks. Journal of Computers 1 (2006)

[7] De Paz, J.F., Rodríguez, S., Bajo, J., Corchado, J.M., Corchado, E.S.: OVACARE: A Multi-Agent System for Assistance and Health Care. In: Setchi, R., Jordanov, I., Howlett, R.J., Jain, L.C. (eds.) KES 2010. LNCS (LNAI), vol. 6279, pp. 318–327. Springer, Heidelberg (2010)

[8] Fok, C., Roman, G., Lu, C.: Mobile agent middleware for sensor networks: An application case study. In: Proc. of the 4th Int. Conf. on Information Processing in Sensor Networks (IPSN 2005), Los Angeles, California, USA, p. 382 (2005)

[9] Hall, D.L., Llinas, J.: An Introduction to Multisensor Data Fusion. Proceedings of the IEEE 85(1) (January 1997)

[10] Hübner, J.F., Sichman, J.S., Boissier, O.: Using the Moise+ for a Cooperative Framework of MAS Reorganisation. In: Bazzan, A.L.C., Labidi, S. (eds.) SBIA 2004. LNCS (LNAI), vol. 3171, pp. 506–515. Springer, Heidelberg (2004)

[11] Marin-Perianu, M., Meratnia, N., Havinga, P., de Souza, L., Muller, J., Spiess, P., Haller, S., Riedel, T., Decker, C., Stromberg, G.: Decentralized enterprise systems: a multiplatform wireless sensor network approach. IEEE Wireless Communications 14, 57–66 (2007)

[12] Preece, H.H., Gray, P.: KRAFT: Supporting virtual organizations through knowledge fusion. In: Artificial Intelligence for Electronic Commerce: Papers from the AAAI-99 Workshop, pp. 33–38. AAAI Press (1999)

[13] Rajagopalan, R., Mohan, C.K., Varshney, P., Mehrotra, K.: Multi-objective mobile agent routing in wireless sensor networks. In: Proc. of IEEE Congress on Evolutionary Comp. (2005)

[14] Rodríguez, S., Pérez-Lancho, B., De Paz, J.F., Bajo, J., Corchado, J.M.: Ovamah: Multiagent-based Adaptive Virtual Organizations. In: 12th International Conference on Information Fusion, Seattle, Washington, USA (2009)

[15] Rodríguez, S., Pérez-Lancho, B., Bajo, J., Zato, C., Corchado, J.M.: Self-adaptive Coordination for Organizations of Agents in Information Fusion Environments. In: Corchado, E., Graña Romay, M., Manhaes Savio, A. (eds.) HAIS 2010. LNCS (LNAI), vol. 6077, pp. 444–451. Springer, Heidelberg (2010) ISBN: 978-3-642-13802-7

[16] Sarangapani, J.: Wireless Ad hoc and Sensor Networks: Protocols, Performance, and Control. CRC (2007)

[17] Sycara, K., Glinton, R., Yu, B., Giampapa, J., Owens, S., Lewis, M., Charles Grindle, L.T.C.: An integrated approach to high-level information fusion. Information Fusion 10(1), 25–50 (2009)

[18] Tapia, D.I., Alonso, R.S., De Paz, J.F., Corchado, J.M.: Introducing a Distributed Architecture for Heterogeneous Wireless Sensor Networks. Distributed Computing, Artificial Intelligence, Bioinformatics, Soft Computing & Ambient Assisted Living, 116–123 (2009)

[19] Tapia, D.I., Alonso, R.S., Rodríguez, S., De Paz, J.F., Corchado, J.M.: Embedding Reactive Hardware Agents into Hetero generous Sensor Networks. In: 13th International Conference on Information Fusion (2010) ISBN: 978-0-9824438-1-1

[20] Tynan, R., O'Hare, G., Ruzzelli, A.: Multi-Agent System Methodology for Wireless Sensor Networks. Multiagent and Grid Systems 2(4), 491–503 (2006)

[21] Volponi, A.: Data Fusion for Enhanced Aircraft Engine Prognostics and Health Management – Task 14: Program Plan Development, Pratt & Whitney Internal Memo (October 2001)

[22] Wooldridge, M.: An Introduction to MultiAgent Systems. Wiley (2009)

[23] Wu, Y., Cheng, L.: A Study of Mobile Agent Tree Routes for Data Fusion in WSN. In: 2009 WRI International Conference on Communications and Mobile Computing, Kunming, Yunnan, China, pp. 57–60 (2009)

Wasp-Like Agents for Scheduling Production in Real-Time Strategy Games

Marco Santos and Carlos Martinho

Instituto Superior Técnico, Universidade Técnica de Lisboa and INESC-ID
Av. Prof. Doutor Aníbal Cavaco Silva, 2744-016 Porto Salvo
{marco.d.santos,carlos.martinho}@ist.utl.pt

Abstract. In this paper, we propose an algorithm inspired in the social intelligence of wasps for scheduling production in real-time strategy games, and evaluate its performance in a scenario developed as a modification of the game Warcraft III The Frozen Throne. Results show that such an approach is well suited for the highly dynamic nature of the environment in this game genre. We also believe such an approach may allow the exploration of new paradigms of gameplay, and provide some examples in the explored scenarios.

Keywords: Real-time Scheduling, Swarm Intelligence, RTS Games.

1 Introduction

For the past years, the real-time strategy (RTS) genre seems confined to a de facto standard; collect resources, construct a base and an army, overcome the enemy and repeat. Some games, such as Company of Heroes (Relic Entertainment, 2006) stand out due to their combat-oriented gameplay, which is achieved at the expense of other key aspects of RTS games, such as economic management. We believe that by offering a mechanism to automate micro-management of certain aspects of a game (for instance, individual unit production or training), we will promote the exploration of novel paradigms of gameplay by allowing the player to choose which aspects to micro-manage while maintaining a macro level control on the others. In this paper, in particular, we explore an approach where the player only specifies the position where an unit is requested, and leave the actual scheduling to the game system. This relieves her from micro-management routines, such as having to specify where each unit is produced. The macro level control of the economic aspect of the game is offered by requiring the player to specify where each production factory is built in the game environment.

To be accepted by the player and not detract her from the game experience, the mechanisms implementing the automation of micro-management tasks has to deal with the unpredictable and unstable environment that characterizes real time strategy games, with efficiency and robustness. To take into consideration the highly dynamic nature of the environment, we explore a solution based on the social behavior of insects, namely wasps, capable of producing complex emergent behaviors as a colony (swarm intelligence), inspired by their ability to respond

L. Antunes and H.S. Pinto (Eds.): EPIA 2011, LNAI 7026, pp. 71–82, 2011.

and adapt to external changes in a decentralized manner. The automated mechanism for scheduling the production of units has already been shown capable of handling intricate engineering problems [3].

This document is organized as follows. First, we discuss how insect behavior and particularly wasp behavior is adequate to tackle the problem of scheduling in real-time the production of units in RTS games. In the following section, we describe WAIST, the wasp-inspired algorithm we implemented in the game Warcraft III The Frozen Throne (Blizzard Entertainment, 2003), as well as the new gameplay concepts present in our evaluation scenario. Then, we present the evaluation methodology and report the results achieved by WAIST in our scenario. Finally, we draw some conclusions and present some possible future directions for our work.

2   Related Work

2.1   Swarm Intelligence

Swarm Intelligence, a term coined by Beni and Wang in 1989 [1], describes the collective emergent behavior resulting from decentralized and self-organized systems. Its roots are the studies of self-organized social insects, such as ants, wasps or termites [2]. In a colony of such insects, there is no central entity or mechanism controlling or even defining objectives, yet these creatures with strict sensory and cognitive limitations manage to perform complex tasks such as food foraging [9], brood clustering [7], nest maintenance and nest construction [6]. As a result, the mechanisms underlying their complex behavior as a whole became subject of great interest and study, resulting in a great wealth of models inspired by Nature [4].

The particular relevance of such models for our work is based from the fact that most problems dealt within a colony, particularly in the case of wasps, are analogous to the scheduling and logistic engineering problems raised when considering unit production for real-time strategy games.

2.2   Wasp Behavior

From their studies of the *polistes dominulus* wasps, Theraulaz and colleagues created a model of dynamic task allocation that successfully emulates the self-organized behavior of wasps [8].

The model consists in a wasp hive in which there are two possible tasks: foraging and brood care. Individuals decide which task to do according to their response threshold and stimulus emitted by the brood. The system has the following main features:

- Tasks have the capacity of emitting stimuli that affects the individuals' task selection decisions (*stimulus*);

- Individuals possess response thresholds that represent their predisposition to perform certain tasks (*response thresholds*);
- Each individual has a force that is taken into account during dominance contests to determine the winner. Dominance contests form a hierarchy within the colony (*force*);
- When an individual performs a task, the respective response threshold is decreased while the other response thresholds associated with other tasks are increased. This means that the more an individual performs a task the more likely he is to do it again, creating task specialists in the society (*specialization*).

These four features guide the model towards both performance and flexibility. The capacity of specialization of each individual leads self-organization towards optimal performance, allowing the whole work force to dynamically adapt to the constantly changing external environment as well as the intrinsic needs of the colony resulting, for instance, from loss of individual, etc. Such characteristics are of importance when considering the production scheduling in RTS games.

2.3 Routing-Wasp

Based on the properties of the natural model created by Theraulaz and colleagues, Cicirello and colleagues proposed an algorithm for dynamic task allocation [3] that later was adapted to Morley's factory problem from General Motors [5], denominated as Routing-Wasp or R-Wasp.

In this algorithm, each different possible task has a type and is capable of emitting stimuli whose strength if proportional to the time the task is unassigned: the longer a task remains unassigned, the stronger the emitted signal will be. Each agent capable of performing certain tasks has a set of response thresholds, one for each task type, that represents its propensity to bid for a task of this same type: the lower the threshold, the higher the propensity to bid for tasks of this type. For each unassigned task, agents stochastically decide to bid or not according to the strength of the emitted stimulus and the response threshold associated to the task type.

After all the agents have decided to bid or not, if more than one candidate exists, a dominance contest occurs. The dominance contest consists in a tournament where duels are made until one last standing agent wins. The winner of a duel is stochastically settled taking into consideration the agent forces, which vary according to the properties that make the agent more or less suited to perform the task it is competing for. The task is assigned to the dominance contest winner.

Finally, response thresholds are updated according to the task being performed. The more an agent performs tasks of a certain type, the more the respective threshold decreases and the more likely it will accept other tasks of the same type. The reverse is also true, i.e. the less an agent performs a certain

type of tasks to the detriment of others, the less likely it will accept tasks of
that type. However, if an agent is not performing any task, all thresholds are
gradually lowered, increasing the propensity to accept any task.

The R-Wasp strength lies in two aspects. Response thresholds allow the creation of task specialists and gives them the ability to bid for the tasks most suited
for them, without however disabling their ability to do other tasks, if needed.
Force permits a fair distribution of the workload trough the agents by taking
into consideration their characteristics for the task when determining the winner.
These two aspects, together, allow the system to self-regulate and dynamically
respond to unexpected events like loss of agents or variations in demand.

Such characteristics are relevant when considering the needs for real-time
scheduling of resources in the RTS game genre. As such, in this work, we extended
and adapted the R-Wasp algorithm to the RTS game domain and evaluate its
adequacy.

3   WAIST: R-Wasp in RTS Games

Before describing the details of our adaptation of the R-Wasp algorithm to the
RTS game genre, we will describe the testbed and scenario used for both the
implementation and evaluation of our approach.

3.1   Scenario

We implemented a modification[1] of the video game Warcraft III The Frozen
Throne (Blizzard Entertainment, 2003) that provides the player with a new
type of interface to request units. The game was chosen because it is one of the
currently most played RTS game and as such an adequate representative of the
target game domain. Using an actual game also allows us to account for other
in-game factors (e.g. unit path finding, in-between unit collisions, unexpected
destruction of factories, etc.) while running the experiment. In the base Warcraft
III game, to build a new unit, the player has first to build a factory and then
issue build orders individually for each unit. Each type of factory is only able to
produce a limited number of units and unit types, and different factories produce
different unit types. As such, the player has to micro-manage unit production
and constantly move between factories, while controlling all the other aspects of
the game simultaneously.

Our scenario provides the player with a macro-level management of unit production and explores a new paradigm of gameplay. The player still decides which
factory type to build and where it is built but, afterwards, only needs to specify
the location where a certain quantity of units of different types are requested,
and the underlying system will devise a near-optimal production schedule accordingly. As such, we ensure there are still meaningful choices for the player to
make regarding resource management, albeit at a completely different level.

---

[1] Also known as *mod* by the online videogame community.

To provide even more interesting choices to the player, since scheduling is automated and does not cognitively hinder the player, we refined the production model and further introduced unit *heterogeneity* in factory production: different factories are able to produce the same unit type but have different production times, and an extra setup time is required when production changes from one type of unit to another.

Before moving to a full game, however, we specifically designed a first scenario to evaluate our approach, in which:

– Two types of units are considered: F(ootman) and R(ifleman);
– Changing from the production of one type of unit to another results in extra setup time before resuming production;
– Five different factories are available to the player (see Fig. 1), each one with different balanced properties:

  • Two specialized factories (F_F and F_R) are efficient at producing one type of unit (F and R respectively) but unable to produce the other.
  • One factory (F_{Hybrid}) is able to produce both unit types but not as efficiently as the specialized factories. Setup time, however, is very high.
  • Two other factories (F_{FR} and F_{RF}) are also able to produce both unit types and have a lower setup time than the hybrid factory. However, one of the unit type has a high production time.

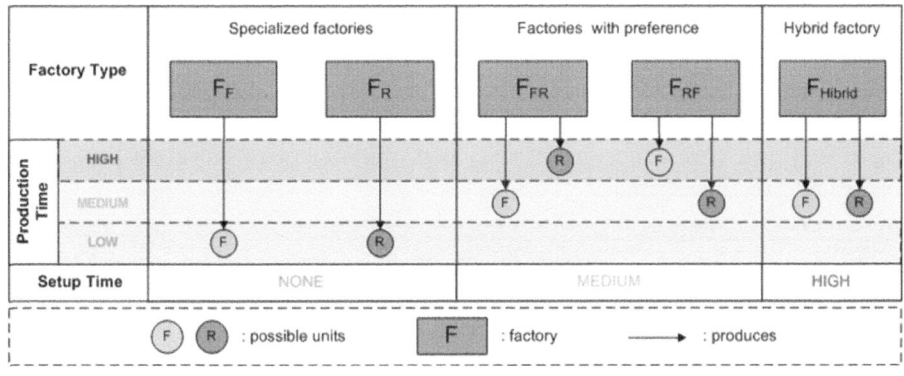

Fig. 1. Representation of the 5 factories considered by our scenario

For the evaluation of the scenario, the objective is to build a determined amount of units in response to a set of requests issued within a certain amount of time, using the available factories. To test the robustness of the approach, several variants are introduced, such as the destruction of factories and different distributions of request orders. The variants will be detailed later in the document.

3.2   Model

We will now describe the algorithm used to implement real-time scheduling in our scenario. Fig 2 depicts the pseudo-code for the algorithm.

```
1   var: list of unassigned orders
2   var: list of factories
3   var: list of candidates = {}
4   for-each unassigned order
5       for-each factory
6           decide to bid or not              (1)
7           if( bid )
8               candidates <- factory
9           dominance contest between candidates (2,3)
10          assign order to the winner
11          update forces                     (4)
12  update thresholds                         (5)
13  update stimuli                            (6)
```

Fig. 2. Pseudo-code of the algorithm. The marked steps are detailed in the main text.

Each unit requested by the player is a task that must be performed. Tasks have a type, which is defined by the type of unit to be produced (in our scenario, Footman or Rifleman), and are capable of emitting stimuli perceptible by all existing factories.

Each factory f capable of producing a certain type of units is conceptually modelled as a *wasp* capable of executing production tasks and has a response threshold $T_{f,t}$ associated with this task t. When a factory f receives a stimulus S_t from a task t, it stochastically decides to bid or not to bid according to the probability defined by Eq. 1.

$$P(bid \mid T_{f,t}, S_t) = \frac{S_t^2}{S_t^2 + T_{f,t}^2} \qquad (1)$$

After all factories have decided to bid or not, three different cases can occur:

- No factory bids for the task: in this case, the task remains unassigned until the next cycle;
- Only one factory bids for the task: in this case, the task is attributed to the single bidder;
- More than one factory bids for the task: in this case, the winner is selected after a tournament of dominance contests.

Dominance contests are performed as follows. First, if an odd number or factories bid, the number of participants is made an even number by moving automatically N factories to the next phase of the tournament according to Eq. 2, in which C stands for the number of competitors.

$$N = 2^{\lceil log_2(C) \rceil} - C \qquad (2)$$

Then, duels are performed. Each dominance contest takes the force of the biding "wasp factories" into consideration. The probability of factory f_1 winning over factory f_2 in a duel is given by Eq. 3, in which F_{f_1} and F_{f_2} represent the forces of each factory respectively.

$$P(f_1 wins \mid F_{f_1}, F_{f_2}) = \frac{F_{f_2}^2}{F_{f_1}^2 + F_{f_2}^2} \tag{3}$$

The force F_f of a factory f is given by Eq. 4, where T_{prd} is the sum of all the production times of the tasks currently in the factory's queue, the task being processed and also the one being disputed; T_{stp} is the sum of all required setups, including the task being disputed; and T_{mov} is the estimated time needed to move the unit from the factory to its target location. Note that the lower the force, the faster the factory will execute the job.

$$F_f = T_{prd} + T_{stp} + T_{mov} \tag{4}$$

In each cycle, the response thresholds of all factories are updated according to the following criteria:

$$T_{f,t} = \begin{cases} T_{f,t} - \delta_1 & \text{if the last task in factory } f \text{ queue is of type } t \\ T_{f,t} + \delta_2 & \text{if the last task in factory } f \text{ queue is not of type } t \\ T_{f,t} - \delta_3 & \text{if factory } f \text{ queue is idle} \\ T_{f,t} - \delta_4 & \text{if factory } f \text{ queue is idle and there is an unassigned} \\ & \text{task of type } t \end{cases} \tag{5}$$

δ_1 acts as a "learning" coefficient, since it diminishes the threshold and encourages the factory to take tasks of the same type, as opposed to δ_2, which acts as an "unlearning" coefficient. These two rules are important to reduce the number of setups. When factories are idle, their response thresholds are slowly decreased by δ_3, if there are no unassigned tasks, or more rapidly decreased by $\delta_4 > \delta_3$ if there are unassigned tasks. Both thresholds ensure factories do not remain inactive if there are still tasks to perform. After being updated, all response thresholds are clamped to a predefined interval.

Task stimuli are also updated every cycle, according to Eq. 6, to ensure that a task does not remain unassigned indefinitely.

$$S_t = S_t + \delta_S \tag{6}$$

This update rule ensures that older requests have higher priorities since they will emit stronger stimuli. This update also allows more specialized factories to bid for other types of tasks, when necessary. Other factors could be easily added to Eq. 6 to create other priority criteria rather than "task unassigned time", such as "unit priority", for example.

4   Evaluation

4.1   Methodology

To evaluate our scenario, we ran a series of automated scripts that, at certain time intervals, requested one or more units at certain points of the map, simulating a typical human player input. To execute the orders, 3 factories were positioned close to each other, implementing the concept of a "base camp". The type of the factories was chosen randomly, with the constraint that the production of both Footman and Rifleman units should be possible within this base camp. Each time a scenario is run, it is run once using each one of the alternative base camp configuration.

To evaluate our approach, five different variants of our base scenario were implemented and tested within the Warcraft III The Frozen Throne World Editor game environment, ensuring that other factors such as path-finding and collisions between units or between the units and the environment were accounted for in the in-game simulation:

- Uniform distribution of requests: at regular time intervals, a new request is issued. The requested unit has 50% chance of being either a Footman or a Rifleman. (*"even"* variant)
- Mirrored distribution of request: at regular time intervals, a new request is issued. The requested unit has 90% chance of being a Footman and 10% chance of being a Rifleman. When half of the desired units have been requested, the distribution is mirrored, i.e. there is 10% chance of the request being a Footman and 90% chance of being a Rifleman. (*"uneven"* variant)
- Destruction and construction of factories: in this variant, two factories are destroyed and two random others are constructed, within a certain time frame. The new factories are selected such as to always allow the construction of the two unit types. Distribution of requests is uniform. (*"destruct"* variant).
- Alternating high intensity and low intensity request: instead of being ordered regularly, units are requested in bursts, i.e. in small time intervals with longer pauses between them. This variant simulates the way that units are usually ordered in RTS games. (*"burst"* variant)
- Additional available factories: this variant increases the number of factories available to produce units to 9. The 9 factories are grouped into 3 base camps spread over the map. Distribution of requests is uniform. (*"more"* variant)

To evaluate the performance of WAIST, we implemented three additional algorithms, and analyzed them comparatively to our approach. The three algorithms were:

- Random attribution: this algorithm randomly selects one of the available factories capable of producing the requested unit that has free slots in its production queue. This scenario serves primarily the purpose of creating a baseline. (*"random"* algorithm)

- Closest-distance attribution: this algorithm attributes production requests to the closest factory with empty slots in its queue. This algorithm simulates the assignment strategy commonly used by "newcomer" players. ("closest" algorithm)
- Global attribution: this algorithm attributes production requests to the expected fastest factory at the moment of the request. It takes into consideration the time needed for the ordered unit to move from the factory to the target rally point in a straight line, and all production and setup times of the requests in its queue, including the one being disputed. This last algorithm represents the centralized optimal choice, using all the available information when decision is taken.

For evaluation purposes, we measured the execution time of a scenario, that is the time elapsed since the start of the scenario until the last unit reaches the destination to which it was requested. Each one of the scenario variant was ran several times:

- "closest" and "global" are deterministic and as such were executed 3 times for each one of the five base camp configurations, to account for the intrinsically unpredictable nature of the game environment;
- "random" and WAIST are intrinsically stochastic, and as such were executed 15 times for each base camp configuration.

In all cases, the final execution time is the average execution time of all runs.

Because our approach requires tuning a considerable amount of parameters, before running the variant scenarios, we performed several iterations of smaller (and faster) runs consisting of 40 random unit requests over the base scenario. After the parameters were set, we performed the five variant scenario with 100 requests each for each one of the base camp configuration and for each one of the four algorithms, to evaluate our approach.

4.2   Results

Fig. 3 compares the total execution time for each algorithm across the five different scenario variants used during the evaluation phase. Both WAIST and "global" algorithms achieved a similar performance, "global" only outperforming WAIST by an average of 0.9% over the five scenarios variants. Both WAIST and "global" performed above "random" and "closest", which shared a same average performance. We consider this an encouraging result for WAIST, taking into consideration that scheduling relies only on local information, while "global" has access to all information in the game environment.

WAIST presented the additional advantage of allowing factories to wait for the best suited requests by using the response threshold mechanism. This is visible in Fig. 4, that compares the total number of setups for each algorithm across the five scenario variants. WAIST outperformed all algorithms by achieving the lowest number of setup times in all but the "more factories" scenario variant.

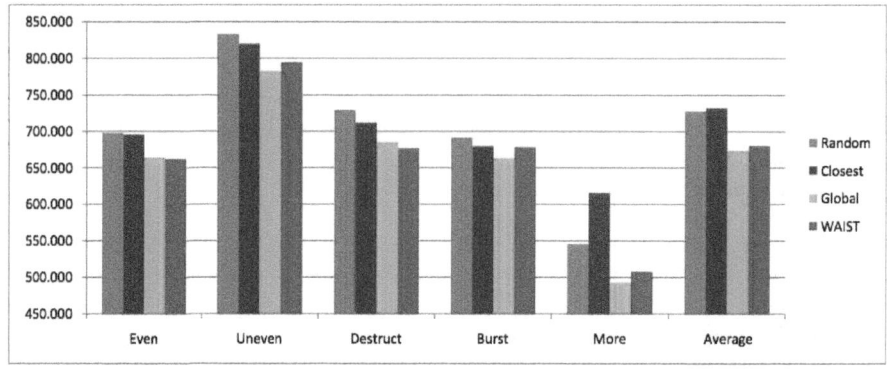

	Even	Uneven	Destruct	Burst	More	Average
Random	697.600	833.067	729.200	691.267	545.867	727.667
Closest	695.667	820.000	712.000	679.333	615.667	732.200
Global	663.667	782.667	684.867	663.000	492.667	673.040
WAIST	661.933	794.600	676.667	678.200	507.667	679.880

Fig. 3. Execution times of the five different scenario variants

A possible explanation for this fact is that, as more factories are available, more idle periods occur which may lead some factories to experience a premature loss of specialization.

A final remark: in this work, WAIST parameters were tuned by hand. We believe the use of an optimization algorithm could provide a better set of parameters, by taking into account the specific characteristic of each particular scenario variant. In future work, we intend to further research this issue.

4.3   WAIST Benefits

Although not the most efficient approach in all possible situations, WAIST demonstrated to be an efficient and versatile approach for unit production scheduling in RTS games, even when compared to a centralized algorithm. WAIST was found to be most appropriate in games with the following characteristics:

- Significant setup times: in this work, setups accounted for approximately 10% of the time needed to train or produce a unit. We expect higher setup times to boost WAIST performance in comparison to the other approaches;
- Congestion of requests: WAIST is most suited for situation with continuous high request rate, which is expected at higher levels of play. Due to the lack of idle times, specialization ensures the adequacy of the algorithm;
- Performance scalability: WAIST is adequate for larger scale scenarios: while the production queue and the capability of all factories are evaluated to select the best possible choice in a centralized solution, bidding and dominance contests require less computation power while achieving a similar performance.

Number of setups

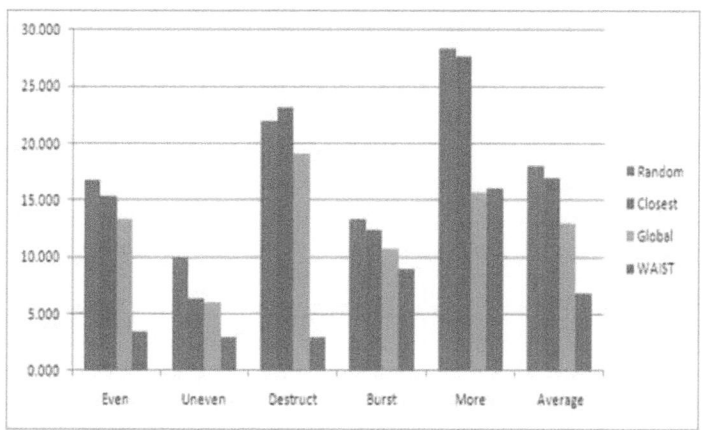

Fig. 4. Number of setups for each algorithm and scenario variant

Another benefit comes from its agent-based nature. In this work, due to the use of the Warcraft III World Editor, the algorithm was executed sequentially. However, this algorithm can take advantage of the parallel computation offered by multi-core processors, increasing its computational efficiency even further.

5   Conclusion

In this paper, we presented WAIST, an algorithm inspired in the social intelligence of wasps for scheduling unit production in real-time strategy games and evaluated its performance with a set of five scenario variants developed as a modification of the game Warcraft III The Frozen Throne. The variants accounted for factors such as: the rate and distribution of the requests issued over the scenario, the number of available factories, and environment changes such as the destruction and construction of factories during the scenario. The performance of WAIST in each scenario variants was compared to three other approaches: random attribution; distance-based attribution, and global attribution, which considers all the information available at the moment from the game environment.

Overall, WAIST performed comparably to the global attribution algorithm (and better than the other two), an encouraging result considering WAIST is a decentralized algorithm that relies on local information while the latter has full global knowledge. As such, we believe WAIST to be an efficient and reliable alternative for real time scheduling in real time strategy games. While WAIST experiences some limitations when dealing with low amounts of requests, it demonstrated good performance in situations of higher congestion of requests, and when setting up from one production type to another has a cost that cannot be ignored. The limitations may be related to fine tuning specialization-related parameters for the specificities of each scenario, and should be researched further. Due to its decentralized and multi-agent nature, WAIST should be able to scale relatively well performance-wise, and take advantage of multi-core processors.

Approaches such as the one described in this paper allows us to start exploring new paradigms of gameplay for real-time strategy games, such as relieving the player from having to micro-manage certain tasks (in this paper, unit production and training) while still having to perform interesting choices at the macro level of play (in this paper, deciding where to build the production or training facility). By relieving the player from certain tasks, we are able introduce her to alternative challenges. In this paper, particularly, we introduced a new feature observable in real factories to the current language of game mechanics related to real-time strategy games: heterogeneity i.e. factories with different characteristics, and extra production costs, either of resources or time, resultant from the change of production. These are some examples of new paradigms of gameplay that could be enabled by the adequate use of artificial intelligence.

References

1. Beni, G., Wang, J.: Swarm intelligence in cellular robotic systems. In: Proc. NATO Advanced Workshop on Robots and Biological Systems, pp. 26–30 (1989)
2. Bonabeau, E., Dorigo, M., Theraulaz, G.: Swarm Intelligence: From Natural to Artificial Systems. Santa Fe Institute Studies on the Sciences of Complexity. Oxford University Press, USA (1999)
3. Cicirello, V., Smith, S.F.: Wasp nests for self-configurable factories. In: Müller, J.P., Andre, E., Sen, S., Frasson, C. (eds.) Proceedings of the Fifth International Conference on Autonomous Agents, pp. 473–480. ACM Press (2001)
4. Engelbrecht, A.: Fundamentals of Computational Swarm Intelligence. Wiley (2006)
5. Morley, D.: Painting trucks at general motors: The effectiveness of a complexity-based approach. In: Embracing Complexity: Exploring the Application of Complex Adaptive Systems to Business, The Ernst and Young Center for Business Innovation, pp. 53–58 (1996)
6. Perna, A., Jost, C., Valverde, S., Gautrais, J., Theraulaz, G., Kuntz, P.: The Topological Fortress of Termites. In: Liò, P., Yoneki, E., Crowcroft, J., Verma, D.C. (eds.) BIOWIRE 2007. LNCS, vol. 5151, pp. 165–173. Springer, Heidelberg (2008)
7. Sendova-Franks, A., Scholes, S., Franks, N., Melhuish, C.: Brood sorting by ants: two phases and differential diffusion. Animal Behaviour 68, 1095–1106 (2004)
8. Theraulaz, G., Goss, S., Gervet, J., Deneubourg, J.L.: Task differentiation in polistes wasp colonies: a model for self-organizing groups of robots. In: Proceedings of the first International Conference on Simulation of Adaptive Behavior - From Animals to Animats. pp. 346–355 (1990)
9. Traniello, J.F.A.: Foraging strategies of ants. Annual Review of Entomology 34, 191–210 (1989)

Operational Problems Recovery in Airlines – A Specialized Methodologies Approach

Bruno Aguiar[1], José Torres[1], and António J.M. Castro[2]

[1] FEUP - Faculty of Engineering, University of Porto,
Rua Dr. Roberto Frias, s/n 4200-465 Porto, Portugal
{bruno.aguiar,ricardo.torres}@fe.up.pt
[2] LIACC, FEUP, Department of Informatics Engineering, University of Porto,
4200-465 Porto, Portugal
antonio.castro@fe.up.pt

Abstract. Disruption management is one of the most important scheduling problems in the airline industry because of the elevated costs associated, however this is relatively new research area comparing for example with fleet and tail assignment. The major goal to solve this kind of problem is to achieve a feasible solution for the airline company minimizing the several costs involved and within time constraints. An approach to solve operational problems causing disruptions is presented using different specialized methodologies for the problems with aircrafts and crewmembers including flight graph based with meta-heuristic optimization algorithms. These approaches were built to fit on a multi-agent system with specialist agents solving disruptions. A comparative analysis of the algorithms is also presented. Using a complete month real dataset we demonstrate an example how the system handled disruption events. The resulting application is able to solve disruption events optimizing costs and respecting operational constraints.

Keywords: disruption management, scheduling in airlines, flight-aircraft recovery, crew rescheduling, genetic algorithms, flight graph.

1   Introduction

Operations control is a critical area for an airline company. Without complying with minimum assignments on the flight schedule, the airline company management could become chaotic and lead to severe losses. In [10], the authors made an overview on disruption management and claim that "(...) due to their frequent occurrences, the ability to dynamically revise the original schedules to suit the newly changed operational environment after disruptions is very important to the airline. Disruption management refers to this process of plan adjustment (...)".

From the informatics and computing scheduling area, the airline scheduling is one of the most challenging problems, and in the specific case of rescheduling it is even more important, because disruptions need a just-in-time answer otherwise aircraft will stay on ground, wasting parking and crew resources, and passengers will not get to their destinations resulting in major financial and reputation losses.

L. Antunes and H.S. Pinto (Eds.): EPIA 2011, LNAI 7026, pp. 83–97, 2011.
© Springer-Verlag Berlin Heidelberg 2011

This problem is usually decomposed in stages and that is also our approach to solve it, separating aircrafts, crew members and passengers because normally these problems are handled by different departments, having also distinct deadlines and needed knowledge to solve them.

Our proposal goes on that direction, adopting separated methodologies representing each department by the problem that it solves and a specialized team for each algorithm that solves the problem in a different way.

This kind of specification has to do with previous works on the disruption management area, regarding a multi-agent system that deals with disruption events with agents organized in a hierarchy structure as proposed by [1]. This specification includes specialist agents to implement different algorithms for each kind of problem and that are the subject of this paper.

The operational control on the OCC, Operational Control Center, is done continuously with the operational plan, tail and crew duty assignments that are prepared within 1 to 6 months for flights and aircrafts and 1 to 2 months for the crew [2],[3]. The passenger flight connections are usually analyzed 1 to 2 days before the day of operations [2]. Our research focus on planning within the day of operations, however the solutions obtained could result in long term changes.

Unfortunately, due to unexpected events, it is very difficult for the airline to keep the original optimized schedule, generated months before. Whether by propagated delays that accumulate until the day of operation or by disruptions occurred within current day, the original schedule becomes unfeasible. The most common issues that lead to disruption are bad weather, unexpected technical problems, with aircraft or the airport logistics, or even paperwork and crew members that do not show for the briefing before duty.

Analyzing the AEA, Association of European Airlines, European punctuality report on the first quarter of 2008 [13], 22,4% of the departures were delayed more than 15 minutes, meaning an increase of 1,9% comparing with 2007.

The main causes for those delays are airport infrastructure and traffic management delays (13,1%), typically the aircraft is ready to leave but some hold-up or authorization is missing, other major source of delays is attributed to pre-flight preparation, i.e. the aircraft is not ready to departure on time for several reasons that might be late loading, pending paperwork or crew members late to duty. On the other hand, bad weather conditions represent less than 2% of the delays.

It is interesting that the busiest airport in Europe, London Heathrow Airport, has the highest percentage of delays, with 44,1% of the departures delayed more than 15 minutes.

In [2] the authors claim that "(...) the delay costs for airlines is estimated between 840 and 1.200 million Euros in 2002 (...) and the delay cost per minute is estimated to be around 72 Euros", showing the importance of dealing well with disruptions in an industry sector that has all five Porter's[1] forces against it.

Recovering from a disruption in airlines involves searching in a very large candidate solutions space looking for a suboptimal solution that allows the recovery from an unfeasible state. This process includes an optimization of the solutions to achieve the lowest possible costs.

---

[1] Michael Eugene Porter, Professor at Harvard Business School.

The work presented here consists on the development of some scheduling methodologies to solve the operational problems in airlines related with delays. The resolution of the problems will be made by specialized approaches to each one of the problems using several meta-heuristics. The approaches will be based on flight graph representation for crew and chromosome for aircraft problems.

In this paper we will focus only on these problems, discarding passenger connection problems.

We believe that using a full month real dataset will make the final application more relevant, as most of the research works on this area are not functional and do not apply to the real size of disruption management.

With this work we expect to have an application capable of dealing with disruptions, generating optimized solutions within time constraints, usually less than 3 minutes. Having different approaches allow us to compare and take some conclusions about the most adequate algorithms for each problem.

The rest of the paper is organized as follows. In section 2 we present some related work on this subject including some of the papers that influenced our work. Sections 3 and 4 present the approaches proposed for each of the sub-problems, aircrafts and crew. These sections include problem statement, constraints, solution, algorithms, results and discussion applied to real disruptions. Section 5 presents the conclusions and finally the remaining section is about future work.

2   State of the Art

Although disruption management is the newest research area regarding airline scheduling [10] there are right now some important papers with approaches to this problem. The related work is divided in two major categories concerning each one of the problems. Passenger recovery is mainly integrated with the other approaches.

Our primary influence, reflected on our work and research, is the work of Castro et al that proposes a multi-agent system to solve disruptions with specialized agents [1], [3], [4], [5]. In our previous work regarding multi-agent systems we have implemented their agent hierarchy. The authors in [1] also propose an algorithm to solve crew and passengers disruptions using quality costs. Most of the papers analyzed during our work are referenced by [5] and we recommend reading them.

Regarding the aircraft recovery problem, one of the most recent papers in this area [6], is an approach to solve aircraft recovery while considering disrupted passengers as part of the objective function cost using an Ant Colony (ACO) algorithm. The solutions are obtained minimizing the recovery scope, returning as soon as possible to the original schedule. According to the authors "(…) the computational results indicate that the ACO can be successfully used to solve the airline recovery problem". Liu et al. [7], solved this problem using a "multi-objective genetic algorithm" that attempts to optimize a multi-objective function that includes several costs associated, according to the authors "(…) our results further demonstrate that the application can yield high quality solutions quickly (…)".

Considering the crew recovery problem, [1] propose a hill climbing algorithm that swaps crewmembers considering penalizations for schedule changes and delay within the objective function that includes the wages costs. The authors compare this approach with the human method performed in a real airline company, they obtained major decreases on costs. Although it is not a recovery oriented approach, Ozdemir et al work [8] on crew scheduling uses the concept of flight graph, combined with a genetic algorithm. The graph includes some of the major constraints to the problem and we based our recovery version with their idea of graph.

In terms of passenger recovery, Castro and Oliveira [5], propose quality costs for the passengers, including compensations, meals and hotel costs. They also present a quantification of this quality costs using passenger profiles. Most of the approaches that take in consideration disrupted passengers embedded their costs in the aircraft and crew recovery objective functions [6], [9].

We consider that the datasets tested within these papers have a reduced size taking into account the number of flights per month on a real airline company. Our work tries to apply some of these approaches with data set representative of a real month flight schedule.

3 Aircraft Recovery Problem

We will focus in the aircraft recovery problem (ARP), which aim is to find a way to get back to the initial schedule by delaying or canceling flights, or by reassign them to other planes (plane swapping), given its disrupted state and the initial planned schedule. Hereupon, the main objectives are to minimize the costs (in terms of delay, cancellation and parking costs of the aircrafts), the difference in average flight times by aircraft, the recovery period and finally the differences between the original schedule and the recovered one.

We can introduce ARP with a small example. In Table 1 we have a schedule for planes p_1 and p_2. When p_1 lands in OPO, it comes up to knowledge that he will not be able to depart for 2 hours, due to an unforeseen task to be performed, such as maintenance work or resolution of a technical malfunction. This delay, d_1, causes a situation of disruption because the schedule cannot be performed as planned (p_1 cannot take off to *BCN* at 10:00, it will only be ready for takeoff at 11:00).

Table 1. The original schedule for two planes

Plane 1	Flight ID	Origin	Destination	Departure time	Arrival time
	F1	LIS	OPO	08:00	09:00
	F2	OPO	BCN	10:00	11:30
	F3	BCN	PAR	12:30	14:15
	F4	PAR	LIS	15:00	17:35
Plane 2	Flight ID	Origin	Destination	Departure time	Arrival time
	F5	MAD	OPO	08:15	09:30
	F6	OPO	FRA	11:20	13:50

For this small example, a possible recovery plan would be to swap the affection of flights *F2*, *F3*, *F4* from *p1* to *p2* and *F6* from *p2* to *p1*. The resulting recovery schedule is presented in Table 2.

Table 2. The recovered schedule for the initial schedule in Table 1

Plane 1	Flight ID	Origin	Destination	Departure time	Arrival time
	F1	LIS	OPO	08:00	09:00
	F6	OPO	FRA	11:20	13:50
Plane 2	Flight ID	Origin	Destination	Departure time	Arrival time
	F5	MAD	OPO	08:15	09:30
	F2	OPO	BCN	10:00	11:30
	F3	BCN	PAR	12:30	14:15
	F4	PAR	LIS	15:00	17:35

3.1 Problem and Solution

Problem Representation. When a disruption happens, the location and the ready time for each aircraft are calculated. This calculation considers past flights and delays inserted. Therefore, with this representation we can easily add delays to the aircrafts, knowing that they will be correctly added to the ready times.

Given the schedule in Table 1 and the delay d_1, at the time dl is inserted the location and ready times will be calculated for each plane, and will be represented as [OPO, 11:00] for $p1$ and [OPO, 09:30] for $p2$, considering 30 minutes rest time between consecutive flights.

Problem Constraints. The aircraft recovery problem involves several constraints, some of them could not be violated, i.e. hard rules, and others that are subject to optimization, i.e. soft rules. Besides the obvious constraints as the aircraft could only do a flight if it is on the departure airport at time of departure and all the flights should have an aircraft allocated, there are other hard rules that must be respected such as rotation time between flights, one hour is the default value, and each aircraft should remain at one of the airline hubs (LIS or OPO), for at least 360 minutes each four days period for maintenance purpose. There are also constraints regarding fleets for every flight, i.e. only narrow body aircrafts could make narrow body flights, the same applies for the wide body ones.

In terms of soft rules, that are subject of optimization, the goals are to minimize flight delays, changes in schedule, parking costs of the aircrafts and achieve an equitable distribution of flight hours among the aircrafts.

Solution Representation. The representation of the tail assignment is held by the use of a chromosome. This chromosome is composed by two connected lists, the first one represents the flights ordered by departure time and the second one represents the planes allocated for the flights list. This kind of representation is also used by [11].

In Table 3 we can see an example of a small chromosome representation, where the first row represents the planes and the second one the flights.

Table 3. A small example of the chromosome representation

CSTNA	CSTNB	CSTOB	CSTNE	CSTTA
114	194	1602	1900	1951

As for the delays, they will be represented as tuples, being those composed by the flight and the respective delay.

With this solution representation we implemented several different meta-heuristic to solve the problem: *hill-climbing, simulated annealing* and *genetic algorithm*.

The *hill climbing* and *simulated annealing* algorithms are very similar and solve the problem iteratively by following the steps:

1. Obtains the flights that are in the time window of the problem. This time window starts at the delayed flight date and ends at the end of the month. These are the only flights that could undergo a change in their aircrafts.
2. While some specific and customizable time or number of iterations has not passed, the steps 3 and 4 are repeated.
3. Generates a successor of the initial or previous solution.
4. Evaluates the fitness value of the generated solution according to Equation 3. If it is better than the fitness value of the current solution, accepts the generated solution as the new current solution. Otherwise, in the case of *hill climbing* algorithm it discards the generated solution. On the other hand *simulated annealing* accepts the generated solution with a small probability given by Equation 1.

$$P = \exp\left(-\Delta E / T\right) \tag{1}$$

Where ΔE is the difference between the fitness values of the generated solution and the current one, and T is given by Equation 2

$$T_{t+1} = \alpha T_t \tag{2}$$

Where α is the cooling factor and lies between 0 and 1. α was given the value of 0.95 and T was given an initial value of 50. T is updated every iteration.

5. Send the information on which flight will be delayed and for how long to the crew solver.

The *genetic algorithm* approach solves the problem iteratively by following the steps:

1. Obtains the flights that are in the time window of the problem. This time window starts at the delayed flight date and ends at the end of the month. These are the only flights that could undergo a change in their aircrafts.
2. Generates randomly an initial population of n chromosomes, all suitable solutions for the problem.
3. While some specific and customizable time or number of iterations has not passed, the steps 4, 5 and 6 are repeated.
4. Evaluates the fitness of each chromosome in the population and if one chromosome is the best so far keep it
5. Create a new population by repeating following steps until the new population is complete
   a. Select two parents chromosomes from the population, using *roulette* or *tournament* selection
   b. Reproduce the two chosen parents to generate the offspring (children)

      c.   With a small mutation probability mutate the new offspring at each position on chromosome

      d.   Place the new offspring in the new population

6.   Replace the previous population by the new one

7.   Send the information on which flight will be delayed and for how long to the crew solver, using the best chromosome saved as the solution.

Solution Generation for *Hill Climbing* and *Simulated Annealing*. The generation of a new solution is made by swapping two aircrafts in their flight assignment. However, the two chosen aircrafts are not randomly selected. The first aircraft selected to swap must be an aircraft that is causing a delay and therefore the second aircraft is found within the set of planes that can prevent the delay or at least decrease it.

Solution Evaluation. Considering the representation of solution previously explained we calculate the value (fitness) of a solution using the Equation 3.

$$\sum_{\substack{i=1 \\ where \\ i \in F; F=\{all\ flights\ in\ solution\}}}^{|F|} D_i * 101 \quad + (SG - WP) * 100/WP \tag{3}$$

Where D is the delay in minutes defined to the flight and SG is the result of the calculation of all secondary goals, defined by Equation 4. WP is the worst possible fitness for the secondary goals. Therefore it is important to point that the slightest improvement in the total delay minutes will be favored over any improvement in the secondary goals.

$$\sum_{\substack{i=1 \\ where \\ i \in F; F=\{all\ flights\ in\ solution\}}}^{|F|} PC_i + \sum_{\substack{j=1 \\ where \\ j \in P; P=\{all\ planes\ in\ solution\}}}^{|P|} \Delta FH_j + SC_j \tag{4}$$

Where PC is the parking costs resulting from the time the plane spent in standby at the airport before the flight, ΔFH is the difference between the flight hours of the plane and the average flight hours from all planes. SC is the number of schedule changes suffered by the plane.

3.2 Results and Discussion

Using information about flights, planes and airports provided by TAP[2], we were able to apply our approaches using real data. We also used an initial schedule calculated by an algorithm created by us. We tested the algorithms with 4 different disrupted instances using 51 planes and 3521 flights during one month. The delays were

---

[2] TAP Portugal - http://www.flytap.com

inserted on airports, meaning that all flights within the time window of the delay could not departure on time, e.g. weather problems, and also aircrafts delays that represent one specific delay, e.g. unexpected maintenance. As said before the data set from TAP has two fleets, Narrow and Wide Body from Airbus[3].

Table 4. Results for some different disruption scenarios. "Affected flights" is the number of flights affected by the delays including the ones affected by propagation. Total delay is the total minutes of all delays.

Disruption	Test1	Test 2	Test 3	Test 4
#Delayed Planes	17	4	11	1
#Affected Flights	41	10	18	2
Total delay (minutes)	7610	2300	3383	630
Solution				
#Flights Delayed	29	6	3	1
Total delay (minutes)	1665	1460	255	600
Run time (seconds)	5,6	4,3	4,1	1,3

The disruption events that generated the results for *Test 1* and *Test 3* were introduced in Lisbon, the primary hub on TAP operation. *Test 2* results from a disruption event in Oporto, Portuguese second largest airport and secondary hub for TAP, and finally *Test 4* on a flight from Funchal (FNC) that is a small airport in Madeira island.

The results presented in Table 4 show that even with big data sets and with relatively high number of delays inserted this kind of approach is reliable and feasible in a short time. The general behavior shows that the flights affected by propagation are the main target of this kind of approaches, because they are more easily influenced by the reallocation.

Analyzing for example *Test 1*, we realize that in the end we are left with only 21.87% of the originals delays, and if considering the 72€ cost per minute of delay proposed by [2] we realize that thanks to the disruption solution we are capable of saving something near 428 040€. In the case of *Test 3* the remaining delays are only 7.54% of the original ones. Such great results are justified by the fact that this two examples have many delayed planes and for a relatively long time and due to such, are overgrown with propagated delays which are the most affected by the planes swapping, and also these are examples of disruption management in a main airport, in this case Lisbon.

Although such a performance shall not always be expected, the remaining results are capable of reaching 63.48%, and 95.24% of the previous delays. As we can see these two examples are unprivileged on propagated delays and these examples are not in main airports, thus decreasing the possibilities of finding replacements for the delayed planes.

Since we have implemented three different algorithms to solve this problem, we present in Fig 1, a chart with the evolution of the delay minutes remaining through time for *Test 1*. As it is visible in the chart, the recommended algorithm would be, as

---

[3] Airbus - http://www.airbus.com/

expected, the genetic algorithm (GA). However, if the execution is intended to be really short, the hill-climbing (HC) or the simulated annealing (SA) could be a better choice, since they are slightly better than the genetic in an initial phase. Thus, there is not really an algorithm that could be excluded.

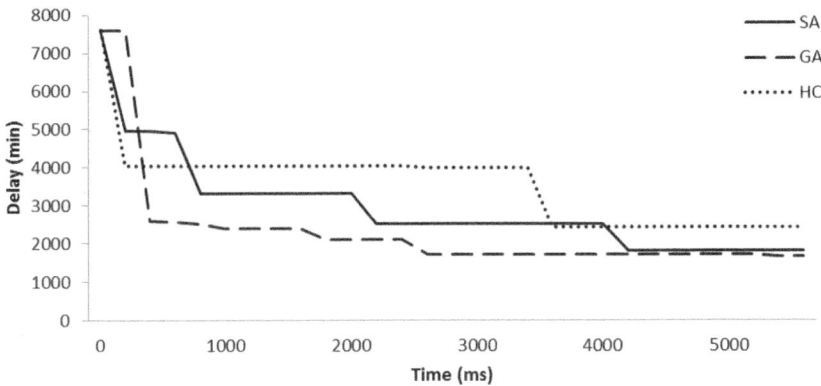

Fig. 1. Chart representing the delay minutes optimization through time using SA, HC and GA

4   Crew Recovery Problem

There is a crew problem whenever a crewmember is not available to perform a flight included in his schedule. This happens when a flight is delayed or the crewmember does not arrive in time for the briefing before the flight. When such events take place another crewmember should be found in order to allow the flight to depart in time. Otherwise the flight would be delayed or cancelled.

Using the schedule example in Table 2, if *F1* was delayed for 45 minutes and if a crewmember *C1* was scheduled to perform *F1* and *F2* there would be a connection problem because *C1* would not be capable of perform *F2* in time. In that case, another crewmember is needed to perform that flight on time. The best solution would be to get a crewmember *C2* that arrived from flight *F5* and allocate him to flight *F2* and all following flights from *C1* schedule. *C1* would then be allocated with the flights from *C2* schedule.

However, when those schedules have a future common flight, it could be possible to maintain the subsequent flights by identifying this flight as the end point of the swapping operation.

If there is not any crewmember available in the airport before the time of departure there will be a need to get a substitute worker that does not have a schedule yet.

4.1 Problem and Solution

Problem Representation. To represent the flight schedule and all the constraint embedded in it, we use a flight graph that will take care of most of the constraints, as suggested by [8]. This graph, instead of using nodes to represent airports and edges to represent flights, uses a much more suitable representation that will embed problem constraints by using nodes to represent flights and edges to represent dependency constraints among flights: an edge that exists from node representing flight X to the node representing flight Y will assure us that flight Y leaves from the destination airport of flight X and that flight Y leaves after a pre-specified rest time following the arrival of flight X, (Fig 2. shows Table 1 example).

With this representation, upon the occurrence of delays, some edges will be eliminated. Those eliminated edges that were used by crewmembers will be the problems to be solved; those edges contain the information about the flight and the delayed crewmembers.

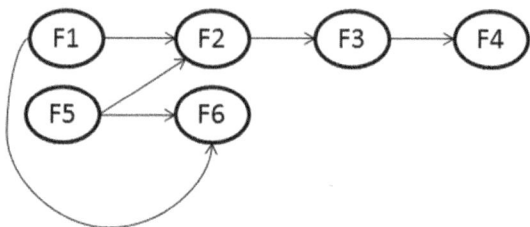

Fig. 2. Directed graph representation for the flight schedule in Table 1 with minimum 30 minutes rest time between consecutive flights

Problem Constraints. The crew recovery problem involves several constraints, some of them could not be violated, i.e. hard rules and others that are subject to optimization, i.e. soft rules. Beyond obvious constraints as the crew member could only take flights accordingly to the airport where it is at time of departure and all the flights should have a complete crew team allocated, i.e. 2 pilots for every flight and 4 cabin crew for narrow body aircrafts and 10 for wide body. There are other hard rules that must be respected, such as rotation time between flights, half an hour is the default value, and duty time limits, i.e. 12 hours daily, 50 hours weekly and 180 hours monthly. As this is mainly an initial allocation constraint, any duty time added during the recovery that does not respect this rule will be added as overtime contributing to the overall cost.

In terms of soft rules, that are subject of optimization, the goals are to minimize crew costs, including wages, overtime and penalizations for adding spare crewmembers. The wages for the 582 crew members were generated following normal distributions between different limits for cabin and pilots.

Solution Representation. For each edge that is considered a problem, using the flight graph we find all the crewmembers that can swap with the ones that are delayed. The solution of this problem will be a set of swap operations to each problem, identifying the crewmembers involved in this swap.

Solution Generation. The generation of a new solution is made by finding a successor that can be obtained by one, and only one of the following operations for each problem:

- Swap two crewmembers in their flight assignment, being those two crewmembers allocated to the flight which the initial crewmembers were delayed.
- Swap two crewmembers in their flight assignment, being one of the crewmembers allocated to the flight which the initial crewmembers were delayed, and the other one of the crewmembers identified as a possible solution.
- Swap two crewmembers in their flight assignment, being one of the crewmembers allocated to the flight which the initial crewmembers were delayed, and the other one an extra worker who does not have any flight assigned.

We implemented different meta-heuristic to solve the problem: *hill-climbing, simulated annealing*.

The *hill climbing* and *simulated annealing* algorithms are very similar and solve the problem iteratively by following the steps:

1. Obtains the flights that are in the time window of the problem. This time window starts at the delayed flight date and ends at the end of the month. These are the only flights that could undergo a change in their crew.
2. While some specific and customizable time or number of iterations has not passed, the steps 3 and 4 are repeated.
3. Generates a successor of the initial or previous solution.
4. Evaluates the cost value of the generated solution according to Equation 5. If it is better than the cost value of the current solution, accepts the generated solution as the new current solution. Otherwise, in the case of *hill climbing* algorithm it discards the generated solution. On the other hand *simulated annealing* accepts the generated solution with a small probability given by Equation 1.
5. Send the information on which flight will be delayed and for how long to the crew solver.

Solution Evaluation. Considering the representation of solution previously explained, we calculate the value (fitness) of a solution using the Equation 5.

$$\sum_{i=1}^{|CM|} Salary_i + Overtime_i + Spare_i \qquad (5)$$

$$where$$
$$i \in CM; CM = \{all \ crew \ members \ involved \ in \ problem \ and \ solution\}$$

The cost includes the salary (*Salary*) costs for all crewmembers involved in the problem or in the possible solutions, it also includes the increased cost in overtime hours (*Overtime*) and the penalization for the addition of spare crewmembers (*Spare*).

4.2 Results and Discussion

As mentioned before, crew connection problems appear on the solution obtained on the aircraft recovery stage, in this phase schedule scenario is analyzed to solve possible crew availability problems.

We tested the algorithms with 3 different disrupted instances using the same number of flights and aircrafts as before and with 473 effective cabin crewmembers and 109 pilots in a full month flights schedule. Note that there are also spare crewmembers within the constraints for workforce needs.

Table 5. Results for some different disruption scenarios. "Delayed flights" is the number of flights delayed and that will cause connection problems for the crewmembers. "Initial Costs Increase" is the increase in the costs for the first solution calculated upon the construction of the graph, while the "Final Costs increase" is the result of the increase in the end of the iterations, we considered this values as monetary units (m.u.).

Disruption	Test1	Test 2	Test 3
#Crew Connection Problem	48	24	12
#Delayed Flights	6	4	1
Solution			
Initial Costs Increase(m.u.)	81450	42204	392
Final Costs Increase(m.u.)	80107	41292	344
Run time (seconds)	13,1	8,7	5,6

As we can see by the values represented in Table 5, the crew recovery is a crucial recovery point due to the costs associated with it. We can easily realize that even a small disruption (with only 4 or 6 flights delayed) can cause large impacts on the budget if it causes problems with crewmembers.

As for the results, it is important to note that all the results do not present any connection problem for any crewmember, having therefore achieved a possible solution for every single one of them. This is important since it allows us to avoid delaying or even canceling any flight.

About the iterative optimization, it is capable of some relevant improvements, improving 1343 m.u. in *Test 1*, 912 m.u. in *Test 2* and 48 m.u. for *Test 3*. It is shown that the higher are the connection problems for crewmembers, the higher will be the increase in wages and therefore higher will be the iterative improvement.

Since we have implemented two different algorithms for the iterative optimization, we present in Fig. 3 a chart with the wages decrease over time for the hill-climbing algorithm (HC) and the simulated annealing (SA) for *Test 3*. It is visible that the simulated annealing is slightly better in the overall.

It is also important to refer that most of the time spent to solve the problem was used for the graph construction (about 4.6 seconds), since with the dataset used the graph would reach 3521 nodes and 1292469 edges. Therefore the time used for the iterative optimization was reduced.

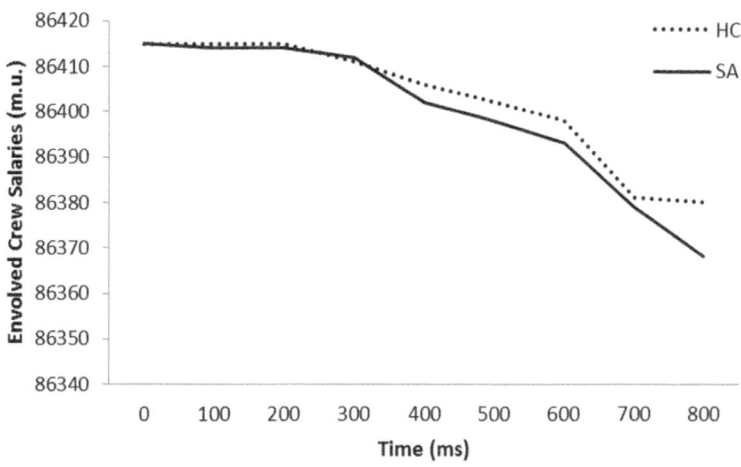

Fig. 3. Chart representing wages optimization through time using SA and HC

5 Conclusions

In this paper while reviewing some related work in disruption management we introduced our approach that deals with a real size dataset using specialized methodologies for the sub-problems.

We have developed an aircraft recovery approach using a chromosome representation for genetic algorithms and applied other meta-heuristics to that representation, the optimization is made through a multi-objective function that optimizes delays and other costs associated to the use of aircrafts. The results were better for GA, although in the initial iterations SA and HC have better results.

Regarding crew recovery we implemented a flight graph based approach that uses HC and SA to generate better solutions. Comparing the results we have shown that this approach can improve wages costs without requiring delay or cancelation of flights.

During this research we came across some difficulties mostly related to our inexperience on this specific domain of airline operation. Modeling the information needed was a complex task along with the data structures to apply the methodologies. Already with the application generating outputs the objective functions had to be adapted so the weights of each factor could reflect on better overall solutions. In terms of results it is hard to compare with other authors due to the specificity of our data set. Nonetheless the number of tests and the performed experiences show that the solutions are acceptable.

We can conclude that the main goals established have been achieved and the final application represents most of reality in airlines and it can solve disruptions taking into account several constraints and costs associated with airlines operation.

6 Future Work

Although the main goals have been achieved, there are several improvements that are possible to be made, some of those we are already working on.

Regarding the methodologies we think that applying a flight graph approach to the aircraft recovery could result on better solutions. Other algorithms are also considered as possibilities, such as ACO proposed by [6] and genetic algorithms on crew. Also a performance analysis of the code could lead to some improvements.

In this paper we did not considered passenger recovery, although we consider that this problem should be treated within two stages. The first one, we have been working on, is a methodology to find solutions for disrupted passengers due to lost connection flights, the approach should include ticket allocation and take into account the class of the passenger, i.e. executive or economic, on the costs. Solutions for disrupted passengers are alternate flights that could be found applying a Dijkstra algorithm [12] on a recovered flight graph. This stage could have better results if we include the number of passenger problems generated within the aircraft recovery objective function.

Another way to improve the work developed is to test the application with other datasets from different airlines and periods, because the results could be polarized by this specific month of operation. A data set of actual disruptions and recovered schedules would be a good way to improve the problem modeling, compare results and possibly show major cost reductions.

We also intent to integrate this methodologies on a multi-agent system, related to previous works, so we can have the computation distributed along several machines leading to a better efficiency. This opens the opportunity to integrate negotiation between approaches as proposed by [1], so the disruptions are not only solved sequentially.

Finally we think that this kind of application should have a statistics module to provide an intuitive interface for understanding the results obtained.

Acknowledgments. The authors are grateful to TAP Portugal for the real data provided, Pedro Torres that contributed to our previous work on this subject and Professor António Pereira that reviewed this paper. The third author is supported by FCT (Fundação para a Ciência e Tecnologia) under research grant SFRH/BD/44109/2008.

References

1. Castro, A.J.M., Oliveira, E.: Disruption Management in Airline Operations Control – An Intelligent Agent-Based Approach. In: Usmani, Z. (ed.) Web Intelligence and Intelligent Agents, pp. 107–132 (2010)
2. Bierlaire, M., et al.: Column Generation Methods for Disrupted Airline Schedules. Presented at: Fifth Joint Operations Research Days, ETHZ, Zürich, August 28 (2007)
3. Castro, A.J.M.: Centros de Controlo Operacional – Organização e Ferramentas, Monograph for Post-graduation in Air Transport Operations. In: ISEC - Instituto Superior de Educação e Ciências, Outubro (2008) (in Portuguese)

4. Mota, A., Castro, A.J.M., Reis, L.P.: Recovering from Airline Operational Problems with a Multi-Agent System: A Case Study. In: Lopes, L.S., Lau, N., Mariano, P., Rocha, L.M. (eds.) EPIA 2009. LNCS, vol. 5816, pp. 461–472. Springer, Heidelberg (2009)

5. Castro, A.J.M., Oliveira, E.: A New Concept for Disruption Management in Airline Operations Control. Proceedings of the Institution of Mechanical Engineers, Part G: Journal of Aerospace Engineering 225(3), 269–290 (2011), doi:10.1243/09544100JAERO864

6. Zegordi, S., Jafari, N.: Solving the Airline Recovery Problem by Using Ant Colonization Optimization. International Journal of Industrial Engineering & Production Research 21(3), 121–128 (2010)

7. Liu, T., Jeng, C., Chang, Y.: Disruption Management of an Inequality-Based Multi-Fleet Airline Schedule by a Multi-Objective Genetic Algorithm. Transportation Planning and Technology 31(6), 613–639 (2008)

8. Ozdemir, H., et Mohan, C.: Flight graph based genetic algorithm for crew scheduling in airlines. Information Sciences—Informatics and Computer Science: An International Journal - Special issue on evolutionary algorithms 133(3-4) (April 2001)

9. Bratu, S., Barhhart, C.: Flight Operations Recovery: New Approaches Considering Passenger Recovery. Journal of Scheduling 9(3), 279–298 (2006)

10. Leung, J.: Handbook of Scheduling: Algorithms, Models, and Performance Analysis, 1st edn (2004)

11. Cheung, A., Ip, W.H., Lu, D., Lai, C.L.: An aircraft service scheduling model using genetic algorithms. Journal of Manufacturing Technology Management 16(1), 109–119 (2005)

12. Dijkstra, E.W.: A note on two problems in connexion with graphs. Numerische Mathematik 1(1), 269271 (1959)

13. AEA – Association of European Airlines, 2008. European Airline Punctuality in 1st quarter 2008. Press release, July 01 (2008)

Solving Heterogeneous Fleet Multiple Depot Vehicle Scheduling Problem as an Asymmetric Traveling Salesman Problem

Jorge Alpedrinha Ramos[1], Luís Paulo Reis[2], and Dulce Pedrosa[3]

[1] DEI - Departamento de Engenharia Informática,
Faculdade de Engenharia da Universidade do Porto
[2] LIACC - Laboratório de Inteligência Artificial e Ciência de Computadores da
Universidade do Porto, Faculdade de Engenharia da Universidade do Porto
{ei06033,lpreis}@fe.up.pt
[3] OPT - Optimização e Planeamento de Transportes
dpedrosa@opt.pt

Abstract. The Vehicle Scheduling Problem is a well-known combinatorial optimization problem that emerges in mobility and transportation sectors. The heterogeneous fleet with multiple depots extension arises in major urban public transportation companies due to different demands throughout the day and some restrictions in the use of different vehicle types. This extension introduces complexity to the problem and makes the known deterministic methods unable to solve it efficiently. This paper describes an approach to create a comprehensive model to represent the Multiple Depot Vehicle Scheduling Problem as an Asymmetric Traveling Salesman Problem. To solve the A-TSP problem an Ant Colony based meta-heuristic was developed. The results achieved on solving problems from a Portuguese major public transportation planning database show the usefulness of the proposed approach.

Keywords: HFMDVSP, Multiple Depot, Vehicle Scheduling Problem, Traveling Salesman Problem, A-TSP, Ant Colony System.

1 Introduction

Public transportation companies try to meet the necessities of their users; ideally they satisfy this demand at the lowest costs possible, providing a better, cheaper and more efficient mobility network to the community they serve. This price/efficiency relation in public transportations serves as an indicator to the life quality in a given community, which with the experienced growth in metropolitan areas, has become a priority.

To provide an efficient mobility network every company shall provide trips that satisfy the flow demand on different day periods, and do it with the lowest expenses possible. One important issue to reach this goal is to use the least expensive vehicle duties possible, and this is where the vehicle scheduling problem

L. Antunes and H.S. Pinto (Eds.): EPIA 2011, LNAI 7026, pp. 98–109, 2011.

becomes important to transportation and mobility companies, but also to our day-to-day life.

Even though the vehicle scheduling problem can be applied in many areas of transportation, the focus throughout the development was urban transportation companies.

The Vehicle Scheduling Problem - VSP - consists in assigning pre-determined timetabled trips to vehicles that will perform them.The goal is to obtain an assignment that reduces the number of vehicles necessary and the total operational costs related. The main sources of resources waste are the empty trips to make connections and the time the vehicle remains still in a bus stop. This problem has been studied and interesting approaches are able to solve it efficiently, but when there are restrictions to the vehicle types that can perform each trip and there is more than one depot, the problem becomes more complex and harder to solve in a considered reasonable time. The multiple vehicle type extension even with a single depot is already a NP-Hard problem [7].

The problem instance is defined by the timetabled trips and the network in which the trips will be performed. In Fig. 1 a graphic representation of the network can be seen as well as a table with trips to be performed.

Timetabled trips				
N⁰	Origin	Departure	Destiny	Arrival
1	Hosp. S. João	07:40	FEUP	07:45
2	FEUP	07:49	ISEP	07:54
3	ISEP	07:58	Cem. Paranhos	08:03
4	ISEP	08:03	Cem. Paranhos	08:08
5	Cem. Paranhos	08:14	Polo Univers.	08:19
6	Cem. Paranhos	08:19	Polo Univers.	08:24
7	Polo Univers.	08:30	U. Fern. Pessoa	08:35
8	Polo Univers.	08:35	U. Fern. Pessoa	08:40
9	U. Fern. Pessoa	08:48	Cem. Paranhos	08:53
10	U. Fern. Pessoa	08:55	Cem. Paranhos	09:00
11	Cem. Paranhos	09:09	FEUP	09:14
12	Cem. Paranhos	09:16	FEUP	09:21
13	ISEP	11:19	U. Fern. Pessoa	11:24
14	U. Fern. Pessoa	11:49	FEUP	11:54

Fig. 1. Network and Time Tabled Trips example

2   Problem Representation

Representing the Vehicle Scheduling Problem has been widely studied; especially when it comes to the simple VSP and Multiple Depot VSP, a good overview of the existing representations is presented by Bunte and Kliewer [1]. The problem is very comprehensible, but there are several ways of representing it, which can be more or less comprehensive.

2.1   Connection Based Network

The Connection Based Network model uses a Network Flow Model to represent layers of the problem, each layer represents a depot, where each trip is represented by two nodes, beginning node and end node, and the depot is represented with two nodes as well, one departure node and one arrival node. The Network flow model can be defined as an integer linear programming model, Eq. 1, where x_{ij} is a boolean variable that defines whether or not a arc between nodes $i, j \in N$ is used and c_{ij} is the cost associated with using that arc. The arcs A are represented as a pair of nodes $ij \in A$, and a subset of them called AT represents the mandatory arcs between the beginning and the end of a given trip.

$$
\begin{aligned}
&\min \sum_{(i,j) \in A} c_{ij} x_{ij} \\
&\text{s.t. } \sum_{i:(i,j) \in A} x_{ij} - \sum_{i:(j,i) \in A} x_{ij} = 0 \qquad \forall n \in N \\
&\qquad 1 \leq x_{ij} \leq 1 \qquad\qquad\qquad\quad \forall (i,j) \in AT \\
&\qquad x_{ij} \geq 0 \qquad\qquad\qquad\qquad\quad x_{ij} \in \mathbb{N}
\end{aligned} \tag{1}
$$

This model uses a multiple layer network where each layer represents a depot. Even though it was proposed to solve the multiple depot problem, using layers to represent a depot/vehicle type pairs it can be used to solve the heterogeneous fleet extension as well. The edges between nodes in Fig. 2 are the possible connections between nodes. These edges can represent trips linking a depot departure to the beginning of a trip, the beginning of a trip to the end of the same trip, the end of a trip to the beginning of another trip and the end of a trip to the depot arrival.

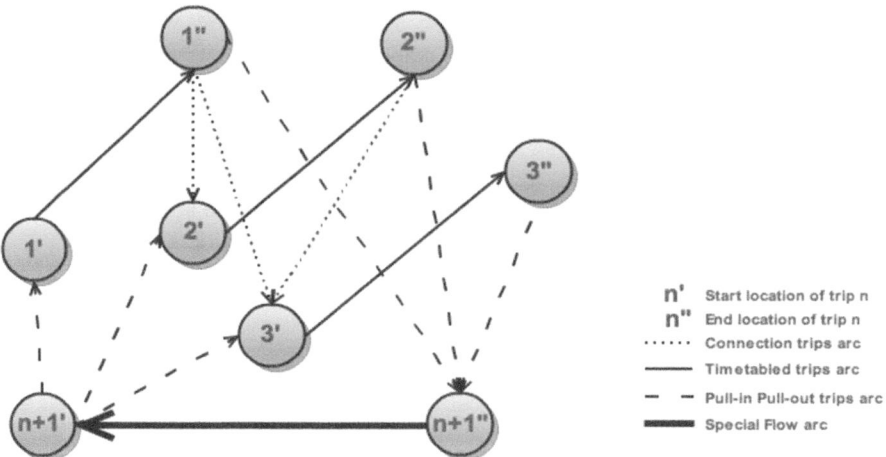

Fig. 2. Representation of a Network Flow Model

In order to link two nodes, there must be a path that can connect the end node i location to the beginning node j location, and it must have a lower duration than the difference between the end node time and the start node time. Also if the location of a node i, representing the end of a trip, that starts after a node j, representing the beginning of a trip, the connection is possible. Links from a beginning i to an end node j, where both are representative of the same trip, are the pre-determined trips links. The links that connect a depot departure to a beginning node, consider only if there is a path between the depot node and the node's location, since the trip can start at any given time. Same logic is followed to connect an end node to a depot arrival. Special links between depot arrivals and depot departure are created to consider the use of a new vehicle.

The subnetworks are modelled as a Network Flow Model, and constraints must be added to Eq. 1 in order to guarantee that each trip is served only in one of the subnetworks (2), and a constraint to ensure that no more than the available vehicles from a depot are used (3). Consider T the timetabled trips and $AT^t \subseteq AT \subset A$ where $t \in T$ as the set of arcs related to trip t in all subnetworks; H is the set of all depots available and d^h represents the number of vehicles in depot $h \in H$ and A^h is the set of circulation flow arcs of depot h.

$$\sum_{(i,j)\in AT^t} x_{ij} = 1 \ \forall t \in T \tag{2}$$

$$\sum_{(i,j)\in A^h} x_{ij} \leq d^h \ \forall h \in H \tag{3}$$

2.2   Objective Function and Costs

The costs related to a work block - a work block represents the trips assigned to a vehicle and will be called that way from now on - can be calculated considering the time the vehicle is performing a pre-determined trip, the time it is idled at any given bus terminal, and the time it spends performing empty connection trips, as well as the costs related to using the vehicle required for that work block.

In order to find a solution with the least time spent on empty trips and idled at a regular stop, penalty values are determined, which will then be multiplied by the time spent at each state and weigh more in the final cost than the pre determined trips cost. The costs related to the pre-determined trips will be considered multiplying the trips time by the used vehicle's cost factor that is characteristic of each vehicle type. This factor is used to benefit the use of the least expensive vehicle type possible in each trip. The costs related to using the vehicle represent the costs related to use an extra vehicle - driver costs, equipment depreciation, among others. This value is determined independently from the vehicle type .

The objective is to minimize the overall cost of the solution, in order to achieve more efficient assignments.

2.3   Asymmetric Traveling Salesman Problem

The Traveling Salesman Problem is a widely studied subject, it is a well-known combinatorial optimisation problem. It is the problem of assigning a route of a traveling salesman that must go through a number of cities with different distances between each other. An instance of the problem with n cities is defined by a distance matrix $M[n][n]$, as the one defined in Tab. 1 and the solution consists of an optimal Hamiltonian tour with the shortest possible length.

Table 1. Traveling Salesman Problem Distance Matrix

	A	B
A	∞	300
B	300	∞

The symmetry of a Traveling Salesman Problem exists if the distance from and to a city C is the same in every case, and otherwise it is considered an Asymmetric Traveling Salesman Problem.

2.4   A-TSP Approach

With the Connection Based Network model, the problem of finding the optimal solution for the assignment can be interpreted as a special case of the Asymmetric Traveling Salesman Problem. The main difference is that in the original A-TSP the traveling salesman only visits each city once, returning to the original city and in this problem the depot is a special node that can be visited more than once, and the links in the model have an infinite cost in the opposite direction.

A simple VSP instance can be represented as an A-TSP graph in Fig. 5 or as a cost matrix as seen in Tab. 2. The links between nodes have a special structure due to the time restrictions that does not allow all nodes to be connected between each other. The connections that are not feasible appear represented with the ∞ symbol in the cost matrix.

Another peculiarity about this case is that the cost of the connections between trips is different whether the vehicle type being used, and given the depot restrictions the connection may not exist for a given depot. The Connection Based Network Model uses a multi layer network to represent the different depots connections that in the case of heterogeneous fleet can be used to represent pairs depot/vehicle type. By doing so every layer can be represented as a costs matrix.

3   Ant Colony Meta-heuristic

The Ant Colony System, initially proposed by Marco Dorigo [4], is a nature inspired meta-heuristic that simulates ant colonies behavior on finding the shortest path to food. It's most known application is to TSP (and A-TSP) problems and was suggested by Dorigo and Gambardella in [3].

The ants search for food starts with an ant wandering and finding food, while laying down a pheromone trail. Other ant, also wandering, eventually finds the pheromone trail and is likely to follow it and reinforce the pheromone trail. This

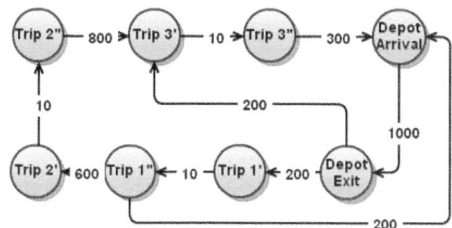

Fig. 3. An A-TSP Graph sample

Table 2. One layer cost matrix

From/To	Depot E.	Depot A.	Trip 1'	Trip 1"	Trip 2'	Trip 2"	Trip 3'	Trip 3"
Depot E.	∞	∞	200	∞	∞	∞	200	∞
Depot A.	1000	∞	∞	∞	∞	∞	∞	∞
Trip 1'	∞	∞	∞	10	∞	∞	∞	∞
Trip 1"	∞	200	∞	∞	600	∞	∞	∞
Trip 2'	∞	∞	∞	∞	∞	10	∞	∞
Trip 2"	∞	∞	∞	∞	∞	∞	800	∞
Trip 3'	∞	∞	∞	∞	∞	∞	∞	10
Trip 3"	∞	300	∞	∞	∞	∞	∞	∞

biological feature by itself and the fact that pheromone trails evaporate with time help this task. The longer the path to food, the higher the probability of evaporation before another ant follows that entire path, the opposite occurs with short paths, because ants will be more likely to follow the trail and reinforce it.

Fig. 4 represents this behavior, in image 1 the first ant finds randomly a path to food and leaves the pheromone trail so the next ant can follow it with a higher probability than any other path. Given the low concentration of pheromones on all paths initially, the ants will eventually travel through the whole search space, as seen on 2, leaving pheromone trails with a concentration related with the distance to food. In the end as it is shown in 3, the ants converge to the shortest path,leaving the shortest path with much more pheromones than any other path.

It can be seen as a GRASP cooperative method, since it uses a greedy random heuristic by preferentially selecting connections with a higher pheromone concentration. Initially the pheromone concentration is based only on the cost of that connection and a calculated factor that balances the number of paths and the distance on each path, promoting exploration. As artificial ants start to find feasible paths, it adapts the connections pheromones based on the total cost of solutions that contain that link. The update process is done in two phases: the first is the local update and the second is the global update.

The local update rule, represented by Eq. 4 and 5, promotes the exploration of different paths, reducing the probability of following the same path repeatedly. This update is done after every artificial ant finds its path, and has a low

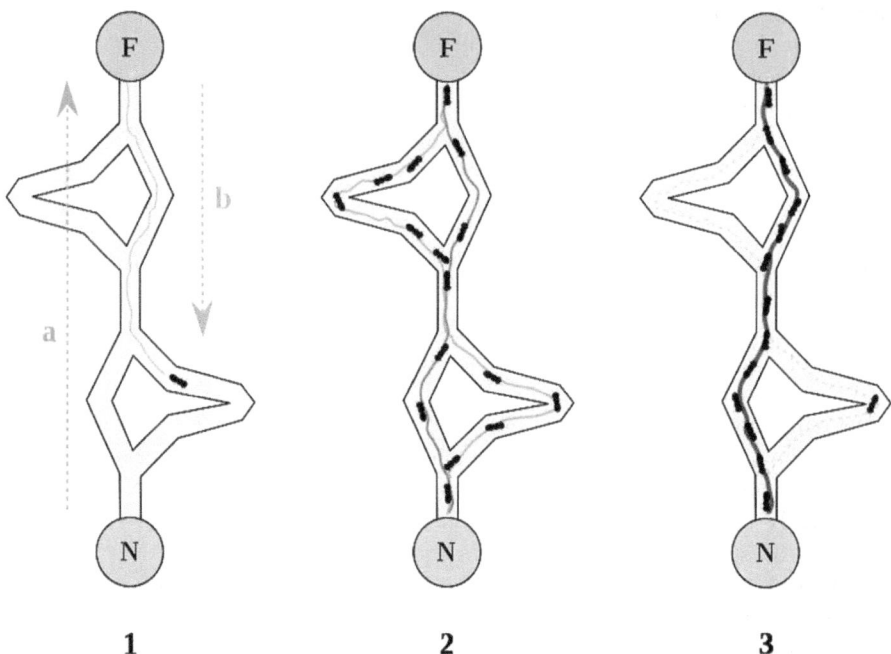

Fig. 4. Ant's search for food behaviour representation

impact in the pheromone concentration. In the formulas presented pheromone concentration is represented as $ph(edge)$ and L_{gb} represents the balanced weight of this link in the global best solution cost.

$$ph(e) = (1 - \alpha) * ph(e) + \alpha * \Delta Lo(e) \tag{4}$$

$$\Delta Lo(e) = \tau_0 \tag{5}$$

The global update, represented by Eq. 6 and 7, is done after a number of artificial ants find their path. Those paths are then evaluated and the best path is selected, the links in the best solution are updated with a high impact on pheromone concentration.

$$ph(e) = (1 - \rho) * ph(e) + \rho * \Delta G(e) \tag{6}$$

$$\Delta G(e) = \begin{cases} (L_{gb})^{-1} & if\, e \in global Best \\ 0 & otherwise \end{cases} \tag{7}$$

The factor that determines the impact on the update rules are α and ρ and in our approach the values used were $\alpha = 0.3$ and $\rho = 0.1$.

The link selection algorithm considers the pheromone density of each link and the costs related to it to calculate the probability of selection of that link. In order to promote exploitation over exploration, the best link is used with a given probability, called the exploitation factor, and otherwise the previously presented probabilities are used in order to select the used link. This selection is formally presented bellow, where σ is a random value, σ_0 is an exploitation factor (defined $\sigma_0 = 0.8$) and S is a random variable selected according to the probability distribution given in the second equation.

$$s = \begin{cases} max \ ph(e) * \eta(e)^{\beta} & if \sigma < \sigma_0 \\ S & otherwise \end{cases}$$

$$p(e) = \begin{cases} \frac{ph(e)*\eta(e)^{\beta}}{\sum ph(e_i)*\eta(e_i)^{\beta}} & e, e_i \in Current Edges \\ 0 & otherwise \end{cases}$$

To define the initial pheromone concentration a static value is used, and then an artificial ant is set to determine a feasible path, and it's solution cost is used to redefine the pheromone concentration. The definition of this initial value is specially important because if there is a huge gap between the initial concentration and the first value used to update the pheromone concentration, it can make those links almost mandatory or irrelevant, so the closer this initial solution is to an optimal solution higher is the causality of that update. This causality improves the learning capacity of the algorithm

The cost of connecting the current node to the depot node only considers the trip required to do so, and no waiting time, which makes this link very likely to be selected, because other links usually have higher costs. In order to avoid the selection of links to depot from any trip, it was considered the cost of using another vehicle in the cost of those links.

3.1   Unlinked Trips Sub Problem

One of the main issues in this approach was to find feasible solutions for the problem. The problem arises mainly due to the fact that, in some problems, nodes are not heavily connected, and the decision tree of building a path may contain leaves that do not correspond to feasible solutions (Figure 5) .

If a trip can be connected to a depot, it will always have an available connection, because depots can be visited more than once, except if the trip is set to another depot, and then this link will no longer be available. Filtering the graph by assigning the nodes that are not connected from or to all depots that they can be assigned is a relaxed problem that ensures that all trips can be connected. This relaxed sub problem allows to avoid unfeasible solutions that can represent a high percentage of the reachable solutions, enhancing the performance of any search heuristic. On the other hand this problem may become very complex, that is the case presented in Figure 6 which has a ratio of 85% unlinked nodes over the total nodes.

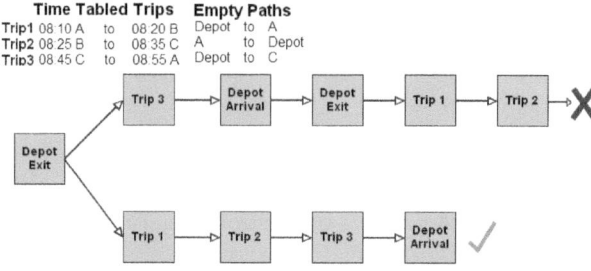

Fig. 5. Example of a path construction tree with two paths, one feasible and one unfeasible

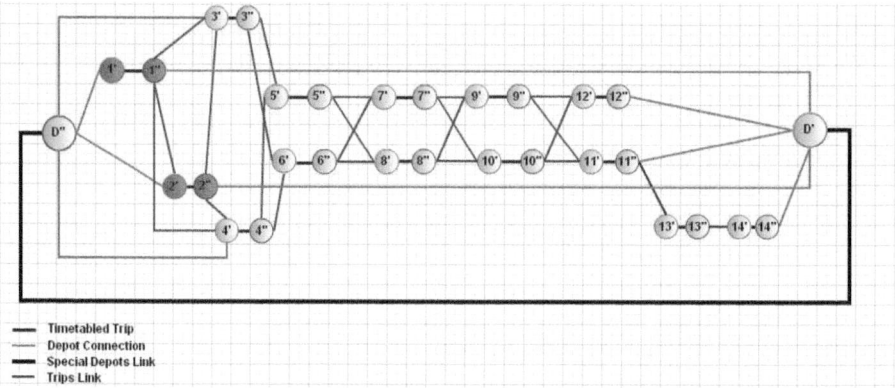

Fig. 6. A CBN layer sample focusing on the unlinked trips sub problem

4   Results

The testing platform was the environment used by many public transportation companies in Portugal to manage their network and planning data, GIST (Integrated Transport Management Systems). The database used was a copy of the one that a major company operating in Porto uses in their daily planning operations. This platform already provides a planning support system that allows the Vehicle Scheduling Problems to be solved automatically, even though it does not support heterogeneous fleets and vehicle type constraints. This method is based on a quasi-assignment algorithm, developed by Paixão and Branco [5].

This method was used as a benchmark for situations in which vehicle type constraints are not defined. These tests allowed verifying the solutions quality, and efficiency.

The Ant Colony System most relevant factor of success is its ability to learn how different connections affect the overall scheduling. To measure its capacity to learn every ant that found a feasible path registered its total cost, which allowed analyzing how the overall progression of solutions occurred. The graphic shown on Figure 7 shows the average costs of ant's solution throughout the process. The schedule for which 330 artificial ants found solutions has 220 trips, and a degree of connectivity around 25% (an average of 50 possible connections per trip). The best cost found for this problem was of 15,5M, while the trend of the solutions evolved from an initial 17,4M to 16,9M a significant evolution.

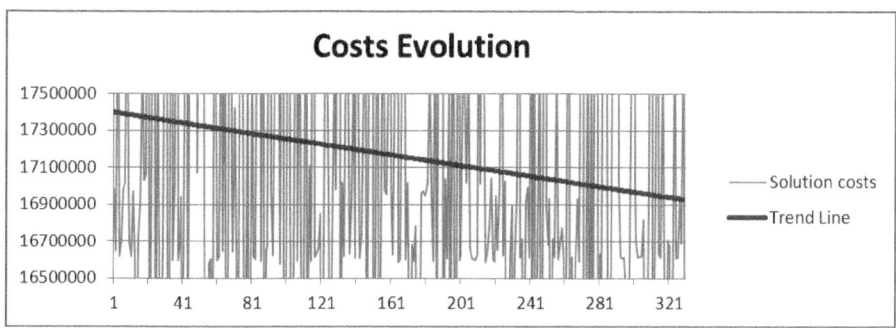

Fig. 7. Solutions cost evolution through algorithm iterations

The variation between solutions is significant as it can be observed in the graphic, which can mislead to think that the learning capacity is not as good as it would be expected, but this is rather necessary in order to avoid local minimums. The exploitation vs. exploration factor was considered and similar values to the ones proposed by Dorigo and Gambardella [3] provided the best learning rate.

This database is a truly heterogeneous database, so it was important to classify different schedules patterns. The parameters used to classify were mainly the amount of trips in the schedule, and the degree of connectivity of the generated graph. The schedules were classified into two classes of trip quantity comprehensively denominated many trips and few trips, and two classes of connectivity lightly connected and heavily connected. Schedules with less than 50 trips were considered with few trips and the remaining with many trips. Schedules with a link average per trip higher than 20% of the number of trips were considered heavily connected, and the remaining lightly connected. The results are shown in Figure 8.

	Few Trips						Many Trips					
Lightly Connected	Trips	Nodes	Links per Trip	Time to Solve	Average Vehicles	Costs	Trips	Nodes	Links per Trip	Time to Solve	Average Vehicles	Costs
	13,25	6,50	1,96	00:00,2	1,17	1233506,67	63,33	4,83	10,48	00:02,6	2,33	2821013,33
Heavily Connected	Trips	Nodes	Links per Trip	Time to Solve	Average Vehicles	Costs	Trips	Nodes	Links per Trip	Time to Solve	Average Vehicles	Costs
	24,67	4,33	7,29	00:00,2	1,67	1860866,67	125,06	7,94	30,41	00:02,7	7,25	8248422,50

Fig. 8. Average results for the different problem classes

5   Conclusions and Future Work

The major constraint found was the construction of feasible solutions with a random factor without using any greedy heuristic that ensured the feasibility throughout the construction process. This forced the use of constraint programming which, even using a first fail heuristic, has a random factor that enhances the probability of failure into exploring the search space.

The weak causality, principle presented by Rechenberg [6], meaning that small changes on a solution may cause significant changes on the overall individual fitness is also found in this problem. Even though every connection has it's related cost, using it makes impossible to use other connections, which is difficult to predict and has a huge impact in the overall fitness. Using a given connection may ruin the rest of the solution, so the solution presented seems to overcome that problem due to its adaptive structure, which has been proven by its learning capacity. The results achieved were really optimistic into adapting known methodologies usually applied to the A-TSP problem into the extended Vehicle Scheduling Problem. The ACS approach used is known as a good metaheuristic for solving the A-TSP problem, and performed well in comparison with known deterministic methods widely used. This approach translates the extended VSP problem into a more comprehensive and widely studied model, which enhances the adoption of methodologies previously used to solve A-TSP problems.

The complexity created with the heterogeneous fleet restrictions makes the traditional VSP solutions difficult to apply. Studies show that allying some metaheuristics search space exploration capacities with more structured and intelligent mathematical methods can be translated into more efficient approaches for the problem. This methods are applied nowadays to A-TSP problems and this way may become easier to apply to the different versions of VSP problems, and bring more interesting approaches to this problem.

The termination criteria for Ant Colony System may be subject of further work, since the relationship between problem complexity and iterations necessary to converge into a solution is not linear and its exploration can turn this method more efficient.

The complexity introduced by Vehicle type groups restrictions may be higher than the tests already performed, since in this cases they appear as ocasional restrictions. Even though the ocasional restrictions are the most common scenario, studying the relationship between aditional restrictions and time to perform may become important in terms of understanding limitations of the approach.

References

1. Bunte, S., Kliewer, N.: An overview on vehicle scheduling models. DSOR Arbeitspapiere / DSOR Working Papers (2006)
2. Freling, R., Paixão, J.P., Wagelmans, A.: Models and Algorithms for Vehicle Scheduling (2004)
3. Dorigo, M., Gambardella, L.M.: Ant Colony System: A Cooperative Learning Approach to the Traveling Salesman Problem. IEEE Transactions on Evolutionary Computation I (April 1997)
4. Dorigo, M.: Optimization, Learning and Natural Algorithms. PhD thesis, Politecnico di Milano. Italie (1992)
5. Paixão, J.P., Branco, I.M.: A quasi-assignment algorithm for bus scheduling. Networks 17, 249–269 (1987), doi:10.1002/net.3230170302
6. Rechenberg, I.: Evolutionsstrategie. Werkstatt Bionik und Evolutionstechnik, vol. 1. Frommann-Holzboog Verlag, Stuttgart (1994)
7. Lenstra, J., Kahn, A.R.: Complexity of vehicle routing and scheduling problems. Networks 11(2), 221–227 (1981)

Evolving Numerical Constants in Grammatical Evolution with the Ephemeral Constant Method

Douglas A. Augusto, Helio J.C. Barbosa,
André M.S. Barreto, and Heder S. Bernardino

Laboratório Nacional de Computação Científica, Petrópolis, RJ, Brazil
{douglas,hcbm,amsb,hedersb}@lncc.br

Abstract. This paper assesses the new numerical-constant generation method called ephemeral constant, which can be seen as a translation of the classical genetic programming's ephemeral random constant to the grammatical evolution framework. Its most distinctive feature is that it decouples the number of bits used to encode the grammar's production rules from the number of bits used to represent a constant. This makes it possible to increase the method's representational power without incurring in an overly redundant encoding scheme. We present experiments comparing ephemeral constant with the three most popular approaches for constant handling: the traditional approach, digit concatenation, and persistent random constant. By varying the number of bits to represent a constant, we can increase the numerical precision to the desired level of accuracy, overcoming by a large margin the other approaches.

Keywords: Constant Creation, Grammatical Evolution, Genetic Programming.

1   Introduction

Perhaps the most distinctive and powerful application of evolutionary computation is the automatic generation of "programs"—be it computer programs in the conventional sense or any other structured set of symbols, such as a mathematical expression or the design of an analog circuit. The first attempt to evolve this type of symbolic structure was probably that of Cramer [4], but it was only after Koza's work that the field achieved maturity and got inexorably associated with the term "Genetic Programming" (GP) [9].

Ryan et al. [16] proposed an alternative way of evolving programs which they called *Grammatical Evolution* (GE). The main difference with respect to GP is that in GE the candidate solutions are regular binary strings which encode programs through a user-specified grammar [13]. Thus, in GE the search and solution spaces are completely dissociated, and, as a result, any metaheuristic can be readily used as the method's search engine. In fact, GE has been successfully combined with genetic algorithms [13], particle swarm optimization [11], differential evolution [10], and artificial immune systems [2].

L. Antunes and H.S. Pinto (Eds.): EPIA 2011, LNAI 7026, pp. 110–124, 2011.

Regardless of the particular metaheuristic used as the optimization tool, the success of any GE method depends crucially on its capability of generating numeric constants within the programs [13]. There are many algorithms specifically developed with this objective in mind, but apparently no consensus yet as to which method is the most effective one [5–7, 12]. In this paper we evaluate the recently developed numerical-constant generation method called *ephemeral constant* ("ec"), which can be seen as a translation of the classical genetic programming's ephemeral random constant to the GE framework [1]. Its most distinctive feature is that it decouples the number of bits used to encode the grammar's production rules from the number of bits used to represent a constant, making it possible to increase the method's representational power without incurring in an overly redundant encoding scheme. A set of experiments is carried out to compare the performance of "ec" with the three most popular algorithms for constant generation in GE: the *traditional approach* [13], *digit concatenation* [12], and *persistent random constant* [5]. We show that, by varying the number of bits to represent a constant, we can increase the precision of "ec" to the desired level of accuracy, overcoming by a large margin the approaches mentioned above.

The paper is organized as follows. Section 2 introduces the field of GE. Common approaches for constant handling in GE are discussed in Section 3, while Section 4 describes "ec" in detail, which is the method under investigation here. In Section 5 experiments are presented to compare the constant handling techniques. Finally, in Section 6 we provide our conclusions and discuss some directions for future work.

2   Grammatical Evolution

Grammatical Evolution (GE) was proposed by Ryan *et al.* [16] and later extended by O'Neill and Ryan [13, 14]. It is an evolutionary computation framework in which a binary genotype is mapped into a program by means of a user-defined grammar. The binary string is in fact a representation of an integer array in which each value is encoded using b bits (usually $b = 8$).

A formal grammar G can be defined as $G = \{N, \Sigma, R, S\}$, where N is a finite set of nonterminals, Σ is a finite set of terminals, R is a finite set of rules (or productions), and $S \in N$ is the start symbol. A context-free grammar, used by standard GE, is one in which the left-hand side of each production rule does not have more than one nonterminal symbol. An example of a context-free grammar that generates mathematical expressions in prefix form is:

$$N = \{<\text{expr}>, <\text{op}>, <\text{uop}>, <\text{const}>\}$$
$$\Sigma = \{sqrt, neg, x, 0, 1, 2, 3, +, -, \times, \div\}$$
$$S = <\text{expr}>$$

with R defined as

$$
\begin{array}{rlr}
\text{<expr>} & ::= & \text{<op> <expr> <expr>} & (0) \\
 & | & \text{<uop> <expr>} & (1) \\
 & | & \text{<const>} & (2) \\
\text{<op>} & ::= & + & (0) \\
 & | & - & (1) \\
 & | & \times & (2) \\
 & | & \div & (3) \\
\text{<uop>} & ::= & sqrt & (0) \\
 & | & neg & (1) \\
\text{<const>} & ::= & 0 & (0) \\
 & | & 1 & (1) \\
 & | & 2 & (2) \\
 & | & 3 & (3)
\end{array}
$$

After decoding the genotype into an integer array, its entries are used to select rules from the grammar via the following expression

$$ rule = i \mod m, $$

where i is the decoded integer and m is the number of rules for the current nonterminal.

Notice that the mapping process might create a phenotype without using the entire integer array. Moreover, it may happen that the decoding of an entire chromosome results in an incomplete program; there are at least three ways to handle this situation: (i) reusing the genes by wrapping around the genotype; (ii) setting the program's fitness to the worst possible value; or (iii) repairing the program [2, 3].

3 Evolving Constants

The grammar presented in Section 2 has an elementary way of generating numerical constants. In particular, only the integers 0, 1, 2, and 3 belong to its language. Of course, there exist more advanced techniques to create numerical constants in GE. In what follows we present some of what we believe to be the most important constant-handling approaches found in the GE literature.

3.1 Traditional Approach

The original method for constant handling in GE, here referred to as the *traditional approach*, operates by defining grammatical rules to allow the mathematical manipulation of single integers (usually digits between 0 and 9). In other words, non-trivial numerical constants are obtained through mathematical expressions having arbitrarily complex arithmetic operations over simple integers.

Thus, considering a grammar containing only the integer values 0, 1, 2, and 3 (like the one presented in Section 2), an example of a program that generates the value 0.5 would be (÷ 1 2). In [12], O'Neill *et al.* only use the integers from 0 to 9, but of course other values can be included in the grammar to facilitate the generation of real values. The traditional approach uses the grammar presented in Section 2 with the following definition for the non-terminal <const>:

$$\text{<const>} \quad ::= \quad 0 \mid 1 \mid 2 \mid 3 \mid 4 \mid 5 \mid 6 \mid 7 \mid 8 \mid 9.$$

3.2 Digit Concatenation

The generation of real values using the traditional approach might require complex mathematical expressions which are not trivial to evolve. The digit concatenation approach was proposed by O'Neill *et al.* [12] as a way to circumvent such difficulty. In this method a numerical constant is generated by the concatenation of decimal digits (including the dot separator for real values). Thus, the non-terminal <const> in the grammar given in Section 2 would be defined as

$$
\begin{aligned}
\text{<const>} \quad &::= \quad \text{<int>.<int>} \\
\text{<int>} \quad &::= \quad \text{<int><int>} \mid \text{<digit>} \\
\text{<digit>} \quad &::= \quad 0 \mid 1 \mid 2 \mid 3 \mid 4 \mid 5 \mid 6 \mid 7 \mid 8 \mid 9.
\end{aligned}
$$

An important characteristic of digit concatenation is that it has a natural bias towards small constants, which may serve as an implicit mechanism for bloat control. The experiments presented by O'Neill *et al.* [12] suggest that, in general, the performance of digit concatenation is better than that of the traditional approach. The same experiments also indicate that there is no clear advantage in having arithmetic operations manipulate the constants generated by the concatenation grammars. In contrast, Dempsey *et al.* [7] obtained their best results when combining digit concatenation with the traditional approach.

3.3 Persistent Random Constant

The use of persistent random constants in GE was first investigated by Dempsey *et al.* [5, 7]. In this method real numbers are randomly generated and included in the production rules at the outset of the search process. Persistent random constants is essentially a variation of the traditional approach, since constants are created through the arithmetic manipulation of a finite number of predefined values. The main difference is that, in this case, the set of fixed constants is sampled from a uniform distribution over a given interval. The production rules for the persistent random constant approach are included in the grammar as

$$\text{<const>} \quad ::= \quad c_1 \mid c_2 \mid \cdots \mid c_\rho,$$

where the $c_i \in [c_{\min}, c_{\max}]$ are not ordered, and $c_{\min}$, $c_{\max}$, and ρ are user-defined parameters. The most commonly-used values for ρ are 100 [6] and 150 [7].

4   Ephemeral Constant

The "ec" method resembles the implementation of GP's ephemeral random constant, in which numerical constants are randomly generated and assigned to a program during its creation [9]. As in GP, the method stores the constants directly in the program's genotype, but, following GE's representation scheme, they are encoded as strings of bits rather than as real values. The implementation of the "ec" method is very simple and, in what concerns the grammar, it only requires a minor modification, which is the introduction of the following production rule:

$$<\text{const}> ::= ephemeral$$

The terminal symbol *ephemeral* carries a special meaning: whenever it is selected during the program's decoding process, the next n bits—counting from the genotype's first unread bit—are decoded into a real number. A way to decode an n-bit string into a real number c in the interval $[a, b]$ is to first convert the binary array to an integer i and then scale it by making $c = a + i \times (b - a)/(2^n - 1)$. Afterwards, the decoding process resumes normally past those n decoded bits. Close to the end of the genotype, there might be less than n bits available for decoding; in this case our implementation reduces the constant precision to the number of remaining bits—if there are no bits available, the decoding procedure returns zero. An example of this decoding process is shown in Figure 1, where the expression + 0.1953 50.0496 is decoded using a precision of 16 bits for the constants in the interval $[0, 100]$.

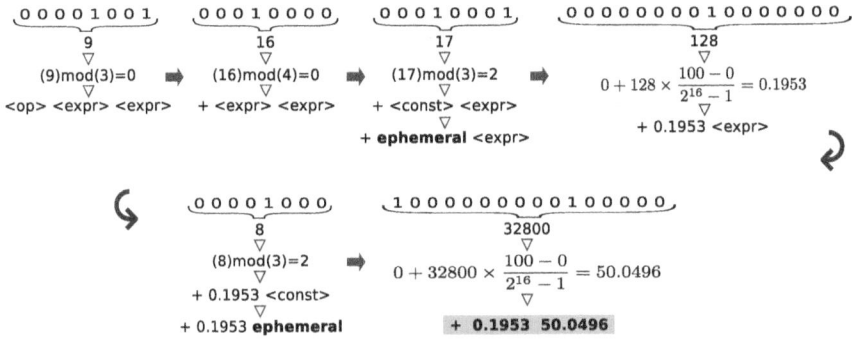

Fig. 1. Example of the ephemeral constant decoding process

There are two user-defined parameters in "ec". The first one is the interval in which the numerical constants will be decoded. The other one is the number of bits n used to represent each constant, that is, the method's numerical precision.

The "ec" method can be seen as a modification of the persistent random constant technique discussed in Section 3.3. The main improvement of the former with respect to the latter is the separation between the representation of *constants* and *production rules*—after all, they are rather different in nature: while one denotes a number, the other denotes a rule. This makes it possible to tune the precision with which the constant are represented without affecting the dynamics of the grammar's decoding process. In principle one could argue that the same effect could be obtained with the persistent random constant method by increasing ρ. However, in this case the number of bits used to encode the production rules would also increase, which could harm the evolutionary process [14]. Besides, since in the persistent random constant approach the constants must be explicitly enumerated in the grammar, increasing ρ may lead to practical difficulties. For example, a 32-bit representation can hold up to 2^{32} constants; using standard double precision to represent such constants would require astonishing 32 gigabytes of physical memory.

Another difference between "ec" and its precursor is the fact that in "ec" the constants are (implicitly) sorted: due to the scaling, as the underlying decoded integer gets bigger, the constant's value increases. This may help the evolutionary search if an appropriate representation scheme, such as a Gray code [15], is adopted. Besides, the constants used by "ec" are evenly distributed over the interval $[c_{\min}, c_{\max}]$, what makes the underlying GE algorithm less susceptible to the contingencies inherent to the persistent random constant's sampling process.

5   Computational Experiments

5.1   The Techniques

Computational experiments with different characteristics were performed to evaluate and to compare the approaches presented above. The methods used in the comparisons are labeled as "trad" (traditional approach), "dc" (digit concatenation), and "prc" (persistent random constant) to facilitate the visualization of the information. Four variants of the "ec" method were analyzed, corresponding to the use of 8, 16, 24, and 32 bits for constant encoding. They are denoted by "ec-8bits", "ec-16bits", "ec-24bits", and "ec-32bits", respectively (as a reference, the number of bits used by the "trad" and "prc" techniques to encode an ordinary numerical value is 8, while "dc" uses at least 16 bits).

5.2   Parameter Setting

The standard search mechanism adopted in our GE is a simple genetic algorithm. Although there is no consensus on the effect of using a Gray code [15] instead of a standard binary representation within GE, we used this scheme due to its capacity of preserving the notion of neighborhood in mapping a binary string into an integer. This can be exploited by evolutionary methods, since in this case small variations in Hamming space lead to small variations in the constant value. Also, there is no evidence that the use of a Gray code can harm the performance of the other techniques. A hundred independent runs were performed with the GE parameters set to: population size 500, chromosome length 800, mutation rate 1/(chromosome length), one-point crossover with probability 0.9, tournament selection with two individuals, and each integer encoded in 8 bits. The search mechanism and the user-defined parameters are the same for all four constant handling approaches considered in the experiments.

5.3   The Test Problems

The objective of the test problems is to evolve a single real constant. Six target constants of different magnitude and number of digits were considered here, namely: 1.23, 123000, 0.000123, 45.60099, 45600.99, and 0.0004560099. The fitness, to be maximized, was defined as $f(x) = 1/(1 + |target - x|)$.

5.4   Results

The results are summarized in Tables 1 to 6, which present the relative error and mean expression size for each technique; the symbol $\pm$ denotes the *standard deviation*. The relative error is defined as $|target - x|/target$ and the expression size as the total number of arithmetic operators and numerical constants appearing in the final expression. For instance, $size\{1.2 + 0.2\} = size\{1 - 1\} = 3$. The techniques are listed according to increasing mean relative error.

Table 1. Results for constant 1.23

Method	Relative error ($\times 10^2$)			Mean size
	Mean	Min.	Max.	
ec-32bits	0.000 ± 0.00	0.000	0.000	4.4 ± 2.5
ec-24bits	0.000 ± 0.00	0.000	0.000	4.2 ± 2.8
dc	0.001 ± 0.01	0.000	0.121	2.0 ± 1.2
ec-16bits	0.003 ± 0.01	0.000	0.024	4.3 ± 2.6
ec-8bits	0.048 ± 0.09	0.000	0.693	8.5 ± 6.3
trad	0.089 ± 0.21	0.000	1.626	17.2 ± 8.9
prc	0.150 ± 0.26	0.000	1.293	10.9 ± 8.1

To summarize the results shown above we use performance profiles [8]. By testing all techniques against all problems and measuring the performance $t_{p,v}$ of technique v when applied to problem p, a performance ratio can be defined with respect to the best performing technique in each problem as $r_{p,v} = \frac{t_{p,v}}{\min_v\{t_{p,v}\}}$. The performance indicators considered here are: (i) the mean relative error, and (ii) the mean expression size found by technique v in problem p. The relative performance of the techniques on the whole set of problems can be plotted using

$$\rho_v(\tau) = |\{p : r_{p,v} \leq \tau\}| /n_p$$

where $|.|$ denotes the cardinality of a set and n_p is the number of test problems. Then $\rho_v(\tau)$ is the probability that the performance ratio $r_{p,v}$ of technique v is

Table 2. Results for constant 123000

| Method | Relative error ($\times 10^2$) | | | Mean size |
	Mean	Min.	Max.	
ec-32bits	0.000 ± 0.00	0.000	0.000	11.0 ± 2.8
ec-16bits	0.000 ± 0.00	0.000	0.001	12.7 ± 4.8
ec-24bits	0.000 ± 0.00	0.000	0.000	11.6 ± 2.9
ec-8bits	0.011 ± 0.04	0.000	0.356	15.2 ± 5.7
prc	0.061 ± 0.15	0.000	1.036	16.4 ± 7.3
trad	0.075 ± 0.11	0.000	0.429	23.1 ± 7.1
dc	0.637 ± 3.23	0.000	18.699	2.2 ± 2.1

Table 3. Results for constant 0.000123

| Method | Relative error ($\times 10^2$) | | | Mean size |
	Mean	Min.	Max.	
ec-32bits	0.000 ± 0.00	0.000	0.000	10.8 ± 3.1
ec-16bits	0.000 ± 0.00	0.000	0.001	11.5 ± 3.9
ec-24bits	0.000 ± 0.00	0.000	0.000	11.4 ± 3.3
dc	0.001 ± 0.01	0.000	0.116	4.0 ± 1.5
prc	0.048 ± 0.14	0.000	1.278	17.3 ± 6.8
ec-8bits	20.012 ± 40.20	0.000	100.000	12.6 ± 8.1
trad	98.029 ± 13.86	0.046	100.000	4.5 ± 7.2

Table 4. Results for constant 45.60099

| Method | Relative error ($\times 10^2$) | | | Mean size |
	Mean	Min.	Max.	
ec-32bits	0.000 ± 0.00	0.000	0.000	1.7 ± 1.9
ec-24bits	0.000 ± 0.00	0.000	0.000	1.8 ± 2.4
dc	0.001 ± 0.00	0.000	0.013	1.4 ± 0.9
ec-16bits	0.001 ± 0.00	0.000	0.001	2.1 ± 3.0
ec-8bits	0.131 ± 0.12	0.000	0.243	5.6 ± 7.3
prc	0.161 ± 0.26	0.000	1.215	10.9 ± 10.5
trad	0.259 ± 0.43	0.000	1.318	17.0 ± 8.7

Table 5. Results for constant 45600.99

| Method | Relative error ($\times 10^2$) | | | Mean size |
	Mean	Min.	Max.	
ec-32bits	0.000 ± 0.00	0.000	0.000	8.9 ± 2.8
ec-16bits	0.000 ± 0.00	0.000	0.001	10.3 ± 3.7
ec-24bits	0.000 ± 0.00	0.000	0.000	10.2 ± 3.8
dc	0.001 ± 0.00	0.000	0.002	1.3 ± 0.9
ec-8bits	0.017 ± 0.03	0.000	0.250	12.6 ± 5.5
prc	0.068 ± 0.17	0.000	1.345	15.9 ± 7.6
trad	0.109 ± 0.17	0.000	0.715	21.3 ± 7.1

Table 6. Results for constant 0.0004560099

| Method | Relative error $(\times 10^2)$ | | | Mean size |
	Mean	Min.	Max.	
ec-32bits	0.000 ± 0.00	0.000	0.002	10.5 ± 3.0
ec-16bits	0.000 ± 0.00	0.000	0.001	11.5 ± 3.6
ec-24bits	0.000 ± 0.00	0.000	0.000	10.8 ± 3.0
dc	0.002 ± 0.01	0.000	0.093	3.9 ± 1.5
prc	0.078 ± 0.27	0.000	2.572	18.2 ± 8.2
ec-8bits	11.018 ± 31.44	0.000	100.000	13.4 ± 7.4
trad	92.027 ± 27.17	0.000	100.000	5.9 ± 9.0

within a factor $\tau \geq 1$ of the best possible solution. As a result, the curve of the best-performing algorithm should always stay above all other curves.

From Figures 2 and 3 it is clear that "dc" presents the best overall behavior among the proposals from the literature; however, when considering the mean relative error, it is outperformed by the "ec-16bits" variant. The variants "ec-24bits" and "ec-32bits" were omitted because their better performance would render the plots unreadable. They are, however, analyzed separately in Figures 4 and 5. Figure 4 clearly indicates that the mean relative error of the "ec-24bits" and "ec-32bits" variants are even smaller, as expected. Finally, Figure 5 indicates that as the number of bits used to encode numerical values in the "ec" variants increase the expression size decreases. However, the "dc" technique produced the shortest expressions in all cases tested.

For the mean error profiles in Figures 2 and 4 the range of τ varies by 3 orders of magnitude in the relative error with respect to the best average solution obtained in all test problems. For the expression size profiles in Figures 3 and 5 the range of τ, $[0, \tau_{max}]$, is such that, for all v, $\rho_v(\tau) = 1$ if $\tau \geq \tau_{max}$.

Figures 6 to 11 present the evolution of the mean relative error (log scale) in approximating the target constants (again, the "ec-24bits" and "ec-32bits" variants are omitted). As early as the 100th generation, "ec-16bits" presents the best results in all 6 cases. The "dc" technique does not present a uniform relative performance as it has a good performance for the constants 1.23 and 45.60099 (which, when compared with the other constants, are "closer" to its initial derivation <int>.<int>) while being outperformed by the "trad" technique (which is the weakest one) for the constant 123000. It seems that there is a favorable bias in "dc" for evolving high magnitude constants with large digits, and low magnitude constants with small digits. It can be noticed that "ec-8bits" is always better than the traditional approach ("trad"). Although "ec-8bits" is not always better than "prc", "ec-16bits" outperforms "prc" in all cases. Indeed, for the constant 1.23, "prc" is worse than the "trad" technique.

6   Conclusions

This paper assessed a novel approach for constant-handling in GE called ephemeral constant. By dissociating the number of bits used to represent a constant from the

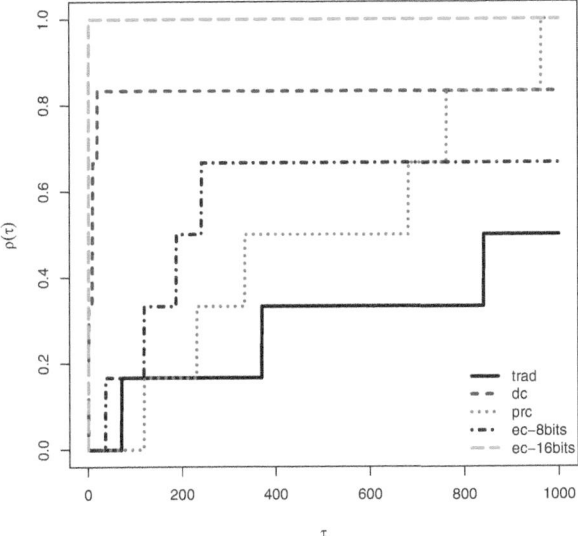

Fig. 2. Performance Profiles for the mean error

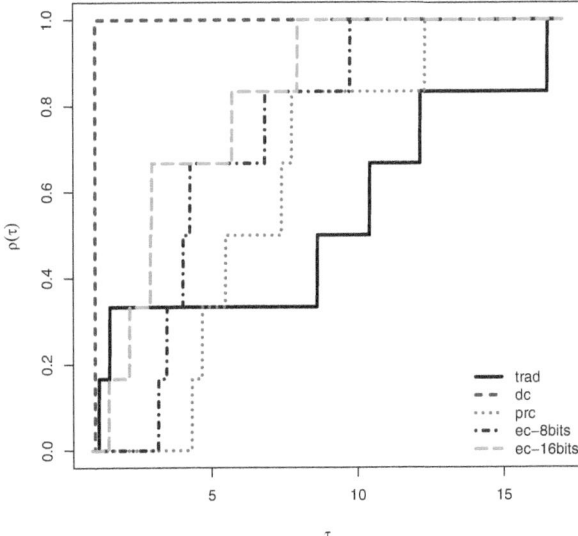

Fig. 3. Performance Profiles for expression size

number of bits used to index a production rule, "ec" provides extra flexibility in the definition of GE's genotypes. In particular, the new approach makes it possible to adjust the numeric precision of the representation of constants without changing the grammar's encoding scheme.

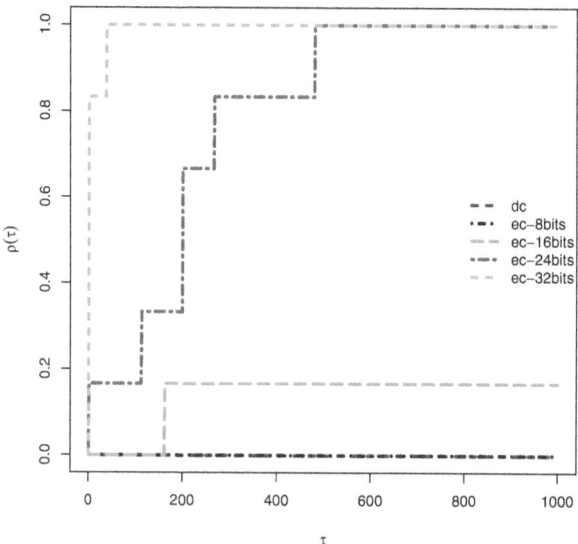

Fig. 4. Performance Profiles for mean error in "ec" variants

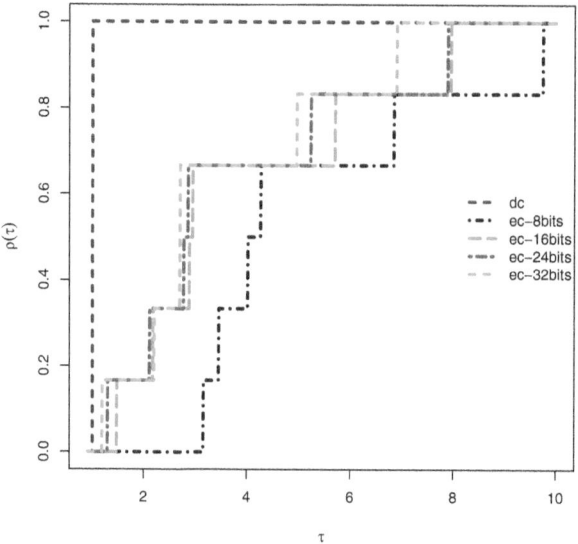

Fig. 5. Performance Profiles for expression size in "ec" variants

The computational experiments presented indicate that an increase on the number of bits used in the constants' representation has two desired effects. First, the level of accuracy of the approximation of constants increases, as expected. Second, and in part as a consequence of the first, the size of the final

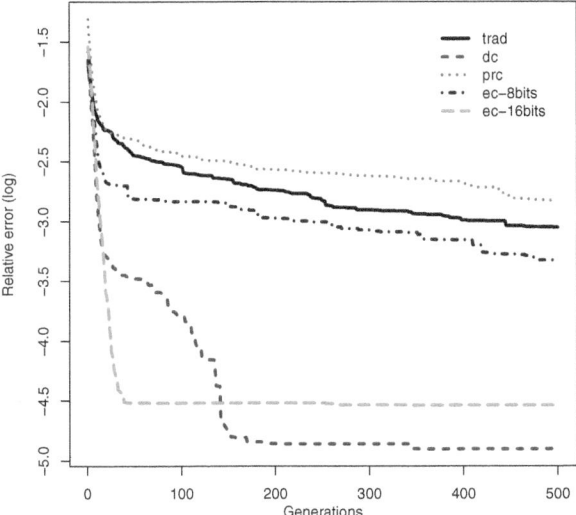

Fig. 6. Mean relative error for the constant 1.23

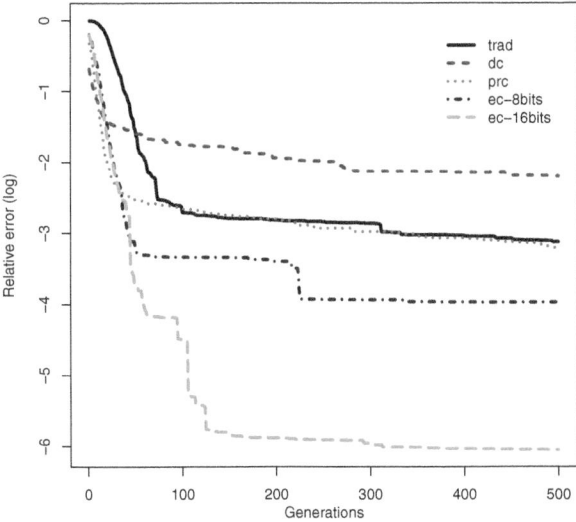

Fig. 7. Mean relative error for the constant 123000

symbolic expressions evolved by GE tends to decrease. We showed that, by increasing the numeric precision, we can easily overcome the most popular alternative approaches with respect to approximation accuracy. As for the complexity of the final solutions evolved, our experiments suggest that digit concatenation generates expressions that are slightly simpler than those returned by "ec".

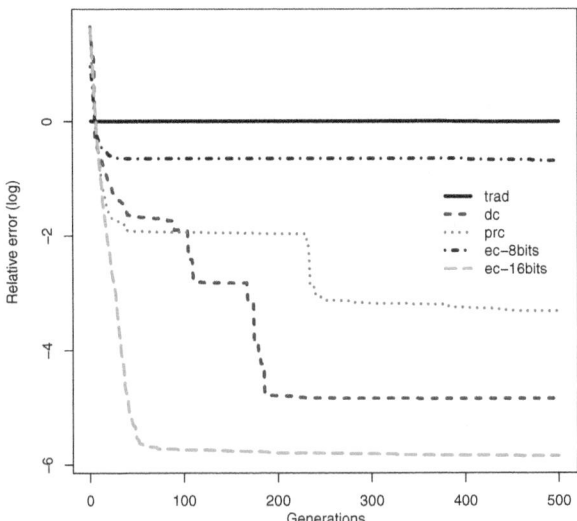

Fig. 8. Mean relative error for the constant 0.000123

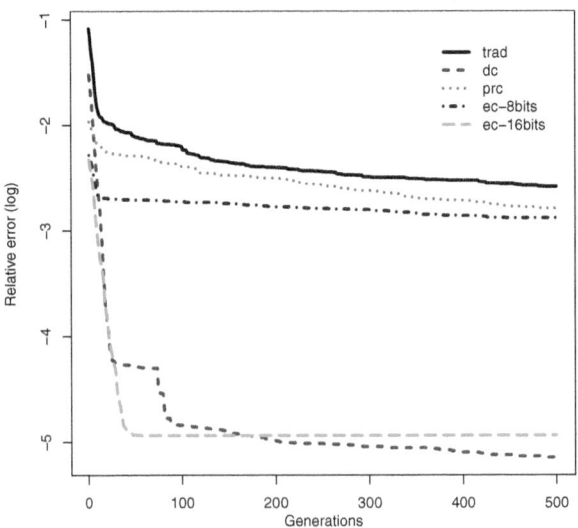

Fig. 9. Mean relative error for the constant 45.60099

As a final remark, it should be mentioned that the experiments presented here are only a simplified version of the real scenario for which GE was designed. In general, the evolution of programs involve the simultaneous approximation of several constants, and the evaluation of each candidate solution is invariably corrupted by noise (think of symbolic regression, for example). In this case it

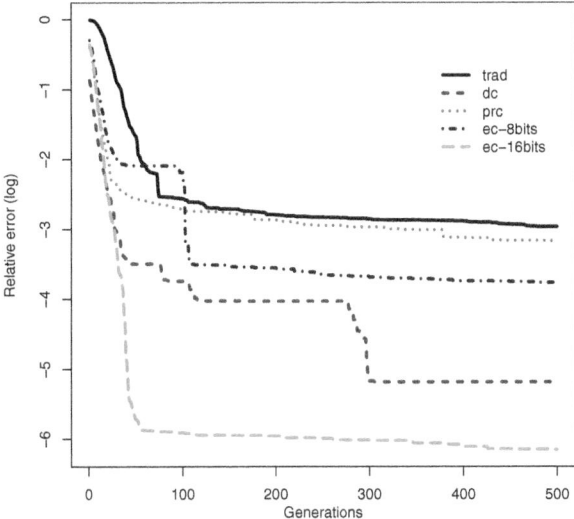

Fig. 10. Mean relative error for the constant 45600.99

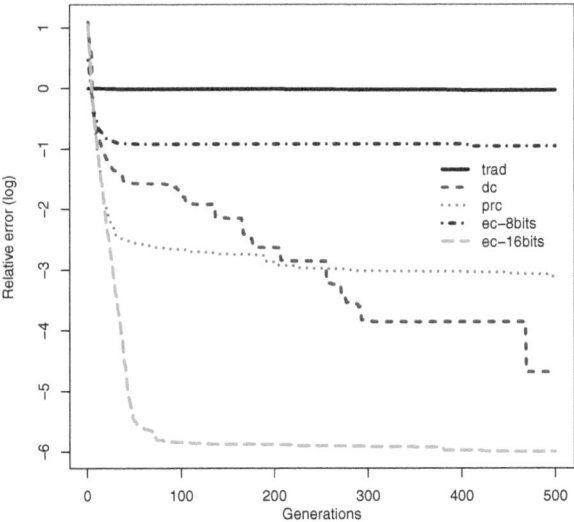

Fig. 11. Mean relative error for the constant 0.0004560099

is not so clear whether increasing the numerical precision of the representation scheme will necessarily lead to better performance. In principle this should not be a problem for the ephemeral-constant method, since it is always possible to decrease the number of bits it uses to represent constants. However, more experimentation is needed to verify whether the "ec" method also outperforms its counterparts when the solutions being evolved are complete programs.

Acknowledgments. The authors thank the support from CAPES, CNPq (308317/2009-2) and FAPERJ (grants E-26/102.825/2008, E-26/102.025/2009 and E-26/100.308/2010).

References

1. Augusto, D.A., Barreto, A.M., Barbosa, H.J., Bernardino, H.S.: A new approach for generating numerical constants in grammatical evolution. In: Proc. of the Conf. on Genetic and Evolutionary Computation (2011) Extended Abstract
2. Bernardino, H.S., Barbosa, H.J.: Grammar-Based Immune Programming for Symbolic Regression. In: Andrews, P.S., Timmis, J., Owens, N.D.L., Aickelin, U., Hart, E., Hone, A., Tyrrell, A.M. (eds.) ICARIS 2009. LNCS, vol. 5666, pp. 274–287. Springer, Heidelberg (2009)
3. Bernardino, H.S., Barbosa, H.J.: Grammar-based immune programming. Natural Computing, 1–33 (2010)
4. Cramer, N.L.: A representation for the adaptive generation of simple sequential programs. In: Proc. 1st Intl. Conf. on Genetic Algorithms, Hillsdale, NJ, USA, pp. 183–187 (1985)
5. Dempsey, I., O'Neill, M., Brabazon, A.: Grammatical constant creation. In: Proc. Conf. on Genetic and Evolutionary Computation, pp. 447–458 (2004)
6. Dempsey, I., O'Neill, M., Brabazon, A.: Meta-grammar constant creation with grammatical evolution by grammatical evolution. In: Proc. Conf. on Genetic and Evolutionary Computation, pp. 1665–1671. ACM, New York (2005)
7. Dempsey, I., O'Neill, M., Brabazon, A.: Constant creation in grammatical evolution. Int. J. Innov. Comput. Appl. 1, 23–38 (2007)
8. Dolan, E., Moré, J.J.: Benchmarking optimization software with performance profiles. Math. Programming 91(2), 201–213 (2002)
9. Koza, J.R.: Genetic Programming: On the Programming of Computers by Means of Natural Selection. MIT Press, Cambridge (1992)
10. O'Neill, M., Brabazon, A.: Grammatical differential evolution. In: Proc. Intl. Conf. on Artificial Intelligence, Las Vegas, NV, USA, pp. 231–236 (2006)
11. O'Neill, M., Brabazon, A., Adley, C.: The automatic generation of programs for classification problems with grammatical swarm. In: Proc. of the Congress on Evolutionary Computation, vol. 1, pp. 104–110 (2004)
12. O'Neill, M., Dempsey, I., Brabazon, A., Ryan, C.: Analysis of a Digit Concatenation Approach to Constant Creation. In: Ryan, C., Soule, T., Keijzer, M., Tsang, E.P.K., Poli, R., Costa, E. (eds.) EuroGP 2003. LNCS, vol. 2610, pp. 173–182. Springer, Heidelberg (2003)
13. O'Neill, M., Ryan, C.: Grammatical evolution. IEEE Trans. Evol. Comput. 5(4), 349–358 (2001)
14. O'Neill, M., Ryan, C.: Grammatical Evolution: Evolutionary Automatic Programming in an Arbitrary Language. Kluwer Academic Publishers (2003)
15. Rowe, J., Whitley, D., Barbulescu, L., Watson, J.-P.: Properties of gray and binary representations. Evol. Comput. 12, 47–76 (2004)
16. Ryan, C., Collins, J.J., Neill, M.O.: Grammatical Evolution: Evolving Programs for an Arbitrary Language. In: Banzhaf, W., Poli, R., Schoenauer, M., Fogarty, T.C. (eds.) EuroGP 1998. LNCS, vol. 1391, pp. 83–95. Springer, Heidelberg (1998)

The Evolution of Foraging
in an Open-Ended Simulation Environment

Tiago Baptista and Ernesto Costa

CISUC, Department of Informatics Engineering, University of Coimbra
P-3030-290 Coimbra, Portugal
{baptista,ernesto}@dei.uc.pt

Abstract. Throughout the last decades, Darwin's theory of natural selection has fueled a vast amount of research in the field of computer
science, and more specifically in artificial intelligence. The majority of
this work has focussed on artificial selection, rather than on natural selection. In this paper we study the evolution of agents' controllers in an
open-ended scenario. To that end, we set up a multi-agent simulation
inspired by the ant foraging task, and evolve the agents' brain (a rule
list) without any explicit fitness function. We show that the agents do
evolve sustainable foraging behaviors in this environment, and discuss
some evolutionary conditions that seem to be important to achieve these
results.

Keywords: artificial life, open-ended, evolution, multi-agent, ant
foraging.

1 Introduction

Today the process of evolution of species is well understood thanks to the natural selection theory proposed by Darwin, and complemented by the laws of
inheritance discovered by Mendel. In the twentieth century these theories were
deepened with a new comprehension supported by the advances in molecular
genetics. Evolution is a slow, time consuming process, that started more than
3000 million years ago. In the last years Darwins theory and the molecular biology central dogma ("DNA made proteins, and proteins made us") has been
under criticism and revision, with some researchers pointing out the importance
of other dimensions (e.g., epigenetic, behavioral, symbolic) to the process of natural evolution [7]. Notwithstanding, nobody denies that there is no goal, no plan
and no end in evolution: it is an open-ended process. An established definition
of open-ended evolution has not yet surfaced, however, most authors consider
that one of the major requirements is the absence of an explicit fitness function. In other words, to have open-ended evolution, a system should be based
on Natural Selection rather than Artificial Selection, as is done in Evolutionary
Computation [4].

The main goal of this paper is to study if, and how, it is possible to evolve
an agents controller, i.e., its brain, in an open-ended scenario. To that end we

L. Antunes and H.S. Pinto (Eds.): EPIA 2011, LNAI 7026, pp. 125–137, 2011.

set up a simulation in a simple, two dimensional world, involving agents — artificial ants, and a task — foraging. Moreover, rather than using the known ant foraging behaviors to create new algorithms, we are interested in investigating the emergence of such behaviors. Namely, we are interested in evolving complex behaviors in an open-ended evolution simulation, and use the example of the foraging behaviors of ants as an inspiration. By implementing a multi-agent simulation whose conditions resemble those found by ants on a foraging task, we hope to be able to evolve behaviors that themselves resemble those found on nature. In Nature, the foraging task, is of vital importance to all organisms. We can find a great variety of different behaviors associated with this task. For example, social insects like bees and ants, have evolved complex collective foraging behaviors to cope with this task. The study of these collective behaviors has crossed the discipline of biology and inspired a number of computational algorithms generally known as swarm intelligence [3]. Specifically, in ants, both the use of a random walk and of pheromone as a tool to communicate and coordinate the foraging task, has fueled a vast number of ant based algorithms like the Ant Colony Optimization algorithms [6].

Although most research on these issues is based on the implementation of ant algorithms, using the known biological behaviors, some research onto the emergence of these behaviors also exists [5] [8] [9]. In those papers, the authors use standard genetic algorithms to evolve populations of agents, controlled by artificial neural networks. In our research, however, we are mostly concerned with open-ended evolution of these behaviors. On a previous paper, the open-ended evolution of a circadian rhythm in an environment with a varying light level was investigated [1]. In that case, the agents had to first evolve a foraging behavior, to then be able to adapt to the varying level of light. Also, in Yaeger's seminal paper about the Polyworld simulation [11], agents evolve foraging in an open-ended evolution scenario. However, in both those papers, the food is distributed throughout all of the environment, whereas in the experiments presented here, the food is placed in a specified zone of the environment, away from the vision range of the agents.

The simulations described in this paper were implemented using the BitBang framework [2]. Implementing a modern autonomous agent model [10], this framework has roots in Artificial Life systems and Complexity Science. The simulated world is composed of entities. These can either be inanimate objects which we designate as *things*, or entities that have reasoning capabilities and power to perceive and affect the world—the *agents*. Both have traits that characterize them, such as color, size, or energy—the *features*. The agents communicate with, and change the environment using *perceptions* and *actions*, taking decisions using the *brain*. In this model, there is no definition of a simulation step, as we will not have any type of centralized control. As such, the simulation is asynchronous. The agents will independently perceive, decide, and act. Moreover, there is no evolutionary mechanism included in the definition of the model, since evolution is implemented as an action. That is accomplished by giving the agents the capability of reproduction. Again, there is no central control bound to the process of reproduction. The agents choose when to reproduce by choosing to execute the reproduction action. In addition, there is no explicit fitness function. The agents die due to lack of resources, predators, age, or any other mechanism implemented in the world. Thus, using this model we can implement open-ended evolution. In the next section we will describe the simulation world developed, detailing the agents, things, brain architecture, evolutionary process, and environmental settings. We will then present the configuration parameters used in the experiments. Finally, we present the results obtained from the simulation runs and end with some conclusions.

2   The Ant World

In this section we will set out all the implementation details and architecture of the simulations. As mentioned above, these simulations were implemented using the BitBang framework, and therefore we will present the architecture according to the framework's specifications. We begin by describing the simulation environment, then detail the agents' architecture (features, perceptions, actions, and brain), and than we present the *things* defined in this world, and finally we present the architecture of brain used in these experiments.

2.1   The Environment

Our world is a two dimensional world where agents and resources are placed (see Fig. 1). The terrain is a bounded square. Inside this arena we define a nest (the place where all the agents are born), and a feeding zone. This feeding zone is a circle of a given radius where all food items are placed. At startup, we fill the zone with food items, and these are periodically replenished so that the total food count is maintained. The number of resources available is configurable to be able to fine tune the system so as to allow agents to survive but also provide enough evolutionary pressure.

On initialization, the world is populated with randomly generated agents, placed at the nest. At this time, as their brain was generated randomly, it is highly probable that the agents will not execute the reproduction action, either by not choosing it, or because they do not have enough energy to reproduce. To keep the population alive, whenever the number of agents in the world falls bellow a given threshold, new agents are created. If there are live agents in the environment, one will be picked for reproduction, otherwise a new random agent is created. Note that, as stated before, there is no explicit fitness function, so the agent chosen for reproduction will be randomly selected from the population. When, and if, the agents evolve to reproduce by themselves (by choosing the reproduction action), this process of introducing new agents in the population stops.

To accommodate the placement and evaporation of pheromone, we created an influence map in the environment. The map divides the continuos environment into a grid. Each cell will have a value indicating the amount of pheromone it

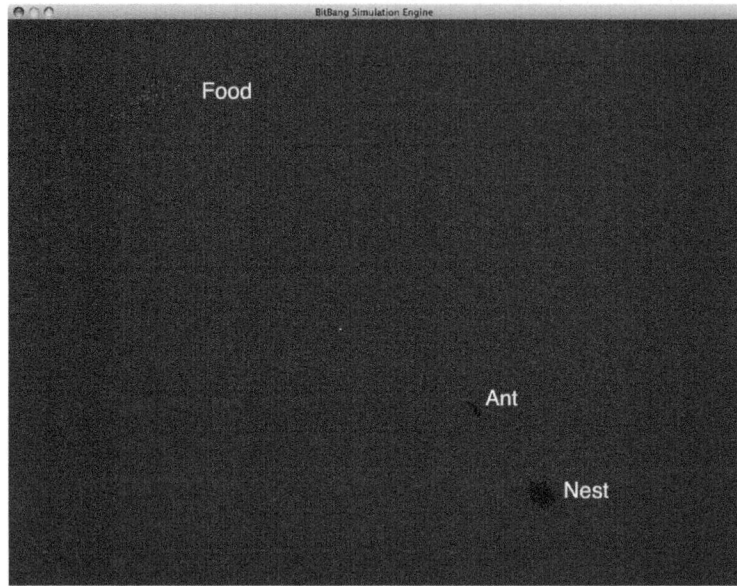

Fig. 1. Screenshot of a running simulation. At the top left corner there is a feeding zone with food items and at the bottom right we see the nest and an ant agent walking away from the nest.

contains. Agents can deposit pheromone in the environment, and once placed, the pheromone will evaporate linearly at a constant rate. Each cell has a maximum value of pheromone it can contain. If an agent tries to place pheromone into a cell that already has the maximum amount, no more pheromone will be deposited.

2.2 The Agents

In this simulation only one type (species) of agent exists, and has the following architecture:

- **Features:** energy, metabolic rate, birth date, and placing pheromone.
- **Perceptions:** energy, resource location, reach resource, pheromone location, placing pheromone, and random number.
- **Actions:** move front, turn left, turn right, eat, reproduce, and place pheromone.
- **Brain:** rule list (see section 2.4).

We will now describe each one of these components.

Features

Energy. This feature represents the current energy level of the agent. When this feature reaches zero, the agent dies. The feature is initialized with a predetermined value at agent birth. For these simulations, the agents are initialized with 10 energy units.

Metabolic Rate. The metabolic rate is the amount of energy the agent consumes per time unit.

Birth Date. This feature is set to the current time at birth and remains constant. It is used to calculate the agent's age. When the agent reaches a given maximum age, it dies. This procedure allows the evolution to continue past the moment when the agents have developed good navigation and eating capabilities, whilst maintaining an asynchronous and open-ended simulation. The maximum age of the agents is configurable.

Placing Pheromone. This boolean feature indicates if the agent is currently placing pheromone on the environment. Whenever this feature is true, the agent will place pheromone at a given rate.

Perceptions

Energy. This is a self-referencing perception on the agent's current energy level. This perception is tied to the corresponding feature.

Resource Location. This is the agent's perception of vision, representing the position of the nearest resource, relative to the agent's position and orientation. The agent's vision field is defined by a given range and angle. An object is within the vision field of an agent if its distance to the agent is less than or equal to the vision range, and the relative angle to the agent is within the defined vision angle. This is a numerical perception with possible values 0, 1, 2, and 3. The value 0 means no resource is visible. The value 1 means there is a resource to the left. The value 2 means there is a resource directly in front of the agent. The value 3 means there is a resource to the right.

Reach Resource. This is a boolean perception that evaluates to true whenever the agent has a resource within its reach.

Pheromone Location. This perception is similar in values to the *Resource Location* perception, but instead of identifying resources, it identifies pheromone values in the vicinity of the agent. To determine the value of this perception we will look to the nearby cells (1 cel neighborhood) in the pheromone influence map, and find the one with the maximum pheromone value. Cells are processed starting from the one at the front and left of the agents. In the case of a tie, the first processed cell will be selected. If none of the neighboring cells have pheromone, the value of the perception will be 0. If the maximum is to the left of the agent (the agent's orientation is taken into account), the value of the perception will be 1. If it is to the right, the value is 3. And if it is in front of the agent, the value is 2.

Placing Pheromone. This is a self-referencing perception that takes the same value as the corresponding feature. With this perception, the agent can perceive if it is currently placing pheromone.

Random Number. This perception provides a source of randomization to the agent. The perception will take a new random number each time it is evaluated. The number is selected from the range 0 to 3. We use this range, as it is the same range of values that the location perceptions can take.

Actions

Movement. We define three actions for movement. One to walk forward, one to turn left, and one to turn right.

Eat. This action enables the agent to eat a resource within its range. If no resource is in range when the action is executed, nothing happens. This action will add a configured amount of energy to the agent's energy feature.

Reproduce. This action allows the agent to reproduce itself. The reproduction implemented is asexual. When the action is executed, a new agent is created and placed in the world. The new agent will be given a brain that is a mutated version of its parent's brain. Note that, as each mutation operator has a given probability of being applied, the child's brain can be a perfect clone of its parent's brain. The action will also transfer energy from the parent to the offspring. The amount of energy consumed in the action is the sum of the initial energy for the new agent and a configurable fixed cost. It's important to have a cost of reproduction higher than the initial energy of an agent, so as to provide evolutionary pressure.

Place Pheromone. This action toggles the value of the *Placing Pheromone* feature, turning it on if is off, and off if it is on.

Brain. The agents' brain used in these experiments is a rule list. The architecture of this system is explained in section 2.4. On initial creation of an agent, the brain is randomly initialized. This initialization conforms to some configurable parameters: the maximum and minimum number of rules, and the maximum number of conditions per rule. Other configured values are the mutation probabilities used in the reproduction action.

2.3   The Things

Only one type of thing is defined in this world, representing the food items that the agents can eat to acquire energy. These things have no associated features.

2.4   The RuleList Brain

The Rule List brain is composed of an ordered list of rules. The reasoning process is straightforward. The rules are evaluated in order, and the first one whose conditions are all true, is selected. Each rule is composed of a conjunction of conditions and an action. The syntax of a rule is shown in Listing 1. Next, to illustrate, in Listing 2 we provide an example of a rule.

The operators available are dependent on the type of perceptions available to the agents. In these experiments, we use boolean and numeric perceptions. Boolean perceptions have two possible operators: TRUE and FALSE. Numeric perceptions have three possible operators: $=$, $<$, and $>$.

Listing 1. Syntax of a rule in the RuleList brain

```
<rule> ::= IF <cond-list> THEN <action>
<cond-list> ::= <condition>
<cond-list> ::= <condition> AND <cond-list>
<condition> ::= <percept> <operator> <percept>
<condition> ::= <percept> <operator>
```

The use of this brain architecture has the added benefit of readability. It is easy to understand the reasoning process by looking at the agent's rule list. This feature will permit a better analysis of the results.

Listing 2. Example of a rule

```
IF energy < 10 AND reach_resource TRUE THEN eat
```

To be able to evolve this brain architecture we need to define its equivalent to the genome, and the operators that modify it on reproduction. The brain's genome is the rule list itself, no translation is applied. To alter it we defined only mutation operators. These operators are show in Table 1.

Table 1. Mutation operators of the RuleList brain

Operator	Description
Mutate List	This operator iterates through the rule list, and replaces a rule with a new random one.
Mutate Rules	This is the lowest level operator. It drills down to the perceptions on the conditions and mutates both the perceptions and their operators. It also mutates the action of the rules.
Mutate Order	This operator iterates through the rule list and moves a rule one position towards the top.

3 Experimental Setup

In this section we present the main configuration values used in the experiments whose results are shown in the next section. Table 2 shows all the parameters

Table 2. Configuration values

Parameter	Values
Terrain Size	1000 x 1000
Time Limit	100 000; 500 000; 5 000 000
Nest Location	100, 100
Food Location	800, 800
Minimum Food	100; 200
Food Energy Content	3
Minimum Agents	20
Vision Range	100
Vision Angle	100 °
Agent Reach	20
Agent Initial Energy	10
Metabolic Rate	0.1
Maximum Age	500
Reproduction Cost	2
Minimum Rules	15; 20; 30
Maximum Rules	20; 30; 40
Maximum Conditions	2; 3
Mutation Probability	0.01
Pheromone Deposit Rate	50
Pheromone Evaporation Rate	10

used, and their configured values. If, for a given parameter, several values were tested, we present them separated by a semicolon. Most of the values used for these parameters are the result of previous experimentation.

The choice of the parameters to test was focussed on the evaluation of three main aspects. As introduced earlier, we know that the evolutionary process is very time consuming, especially when considering open-ended evolution. So, we tested three different values for the time limit of the simulations, one small, one medium, and one large. Secondly, we wanted to test if the size of the brain would influence the behaviors that emerge. So, we tested different values for the parameters that govern the creation of the brains. And lastly, as discussed earlier, the configuration of the number of food items available is important to provide enough evolutionary pressure, but also creating a world where life is at all possible. In that regard, we also tested two different values for the number of food items in the environment.

For each tested configuration, we ran 30 independent simulations. The combinations of tested configuration values are presented in Table 3.

4 Results

In this section we will present and analyze the results of the several experiments. As explained in the introduction, in these experiments we are interested in finding out if the agents evolve good foraging behaviors in this environment. Namely

Table 3. Experiments Configuration

Exp. Number	Time Limit	Min. Rules	Max. Rules	Max. Conditions	Food
1	100 000	15	20	2	100
2	500 000	15	20	2	100
3	500 000	20	30	3	100
4	500 000	30	40	3	100
5	5 000 000	15	20	2	100
6	500 000	15	20	2	200

we are looking for the random walk that characterizes the foraging behavior of some species of ants. To that end we will first show some plots of an example typical simulation run. In Fig. 2 we show the data from a run of experiment 2.

By inspecting the plots shown on Fig. 2, it's clear that the agents evolve foraging capabilities. We can see that at about 80 000 time units, both the average age and average energy of the agents rises, and in the case of the energy, it keeps improving up to time 150 000. At that time, we see that the agents start reproducing by themselves, and the population size quickly goes to about 250 agents. With the increase in population, the average energy per agent drops, because now there is greater competition for the same total amount of food.

Next, in Fig. 3, we show another group of plots from the same simulation run, showing the evolution of the average brain size and average number of used rules. We define used rules as the ones that have been selected at least once during the lifetime of the agent. The results shown are from a run of experiment 2, and thus the number of rules of an agent is between 15 and 20. We can see from the plot that, at the start of the simulation, the average brain size varies between these two values. Then, after about 80 000 time units (the time when agents start gathering food, as seen on above), the average brain size stabilizes. From the analysis of all the simulations, we found that from the point where an agent starts to have a good foraging behavior, most of the subsequent agents will be from the lineage of this first ancestor. Moreover, considering that the mutation operators do not change the number of rules, the brain size will consequently stabilize.

However, if we examine the second plot, we see that the number of used rules does not completely stabilize. Thus, having a stable brain size, and an increasing number of used portions of that brain, indicates that there is an increase in the complexity of the agents and their behaviors. It is important to note that within the population of about 250 agents, there are some that make it to the feeding zone, but also a large number that do not, pulling the average down. In the simulation run shown here, towards the end, the number of used rules for agents that do make it to the feeding zone is 7.

The analysis of the previous plots tells us that the agents do develop foraging capabilities, but does not show what kind of behaviors are emerging. These could be better observed by watching the running simulation, but that is not possible to show in the paper. In any case, as the brain algorithm used in this experiments

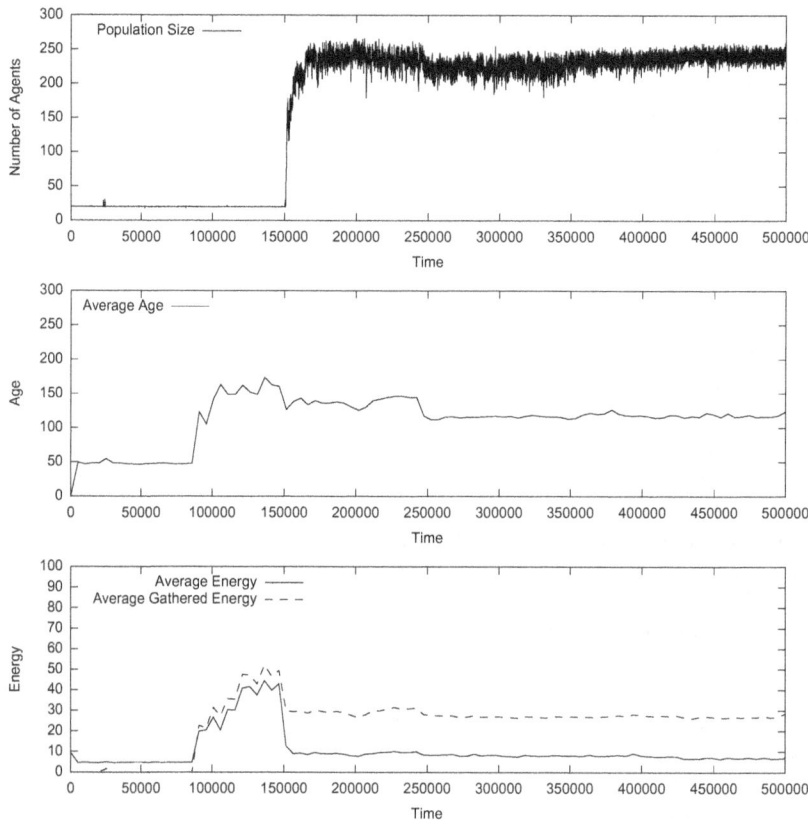

Fig. 2. Plot of the evolution of the number of agents, their average age, average energy, and average gathered energy, over the course of one simulation run

is fairly readable, we present one example of a brain taken from the population of agents from the same run as the plots shown. We show in Listing 3 an example of a brain taken from an agent living at the end of the simulation.

The example brain shown, allows us to decode some of the agent's behavior from the rule list presented. From rule 1, we see that the agent has as first priority reproduction, and reproduces whenever it has more than 18.5 energy. And, from rule 2, the agent eats whenever a food item is within its reach. The rest of the behavior can be divided into two parts: the movement behavior when either there is some, or no food in the vision range. In other words, the behavior is different when the agent finds the feeding zone. Examining the rules 3, 4, and 5, we can see that the agent will move forward if there is a food item in front or to the right, and turn left if there is a food item to the left. This is not the best possible behavior because the agent does not turn right when the food is on its right, but it is a good enough behavior. If there is no food item in the vision range, the value of the Resource Location perception is 0. Thus, if we consider

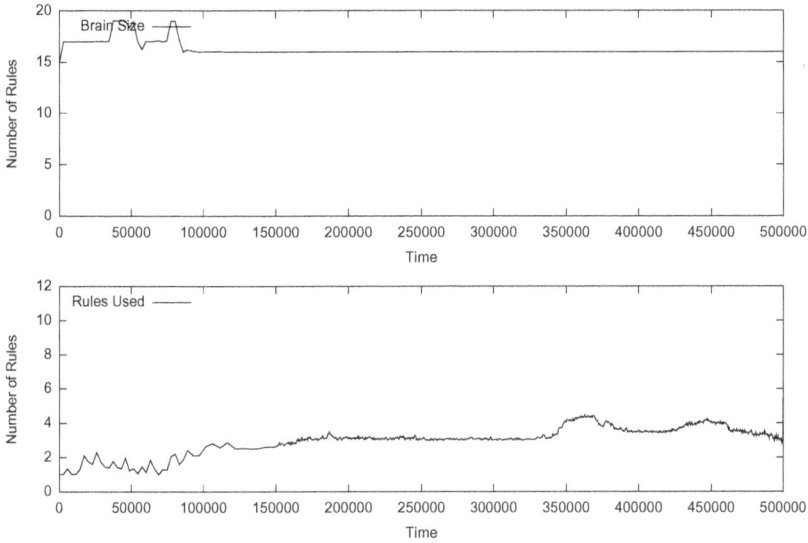

Fig. 3. Plot of the evolution of the average brain size and average number of used rules, over the course of one simulation run

the rules 7 and 12, we can see that the agent has a random walk behavior. The agent will move forward if the Random Number perception greater than 0 (probability of 0.75), and will turn left otherwise. Again, this is not a perfect random walk, but it is good enough.

The data shown above is from one simulation run of one of the experiments conducted. Although the results are typical of all the runs where the agents evolve good foraging behaviors, we need to analyze global results from all the simulations. To that end we show in Table 4, for each experiment defined (see Table 3), the number of runs that were successful. We define a run as being successful if the agents evolve foraging and reproductive behaviors. This will give us an indication on how the conditions set by the parameters in each experiment influence the evolution of the agents.

Table 4. Number of successful runs

Exp. Number	Total Runs	Successful
1	30	0
2	30	13
3	30	11
4	30	10
5	30	30
6	30	27

Listing 3. Example brain of an agent that evolved good foraging capabilities. This agent was born at time 499 211. Used rules are set in bold.

1. **IF Feature energy > 18.4609 THEN reproduce**
2. **IF istrue(Reaching Resource) THEN eat**
3. **IF Resource Location > 1 THEN go front**
4. **IF istrue(Feature placing pheromone) THEN turn left**
5. **IF Resource Location = 1 THEN place pheromone**
6. IF istrue(Feature placing pheromone) THEN place pheromone
7. **IF Resource Location < Random Number THEN go front**
8. IF Feature energy > 25.0065 THEN go front
9. IF Random Number = 2 THEN reproduce
10. IF istrue(Reaching Resource) THEN turn right
11. IF Random Number = Feature energy THEN go front
12. **IF Resource Location = 0 THEN turn left**
13. IF Feature energy > 34.5193 THEN turn right
14. IF istrue(Feature placing pheromone) THEN eat
15. IF Feature energy < Feature energy THEN turn right
16. IF istrue(Feature placing pheromone) THEN reproduce

To analyze the data from Table 4, we recall the three aspects considered for the definition of experiments: time limit, brain size, and environment's resources. Considering the value of the time limit we will look at experiments 1, 2, and 5. The results supports the hypothesis that time plays a fundamental role in this open-ended evolutionary process. When given enough time (5 000 000), all runs were successful. To analyze the effect of the brain size, we look at experiments 2, 3, and 4. In these experiments there is no significant difference in the number of successful runs. That seems to indicate that the lower value used (15 to 20 rules, with 1 or 2 conditions) is sufficient to accommodate the evolved behaviors. Finally, considering the amount of available food in the environment, we'll examine experiments 2 and 6. Again, we see that, as was the case for the time limit, the availability of resources has a significant influence in the evolutionary process. By incrementing the food count from 100 to 200, the number of successful runs raised form 13 to 27. This result supports our belief that the environment conditions are important to provide the necessary evolutionary pressure.

5   Conclusion

We set out to show that it is possible to evolve agents' controllers in an open-ended simulation scenario. Using the ant foraging task as an inspiration to develop a simulation with the BitBang framework, we showed that the agents do evolve behaviors capable of surviving in this world, from initial random brains. We also verify that the population evolves in two different steps. First the agents evolve foraging capabilities, and then start reproducing by themselves, creating a sustainable population. Moreover, through the analyses of the evolved brains, we find that the behaviors are similar to those found in Nature. The random walk is used by some species of ants to effectively cover the terrain in search for food. Additionally, we've shown that both the time limit and the environmental conditions are crucial to the success of the simulations.

As future work, we plan to further develop the environment to hopefully evolve more complex behaviors. Namely, we hope to construct a world where agents evolve stigmergic communication using the available pheromone tool.

References

1. Baptista, T., Costa, E.: Evolution of a multi-agent system in a cyclical environment. Theory in Biosciences 127(2), 141–148 (2008)
2. Baptista, T., Menezes, T., Costa, E.: Bitbang: A model and framework for complexity research. In: Proceedings of the European Conference on Complex Systems (September 2006)
3. Bonabeau, E., Theraulaz, G., Dorigo, M.: Swarm Intelligence: From Natural to Artificial Systems, 1st edn. Oxford University Press, USA (1999)
4. Channon, A.: Three evolvability requirements for open-ended evolution. In: Artificial Life VII Workshop Proceedings (2000)
5. Collins, R.J., Jefferson, D.R.: AntFarm: Towards Simulated Evolution. In: Artificial Life II, University of California, pp. 1–23 (May 1991)
6. Dorigo, M., Stutzle, T.: Ant Colony Optimization (Bradford Books). The MIT Press, Cambridge (2004)
7. Jablonda, E., Lamb, M.: Evolution in Four Dimensions: Genetic, Epigenetic, Behavioral, and Symbolic Variation in the History of Life. MIT Press, MA (2006)
8. Kawamura, H., Ohuchi, A.: Evolutionary Emergence of Collective Intelligence with Artificial Pheromone Communication. In: IEEE International Conference on Industrial Electronics, Control and Instrumentation, pp. 2831–2836. IEEE, Graduate School of Eng. (2000)
9. Nakamichi, Y.: Effectiveness of emerged pheromone communication in an ant foraging model. In: Proceedings of the Tenth International Symposium... (2005)
10. Russell, S., Norvig, P.: Artificial Intelligence: A Modern Approach, 2nd edn. Prentice Hall, Englewood Cliffs (2002)
11. Yaeger, L.: Computational genetics, physiology, metabolism, neural systems, learning, vision, and behavior or Poly World: Life in a new context. In: Proceedings of the Artificial Life III Conference, pp. 263–298 (1994)

A Method to Reuse Old Populations in Genetic Algorithms

Mauro Castelli, Luca Manzoni, and Leonardo Vanneschi

Dipartimento di Informatica, Sistemistica e Comunicazione
Università degli Studi di Milano - Bicocca
Viale Sarca 336, 20126 Milano, Italy
{mauro.castelli,luca.manzoni,vanneschi}@disco.unimib.it

Abstract. In this paper a method to increase the optimization ability of genetic algorithms (GAs) is proposed. To promote population diversity, a fraction of the worst individuals of the current population is replaced by individuals from an older population. To experimentally validate the approach we have used a set of well-known benchmark problems of tunable difficulty for GAs. Standard GA with and without elitism and steady state GA have been augmented with the proposed method. The obtained results show that the algorithms augmented with the proposed method perform better than the not-augmented algorithms or have the same performances. Furthermore, the proposed method depends on two parameters: one of them regulates the size of the fraction of the population replaced and the other one decides the "age" of the population used for the replacement. Experimental results indicate that better performances have been achieved with high values of the former parameter and low values of the latter one.

1 Introduction

The goal of this paper is to define a simple-to-implement method to improve the optimization ability of Genetic Algorithms (GAs) [6,4] that can be designed over other variation of the standard GAs. The idea is to re-use genetic material from older generations. Even though similar to the concept of "short-term" memory, which is typical of Tabu Search [3], and that has already been employed in evolutionary computation (see Section 2 for a discussion of previous and related work), the method we propose is new. We allow a set of individuals belonging to "old" generations to take part again in the selection process at given time intervals, thus giving them a second chance to be involved in mating. For this reason, we call the proposed method "second chance GA". In some senses, our method simulates the idea that partners do not need to have all the same age, but partners of different ages can mate and produce offspring. To experimentally validate this idea, we use a set of well-known GA benchmarks, including the one-max problem [11] and NK landscapes [8,7]. On these problems, we investigate the suitability of second chance GA for various values of some important parameters of the method and we compare the quality of the solutions returned by second chance GA when used to augment standard GA, GA with elitism and steady state GA.

L. Antunes and H.S. Pinto (Eds.): EPIA 2011, LNAI 7026, pp. 138–152, 2011.

The rest of the paper is organized as follows: in Section 2 we review some of the most relevant previous and related contributions, pointing out, for each of them, similarities and differences with the work presented here. Section 3 introduces the second chance GA method. In Section 4 we present the test problems used in the experimental study (one-max and NK landscapes) and we describe in detail the used parameter setting. Section 5 contains a presentation and a discussion of the obtained experimental results. Finally, Section 6 concludes the paper and suggests ideas for future research.

2   Previous and Related Work

The idea of using a "memory", for instance to store individuals belonging to previous generations, is not new in GAs. For instance, this idea has been exploited in [14,1], where a hybrid memory and random immigrants scheme in dynamic environments was proposed. In that approach, the best solution in memory is retrieved and acts as the base to create random immigrants to replace the worst individuals in the population. In this way, not only can diversity be maintained, but it is more efficient to adapt the GA to the changing environment. Even though interesting, the approach differs from the one presented here in at least the following points: (1) only one solution is kept in memory at each generation, i.e. either a random one (useful in case of modification of the target function) or the one with the best fitness value; (2) the goal of that work is to make GAs more effective in the presence of dynamic optimization environments, i.e. where the target function is subject to modifications in time. On the other hand, as it will be clear in the following, the focus here is on static problems. Furthermore, we use a larger memory than in [14,1], storing a pool of "good" individuals, instead of just the best one, and the storage happens at prefixed time steps (regulated by an apposite parameter) instead of happening at each generation. Finally, contrarily to [14,1], no random genetic material is ever re-injected in the population.

More similar to the present approach is the idea proposed in [9], where individuals are periodically stored in a database during a GA run. Nevertheless, contrarily to the main idea followed in this article, the focus of [9] is on the issue of generalization: stored individuals are good solutions for a slightly different problem instance than the current one. Furthermore the idea proposed in [9] differs from the one proposed here since it considers a "long term" memory. In other words, in [9] each GA run, instead of starting from scratch, uses an initial population seeded with the stored individuals, that previously solved "similar" problems. On the other hand, in the present work, only individuals from the same run are used (and in this sense we can say that we use a "short term" memory). Different from the method proposed here, although slightly related, is the approach proposed in [12], where GAs are used as a bootstrapping method for a further learning technique (i.e., memory-based reasoning). Another typical use of the concept of memory, that is different to, but perfectly integrable with, the one used here is the use of memory to avoid re-sampling already considered individuals, as, for instance, in [13].

Others similar concepts are common in the field of Artificial Immune Systems (AISs) [5]. As an example, it is possible to consider the work proposed in [2], where the clonal selection algorithm was defined to mimic the basic features of an immune response to an antigenic stimulus. It models the idea that only those cells that recognize the antigens are selected to proliferate and, before selection, they can be cloned in many copies. Differences between that approach and the one proposed here are numerous: besides the obvious differences between GAs and AISs (such as, for instance, the fact that GAs use crossover while AISs only use mutation, a different selection strategy, etc.), it is interesting to remark that, in the AISs clonal selection algorithm, only copies of solutions belonging to the same generation are considered, while the approach proposed here involves the use of solutions belonging to previous generations.

3   Second Chance GA

The main idea of the proposed "second chance" method is to insert genetic material from older populations into the current one, replacing the worst individuals in the current population. To accomplish this goal, every k generations (where $k \in \mathbb{N}$ and $k > 1$), the worst $p_r\%$ individuals in the population (where we call p_r the *replacement pressure*) are replaced. The individuals that replace them are extracted from the population of k generations before the current one (for this reason, we call k the *refresh rate*), and they are chosen from that population using exactly the same selection method used by the standard algorithm. The name "second chance" is inspired by the fact that, in this way, an individual can participate in at least two selection phases, increasing its probability of being selected. The pseudo code of the "second chance GA" method is given in Algorithm 1, where the "merge" procedure can be implemented as in Algorithm 2.

The algorithm as presented is not dependent on the specific variation of GA in use. It can be applied to standard GA as well as elitist GA or other variations. In other words, the proposed method can be "plugged" into an existing GA system in order to try to improve its performance.

Our motivations for introducing this method are the following: first of all we hypothesize that, generation by generation, the GA individuals become more and more specialized (that is, close to the optimum). Even though this behavior can be a good one in some cases, in others it can also cause a stagnation of the algorithm into a local optimum. The insertion of earlier - and probably less specialized - individuals with a good fitness should, in our intention, allow the population diversity to increase and facilitate the algorithm to escape from local optima. The second idea is that the processes of selection, crossover and mutation of GAs can discard (the former one) or disrupt (the latter ones) good genetic material. Allowing a reinsertion increases the probability of good genetic material to be propagated.

The presence of two different parameters in the proposed algorithm (the *replacement pressure p_r* and the *refresh rate k*), forces us to investigate their effect on the effectiveness of the GA. Informally, we could say that when the k parameter is low, reinserted individuals are not very "old". This can be beneficial

Algorithm 1. The *second chance* algorithm

```
initialize_GA();
number_of_generations := 0;
saved_population := ∅;
while ¬termination_criteria() do
      execute_ga_population();
      number_of_generations := number_of_generations + 1;
      if mod(number_of_generations, k) = 0
        then
                tmp_population := current_population;
                merge(current_population, saved_population);
                saved_population := tmp_population;
      fi
od
```

since, if fitness has not greatly improved during the last k generations, probably the previous populations still contain good individuals that might have been removed by selection, mutation or crossover, but that might have been useful. But it is also true that using a low value of k may prevent us from reinjecting the necessary amount of diversity in the current population. On the other hand, a big value of k can increase the diversity but may also be unable to supply "good enough" individuals (simply because they are too "old"), able to meaningfully influence the evolution of the population. The value of p_r can also have different effects. A low value of p_r can make the proposed method useless since the individuals inserted are not numerous enough to influence the search process. On the other hand, a high value of p_r can harm the search process by removing too many individuals with old, and thus possibly less good, ones.

Another choice that can influence the behavior of the method is *how* individuals are inserted and removed from the population. We have decided that at every removal of the worst individual from the current population, a new individual selected from the old population is inserted. With this method an individual with a bad fitness, that has eventually been selected from the old population, could be removed at the subsequent step (see Algorithm 2).

4   Test Problems and Experimental Setting

In all the experiments discussed later, the selection method used was tournament with size equal to 4. We used one-point crossover [6,4] with a crossover rate equal to 0.95 and standard GA mutation [6,4] with a mutation rate equal to 0.05. The tests were performed using both standard GA with and without elitism (i.e. unchanged copy of the best individual in the next population at each generation) and steady state GA. For the tests with the steady state GA at every step one quarter of the individuals in the current generation where replaced. In the results we considered four steps of the steady state GA as one generation (to obtain a number of fitness evaluations equal to the standard GA). The experiments were

Algorithm 2. The replacement and insertion phase of the algorithm

```
proc merge(P₁, P₂)
    for i := 0 to Population_Size * pᵣ do
        remove_worst(P₁);
        x = selection(P₂);
        add_individual(P₁, x);
    od
```

run using different individuals lengths and population sizes. In the continuation, we define *small runs* the ones where individuals length was 32 and population size was 64 and *medium runs* the ones where individuals length was 64 and population size was 128.

The second chance method was tested with a *replacement pressure p_r* of 0.75, 0.5 and of 0.25 and with a *refresh rate k* of 2, 5, 10, 15, 20 and 25. Every test was composed by 1000 runs, each of which was executed for 100 generations. The test functions used were the one-max problem [11] and the NK landscapes [8,7]. For all these problems binary genomes were considered (i.e. individuals are fixed length strings from the alphabet $\{0, 1\}$). The second chance method has been used to augment standard GA, standard GA with elitism (in the results discussion abbreviated with elitist GA) and steady state GA. The names of the resulting augmented algorithms are respectively second chance GA, elitist second chance GA and steady state second chance GA. Every GA variation is compared with the corresponding augmented method.

The one-max function, also called the counting-ones function, considers as a fitness of an individual the number of its bits that are equal to 1. In this case the global optimum is clearly the individual composed by only 1s.

The NK landscapes is a different family of tunable difficult functions. The difficulty of the fitness landscape can be modified by changing the value of a parameter $K_{NK} \in \{0, \ldots, N-1\}$ (where N is the length of the individual), that influences the ruggedness of the resulting fitness landscape. Here we use the notation K_{NK}, instead of the more usual K, to avoid confusing this parameter with the refresh rate of the second chance method, indicated by k. The fitness of an individual is given by the average contributions of its genes. The contribution of a gene in position i is determined by a function F_i. F_i has the gene in position i and other K_{NK} genes as its arguments. This means that the variation of a gene can modify the contribution of that gene, but also the contribution of the other ones. High values of K_{NK} produce rugged landscapes while low values reduce ruggedness. In particular, for $K_{NK} = 0$ the landscape has only one optimum while for $K_{NK} = N - 1$ the landscape is random. The experiments on NK landscapes were performed using 2, 4, 6, 8 and 10 as the values for K_{NK}.

For all the experiments the *average best fitness* (ABF) for all the generations has been recorded and normalized into the $[0, 1]$ range. For every method, the set of the best fitnesses at generation 100 has been recorded. The results obtained using the standard and elitist GA method and the ones obtained with the proposed "second chance" method have been compared. Statistical significance of the different results was investigated using a t-test [10] with a significance level $\alpha = 0.05$ and under the alternative hypothesis that average fitness of the proposed method was greater than the average fitness of the standard (or elitist) method. Since t-test requires normal distributions, a Kolmogorov-Smirnov test has been performed to check for this requirement.

5   Experimental Results

In all the tables showed here, the best ABF between all the tested methods is reported in **bold**, while the worst is reported in *italic*. In all the experiments, a larger value corresponds to a better fitness. For the sake of compactness only the results for the *onemax* problem and *NK landscapes* with K_{NK} equals to 2, 6 and 10 are reported.

5.1   Onemax

The results for $p_r = 0.25$ on the *onemax* problem, both for "small" and "medium" populations and individuals size are summarized in Table 1. It is possible to see that for standard GA the augmented method performs better in every case. The difference is also statistically significant. The elitist GA performs better than the corresponding augmented algorithm but the difference, in particular for $k = 2$, is quite small. For the steady state GA, considering the small populations case, the algorithm always finds the optimum. For the other cases the augmented method performs better and the difference in terms of performances is statistically significant except for $k = 25$.

Table 1. The ABF at generation 100 for the *one-max* problem with $p_r = 0.25$

	Standard GA		Elitist GA		Steady State GA	
	small	medium	small	medium	small	medium
Standard	*0.96875*	*0.88328*	**0.98269**	**0.90877**	*1.00000*	*0.98198*
$k = 2$	**0.99187**	**0.92470**	0.97978	0.90189	1.00000	**0.98780**
$k = 5$	0.98738	0.91461	0.97459	0.89070	1.00000	0.98591
$k = 10$	0.98566	0.91198	0.97056	0.88750	1.00000	0.98369
$k = 15$	0.98466	0.91281	0.97066	0.88569	1.00000	0.98297
$k = 20$	0.98381	0.90936	0.96941	0.88456	1.00000	0.98266
$k = 25$	0.98456	0.91016	*0.96931*	*0.88342*	1.00000	0.98241

When the *replacement pressure* grows to 0.5 (see Table 2) the results are similar for standard GA, but for elitist GA the situation changes. In fact, when $k = 2$, the augmented method shows to be better both for small and medium populations, with a statistically significant difference. For the steady state GA the results are similar to the ones for $p_r = 0.25$, except for the fact that the proposed method is not statistically better than the steady state GA for $k = 20$ (other than $k = 25$).

Table 2. The ABF at generation 100 for the *one-max* problem with $p_r = 0.5$

	Standard GA		Elitist GA		Steady State GA	
	small	medium	small	medium	small	medium
Standard	*0.96875*	*0.88328*	0.98269	0.90877	*1.00000*	*0.98198*
$k = 2$	**0.99503**	**0.93322**	**0.98684**	**0.91370**	1.00000	**0.98684**
$k = 5$	0.99019	0.91981	0.97728	0.89630	1.00000	0.98666
$k = 10$	0.98706	0.91467	0.97209	0.88997	1.00000	0.98452
$k = 15$	0.98525	0.91320	0.97066	0.88714	1.00000	0.98339
$k = 20$	0.98528	0.91152	0.97062	0.88680	1.00000	0.98288
$k = 25$	0.98450	0.91077	*0.96909*	*0.88506*	1.00000	0.98256

In Table 3 results for $p_r = 0.75$ are reported. Besides the ABF at generation 100, also the evolution of the ABF generation by generation is presented (see Figure 1). In this case the situation is almost identical to the one for $p_r = 0.5$.

For both standard GA and elitist GA results confirm that the best performances are obtained with $k = 2$. For steady state GA the proposed method with $k = 2$ still performs better than the non-augmented algorithm but the best performances are obtained with $k = 5$.

Table 3. The ABF at generation 100 for the *one-max* problem with $p_r = 0.75$

	Standard GA		Elitist GA		Steady State GA	
	small	medium	small	medium	small	medium
Standard	*0.96875*	*0.88328*	0.98269	0.90877	*1.00000*	*0.98198*
$k = 2$	**0.99803**	**0.94059**	**0.99175**	**0.92220**	1.00000	0.98656
$k = 5$	0.99134	0.92439	0.97897	0.89973	1.00000	**0.98705**
$k = 10$	0.98703	0.91687	0.97431	0.89238	1.00000	0.98553
$k = 15$	0.98628	0.91369	0.97294	0.88844	1.00000	0.98389
$k = 20$	0.98597	0.91323	0.97122	0.88748	1.00000	0.98309
$k = 25$	0.98500	0.91194	*0.96962*	*0.88548*	1.00000	0.98261

In conclusion, for the *onemax* problem the best performances are obtained with high values of p_r and low values of k. The effects of the proposed method seem to diminish when the base method performs already well.

5.2   NK-Landscapes

$K_{NK} = 2$. The results for the smallest K_{NK} tested for the *NK landscapes* problem when $p_r = 0.25$ is presented in Table 4. In this case the proposed method performs better than the standard GA for all the considered k values with a statistically significant difference. The results are inverted for the elitist GA case, where the non-augmented method is the best performer. For the steady state GA case it is necessary to make a distinction between "small" and "medium" populations. For small populations the base algorithm is the best performer. When the individuals size (and, consequently, the difficulty of the problem) increases,

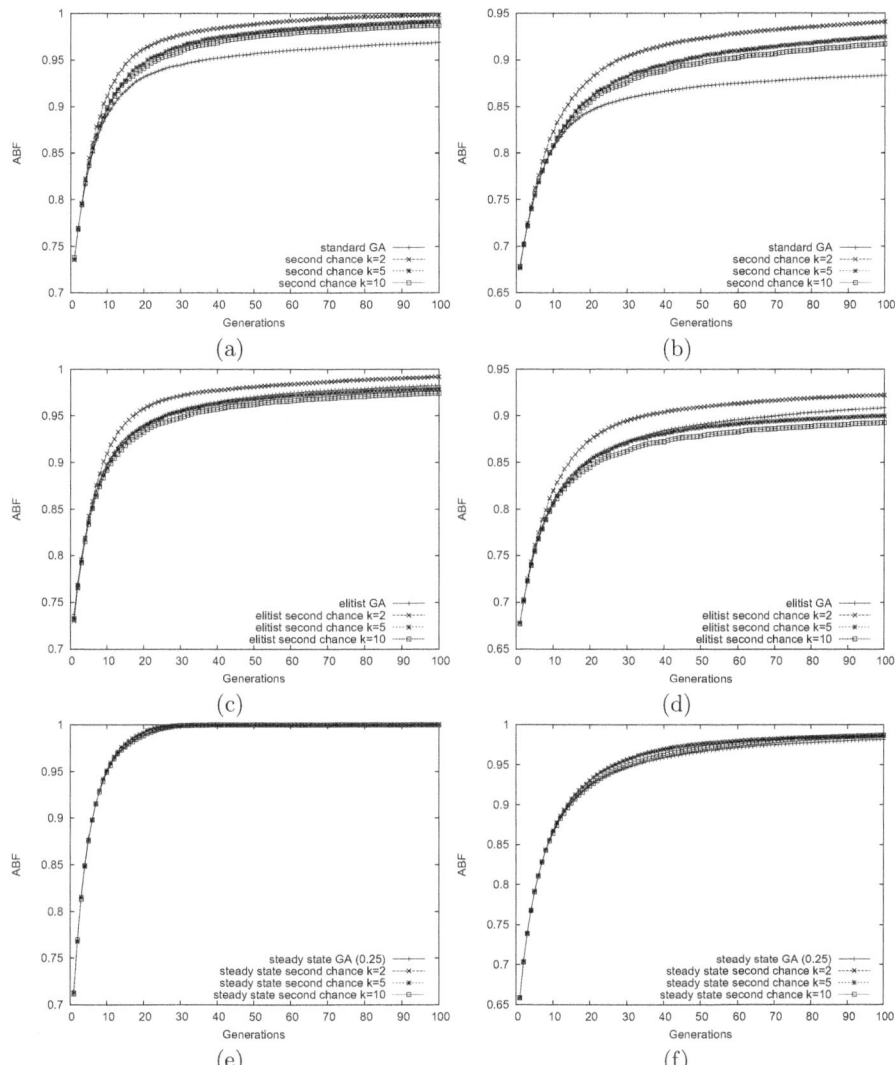

Fig. 1. The ABF for the *onemax* problem with replacement pressure $p_r = 0.75$. Plots *a*, *c* and *e* use small population size while the others use medium population size.

the proposed method performs better than the base one with a statistically significant difference for all considered k values.

When $p_r = 0.5$ (see Table 5) the proposed method always performs better than the base algorithms for $k = 2$. The difference between the different methods is quite small and not statistically significant, except when comparing with standard GA.

Table 4. The ABF at generation 100 for the *NK landscapes* problem with $K_{NK} = 2$ and $p_r = 0.25$

	Standard GA		Elitist GA		Steady State GA	
	small	medium	small	medium	small	medium
Standard	*0.54536*	*0.48216*	**0.55709**	**0.49964**	**0.56806**	*0.54715*
$k = 2$	**0.56275**	**0.50878**	0.55347	0.49407	*0.56608*	**0.56856**
$k = 5$	0.55994	0.50374	0.54978	0.48684	0.56650	0.56708
$k = 10$	0.55904	0.50155	0.54760	0.48376	0.56664	0.56609
$k = 15$	0.55811	0.50106	0.54751	0.48370	0.56682	0.56535
$k = 20$	0.55712	0.49970	*0.54695*	0.48332	0.56692	0.56500
$k = 25$	0.55750	0.49984	0.54727	*0.48242*	0.56674	0.56507

Table 5. The ABF at generation 100 for the *NK landscapes* problem with $K_{NK} = 2$ and $p_r = 0.5$

	Standard GA		Elitist GA		Steady State GA	
	small	medium	small	medium	small	medium
Standard	*0.54536*	*0.48216*	0.55709	0.49964	*0.56806*	0.54715
$k = 2$	**0.57262**	**0.51697**	**0.56603**	**0.50326**	0.57162	0.54845
$k = 5$	0.56925	0.50863	0.55929	0.49286	0.57220	**0.54930**
$k = 10$	0.56708	0.50539	0.55629	0.48856	0.57243	0.54810
$k = 15$	0.56684	0.50391	0.55618	0.48760	**0.57247**	0.54728
$k = 20$	0.56495	0.50332	0.55532	0.48641	0.57245	0.54728
$k = 25$	0.56591	0.50329	*0.55448*	*0.48594*	0.57244	*0.54694*

When $p_r = 0.75$ (see Table 6) the results show that the proposed method performs always better than the base algorithm. In particular the difference is statistically significant for standard GA and steady state GA with "medium" population size. It is interesting to note that except for steady state GA the best results are obtained for small values of k. In the remaining cases the differences are small but the best performer can have an high k.

Table 6. The ABF at generation 100 for the *NK landscapes* problem with $K_{NK} = 2$ and $p_r = 0.75$

	Standard GA		Elitist GA		Steady State GA	
	small	medium	small	medium	small	medium
Standard	*0.54536*	*0.48216*	0.55709	0.49964	*0.56806*	*0.54715*
$k = 2$	**0.56980**	**0.51950**	**0.56600**	**0.50581**	0.57983	0.56808
$k = 5$	0.56603	0.50927	0.55708	0.49331	0.58043	**0.56900**
$k = 10$	0.56351	0.50453	0.55386	0.48742	0.58085	0.56781
$k = 15$	0.56277	0.50253	0.55104	0.48641	0.58097	0.56711
$k = 20$	0.56160	0.50176	*0.55078*	0.48526	0.58073	0.56609
$k = 25$	0.56237	0.50130	0.55126	*0.48441*	**0.58108**	0.56618

Table 7. The ABF at generation 100 for the *NK landscapes* problem with $K_{NK} = 6$ and $p_r = 0.25$

	Standard GA		Elitist GA		Steady State GA	
	small	medium	small	medium	small	medium
Standard	*0.60953*	*0.52223*	**0.63078**	**0.54638**	**0.64841**	**0.61586**
$k = 2$	**0.63867**	**0.55849**	0.62522	0.53742	*0.63426*	0.61056
$k = 5$	0.63497	0.55119	0.61589	0.52855	0.63556	0.60837
$k = 10$	0.63195	0.54851	0.61299	0.52575	0.63600	0.60729
$k = 15$	0.63167	0.54820	0.61256	0.52463	0.63558	0.60646
$k = 20$	0.63110	0.54545	0.61083	0.52343	0.63599	0.60625
$k = 25$	0.63094	0.54667	*0.61081*	*0.52337*	0.63567	*0.60575*

Table 8. The ABF at generation 100 for the *NK landscapes* problem with $K_{NK} = 6$ and $p_r = 0.5$

	Standard GA		Elitist GA		Steady State GA	
	small	medium	small	medium	small	medium
Standard	*0.60953*	*0.52223*	**0.63078**	**0.54638**	**0.64841**	**0.61586**
$k = 2$	**0.63500**	**0.55829**	0.62267	0.53890	*0.63798*	0.60584
$k = 5$	0.62843	0.54640	0.61189	0.52518	0.64046	0.60515
$k = 10$	0.62538	0.54125	0.60738	0.52050	0.64099	0.60230
$k = 15$	0.62286	0.54094	0.60547	0.51896	0.64098	0.60236
$k = 20$	0.62317	0.53817	0.60445	0.51731	0.64152	0.60222
$k = 25$	0.62345	0.53808	*0.60388*	*0.51695*	0.64133	*0.60090*

Concluding the discussion for $K_{NK} = 2$, there are some interesting remarks that can be done. First of all, the results show that the replacement pressure is an important factor, in particular when comparing with more sophisticated algorithms. High values of p_r seem to provide the best performances. The value of k is also important since using populations that are "too old" seems to be detrimental to the learning process. Rather unexpectedly, the previous consideration is not entirely valid when the augmentation of steady state GA is considered.

$K_{NK} = 6$. When K_{NK} is equal to 6 the resulting problem has an intermediate difficulty between the ones tested. When $p_r = 0.25$ the proposed method is better than the base one only for standard GA with a difference that is statistically significant.

When the *replacement pressure* is equal to 0.5 (see Table 8) the proposed method still outperforms standard GA but its performances are worse than the non-augmented algorithm for both elitist and steady state GA.

When $p_r = 0.75$ (see Table 9) the situations changes. In fact, the proposed method performs better than both standard and elitist GA. The difference between the proposed method and standard GA is statistically significant for all values of k. Concluding the results discussion for $K_{NK} = 6$, it is interesting to note that standard GA remains the worst performer while other methods can obtain better performances than the augmented method (even if the difference

Table 9. The ABF at generation 100 for the *NK landscapes* problem with $K_{NK} = 6$ and $p_r = 0.75$

	Standard GA		Elitist GA		Steady State GA	
	small	medium	small	medium	small	medium
Standard	*0.60953*	*0.52223*	0.63078	0.54638	**0.64841**	**0.61586**
$k = 2$	**0.64413**	**0.57282**	**0.63747**	**0.55535**	*0.64038*	0.60481
$k = 5$	0.63830	0.55831	0.62244	0.53797	0.64282	0.60575
$k = 10$	0.63423	0.55128	0.61697	0.52967	0.64351	0.60363
$k = 15$	0.63369	0.54892	0.61368	0.52734	0.64334	0.60237
$k = 20$	0.63146	0.54787	0.61354	0.52630	0.64372	*0.60108*
$k = 25$	0.63055	0.54811	*0.61136*	*0.52503*	0.64324	0.60140

Table 10. The ABF at generation 100 for the *NK landscapes* problem with $K_{NK} = 10$ and $p_r = 0.25$

	Standard GA		Elitist GA		Steady State GA	
	small	medium	small	medium	small	medium
Standard	*0.60764*	*0.50900*	**0.63787**	**0.53214**	**0.66529**	*0.61123*
$k = 2$	**0.64905**	**0.54800**	0.62518	0.52564	*0.65915*	**0.62795**
$k = 5$	0.64202	0.53862	0.61397	0.51529	0.66058	0.62586
$k = 10$	0.64087	0.53558	0.61090	0.51197	0.66179	0.62422
$k = 15$	0.63706	0.53486	0.61001	0.51124	0.66209	0.62371
$k = 20$	0.63952	0.53290	0.60752	0.51073	0.66169	0.62313
$k = 25$	0.63484	0.53262	*0.60735*	*0.50984*	0.66227	0.62251

in performances is generally quite small). The best performances are obtained with small values of k and high values of p_r.

$\mathbf{K_{NK} = 10}$. The last considered case for the *NK landscapes* problem is also the most difficult among the ones that we have studied. When the *replacement pressure* is equal to 0.25 (see Table 10) the proposed method is the best performer when compared to standard GA (the difference is statistically significant). When compared with elitist GA, the base algorithm performs better and when compared with steady state GA the situation needs to be discussed further. For "small" populations the base algorithm performs better than the second chance one. For "medium" sized populations the situation is inverted. In fact, the difference is statistically significant for all considered k values.

The results for $p_r = 0.5$ are presented in Table 11. In this case only the results of the comparison with standard GA do not change significantly when the population size varies. For elitist GA the proposed method performs better (in a statistically significant way) only for "medium" sized populations and individuals and only for $k = 2$. The comparison with the steady state GA shows that the proposed method performs better than the non-augmented algorithm but the difference is statistically significant only for "medium" sized populations and individuals. It is interesting to note that in this case the proposed method performs better for all values of k.

Table 11. The ABF at generation 100 for the *NK landscapes* problem with $K_{\mathrm{NK}} = 10$ and $p_r = 0.5$

	Standard GA		Elitist GA		Steady State GA	
	small	medium	small	medium	small	medium
Standard	*0.60764*	*0.50900*	**0.63787**	0.53214	0.66529	*0.61123*
$k = 2$	**0.65451**	**0.56027**	0.63606	**0.54014**	*0.66095*	0.62602
$k = 5$	0.64459	0.54607	0.61737	0.52393	0.66590	**0.62725**
$k = 10$	0.64058	0.54008	0.61473	0.51669	0.66670	0.62481
$k = 15$	0.63963	0.53910	0.61099	0.51515	**0.66741**	0.62311
$k = 20$	0.63593	0.53752	0.60872	*0.51399*	0.66689	0.62321
$k = 25$	0.63655	0.53650	*0.60662*	0.51467	0.66720	0.62285

The results for $p_r = 0.75$ are presented in Table 12. The evolution of the ABF generation by generation is presented in Figure 2. As a confirmation of the previous results, the proposed method performs better than standard GA. Also for elitist GA the best performer is the second chance method but the results change when population and individual size varies. In particular, for "small" populations the difference is not statistically significant. For "medium" populations the difference is statistically significant for $k = 2$ but also the other values of k performs worse than the base algorithm. This is a confirmation of the importance of the parameter k for the performances of the second chance method. The comparison with steady state GA also allows to take different conclusions when considering differently sized populations. For "small" populations the proposed method is the best performer but with a difference that is not statistically significant. Interestingly, the worst performance is obtained with $k = 2$. For "medium" sized populations the proposed method performs better, but not with a statistically significant difference, than the base algorithm for all k.

In conclusion, the setting of the two parameters has a great influence in the behaviour of the proposed method. In particular, the best combinations are confirmed to be those with a low value of k and an high value of p_r.

It is interesting to note that in almost all tests the elitist second chance GA performs worse than the second chance GA. This means that the presence of elitism hurts the search process when reusing old individuals.

5.3   Results Summary

The presented results indicate that second chance GA can perform better than standard, elitist and steady state GA and, when it does not perform better the performances are generally comparable. An essential part of the method is the setting of the parameters p_r and k:

– p_r needs to be at least 0.5 in order to allow the proposed method to perform better than elitist and steady state GA. Values higher than 0.5 seem to work even better. In fact, $p_r = 0.75$ seems to be the best setting of the parameter among the tested ones.

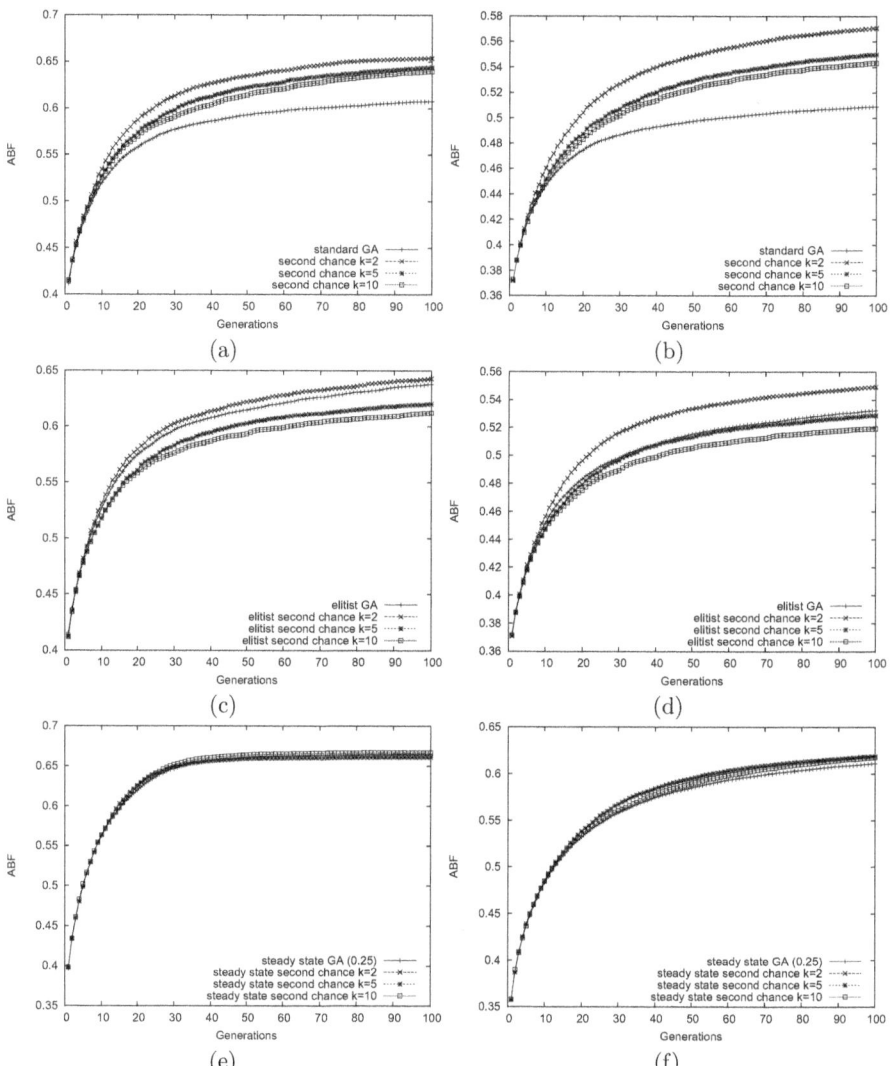

Fig. 2. The ABF for the *NK landscapes* problem with $K_{NK} = 10$ with replacement pressure $p_r = 0.75$. Plots *a*, *c* and *e* use small population size while the others use medium population size.

- k plays a role that is of primary importance. Only small values seem to be of interest. In fact, for high values of k, second chance GA performances are often inferior to second chance GA with low k. In particular, the value $k = 2$ seems to provide the best performances.

It is interesting to note that the different behaviour obtained by changing the values of the two parameters was consistent for all the studied benchmark problems. Thus, we hypothesize that the observed behaviour may be problem independent.

Table 12. The ABF at generation 100 for the *NK landscapes* problem with $K_{\mathrm{NK}} = 10$ and $p_r = 0.75$

	Standard GA		Elitist GA		Steady State GA	
	small	medium	small	medium	small	medium
Standard	*0.60764*	*0.50900*	0.63787	0.53214	0.66529	*0.61123*
$k = 2$	**0.65372**	**0.57059**	**0.64288**	**0.54913**	*0.66119*	**0.61874**
$k = 5$	0.64412	0.54995	0.62032	0.52874	0.66690	0.61772
$k = 10$	0.63966	0.54334	0.61222	0.51928	0.66854	0.61602
$k = 15$	0.63863	0.54192	0.60931	0.51836	0.66898	0.61472
$k = 20$	0.63548	0.53955	0.60882	0.51505	0.66881	0.61445
$k = 25$	0.63497	0.53886	*0.60595*	*0.51424*	**0.66949**	0.61366

It is also useful to remark that elitism seems to degrade performances of second chance GA. Even if the idea of having a GA with both a population with a good variety and the possibility to retain good solutions seems intriguing, the experimental results show that, for the proposed method, this idea is not beneficial.

6 Conclusions and Future Work

A new method for improving the optimization ability of Genetic Algorithms (GAs) has been presented in this paper. It is based on the idea of re-using "good" but "old" individuals in the current population, giving them a second chance to survive and mate (hence the name of the proposed method: second chance GA). It is inspired by well-known concepts such as various forms of short-term memory scheme. The main motivation for introducing this method has been that we expect re-injection of old genetic material to be beneficial in case of premature convergence of the standard algorithm towards local optima. The proposed method depends on two new parameters that we have called refresh rate and replacement pressure. In order to experimentally validate the approach, we have used a set of benchmarks including the one-max problem and NK landscapes. The obtained results have indicated that second chance GA is able to outperform standard GA with no elitism and it has performances better or comparable to GA with elitism and steady state GA. In particular, "low" values of the refresh rate and "high" values of the replacement pressure seems to be the best choice for all the considered test problems.

Future work includes the experimental validation of second chance GA on a wider set of test problems of different nature, including real-life applications, and the attempt to theoretically justify the suitability of the parameter values that have been indicated by the experimental results presented here.

References

1. Cao, Y., Luo, W.: Novel associative memory retrieving strategies for evolutionary algorithms in dynamic environments. In: Cai, Z., Li, Z., Kang, Z., Liu, Y. (eds.) ISICA 2009. LNCS, vol. 5821, pp. 258–268. Springer, Heidelberg (2009)
2. De Castro, L.N., Von Zuben, F.J.: The clonal selection algorithm with engineering applications. In: GECCO 2002 - Workshop Proceedings, pp. 36–37. Morgan Kaufmann (2002)

3. Glover, F., Laguna, M.: Tabu Search. Kluwer Academic Publishers, Norwell (1997)
4. Goldberg, D.E.: Genetic Algorithms in Search, Optimization and Machine Learning. Addison-Wesley, Reading (1989)
5. Greensmith, J., Whitbrook, A.M., Aickelin, U.: Artificial immune systems (2010). Computing Research Repository (CoRR), abs/1006.4949 (2010), http://arxiv.org/abs/1006.4949
6. Holland, J.H.: Adaptation in Natural and Artificial Systems. The University of Michigan Press, Ann Arbor (1975)
7. Kauffman, S.A.: Adaptation on rugged fitness landscapes. In: Stein, D. (ed.) Lectures in the Sciences of Complexity, Redwood City. SFI Studies in the Sciences of Complexity, Lecture, vol. I, Addison-Wesley (1989)
8. Kauffman, S., Levin, S.: Towards a general theory of adaptive walks on rugged landscapes. J. Theoret. Biol. 128(1), 11–45 (1987)
9. Louis, S., Li, G.: Augmenting genetic algorithms with memory to solve traveling salesman problems. In: Proceedings of the Joint Conference on Information Sciences, pp. 108–111. Duke University Press (1997)
10. Moore, D.S.: The Basic Practice of Statistics, 2nd edn. W. H. Freeman & Co., New York (1999)
11. Schaffer, J.D., Eshelman, L.J.: On crossover as an evolutionary viable strategy. In: Belew, R.K., Booker, L.B. (eds.) Proceedings of the 4th International Conference on Genetic Algorithms, pp. 61–68. Morgan Kaufmann, San Francisco (1991)
12. Sheppard, J., Salzberg, S.: Combining genetic algorithms with memory based reasoning. In: Proceedings of the Sixth International Conference on Genetic Algorithms, pp. 452–459. Morgan Kaufmann (1995)
13. Wiering, M.: Memory-based memetic algorithms. In: Nowe, A., Lenaerts, T., Steenhout, K. (eds.) Benelearn 2004: Proceedings of the Thirteenth Belgian-Dutch Conference on Machine Learning, pp. 191–198 (2004)
14. Yang, S.: Genetic algorithms with memory-and elitism-based immigrants in dynamic environments. Evol. Comput. 16, 385–416 (2008)

Towards Artificial Evolution of Complex Behaviors Observed in Insect Colonies

Miguel Duarte, Anders Lyhne Christensen, and Sancho Oliveira

Instituto de Telecomunicações
Instituto Universitário de Lisboa (ISCTE-IUL)
Lisbon, Portugal
mail@miguelduarte.org, {anders.christensen,sancho.oliveira}@iscte.pt

Abstract. Studies on social insects have demonstrated that complex, adaptive and self-organized behavior can arise at the macroscopic level from relatively simple rules at the microscopic level. Several past studies in robotics and artificial life have focused on the evolution and understanding of the rules that give rise to a specific macroscopic behavior such as task allocation, communication or synchronization. In this study, we demonstrate how colonies of embodied agents can be evolved to display multiple complex macroscopic behaviors at the same time. In our evolutionary model, we incorporate key features present in many natural systems, namely energy consumption, birth, death and a long evaluation time. We use a generic foraging scenario in which agents spend energy while they move and they must periodically recharge in the nest to avoid death. New robots are added (born) when the colony has foraged a sufficient number of preys. We perform an analysis of the evolved behaviors and demonstrate that several colonies display multiple complex and scalable macroscopic behaviors.

1 Introduction

In nature, millions of years of evolution generated complex behavior across different types of insects. Fireflies, for instance, rely on the synchronization of light emission in order to attract mates. Other more sophisticated techniques are used to convey information to peers: ants use pheromones to mark paths, while bees use a waggle dance to indicate the direction and distance of food sources. Bees also display task allocation: some worker bees remain in the hive to produce honey, while others scout and forage pollen.

Previous studies in evolutionary robotics and artificial life have mainly focused on evolving controllers that exhibit a single macroscopic behavior observed in insect colonies [19,1,2,11,5,15,24,21]. In this study, we approach the evolution of complex behavior at the macroscopic level differently: we propose to include a set of nature-inspired conditions in the evolutionary model in order to allow for the evolution of colonies that display multiple different complex macroscopic behaviors. More specifically, we incorporate energy consumption, addition (birth) of new robots during evaluation, death and a long evaluation time into our

L. Antunes and H.S. Pinto (Eds.): EPIA 2011, LNAI 7026, pp. 153–167, 2011.

evolutionary model. We go on to demonstrate that colonies that display the abovementioned behaviors can be evolved under these conditions.

We use a multirobot foraging task [23] in which simple robots must carry or push preys to a nest. We designed the experiment in such a way that robots have to overcome several challenges. Preys are located in clusters which may be hard to find due to the robots' limited sensing range. The difficulty in finding the prey clusters makes their location important information to share between group members. The robots spend energy when they are outside of the nest and recharge when they are in the nest. If a robot spends all its energy, it dies. Each time a certain number of preys have been foraged, a new robot is added and it can participate in the foraging. In addition to these features, a relatively long evaluation time (12000 simulation steps or 20 minutes of virtual time in each simulation) has been imposed upon the robots, which means that there is an evolutionary pressure towards sustainable behaviors. We use JBotEvolver, an open source neuro-evolution framework for our experiments. The simulator can be downloaded at the following location: `http://sourceforge.net/projects/jbotevolver`.

This research is a first step towards the study of the conditions necessary for the evolution of complex macroscopic behaviors such as those observed in social insects. The contributions of the present work are: i) to propose and include a set of bio-inspired conditions, namely energy consumption, birth of new robots during simulation, death and a long evaluation time in the evolutionary setup, and ii) to show that under these conditions, simple controllers that display multiple different complex macroscopic behaviors can be evolved.

2   Related Work and Macroscopic Behaviors

2.1   Task Allocation

Task allocation refers to the way that tasks are chosen, assigned, subdivided and coordinated among different members of a group within the same environment (in some studies, the term *division of labor* is used to denote this process).

Task allocation is a widely studied field in multirobot systems [6]. Several approaches to task allocation in tightly-coupled multirobot systems, often with centralized coordination, have been proposed ranging from mathematically modeled motivations [18] to market-based algorithms [4]. Our aim is to evolve complex, insect-like behavior. In our discussion of related work on task allocation, we will therefore focus on studies on task allocation in loosely coupled and decentralized systems.

Krieger et al. [8] hand coded the behavioral control for a group of robots performing a foraging task. The authors demonstrated that recruitment and knowledge transfer as a means for task allocation could improve the performance of a group when the virtual food items were clustered.

Labella et al. [9] studied task allocation in a group of relatively simple robots. Each robot would perform one of two tasks: rest in the nest or forage. Each robot would choose which task to perform based on an individually learned parameter reflecting its continuous success in finding and retrieving preys. The more successful a robot was at foraging preys, the more likely it would be to forage instead of resting in the nest. The authors showed that such a simple parameter adaptation brings forth task allocation and that it is an effective way to improve the efficiency of the group. Liu et al. [10] took a similar approach but in which individuals also took environmental cues (collision with teammates) and social cues (teammate success in food retrieval) into account.

In evolutionary robotics, focus has mainly been on task allocation in groups of relatively few robots (often referred to as *role allocation*). For instance: Quinn et al. [19] evolved controllers for a team of three simple robots. The robots had to first aggregate and then to travel one meter as a coherent group. The team members dynamically adopted different roles and moved in a line formation. The robot that assumed the leader role moved backward in order to perceive the middle robot. The middle robot and the rear robot, on the other hand, moved forward. Ampatzis et al. [1] evolved homogeneous controllers for two real robots that allowed them to physically connect to one another. The robots first had to allocate roles so that one would be the gripping robot, while the other would be the gripped robot. The roles were allocated during a series of oscillating movements while circling around one another until one would approach and perform the grip.

In this study, we show how task allocation often plays a role in the solutions evolved. In many cases, the colony of homogeneous robots divides into subgroups that perform different tasks, such as foraging, scouting or resting.

2.2   Communication

Communication, either explicit or implicit, relates to the transmission of information from one member of a group to another. This is a fundamental behavior in most organisms, particularly in social species [12,22]. Although significant progress has been made towards the understanding of the neurophysiological processing of signals involved in the communication process, the conditions that allow for the evolution of reliable communication systems are still largely unknown. Most studies in evolutionary robotics on the emergence of communication have focused on relatively simple systems.

In [2], Baldassarre et al. studied robots with actuators and sensors that allowed for explicit communication through sound. In their experiment, the robots had to aggregate and move together. The evolved solutions exploited the robots' simple sound communication capabilities to locate teammates.

In a slightly more complex collective navigation problem, Marroco et al. [11] showed that an effective communication system can arise from a collective of initially non-communicating agents through an evolutionary process. The evolved agents communicated by producing and detecting five different signals that affected both their motor behavior and their signaling behavior.

Floreano et al. [5] evolved communication patterns for colonies of 10-wheeled robots. The robots were placed in an arena with a healthy food source and a poisonous food source. The food sources were detectable by the robots from anywhere in the arena, but they were only distinguishable at short distances. The robots were also equipped with a controllable light emitter. When illuminated, a light from one robots could be detected by the others from anywhere in the arena. The authors managed to evolve both cooperative communication and deceptive signaling. Using a similar setup, Mitri et al. [15] studied the role of genetic relatedness between signalers and receivers. They found a strong positive correlation between relatedness and the reliability of the evolved signaling. Unrelated robots produced unreliable signals, whereas highly related robots produced signals that reliably indicated the location of the food source.

In this study, we find that communication evolves and becomes part of an integrated solution to the foraging task. In some of the solutions evolved, the robots convey information about the position of prey clusters.

2.3   Self-organized Synchronization

Synchronization refers to the coordinated action of several individuals. It can be expressed by a simultaneous change in behavior or by a continuously coordinated group behavior. Synchronization is a pervasive phenomenon that can be found in many different places: from self-synchronization of cardiac cells [7], to synchronous choruses of grasshoppers [20], self-synchronization of female menstrual cycles [14], and collective, synchronous clapping in theaters [16].

Synchronization as a means to coordination has been successfully used in a wide range of applications. Nijmeijer and Rodriguez-Angeles [17] analyzed the use of synchronization to control multiple manipulators. Mazzapioda and Nolfi [13] evolved neural controllers for a hexapod robot that manage to synchronize its legs in a gait that adjusted to the configuration of the ground.

In evolutionary robotics, synchronization has also been studied. Wischmann et al. [24] have evolved synchronous behaviors that reduce the interference among communicating robots. Trianni and Nolfi [21] evolved robot controllers, which proved to be capable of synthesizing minimal synchronization strategies based on the dynamical coupling between robots and environment.

In this study, synchronous movement proved an efficient scouting mechanism. The robots could maximize the coverage of the environment while minimizing the energy needed for the colony to locate prey clusters by moving synchronously.

3   Robot Model

We used a circular differential drive robot model with a diameter of 10 cm. Each robot is equipped with a variety of sensors and actuators. The set of actuators is composed of two wheels, that enables the robot to move at speeds of up to 50 cm/s, and a prey carry mechanism. The prey carry mechanism allows a robot to pick up one prey within 5 cm. A robot is only able to carry a prey at the time. If a robot that is carrying a prey collides with another robot, it loses the prey, and the prey is returned to its original location.

The robots are equipped with several sensors that allow them to perceive: i) whether they are currently carrying a prey or not (prey-carried sensor), ii) the direction they are facing (compass sensor), iii) their current energy level (energy sensor), and iv) the presence of nearby objects: eight nest sensors, eight prey sensors, and eight robot sensors.

The distance-related sensors are located in the center of each robot. The sensors' orientations are distributed evenly at intervals of 45° providing each robot with omnidirectional sensing capabilities. A sensor only registers objects within a certain distance and angle with respect to its orientation. All sensors have an opening angle of 90°, but the maximum range varies: prey sensors have a range of 1 m, nest sensors have a range of 30 m, and robot sensors have a range of 20 m.[1] If there are no sources within a sensor's range and opening angle, its reading is 0. Otherwise, the reading is based on distance to the closest source according to the following equation:

$$s = \frac{range - d_c}{range} \tag{1}$$

where $range$ is the sensor's detection range and d_c is the distance between the closest source c and the robot.

4 The Foraging Task

The robots must perform a foraging task in a two dimensional, circular environment. An illustration of the environment can be seen in Figure 1. The foraging area has a radius of 13 m, and a nest with a diameter of 1 m occupies the center of the environment. An intermediate zone surrounds the environment. A colony is penalized if robots are in this zone. A higher penalty is assigned if robots move beyond the intermediate zone. A cluster of 30 preys is positioned in the foraging area. A new cluster is created once all the preys in the previous cluster have been foraged. In the beginning of each simulation, 5 robots are created and positioned in the nest. Their initial energy level is 50 units out of a maximum of 100 units. The robots can recharge energy in the nest at a rate of 0.5 units every second (10 time steps). They lose energy at a rate of 0.3 units every second whenever they are outside the nest. To avoid dying, the robots thus need to return to the nest and recharge from time to time.

---

[1] Since the focus of this study is on whether multiple different macroscopic behaviors can evolve under a fixed set of nature-inspired conditions, we chose to simplify the robot model at the microscopic level. We allowed the robots to sense the nest and other robots in most of the arena in order to avoid that individual skills such as odometry (to be able to navigate back to the nest) or the capacity to remember or infer other robots' positions and so on, would have to be learned before macroscopic behaviors would evolve. In future studies, we intend to study how the abilities of the individual robots affect the solutions evolved – especially when sensors are more similar to the sensory apparatus of specific species of insects.

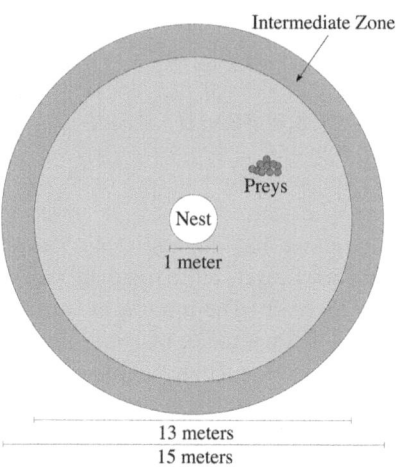

Fig. 1. The foraging environment. The nest is positioned at the center and has a diameter of 1 m. Preys are positioned in a cluster at varying distances from the center of the nest (4 m to 10 m). A colony of robots is penalized if any robot is in the intermediate zone or beyond.

5   Behavioral Control and the Evolutionary Algorithm

Each robot is controlled by an artificial neural network (ANN). We use continuous time recurrent neural networks [3] with a reactive layer of input neurons, one layer of hidden neurons, and one layer of output neurons. The input layer is fully connected to the hidden layer, which, in turn, is fully connected to the output layer. The input layer has one neuron for each sensor and the ouput layer has one neuron for each actuator. The neurons in the hidden layer are fully connected and governed by the following equation:

$$\tau_i \frac{dH_i}{dt} = -H_i + \sum_{j=1}^{27} \omega_{ji} I_i + \sum_{k=1}^{3} \omega_{ki} Z(H_k + \beta_k), \qquad (2)$$

where τ_i is the decay constant, H_i is the neuron's state, ω_{ji} the strength of the synaptic connection from neuron j to neuron i, β the bias terms, and $Z(x) = (1 + e^{-x})^{-1}$ is the sigmoid function. β, τ, and ω_{ji} are genetically controlled network parameters. The possible ranges of these parameters are: $\beta \in [-10, 10]$, $\tau \in [0.1, 32]$ and $\omega_{ji} \in [-10, 10]$. Circuits are integrated using the forward Euler method with an integration step-size of 0.2 and cell potentials are set to 0 when the network is initialized.

Each generation is composed of 100 genomes, and each genome corresponds to an ANN with the topology described above. The fitness of a genome is sampled 20 times and the average fitness is computed. Each sample lasts 12000 time steps unless all the robots die and unless the robots do not find any prey within the first 2000 time steps. A prey cluster is placed at different distances from the nest in every sample. The distance of the prey cluster from the center of the nest ranges from 4 m to 10 m.

After all the genomes have been evaluated, an elitist approach is used: the top 5 are chosen to populate the next generation. Each of the top 5 genomes becomes the parent of 19 offspring. An offspring is created by applying a Gaussian noise to each gene with a probably of 15%. The 95 mutated offspring and the original 5 genomes constitute the next generation. For every 5 preys that are foraged, a new robot is added to the colony. The new robot is identical to the ones that are already foraging.

The fitness function is composed of a main component and a bootstrapping component. The main component reflects the dominant characteristics of the fitness function, which are calculated at the end of the simulation. The bootstrapping component is calculated at each time step and is the sum of a rewarding subcomponent and a penalizing subcomponent. The fitness function is defined as follows:

$$F(i) = \underbrace{100P_i - 50D_i}_{\text{main}} + \underbrace{\sum_{s=1}^{\text{time-steps}} (p_{i,s} + r_{i,s})}_{\text{bootstrap}} \tag{3}$$

where i is the genome being evaluated, P_i is the number of preys foraged, D_i is the number of robots that died during the current simulation, $p_{i,s}$ is the bootstrapping subcomponent that penalizes the colony, and $r_{i,s}$ is the bootstrapping subcomponent that rewards the colony.

The term $p_{i,s}$ penalizes robots for moving beyond the foraging area (intermediate zone and beyond). It is computed using the formula:

$$p_{i,s} = -\frac{10^{-4}L_{i,s} + 10^{-1}F_{i,s}}{T_{i,s}} \tag{4}$$

where $L_{i,s}$ is the number of robots in the intermediate zone, $F_{i,s}$ is the number of robots beyond the intermediate zone, and $T_{i,s}$ is the total number of robots.

The term $r_{i,s}$ rewards robots that are near preys and robots that are carrying preys. It also gives a reward that is inversely proportional to the average distance of all the preys to the center of the nest. The formula for $r_{i,s}$ is the following:

$$r_{i,s} = \frac{10^{-4}N_{i,s} + 10^{-2}C_{i,s}}{T_{i,s}} + \sum_{j=1}^{\text{n}^\circ \text{ preys}} \frac{limit - dist(p_j, \text{nest})}{limit} \tag{5}$$

where $N_{i,s}$ is the number of robots near preys, $C_{i,s}$ is the number of robots carrying preys, $limit$ is the distance from the nest to the intermediate zone, and $dist(p_j, \text{nest})$ represents the distance of prey p_j to the center of the nest. The bootstrapping component should allow for a faster convergence towards good solutions, in which the robots forage.

If the robots are unable to encounter any prey during the first 2000 steps, a fitness of -1000 is awarded and the evaluation is stopped. The evaluation is also stopped if all the robots die. The parameters described in this section were fine-tuned through a trial-and-error process.

6   Results and Discussion

The purpose of this study is to demonstrate how colonies of embodied agents can evolve complex macroscopic behaviors akin to those found in colonies of social insects. In order to evolve these types of behaviors, we create two experimental setups, A and B. In setup A, the robots can only communicate implicitly, while in setup B, we give the robots the capacity to communicate explicitly. Below, we analyze behaviors that evolved in our experiments and take a closer look at some of the individual solutions.

6.1   Experimental Setup A: Implicit Communication

We conducted a total of 25 evolutionary runs in experimental setup A, each with a unique initial random seed. In all runs, the best solutions evolved displayed foraging behaviors. However, in terms of fitness and behavior, the solutions evolved were quite different. The fitness of the best genomes from the 25 runs ranged from 1772 to 15520. Different types of macroscopic behaviors evolved: communication evolved in 2 runs, task allocation evolved in 6 runs, and in synchronization evolved 1 run. Below, we describe the macroscopic behaviors in more detail.

6.2   Behavioral Patterns

Communication: Robots that have been evolved in experimental setup A have no way of explicitly conveying information, so they have to rely on implicit communication. Implicit communication means that a robot must interpret cues from the environment and other robots in order to extract information that is not directly expressed.

Implicit communication has been observed in two of the evolutionary runs (A1 and A23). In these runs, a robot that encounters a prey cluster will remain near the cluster for other robots to be able to locate and forage the preys. This *signal-by-waiting* approach is efficient because the prey sensor range is limited (1 m) compared to the range of the robot distance sensors (20 m). When another robot arrives, the signaler abandons its position and returns to the nest with a prey. The arriving robot then remains close to the preys and thereby assumes the role as the (implicit) signaler until a third robot arrives and so forth. The *signal-by-waiting* behavior allows the colony to "remember" the location of a prey cluster, which improves their foraging efficiency.

Task allocation: In the behaviors evolved in 6 evolutionary runs, most of the robots stay in the nest, while only a fixed number of robots scout and forage.

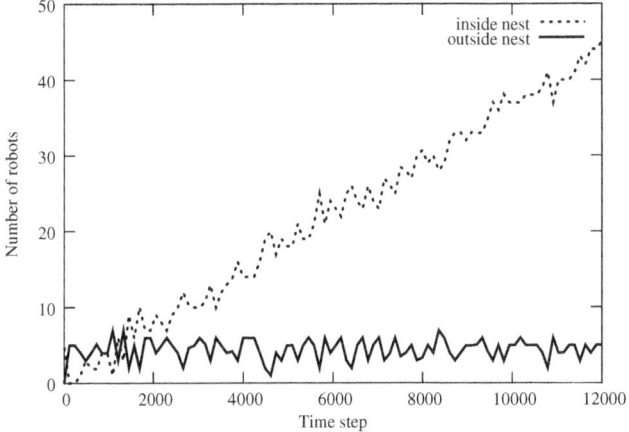

Fig. 2. Task allocation in evolutionary run A6. Even though the number of robots grows during the simulation, only a fixed number of robots scout for preys.

To illustrate this type of task allocation, we have plotted the number of robots inside and outside the nest for each time step in one simulation in Figure 2. As it can be seen in the figure, the number of scouting robots outside of the nest remains between 3 and 7, even as the total number of robots in the colony grows to 50.

Synchronization: In one of the evolutionary runs (A1), the robots learned how to synchronize their movements in order to scout for preys. The robots gather in the nest and simultaneously exit in an outwards spiraling fashion. An example is shown in Figure 3. The robots remain equally spaced as they spiral away from the nest. The robots continually influence each others' trajectories, which means that when one robot encounters the prey cluster, the robot after it moves towards the cluster. This effect propagates backwards to all robots and thus leads them all to the prey cluster. The maximum distance between the arms of the spiral path is less than the prey sensor distance (1 m), which guarantees a total coverage of the foraging area. After the colony has located a prey cluster, they stop their synchronized spiraling and start foraging.

6.3 Behavior Analysis

The macroscopic behaviors described above were observed in 8 out of 25 evolutionary runs. Although the other 17 runs did not show signs of communication, task allocation or synchronization, some of them produced good foraging solutions. In the majority of these 17 runs, the observed behavior consisted of the robots moving in a circle around the nest, while following one another. If the prey cluster was found by one robot, the other robots would often find it as well. This was typically a better approach than robots that searched for preys

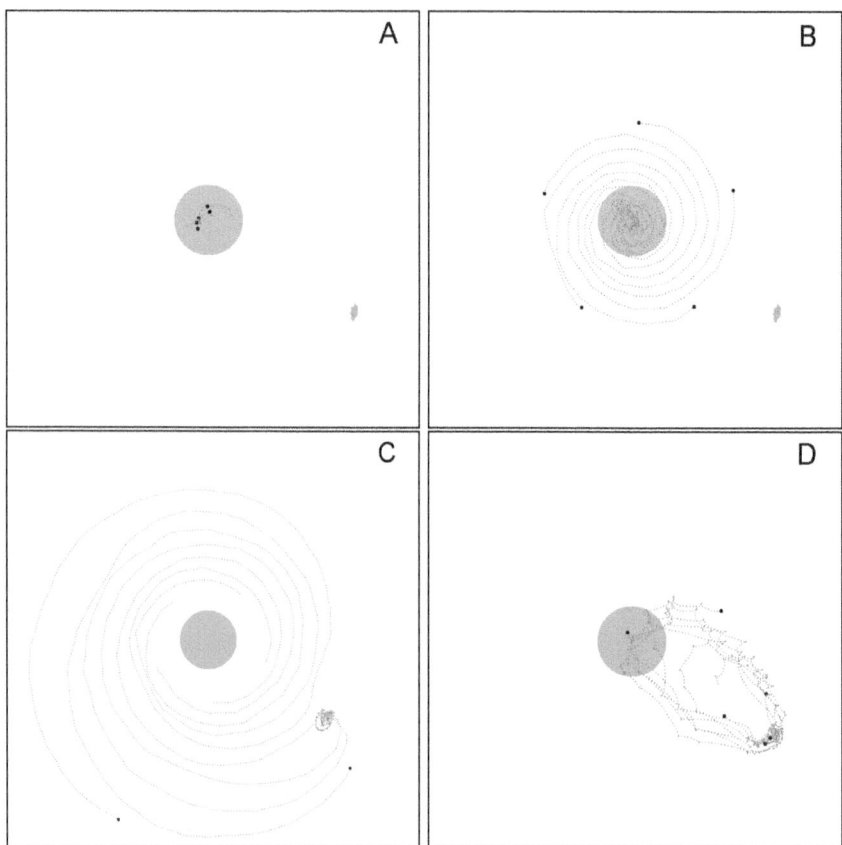

Fig. 3. An example of synchronization behavior from run A1 (four screenshots from the same run). A: the robots gather in the nest. B: the robots start scouting simultaneously in an outwards spiraling motion. C: when the prey cluster is found, other robots are influenced by the finding robot's change of course and they all find the cluster quickly. D: from this point on, the robots are able to go back and forth between the nest and the preys, foraging them efficiently.

individually. The "following" behavior was only worse in the case where one robot would overshot the preys' location causing all following robots to overshoot it as well.

We observed different strategies in terms of the robots' energy management (or lack thereof). In 11 of the 25 runs, the robots did not learn how to recharge in the nest. This meant that they would die gradually as they foraged. If the foraging behavior was not efficient, the robots would eventually all die before the end of a simulation, since they did not manage to forage enough preys to generate a new robot before running out of energy.

Solutions found by evolution proved to be scalable. Colonies with over 100 robots still managed to exhibit these macroscopic behaviors.

6.4   Experimental Setup B: Explicit Communication

We setup a new series of experiments (setup B) to study if the robots could benefit from being able to communication explicitly. We gave the robots the capacity to change the luminosity of their bodies. For each distance sensor, we also added a luminosity sensor. While each distance sensor would provide the robot with information about the distance to the closest robot, the matching luminosity sensor would provide information about the luminance of its body. We increased the number of preys that needed to be foraged before a new robot was added to the colony from 5 to 10. This change was motivated by the overcrowding (more than 100 robots) observed in some of the simulations of colonies evolved in experimental setup A. A large number of foraging robots has some disadvantages: i) if the robots heavily outnumber the number of preys, there is little or no benefit in communicating, and ii) often most of the robots need not participate in the foraging as there may be fewer preys in a cluster (30) than robots in a colony.

We ran 25 evolutionary runs in setup B. In 15 of the 25 runs, the colonies evolved energy recharging behaviors that enabled the colonies to reach and maintain high numbers of robots. In 2 of the 25 runs, no foraging behavior evolved. We identified complex macroscopic behaviors in 11 of the 25 evolved solutions. In 4 runs of the 25 runs conducted in experimental setup B, we observed communication. However, only in one of those runs (B14) did the robots communicate explicitly by changing the luminosity of their bodies to signal that a prey cluster had been found.

In order to confirm that explicit communication was used in run B14, we measured the average luminosity of the robots with prey clusters placed at distances of 4 m, 7 m and 10 m from the center of the nest. We then ran a set of experiments for each prey distance in which we ignored the robots' signals to change the luminosity of their body and instead fixed the luminosity to the average value measured. For a prey distance of 4 m, both the robots with the fixed luminosity and the robots that could control their luminosity performed well and found the preys in every sample. For a distance of 7 m, the performance of the robots with fixed body luminosity declined considerably. They were only able to find preys in 65 out of 100 samples, while the robots that could control their body luminosity still managed to find the prey cluster and forage in all the samples. For a prey distance of 10 m, robots with the fixed body luminosity did not find preys in any of the samples, while the robots with control over their body luminosity foraged preys in 82 out of 100 samples. In B14, explicit communication is thus important for coordinating the foraging effort, especially when prey clusters are located far from the nest.

Body luminosity was used to convey information about the localization of prey clusters. The robots' normal luminosity is set at around 40% of intensity when they are scouting but as soon as they find a prey cluster, the luminosity is set to 0%. Other robots in the environment immediately change behavior when they see this: the robots that are far away head towards the nest where they first gather and then synchronously head towards the prey cluster, while the ones that are closer head directly to the cluster. Figure 4 illustrates this change of

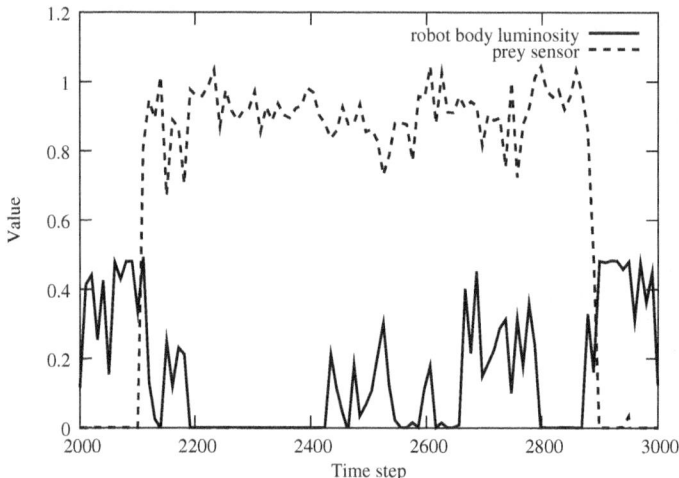

Fig. 4. Communication in run B14. Both lines represent internal values from a single robot. When preys are within range (1 m), the prey sensor's value increases towards the value of 1. At the same time, the robot's luminosity decreases in intensity towards the value of 0. This behavior has an effect on other robots: they head towards the signaling robot's position. Hence, the robot signals by decreasing its body luminosity. The luminosity value fluctuates when other robots arrive to forage. Around step 2900, the robot exchanges role with another robot and forages a prey, returning its luminosity to the normal value (around 40%).

luminosity of a robot that is near the prey cluster. Figure 5 contains screenshots from a simulation in which the robots display the communication behavior.

Three evolutionary runs (B5, B21 and B25) evolved the implicit communication behavior *signal-by-waiting*, similar to the behavior evolved in A1 and A23. One robot remains near the cluster signaling its position until other robots arrive. After some of the robots have foraged, the signaler exchanges its role with another robot.

The task allocation behavior where most of the robots scout while the others rest was observed in 4 out of 25 runs conducted in experimental setup B. Synchronization evolved in 7 runs, with similar behavior to what had been observed in one run in experimental setup A.

We compared the fitness of the two best runs from experimental setup B (B5 and B14). In run B5, the robots evolved to use implicit communication while in run B14, the robots evolved to use explicit communication. In order to compare the two solutions, we evaluated their performance with prey clusters at distances of 4 m, 7 m, and 10 m from the center of the nest, respectively. 100 samples were run for each of the three distances. The fitness obtained by solution evolved in B5 was similar to the fitness obtained by the solution evolved in B14 when preys were placed 4 m and for 7 m from the center of the nest. When the preys are placed 10 m from the center of the nest, however, the solution evolved in

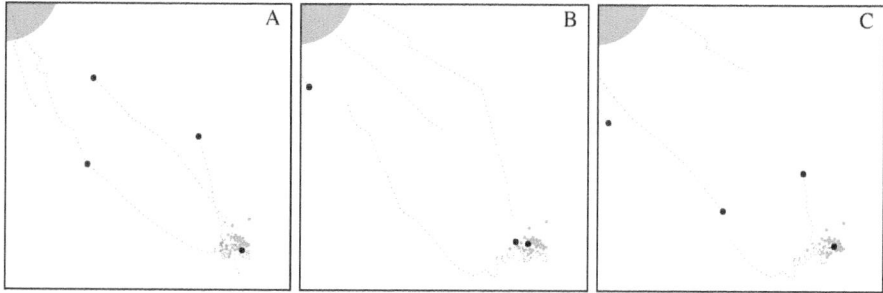

Fig. 5. An example of communication in evolutionary run B14 (three screenshots from the same run). A: after finding a prey cluster, a robot stays near the preys to signal the cluster's location. In order to signal, the robot reduces its luminosity to attract other robots. B: other robots quickly find the prey cluster. C: the signaler stays near the cluster while the other robots forage. The nest can be seen at the top-left corner of the figures. In this example, the prey cluster is positioned at a distance of 4 m from the center of the nest.

B14 (uses explicit communication) obtained a higher fitness (Mann-Whitney U, $p < 0.05$) than the solution evolved in B5 (uses implicit communication). The solution evolved in B14 on average obtains a fitness that is 29% higher than the fitness obtained by the solution evolved in B5 with preys are placed 10 m from the center of the nest.

Table 1. Summary of the macroscopic behaviors observed in the different evolutionary runs

	A1	A6	A8	A10	A18	A19	A21	A23	B4	B5	B7	B10	B11	B12	B14	B16	B21	B24	B25
Communication	•							•		•					•		•		•
Synchronization	•									•		•	•		•	•		•	•
Task allocation		•	•	•	•	•	•		•		•			•		•			

In Table 1, we have summarized the results obtained in experimental setup A and in experimental setup B. In the table, we indicate which evolutionary runs evolved which macroscopic behaviors. The behaviors from the best runs can be seen on the following page: `http://home.iscte-iul.pt/~alcen/epia2011/`

7   Conclusions

In this paper, we proposed to include four nature-inspired conditions in an evolutionary setup, namely energy consumption, birth of new robots, death, and a long evaluation time. We implemented these conditions and demonstrated that multiple complex macroscopic behaviors, such as communication, task allocation and synchronization evolved.

We conducted evolutionary runs in two different setups: in the first setup (A), the robots had no means to communicate explicitly while in the second setup (B), we gave the robots the capacity to change their body luminosity and to detect the body luminosity of other robots. We found that complex macroscopic behaviors evolved in both setups. Implicit communication enabled a robot that found the location of a prey cluster to stop and thereby inform the rest of the colony of its location. Only in one of the evolutionary runs conducted in setup B did the robots evolve to use their capacity to change body luminosity for explicit communication. Although explicit communication only evolved in a single run, it proved more efficient than any of the other behaviors evolved in setup B – including those that rely on implicit communication.

In this study, we have shown how simple controllers can be evolved to display several complex and scalable macroscopic behaviors similar to those observed in colonies of social insects. In our ongoing work, we are studying how to specialize evolutionary setups in order to more closely mimic the conditions and capabilities of specific species of social insects. Future experimental setups may include the concept of age, more complex environments, and/or biologically plausible sensors. We expect that such studies may shed light on how complex colony level behaviors may have evolved in nature.

References

1. Ampatzis, C., Tuci, E., Trianni, V., Christensen, A.L., Dorigo, M.: Evolving self-assembly in autonomous homogeneous robots: experiments with two physical robots. Artificial Life 15(4), 465–484 (2009)
2. Baldassarre, G., Nolfi, S., Parisi, D.: Evolving mobile robots able to display collective behaviours. Artificial Life 9, 255–267 (2002)
3. Beer, R.D., Gallagher, J.C.: Evolving dynamical neural networks for adaptive behavior. Adaptive Behavior 1, 91–122 (1992)
4. Dias, M.B., Zlot, R., Kalra, N., Stentz, A.: Market-based multirobot coordination: A survey and analysis. Proceedings of the IEEE 94(7), 1257–1270 (2006)
5. Floreano, D., Mitri, S., Magnenat, S., Keller, L.: Evolutionary conditions for the emergence of communication in robots. Current Biology 17(6), 514–519 (2007)
6. Gerkey, B.P., Matarić, M.J.: A formal analysis and taxonomy of task allocation in multi-robot systems. International Journal of Robotics Research 23(9), 939–954 (2004)
7. Glass, L.: Synchronization and rhythmic processes in physiology. Nature 410, 277–284 (2001)
8. Krieger, M.J.B., Billeter, J.-B., Keller, L.: Ant-like task allocation and recruitment in cooperative robots. Nature 406, 992–995 (2000)
9. Labella, T.H., Dorigo, M., Deneubourg, J.-L.: Division of labor in a group of robots inspired by ants' foraging behavior. ACM Transactions on Autonomous and Adaptive Systems 1(1), 4–25 (2006)
10. Liu, W., Winfield, A.F.T., Sa, J., Chen, J., Dou, L.: Towards energy optimization: Emergent task allocation in a swarm of foraging robots. Adaptive Behavior 15(3), 289–305 (2007)

11. Marocco, D., Nolfi, S.: Emergence of communication in teams of embodied and situated agents, in the evolution of language. In: Proceedings of the 6th Internation Conference on the Evolution of Language, pp. 198–205. World Scientific Publishing Ltd., Toh Tuck Link (2006)
12. Maynard-Smith, J., Szathmary, E.: The Major Transitions in Evolution. Oxford University Press, New York (1997)
13. Mazzapioda, M., Nolfi, S.: Synchronization and Gait Adaptation in Evolving Hexapod Robots. In: Nolfi, S., Baldassarre, G., Calabretta, R., Hallam, J.C.T., Marocco, D., Meyer, J.-A., Miglino, O., Parisi, D. (eds.) SAB 2006. LNCS (LNAI), vol. 4095, pp. 113–125. Springer, Heidelberg (2006)
14. McClintock, M.K.: Menstrual synchrony and suppression. Nature 229(5282), 244–245 (1971)
15. Mitri, S., Floreano, D., Keller, L.: Relatedness influences signal reliability in evolving robots. In: Proceedings of The Royal Society / Biological Sciences, vol. 278(1704), pp. 378–383 (2010)
16. Néda, Z., Ravasz, E., Brechet, Y., Vicsek, T., Barabási, A.L.: Self-organizing processes: The sound of many hands clapping. Nature 403(6772), 849–850 (2000)
17. Nijmeijer, H., Rodriguez-Angeles, A.: Synchronization of mechanical systems. World Scientific Publishing Ltd., Toh Tuck Link (2003)
18. Parker, L.E.: ALLIANCE: an architecture for fault tolerant multirobot cooperation. IEEE Transactions on Robotics and Automation 14(2), 220–240 (1998)
19. Quinn, M., Smith, L., Mayley, G., Husbands, P.: Evolving controllers for a homogeneous system of physical robots: Structured cooperation with minimal sensors. Philosophical Transactions of the Royal Society of London. Series A: Mathematical, Physical and Engineering Sciences 361(1811), 2321–2343 (2003)
20. Snedden, W.A., Greenfield, M.D., Jang, Y.: Mechanisms of selective attention in grasshopper choruses: who listens to whom? Behavioral Ecology & Sociobiology 43(1), 59–66 (1998)
21. Trianni, V., Nolfi, S.: Self-organising sync in a robotic swarm. a dynamical system view. IEEE Transactions on Evolutionary Computation, 1–21 (2009)
22. Wilson, E.O.: Sociobiology: The New Synthesis. Belknap Press, Cambridge (1975)
23. Winfield, A.F.T.: Foraging Robots. In: Encyclopedia of Complexity and Systems Science, pp. 3682–3700. Springer, Berlin (2009)
24. Wischmann, S., Pasemann, F.: The Emergence of Communication by Evolving Dynamical Systems. In: Nolfi, S., Baldassarre, G., Calabretta, R., Hallam, J.C.T., Marocco, D., Meyer, J.-A., Miglino, O., Parisi, D. (eds.) SAB 2006. LNCS (LNAI), vol. 4095, pp. 777–788. Springer, Heidelberg (2006)

Network Regularity and the Influence of Asynchronism on the Evolution of Cooperation

Carlos Grilo[1,2] and Luís Correia[2]

[1] Dep. Eng. Informática, Escola Superior de Tecnologia e Gestão,
Instituto Politécnico de Leiria Portugal
[2] LabMag, Dep. Informática, Faculdade Ciências da Universidade de Lisboa, Portugal
carlos.grilo@ipleiria.pt, luis.correia@di.fc.ul.pt

Abstract. In a population of interacting agents, the update dynamics defines the temporal relation between the moments at which agents update the strategies they use when they interact with other agents. The update dynamics is said to be synchronous if this process occurs simultaneously for all the agents and asynchronous if this is not the case. On the other hand, the network of contacts defines who may interact with whom. In this paper, we investigate the features of the network of contacts that play an important role in the influence of the update dynamics on the evolution of cooperative behaviors in a population of agents. First we show that asynchronous dynamics is detrimental to cooperation only when 1) the network of contacts is highly regular and 2) there is no noise in the strategy update process. We then show that, among the different features of the network of contacts, network regularity plays indeed a major role in the influence of the update dynamics, in combination with the temporal scale at which clusters of cooperator agents grow.

1 Introduction

Why cooperative behaviors do exist in nature? How can we promote this type of behaviors in human and artificial societies? In light of the evolution theory, there seems to exist a contradiction between the existence of altruistic behaviors in nature and the fact they seem apparently less advantageous from an evolutionary point of view [21], hence the first question. The second question is more relevant in social sciences and informatics, for example. In these cases, besides explaining observed phenomena, the goal is to identify mechanisms that promote the emergence and maintenance of cooperative behaviors.

Evolutionary games [21] have been one of the main tools used to help answering these questions. In these models there is a population of agents interacting with each other during several time steps through a given game that is used as a metaphor for the type of interaction that is being studied. The structure which defines who may interact with whom is called the *network of contacts*. On each iteration, the agents may update the strategy they use to play the game using a so called *transition rule*. The *update dynamics* defines the temporal relation between the moments at which agents update their strategy. If this process

L. Antunes and H.S. Pinto (Eds.): EPIA 2011, LNAI 7026, pp. 168–181, 2011.

is modeled as if occurring simultaneously for all the agents for all time steps, one says that the system is under a *synchronous dynamics*. If only a subset of the agents (simultaneously) update their strategies, the system is under an *asynchronous dynamics*. In this paper we analyze the features of the network of contacts that are determinant on the influence of the update dynamics on the evolution of cooperation.

There are several aspects whose influence on the evolution of cooperation has been studied, among which, the network of contacts [12, 1, 7, 19], the presence of noise in the strategy update process [22] and direct and indirect reciprocity [2, 13], to name just a few. The update dynamics is also among the studied aspects. The results reported in most of previous studies vary with the conditions used [9, 11, 25, 10, 17, 18, 5]. That is, depending on the conditions, asynchronous dynamics can be beneficial, detrimental or innocuous. In [5] we tested a broad number of conditions, namely different networks of contacts, transition rules with tunable noise levels and intermediate levels of asynchronism. We have confirmed the results of previous works where the conditions coincide. However, the broad number of conditions allowed us to show that asynchronous dynamics is detrimental to cooperation only when there is no noise involved in the strategy update process and only for highly regular networks. That is, in general, an asynchronous dynamics supports more cooperation than the synchronous counterpart. We identified also the features of the transition rules that play an important role in the influence of the update dynamics. On the other hand, the variety of the networks we used is not enough to completely identify network features that also play a relevant role in this influence. Identifying these features is the subject of this paper.

The paper is structured as follows: in Section 2 we describe the model used in the simulations and the experimental setup. In Section 3 we show the influence of the update dynamics when the games are played on different networks of contacts and with two transition rules which model distinct noise levels. In Section 4 we analyze the network features that determine the type of influence of the update dynamics. Finally, in the last section some conclusions are drawn and future work is proposed.

2   The Model

2.1   The Games

Symmetric 2-player games are among the most studied games in evolutionary game theory. These games can be described by the payoff matrix

$$
\begin{array}{c}
\quad C \quad D \\
\begin{array}{c} C \\ D \end{array}
\begin{pmatrix} R & S \\ T & P \end{pmatrix}
\end{array}
\tag{1}
$$

where C (cooperate) and D (defect) are the possible actions for each player. Each element of the matrix represents the payoff received by the row-player when it plays the game against the column-player. Let us consider $R = 1$ and $P = 0$, and restrict S and T to the intervals $-1 < S < 1$, $0 < T < 2$ [6, 17]. The $S > 0$, $T < 1$ region corresponds to the Harmony game where the rational action for both players is to play C in a one shot game. The famous Prisoner's Dilemma game [2] corresponds to the region $S < 0$, $T > 1$. In this game there is a strong temptation to play D, which is the rational choice. However, if both players play D, they receive a smaller payoff than if they both play C, hence the dilemma. In the Snowdrift game [7], $S > 0$, $T > 1$, the best action depends on the opponent's decision: it is better to play C if the other player plays D and vice-versa. Finally, the region $S < 0$, $T < 1$ corresponds to the Stag-Hunt game [20]. In this game there is a dilemma between playing the potentially more profitable but risky action C and the less profitable but less risky action D.

2.2   Network of Contacts

We use two types of network models: the *small-world networks* model of Watts-Strogatz [27] and the *scale-free networks* model of Barabási-Albert [3]. In order to build small-world networks, first a toroidal regular 2D grid is built so that each node is linked to its 8 surrounding neighbors by undirected links (this is called the *Moore neighborhood*); then, with probability ϕ, each link is replaced by another one linking two randomly selected nodes. Self links, repeated links and disconnected graphs are not allowed. Two measures are often used to characterize networks of contacts: The *average path length L* and the *clustering coefficient C*. L measures the average (smallest) distance between two nodes in the network. C measures the average probability that the neighbors of a node are also connected. In general, regular networks ($\phi = 0$) have both large L and C. Random networks ($\phi = 1$) have both very small L and C. Since L decreases at a faster rate with ϕ than C, there are networks with low L and large C between regular and random networks ($0.01 \leq \phi \leq 0.1$). These are the so called small-world networks. The values used in the simulations are $\phi = \{0, 0.01, 0.05, 0.1, 1\}$.

Scale-free networks are built in the following way: the network is initialized with m fully connected nodes. Then, new nodes are added, one at a time, until the network has the desired size. Each added node is linked to m already existing nodes so that the probability of creating a link to some existing node i is equal to $k_i / \sum_j k_j$, where k_i is the degree of i, which is defined as the number of nodes to which it is connected. This method of link creation, named *preferential attachment*, leads to a power law degree distribution $P(k) \sim k^{-\gamma}$ that is very common in real social networks. Scale-free networks built with this model have very low L and C values. Unless stated otherwise, all the networks for which results are presented have average degree $\overline{k} = 8$ (equivalent to $m = 4$ in scale-free networks).

2.3   Dynamics

In the synchronous model, at each time step all the agents play a one round game with all their neighbors and collect the payoffs resulting from these games, forming an *aggregated payoff*. After this, they all simultaneously update their strategies using the transition rule (see below). In the asynchronous model, at each time step, one agent x is randomly selected; x and its neighbors play the game with their neighbors and, after this, x updates its strategy. This is an extreme case of asynchronism, named *sequential dynamics*, in which only one agent is updated at each time step. In our opinion, both synchronous and sequential dynamics are artificial ways of modeling the update dynamics of real social systems. Synchronous dynamics is artificial because there is no evidence that behavior updating occurs simultaneously for all individuals in a population. We consider that sequential dynamics is artificial because it presupposes that 1) events are instantaneous, 2) events never occur simultaneously and 3) the information resulting from an event becomes immediately accessible by other members of the population. In [5] we used an update method which allows us to model intermediate levels of asynchronism. We verified that, in general, the results change monotonically as we go from synchronous to sequential dynamics. This means that we can evaluate the maximum influence of the update dynamics by using these two extreme methods and that is why we use them here.

We use two imitation transition rules: the *best-takes-over* rule and the *Moran* rule. With the *best-takes-over* rule each agent x always imitates its most successful neighbor y, provided y's payoff is larger than x's payoff. The Moran rule is defined in the following way: let G_x be the aggregated payoff earned by agent x in the present time-step; let $N_x^* = N_x \cup x$, where N_x is the set of x's neighbors, with $k_x = |N_x|$. According to this rule, the probability that an agent x, with strategy s_x, imitates agent y, with strategy s_y, is given by

$$p(s_x \rightarrow s_y) = \frac{G_y - \Psi}{\sum_i (G_i - \Psi)}, \quad y, i \in N_x^*. \tag{2}$$

The constant Ψ is subtracted from G_i because payoffs in the Stag-Hunt and the Prisoner's Dilemma games can be negative. If G_x is set to the accumulated payoffs gained by agent x in the games played in one time step, then $\Psi = \max_{i \in N_x^*}(k_i)\min(0, S)$. If the average of the payoffs gained in one time step is considered instead, then $\Psi = \min(0, S)$. For small-world networks, the results obtained with the two approaches are similar since all the agents have approximately the same k (k is the same for all agents when $\phi = 0$). Major differences appear for scale-free networks due to the large degree heterogeneity. Average payoffs are intended to model the fact that agents have limitations in the number of interactions they can sustain simultaneously and also that relationships are costly [26, 24]. Finally, we note that this rule models the presence of noise in the decision process since it allows agents to imitate neighbors less successful than themselves, contrary to what happens with the *best-takes-over* rule.

2.4 Experimental Setup

The charts presented were obtained with populations of $n = 10^4$ agents. We let the system run during 10^4 time steps for the synchronous model and $10^4 \times n$

time steps for the sequential model, which is enough for the system to converge to homogeneous populations of cooperators or defectors, or to stabilize around a ρ value (we confirmed that the results do not change if we use larger evolution periods). The steady state ρ value is computed as the average proportion of cooperators in the last 10^3 time steps for the synchronous model and in the last $10^3 \times n$ time steps for the sequential model. Populations are randomly initialized with $\rho^0 = 0.5$. Each point in the charts presented is an average of 50 independent simulations. For each simulation a new network is generated, which is kept static during the evolutionary process. As in [17], we use the average of the ρ values corresponding to the region of each game in the ST-plane as a global measure of the cooperation level obtained with that game. This average is presented next to the quadrant of each game in Figures 1-5.

3   The Influence of the Update Dynamics

The results obtained with regular grids (Fig. 1) can be summarized in the following way: sequential dynamics supports less cooperation for the best-takes-over rule, with the exception of the Stag-Hunt game; sequential dynamics supports more cooperation for the Moran rule. With the best-takes-over rule, the main differences appear in the Snowdrift game, while there are no large differences in the Stag-Hunt and Prisoner's Dilemma games. The main differences appear for the Moran rule, especially in the Snowdrift and Stag-Hunt games. The influence in the Prisoner's Dilemma game is limited to a small region. However, in this region, synchronous dynamics leads to uniform populations of defectors, while sequential dynamics leads to populations strongly dominated by cooperators or even states where $\rho = 1$. This is also the case of the Stag-Hunt and Snowdrift games for a noticeable portion of the space.

The main differences between the results obtained with regular and small-world networks exist for the best-takes-over rule: in the Snowdrift and Prisoner's Dilemma games, sequential dynamics becomes progressively beneficial to cooperation as ϕ is increased. For $\phi = 0.05$, sequential dynamics already supports more cooperation on average than synchronous dynamics when the best-takes-over rule is used (Fig. 2). When the Moran rule is used the results obtained with small-world networks are similar to the ones obtained with regular grids.

When scale-free networks are used there are relevant differences for both rules. The differences are larger for accumulated payoffs (Fig. 3) than for average payoffs (Fig. 4). In the first case, and for the best-takes-over rule, cooperation completely dominates the whole quadrant corresponding to the Snowdrift game when sequential dynamics is used. For the Stag-Hunt and Prisoner's Dilemma games, sequential dynamics leads to a significant increment of cooperation in large portions of the space. We note also that with these networks, when the update dynamics has some influence over ρ, sequential updating is always beneficial to cooperation when accumulated payoffs are used, with only a few exceptions to this behavior when average payoffs are used.

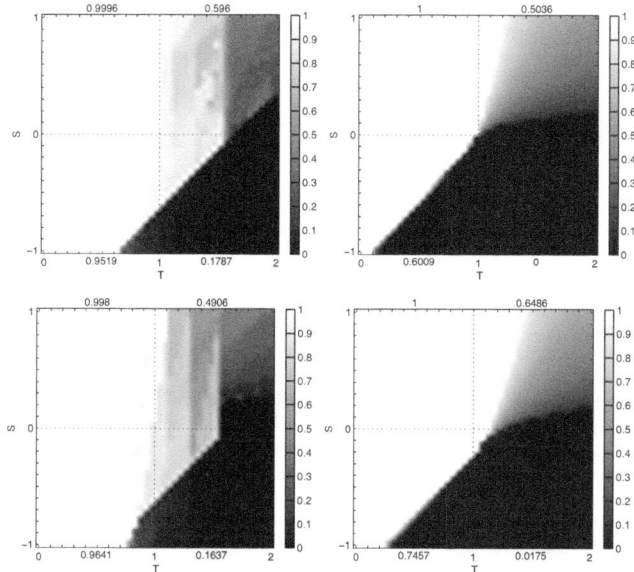

Fig. 1. Proportion of cooperators ρ in regular grids ($\phi = 0$), with synchronous dynamics (upper row) and sequential dynamics (lower row). Left column: best-takes-over rule; Right column: Moran rule. The games are the Harmony game (upper left quadrant), the Snowdrift game (upper right quadrant), the Stag-Hunt game (lower left quadrant) and the Prisoner's Dilemma game (lower right quadrant). The numbers, respectively, above the Harmony and Snowdrift games, and below the Stag-Hunt and the Prisoner's Dilemma games, are the average values of the corresponding quadrant. The S and T parameters are varied in steps of 0.05.

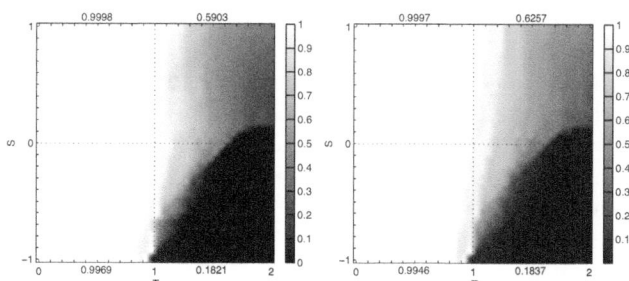

Fig. 2. Proportion of cooperators ρ on small-world networks ($\phi = 0.05$) with the best-takes-over rule. Left column: synchronous dynamics; Right column: sequential dynamics.

These results suggest that asynchronous updating is beneficial to the emergence of cooperation more often than it is detrimental. More specifically, they suggest that asynchronism is detrimental to cooperation only for networks with a high degree of regularity and for low or no noise. We studied the role of noise in

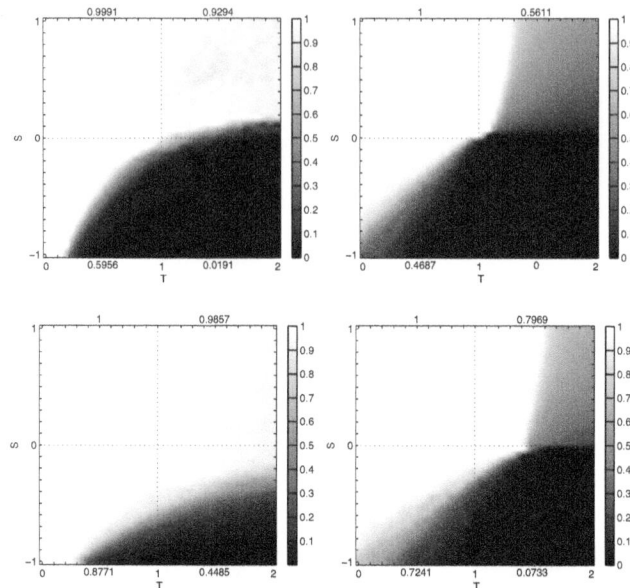

Fig. 3. Proportion of cooperators ρ in scale-free networks ($m = 4$), with synchronous dynamics (upper row) and sequential dynamics (lower row) using accumulated payoffs. Left column: best-takes-over rule; Right column: Moran rule.

[5]. In the next section we investigate the network features that are responsible for the detrimental effect of asynchronous dynamics when the games are played on regular grids with the best-takes-over rule.

4   The Role of the Network of Contacts

The results in the previous section show that the influence of the update dynamics depends on the network of contacts, mainly when the best-takes-over rule is used. We now derive some conclusions about the network features that may determine this influence. We first recall that sequential dynamics becomes beneficial to cooperation above a certain ϕ value in small-world networks when the best-takes-over rule is used. This means that degree heterogeneity, which occurs in scale-free networks, is not a necessary condition for sequential dynamics to become beneficial to cooperation, though it may potentiate this effect.

The detrimental effect of sequential dynamics for both $\phi = 0$ (regular grid) and $\phi = 0.01$ (small-world networks) indicates that the mean path length L has no determinant role also. This conclusion comes from the significant drop of L when we change ϕ from 0 to 0.01 [27].

Concerning the clustering coefficient C, we note that the regular grid with a Moore neighborhood has a large C value ($C \approx 0.428$), while it is very low for scale-free networks built with the Barabási-Albert model [3] described in Section 2.2. Considering that the influence of the update dynamics is different

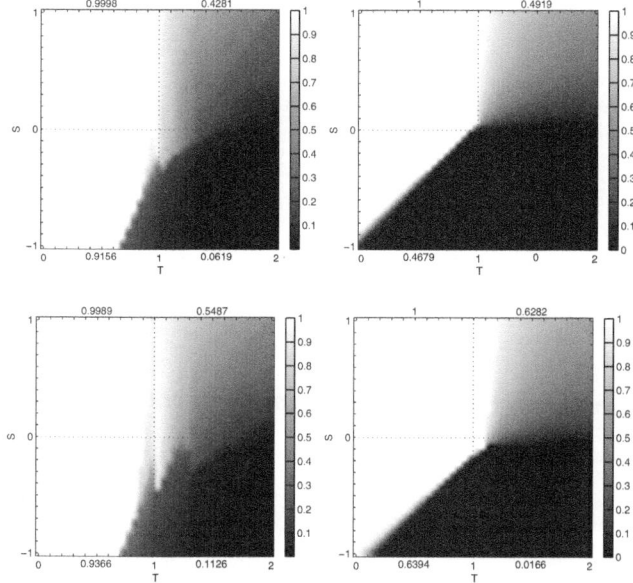

Fig. 4. As in Fig. 3 but with average payoffs

for these two types of network when the best-takes-over rule is used, nothing can be concluded about the role of C based on the results described in the previous section. We note that the beneficial and detrimental effect of the sequential dynamics for $\phi < 0.05$ and $\phi \geq 0.05$, respectively, indicates that this property also does not play a determinant role in the influence of the update dynamics. In order to verify this, we have done simulations with both regular grids with a von Neumann neighborhood, where $C = 0$, and with the scale-free networks model of Holme-Kim [8], which allows us to tune the value of C. These simulations were done only with the best-takes-over rule since the results obtained with the Moran rule are coherent for all networks of contacts: asynchronous dynamics benefits cooperation.

von Neumann grids are built so that each agent is linked to the four closest agents located in the four main cardinal directions. We note that, with this type of neighborhood, agents have no common neighbors. Holme-Kim networks are built in the following way: the network is initialized with m fully connected nodes. Then, new nodes are added one at a time until the network has the desired size. Each added node is linked to m already existing nodes. The first link between a new node v and an already existing node w is added using preferential attachment as in the Barabási-Albert model. The remaining $m - 1$ links are created using two different processes: (i) with probability p, v is linked to a randomly chosen neighbor of w and, (ii) with probability $1 - p$, preferential attachment is used. This model builds a scale-free network where the value of C depends on p: When $p = 0$, it generates Barabási-Albert scale-free networks with a very low C value. For $p > 0$, C grows with p.

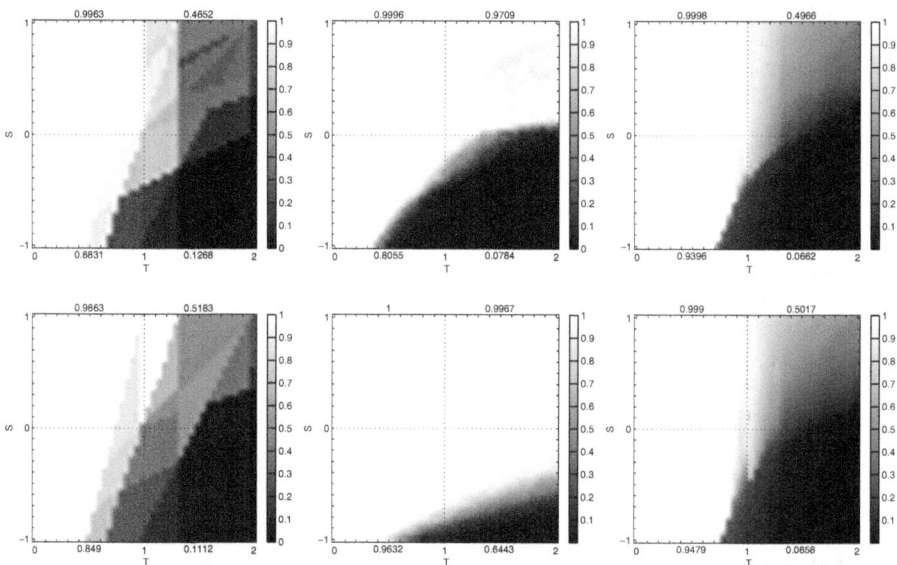

Fig. 5. Proportion of cooperators ρ in von Neumann grids (left column) and Holme-Kim networks with accumulated payoffs (middle column) and average payoffs (right column). Upper row: synchronous dynamics; Lower row: sequential dynamics. The transition rule is the best-takes-over rule.

Considering that $\bar{k} = 4$ in von Neumann grids, the simulations with Holme-Kim scale-free networks were done with both $\bar{k} = 4$ and $\bar{k} = 8$, corresponding to $m = 2$ and $m = 4$, respectively. However, we only present results for $\bar{k} = 8$, since the results are qualitatively similar. We used $p = 0.871$ for $\bar{k} = 8$ and $p = 0.582$ for $\bar{k} = 4$. Both values lead do networks where $C \approx 0.428$ (C value for Moore grids).

Fig. 5 shows that, concerning the influence of the update dynamics, the results obtained with the Holme-Kim model are qualitatively similar to the ones obtained with the Barabási-Albert model. This means that, independently of C, sequential dynamics benefits cooperation in scale-free networks. This is a strong evidence that the clustering coefficient plays no determinant role in the influence of the update dynamics.

Fig. 5 also shows that the results obtained with von Neumann grids qualitatively coincide with the ones obtained with Moore grids for Stag-Hunt and Prisoner's Dilemma but that they differ for Snowdrift. For this game, on average, sequential dynamics is beneficial to cooperation in von Neumann grids but detrimental in Moore grids. This result casts some doubts concerning the role of C in the influence of the update dynamics. However, as we will show next, the main role seems to be played by network regularity and less by the clustering coefficient.

We first note that, in regular grids, cooperator agents often form clusters with straight boundaries when a deterministic transition rule is used, as is the case of

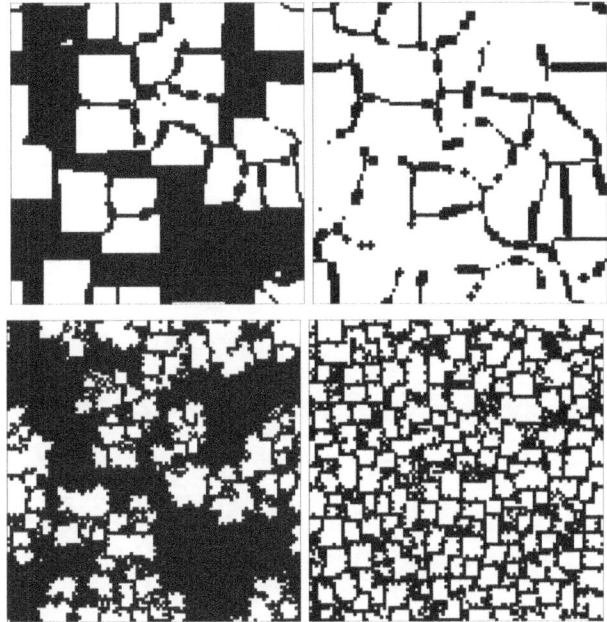

Fig. 6. Population states on Moore grids for the Prisoner's Dilemma game with $S = -0.05$ and $T = 1.35$ during the transient phase (left column) and on equilibrium (right column). Upper row: synchronous dynamics; Lower row: sequential dynamics.

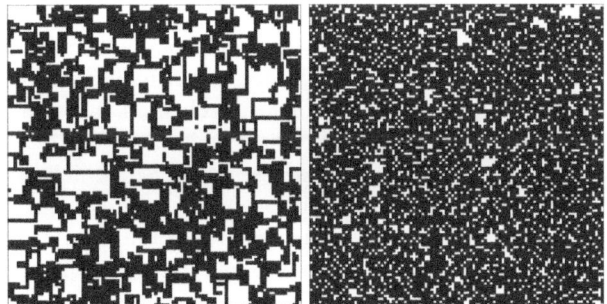

Fig. 7. Equilibrium population states on Moore grids for the Snowdrift game with $S = 0.6$ e $T = 1.6$. Left column: synchronous dynamics; Right column: sequential dynamics.

the best-takes-over rule. Let us start by analyzing the situations where cooperator clusters grow through their straight boundaries, which occur mainly when Moore grids are used. In this case, any mechanism that hampers the formation of straight boundaries hampers also the growth of cooperator clusters. This may happen, for example, if we use a stochastic transition rule [18] or an asynchronous dynamics. Fig. 6 shows an example for the Prisoner's Dilemma game

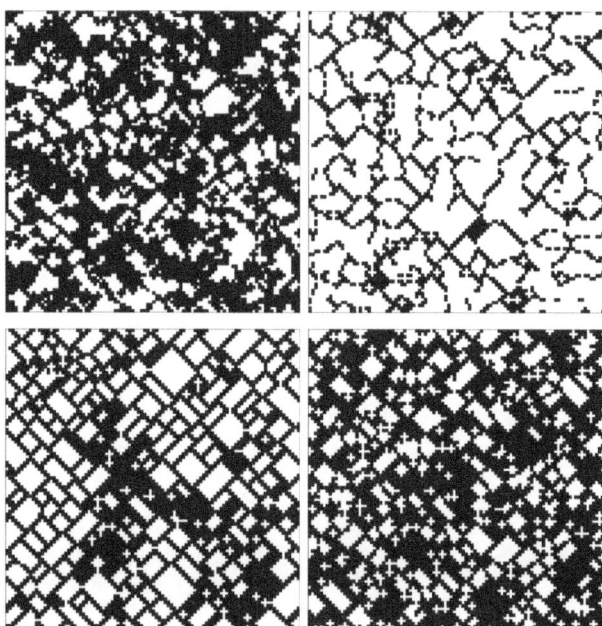

Fig. 8. Equilibrium states on von Neumann grids for the Prisoner's Dilemma with $S = -0.05$ e $T = 1.4$ (left column) and $S = -0.4$ e $T = 1.15$ (right column). Upper row: synchronous dynamics; Lower row: sequential dynamics.

where, under synchronous dynamics (upper row), cooperators form clusters that quickly grow through their straight boundaries, eventually forming a single large cluster which coexists with filament like defector clusters. The same figure (lower row) also shows that, when sequential dynamics is used, clusters have more irregular boundaries. Due to the local processes involved, which depend strongly on the relative value of game parameters S and T, irregular boundaries reduce the timescale at which clusters grow. This leads to equilibrium states with many small cooperator clusters that are not able to join due to the presence of defector agents between them. Fig. 7 shows an example for the Snowdrift game where cooperators are not able to form compact clusters when the population is under an asynchronous dynamics.

Let us now concentrate on situations for which sequential dynamics does not prevent the formation of straight boundaries. These situations occur mainly on von Neumann grids. When the games are played on these networks, there are many combinations of S and T for which straight boundaries remain fixed once formed, unless another cluster "collides" with them. That is, contrary to what happens in Moore grids (Fig. 6), in von Neumann grids clusters grow mainly through their irregular boundaries. In these cases, the influence of the update dynamics depends on how quickly straight boundaries are formed.

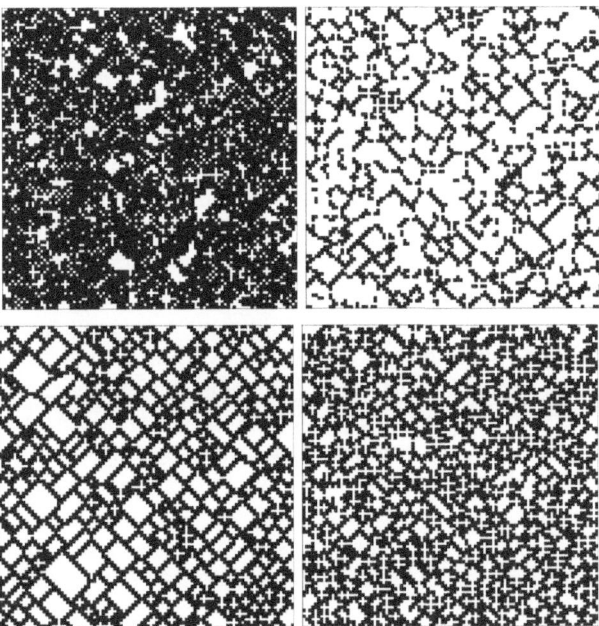

Fig. 9. Equilibrium states on von Neumann grids for the Snowdrift with $S = 0.6$ e $T = 1.7$ (left column) and $S = 0.2$ e $T = 1.2$ (right column). Upper row: synchronous dynamics; Lower row: sequential dynamics.

Figures 8 and 9 show typical equilibrium population states for the Prisoner's Dilemma and Snowdrift games, respectively. They allow us to understand the relation between straight boundaries and the influence of the update dynamics. The left side images depict situations where sequential dynamics is beneficial for cooperation. In these cases, synchronous updating leads to a chaotic dynamics where cooperators are not able to form compact clusters. With sequential updating, the boundaries advance slowly and cluster growth is interrupted only when (diagonal) straight boundaries are finally formed, which happens when clusters have already a significant size.

The right side images depict the inverse situation. In these cases, with synchronous updating, straight boundaries are formed only when the population is already dominated by a big cluster of cooperators. On the other hand, with sequential updating, straight boundaries are formed at an early phase of the evolutionary process, preventing cooperator clusters from growing. These cases show the role that network regularity (which allows the formation of straight cluster boundaries), in combination with the timescale at which clusters grow, has on the influence of the update dynamics. They also explain why sequential dynamics is beneficial to the evolution of cooperation in some cases and detrimental in others when highly regular networks are used.

5   Conclusions and Future Work

We have shown that, when the update dynamics has some influence on the evolution of cooperation, asynchronous dynamics is beneficial to cooperation in the case of the Stag-Hunt game. For the Prisoner's Dilemma and Snowdrift games, asynchronous dynamics is detrimental for cooperation only when these games are played on strongly regular networks and only for the best-takes-over rule, which models the absence of noise in the strategy update process. Moreover, we verify a strong increment of cooperation on scale-free networks when we change from a synchronous to an asynchronous dynamics, mainly when accumulated payoffs are used. An analysis of these results, taking into account the features of the networks, indicates that network regularity, in combination with the temporal scale at which clusters grow, plays the main role concerning the positive or negative influence of the asynchronous dynamics. Given that both regular networks and noise free environments seldom exist in real systems, this is a strong evidence that an asynchronous dynamics is, in general, beneficial to the evolution of cooperation.

In this paper we used a stochastic asynchronous update method. A future direction for this work will be to explore deterministic asynchronous update methods in order to verify to what extent the observed behaviors are due to the stochastic nature of the update dynamics [4].

The network of contacts rarely is a static structure: agents continuously enter and leave the population; moreover, agents continuously establish new connections and break existing ones. Several works have shown that it is possible for cooperation to strive in such scenarios [14, 15, 16, 28, 23]. Another possible future direction for this work is, thus, to analyze the interplay between dynamic models of networks and the influence of the update dynamics.

Acknowledgements. This work is partially supported by FCT/MCTES grant SFRH/BD/37650/2007.

References

[1] Abramson, G., Kuperman, M.: Social games in a social network. Physical Review E 63, 030901 (2001)

[2] Axelrod, R.: The Evolution of Cooperation. Penguin Books (1984)

[3] Barabási, A.-L., Albert, R.: Emergence of scaling in random networks. Science 286(5439), 509–512 (1999)

[4] Gershenson, C.: Classification of random boolean networks. In: Artificial Life VIII: Proceedings of the Eighth International Conference on Artificial Life, pp. 1–8. The MIT Press, Cambridge (2002)

[5] Grilo, C., Correia, L.: Effects of asynchonism on evolutionary games. Journal of Theoretical Biology 269(1), 109–122 (2011)

[6] Hauert, C.: Effects of space in 2x2 games. International Journal of Bifurcation and Chaos 12(7), 1531–1548 (2002)

[7] Hauert, C., Doebeli, M.: Spatial structure often inhibits the evolution of cooperation in the snowdrift game. Nature 428, 643–646 (2004)

[8] Holme, P., Kim, B.J.: Growing scale-free networks with tunable clustering. Physical Review E 65(2), 026107 (2002)

[9] Huberman, B., Glance, N.: Evolutionary games and computer simulations. Proceedings of the National Academy of Sciences of the United States of America 90(16), 7716–7718 (1993)

[10] Newth, D., Cornforth, D.: Asynchronous spatial evolutionary games. BioSystems 95, 120–129 (2009)

[11] Nowak, M., Bonhoeffer, S., May, R.M.: More spatial games. International Journal of Bifurcation and Chaos 4(1), 33–56 (1994)

[12] Nowak, M., May, R.M.: Evolutionary games and spatial chaos. Nature 359, 826–829 (1992)

[13] Nowak, M., Sigmund, K.: Evolution of indirect reciprocity. Nature 437, 1291–1298 (2005)

[14] Pacheco, J.M., Traulsen, A., Nowak, M.A.: Active linking in evolutionary games. Journal of Theoretical Biology 243, 437–443 (2006)

[15] Pacheco, J.M., Traulsen, A., Nowak, M.A.: Co-evolution of strategy and structure in complex networks with dynamical linking. Physical Review Letters 97(25), 258103 (2006)

[16] Poncela, J., Gómez-Gardeñes, J., Floría, L.M., Sánchez, A., Moreno, Y.: Complex cooperative networks from evolutionary preferential attachment. PLoS ONE 3(6), e2449 (2008)

[17] Roca, C.O., Cuesta, J.A., Sánchez, A.: Effect of spatial structure on the evolution of cooperation. Physical Review E 80(4), 046106 (2009)

[18] Roca, C.O., Cuesta, J.A., Sánchez, A.: Imperfect imitation can enhance cooperation. Europhysics Letters 87, 48005 (2009)

[19] Santos, F.C., Pacheco, J.M.: Scale-free networks provide a unifying framework for the emergence of cooperation. Physical Review Letters 95(9), 098104–+ (2005)

[20] Skyrms, B.: The Stag Hunt and the Evolution of Social Structure. Cambridge University Press, Cambridge (2004)

[21] Smith, J.M.: Evolution and the Theory of Games. Cambridge University Press, Cambridge (1982)

[22] Szabó, G., Tóke, C.: Evolutionary prisoner's dilemma game on a square lattice. Physical Review E 55(1), 69–73 (1998)

[23] Szolnoki, A., Perc, M., Danku, Z.: Making new connections towards cooperation in the prisoner's dilemma game. EPL 84(5), 50007 (2008)

[24] Szolnoki, A., Perc, M., Danku, Z.: Towards effective payoffs in the prisoner's dilemma game on scale-free networks. Physica A: Statistical Mechanics and its Applications 387, 2075–2082 (2008)

[25] Tomassini, M., Luthi, L., Giacobini, M.: Hawks and doves on small-world networks. Physical Review E 73(1), 016132 (2006)

[26] Tomassini, M., Luthi, L., Pestelacci, E.: Social dilemmas and cooperation in complex networks. International Journal of Modern Physics C 18, 1173–1185 (2007)

[27] Watts, D., Strogatz, S.H.: Collective dynamics of small-world networks. Nature 393, 440–442 (1998)

[28] Zimmermann, M.G., Eguíluz, V.M.: Coevolution of dynamical states and interaction in dynamic networks. Physical Review E 69, 065102(R) (2004)

The Squares Problem and a Neutrality Analysis with ReNCoDe

Rui L. Lopes and Ernesto Costa

Center for Informatics and Systems of the University of Coimbra
Polo II - Pinhal de Marrocos 3030-290 Coimbra, Portugal
{rmlopes,ernesto}@dei.uc.pt

Abstract. Evolutionary Algorithms (EA) are stochastic search algorithms inspired by the principles of selection and variation posited by the theory of evolution, mimicking in a simple way those mechanisms. In particular, EAs approach differently from nature the genotype - phenotype relationship, and this view is a recurrent issue among researchers. Moreover, in spite of some performance improvements, it is a true fact that biology knowledge has advanced faster than our ability to incorporate novel biological ideas into EAs. Recently, some researchers start exploring computationally our new comprehension about the multitude of the regulatory mechanisms that are fundamental in both processes of inheritance and of development in natural systems, trying to include those mechanism in the EA. One of the first successful proposals is the Artificial Gene Regulatory (ARN) model, by Wolfgang Banzhaf. Soon after some variants of the ARN with increased capabilities were tested. In this paper, we further explore the capabilities of one of those, the Regulatory Network Computational Device, empowering it with feedback connections. The efficacy and efficiency of this alternative is tested experimentally using a typical benchmark problem for recurrent and developmental systems. In order to gain a better understanding about the reasons for the improved quality of the results, we undertake a preliminary study about the role of neutral mutations during the evolutionary process.

Keywords: regulation, network, evolution, development, neutrality.

1 Introduction

Nature-inspired algorithms are used today extensively to solve a multitude of learning, design and optimization problems, giving rise to a new research area called Evolutionary Computation (EC) [6]. Along time many variants of a basic algorithm especially tuned for some problems and/or situations were proposed (e.g., algorithms for dealing with noisy, uncertain or dynamic environments, for evolving rather than designing the algorithm's parameters or some of its components, algorithms with local search operators or for multi-objective optimization). The standard evolutionary algorithm can be described in simple terms:

L. Antunes and H.S. Pinto (Eds.): EPIA 2011, LNAI 7026, pp. 182–195, 2011.

(1) randomly define an initial population of solution candidates; (2) select, according to fitness, some individuals for reproduction with variation; (3) define the survivors for the next generation; (4) repeat steps (2) and (3) until some condition is fulfilled. Typically, the objects manipulated by the algorithms are represented at two different levels. At a low level, the genotype, the representations are manipulated by the variation operators; at a high level, the phenotype, the objects are evaluated to determine their fitness and are selected accordingly. Because of that, we need a mapping between these two levels. The issue of the relationship between the genotype and the phenotype is as old as the area itself, many claiming that the standard approach is too simplistic. For example, typically in an EA, the two phases of transcription and translation are merged into just one and the regulatory processes are missing. At a larger scale, we could add the lack of epigenetic phenomena that contribute to the evolution and all the mechanisms involved in the construction of an organism (Figure 1).

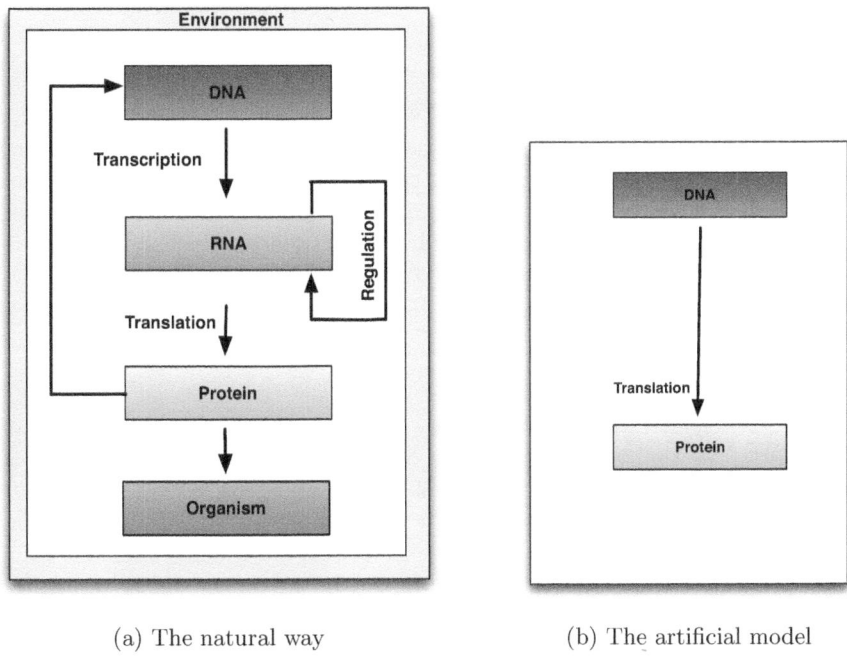

(a) The natural way   (b) The artificial model

Fig. 1. From genes to an organism

In this work we will be focused on regulation. Today we are aware of the crucial role of regulation in the evolution and development of complex and well adapted creatures [3]. Regulation is a consequence of external (i.e., epigenetic) or internal (i.e., genetic) influences. Some computational explorations have been proposed to deal with the inclusion of regulatory mechanisms into the standard evolutionary algorithm [5,2,12]. Wolfgang Banzhaf [1] proposed an artificial gene regulatory (ARN) model and showed how it could be used computationally in different settings [4,11]. More recently [8] presented a variant of the ARN aimed at solving some weaknesses of that model and enlarge its applicability, by transforming the regulatory gene network into a computable tree-like expression as we do in standard GP. In [9] we complement and enhance this latter proposal by making possible the use of feedback edges (equivalent to delays in an iterative process). In the present work we further explore the capabilities of this method by experimentally testing it with another standard benchmark problem. We clearly establish the generalization capabilities of the solutions produced and we present a preliminary analysis of neutrality in the system.

The paper is organized as follows. Section 2 describes the ARN model as proposed by W. Banzhaf. Then, Section 3 describes an extension of that model, elucidating how we can extract a program from a network. In Section 4 we briefly refer to the problem used and the experimental setup. The results are presented and analyzed in Section 5. Finally, in Section 6 we draw some conclusions and present some ideas for future work.

2   Banzhaf's ARN

The Artificial Regulatory Network (ARN) [1] is an attempt to incorporate regulatory mechanisms between the genotype and the phenotype. There is no other products, i.e. DNA, and processes in between these two levels. The genome has fixed length and is constructed by simple duplication with mutation events. Regulation between genes is a mediated process, and is achieved by means of a binding process between proteins (i.e., transcription factors) and special zones in the genome that appear upstream the promoter of a gene. Let's describe this in more detail.

Genome. At the center of the model is a binary genome and proteins. The genome can be generated randomly or by a process of duplication with mutation, also called *DM*, that is considered the driving force for creating new genes in biological genomes and as an important role in the growth of gene regulatory networks [14]. In the latter case we start with a random 32-bit binary sequence, that is followed by several duplication episodes. As we will see later the number of duplications is an important parameter. The mutation rate is typically of 1%. So, if we have 10 duplication events then the final length of the genome is $2^5 \times 2^{10} = 32768$. The genome is divided in several regions, namely a regulatory site, the promoter and the gene itself. The first 32 bits of the regulation zone are the enhancer site, while the following 32 bits are the inhibitory site. The promoter is located downstream and has the form $XYZ01010101$. This means that only the last 8 bits are fixed. A gene is a five 32-bit long sequence, i.e., a 160-bit string.

Gene expression. The genotype - phenotype mapping is defined by expressing each 160-bit long gene, resulting in a 32-bit protein. This correspondence is based on using a majority rule: if we consider the gene divided into 5 parts of size 32

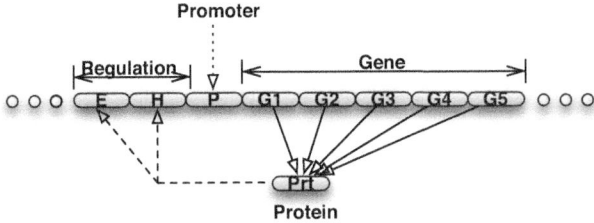

Fig. 2. Artificial Regulatory Network, after W. Banzhaf

each, at position i, say, the protein's bit will have a value corresponding to the most frequent value in each of these 5 parts, at the same position. Figure 2 gives an idea of the representation.

Regulation. The proteins can bind to the regulatory region. The strength of the binding is computed by calculating the degree of complementarity between the protein and each of the regulatory regions, according to formula 1:

$$x_i = \frac{1}{N} \sum_{j=1}^{N} c_j e^{\beta(\mu_j - \mu_{max})} \tag{1}$$

where x_i can the the enhancer or the inhibitory region, N is the number of proteins, c_j the concentration of protein j, μ_j is the number of bits that are different in each position of the protein and of the regulation site, μ_{max} is the maximum match achievable, and β is a scaling factor. The production of a protein along time depends on its concentration, which in turn is a function of the way it binds to the regulatory regions. It is defined by the differential equation

$$\frac{dc_i}{dt} = \delta(e_i - h_i)c_i \ .$$

Genes interact mediated by proteins. If, say, gene **A** expresses protein p_A and that protein contributes to the activation of gene **B**, we say that gene **A** regulates **B** (see Fig. 3).

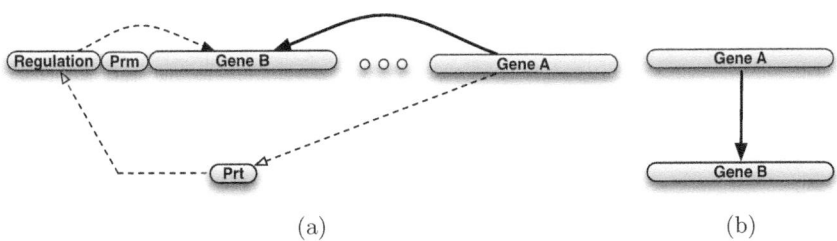

(a)   (b)

Fig. 3. Gene - Protein - Gene interaction

Notice that in order for a link to exist between any two genes, the concentration of the corresponding protein must attain a certain level, and that depends on the strength of the binding.

3   Regulatory Network Computational Device (ReNCoDe)

Based on this model we can build for each genome the corresponding gene regulatory network. These networks can be studied in different ways. We can be concerned by topological aspects (i.e., to study the degrees distribution, the clustering coefficient, small world or scale free, and the like) or the dynamics of the ARNs (i.e., attractors, influence of the protein-gene binding threshold) [10,4]. This is interesting, but from a problem-solving perspective what we want is to see how the model can be used as a computational device. That was already done. As a matter of fact, in [8] the ARN architecture was used as a genotypical representation for a new computational model. The main idea is to simplify the ARN to produce a computable tree-like expression similar to a GP tree. Besides that, in order to greatly increase the efficacy and the performance of the system, special variation operators, besides mutation, were also introduced. This model was tested successfully with several benchmark problems. Later, the model was further extended by allowing feedback connections during the transformation of the regulatory network into a computational circuit [9]. In the latter case, instead of a tree we now have to deal with a cyclic graph.

3.1   Extracting Circuits from ARNs

The networks resultant from ARN genomes are very complex, composed of multiple links (inhibition and excitation) between different nodes (genes). In order to extract a circuit from these networks they must first be reduced, input and output nodes must de identified, and we must ascribe a semantic to its nodes. In the simplest cases, the final product will be an executable feed-forward circuit, although it is also possible to use feedback connections, creating cyclic graphs that can be iterated and allow to keep state information.

Listing 1.1 shows the pseudocode for the reduction algorithm. We start by transforming every pair of connections, excitation (e) and inhibition (h), into one single connection with strength equal to the difference of the originals (e-h) (step 3). Every connection with negative or null strength will be discarded (steps 4-5). Then a directed graph is built adding only the nodes with active edges and the strongest edge between each pair of nodes (steps 6-10). This process is illustrated in Figure 4a (the network) and 4b (the reduced network).

Next, the node with the highest connectivity is chosen as the output. After this, the circuit is built backwards from the output function until the terminals (nodes without input edges) are reached. If, at any point of this process, there is a deadlock (every function is input to some other), again the gene with highest connectivity is chosen. To solve problems like symbolic regression and the santa-fe ant trail, where there is no need for feedback, each time a node is added to the

Listing 1.1. The reduction algorithm.

```
1  def reduce(network):
2    for each gene in network:
3      replace bindings by edge (e-h)
4        if (e - h ) <= 0:
5          remove edge
6    for each edge(i,j) in new network:
7      if  edge(i,j) < edge(j,i):
8        remove edge(i,j)
9      else:
10        remove edge(j,i)
```

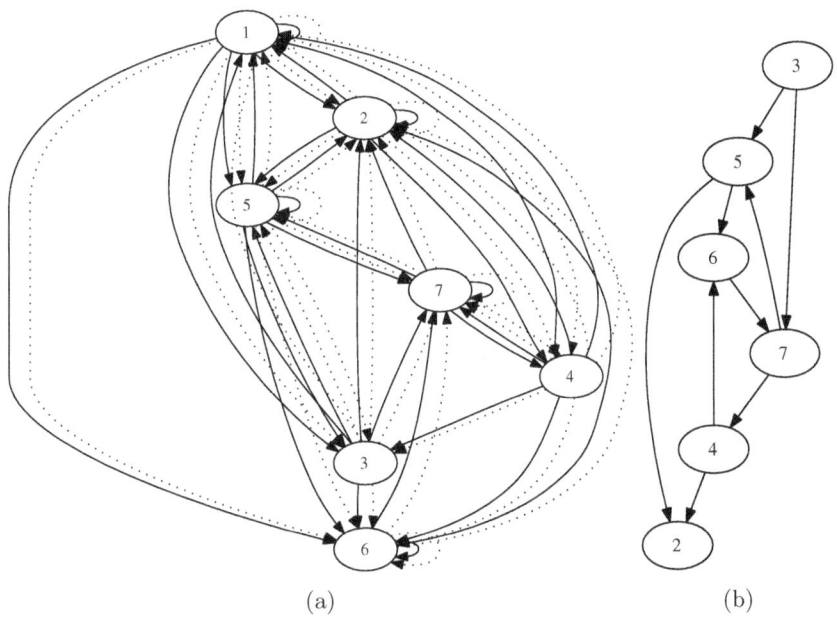

(a)   (b)

Fig. 4. a) ARN example. The dotted and full edges represent, respectively, inhibition and excitation relationships. The numbers are just identifiers. b) The same network after application of the algorithm in listing 1.1.

circuit, the inputs from nodes already added are simply discarded (Fig. 5a). This results in state-less feed-forward circuits, which are not adequate for problems where some kind of memory is needed. On the other side, when a node takes input from one already in the circuit, it is possible to use it as feedback (instead of discarding it), resulting in a state-full feed-forward circuit (please refer to Fig. 5b for an example), similarly to a cyclic graph. This one-level of indirection allows to save information from the previous state(s) when the circuit is iterated.

To complete the process, a mapping is needed linking nodes (i.e., genes) to functions and terminals. To that end we use the gene-protein correspondence using the protein's signature to obtain the function/terminal by a majority vote process. As an example, to code the function set { +, -, *, / } only two bits are necessary. These are obtained by splitting the 32-bit protein into sixteen 2-bit chunks and applying the majority rule in each position of the chunks. This will provide us with an index to the function set. If some determined problem has more than one input, for instance four, then the majority rule is applied again over the terminal protein signature (its binary stream) to define to which input it corresponds.

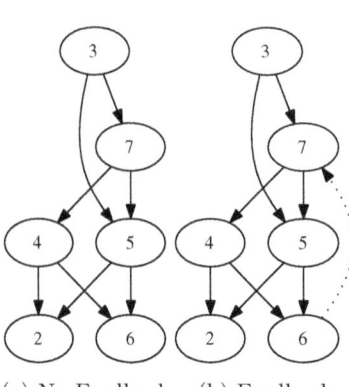

(a) No Feedback   (b) Feedback

Fig. 5. The resulting circuit from the reduced network depicted in Fig. 4b. This circuit is obtained by choosing node 3 as the output and then successively adding the inputs of each node until the terminal nodes are reached. Note that one of the cycles in Fig. 4b is composed of the nodes 4, 6, and 7. There are two possible ways to break the cycle: a) the connection from 6 to 7 is discarded; b) the connection from 6 to 7 is used as feedback, since the node 7 was already in the circuit. Functions take their inputs from the nodes where their out-edges point to. The dotted edges represent feedback connections.

Finally, in order to improve the evolvability of the genomes, in [8] the authors proposed also variation operators inspired by the concepts of transposons and non-coding DNA, which can copy a part of the genome (*transposon-like*, see Fig. 6), or introduce non-coding (*junk*) genetic material (streams with 0s). Also, a *delete* operator is used to allow the genomes to be shrunk if beneficial. Every iteration one of these operators may be applied with a determined probability. The results reported show an efficiency increase when using any of the operators with small lengths (versus fixed size genomes with only the mutation operator). The average results also show that the *transposon-like* operator is more *stable*.

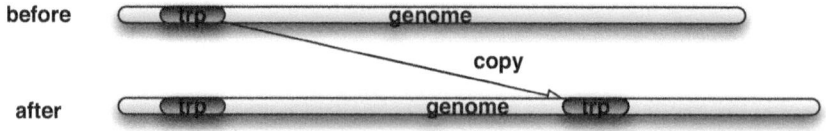

Fig. 6. The effect of using a transposon-like operator on modifying the genome: the copy example

4   Experimental Setup

In order to study the performance of the new approach we setup an experiment involving a benchmark problem knowing that, in this case, the use of feedback connections is crucial.

The Problem. The sequence of squares is a sequential regression problem, where the target function x^2, over the non-negative integers, is to be evolved. The particularity of this problem is that the function set is composed of only $\{+,-\}$. As pointed out in [13] this limited function set only allows the regression of linear functions, and so this would be impossible to solve with traditional GP methods. Based on the use of self-modification functions in the phenotype, different developmental approaches have been proposed that solve this problem with success [13,7].

In this work we take advantage of the feedback connections in ReNCoDe to obtain circuits that when iterated will produce the correct squared value for that iteration. Similarly to [7] the evolutionary process tries to find a circuit that generates correctly the first ten terms of the sequence. Once a correct solution is reached we test the circuit over the first hundred terms of the sequence to assess the generalization capabilities of this approach.

A standard Evolution Strategy was used for the experiments: $(10 + 100) - ES$. As in previous implementations of ReNCoDe, there is no crossover but a *transposon-like* and a *delete* operator are used, which can, respectively, copy or delete a portion of the genome. As shown in Fig. 6, the application of the *transposon-like* operator will render a variable length genome. This operator copies one part of the genome and adds it at another point of the stream while the *delete* operator allows the removal of detrimental sections. The mutation operator is always applied, while the remaining variation operators are mutually-exclusively applied, according to the probabilities defined in Table 1. The number of runs and their parameterizations are summarized in Table 2. The parameters were chosen by trial and error during testing. No sensitivity tests nor parameter optimization were realized.

Moreover, we will be interested in analyzing the effect of using the new variation operators. In previous studies it was clear that these operators were fundamental for the high quality of the results. In [8] the authors hypothesized that this is due to the fact that these operators are responsible for the existence of neutral mutations.

Parameter	Value
Operator Length	80
P_{Idle}	0.5
$P_{Transposon}$	0.3
P_{Delete}	0.2

Table 1. Parameterization of the variation operators. The probability of applying each operator is referred by $P_{OperatorName}$, with P_{Idle} being the probability of not applying any of them.

Table 2. Evolutionary Parameters Values

Parameter	Value
Number of Runs	100
Initial Population	100
Max. Evaluations	10^6
Number of DMs	6
DM Mutation Rate	0.01
Mutation Operator Rate	0.01
Protein Bind Threshold	16
Genome Length	Variable
Operator Type	Transposon
Operator Length	80

5   Results and Analysis

In Table 3 we can see the summary of the results for the given parameterization. Similarly to previous applications of this model, the achieved performance (in number of evaluations to reach a solution) is encouraging and there is not many bloat in the final solutions, as shown by the average number of functions and connections. Moreover, all the solutions found generalize correctly for the first hundred elements of the sequence. The distribution of the effort values for all the runs can be observed in Fig. 7. We can see that most of the results are better than the average value found (visible also by the median value), an effect of the presence of some outliers in the data.

One possible solution, the smallest found, is presented in Fig. 8. In this picture, the dotted edges represent the feedback connections, taking the value of that function in the previous iteration. The full edges represent the regular connections where a node takes its inputs from.

Table 3. Summary of the squares' sequence results for the 100 runs

% of Successful Runs	99
% of General Solutions	100
Min. Numb. of Evaluations	15500
Median Numb. of Evaluations	162200
Avg. Numb. of Evaluations (Std. Dev.)	254059 (247206)
Min. Numb. of Functions/Connections/Feedbacks	6/13/3
Avg. Numb. of Functions/Connections/Feedbacks	8/18/5

5.1   Generality

The generality of the solutions found is not very clear to analyze, especially when the number of feedback connections increases. For the following demonstration

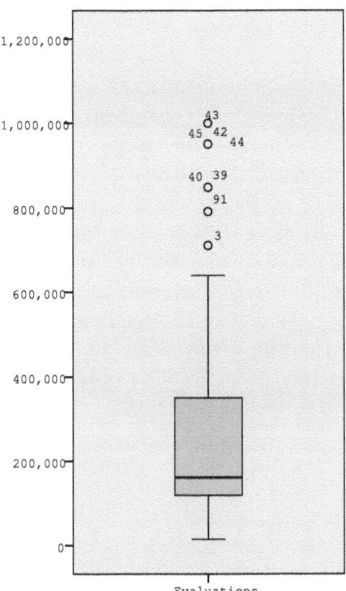

Fig. 7. Distribution of the effort values for the hundred runs with identification of outliers

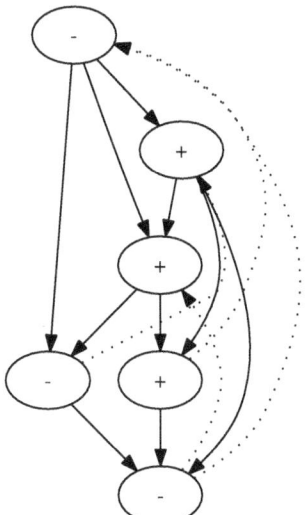

Fig. 8. Circuit with the smallest number of functions found that generates the correct sequence. The nodes represent arithmetic operators. The output of the circuit is taken from the top. Each iteration the nodes are updated from bottom to top and the result is taken as an element of the sequence. Functions take their inputs from the nodes where their out-edges point to. The dotted edges represent feedback connections.

we used a circuit with the minimum feedback connections (3) and removed any node that was just a duplication of some other node. This solution is depicted in Fig. 9.

Based on this we have done a simulation of the resulting circuit over the first few iterations, as shown in Fig. 10. The nodes are represented in the column "ID" by the same numerical identifications used inside squared brackets in Fig. 9, with the corresponding function (and input connections). All the nodes are

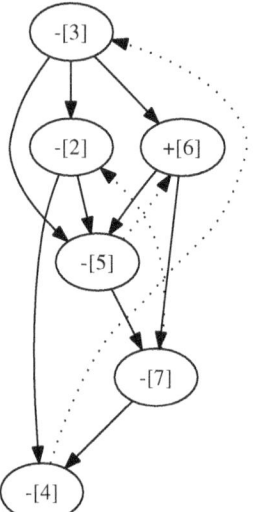

Fig. 9. Circuit found with the least feedback connections, after removal of nodes duplicating others. The nodes represent arithmetic operators with a label containing the identifier of the function inside brackets. The output of the circuit is taken from the top. Each iteration the nodes are updated from bottom to top and the result is taken as an element of the sequence. Functions take their inputs from the nodes where their out-edges point to. The dotted edges represent feedback connections.

ID		Init / k	0	1	2	3	4	5	6	7	8	9	10	Equivalence
3	= 5 - 2 - 6	1	0	1	4	9	16	25	36	49	64	81	100	k^2
2	= 4 - 5	1	0	-1	-2	-3	-4	-5	-6	-7	-8	-9	-10	$-k$
6	= 5 + 7	1	1	1	1	1	1	1	1	1	1	1	1	1
5	= 6* - 7	1	1	1	3	7	13	21	31	43	57	73	91	$1 + k^2 - k$
7	= 2* - 4	1	0	0	-2	-6	-12	-20	-30	-42	-56	-72	-90	$-k^2 + k$
4	= 3*	1	1	0	1	4	9	16	25	36	49	64	81	$[k-1]^2$

Fig. 10. A simulation of the circuit depicted in Fig. 9 for the first iterations. The nodes are initialized with 1 and then the consecutive values are propagated from bottom to top (similarly to the circuit graph). The '*' denote feedback connections. The invariant representation of some nodes has been identified in *Equivalence*.

initialized with the value 1 (column "Init"). The simulation of the circuit for the first 11 iterations is presented in the following columns, numbered from 0 to 10. It is updated from bottom to top, similarly to the circuits presented before. If a node has an input from a recurrent connection that value is taken from the previous column. In the final column the equivalence is deduced from the consecutive values outputted by each node.

Finally, we replace the output node by its mathematical function, based on the invariant values identified by *Equivalence* in Fig. 10, and show that it is equivalent to k^2.

$$[3] = [5] - [2] - [6]$$
$$= 1 + k^2 - k - (-k) - 1$$
$$= k^2 \qquad (2)$$

5.2   Preliminary Results on Neutrality

In previous studies it was shown that the introduction of biologically inspired variation operators is fundamental for the high quality of the results. In [8] the authors have hypothesized that this is due to the fact that these operators are responsible for the existence of a bigger amount of neutral mutations accumulated in the genomes.

In this section we analyze the amount of *effective* and *neutral* mutations accumulated in the population throughout the evolutionary process. Typical runs of the *Squares* problem will be used, selected within the 95% confidence interval of the mean number of evaluations. Moreover, we compare the dynamics of the mutations with those from runs where no variation operator was used (fixed length genomes).

We analyzed the neutral mutations only at the genotype level. If a mutation occurs at either a binding site, the coding part of a promoter, or at a gene it is taken as *effective*. Otherwise, it is accounted for as a *neutral* mutation. Figure 11 shows the accumulation of mutations in the population throughout the evolutionary process of typical runs selected from the results presented in Section 5. We can observe that the number of accumulated neutral mutations is increasing more than the effective during the evolutionary process.

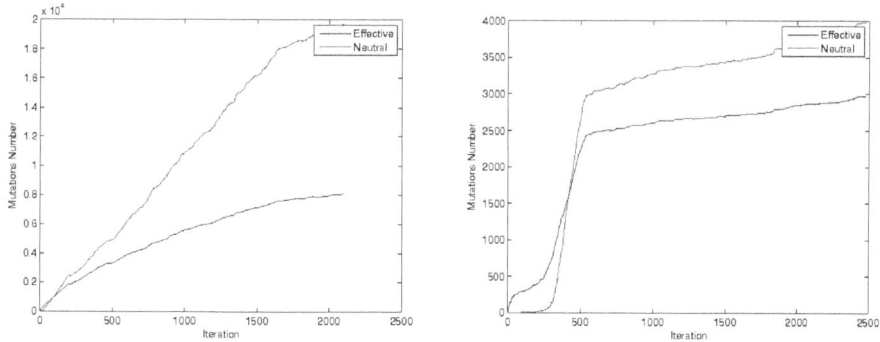

Fig. 11. Number of accumulated mutations in the population, effective and neutral, during two typical runs solving the Squares problem using the *transposon-like* operator

In order to know if the operators are responsible for the accumulation of neutral mutations we analyzed also the behavior of the algorithm without the use of the variation operators, when applied to the same problem. The typical mutational dynamic found in the results is shown in Fig. 12. In this case, however, it is clear that the number of accumulated effective mutations is increasing more than the neutral during the evolutionary process.

These preliminary results on the amount of mutations in the population tend to confirm that indeed it is the variation operators that allow the accumulation of neutral mutations, by allowing the genome to expand with non-coding DNA.

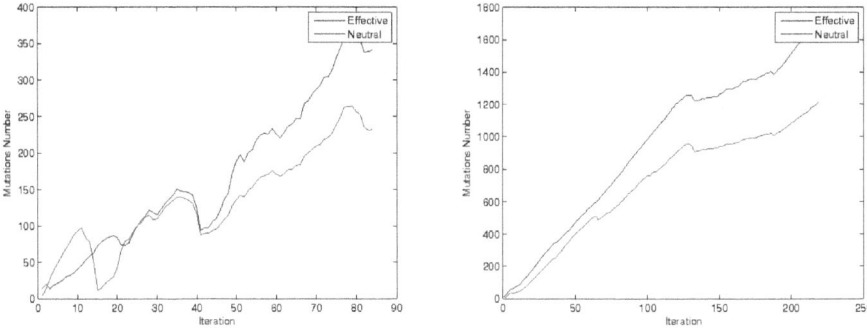

Fig. 12. Number of accumulated mutations in the population, effective and neutral, during two typical runs solving the Squares problem, when the variation operators are not used

6   Conclusions and Future Work

The ReNCoDe is a biologically inspired technique aimed at solving different classes of problems, which uses as genomic representation a well established computational biology model, the ARN. In previous work [8] it was shown not only that it is effective in solving a set of GP benchmark problems but also that the introduction of new variation operators (also based on biological concepts) could improve the performance of the algorithm. In later work [9], the model has been extended to allow the inclusion of feedback connections, extending the applicability of the model to a different class of problems, where there is recursion. Examples of problems tackled using this algorithm include the Cart-Centering and Pole-Balancing, the N-bit Parity and the Fibonacci sequence, obtaining good results in both performance, and success and generalization rates.

In this paper we confirmed these results by showing how it can be applied to another problem, the Squares Sequence, a typical benchmark for developmental and recurrent systems. The performance indicators are good, and both the success and generalization rates improve on known results from the literature [7]. Moreover, we did a preliminary study on the accumulation of mutations in the evolved population. This revealed that indeed the variation operators allow a bigger amount of neutral mutations than effective, opposed to the results without operator, where the number of effective mutations is bigger than the neutrals.

Future work on ReNCode should aim at a better understanding of the neutrality in the model and its relation to the performance. Also, the model could be extended to deal with dynamic environments.

References

1. Banzhaf, W.: Artificial regulatory networks and genetic programming. Genetic Programming Theory and Practice, 43–62, 2003
2. Bongard, J.: Evolving modular genetic regulatory networks. In: IEEE 2002 Congress on Evolutionary Computation (CEC 2002), pp. 1872–1877. IEEE Press (2002)
3. Davidson, E.H.: The regulatory genome: gene regulatory networks in development and evolution. Academic Press (2006)
4. Dwight Kuo, P., Banzhaf, W., Leier, A.: Network topology and the evolution of dynamics in an artificial genetic regulatory network model created by whole genome duplication and divergence. Bio Systems 85(3), 177–200 (2006)
5. Eggenberger, P.: Evolving morphologies of simulated 3D organisms based on differential gene expression. In: Husbands, P., Harvey, I. (eds.) Fourth European Conference of Artificial Life. MIT Press, Cambridge (1997)
6. Eiben, A.E., Smith, J.E.: Introduction to Evolutionary Computing. Springer, Heidelberg (2003)
7. Harding, S., Miller, J., Banzhaf, W.: Self modifying cartesian genetic programming: Fibonacci, squares, regression and summing. Genetic Programming, 133–144 (2009)
8. Lopes, R.L., Costa, E.: ReNCoDe: A Regulatory Network Computational Device. Genetic Programming 6621(EuroGP 2011), 142–153 (2011)
9. Lopes, R.L., Costa, E.: Using Feedback in a Regulatory Network Computational Device. In: GECCO 2011: Proceedings of the 13th Annual Conference on Genetic and Evolutionary Computation (to be published in, 2011)
10. Nicolau, M., Schoenauer, M.: Evolving specific network statistical properties using a gene regulatory network model. In: Raidl, G., et al. (eds.) GECCO 2009: Proceedings of the 11th Annual Conference on Genetic and Evolutionary Computation, pp. 723–730. ACM, Montreal (2009)
11. Nicolau, M., Schoenauer, M., Banzhaf, W.: Evolving Genes to Balance a Pole. In: Esparcia-Alcázar, A.I., Ekárt, A., Silva, S., Dignum, S., Uyar, A.Ş. (eds.) EuroGP 2010. LNCS, vol. 6021, pp. 196–207. Springer, Heidelberg (2010)
12. Roggen, D., Federici, D., Floreano, D.: Evolutionary morphogenesis for multicellular systems. Genetic Programming and Evolvable Machines 8(1), 61–96 (2006)
13. Spector, L., Stoffel, K.: Ontogenetic programming. In: Proceedings of the First Annual Conference on Genetic Programming, pp. 394–399. MIT Press, Cambridge (1996)
14. Teichmann, S.a., Babu, M.M.: Gene regulatory network growth by duplication. Nature Genetics 36(5), 492–496 (2004)

Particle Swarm Optimization for Gantry Control: A Teaching Experiment

Paulo B. de Moura Oliveira[1], Eduardo J. Solteiro Pires[2], and José Boaventura Cunha[3]

[1] CIDESD - Departamento de Engenharias
[2] CITAB – Departamento de Engenharias
[3] CIDESD - Departamento de Engenharias, Universidade de Trás-os-Montes e Alto Douro,
5000–911 Vila Real, Portugal
{oliveira,epires,jboavent}@utad.pt

Abstract. The particle swarm optimization algorithm is proposed as a tool to solve the Posicast input command shaping problem. The design technique is addressed, in the context of a simulation teaching experiment, aiming to illustrate second-order system feedforward control. The selected experiment is the well known suspended load or gantry problem, relevant to the crane control. Preliminary simulation results for a quarter-cycle Posicast shaper, designed with the particle swarm algorithm are presented. Illustrating figures extracted from an animation of a gantry example which validate the Posicast design are presented.

Keywords: Particle swarm optimization, Gantry-crane control, Posicast shaping.

1 Introduction

Second order system dynamics is a very important topic under any curricula for feedback control systems. The applications of second order systems in Engineering are huge, ranging from electronic circuit analysis and design to mechanical engineering vibration control. Teaching second-order systems time response and control to undergraduate students can beneficiate greatly with the support of computer-aided design examples, to illustrate the use and application of the associated techniques. A system which can be modeled by second-order dynamics is a gantry crane, and has been used with success for teaching purposes [11]. Cranes are used in real engineering applications, to carry heavy loads, as in shipyards, building construction sites, factories, among others. The overall control objective is to move the crane trolley fast (enough) and to avoid the suspended load oscillation (swing), at the specified final position. The automatic control of cranes is a research topic which has been studied using different techniques [1,10]. However their control design can beneficiate from using computational intelligence techniques.

A pioneering work to control a pendulum with a suspended load, was the Posicast, proposed by Smith (1957) [2]. Posicast is an open-loop feedforward control technique which can be taught simultaneously with second-order systems, as a simple technique to avoid overshooting the final position. Indeed, the Posicast concept originated most

L. Antunes and H.S. Pinto (Eds.): EPIA 2011, LNAI 7026, pp. 196–207, 2011.

of the flexible structure vibration techniques in use currently [3,4]. As many other control engineering design applications, Posicast control, can be designed using optimization techniques such as biological and natural inspired meta-heuristics.

One of the most successful search and optimization algorithms is the particle swarm optimization algorithm (PSO), proposed by [5], inspired in the collective behaviour of animal groups. The PSO success is due to several reasons: the beauty associated with the collective behaviors of swarms such as bird flocks and fish schools which capture very strongly human interest; the extreme simplicity of the standard algorithm and finally the success proved in solving a myriad of search and optimization problems [12,13]. As an example, while the popular genetic algorithm, in the standard form, is a very simple algorithm, its implementation is more complex than the standard PSO. This makes the PSO algorithm a good candidate to be taught to undergraduates engineering students. Indeed, a successful application of PSO within an open-loop system identification teaching experiment, to undergraduate students, was reported in [14].

In this paper a particle swarm optimization algorithm is proposed to solve the gantry control input command shaping problem. Simulation results are presented through the proposal of a teaching experiment.

2   Particle Swarm Optimization: Elementary Notions

The particle swarm optimization can be classified as an animal inspired algorithm, as it's based in some swarms (bird flocks, fish schools) capability to move in an organized form. From this synchronized motion, it emerges an intelligent collective behavior. The advantage of sharing social information within the swarm clearly compensates the disadvantages of competition. Since the pioneering particle swarm optimization (PSO) proposal by [5], its algorithm has been subjected to significant research efforts resulting in some refinements [6,7] and extensions for multiple-objective optimization [8]. In this work a standard PSO algorithm [9] is deployed to optimize Posicast input command shaping.

Each swarm population element represents a potential solution for the search and optimization problem, and it is represented within the PSO algorithm by two fundamentals n-dimensional variables: a position and velocity vectors, x and v, respectively. The first algorithm step consists in initializing the swarm position and velocity, within a predefined search space interval, usually using a random procedure. After evaluating swarm elements fitness in terms of the problem to be solved, using a specific objective function, swarm particles position and velocity, are updated in each iteration t, by the following two equations:

$$v_{id}(t+1) = \omega v_{id}(t) + c_1\varphi_c.(b_{id}(t) - x_{id}(t)) + c_2\varphi_s.(g(t) - x_{id}(t)) \tag{1}$$

$$x_{id}(t+1) = x_{id}(t) + v_{id}(t+1) \tag{2}$$

with: $1 \leq i \leq m$ and $1 \leq d \leq n$. In (1), b_i represents the best position achieved by particle i and g represents the best global position achieved within a specified neighborhood, which, in this paper, considers the entire swarm in a fully connected topology; In the same expression c_1 and c_2 are known as the cognitive and social constants, respectively, and φ_c and φ_s represent uniformly randomly generated numbers within interval [0,1]. While the originally proposed value for c_1 and c_2 is 2, other values can be used. Parameter ω, called inertia weight, is a modification to the original algorithm [15] to help controlling the convergence of the PSO algorithm. A high value assigned to the inertia weight (near one) promotes exploitation of the search space, making the search global, while a small value, promotes imploitation making the search local. Thus, usually it is common [9] to linearly decrease this weight from a high initial value, ω_{init} to a small value final value, ω_{fin}, throughout the search procedure.

Alternative versions of the PSO algorithm include a constriction coefficient, χ, which has been proved to guarantee swarm convergence [6], but with the appropriate parameter selection can be equivalent to equation (1). While not mandatory it is usual to limit the velocity to a maximum value, V_{max}. Thus, in each iteration, the velocity may vary within the interval defined by $[-V_{max}, V_{max}]$. The value of V_{max} is often a fraction of the range defined for the search space interval: $[x_{min}, x_{max}]$ for all dimensions. The structure of a standard PSO algorithm it is presented in Fig. 1.

$t = 0$
initialize swarm $X(t)$
evaluate $X(t)$
while(!(termination criterion))
 $t = t + 1$
 update local bests and global best(s)
 update particles velocity
 update particles position
 generate $X(t + 1)$
 evaluate $X(t + 1)$
end

Fig. 1. Standard PSO algorithm

3   The Gantry-Crane Control Problem

In this section the gantry control problem is described using elementary mathematical modelling technique, as we are dealing with a didactic experience. Consider the simplified representation of a trolley crane mechanism represented in Figure 2. In this figure: x and x_m represent the trolley position and the suspended load (payload), respectively, relative to a predefined referential, m_1 and m_2 are the load and trolley masses, respectively, l the cable length, θ the swing angle and F the applied force to the trolley.

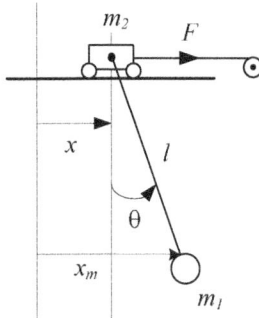

Fig. 2. Schematic representation of the trolley crane mechanism

The mathematical model for this system can be derived from the Lagrange equations based on the kinetic and potential energy [10]. Neglecting the cable elongation due to the tension force and considering x and θ generalized coordinates, the key dynamic equations are (3-7), with d representing the damping coefficient and g is the gravitational constant.

$$(m_1 + m_2)\ddot{x} + m_1 l(\ddot{\theta}\cos\theta - \dot{\theta} - \dot{\theta}^2 \sin\theta) = F \tag{3}$$

$$l\theta + d\theta + \ddot{x}\cos\theta + g\sin\theta = 0 \tag{4}$$

$$l\theta + d\theta + \ddot{x} + g\theta = 0 \tag{5}$$

$$x_m = x + l\sin\theta \tag{6}$$

$$y_m = -l\cos\theta \tag{7}$$

As this study aims to control the payload position by actuating in the trolley position, it is important to derive a transfer function relating its position with the cable swing angle. Equation (4) can be simplified to (5) assuming that the swing angle is kept small and thus: $\sin\theta \cong \theta$ and $\cos\theta \cong 1$. Representing equation (5) in the Laplace complex domain equation (8) it is simple to derive the transfer function represented by (9), which allows to evaluate the swing angle from the trolley position.

$$\Theta(s)(ls^2 + ds + g) = -s^2 X(s) \tag{8}$$

$$\frac{\Theta(s)}{X(s)} = \frac{-\dfrac{1}{l}s^2}{s^2 + \dfrac{d}{l}s + \dfrac{g}{l}} \tag{9}$$

Equation (9) can be rewritten using the standard second order transfer function notation, represented by (10), where ω_n (11), represents the undamped natural frequency, and ζ the damping factor. The suspended load Cartesian coordinates can be evaluated using (6,7), or any approximated expression.

$$\frac{\Theta(s)}{X(s)} = \frac{-\left(\frac{\omega_n^2}{g}\right)s^2}{s^2 + 2\zeta\omega_n s + \omega_n^2} \qquad (10)$$

$$\omega_n^2 = \frac{g}{l} \quad 2\zeta\omega_n = \frac{d}{l} \qquad (11)$$

One of the control objectives associated to the system illustrated in Figure 2, is to move the suspended load from an initial position to a final position. The optimal control methodology should allow the suspended load to be moved as fast as possible and without overshooting the final position. This is a set-point tracking problem, which can be controlled using an open-loop feedforward strategy, as illustrated by Figure 3 a), originally proposed by Smith [2], known as Posicast Control. The former technique can be integrated within a two-degrees control configuration as illustrated in Figure 3 b), in which the feedback loop takes care of the disturbance rejection and model uncertainty mismatch. In Figure 3, x_r represent the input reference signal, u the control signal and x_m the payload position. This study concerns the design of the feedforward Posicast shaper using a PSO algorithm.

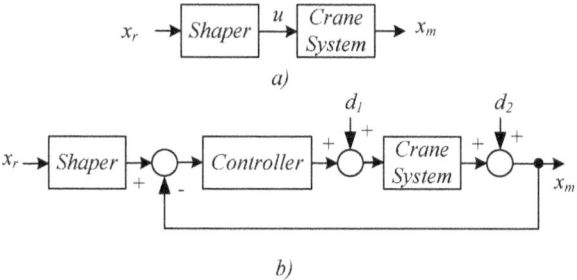

Fig. 3. Crane payload position control: a) using a open-loop feedforward approach, b) using a two-degree of freedom closed-loop configuration

4   Posicast Input Command Shaping

Consider the illustration presented in Figure 4 a), where it is shown a crane in the initial and final position, I and F, respectively. Considering the trolley moving to the right (forward direction), from the initial position to the final position, fast enough, when it stops the suspended load is prone to oscillate, as illustrated by Figure 4, b). The input command which represents the crane position change is a step with amplitude normalized to one, for analysis convenience. Smith, had the original idea to split the input step command in two components. The first input component puts the trolley in an intermediate position, H, between I and F, as illustrated by Figure 4, c). The payload reaches the final position by inertia, and the second input step component puts the trolley in the final position, without oscillation and avoiding overshoot. This type of control was termed half-cycle Posicast.

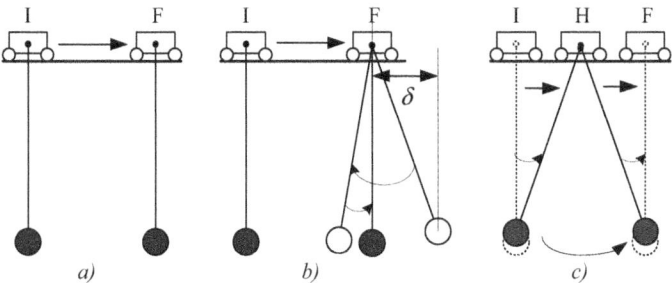

Fig. 4. Half-cycle analogy using a crane with a suspended load: a) Initial (I) and final (F) positions without load oscillation, b) Final position reached with load oscillation, c) Half-cycle control without load oscillation

Assuming, as an illustrative example, that the gantry has a standard second order system dynamics (12), the unitary open-loop underdamped step response is presented in Figure 5. For this underdamped second order response the first overshoot time, or peak time is evaluated by (13), which is equal to half the underdamped time period as represented by (14).

$$G_1(s) = \frac{Y(s)}{U(s)} = \frac{4}{s^2 + 1.4s + 4} = \frac{\omega_n^2}{s^2 + 2\zeta\omega_n + \omega_n^2} \tag{12}$$

$$\delta = e^{\left(-\frac{\zeta\pi}{\sqrt{1-\zeta^2}}\right)} = 0.31 \tag{13}$$

$$T_p = \frac{T_d}{2} = \frac{\pi}{\omega_n\sqrt{1-\zeta^2}} = 1.68s \tag{14}$$

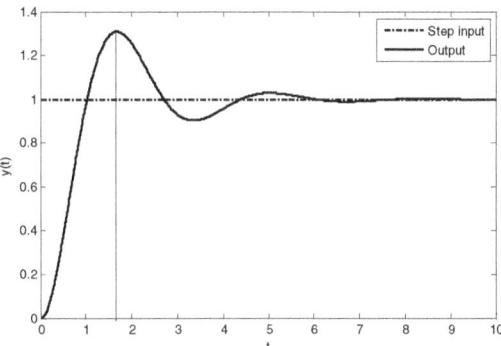

Fig. 5. Unit step open-loop step response for the second-order system (12)

The way Posicast half-cycle splits the unit input step in two is represented by (15-18), in which the second step component is delayed, in relation to the first component, by half the underdamped time period, T_d. In equations (16-17), r, represents the input command reference signal. The half-cycle Posicast input shaping can be represented by a block diagram shown in Figure 6, and the transfer function, P, representing the half-cycle Posicast is represented by (19).

$$u(t) = u_1(t) + u_2(t) \tag{15}$$

$$u_1(t) = \frac{1}{1+\delta} r(t) = A_1 = \frac{1}{1+0.31} = 0.7634 \tag{16}$$

$$u_2(t) = \frac{\delta}{1+\delta} r(t) = \frac{0.31}{1+0.31} = 0.2366 \quad t \geq \frac{T_d}{2} \tag{17}$$

$$A_1 + A_2 = 1 \tag{18}$$

$$P_{hc}(s) = \frac{U(s)}{R(s)} = A_1 + A_2 e^{-\frac{T_d}{2}s} = A_1 + (1-A_1)e^{-\frac{T_d}{2}s} \tag{19}$$

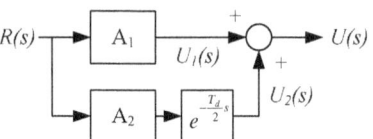

Fig. 6. Half-cycle open-loop block diagram

Applying the two step components to system (12) separately, evaluating the system response as the sum of each output (as the system is linear), results in the Posicast response illustrated in Figure 7, without overshoot.

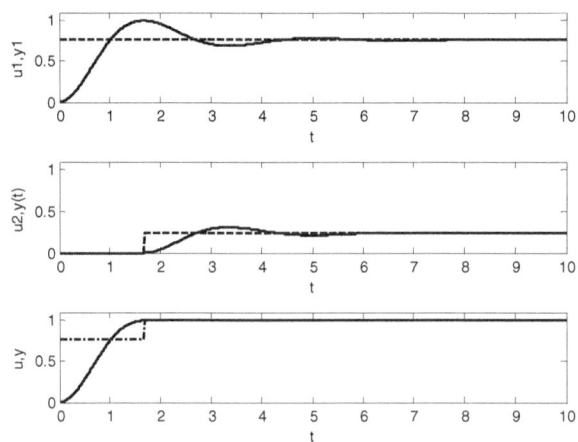

Fig. 7. Unit step open-loop step response input for the second-order system with half-cycle Posicast command shaping

Smith [2] also proposed a more elaborated Posicast shaper, using three steps, termed quarter-cycle Posicast, which is not so very well known and used as the half-cycle case. Consider the example illustrated in Figure 4. In this case the trolley is moved forward to a position with passes over the final desired position, F, moved back to an intermediate position, H, and then moved forward again to the final position. The model is represented by (20), in which the transition times are tr_1 and tr_2.

$$P_{qc}(s) = A_1 + A_2 e^{-tr_1 s} + A_3 e^{-tr_2 s} \tag{20}$$

$$A_1 + A_2 + A_3 = 1 \tag{21}$$

As the original Posicast input command shaping technique was designed for pure second order system, its design parameters, the steps amplitudes and the transition time are prone not to work for other system dynamics, or when there are model mismatches. In this application as the system is second-order the classical half-cycle shaping works quite well and there is no benefit in applying the PSO algorithm. However for the quarter cycle case, the PSO can be used with benefit, not only to select the amplitude of the steps, as well as their transition instants.

5   Teaching Experiment

In this section the PSO algorithm is deployed as an optimization tool to solve the half-cycle Posicast input command shaping problem. While issues concerning the PSO algorithm heuristics are very important to be taught within this type of experience, this paper focus is in the automatic control aspects, with a standard PSO being used as an optimization tool. The simulation experiment main objective is to design input command shapers for the gantry control problem described. The main learning outcomes are:

   i)   To design a half-cycle Posicast controller using the Smith technique.
   ii)   To design a quarter-cycle Posicast controller using the PSO technique.
   iii) To simulate the movement of the crane through an animated Matlab program
         in order to see the design performed with both methods.

The gantry system to be simulated assumes that the cable length is $l=0.6m$ a damping coefficient which gives an approximate transfer function expressed by:

$$G_2(s) = \frac{\Theta(s)}{X(s)} = \frac{-\left(16/9.8\right)s^2}{s^2 + 1.6s + 16} \tag{22}$$

The half-cycle Posicast design using an approximated normalized second order model, with $\omega_n=2$ rad/s and $\zeta=0.2$ results in the design parameters of $A_1=0.66$, $A_2=0.34$ and the delay time for the second step, or transition time between steps, $t_r=0.8s$. If these settings are applied to system G_2, the plots presented in Figure 8 are obtained for the payload position and modified input reference signal, for the trolley position. It is important to state, that no physical limitation was imposed to the trolley velocity. Thus, in the presented simulations, the trolley can ideally be placed instantaneously in a specified position, with large swing angles, which would not be obtained otherwise.

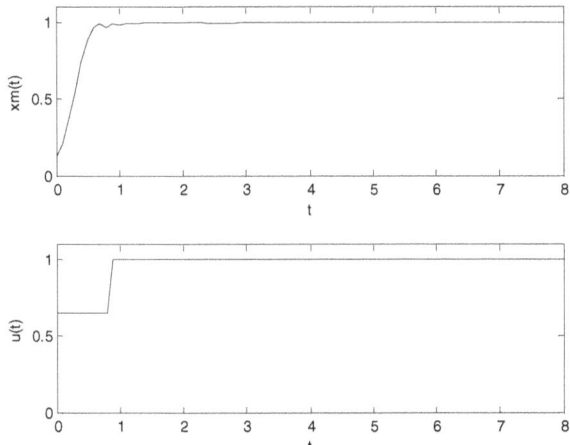

Fig. 8. Step response obtained for the crane suspended load and modified half-cycle input

Figure 9 illustrates part of the gantry system simulation for the half-cycle control. The trolley is moved first forward to the intermediate position, H, corresponding to 66% of the final destination. It rests in that position during 0.8 seconds, and when the suspended load reaches the final position by inertia, moves the trolley to the F position. There is a tiny oscillation in the final position, as the system is not purely second order.

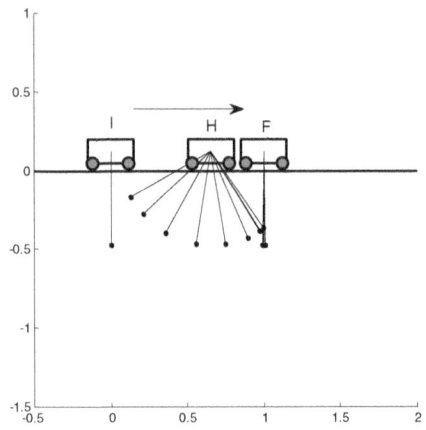

Fig. 9. Simulation of the half-cycle Posicast crane pendulum movement. No physical constraints were imposed on the trolley movement simulation.

A PSO algorithm is now used to design a quarter cycle Posicast shaper. The optimization criterion is the integral of time multiplied by the absolute value of the error, (ITAE) for a step input. The four quarter-cycle design parameters: $\{A_1, A_2, A_3, t_{r2}\}$ are reduced to three parameters.: $\{d_1, d_2, t_{r2}\}$, as $t_{r1}=t_{r2}/2$. The significance and relation of d_1 and d_2 with A_1, A_2 and A_3 are represented in following equation.

$$P_{qc}(s) = A_1 + A_2 e^{-t_{r1}s} + A_3 e^{-t_{r2}s} = (1+d_1) - d_2 e^{-t_{r1}s} + (d_2 - d)e^{-t_{r2}s} \quad (23)$$

A swarm of size 30 was used, randomly initialized considering the following parameter search space [0.1 1], [0.1 1.0], [0.2 1.1] for $\{d_1, d_2, t_{r2}\}$ respectively. The PSO algorithm termination criterion was 100 iterations. The inertia weight was linearly decreased, through the 100 iterations, with $\omega_{init}=0.9$ and $\omega_{fin}=0.4$. The PSO converge for $\{d_1=1.1, d_2=0.834, t_{r2}=0.58\}$. The response obtained with these quarter-cycle Posicast settings is illustrated in Figure 10.

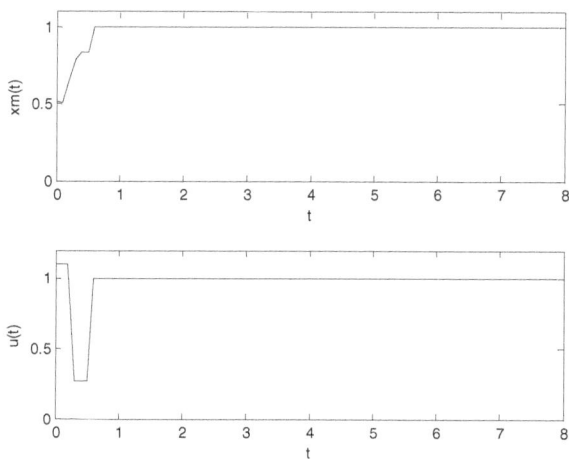

Fig. 10. Step response obtained for the crane suspended load and modified quarter-cycle input. Design using the PSO algorithm.

Figure 11 illustrates the ideal movement achieved with the PSO designed quarter-cycle Posicast input command shaper. The trolley is first moved forward to a position past the final desired position, Q, moved back to and intermediate position, H, and then moves forward again to the final position.

As it can be seen from the animation plot presented in Figure 11, while the oscillation in the playload x-axis, x_m, it is in accordance to the step-point tracking response illustrated by Figure 10, the swing presented in the y-axis variable is not acceptable for practical crane-control. This can be avoided incorporating the maximum swing angle variation and the y_m variable within the design, and considering the trolley physical movement constraints. However, as a didactic experience purpose the results are very interesting. From the comparison between the half-cycle and quarter-cycle Posicast case, the simulation results indicate a possible reason why the latest, it is not so well known. The reason may be associated with the aggressiveness of the quarter-cycle control, which while acceptable for some control applications, may compromise its application for crane-control.

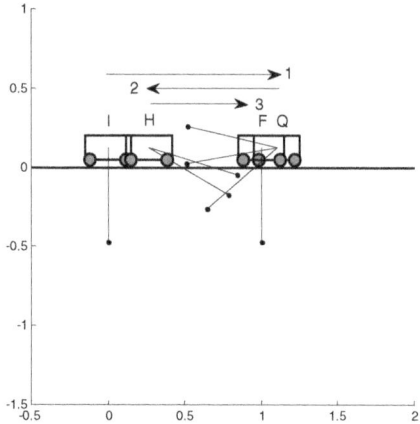

Fig. 11. Simulation of the PSO designed quarter-cycle Posicast crane pendulum movement. No physical constraints were imposed on the trolley movement simulation.

6 Conclusion

In this paper a particle swarm optimization algorithm was proposed and deployed in designing Posicast feedforward controllers for the gantry problem. Preliminary results were presented which indicate that the PSO can successfully design quarter-cycle Posicast controllers.

The crane suspended load example presented, controlled by Posicast input command shaper, validated with the simulation of its movement to prove that the final position overshoot is eliminated, works very well as a didactic experience to teach second-order systems control.

Further work is necessary by carrying out a battery of tests in order to consolidate the results obtained, and to refine the PSO design as well as to fully propose an experiment protocol. The physical constraints related to the trolley velocity and swing payload angles must be incorporated within the design and animation script.

References

1. Stergiopoulos, J., Manesis, S., Tzes, A., Nikolakopoulos, G.: Control via Input Shaping of a Pneumatic Crane System. In: 2005 American Control Conference, Portland, OR, USA, pp. 545–550 (2005)
2. Smith, O.J.M.: Posicast Control of Damped Oscillatory Systems. Proc. IRE 45(9), 1249–1255 (1957)
3. Singhose, W.: Command Shaping for Flexible Systems: A Review of the First 50 Years. Int. Journal of Precision Eng. and Manufacturing 10(4), 153–168 (2009)
4. Hung, H.Y.: Posicast Control Past and Present. IEEE Multidisciplinary Engineering Education Magazine 2(1), 7–11 (2007)
5. Kennedy, J., Eberhart, R.C.: Particle swarm optimization. In: Proc. IEEE Int'l. Conf. on Neural Networks, IV, pp. 1942–1948. IEEE Service Center, Piscataway (1995)

6. Clerc, M., Kennedy, J.: The particle swarm-explosion, stability, and convergence in a multidimensional complex space. IEEE Transactions on Evolutionary Computation 6(1), 58–73 (2002)
7. Banks, A., Vincent, J., Anyajoaha: A review of particle swarm optimization. Part I: background and development. Natural Computing 6, 467–484 (2007)
8. Reyes-Sierra, M., Coello Coello, A.A.: Multi-Objective Particle Swarm Optimiz-ers: A Survey of the State-of-the-Art. International Journal of Computational Intelligence Research 2(3), 287–308 (2006)
9. Bratton, D., Kennedy, J.: Defining a Standard for Particle Swarm Optimization. In: Proc. of the 2007 IEEE Swarm Intelligence Symposium, pp. 120–127 (2007)
10. Solihin, M.I., Wahyudi: Sensorless Anti-swing Control for Automatic Gantry Crane System: Model-based Approach. International Journal of Applied Engineering Re-search 2(1), 147–161 (2007) ISSN 0973-4562
11. Reynolds, M.C., Meckl, P.H., Yao, B.: The Educational Impact of a Gantry-Crane Project in an Undergraduate Controls Class. In: Proceedings of IMECE 2002, ASME International Mechanical Engineering Congress & Exposition, New Orleans, Louisiana (November 2002)
12. Coelho, P.C., De Moura Oliveira, P.B., Boaventura Cunha, J.: Greenhouse Air Temperature Control using the Particle Swarm Optimization Algorithm. Computers and Electronics in Agriculture 49, 330–344 (2005)
13. Azevedo, F., Vale, Z.A., Moura Oliveira, P.B., Khodr, H.M.: A long-term management tool for electricity markets using swarm intelligence. Electric Power Systems Research 80(4), 380–389 (2009)
14. Moura Oliveira, P.B., Vrančić, D., Boaventura Cunha, J., Solteiro Pires, E.J.: Teaching particle swarm optimization through an open-loop system identification project. In: Computer Applications in Engineering Education. Wiley (article first published online, 2011), doi:10.1002/cae.20549
15. Shi, Y., Eberhart, R.C.: A Modified Particle Swarm Optimizer. In: Proceedings of the IEEE Congress of Evolutionary Computation, pp. 69–73 (1998)

Evolving Reaction-Diffusion Systems on GPU

Lidia Yamamoto[1], Wolfgang Banzhaf[2], and Pierre Collet[1]

[1] LSIIT-FDBT, University of Strasbourg, France
{Lidia.Yamamoto,Pierre.Collet}@unistra.fr
[2] Computer Science Department, Memorial University of Newfoundland, Canada
banzhaf@mun.ca

Abstract. Reaction-diffusion systems contribute to various morphogenetic processes, and can also be used as computation models in real
and artificial chemistries. Evolving reaction-diffusion solutions automatically is interesting because it is otherwise difficult to engineer them to
achieve a target pattern or to perform a desired task. However most of
the existing work focuses on the optimization of parameters of a fixed
reaction network. In this paper we extend this state of the art by also
exploring the space of alternative reaction networks, with the help of
GPU hardware. We compare parameter optimization and reaction network optimization on the evolution of reaction-diffusion solutions leading
to simple spot patterns. Our results indicate that these two optimization
modes tend to exhibit qualitatively different evolutionary dynamics: in
the former, the fitness tends to improve continuously in gentle slopes,
while the latter tends to exhibit large periods of stagnation followed by
sudden jumps, a sign of punctuated equilibria.

1 Introduction

In 1952 Alan Turing [27] proposed reaction-diffusion (RD) as a possible mathematical explanation for morphogenetic processes in nature, especially the formation of patterns on the skin of animals, such as zebra stripes and leopard spots.
In a reaction-diffusion system (RDS), a number of chemicals (morphogens) diffuse and react in two or three dimensions within a chemical medium. An RDS
is a dynamical system usually described by a set of partial differential equations
(PDEs) that quantify the speed of diffusion of chemicals and their reactions. Under some conditions, the equilibrium instability in an RDS may lead to spatial
patterns such as Turing patterns. Other well-known RD patterns include circles
and spirals in the Belousov-Zhabotinsky reaction, self-replicating spots in the
Gray-Scott system [10, 23], and diverse patterns on the surface of sea shells [18].
More complex morphogenetic processes such as the formation of net patterns in
leafs or veins [7], the formation of insect eyes and various body parts, are only
partially explained by reaction-diffusion processes, nevertheless RD remains an
important component of morphogenesis in general.

Several applications of reaction-diffusion systems exist in the literature: first,
as abstract models of biological pattern formation [6, 17, 18, 21]: feathers, leaves,
veins, nerves, body segments, and so on. More recently, they are considered as

L. Antunes and H.S. Pinto (Eds.): EPIA 2011, LNAI 7026, pp. 208–223, 2011.

new substrates for unconventional computation on real chemical media, such as reaction-diffusion computers [1]: algorithms such as image processing *in vitro*, shortest path, robot controllers, and Voronoi diagrams have been proposed on top of such chemical computers. The potential of RD in micro- and nanotechnology is reviewed in [11], such as the fabrication of structures and devices at very small scales. RD can also be used in models of distributed computation inspired by chemistry, with applications to sensor networks (role assignment [22], routing [15]), and robotics (robot controllers [5], swarm robotics [25]).

Although there is a vast literature on the mathematical analysis of various reaction-diffusion systems, their overall design space remains poorly understood: even very simple systems like the Gray-Scott system (composed essentially of two chemicals and one autocatalytic reaction between them) lead to complex analytical solutions and a very narrow region of the parameter space in which interesting patterns occur. Therefore in many cases it makes sense to sweep RD parameter spaces using search heuristics such as evolutionary algorithms.

There are more reasons to evolve reaction-diffusion systems. Given a desired pattern, it is difficult to find a set of chemicals, their reactions and corresponding parameters (speed of diffusion and reaction) that lead to the target pattern. In addition, in the literature sometimes only the PDEs of an RDS are given: deriving the corresponding chemical reactions is straightforward in some cases, but difficult in others. Beyond parameter sweeping with a Genetic Algorithm (GA) or other techniques, it is worth exploring the RDS design space by finding the appropriate set of chemical reactions leading to a given target pattern. This is analogous to finding a program (here, a chemical program) that solves a given problem, therefore it can be considered as a form of genetic programming (GP). A GP-based approach to RD evolution covers a much larger search space than a GA-based one. Whereas the numeric integration of RD PDEs is already computationally expensive, evolving a population of these PDEs is even more demanding. Luckily, both GP and RD numeric integration are well suited for parallelization on top of GPU (Graphics Processing Unit) hardware [2, 16, 20, 24].

In this paper we compare (experimentally) parameter optimization (GA analog) with reaction network optimization (GP analog) to explore the space of RD solutions forming spot patterns. Such comparison is enabled by a hybrid CPU-GPU evolutionary algorithm that makes the problem tractable with a modest investment in computation resources. The paper is structured as follows: Section 2 provides background on RDS, their evolution and parallelization. Section 3 presents our approach to evolving RDS. Section 4 reports our evolution experiments and discusses their results. Section 5 concludes the paper.

2   Reaction-Diffusion Systems

An RDS is a chemical reaction system in which substances react and diffuse in space. The movement of molecules, their collisions and reactions are stochastic processes at the microscopic level. At the macroscopic level though (for large numbers of molecules) the system can be expressed as a set of PDEs describing the change in concentrations of substances caused by both reaction and diffusion effects combined:

$$\frac{\partial s_i(\boldsymbol{p}, t)}{\partial t} = f_i(s_i(\boldsymbol{p}, t)) + D_i \nabla^2 s_i(\boldsymbol{p}, t) \tag{1}$$

where s_i is the concentration level at time t of each chemical S_i at position $\boldsymbol{p} = (x, y, z)$. The reaction term $f_i(s_i(\boldsymbol{p}, t))$ describes the reaction kinetics for chemical S_i at each point $\boldsymbol{p}$. The diffusion term $D_i \nabla^2 s_i(\boldsymbol{p}, t)$ tells how fast each chemical substance diffuses in space. D_i is the diffusion coefficient of S_i (a constant scalar in the case of isotropic diffusion), and ∇^2 is the Laplacian operator.

In the simplest case, the dynamics of the reaction term (described by the set of functions f_i) follows the *Law of Mass Action*: it states that, in a well-stirred reactor, the average speed (or rate) of a chemical reaction is proportional to the product of the concentrations of its reactants. For n chemicals and m reactions, the system of differential equations for the reaction terms can be described in matrix notation as:

$$\frac{d\boldsymbol{s}(t)}{dt} = \mathbf{M}\boldsymbol{v}(t) \tag{2}$$

where $\mathbf{M}$ is the stoichiometric matrix of the system, which expresses the net changes in number of molecules for each species $S_i, 1 \leq i \leq n$ in each reaction $R_j, 1 \leq j \leq m$; and $\boldsymbol{v}(t) = \{v_1, ... v_j, ... v_m\}$ is a vector of rates for each reaction, for instance, following the Law of Mass Action:

$$v_j = k_j \prod_{1 \leq l \leq n} s_l^{\mathbf{M}_{e(l,j)}} \tag{3}$$

where k_j is kinetic or rate coefficient of reaction R_j (constant for our purposes), and $\mathbf{M}_e$ is the educt stoichiometric matrix where each element $\mathbf{M}_{e(l,j)}$ expresses the consumption of molecules of S_l in reaction R_j. This provides a simple and automatic way to obtain the system of PDEs from a given set of chemical reactions. The system can then be integrated numerically by discretizing the equation terms in space and time.

2.1   Activator-Inhibitor Models

Activator-inhibitor models [17] are among the simplest reaction-diffusion systems known. The experiments described in this paper focus on the automatic evolution of solutions that fall into this class. The activation-inhibition effect can be achieved by two interacting morphogens: an activator catalyzes its own production and produces an inhibitor that inhibits the autocatalytic action of the activator. Moreover the inhibitor travels faster than the activator, leading to a short-range activation and long-range inhibition effect that under some conditions [21] results in the formation of spot and stripe patterns.

Numerous alternative activator-inhibitor models are available in the literature. A lot is known about specific cases, however much remains to be explored in the vast design space of reaction-diffusion systems that form given patterns.

A well-known activator-inhibitor model is the one by Gierer and Mein-hardt [17], described by the following equations:

$$\frac{\partial a}{\partial t} = \frac{\sigma a^2}{h} - \mu_a a + \rho_a + D_a \nabla^2 a \tag{4}$$

$$\frac{\partial h}{\partial t} = \sigma a^2 - \mu_h h + \rho_h + D_h \nabla^2 h \tag{5}$$

where a and h are the concentrations of activator (A) and inhibitor (H), respectively; μ_a and μ_h are the decay rates for each substance; ρ_a and ρ_h are their inflow rates; D_a and D_h are their respective diffusion coefficients.

Note that the inhibition factor $1/h$ does not stem directly from the Law of Mass Action. In [28] the corresponding chemical reactions are derived for the above equations, by introducing a catalyst C that is needed in the autocatalysis of A but is consumed by H, leading to the expected inhibition effect:

$$2A \xrightarrow{\sigma} 2A + H \tag{6}$$

$$C + 2A \xrightarrow{k_1} C + 3A \tag{7}$$

$$C + H \xrightarrow{k_2} H \tag{8}$$

The factor $1/h$ is obtained by assuming that the catalyst C is in steady state with constant concentration, and by setting $\sigma = k_1 \rho_c / k_2$ (where ρ_c is the injection rate of C). C neither decays ($\mu_c = 0$) nor diffuses ($D_c = 0$).

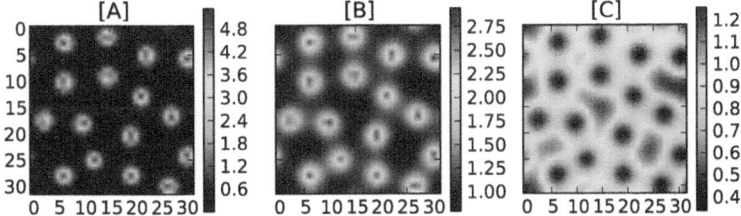

Fig. 1. Activator-inhibitor pattern

Figure 1 shows a typical pattern formed using our implementation of this model, using the parameters from [13, 28], namely: $\sigma = 0.02$, $\rho_a = \rho_h = 0$, $\rho_c = 0.1$, $\mu_a = 0.01$, $\mu_h = 0.02$, $\mu_c = 0$, $k_1 = 0.01$, $k_2 = 0.1$, $D_a = 0.005$, $D_c = 0$, $D_h = 0.2$. [A], [B] and [C] represent the concentrations of activator, inhibitor and catalyst, respectively, plotted as heatmaps (colors indicate concentration levels at each point in space, according to the colorbars to the right of each plot). One can see that the activator peaks coincide with the inhibitor peaks, which are smoother. Moreover the activator/inhibitor peaks coincide with the catalyst valleys, consistent with the fact that the inhibitor consumes such catalyst.

2.2   Evolving Reaction-Diffusion Solutions

Most interesting RDSs are open systems: they rely on a constant inflow and out-flow of substances that drives the system away from equilibrium. Moreover they rely heavily on positive and negative feedback (e.g. activation and inhibition), and often result in non-linear PDEs. Combined with the high computation demands on their numeric integration, evolving them automatically becomes very challenging.

The evolution of spot and stripe patterns is presented in [9], via the optimization of parameters of transition rules that emulate Turing patterns on a binary Cellular Automaton (CA). Such CA emulation is simpler and faster than full RD, but sheds no light on the chemical processes behind the observed patterns.

RD parameter sweeping for biology has been demonstrated in [12] using the covariance matrix adaptation evolution strategy (CMA-ES), a variant of evolution strategy suitable for sweeping parameters on difficult fitness landscapes. CMA-ES has also been used for tuning parameters in self-configuring morphogenetic modular robots [19] (a system that is loosely inspired by chemistry), and in the numerical simulation of turbulent fluids [8].

RDS solutions for one-dimensional segmentation (head-tail pattern) were evolved in [26]. Four models were compared: three parameter sweeping variants and the evolution of arbitrary differential equations via GP. The equations obtained by GP performed best, however their actual chemical implementation was not covered.

Stochastic particle-based (discrete) simulations of RDS have been used in [3] to study the molecular evolution of replicators. In [3] molecules are strings that may move in space and react due to complementary substrings, leading to an emergent evolutionary dynamics in space.

RD controllers for robots were evolved in [5], based on the Gray-Scott system in one dimension (ring). Chemicals in the RD ring were used to activate the robot's motors in response to sensor input. The wiring from the sensors to the RD ring were evolved (not the actual RDS).

To the best of our knowledge, the use of evolutionary optimization to discover new chemical reaction networks implementing desired patterns has not been reported so far in the literature (most cases [8, 12, 19] are restricted to parameter sweeping; novel reaction networks may emerge in [3] but with no assigned computation task; arbitrary equations are evolved in [26] but with no guaranteed chemical analog). Our contribution represents a step in this direction.

2.3   Reaction-Diffusion on GPU

GPU approaches to parallel PDE solving for advection-reaction-diffusion equations are discussed and evaluated in [24]. Besides reaction and diffusion, these systems take into account advection phenomena in which chemicals also move by fluid transport, such as pollutants in the air or water. An implementation of reaction-diffusion in three dimensions is reported in [20]. Accompanying source code is publicly available for both [20, 24], however both are far more elaborate than what we needed for our own experiments. Moreover, none of these implementations covered the execution and fitness evaluation of multiple reaction-diffusion individuals in parallel on the same GPU card. Therefore we have decided to reimplement our own solution from scratch. To the best of our knowledge, the evolution of RDS on GPUs has not been reported so far.

3   RD Evolutionary Algorithm

In order to cope with the high computation demands of evolving RDS, we have developed an evolutionary algorithm that runs on the CPU and evaluates the population of candidate solutions (individuals) on the GPU. Two levels of parallelism are exploited: population (multiple individuals on the GPU) and individual (parallel PDE integration on a 2D surface).

As a developmental system, an RDS implies a separation of genotype (the set of reactions that is evolved) and phenotype (the resulting pattern on the surface). The genotype and phenotype encoding are described in Section 3.1, the genetic operations applied to the genotype in Section 3.2, and the fitness evaluation process in Section 3.3.

3.1   Genotype and Phenotype

An individual in the population consists of a genotype, a phenotype and a fitness value. The genotype or genome encodes the reactions and parameters of the RDS. The genome of the individuals is shown in Fig. 2(a). It consists of four elements: the vector of kinetic coefficients $k = \{k_1, ...k_m\}$ for each reaction; the vector of diffusion coefficients $D = \{D_1, ...D_n\}$ for each substance; the educt stoichiometric matrix $\mathbf{M}_e$; and the product stoichiometric matrix $\mathbf{M}_p$. Both matrices have size $n \times m$. An element $\mathbf{M}_e(i, j)$, respectively $\mathbf{M}_p(i, j)$, tells how many molecules of species i are consumed (resp. produced) in reaction j. For example, reaction R_2 in Fig. 2(a) takes two molecules of S_1 and one of S_2 to produce 3 molecules of the last species S_n. The matrices $\mathbf{M}_p$ and $\mathbf{M}_e$ together describe all the reactions that occur in the system (without their rates k_j). The net stoichiometric matrix $\mathbf{M}$ (needed for the PDE integration as shown in Eq. 2) can then be simply calculated as $\mathbf{M} = \mathbf{M}_p - \mathbf{M}_e$.

The phenotype is the pattern that results from numerically integrating the PDE of the corresponding genome on a 2D surface of $N_x \times N_y$ points (cells) arranged on a regular grid. More specifically, it corresponds to $s(p, t)$, the set of n matrices of size $N_x \times N_y$ containing the concentrations of each of the n species at each point p on the surface, at an observed time t. The phenotype changes in time as the pattern takes shape starting from the initial conditions. After allowing some initial time for a potential pattern to form, the fitness evaluation of the individual starts in periodic rounds. At the end of the evaluation period, the average of all the periodic evaluations is computed, and the obtained fitness value is assigned to the corresponding individual. At every generation, all the individuals are evaluated under new, random initial conditions. The same initial

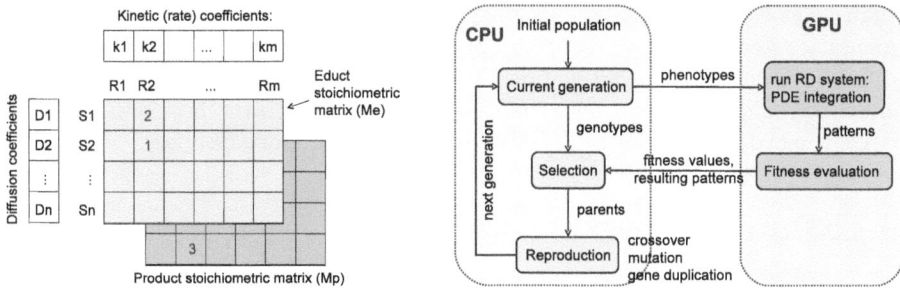

(a) genome of RD individuals   (b) CPU-GPU division of roles

Fig. 2. Elements of the reaction-diffusion CPU-GPU evolutionary algorithm

conditions are applied to all individuals of the same generation, like exposing them to the same environment. Even if an individual may survive from one generation to the next, its fitness is re-evaluated under new conditions. Therefore only those individuals who have a good fitness under all faced situations can survive in the long run.

3.2   Genetic Operators

Three types of genetic operators are used: mutation, crossover and gene duplication. The maximum number of chemicals (n) and reactions (m) is constant, for both parameter and network optimization cases.

For the case of parameter sweeping, only the coefficients in k may be modified by mutation and crossover (gene duplication is disabled). Since these coefficients typically vary by several orders of magnitude, they are expressed in scientific notation, with integer mantissa and exponent fields that may be mutated independently. A mutation increments or decrements one of these fields. A circular crossover is used, in which the vector of coefficients is treated as a ring; two points are chosen on each parent's ring and their segments are swapped.

For the case of reaction network evolution, in addition to k, the matrices $\mathbf{M}_e$ and $\mathbf{M}_p$ may be modified too. A mutation may increment or decrement one element of $\mathbf{M}_e$ or $\mathbf{M}_p$, which results in modifying the corresponding chemical reaction. For instance, incrementing $\mathbf{M}_e(i,j)$ (resp. $\mathbf{M}_p(i,j)$) means that now an additional molecule of S_i is needed (resp. is produced) in reaction R_j. Circular crossover can also be applied: in this case, individuals swap the full description of the affected reaction ($\mathbf{M}_e$ and $\mathbf{M}_p$ columns, plus corresponding rate coefficient).

For network evolution, the population is initialized with random genomes containing a small number of reactions. Gene duplication may then be used to create new reactions from existing ones. In this case, the child individual will contain two identical reactions that may diverge in subsequent mutations.

Currently the diffusion coefficients are taken as fixed, i.e. they are not evolved, assuming that the diffusion speed is an inherent property of the mobility of the chemical in the medium due to its shape or size.

3.3   Fitness Evaluation

Since we are interested in the computation properties of RDS, a task-based fitness function was chosen. This function measures the capacity of the RDS to perform a desired computation task. The goal is then to maximize the individual's performance on this task. The RDS is then regarded as a chemical algorithm performing a distributed computation. The goal task chosen for this paper is a distributed cluster head election as described in [28]: the cells where activator peaks are found in spot patterns are "elected" as "cluster heads" (local leaders) of their region in space.

The cluster head evaluation function works as follows. First, four cell types are distinguished: *invalid* (with concentrations outside bounds, inf (infinity), or nan ("not a number" numerical error)); *peak* (valid cell with activator concentration above a threshold); *alive*: valid non-peak cell sufficiently close to a peak; *dead* (a valid non-peak cell without any nearby peaks). The fitness Φ of a single cell is then computed by respecting the following relations:

$$\Phi(\text{invalid}) < \Phi(\text{dead}) < \Phi(\text{peak}) < \Phi(\text{alive}) \tag{9}$$

Therefore a pattern containing the maximum number of alive cells has the maximum fitness. However, alive cells rely on peak cells, therefore peaks must be present too. Dead cells must be penalized with a low fitness, but invalid cells are even worse, so they should receive a larger penalty. The global fitness of the pattern (individual) is then the sum of the fitness values of all cells on the surface. The actual Φ values adopted in the experiments are reported in Sec. 4.

In RDS, there is a vast region of the search space in which no patterns occur [12], or which lead to misbehaving solutions, for instance, solutions that include autocatalytic reactions that quickly exhaust the available resources and produce infinity values. In order to eliminate these invalid individuals as soon as possible, an incremental fitness evaluation approach is adopted, with three stages: First, the individual is tested for having concentrations within valid bounds. Second, it is tested for the presence of any non-homogeneous pattern. At the third and last stage comes the evaluation against the real task according to the above relations (9). Only when the first stage is overcome can the individual move to the second stage, and so on.

3.4   CPU-GPU Evolutionary Algorithm

In order to cope with the intensive computation requirements of RD evolution, the PDE integration and most of the fitness evaluation are delegated to a GPU card. The resulting hybrid CPU-GPU evolutionary algorithm is shown in Fig. 2(b).

The initial population is generated on the host computer, together with the initial conditions for the patterns. The phenotypes are then transferred to the GPU where they are integrated for pattern formation. PDE integration is implemented as kernel routine on the GPU, where each thread is responsible for numerically integrating Eq. 1 for one point on the grid of cells.

The fitness of the obtained patterns is then evaluated on the GPU, as follows: First, each thread computes the local fitness value of its cell by observing its concentration values and the concentration values in the nearby cells. After all threads have completed this step, the first thread of each block computes the sum of the fitness values of all the cells in the block. These values are then returned to the CPU, where the fitness of all the blocks is summed up to obtain the global fitness of the pattern. This process is repeated for a number of evaluation rounds, at the end of which the fitness of the individual is computed as the average fitness of all the evaluations.

After all the fitness values are computed, a selection process takes place, during which individuals compete in tournaments. The winners of the tournaments gain the right to reproduce via mutation, crossover and/or gene duplication, in order to produce the next generation, which is then re-evaluated on the GPU, and so on, until the maximum amount of generations is reached.

The PDE integration routine deserves special attention since it is the most computationally demanding part of the evolutionary algorithm. It typically consumes over 99% of the total GPU plus CPU computation time. Therefore it is important to optimize its performance. For this purpose, the stoichiometric matrices, reaction and diffusion coefficients are placed in the GPU block's shared memory. This results in considerable performance improvements, as frequently reported in the GPU literature. The experiments were run on a host containing three NVIDIA GTX 480 GPU devices. For the pure PDE integration without evolution, the typical speedups obtained with the GPU implementation range from 100 to 180 times the CPU runtime, for one RD individual filling the whole GPU card (size 128x120 points). A typical evolution run with a population of 150 individuals of 32x32 points over 100 generations still takes around 20 to 50 minutes to complete using the hybrid CPU-GPU evolutionary algorithm presented (against more than 30 hours on a single CPU), depending on the size of the reaction networks involved. Since several runs are usually needed in order to explore the RD design space, such experiments would have been infeasible on a single CPU.

4   Evolution Experiments

We start by sweeping parameter space for spot patterns using the activator-inhibitor system shown in Sec. 2.1. We then explore the space of possible reaction networks leading to such patterns. The results are presented in Secs. 4.1 and 4.2 respectively. Table 1 lists the parameters of the experiments. The population size, number of generations and grid surface were set to the minimum values that still showed some evolutionary behavior, while keeping the runtimes short enough (below one hour per run). The mutation and crossover probabilities were set to the typical ranges used in GP, whereas the gene duplication probability was set low enough according to the results in [4]. The tournament size was set to the minimum, in order to reduce the selection pressure in an attempt to prevent premature convergence, which was very commonly observed in these

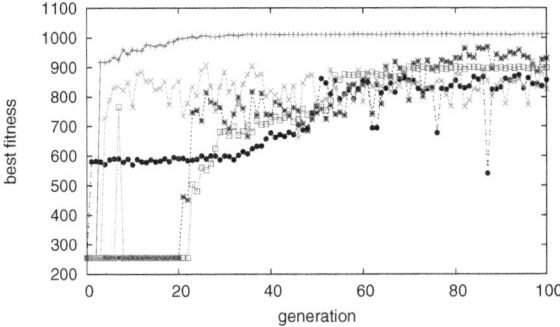

Fig. 3. Selected parameter sweeping runs (each curve corresponds to one run)

experiments. Indeed, we have observed that increasing the selection pressure (up to a tournament of size 7) led more often to premature convergence to non-optimum solutions.

The global fitness of a pattern (sum of the fitness values of each cell) ranges from zero to the total number of grid cells, i.e. 32x32=1024 for the given setup. Peaks are points of maximum activator concentration, above a peak threshold of $a_{peak} = 3.0$. The distance threshold for a cell to be considered as alive is $d_{max} = 6$ cells from the nearest peak. For each cell type, the values of the task-based fitness function per cell are: $\Phi(\text{invalid}) = 0$; $\Phi(\text{dead}) = 0.25$; $\Phi(\text{peak}) = 0.5$ if there is another peak at distance less than d_{max} from itself; $\Phi(\text{peak}) = 0.75$ otherwise; $\Phi(\text{alive}) = 1$.

Table 1. Parameters of the evolution experiments

Evolution parameters		Phenotype parameters	
population size	150 individuals	grid surface	32x32 points
generations	100	PDE integration timestep	$\Delta t = 0.1$ s
selection method	tournament	activator peak threshold	$a_{peak} = 3.0$
tournament size	2 individuals	max. dist. to nearest peak	$d_{max} = 6.0$
mutation prob.	0.1 per reaction	max. valid concentration	$s_{max} = 100$
mutation prob.	0.2 per mating event	Fitness evaluation parameters	
crossover prob.	0.8 per mating event	n. evaluation rounds	10 per indiv.
gene dup. prob.	0.04 per mating event	evaluation start time	at t=2000 s
		evaluation interval	every 200 s

4.1 Sweeping Parameter Spaces

In this set of experiments, we take Reactions 6 to 8 from Sec. 2.1 as given, and let evolution find matching rate coefficients that lead to spot patterns similar to those on Fig. 1. The coefficients are allowed to vary within the interval $0 \leq k < 1000$.

Figure 3 shows the best fitness for the 5 runs that gave the best results. Each curve corresponds to one run. The other runs got stuck at poor fitness values near the bottom, and are not shown for better readability of the plots. These poor runs usually overcome the first fitness stage but either reach flat homogeneous surfaces with no patterns, unstable patterns, or patterns that do not match the fitness criteria well enough.

Note that the fitness curves vary widely across runs, hence showing averages over several runs would make little sense. This is why we have selected a few runs to plot such as to achieve a compromise between the readability of the plots and the amount of information conveyed by them.

The best evolved patterns are shown in Fig. 4, where [A], [B] and [C] represent the concentrations of activator, inhibitor and catalyst, respectively. The rate coefficients for the best solutions found are shown in Table 2, compared to the human-designed parameter values shown in Sec. 2.1. The solution producing Fig. 4 presents tall and narrow activator peaks, and depletes the catalyst almost entirely. It does that by increasing the rate of autocatalytic production of A, and increasing the rate of consumption of C by the inhibitor, when compared to the hand-made case. These narrow peaks are very good for the cluster head task, since the number of alive cells is maximized. This solution is probably more efficient than the human-designed case.

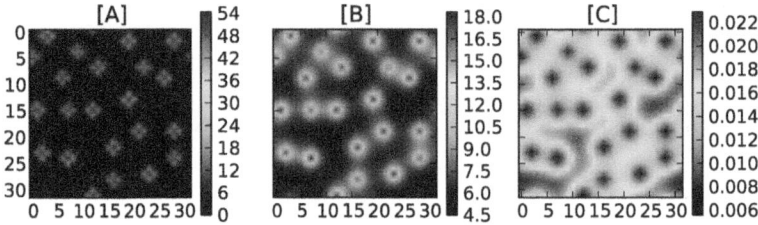

Fig. 4. Evolved patterns for parameter sweeping

Table 2. Evolved vs. hand-made rate coefficients

Reactions	hand-made	Fig. 4
$2A + C \rightarrow 3A + C$	1.0e-2	9.4e-2
$B + C \rightarrow B$	1.0e-1	3.0e-1
$2A \rightarrow 2A + B$	2.0e-2	1.9e-3

Reactions	hand-made	Fig. 4
$\emptyset \rightarrow C$	1.0e-1	3.1e-2
$A \rightarrow \emptyset$	1.0e-2	1.5e-2
$B \rightarrow \emptyset$	2.0e-2	2.0e-2

4.2   Exploring New Reaction Networks

We now let the reaction networks evolve, for a set of $n = 3$ chemicals and up to $m = 4n = 12$ reactions, including inflow and decay reactions. All the individuals have at least n inflow reactions and n decay reactions (one inflow and one decay reaction for each chemical). These reactions are statically defined and cannot be added, deleted nor mutated. Only their rate coefficients may change during evolution. Some coefficients may be imposed. For instance, we disable the injection of activator and inhibitor by setting their coefficients to zero, in order to avoid solutions that "cheat" by artificially injecting these chemicals. We refer to the remaining up to $2n$ reactions as the set of *evolvable* reactions. In order to avoid unfeasible high-order reactions, when mutating evolvable reactions, the maximum stoichiometric coefficient per molecule is kept at 3: $\mathbf{M}_e(i,j) \leq 3$ and $\mathbf{M}_p(i,j) \leq 3 \; \forall i,j$; moreover the maximum number of molecules that participate in a reaction (on either side) is 4: $\sum_i \mathbf{M}_e(i,j) \leq 4$ and $\sum_i \mathbf{M}_p(i,j) \leq 4$ for all reactions R_j.

In the initial population, each individual is initialized with a genome containing a single evolvable reaction (plus the default $2n$ static inflow and decay reactions). More reactions may appear later as genes duplicate and diverge. In this way, we start by exploring the search space of smaller, parsimonious solutions, and then let evolution grow more complex solutions if they happen to be fitter.

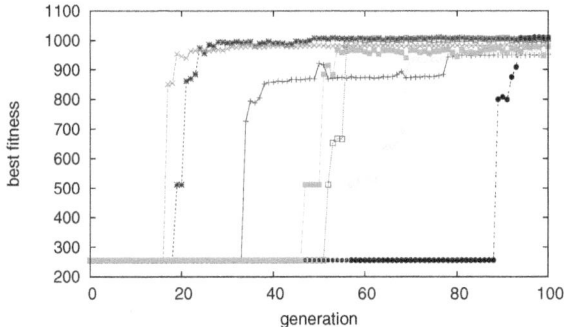

Fig. 5. Selected reaction network evolution runs (each curve corresponds to one run)

Again, a set of 10 runs were performed. Figure 5 shows the best fitness over the generations, for the 7 runs that made progress. The other 3 runs got stuck at the bottom of the plot and have been omitted. Compared to parameter evolution (Fig. 3) a remarkably different qualitative behavior can be noticed: Here the fitness exhibits periods of stagnation followed by steep jumps. Such profile is characteristic of indirect phenotype-genotype-fitness encodings, of which RD is an example, so it should come at no surprise. However, it seems to become significantly more apparent in network evolution than in parameter evolution. A tentative explanation for this might be that network evolution has a much larger range of possibilities of neutral mutations (mutations that have no impact in the fitness) than parameter evolution. These neutral mutations may accumulate over time, until they start to play a role in the fitness. This topic deserves a deeper investigation.

Two of the best evolved patterns are shown in Fig. 6. Their peaks are sharper than those in the hand-made solution of Fig. 1, and even sharper than in the case of Fig. 4, consistent with the fitness function that rewards for a maximum amount of alive cells that must nonetheless be close to a peak. Note that in these

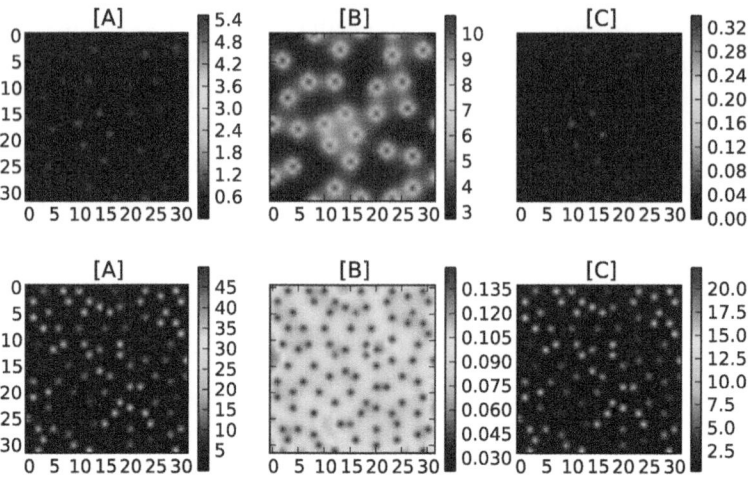

Fig. 6. Patterns stemming from reaction network evolution

Table 3. Two evolved reaction network solutions

Fig. 6 (top)

Reactions	k
$A + C \rightarrow 2A + B$	6.2
$2A \rightarrow A + 2C$	6.4e-2
$2A + B + C \rightarrow B + C$	2.3e-2
$A + B + C \rightarrow \emptyset$	1.0e-1
$A + B \rightarrow \emptyset$	2.0e-2
$B + 2C \rightarrow A + B + 2C$	1.0

Fig. 6 (bottom)

Reactions	k
$A + 2B + C \rightarrow A + B$	5.0e-3
$2C \rightarrow B + C$	0.0
$B + 2C \rightarrow A + C$	0.0
$2B \rightarrow A + B + C$	1.0
$B + 2C \rightarrow A + 2B + C$	1.0
$2A + B \rightarrow 2A + C$	2.0e-1

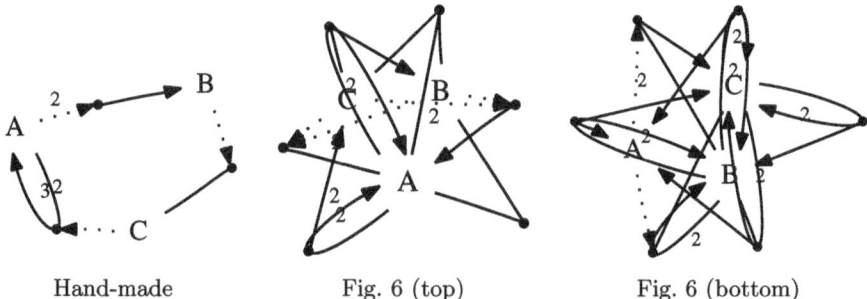

Hand-made   Fig. 6 (top)   Fig. 6 (bottom)

Fig. 7. Best reaction networks evolved (center, right) vs. hand-made solution (left)

solutions the substance C no longer plays the role of catalyst consumed by an inhibitor, since the peaks of C now coincide with those of A.

The corresponding genomes for the patterns of Fig. 6 are listed in Table 3. The respective reaction networks are depicted in Fig. 7, compared with the hand-made case. Only the evolvable reactions are shown. In this figure, a dot indicates a reaction, continuous lines indicate educts, continuous arrows indicate products, dotted arrows indicate catalysis, and numbers indicate stoichiometric coefficients. We can see that the evolved networks are densely connected and redundant, i.e. there are many ways to produce and consume each substance. This is consistent with the fact that the evolved networks must be robust to mutations in the network topology in order to survive to the next generation.

5   Conclusions

RD evolution is a difficult task with huge computation demands. In this paper we have presented experimental results that compare parameter sweeping and reaction network evolution on the activator-inhibitor model. The experiments were made feasible by the parallelization of fitness evaluation and PDE integration on GPU hardware. The outcome of our experiments indicates that the evolutionary dynamics of RD network evolution seems to present remarkable qualitative differences when compared to RD parameter sweeping. Steep jumps in fitness are very frequently observed during network evolution, while similar jumps are only rarely observed in the case of parameter sweeping. This could be an evidence of punctuated equilibria in the evolution of RD networks, and is a topic that deserves further investigation.

The present paper covered only very simple spot patterns. The evolution of more complex patterns will require a number of extensions to our system. Further investigation into the genotype and phenotype structures is needed, moving towards Genetic Regulatory Networks (GRN), and looking at how modularity could emerge in the corresponding chemical reaction networks [4]. Another envisaged extension is to impose mass and energy conservation constraints to reduce the search space to resource-saving and physically plausible solutions. In order to evolve vein and leaf patterns [7, 18], the active transport of chemicals is desirable, beyond passive diffusion. In order to improve convergence speed and escape more easily from local optima, another research topic would be the adaptation of more sophisticated algorithms, such as CMA-ES [12, 19] or the nested evolution algorithm from [14], to the evolution of reaction networks in the RDS context.

Acknowledgments. This work was supported by the French Region Alsace through the EVOL grant.

References

1. Adamatzky, A., Costello, B.D.L., Asai, T.: Reaction-Diffusion Computers. Elsevier Science Inc., New York (2005)
2. Banzhaf, W., Harding, S., Langdon, W.B., Wilson, G.: Accelerating Genetic Programming through Graphics Processing Units. In: Genetic Programming Theory and Practice VI, pp. 1–19. Springer, US (2009)
3. Breyer, J., Ackermann, J., McCaskill, J.: Evolving Reaction-Diffusion Ecosystems with Self-Assembling Structures in Thin Films. Artificial Life 4(1), 25–40 (1998)

4. Calabretta, R., Nolfi, S., Parisi, D., Wagner, G.P.: Duplication of Modules Facilitates the Evolution of Functional Specialization. Artificial Life 6(1), 69–84 (2000)
5. Dale, K., Husbands, P.: The Evolution of Reaction-Diffusion Controllers for Minimally Cognitive Agents. Artificial Life 16(1), 1–20 (2010)
6. Deutsch, A., Dormann, S.: Cellular automaton modeling of biological pattern formation: characterization, applications, and analysis. Birkhäuser (2005)
7. Fujita, H., Mochizuki, A.: The Origin of the Diversity of Leaf Venation Pattern. Developmental Dynamics 235(10), 351–361 (2006)
8. Fukagata, K., Kern, S., Chatelain, P., Koumoutsakos, P., Kasagi, N.: Evolutionary optimization of an anisotropic compliant surface for turbulent friction drag reduction. Journal of Turbulence 9(35), 1–17 (2008)
9. Graván, C.P., Lahoz-Beltra, R.: Evolving morphogenetic fields in the zebra skin pattern based on Turing's morphogen hypothesis. Int. J. Appl. Math. Comp. Sci. 14(3), 351–361 (2004)
10. Gray, P., Scott, S.: Chemical Oscillations and Instabilities: Nonlinear Chemical Kinetics. Oxford Science Publications, Oxford (1990)
11. Grzybowski, B.A., Bishop, K.J.M., Campbell, C.J., Fialkowski, M., Smoukov, S.K.: Micro-and nanotechnology via reaction-diffusion. Soft Matter 1, 114–128 (2005)
12. Hohm, T., Zitzler, E.: A Hierarchical Approach to Model Parameter Optimization for Developmental Systems. BioSystems 102, 157–167 (2010)
13. Koch, A.J., Meinhardt, H.: Biological pattern formation: from basic mechanisms to complex structures. Reviews of Modern Physics 66 (1994)
14. Lenser, T., Hinze, T., Ibrahim, B., Dittrich, P.: Towards Evolutionary Network Reconstruction Tools for Systems Biology. In: Marchiori, E., Moore, J.H., Rajapakse, J.C. (eds.) EvoBIO 2007. LNCS, vol. 4447, pp. 132–142. Springer, Heidelberg (2007)
15. Lowe, D., Miorandi, D., Gomez, K.: Activation- inhibition-based data highways for wireless sensor networks. In: Proc. Bionetics. ICST (2009)
16. Maitre, O., Baumes, L.A., Lachiche, N., Corma, A., Collet, P.: Coarse grain parallelization of evolutionary algorithms on GPGPU cards with EASEA. In: Proc. GECCO, pp. 1403–1410 (2009)
17. Meinhardt, H.: Models of biological pattern formation. Academic Press, London (1982)
18. Meinhardt, H.: The Algorithmic Beauty of Sea Shells, 4th edn. Springer, Heidelberg (2009)
19. Meng, Y., Zhang, Y., Jin, Y.: Autonomous self-reconfiguration of modular robots by evolving a hierarchical mechanochemical model. IEEE Computational Intelligence Magazine 6(1), 43–44 (2011)
20. Molnár Jr., F., Izsák, F., Mészáros, R., Lagzi, I.: Simulation of reaction-diffusion processes in three dimensions using CUDA. ArXiv e-prints (April 2010)
21. Murray, J.D.: Mathematical Biology: Spatial Models and Biomedical Applications, vol. 2. Springer, Heidelberg (2003)
22. Neglia, G., Reina, G.: Evaluating activator-inhibitor mechanisms for sensors coordination. In: Proc. Bionetics. ICST (2007)
23. Pearson, J.E.: Complex patterns in a simple system. Science 261(5118), 189–192 (1993)
24. Sanderson, A.R., Meyer, M.D., Kirby, R.M., Johnson, C.R.: A framework for exploring numerical solutions of advection-reaction-diffusion equations using a GPU-based approach. Computing and Visualization in Science 12(4), 155–170 (2009)

25. Shen, W.M., Will, P., Galstyan, A., Chuong, C.M.: Hormone-inspired self-organization and distributed control of robotic swarms. Autonomous Robots 17(1), 93–105 (2004)
26. Streichert, F., Spieth, C., Ulmer, H., Zell, A.: How to evolve the head-tail pattern from reaction-diffusion systems. In: NASA/DoD Conference on Evolvable Hardware, pp. 261–268. IEEE Computer Society Press, Los Alamitos (2004)
27. Turing, A.M.: The chemical basis of morphogenesis. Phil. Trans. Royal Soc. London B 327, 37–72 (1952)
28. Yamamoto, L., Miorandi, D.: Evaluating the Robustness of Activator-Inhibitor Models for Cluster Head Computation. In: Dorigo, M., Birattari, M., Di Caro, G.A., Doursat, R., Engelbrecht, A.P., Floreano, D., Gambardella, L.M., Groß, R., Şahin, E., Sayama, H., Stützle, T. (eds.) ANTS 2010. LNCS, vol. 6234, pp. 143–154. Springer, Heidelberg (2010)

Optimal Divide and Query*

David Insa and Josep Silva

Universidad Politécnica de Valencia
Camino de Vera s/n, E-46022 Valencia, Spain
{dinsa,jsilva}@dsic.upv.es

Abstract. Algorithmic debugging is a semi-automatic debugging technique that allows the programmer to precisely identify the location of bugs without the need to inspect the source code. The technique has been successfully adapted to all paradigms and mature implementations have been released for languages such as Haskell, Prolog or Java. During three decades, the algorithm introduced by Shapiro and later improved by Hirunkitti has been thought optimal. In this paper we first show that this algorithm is not optimal, and moreover, in some situations it is unable to find all possible solutions, thus it is incomplete. Then, we present a new version of the algorithm that is proven optimal, and we introduce some equations that allow the algorithm to identify all optimal solutions.

Keywords: Algorithmic Debugging, Strategy, Divide & Query.

1 Introduction

Debugging is one of the most important but less automated (and, thus, time-consuming) tasks in the software development process. The programmer is often forced to manually explore the code or iterate over it using, e.g., breakpoints, and this process usually requires a deep understanding of the source code to find the bug. *Algorithmic debugging* [17] is a semi-automatic debugging technique that has been extended to practically all paradigms [18]. Recent research has produced new advances to increase the scalability of the technique producing new scalable and mature debuggers. The technique is based on the answers of the programmer to a series of questions generated automatically by the algorithmic debugger. The questions are always whether a given result of an activation of a subcomputation with given input values is actually correct. The answers provide the debugger with information about the correctness of some (sub)computations of a given program; and the debugger uses them to guide the search for the bug until a buggy portion of code is isolated.

Example 1. Consider this simple Haskell program inspired in a similar example by [6]. It wrongly (it has a bug) implements the sorting algorithm *Insertion Sort*:

---

* This work has been partially supported by the Spanish *Ministerio de Ciencia e Innovación* under grant TIN2008-06622-C03-02 and by the *Generalitat Valenciana* under grant PROMETEO/2011/052.

L. Antunes and H.S. Pinto (Eds.): EPIA 2011, LNAI 7026, pp. 224–238, 2011.

```
main = insort [2,1,3]

insort [] = []
insort (x:xs) = insert x (insort xs)

insert x [] = [x]
insert x (y:ys) = if x>=y then (x:y:ys)
                          else (y:(insert x ys))
```

An algorithmic debugging session for this program is the following (YES and NO answers are provided by the programmer):

```
Starting Debugging Session...
(1)  insort [1,3] = [3,1]? NO
(2)  insort [3] = [3]? YES
(3)  insert 1 [3] = [3,1]? NO
(4)  insert 1 [] = [1]? YES

Bug found in rule:
insert x (y:ys) = if x>=y then _ else (y:(insert x ys))
```

The debugger points out the part of the code that contains the bug. In this case x>=y should be x<=y. Note that, to debug the program, the programmer only has to answer questions. It is not even necessary to see the code.

Typically, algorithmic debuggers have a front-end that produces a data structure representing a program execution—the so-called *execution tree* (ET) [15]—; and a back-end that uses the ET to ask questions and process the answers of the programmer to locate the bug. For instance, the ET of the program in Example 1 is depicted in Figure 1.

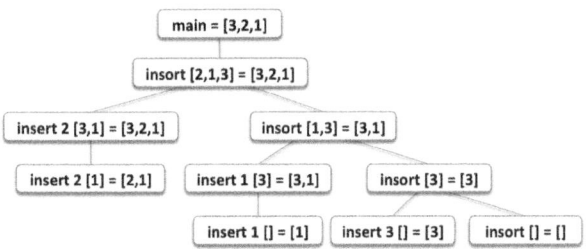

Fig. 1. ET of the program in Example 1

The strategy used to decide what nodes of the ET should be asked is crucial for the performance of the technique. Since the definition of algorithmic debugging, there have been a lot of research concerning the definition of new strategies trying to minimize the number of questions [18]. We conducted several experiments to measure the performance of all current algorithmic debugging strategies. The results of the experiments are shown in Figure 2, where the first column contains the names of the benchmarks; column `nodes` shows the number of nodes in the ET associated with each benchmark; and the other columns represent algorithmic debugging strategies [18] that are ordered according to their performance: Optimal Divide & Query (`D&QO`), Divide & Query by Hirunkitti (`D&QH`), Divide & Query by Shapiro (`D&QS`), Divide by Rules & Query (`DR&Q`), Heaviest First (`HF`), More Rules First (`MRF`), Hat Delta Proportion (`HD-P`), Top-Down (`TD`), Hat Delta YES (`HD-Y`), Hat Delta NO (`HD-N`), Single Stepping (`SS`).

Benchmark	Nodes	D&QO	D&QH	D&QS	DR&Q	HF	MRF	HD-P	HD-Y	TD	HD-N	SS	Average
NumReader	12	28,99	28,99	31,36	29,59	44,38	44,38	49,70	49,70	49,70	49,70	53,25	41,80
Orderings	46	12,04	12,09	12,63	14,40	17,16	17,29	21,05	20,60	20,82	19,60	51,02	19,88
Factoricer	62	9,83	9,83	9,93	20,03	12,55	12,55	15,04	15,04	12,55	18,29	50,77	16,94
Sedgewick	12	30,77	30,77	33,14	30,77	34,91	34,91	43,79	43,79	43,20	43,79	53,25	38,46
Clasifier	23	19,79	20,31	22,40	21,88	22,92	23,26	32,12	32,12	31,94	34,55	51,91	28,47
LegendGame	71	8,87	8,87	8,95	16,72	11,15	11,23	14,68	14,68	13,37	16,94	50,68	16,01
Cues	18	31,58	32,41	32,41	32,41	33,24	34,63	39,06	39,06	42,11	44,32	52,35	37,60
Romanic	123	6,40	10,84	11,23	13,56	7,44	11,88	13,29	13,29	13,41	13,30	50,40	15,00
FibRecursive	4.619	0,27	0,27	0,28	1,20	0,33	0,41	3,92	3,92	0,46	0,48	50,01	5,59
Risk	33	16,78	16,78	18,08	19,38	18,69	18,69	24,31	24,31	31,14	32,79	51,38	24,76
FactTrans	198	3,89	3,89	3,93	6,22	6,58	6,58	7,37	7,24	7,16	7,50	50,25	10,06
RndQuicksort	72	8,73	8,73	8,73	11,41	12,03	12,23	13,62	12,93	13,51	14,54	50,67	15,19
BinaryArrays	128	5,52	5,52	5,71	7,13	7,75	7,94	7,90	8,15	8,59	8,71	50,38	11,21
FibFactAna	351	2,44	2,44	2,45	5,38	7,61	7,71	6,40	7,39	8,57	5,99	50,14	9,68
NewtonPol	7	39,06	39,06	43,75	39,06	43,75	43,75	45,31	45,31	45,31	45,31	54,69	44,03
RegresionTest	18	23,27	23,27	25,21	25,21	26,87	26,87	32,96	32,96	32,96	32,96	52,35	30,45
BoubleFibArrays	171	4,40	4,41	4,57	11,40	5,95	6,96	24,50	24,87	6,96	6,96	50,29	13,75
ComplexNumbers	60	10,02	10,02	10,32	11,31	11,39	11,39	15,78	15,80	15,75	19,19	50,79	16,53
Integral	5	44,44	44,44	47,22	44,44	50,00	50,00	50,00	50,00	50,00	50,00	55,56	48,74
TestMath	48	11,91	11,91	12,16	12,99	15,95	16,28	22,41	23,87	24,20	22,37	50,98	20,46
TestMath2	228	3,51	3,51	3,51	9,73	10,55	10,81	12,29	13,24	28,56	14,37	50,22	14,57
Figures	113	6,72	6,75	6,79	8,09	7,68	7,79	10,17	10,16	10,60	10,76	50,43	12,36
FactCalc	59	10,11	10,14	10,42	11,53	13,69	14,22	20,47	20,47	18,50	20,69	50,81	18,28
SpaceLimits	127	12,95	16,07	19,15	21,74	13,68	16,80	22,87	22,86	22,78	26,15	50,38	22,31
Average	275,17	14,68	15,06	16,01	17,73	18,18	18,69	22,87	22,99	23,01	23,30	51,37	22,17

Fig. 2. Performance of algorithmic debugging strategies

For each benchmark, we produced its associated ET and assumed that the buggy node could be any node of the ET (i.e., any subcomputation in the execution of the program could be buggy). Therefore, we performed a different experiment for each possible case and, hence, each cell of the table summarizes a number of experiments that were automatized. In particular, benchmark *Factoricer* has been debugged 62 times with each strategy; each time, the buggy node was a different node, and the results shown are the average number of questions performed by each strategy with respect to the number of nodes (i.e., the mean percentage of nodes asked). Similarly, benchmark *Cglib* has been debugged 1216 times with each strategy, and so on.

Observe that the best algorithmic debugging strategies in practice are the two variants of Divide and Query (ignoring our new technique D&QO). Moreover, from a theoretical point of view, this strategy has been thought optimal in the worst case for almost 30 years, and it has been implemented in almost all current algorithmic debuggers (see, e.g., [4,5,8,16]). In this paper we show that current algorithms for D&Q are suboptimal. We show the problems of D&Q and solve them in a new improved algorithm that is proven optimal. Moreover, the original strategy was only defined for ETs where all the nodes have an individual weight of

1. In contrast, we allow our algorithms to work with different individual weights that can be integer, but also decimal. An individual weight of zero means that this node cannot contain the bug. A positive individual weight approximates the probability of being buggy. The higher the individual weight, the higher the probability. This generalization strongly influences the technique and allows us to assign different probabilities of being buggy to different parts of the program. For instance, a recursive function with higher-order calls should be assigned a higher individual weight than a function implementing a simple base case [18].

We show that the original algorithms are inefficient with ETs where nodes can have different individual weights in the domain of the positive real numbers (including zero) and we redefine the technique for these generalized ETs.

The rest of the paper has been organized as follows. In Section 2 we recall and formalize the strategy D&Q and we show with counterexamples that it is suboptimal and incomplete. Then, in Section 3 we introduce two new algorithms for D&Q that are optimal and complete. Each algorithm is useful for a different type of ET. Finally, Section 4 concludes. Proofs of technical results can be found in [9].

2   D&Q by Shapiro vs. D&Q by Hirunkitti

In this section we formalize the strategy D&Q to show the differences between the original version by Shapiro [17] and the improved version by Hirunkitti [7]. We start with the definition of *marked execution tree*, that is an ET where some nodes could have been removed because they were marked as correct (i.e., answered YES), some nodes could have been marked as wrong (i.e., answered NO) and the correctness of the other nodes is undefined.

Definition 1 (Marked Execution Tree). *A marked execution tree (MET) is a tree $T = (N, E, M)$ where N are the nodes, $E \subseteq N \times N$ are the edges, and $M : N \to V$ is a marking total function that assigns to all the nodes in N a value in the domain $V = \{Wrong, Undefined\}$.*

Initially, all nodes in the MET are marked as *Undefined*. But with every answer of the user, a new MET is produced. Concretely, given a MET $T = (N, E, M)$ and a node $n \in N$, the answer of the user to the question in n produces a new MET such that: (i) if the answer is YES, then this node and its subtree is removed from the MET. (ii) If the answer is NO, then, all the nodes in the MET are removed except this node and its descendants.[1]

Therefore, the size of the MET is gradually reduced with the answers. If we delete all nodes in the MET then the debugger concludes that no bug has been found. If, contrarily, we finish with a MET composed of a single node marked as wrong, this node is called *buggy node* and it is pointed as responsible of the bug of the program.

---

[1] It is also possible to accept *I don't know* as an answer of the user. In this case, the debugger simply selects another node [8]. For simplicity, we assume here that the user only answers *Correct* or *Wrong*.

All this process is defined in Algorithm 1 where function *selectNode* selects a node in the MET to be asked to the user with function *askNode*. Therefore, *selectNode* is the central point of this paper. In the rest of this section, we assume that *selectNode* implements D&Q. In the following we use E^* to refer to the reflexive and transitive closure of E and E^+ for the transitive closure.

Algorithm 1. General algorithm for algorithmic debugging

Input: A MET $T = (N, E, M)$
Output: A buggy node or $\perp$ if no buggy node exists
Preconditions: $\forall n \in N, M(n) = Undefined$
Initialization: buggyNode $= \perp$

begin
(1) **do**
(2)   node = selectNode(T)
(3)   answer = askNode(node)
(4)   **if** (answer = *Wrong*)
(5)   **then** M(node) = *Wrong*
(6)   buggyNode = node
(7) $N = \{n \in N \mid (\text{node} \to n) \in E^*\}$
(8)   **else** $N = N \backslash \{n \in N \mid (\text{node} \to n) \in E^*\}$
(9) **while** ($\exists n \in N, M(n) = Undefined$)
(10) **return** buggyNode
end

Both D&Q by Shapiro and D&Q by Hirunkitti assume that the individual weight of a node is always 1. Therefore, given a MET $T = (N, E, M)$, the weight of the subtree rooted at node $n \in N$, w_n, is defined as its number of descendants including itself (i.e., $1 + \sum \{w_{n'} \mid (n \to n') \in E\}$).

D&Q tries to simulate a dichotomic search by selecting the node that better divides the MET into two subMETs with a weight as similar as possible. Therefore, given a MET with n nodes, D&Q searches for the node whose weight is closer to $\frac{n}{2}$. The original algorithm by Shapiro always selects:

– the heaviest node n' whose weight is as close as possible to $\frac{n}{2}$ with $w_{n'} \leq \frac{n}{2}$

Hirunkitti and Hogger noted that this is not enough to divide the MET by the half and their improved version always selects the node whose weight is closer to $\frac{n}{2}$ between:

– the heaviest node n' whose weight is as close as possible to $\frac{n}{2}$ with $w_{n'} \leq \frac{n}{2}$,
  or
– the lightest node n' whose weight is as close as possible to $\frac{n}{2}$ with $w_{n'} \geq \frac{n}{2}$

Because it is better, in the rest of the article we only consider Hirunkitti's D&Q and refer to it as D&Q.

2.1   Limitations of D&Q

In this section we show that D&Q is suboptimal when the MET does not contain a wrong node (i.e., all nodes are marked as undefined).[2] Moreover, we show that if the MET contains a wrong node, then D&Q is correct (all nodes found divide the MET optimally), but it is incomplete (it cannot find some nodes that optimally divide the MET). The intuition beyond these limitations is that the objective of D&Q is to divide the tree by two, but the real objective should be to reduce the number of questions to be asked to the programmer. For instance, consider the MET in Figure 3 (left) where the black node is marked as wrong and D&Q would select the gray node. The objective of D&Q is to divide the 8 nodes into two groups of 4. Nevertheless, the real motivation of dividing the tree should be to divide the tree into two parts that would produce the same number of remaining questions (in this case 3).

The problem comes from the fact that D&Q does not take into account the marking of wrong nodes. For instance, observe the two METs in Figure 3 (center) where each node is labeled with its weight and the black node is marked as wrong. In both cases D&Q would behave exactly in the same way, because it completely ignores the fact that some nodes are marked as wrong. Nevertheless, it is evident that we do not need to ask again for a node that is already marked as wrong to determine whether it is buggy. However, D&Q counts the nodes marked as wrong as part of their own weight, and this is a source of inefficiency.

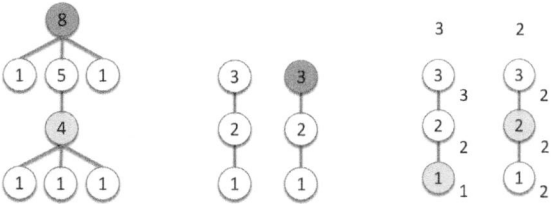

Fig. 3. Behavior of Divide and Query

In the METs of Figure 3 (center) D&Q would select either the node with weight 1 or the node with weight 2 (both are equally close to $\frac{3}{2}$). However, we show in Figure 3 (right) that selecting node 1 is suboptimal, and the strategy should always select node 2. Considering that the gray node is the first node selected by the strategy, then the number at the side of a node represents the number of questions needed to find the bug if the buggy node is this node. The number at the top of the figure represents the number of questions needed to determine that there is not a bug. Clearly, as an average, it is better to select first the node with weight 2 because we would perform less questions ($\frac{8}{4}$ vs. $\frac{9}{4}$ considering all four possible cases).

---

[2] Modern debuggers [8] allow the programmer to debug the MET while it is being generated. Thus the root node of the subtree being debugged is not necessarily marked as *Wrong*.

Therefore, D&Q returns a set of nodes that contains the best node, but it is not able to determine which of them is the best node, thus being suboptimal when it is not selected. In addition, the METs in Figure 4 show that D&Q is incomplete. Observe that the METs have 5 nodes, thus D&Q would always select the node with weight 2. However, the node with weight 4 is equally optimal (both need $\frac{16}{6}$ questions as an average to find the bug) but it will be never selected by D&Q because its weight is far from the half of the tree $\frac{5}{2}$.

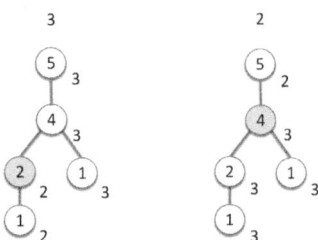

Fig. 4. Incompleteness of Divide and Query

Another limitation of D&Q is that it was designed to work with METs where all the nodes have the same individual weight, and moreover, this weight is assumed to be one. If we work with METs where nodes can have different individual weights and these weights can be any value greater or equal to zero, then D&Q is suboptimal as it is demonstrated by the MET in Figure 5. In this MET, D&Q would select node n_1 because its weight is closer to $\frac{21}{2}$ than any other node. However, node n_2 is the node that better divides the tree in two parts with the same probability of containing the bug.

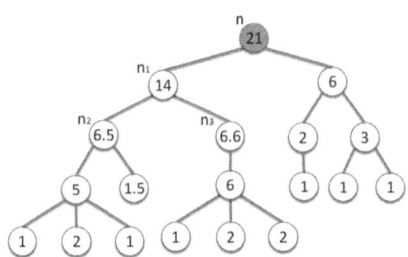

Fig. 5. MET with decimal individual weights

In summary, (1) D&Q is suboptimal when the MET is free of wrong nodes, (2) D&Q is incomplete when the MET contains wrong nodes, (3) D&Q is correct when the MET contains wrong nodes and all the nodes of the MET have the same weight, but (4) D&Q is suboptimal when the MET contains wrong nodes and the nodes of the MET have different individual weights.

3   Optimal D&Q

In this section we introduce a new version of D&Q that tries to divide the MET into two parts with the same probability of containing the bug (instead of two parts with the same weight). We introduce new algorithms that are correct and complete even if the MET contains nodes with different individual weights. For this, we define the *search area* of a MET as the set of undefined nodes.

Definition 2 (Search area). *Let* $T = (N, E, M)$ *be a MET. The* search area *of* T, $Sea(T)$, *is defined as* $\{n \in N \mid M(n) = Undefined\}$.

While D&Q uses the whole T, we only use $Sea(T)$, because answering all nodes in $Sea(T)$ guarantees that we can discover all buggy nodes [10]. Moreover, in the following we refer to the individual weight of a node n with wi_n; and we refer to the weight of a (sub)tree rooted at n with w_n that is recursively defined as:

$$w_n = \begin{cases} \sum \{w_{n'} \mid (n \to n') \in E\} & \text{if } M(n) \neq Undefined \\ wi_n + \sum \{w_{n'} \mid (n \to n') \in E\} & \text{otherwise} \end{cases}$$

Note that, contrarily to standard D&Q, the definition of w_n excludes those nodes that are not in the search area (i.e., the root node when it is wrong). Note also that wi_n allows us to assign any individual weight to the nodes. This is an important generalization of D&Q where it is assumed that all nodes have the same individual weight and it is always 1.

3.1   Debugging ETs Where All Nodes Have the Same Individual Weight $wi \in \mathcal{R}^+$

For the sake of clarity, given a node $n \in Sea(T)$, we distinguish between three subareas of $Sea(T)$ induced by n: (1) n itself, whose individual weight is wi_n; (2) descendants of n, whose weight is

$$Down(n) = \sum \{wi_{n'} \mid n' \in Sea(T) \land (n \to n') \in E^+\}$$

and (3) the rest of nodes, whose weight is

$$Up(n) = \sum \{wi_{n'} \mid n' \in Sea(T) \land (n \not\to n') \in E^*\}$$

Example 2. Consider the MET in Figure 6. Assuming that the root n is the only node marked as wrong and all nodes have an individual weight of 1, then $Sea(T)$ contains all nodes except n, $Up(n') = 4$ (total weight of the gray nodes), and $Down(n') = 3$ (total weight of the white nodes).

Clearly, for any MET whose root is n and a node n', $M(n') = Undefined$, we have that:

$$w_n = Up(n') + Down(n') + wi_{n'} \qquad \text{(Equation 1)}$$
$$w_{n'} = Down(n') + wi_{n'} \qquad \text{(Equation 2)}$$

Intuitively, given a node n, what we want to divide by the half is the area formed by $Up(n) + Down(n)$. That is, n will not be part of $Sea(T)$ after it has been

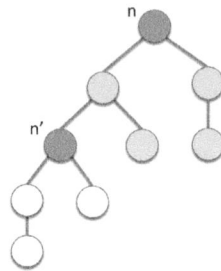

Fig. 6. Functions Up and Down

answered, thus the objective is to make $Up(n)$ equal to $Down(n)$. This is another important difference with traditional D&Q: wi_n should not be considered when dividing the MET. We use the notation $n_1 \gg n_2$ to express that n_1 divides $Sea(T)$ better than n_2 (i.e., $|Up(n_1) - Down(n_1)| < |Up(n_2) - Down(n_2)|$). And we use $n_1 \equiv n_2$ to express that n_1 and n_2 equally divide $Sea(T)$. If we find a node n such that $Up(n) = Down(n)$ then n produces an optimal division, and should be selected by the strategy. If an optimal solution cannot be found, the following theorem states how to compare the nodes in order to decide which of them should be selected.

Theorem 1. *Given a MET $T = (N, E, M)$ whose root is $n \in N$, where $\forall n, n' \in N, wi_n = wi_{n'}$ and $\forall n \in N, wi_n > 0$, and given two nodes $n_1, n_2 \in Sea(T)$, with $w_{n_1} > w_{n_2}$, if $w_n > w_{n_1} + w_{n_2} - wi_n$ then $n_1 \gg n_2$.*

Proposition 1. *Given a MET $T = (N, E, M)$ whose root is $n \in N$, where $\forall n, n' \in N, wi_n = wi_{n'}$ and $\forall n \in N, wi_n > 0$, and given two nodes $n_1, n_2 \in Sea(T)$, with $w_{n_1} > w_{n_2}$, if $w_n = w_{n_1} + w_{n_2} - wi_n$ then $n_1 \equiv n_2$.*

Theorem 1 is useful when one node is heavier than the other. In the case that both nodes have the same weight, then the following theorem guarantees that they both equally divide the MET in all situations.

Theorem 2. *Let $T = (N, E, M)$ be a MET where $\forall n, n' \in N, wi_n = wi_{n'}$ and $\forall n \in N, wi_n > 0$, and let $n_1, n_2 \in Sea(T)$ be two nodes, if $w_{n_1} = w_{n_2}$ then $n_1 \equiv n_2$.*

Corollary 1. *Given a MET $T = (N, E, M)$ where $\forall n, n' \in N, wi_n = wi_{n'}$ and $\forall n \in N, wi_n > 0$, and given a node $n \in Sea(T)$, then n optimally divides $Sea(T)$ if and only if $Up(n) = Down(n)$.*

While Corollary 1 states the objective of optimal D&Q (finding a node n such that $Up(n) = Down(n)$), Theorems 1 and 2 provide a method to approximate this objective (finding a node n such that $|Up(n) - Down(n)|$ is minimum in $Sea(T)$).

An Algorithm for Optimal D&Q. Theorem 1 and Proposition 1 provide equation $w_n \geq w_{n_1} + w_{n_2} - wi_n$ to compare two nodes n_1, n_2 by efficiently determining $n_1 \gg n_2$, $n_1 \equiv n_2$ or $n_1 \ll n_2$. However, with only this equation, we should compare all nodes to select the best of them (i.e., n such that $\nexists n', n' \gg n$). Hence, in this section we provide an algorithm that allows us to find the best node in a MET with a minimum set of node comparisons.

Given a MET, Algorithm 2 efficiently determines the best node to divide $Sea(T)$ by the half (in the following the *optimal node*). In order to find this node, the algorithm does not need to compare all nodes in the MET. It follows a path of nodes from the root to the optimal node which is closer to the root producing a minimum set of comparisons.

Algorithm 2. Optimal D&Q (SelectNode)

Input: A MET $T = (N, E, M)$ whose root is $n \in N$,
    $\forall n_1, n_2 \in N, wi_{n_1} = wi_{n_2}$ and $\forall n_1 \in N, wi_{n_1} > 0$
Output: A node $n' \in N$
Preconditions: $\exists n \in N, M(n) = Undefined$

begin
(1) Candidate $= n$
(2) **do**
(3)   Best $=$ Candidate
(4)   Children $= \{m \mid (\text{Best} \rightarrow m) \in E\}$
(5)   **if** (Children $= \emptyset$) **then break**
(6)   Candidate $= n' \in$ Children $\mid \forall n'' \in$ Children, $w_{n'} \geq w_{n''}$
(7) **while** ($w_{Candidate} > \frac{w_n}{2}$)
(8) **if** ($M(\text{Best}) = Wrong$) **then return** Candidate
(9) **if** ($w_n \geq w_{Best} + w_{Candidate} - wi$) **then return** Best
(10)   **else return** Candidate
end

Example 3. Consider the MET in Figure 7 where $\forall n \in N, wi_n = 1$ and $M(n) = Undefined$. Observe that Algorithm 2 only needs to apply the equation in Theorem 1 once to identify an optimal node. Firstly, it traverses the MET top-down from the root selecting at each level the heaviest node until we find a node whose weight is smaller than the half of the MET ($\frac{w_n}{2}$), thus, defining a

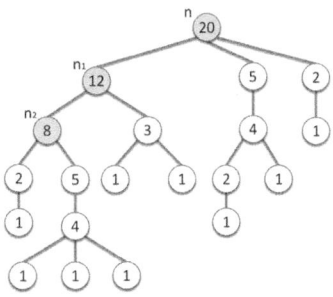

Fig. 7. Defining a path in a MET to find the optimal node

path in the MET that is colored in gray. Then, the algorithm uses the equation $w_n \geq w_{n_1} + w_{n_2} - w_{in}$ to compare nodes n_1 and n_2. Finally, the algorithm selects n_1.

In order to prove the correctness of Algorithm 2, we need to prove that (1) the node returned is really an optimal node, and (2) this node will always be found by the algorithm (i.e., it is always in the path defined by the algorithm).

The first point can be proven with Theorems 1 and 2. The second point is the key idea of the algorithm and it relays on an interesting property of the path defined: while defining the path in the MET, only four cases are possible, and all of them coincide in that the subtree of the heaviest node will contain an optimal node.

In particular, when we use Algorithm 2 and compare two nodes n_1, n_2 in a MET whose root is n, we find four possible cases:

Case 1: n_1 and n_2 are brothers.
Case 2: $w_{n_1} > w_{n_2} \wedge w_{n_2} > \frac{w_n}{2}$.
Case 3: $w_{n_1} > \frac{w_n}{2} \wedge w_{n_2} \leq \frac{w_n}{2}$.
Case 4: $w_{n_1} > w_{n_2} \wedge w_{n_1} \leq \frac{w_n}{2}$.

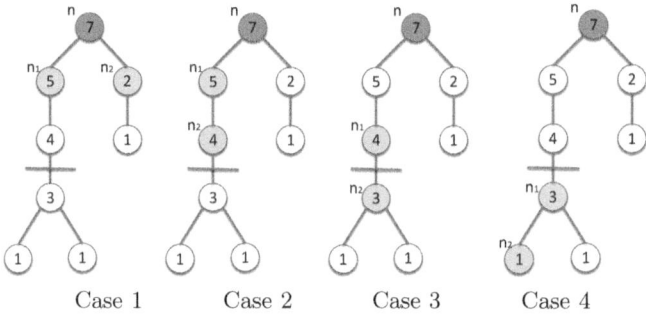

Case 1   Case 2   Case 3   Case 4

Fig. 8. Determining the best node in a MET (four possible cases)

We have proven—the individual proofs are part of the proof of Theorem 3—that in cases 1 and 4, the heaviest node is better (i.e., if $w_{n_1} > w_{n_2}$ then $n_1 \gg n_2$); In case 2, the lightest node is better; and in case 3, the best node must be determined with the equation of Theorem 1. Observe that these results allow the algorithm to determine the path to the optimal node that is closer to the root. For instance, in Example 3 case 1 is used to select a child, e.g., node 12 instead of node 5 or node 2, and node 8 instead of node 3. Case 2 is used to go down and select node 12 instead of node 20. Case 4 is used to stop going down and stop at node 8 because it is better than all its descendants. And it is also used to determine that node 2, 3 and 5 are better than all their descendants. Finally, case 3 is used to select the optimal node, 12 instead of 8. Note that D&Q could have selected node 8 that is equally close to $\frac{20}{2}$ than node 12; but it is suboptimal because $Up(8) = 12$ and $Down(8) = 7$ whereas $Up(12) = 8$ and $Down(12) = 11$.

The correctness of Algorithm 2 is stated by the following theorem.

Theorem 3 (Correctness). *Let $T = (N, E, M)$ be a MET where $\forall n, n' \in N, win = win'$ and $\forall n \in N, win > 0$, then the execution of Algorithm 2 with T as input always terminates producing as output a node $n \in Sea(T)$ such that $\nexists n' \in Sea(T) \mid n' \gg n$.*

Algorithm 2 always returns a single optimal node. However, the equation in Theorem 1 in combination with the equation in Proposition 1 can be used to identify all optimal nodes in the MET. In particular, we could add a new line between lines (7) and (8) of Algorithm 2 to collect all candidates instead of one (see Theorem 2):

$$\text{Candidates} = \{n' \in \text{Children} \mid \forall n'' \in \text{Children} , w_{n'} \geq w_{n''}\}$$

then, in line (8) we could replace Candidate by Candidates, and we could modify lines (9) and (10) to return both Best and Candidates when the equation is an equality (see Proposition 1):

if $(w_n > w_{Best} + w_{Candidate} - win)$ **then return** {Best}
if $(w_n = w_{Best} + w_{Candidate} - win)$ **then return** {Best} $\cup$ Candidates
                                                else return Candidates

With this modifications the algorithm is complete, and it returns nodes 2 and 4 in the MET of Figure 4 where D&Q can only detect node 2 as optimal.

3.2 Debugging METs Where Nodes Can Have Different Individual Weights in $\mathcal{R}^+ \cup \{0\}$

In this section we generalize divide and query to the case where nodes can have different individual weights and these weights can be any value greater or equal to zero. As shown in Figure 5, in this general case traditional D&Q fails to identify the optimal node (it selects node n_1 but the optimal node is n_2). The algorithm presented in the previous section is also suboptimal when the individual weights can be different. For instance, in the MET of Figure 5, it would select node n_3. For this reason, in this section we introduce Algorithm 3, a general algorithm able to identify an optimal node in all cases. It does not mean that Algorithm 2 is useless. Algorithm 2 is optimal when all nodes have the same weight, and in that case, it is more efficient than Algorithm 3. Theorem 4 ensures the finiteness and correctness of Algorithm 3.

Theorem 4 (Correctness). *Let $T = (N, E, M)$ be a MET where $\forall n \in N$, $win \geq 0$, then the execution of Algorithm 3 with T as input always terminates producing as output a node $n \in Sea(T)$ such that $\nexists n' \in Sea(T) \mid n' \gg n$.*

3.3 Debugging METs Where Nodes Can Have Different Individual Weights in $\mathcal{R}^+$

In the previous section we provided an algorithm that optimally selects an optimal node of the MET with a minimum set of node comparisons. But this algorithm is

---

Algorithm 3. Optimal D&Q General (SelectNode)

Input: A MET $T = (N, E, M)$ whose root is $n \in N$ and $\forall n_1 \in N, wi_{n_1} \geq 0$
Output: A node $n' \in N$
Preconditions: $\exists n \in N, M(n) = Undefined$

begin

(1) Candidate $= n$
(2) **do**
(3)   Best $=$ Candidate
(4)   Children $= \{m \mid (\text{Best} \rightarrow m) \in E\}$
(5)   **if** (Children $= \emptyset$) **then break**
(6)   Candidate $= n' \mid \forall n''$ with $n', n'' \in$ Children, $w_{n'} \geq w_{n''}$
(7) **while** $(w_{Candidate} - \frac{wi_{Candidate}}{2} > \frac{w_n}{2})$
(8) Candidate $= n' \mid \forall n''$ with $n', n'' \in$ Children, $w_{n'} - \frac{wi_{n'}}{2} \geq w_{n''} - \frac{wi_{n''}}{2}$
(9) **if** $(M(\text{Best}) = Wrong)$ **then return** Candidate
(10) **if** $(w_n \geq w_{Best} + w_{Candidate} - \frac{wi_{Best}}{2} - \frac{wi_{Candidate}}{2})$ **then return** Best
(11)   **else return** Candidate

end

---

not complete due to the fact that we allow the nodes to have an individual weight of zero. For instance, when all nodes have an individual weight of zero, Algorithm 3 returns a single optimal node, but it is not able to find all optimal nodes.

Given a node, the difference between having an individual weight of zero and having a (total) weight of zero should be clear. The former means that this node

---

Algorithm 4. Optimal D&Q General (SelectNode)

Input: A MET $T = (N, E, M)$ whose root is $n \in N$ and $\forall n_1 \in N, wi_{n_1} > 0$
Output: A set of nodes $O \subseteq N$
Preconditions: $\exists n \in N, M(n) = Undefined$

begin

(1) Candidate $= n$
(2) **do**
(3)   Best $=$ Candidate
(4)   Children $= \{m \mid (\text{Best} \rightarrow m) \in E\}$
(5)   **if** (Children $= \emptyset$) **then break**
(6)   Candidate $= n' \mid \forall n''$ with $n', n'' \in$ Children, $w_{n'} \geq w_{n''}$
(7) **while** $(w_{Candidate} - \frac{wi_{Candidate}}{2} > \frac{w_n}{2})$
(8) Candidates $= \{n' \mid \forall n''$ with $n', n'' \in$ Children, $w_{n'} - \frac{wi_{n'}}{2} \geq w_{n''} - \frac{wi_{n''}}{2}\}$
(9) Candidate $= n' \in$ Candidates
(10) **if** $(M(\text{Best}) = Wrong)$ **then return** Candidates
(11) **if** $(w_n > w_{Best} + w_{Candidate} - \frac{wi_{Best}}{2} - \frac{wi_{Candidate}}{2})$ **then return** $\{\text{Best}\}$
(12) **if** $(w_n = w_{Best} + w_{Candidate} - \frac{wi_{Best}}{2} - \frac{wi_{Candidate}}{2})$ **then**
                                            return $\{\text{Best}\} \cup$ Candidates
(13)   **else return** Candidates

end

---

did not cause the bug, the later means that none of the descendants of this node (neither the node itself) caused the bug. Surprisingly, the use of nodes with individual weights of zero has not been exploited in the literature. Assigning a (total) weight of zero to a node has been used for instance in the technique called *Trusting* [11]. This technique allows the user to trust a method. When this happens all the nodes related to this method and their descendants are pruned from the tree (i.e., these nodes have a (total) weight of zero).

If we add the restriction that nodes cannot be assigned with an individual weight of zero, then we can refine Algorithm 3 to ensure completeness. This refined version is Algorithm 4.

4 Conclusion

During three decades, D&Q has been the more efficient algorithmic debugging strategy. On the practical side, all current algorithmic debuggers implement D&Q [1,3,5,8,12,13,14,15,16], and experiments [2,19] (see also http://users.dsic.upv.es/~jsilva/DDJ/#Experiments) demonstrate that D&Q performs on average 2-36% less questions than other strategies. On the theoretical side, because D&Q intends a dichotomic search, it has been thought optimal with respect to the number of questions performed, and thus research on algorithmic debugging strategies has focused on other aspects such as reducing the complexity of questions.

The main contribution of this work is a new algorithm for D&Q that is optimal in all cases; including a generalization of the technique where all nodes of the ET can have different individual weights in $\mathcal{R}^+ \cup \{0\}$. The algorithm has been proved terminating and correct. And a slightly modified version of the algorithm has been provided that returns all optimal solutions, thus being complete.

We have implemented the technique and experiments show that it is more efficient than all previous algorithms (see column D&QO in Figure 2). The implementation—including the source code—and the experiments are publicly available at: http://users.dsic.upv.es/~jsilva/DDJ.

References

1. Braßel, B., Huch, F.: The Kiel Curry system KiCS. In: Proc of 17th International Conference on Applications of Declarative Programming and Knowledge Management (INAP 2007) and 21st Workshop on (Constraint) Logic Programming (WLP 2007), pp. 215–223. Technical Report 434, University of Würzburg (2007)
2. Caballero, R.: A Declarative Debugger of Incorrect Answers for Constraint Functional-Logic Programs. In: Proc. of the 2005 ACM SIGPLAN Workshop on Curry and Functional Logic Programming (WCFLP 2005), pp. 8–13. ACM Press, New York (2005)
3. Caballero, R.: Algorithmic Debugging of Java Programs. In: Proc. of the 2006 Workshop on Functional Logic Programming (WFLP 2006). Electronic Notes in Theoretical Computer Science, pp. 63–76 (2006)

4. Caballero, R., Martí-Oliet, N., Riesco, A., Verdejo, A.: A Declarative Debugger for Maude Functional Modules. Electronic Notes in Theoretical Computer Science 238, 63–81 (2009)
5. Davie, T., Chitil, O.: Hat-delta: One Right Does Make a Wrong. In: Seventh Symposium on Trends in Functional Programming, TFP 2006 (April 2006)
6. Fritzson, P., Shahmehri, N., Kamkar, M., Gyimóthy, T.: Generalized Algorithmic Debugging and Testing. LOPLAS 1(4), 303–322 (1992)
7. Hirunkitti, V., Hogger, C.J.: A Generalised Query Minimisation for Program Debugging. In: Adsul, B. (ed.) AADEBUG 1993. LNCS, vol. 749, pp. 153–170. Springer, Heidelberg (1993)
8. Insa, D., Silva, J.: An Algorithmic Debugger for Java. In: Proc. of the 26th IEEE International Conference on Software Maintenance, pp. 1–6 (2010)
9. Insa, D., Silva, J.: Optimal Divide and Query (extended version). Available in the Computing Research Repository (July 2011), http://arxiv.org/abs/1107.0350
10. Lloyd, J.W.: Declarative Error Diagnosis. New Gen. Comput. 5(2), 133–154 (1987)
11. Luo, Y., Chitil, O.: Algorithmic debugging and trusted functions. Technical report 10-07, University of Kent, Computing Laboratory, UK (August 2007)
12. Lux, W.: Münster Curry User's Guide (release 0.9.10 of May 10, 2006), http://danae.uni-muenster.de/~lux/curry/user.pdf
13. MacLarty, I.: Practical Declarative Debugging of Mercury Programs. PhD thesis, Department of Computer Science and Software Engineering, The University of Melbourne (2005)
14. Naish, L., Dart, P.W., Zobel, J.: The NU-Prolog Debugging Environment. In: Porto, A. (ed.) Proceedings of the Sixth International Conference on Logic Programming, Lisboa, Portugal, pp. 521–536 (June 1989)
15. Nilsson, H.: Declarative Debugging for Lazy Functional Languages. PhD thesis, Linköping, Sweden (May 1998)
16. Pope, B.: A Declarative Debugger for Haskell. PhD thesis, The University of Melbourne, Australia (2006)
17. Shapiro, E.: Algorithmic Program Debugging. MIT Press (1982)
18. Silva, J.: A Comparative Study of Algorithmic Debugging Strategies. In: Puebla, G. (ed.) LOPSTR 2006. LNCS, vol. 4407, pp. 143–159. Springer, Heidelberg (2007)
19. Silva, J.: An Empirical Evaluation of Algorithmic Debugging Strategies. Technical Report DSIC-II/10/09, UPV (2009), http://www.dsic.upv.es/~jsilva/research.htm#techs

A Subterm-Based Global Trie for Tabled Evaluation of Logic Programs

João Raimundo and Ricardo Rocha

CRACS & INESC TEC, Faculty of Sciences, University of Porto
Rua do Campo Alegre, 1021/1055, 4169-007 Porto, Portugal
{jraimundo,ricroc}@dcc.fc.up.pt

Abstract. Tabling is an implementation technique that overcomes some limitations of traditional Prolog systems in dealing with redundant sub-computations and recursion. A critical component in the implementation of an efficient tabling system is the design of the table space. The most popular and successful data structure for representing tables is based on a two-level trie data structure, where one trie level stores the tabled subgoal calls and the other stores the computed answers. The Global Trie (GT) is an alternative table space organization designed with the intent to reduce the tables's memory usage, namely by storing terms in a global trie, thus preventing repeated representations of the same term in different trie data structures. In this paper, we propose an extension to the GT organization, named *Global Trie for Subterms (GT-ST)*, where compound subterms in term arguments are represented as unique entries in the GT. Experimental results using the YapTab tabling system show that GT-ST support has potential to achieve significant reductions on memory usage, for programs with increasing compound subterms in term arguments, without compromising the execution time for other programs.

Keywords: Logic Programming, Tabling, Table Space, Implementation.

1 Introduction

Tabling [1] is an implementation technique that overcomes some limitations of traditional Prolog systems in dealing with redundant sub-computations and recursion. Tabling became a renowned technique thanks to the leading work in the XSB-Prolog system and, in particular, in the SLG-WAM engine [2]. A critical component in the implementation of an efficient tabling system is the design of the data structures and algorithms to access and manipulate the *table space*. The most popular and successful data structure for representing tables is based on a two-level *trie data structure*, where one trie level stores the tabled subgoal calls and the other stores the computed answers [3].

Tries are trees in which common prefixes are represented only once. The trie data structure provides complete discrimination for terms and permits look up and possibly insertion to be performed in a single pass through a term, hence resulting in a very efficient and compact data structure for term representation.

L. Antunes and H.S. Pinto (Eds.): EPIA 2011, LNAI 7026, pp. 239–253, 2011.

Despite the good properties of tries, one of the major limitations of tabling, when used in applications that pose many queries and/or have a large number of answers, is the overload of the table space memory [4].

The *Global Trie (GT)* [5,6] is an alternative table space organization where tabled subgoal calls and tabled answers are represented only once in a *global trie* instead of being spread over several different trie data structures. The major goal of GT's design is to save memory usage by reducing redundancy in the representation of tabled calls/answers to a minimum.

In this paper, we propose an extension to the GT organization, named *Global Trie for Subterms (GT-ST)*, where compound subterms in term arguments are represented as unique entries in the GT. Our new design extends a previous design, named *Global Trie for Terms (GT-T)* [6], where all argument and substitution compound terms appearing, respectively, in tabled subgoal calls and tabled answers are already represented only once in the GT. Experimental results, using the YapTab tabling system [7], show that GT-ST support has potential to achieve significant reductions on memory usage for programs with increasing compound subterms in term arguments, when compared with the GT-T design, without compromising the execution time for other programs.

The remainder of the paper is organized as follows. First, we introduce some background concepts about tries and the original table space organization in YapTab. Next, we present the previous GT-T design. Then, we introduce the new GT-ST organization and describe how we have extended YapTab to provide engine support for it. At last, we present some experimental results and we end by outlining some conclusions.

2   YapTab's Original Table Space Organization

The basic idea behind a tabled evaluation is, in fact, quite straightforward. The mechanism basically consists in storing, in the table space, all the different tabled subgoal calls and answers found when evaluating a program. The stored subgoal calls are then used to verify if a subgoal is being called for the first time or if it is a repeated call. Repeated calls are not re-evaluated against the program clauses, instead they are resolved by consuming the answers already stored in the table space. During this process, as further new answers are found, they are stored in their tables and later returned to all repeated calls.

The table space may thus be accessed in a number of ways: (i) to find out if a subgoal is in the table and, if not, insert it; (ii) to verify whether a newly found answer is already in the table and, if not, insert it; and (iii) to load answers from the tables to the repeated subgoals. With these requirements, a correct design of the table space is critical to achieve an efficient implementation. YapTab uses *tries* which is regarded as a very efficient way to implement the table space [3].

A trie is a tree structure where each different path through the *trie nodes* corresponds to a term described by the tokens labelling the nodes traversed. Two terms with common prefixes will branch off from each other at the first distinguishing token. For example, the tokenized form of the term $f(X, g(Y, X), Z)$ is the sequence of 6 tokens: $f/3$, VAR_0, $g/2$, VAR_1, VAR_0 and VAR_2, where each variable is represented as a distinct VAR_i constant [8]. YapTab's original table design implements tables using two levels of tries, one level stores the tabled subgoal calls and the other stores the computed answers.

More specifically, each tabled predicate has a *table entry* data structure assigned to it, acting as the entry point for the predicate's *subgoal trie*. Each different subgoal call is then represented as a unique path in the subgoal trie, starting at the predicate's table entry and ending in a *subgoal frame* data structure, with the argument terms being stored within the path's nodes. The subgoal frame data structure acts as an entry point to the *answer trie*. Each different subgoal answer is then represented as a unique path in the answer trie. Contrary to subgoal tries, answer trie paths hold just the substitution terms for the free variables which exist in the argument terms of the corresponding subgoal call [3]. Repeated calls to tabled subgoals load answers by traversing the answer trie nodes bottom-up. An example for a tabled predicate t/2 is shown in Fig. 1.

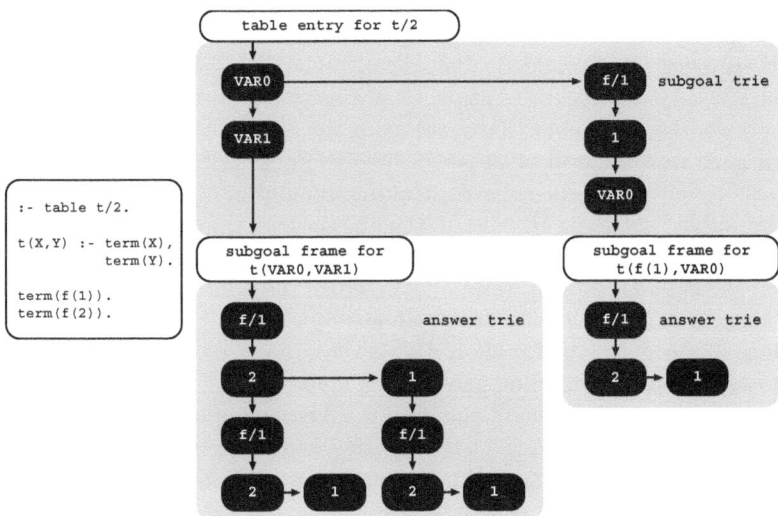

Fig. 1. YapTab's original table space organization

Initially, the subgoal trie is empty. Then, the subgoal t(f(1),Y) is called and three trie nodes are inserted: one for functor f/1, a second for integer 1 and one last for variable Y (VAR0). The subgoal frame is inserted as a leaf, waiting for the answers. Next, the subgoal t(X,Y) is also called. The two calls differ in the first argument, so tries bring no benefit here. Two new trie nodes, for variables X (VAR0) and Y (VAR1), and a new subgoal frame are inserted. Then, the answers for each subgoal are stored in the corresponding answer trie as their values are computed. Subgoal t(f(1),Y) has two answers, Y=f(1) and Y=f(2), so we need three trie nodes to represent both: a common node for functor f/1 and two nodes for integers 1 and 2. For subgoal t(X,Y) we have four answers, resulting from the combination of the answers f(1) and f(2) for variables X and Y, which requires nine trie nodes to represent them. Note that, for this particular example, the completed answer trie for t(X,Y) includes in its representation the completed answer trie for t(f(1),Y).

3 Global Trie

In this section, we introduce the new *Global Trie for Subterms (GT-ST)* design. Our new proposal extends a previous design named *Global Trie for Terms (GT-T)* [6]. We start by briefly presenting the GT-T design and then we discuss in more detail how we have extended and optimized it to our new GT-ST approach.

3.1 Global Trie for Terms

The GT-T was designed in order to maximize the sharing of tabled data which is structurally equal. In GT-T, all argument and substitution compound terms appearing, respectively, in tabled subgoal calls and tabled answers are represented only once in the GT, thus preventing situations where argument and substitution terms are represented more than once as in the example of Fig. 1.

Each path in a subgoal or answer trie is composed of a fixed number of trie nodes, representing, in the subgoal trie, the number of arguments for the corresponding tabled subgoal call, and, in the answer trie, the number of substitution terms for the corresponding answer. More specifically, for the subgoal tries, each node represents an argument term arg_i in which the node's token is used to store either arg_i, if arg_i is a *simple term* (an atom, integer or variable term), or the reference to the path's leaf node in the GT representing arg_i, if arg_i is a *compound (non-simple) term*. Similarly for the answer tries, each node represents a substitution term $subs_i$, where the node's token stores either $subs_i$, if $subs_i$ is a simple term, or the reference to the path's leaf node in the GT representing $subs_i$, if $subs_i$ is a compound term. Figure 2 uses the same example from Fig. 1 to illustrate how the GT-T design works.

Initially, the subgoal trie and the GT are empty. Then, the subgoal t(f(1),Y) is called and the argument compound term f(1) (represented by the tokens f/1 and 1) is first inserted in the GT. The two argument terms are then represented in the subgoal trie (nodes arg1 and VAR0), where the node's token for arg1 stores the reference to the leaf node of the corresponding term representation inserted in the GT. For the second subgoal call t(X,Y), the argument terms VAR0 and VAR1, representing respectively X and Y, are both simple terms and thus we simply insert two nodes in the subgoal trie to represent them.

When processing answers, the procedure is similar to the one described above for the subgoal calls. For each substitution compound term f(1) and f(2), we also insert first its representation in the GT and then we insert a node in the corresponding answer trie (nodes labeled subs1 and subs2 in Fig. 2) storing the reference to its path in the GT. As f(1) was inserted in the GT at the time of

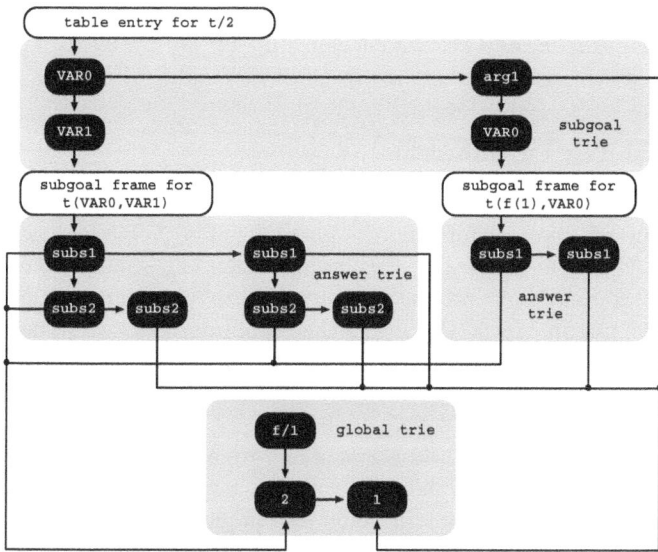

Fig. 2. GT-T's table space organization

the first subgoal call, we only need to insert f(2) (represented by the nodes f/1 and 2), meaning that in fact we only need to insert the token 2 in the GT, in order to represent the full set of answers. So, we are maximizing the sharing of common terms appearing at different arguments or substitution positions. For this particular example, the result is a very compact representation of the GT, as most subgoal calls and/or answers share the same term representations.

On completion of a subgoal, a strategy exists that avoids loading answers from the answer tries using bottom-up unification, performing instead what is called a *completed table optimization* [3]. This optimization implements answer recovery by top-down traversing the completed answer trie and by executing specific WAM-like instructions from the answer trie nodes. In the GT-T design, the difference caused by the existence of the GT is a new set of WAM-like instructions that, instead of working at the level of atoms/terms/functors/lists as in the original design [3], work at the level of the substitution terms. Consider, for example, the loading of four answers for the call t(X,Y). One has two choices for variable X and, to each X, we have two choices for variable Y. In the GT-T design, the answer trie nodes representing the choices for X and for Y (nodes subs1 and subs2 respectively) are compiled with a WAM-like sequence of trie instructions, such as try_subs_compound (for first choices) and trust_subs_compound (for second/last choices). GT-T's compiled tries also include a retry_subs_compound instruction (for intermediate choices), a do_subs_compound instruction (for single choices) and similar variants for simple (non-compound) terms: do_subs_simple, try_subs_simple, retry_subs_simple and trust_subs_simple.

Regarding space reclamation, GT-T uses the child field of the leaf nodes (that is always NULL for a leaf node in the GT) to count the number of references to the path it represents. This feature is of utmost importance for the deletion process of a path in the GT, which can only be performed when there is no reference to it, this is true when the leaf node's child field reaches zero.

3.2   Global Trie for Subterms

The GT-ST was designed taking into account the use of tabling in problems where redundant data occurs more commonly. The GT-ST design maintains most of the GT-T features, but tries to optimize GT's memory usage by representing compound subterms in term arguments as unique entries in the GT. Therefore, we maximize the sharing of the tabled data that is structurally equal at a *second level*, by avoiding the representation of equal compound subterms, and thus preventing situations where the representation of those subterms occur more than once.

Although GT-ST uses the same tree structure as GT-T for implementing the GT, every different path in the GT can now represent a complete term or a subterm of another term, but still being an unique term. Consider, for example the insertion of the term f(g(1)) in the GT. After storing the node representing functor f/1, the process is suspended and the subterm g(1) is inserted as an individual term in the GT. After the complete insertion of subterm g(1) in the GT, the insertion of the main term is resumed by storing a node referencing the g(1) representation in the GT, i.e., by storing a node referencing the leaf node that represents g(1) in the GT.

Despite these structural differences in the GT design, all the remaining data structures remain unaltered. In particular, the GT-T's structure for the subgoal and answer tries, where each path is composed by a fixed number of nodes representing, respectively, the number of arguments for table subgoal calls and the number of substitution terms for tabled answers, is used without changes. Moreover, features regarding the subgoal frame structure used to maintain the chronological order of answers and to implement answer recovery, also remain unchanged. Figure 3 shows an example of how the GT-ST design works by illustrating the resulting data structures for a tabled program with compound subterms.

Initially, the subgoal trie and the GT are empty. Then, a first subgoal call occurs, t(f(g(1),g(1)),Y), and the two argument terms for the call are inserted in the subgoal trie with the compound term being first inserted in the GT. Regarding the insertion of the compound term f(g(1),g(1)) in the GT, we next emphasize the differences between the GT-ST and the GT-T designs.

At first, a node is inserted to represent the functor f/2, but then the insertion of the first subterm g(1) is suspended, since g(1) is a compound term. The compound term g(1) is then inserted as a distinct term in the GT and two nodes, for functor g/1 and integer 1, are then inserted in the GT with the node for functor g/1 being a sibling of the already stored node for functor f/2. After storing g(1) in the GT, the insertion of the main term f(g(1),g(1)) is resumed and a new node, referencing the leaf node of g(1), is inserted as a child node of the node for functor f/2. The construction of the main term then continues

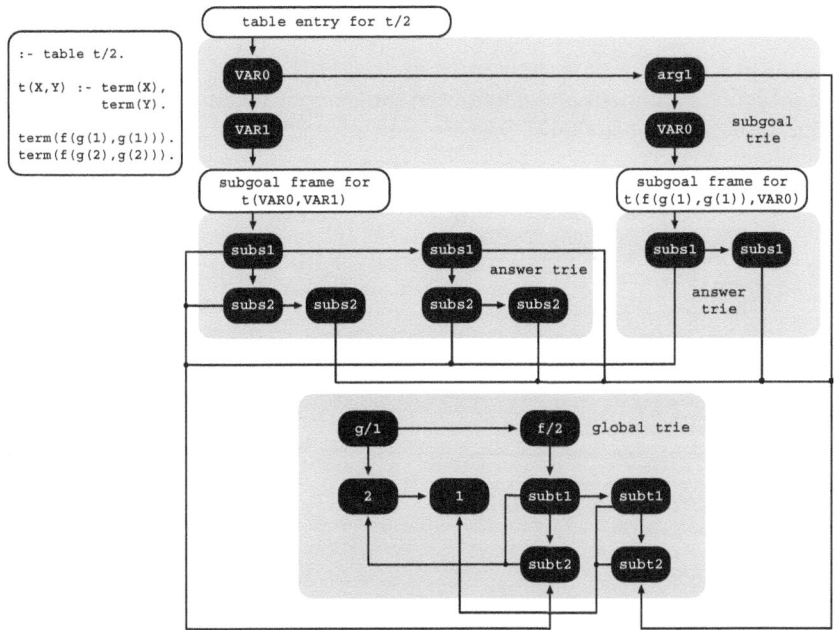

Fig. 3. GT-ST's table space organization

by applying an analogous procedure to its second argument, `g(1)`. However, the term `g(1)` is already stored in the GT, therefore it is only required the insertion of a new node referencing again the leaf node of `g(1)`.

As for the GT-T design, for the second subgoal call `t(X,Y)`, we do not interact with the GT. Both arguments are simple terms and thus we simply insert two nodes, `VAR0` and `VAR1`, in the subgoal trie to represent them.

The procedure used for processing answers is similar to the one just described for the subgoal calls. For each substitution compound term, we first insert the term in the GT and then we insert a node in the corresponding answer trie storing the reference to its path in the GT (nodes labeled `subs1` and `subs2` in Fig. 3). The complete set of answers for both subgoal calls is formed by the substitution terms `f(g(1),g(1))` and `f(g(2),g(2))`. Thus, as `f(g(1),g(1))` was already inserted in the GT when storing the first subgoal call, only `f(g(2),g(2))` needs to be stored in order to represent the whole set of answers. As we are maximizing the sharing of common subterms appearing at different argument or substitution positions, for this particular example, this results in a very compact representation of the GT.

Regarding the completed table optimization and space reclamation, the GT-ST design implements the same GT-T's mechanisms described previously. In particular, for space reclamation, the use of the child field of the leaf nodes (that is always NULL for a leaf node in the GT) to count the number of references to the path it represents can be used as before for subterm counting.

4   Implementation

We then describe the data structures and algorithms for GT-ST's table space design. Figure 4 shows in more detail the table organization previously presented in Fig. 3 for the subgoal call t(X,Y).

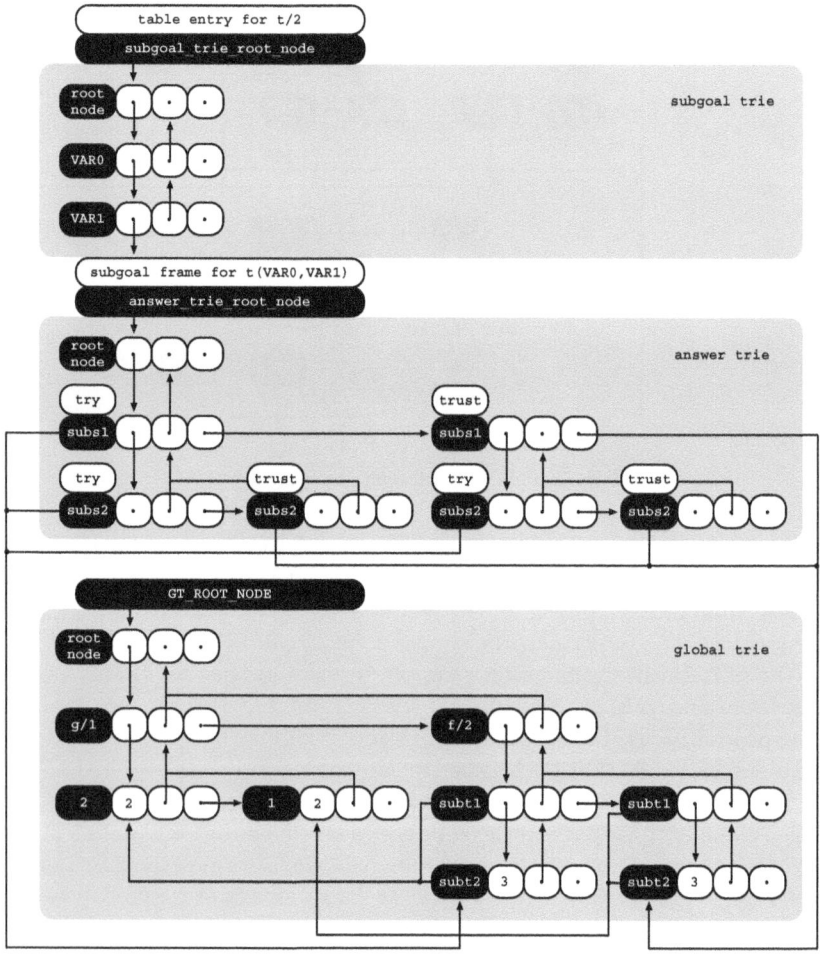

Fig. 4. Implementation of the GT-ST's table space organization

Internally, all tries are represented by a top *root node*, acting as the entry point for the corresponding subgoal, answer or global trie data structure. For the subgoal tries, the root node is stored in the corresponding table entry's subgoal_trie_root_node data field. For the answer tries, the root node is stored in the corresponding subgoal frame's answer_trie_root_node data field. For the GT, the root node is stored in the GT_ROOT_NODE global variable.

Regarding trie nodes, they are internally implemented as 4-field data structures. The first field (`token`) stores the token for the node and the second
(`child`), third (`parent`) and fourth (`sibling`) fields store pointers, respectively,
to the first child node, to the parent node, and to the next sibling node. Remember that for the GT, the leaf node's `child` field is used to count the number
of references to the path it represents. For the answer tries, an additional field
(`code`) is used to support compiled tries.

Traversing a trie to check/insert for new calls or for new answers is implemented by repeatedly invoking a `trie_token_check_insert()` procedure for
each token that represents the call/answer being checked. Given a trie node
`n` and a token `t`, the `trie_token_check_insert()` procedure returns the child
node of `n` that represents the given token `t`. Initially, the procedure traverses
sequentially the list of sibling nodes checking for one representing the given token `t`. If no such node is found, a new trie node is initialized and inserted in the
beginning of the list.

Searching through a list of sibling nodes could be too expensive if we have hun-
dreds of siblings. A threshold value (8 in our implementation) controls whether
to dynamically index the nodes through a hash table, hence providing direct
node access and optimizing search. Further hash collisions are reduced by dynamically expanding the hash tables. For simplicity of presentation, in what
follows, we omit the hashing mechanism.

For YapTab's original table space organization, looking up a term of T tokens
in a trie is thus linear in the number of tokens the term contains plus the number
of sibling nodes visited, i.e., $O(8 \cdot T) = O(T)$. For the GT-T design, we have the
extra cost of inserting a fixed number of A nodes in the original subgoal/answer
tries representing, respectively, the number of A arguments for table subgoal
calls or the number of A substitution terms for tabled answers. Thus, the time
complexity for GT-T is $O(8 \cdot T) + O(A) = O(T)$, since $T > A$. For the GT-ST
design, we have an extra lookup operation in the global trie for each compound
subterm. If S is the number of compound subterms, then the time complexity
for GT-ST is $O(8 \cdot T) + O(A) + O(S) = O(T)$, since $T > S$.

When inserting terms in the table space we need to distinguish two situations:
(i) inserting tabled calls in a subgoal trie structure; and (ii) inserting answers
in a particular answer trie structure. These two situations are handled by the
`trie_subgoal_check_insert()` and `trie_answer_check_insert()` procedures,
respectively. The pseudo-code for the `trie_subgoal_check_insert()` procedure
is shown in Fig. 5. The `trie_answer_check_insert()` procedure works similarly.

For each argument term `t` of the given subgoal call, the procedure first checks
if it is a simple term. If so, `t` is inserted in the current subgoal trie. Otherwise,
`t` is first inserted in the GT and, then, it uses the reference to the leaf node
representing `t` in the GT (`gt_node` in Fig. 5) as the token to be inserted in the
current subgoal trie.

The main difference to the previous GT-T design relies in the insertion of
terms in the GT, and for that we have changed the `trie_term_check_insert()`
procedure in such a way that when a compound term has compound subterms

```
trie_subgoal_check_insert(TABLE_ENTRY te, SUBGOAL_CALL call) {
  sg_node = te->subgoal_trie_root_node
  arity = get_arity(call)
  for (i = 1; i <= arity; i++) {
    t = get_argument_term(call, i)
    if (is_simple_term(t))
      sg_node = trie_token_check_insert(sg_node, t)
    else {                                    // t is a compound term
      gt_node = trie_term_check_insert(GT_ROOT_NODE, t)
      sg_node = trie_token_check_insert(sg_node, gt_node)
    }
  }
  return sg_node
}
```

Fig. 5. Pseudo-code for the `trie_subgoal_check_insert()` procedure

as arguments, the procedure calls itself. Figure 6 shows the pseudo-code for the changes made to the `trie_term_check_insert()` procedure in order to support the new GT-ST design.

As we can see in Fig. 5, the initial call to the `trie_term_check_insert()` procedure is always made with `GT_ROOT_NODE` as the first argument and with a compound term as the second argument (respectively, arguments `gt_node` and t in Fig. 6).

```
trie_term_check_insert(TRIE_NODE gt_node, TERM t) {
  if (is_simple_term(t))
    gt_node = trie_token_check_insert(gt_node, t)
  else {                                      // t is a compound term
    if (gt_node == GT_ROOT_NODE) {
      name = get_name(t)
      arity = get_arity(t)
      gt_node = trie_token_check_insert(gt_node, name)
      for (i = 1; i <= arity; i++) {
        sub_t = get_argument_term(t, i)
        gt_node = trie_term_check_insert(gt_node, sub_t)
      }
    } else {                  // t is a compound subterm of a compound term
      sub_gt_node = trie_term_check_insert(GT_ROOT_NODE, t)
      gt_node = trie_token_check_insert(gt_node, sub_gt_node)
    }
  }
  return gt_node
}
```

Fig. 6. Pseudo-code for the `trie_term_check_insert()` procedure

Initially, the `trie_term_check_insert()` procedure checks if t is a simple term (always false for the initial call) and, if so, t is simply inserted in the GT as a child node of the given `gt_node`. Otherwise, t is a compound term and two situations can occur: (i) if gt_node is GT_ROOT_NODE, then the term's name is inserted in the GT and, for each subterm of t, the procedure is invoked recursively; (ii) if gt_node is not GT_ROOT_NODE, which means that t is a compound subterm of a compound term, the procedure calls itself with GT_ROOT_NODE as the first argument. By doing that, t is inserted as a unique term in the GT. When the procedure returns, the reference sub_gt_node to the leaf node of the subterm's path representation of t in the GT is inserted as a child node of the given gt_node.

Regarding the traversal of the answer tries to consume answers, the GT-ST design follows the same implementation as in the GT-T design, and, in particular, for compound terms, it uses a trie_term_load() procedure to load, from the GT back to the Prolog engine, the substitution term given by the reference stored in the corresponding token field. The main difference to the previous GT-T design is in the cases of subterm references in the GT, where the trie_term_load() procedure calls itself to first load the subterm reference from the GT.

5   Experimental Results

We next present some experimental results comparing YapTab with and without support for the GT-T and GT-ST designs. The environment for our experiments was a PC with a 2.66 GHz Intel(R) Core(TM) 2 Quad CPU and 4 GBytes of memory running the Linux kernel 2.6.24 with YapTab 6.2.0.

To put the performance results in perspective and have a well-defined starting point comparing the GT-T and GT-ST approaches, first we have defined a tabled predicate t/5 that simply stores in the table space terms defined by term/1 facts, and then we used a top query goal test/0 to recursively call t/5 with all combinations of one and two free variables in the arguments. An example of such code for functor terms of arity 1 (1,000 terms in total) is shown next.

```
:- table t/5.
t(A,B,C,D,E) :- term(A), term(B), term(C), term(D), term(E).

test :- t(A,f(1),f(1),f(1),f(1)), fail.          term(f(1)).
...                                               term(f(2)).
test :- t(f(1),f(1),f(1),f(1),A), fail.          term(f(3)).
test :- t(A,B,f(1),f(1),f(1)), fail.
...                                               term(f(998)).
test :- t(f(1),f(1),f(1),A,B), fail.             term(f(999)).
test.                                             term(f(1000)).
```

We experimented the test/0 predicate with 9 different kinds of 1,000 term/1 facts: integers, atoms, functor (with arity 1, 2, 4 and 6) and list (with length 1, 2 and 4) terms. Table 1 shows the table memory usage (column **Mem**), in MBytes, and the execution times, in milliseconds, to store (column **Str**) the tables (first execution) and to load from the tables (second execution) the complete set of answers without (column **Ld**) and with (column **Cmp**) compiled tries for YapTab's original table design (column **YapTab**) and for the GT-T (column

GT-T/YapTab) and GT-ST (column *GT-ST/YapTab*) designs. For GT-T and GT-ST, we only show the ratios over YapTab's original table design. The execution times are the average of five runs.

Table 1. Table memory usage (in MBytes) and store/load times (in milliseconds) comparing YapTab's original table design with the GT-T and GT-ST designs

Terms	YapTab				GT-T/YapTab				GT-ST/YapTab			
	Mem	Str	Ld	Cmp	Mem	Str	Ld	Cmp	Mem	Str	Ld	Cmp
1,000 ints	191	1,270	345	344	1.00	1.05	1.00	1.00	1.00	1.09	1.11	1.07
1,000 atoms	191	1,423	343	406	1.00	1.04	1.01	1.02	1.00	1.04	1.03	1.08
1,000 f/1	191	1,680	542	361	1.00	1.32	1.16	2.10	1.00	1.34	1.17	2.13
1,000 f/2	382	2,295	657	450	0.50	1.10	1.14	1.84	0.50	1.06	1.11	1.88
1,000 f/4	764	3,843	973	631	0.25	0.81	0.98	1.44	0.25	0.78	1.04	1.53
1,000 f/6	1,146	5,181	1,514	798	0.17	0.72	0.72	1.38	0.17	0.66	0.71	1.36
1,000 [ ]/1	382	2,215	507	466	0.50	1.08	1.05	1.61	0.50	1.10	1.02	1.58
1,000 [ ]/2	764	3,832	818	604	0.25	0.80	0.94	1.38	0.25	1.00	1.05	1.48
1,000 [ ]/4	1,528	6,566	1,841	1,066	0.13	0.63	0.54	0.96	0.13	0.89	0.66	1.14
Average					0.53	0.95	0.95	1.42	0.53	0.99	0.99	1.47

The results in Table 1 suggest that both GT designs are a very good approach to reduce memory usage and that this reduction increases proportionally to the length and redundancy of the terms stored in the GT. In particular, for functor and list terms, the results show an increasing and very significant reduction on memory usage, for both GT-T and GT-ST approaches. The results for the special cases of integer and atom terms are also very interesting as they show that the cost of representing only simple terms in the respective tries. Note that, although, integer and atom terms are only represented in the respective tries, it is necessary to check for these types of terms, in order to proceed with the respective store/load algorithm.

Regarding execution time, the results suggest that, in general, GT-ST spends more time in the store and load term procedures than GT-T. Such behaviour can be easily explained by the fact that, the GT-ST's storing and loading algorithms have more sub-cases to process in order to support subterms. These results also seem to indicate that memory reduction for small sized terms, generally comes at a price in storing time (between 4% and 32% more for GT-T and between 4% and 34% more for GT-ST in these experiments). The opposite occurs in the tests where term's length are higher (between 19% and 37% less for GT-T and 11% and 34% less for GT-ST). Note that with GT-T and GT-ST support, we pay the cost of navigating in two tries when checking/storing/loading a term. Moreover, in some situations, the cost of storing a new term in an empty/small trie can be less than the cost of navigating in the GT, even when the term is already stored in the GT. However, our results seem to suggest that this cost decreases proportionally to the length and redundancy of the terms stored in the GT. In particular, for functor and list terms, GT-T and GT-ST support showed to outperform the original YapTab design when we increase the length of the terms stored in the GT.

The results obtained for loading terms also show some gains without compiled tries (around 5% for GT-T and 1% for GT-ST on average) but, when using compiled tries the results show some significant costs on execution time (around 42% for GT-T and 47% for GT-ST on average). We believe that this cost is smaller for GT-T as a result of having less sub-cases in the storing/loading algorithms. On the other hand, we also believe that some cache behaviour effects, reduce the costs on execution time, for both GT designs. As we need to navigate in the GT for each substitution term, we kept accessing the same GT nodes, thus reducing eventual cache misses. This seems to be the reason why for list terms of length 4, GT-T outperforms the original YapTab design, both without and with compiled tries. Note that, for this particular case, both GT-T and GT-ST only consumes 13% of the memory used with the original YapTab design.

Next, we experimented with a new set of tests specially designed to provide more expressive results regarding the comparison between the GT-ST and the GT-T designs. In this tests, we have defined a tabled predicate t/1 that simply stores in the table space terms defined by term/1 facts and then we used a test/0 predicate to call t/1 with a free variable. We experimented the test/0 predicate with 9 different sets of 500,000 term facts of compound terms (with arity 1, 2, 3) where its arguments are also compound subterms (with arity 1, 3, 5). An example of such code for a functor term f/2 with argument subterms g/3 (500,000 terms in total) is shown next.

```
:- table t/1.                      test :- t(A), fail.
t(A) :- term(A).                   test.

term(f(g(1,1,1), g(1,1,1))).
term(f(g(2,2,2), g(2,2,2))).
term(f(g(3,3,3), g(3,3,3))).
...
term(f(g(499998,499998,499998), g(499998,499998,499998))).
term(f(g(499999,499999,499999), g(499999,499999,499999))).
term(f(g(500000,500000,500000), g(500000,500000,500000))).
```

As opposed to the previous experiments, here we just used one free variable for the tabled predicate t/1. This difference is necessary because, when we have more than one free variable and we produce different combinations between those free variables, we are raising the number of nodes represented in the local tries. More precisely, different combinations of free variables raises the number of answers and therefore the number of nodes in the local answer tries.

Table 2 shows the table memory usage (columns *Memory*) composed by two columns, one for total memory (columns *Total*) and the other for GT's memory only (columns *GT*), in MBytes, and the execution times, in milliseconds, to store (columns *Str*) the tables (first execution) and to load from the tables (second execution) the complete set of answers without (columns *Ld*) and with (columns *Cmp*) compiled tries using the GT-T table design (column *GT-T*), and using the GT-ST design (column *GT-ST/GT-T*). For the values referring the GT-ST we only show the ratios over the GT-T design. Since the main purpose of this second set of experiments is to compare the differences between GT-T and GT-ST, we did not include YapTab's original table design in these experiments. The execution times are the average of five runs.

Table 2. Table memory usage (in MBytes) and store/load times (in seconds) comparing the GT-T and GT-ST designs for subterm representation

Terms	GT-T Memory					GT-ST/GT-T Memory				
	Total	GT	Str	Ld	Cmp	Total	GT	Str	Ld	Cmp
f/1										
500,000 g/1	17.17	7.63	126	28	51	1.44	2.00	1.55	1.14	1.00
500,000 g/3	32.43	22.89	198	34	61	1.24	1.33	3.29	1.12	1.25
500,000 g/5	47.68	38.15	293	47	83	1.16	1.20	1.46	1.00	**0.99**
f/2										
500,000 g/1	32.43	22.89	203	38	71	1.00	1.00	1.28	1.13	1.09
500,000 g/3	62.94	53.41	45	60	103	**0.76**	**0.71**	1.18	**0.84**	**0.95**
500,000 g/5	93.46	83.92	438	111	146	**0.67**	**0.64**	1.10	**0.67**	**0.80**
f/3										
500,000 g/1	47.68	38.15	296	50	89	**0.84**	**0.80**	2.87	1.02	1.03
500,000 g/3	93.46	83.92	616	142	164	**0.59**	**0.55**	1.25	**0.80**	**0.85**
500,000 g/5	139.24	129.7	832	197	224	**0.51**	**0.47**	0.96	**0.67**	**0.74**
Average						**0.96**	**0.97**	**0.93**	**0.97**	**0.91**

The results in Table 2 suggest that GT-ST support has potential to outperform GT-T's design with significant reductions on memory usage and execution time for programs with increasing redundancy on compound subterms. However, the results also show that, for some base cases, the storing process can be a very expensive procedure when compared with GT-T's design.

In general, the results suggest three different situations. For f/1 terms, the costs for GT-ST are globally higher. This happens because GT-ST needs to store one extra node for every distinct subterm representation and there is no redundancy in the f/1 subterms. However, the results show that the memory and execution costs can be reduced when the subterm's arity increases from g/1 to g/5. This occurs because the cost of the extra node for each subterm became diluted in the number of nodes represented in the GT.

For f/2 terms, a particular situation occurs for the case of g/1 subterms, where the memory spent is the same for both designs. This happens because the extra node used by GT-ST, to represent the reference to the subterm representation, is balanced by the arity of the functor term f/2. From this point on, for the remaining f/2 terms and all the f/3 terms, the GT-ST always outperforms the GT-T, not only for the system's memory, but also for the execution times with and without compiled tries. These results suggest that, at least for some applications, GT-ST support has potential to achieve significant reductions on memory usage and execution time when compared with GT-T's design.

6   Conclusions

We have presented a new design for the table space organization, named *Global Trie for Subterms (GT-ST)*, that extends the previous *Global Trie for Terms (GT-T)* design. The GT-ST design maintains most of the GT-T features, but tries to optimize GT's memory usage by avoiding the representation of equal compound subterms, thus preventing situations where the representation of those subterms occur more than once and maximizing the sharing of the tabled data that is structurally equal at a second level.

Experimental results, using the YapTab tabling system, show that GT-ST support has potential to achieve significant reductions on memory usage and execution time for programs with increasing compound subterms in term arguments, without compromising the execution time for other programs.

Further work will include seeking real-world applications, that pose many subgoal queries possibly with a large number of redundant answers, thus allowing us to improve and expand the current implementation. In particular, we intend to study how alternative/complementary designs for the table space organization can further reduce redundancy in term representation.

Acknowledgments. This work has been partially supported by the FCT research projects HORUS (PTDC/EIA-EIA/100897/2008) and LEAP (PTDC/EIA-CCO/112158/2009).

References

1. Chen, W., Warren, D.S.: Tabled Evaluation with Delaying for General Logic Programs. Journal of the ACM 43(1), 20–74 (1996)
2. Sagonas, K., Swift, T.: An Abstract Machine for Tabled Execution of Fixed-Order Stratified Logic Programs. ACM Transactions on Programming Languages and Systems 20(3), 586–634 (1998)
3. Ramakrishnan, I.V., Rao, P., Sagonas, K., Swift, T., Warren, D.S.: Efficient Access Mechanisms for Tabled Logic Programs. Journal of Logic Programming 38(1), 31–54 (1999)
4. Rocha, R., Fonseca, N.A., Santos Costa, V.: On Applying Tabling to Inductive Logic Programming. In: Gama, J., Camacho, R., Brazdil, P.B., Jorge, A.M., Torgo, L. (eds.) ECML 2005. LNCS (LNAI), vol. 3720, pp. 707–714. Springer, Heidelberg (2005)
5. Costa, J., Rocha, R.: One Table Fits All. In: Gill, A., Swift, T. (eds.) PADL 2009. LNCS, vol. 5418, pp. 195–208. Springer, Heidelberg (2008)
6. Costa, J., Raimundo, J., Rocha, R.: A Term-Based Global Trie for Tabled Logic Programs. In: Hill, P.M., Warren, D.S. (eds.) ICLP 2009. LNCS, vol. 5649, pp. 205–219. Springer, Heidelberg (2009)
7. Rocha, R., Silva, F., Santos Costa, V.: On applying or-parallelism and tabling to logic programs. Theory and Practice of Logic Programming 5(1 & 2), 161–205 (2005)
8. Bachmair, L., Chen, T., Ramakrishnan, I.V.: Associative Commutative Discrimination Nets. In: Gaudel, M.-C., Jouannaud, J.-P. (eds.) CAAP 1993, FASE 1993, and TAPSOFT 1993. LNCS, vol. 668, pp. 61–74. Springer, Heidelberg (1993)

Intention-Based Decision Making with Evolution Prospection

The Anh Han and Luís Moniz Pereira

Centro de Inteligência Artificial (CENTRIA)
Departamento de Informtica, Faculdade de Cincias e Tecnologia
Universidade Nova de Lisboa, 2829-516 Caparica, Portugal
h.anh@fct.unl.pt, lmp@di.fct.unl.pt

Abstract. We explore a coherent combination, for decision making, of two Logic Programming based implemented systems, Evolution Prospection and Intention Recognition. The Evolution Prospection system has proven to be a powerful system for decision making, designing and implementing several kinds of preferences and useful environment-triggering constructs. It is here enhanced with an ability to recognize intentions of other agents—an important aspect not explored so far. The usage and usefulness of the combined system is illustrated with several extended examples.

Keywords: Decision Making, Evolution Prospection, Preferences, Intention Recognition, Logic Programming.

1 Introduction

Given the important role that intentions play in the way we make decisions [4,24], one would expect intentions to occupy a substantial place in any theory of action. Surprisingly enough, in what is perhaps the most influential theory of action—rational choice theory—which includes the theory of decision making—explicit reference is made to actions, strategies, information, outcomes and preferences but not to intentions.

This is not to say that no attention has been paid to the relationship between rational choice and intentions. Quite the contrary, a rich philosophical literature has developed on the relation between rationality and intentions (see for example [29]). However, to our knowledge, there has been no real attempt to model and implement the role of intentions in decision making, within a rational choice framework.

In this paper, we set forth a coherent Logic Programming (LP) based system for decision making—which extends the existing work on Evolution Prospection (EP) for decision making [18,19]—but taking into consideration now the intentions of other agents. Obviously, when being immersed in a multi-agent system, knowing the intentions of other agents can benefit the agent in a number of ways. It enable the recognizing agents to predict what other agents will do next or might have done before—thereby, being able to plan in advance and taking

L. Antunes and H.S. Pinto (Eds.): EPIA 2011, LNAI 7026, pp. 254–267, 2011.

the best advantage from the prediction, or acting to take remedial action. In addition, an important role of recognizing intentions is to enable coordination of your own actions and in collaborating with others [5,4]. We have recently studied the role of intention recognition in the evolution of cooperative behavior [14,27], showing that intention recognition strongly promotes the emergence of cooperation in populations of self-regarding individuals [11,10].

The Evolution Prospection system is an implemented LP-based system for decision making [19,22] (described in Section 2). An EP agent can prospectively look ahead a number of steps into the future to choose the best course of evolution that satisfies a goal. This is achieved by designing and implementing several kinds of prior and post preferences and several useful environment-triggering constructs for decision making.

In order to take into account the intentions of other agents in decision making processes, we integrate into EP a previously and separately implemented, but also LP-based, intention recognition system [20,22]. Intention recognition can be defined as the process of inferring the intention or goal of one other agent (called *"individual intention recognition"*) or a group of other agents (called *"collective intention recognition"*) through their observable actions or their actions' observable effects [13,26,7,28]. The intention recognition system performs via Causal Bayesian Networks [15] and plan generation techniques. We will briefly recall the system in Section 3.

2   Evolution Prospection

2.1   Preliminary

The implemented EP system has been proven to be useful for decision making [19]. It has been applied for providing appropriate suggestions for elderly people in Ambient Intelligence domain [8,7]. The advance and easiness of expressing preferences in EP [17,19] enable to closely take into account elders' preferences. The EP system is implemented on top of XSB Prolog [31]. We next describe the constructs of EP, to the extent we use them here. A full account can be found in [19].

Language. Let $\mathcal{L}$ be a first order language. A domain literal in $\mathcal{L}$ is a domain atom A or its default negation *not* A. The latter is used to express that the atom is false by default (Closed World Assumption). A domain rule in $\mathcal{L}$ is a rule of the form:

$$A \leftarrow L_1, \ldots, L_t \qquad (t \geq 0)$$

where A is a domain atom and $L_1, \ldots, L_t$ are domain literals. An integrity constraint in $\mathcal{L}$ is a rule with an empty head. A (logic) program P over $\mathcal{L}$ is a set of domain rules and integrity constraints, standing for all their ground instances.

In this paper, we consider solely Normal Logic Programs (NLPs), those whose heads of rules are positive literals, i.e. positive atoms, or empty. We focus furthermore on abductive logic programs, i.e. NLPs allowing for abducibles – user-specified positive literals without rules, whose truth-value is not fixed. Abducibles instances or their default negations may appear in bodies of rules, like any other literal. They stand for hypotheses, each of which may independently be assumed true, in positive literal or default negation form, as the case may be, in order to produce an abductive solution to a query.

Definition 1 (Abductive Solution). *An abductive solution is a consistent collection of abducible instances or their negations that, when replaced by true everywhere in P, affords a model of P that satisfies the query true and the ICs – a so-called abductive model, for the specific semantics being used on P.*

Active Goals. In each cycle of its evolution the agent has a set of active goals or desires. We introduce the *on_observe*/1 predicate, which we consider as representing active goals or desires that, once triggered by the observations figuring in its rule bodies, cause the agent to attempt their satisfaction by launching all the queries standing for them, or using preferences to select them. The rule for an active goal AG is of the form:

$$on\_observe(AG) \leftarrow L_1, ..., L_t \ (t \geq 0)$$

where $L_1,...,L_t$ are domain literals. During evolution, an active goal may be triggered by some events, previous commitments or some history-related information. When starting a cycle, the agent collects its active goals by finding all the $on\_observe(AG)$ that hold under the initial theory without performing any abduction, then finds abductive solutions for their conjunction.

Preferring Abducibles. An abducible A can be assumed only if it is a considered one, i.e. if it is expected in the given situation, and, moreover, there is no expectation to the contrary

$$consider(A) \leftarrow expect(A), \ not \ expect\_not(A), \ A$$

The rules about expectations are domain-specific knowledge contained in the theory of the program, and effectively constrain the hypotheses available in a situation. To express preference criteria among abducibles, we envisage an extended language $\mathcal{L}^\star$. A preference atom in $\mathcal{L}^\star$ is of the form $a \lhd b$, where a and b are abducibles. It means that if b can be assumed (i.e. considered), then $a \lhd b$ forces a to be assumed too if it can. A preference rule in $\mathcal{L}^\star$ is of the form:

$$a \lhd b \leftarrow L_1, ..., L_t \ (t \geq 0)$$

where $L_1, ..., L_t$ are domain literals over $\mathcal{L}^\star$.

A priori preferences are used to produce the most interesting or relevant conjectures about possible future states. They are taken into account when generating possible scenarios (abductive solutions), which will subsequently be preferred amongst each other a posteriori.

A Posteriori Preferences. Having computed possible scenarios, represented by abductive solutions, more favorable scenarios can be preferred a posteriori. Typically, *a posteriori* preferences are performed by evaluating consequences of abducibles in abductive solutions. An *a posteriori* preference has the form:

$$A_i \ll A_j \leftarrow holds\_given(L_i, A_i),\ holds\_given(L_j, A_j)$$

where A_i, A_j are abductive solutions and L_i, L_j are domain literals. This means that A_i is preferred to A_j a posteriori if L_i and L_j are true as the side-effects of abductive solutions A_i and A_j, respectively, without any further abduction when testing for the side-effects. Optionally, in the body of the preference rule there can be any Prolog predicate used to quantitatively compare the consequences of the two abductive solutions.

Evolution Result A Posteriori Preference. While looking ahead a number of steps into the future, the agent is confronted with the problem of having several different possible courses of evolution. It needs to be able to prefer amongst them to determine the best courses from its present state (and any state in general). The *a posteriori* preferences are no longer appropriate, since they can be used to evaluate only one-step-far consequences of a commitment. The agent should be able to also declaratively specify preference amongst evolutions through quantitatively or qualitatively evaluating the consequences or side-effects of each evolution choice.

A *posteriori* preference is generalized to prefer between two evolutions. An *evolution result a posteriori* preference is performed by evaluating consequences of following some evolutions. The agent must use the imagination (look-ahead capability) and present knowledge to evaluate the consequences of evolving according to a particular course of evolution. An *evolution result a posteriori preference* rule has the form:

$$E_i \lll E_j \leftarrow holds\_in\_evol(L_i, E_i), holds\_in\_evol(L_j, E_j)$$

where E_i, E_j are possible evolutions and L_i, L_j are domain literals. This preference implies that E_i is preferred to E_j if L_i and L_j are true as evolution history side-effects when evolving according to E_i or E_j, respectively, without making further abductions when just checking for the side-effects. Optionally, in the body of the preference rule there can be recourse to any Prolog predicate, used to quantitatively compare the consequences of the two evolutions for decision making.

3   Intention Recognition

In [20], a method for individual intention recognition via Causal Bayesian Nets (CBN) and plan generation techniques was presented. The CBN is used to generate conceivable intentions of the intending agent and compute their likelihood conditional on the initially available observations, and so allow to filter out the much less likely ones. The plan generator thus only needs to deal with the remaining more relevant intentions, because they are more probable or credible, rather than all conceivable intentions. In this work we do not need the network causal property; hence, only background of the naive Bayesian Networks is recalled. Note that the first component of the intention recognition system is implemented based on P-log [12,1]—a probabilistic logic framework—implemented on top of XSB Prolog [31]. The second component is also implemented on top of XSB Prolog. This allows us an appropriate and coherent integration of EP and the intention recognition system.

Definition 2. *A Bayes Network is a pair consisting of a directed acyclic graph (DAG) whose nodes represent variables and missing edges encode conditional independencies between the variables, and an associated probability distribution satisfying the Markov assumption of conditional independence, saying that variables are independent of non-descendants given their parents in the graph [16,15].*

Definition 3. *Let G be a DAG that represents causal relations between its nodes. For two nodes A and B of G, if there is an edge from A to B (i.e. A is a direct cause of B), A is called a parent of B, and B is a child of A. The set of parent nodes of a node A is denoted by* parents(A). *Ancestor nodes of A are parents of A or parents of some ancestor nodes of A. If node A has no parents (parents(A) = ∅), it is called a* top *node. If A has no child, it is called a* bottom *node. The nodes which are neither top nor bottom are said* intermediate. *If the value of a node is observed, the node is said to be an* evidence *node.*

In a BN, associated with each intermediate node of its DAG is a specification of the distribution of its variable, say A, conditioned on its parents in the graph, i.e. $P(A|parents(A))$ is specified. For a top node, the unconditional distribution of the variable is specified. These distributions are called Conditional Probability Distribution (CPD) of the BN.

Suppose nodes of the DAG form a causally sufficient set, i.e. no common causes of any two nodes are omitted, the joint distribution of all node values of a causally sufficient can be determined as the product of conditional probabilities of the value of each node on its parents $P(X_1, ..., X_N) = \prod_{i=1}^{N} P(X_i|parents(X_i))$, where $V = \{X_i | 1 \leq i \leq N\}$ is the set of nodes of the DAG.

Suppose there is a set of evidence nodes in the DAG, say $O = \{O_1, ..., O_m\} \subset V$. We can determine the conditional probability of a variable X given the observed value of evidence nodes by using the conditional probability formula

$$P(X|O) = \frac{P(X,O)}{P(O)} = \frac{P(X, O_1, ..., O_m)}{P(O_1, ..., O_m)} \tag{1}$$

where the numerator and denominator are computed by summing up the joint probabilities over all absent variables with respect to V (see [20] for details).

In short, to define a BN, one needs to specify the structure of the network, its CPD and the prior probability distribution of the top nodes.

Network Structure for Intention Recognition. The first phase of the intention recognition system is to find out how likely each conceivable intention is, based on current observations such as observed actions of the intending agent or the effects of its actions had in the environment. A conceivable intention is the one having causal relations to all current observations. It is brought out by using a CBN with nodes standing for binary random variables that represent causes, intentions, actions and effects.

Intentions are represented by those nodes whose ancestor nodes stand for causes that give rise to intentions. Intuitively, we extend Heinze's tri-level model [13,22] with a so-called pre-intentional level that describes the causes of intentions, used to estimate prior probabilities of the intentions. However, if these prior probabilities can be specified without considering the causes, intentions are represented by top nodes (i.e. nodes that have no parents). These reflect the problem context or the intending agent's mental state.

Observed actions are represented as children of the intentions that causally affect them. Observable effects are represented as bottom nodes (having no children). They can be children of observed action nodes, of intention nodes, or of some unobserved actions that might cause the observable effects that are added as children of the intention nodes.

The causal relations among nodes of the CBNs (e.g. which causes give rise to an intention, which intentions trigger an action, which actions have an effect), as well as their Conditional Probability Distribution tables and the distribution of the top nodes, are specified by domain experts. However, they might be learnt mechanically from plan corpora [2,9]. In addition, as it is usually not easy to create the whole BN, we have recently provided a method to incrementally construct it from simple, easily maintained, small fragments of Bayesian Networks [9]. It would enable an easy deployment of the method for real application domains.

Example 1 (Fox-Crow). Consider Fox-Crow story, adapted from Aesop's fable. There is a crow, holding a cheese. A fox, being hungry, approaches the crow and praises her, hoping that the crow will sing and the cheese will fall down near him. Unfortunately for the fox, the crow is very intelligent, having the ability of intention recognition.

The Fox's intentions CBN is depicted in the Figure 1. The initial possible intentions of Fox that Crow comes up with are: Food - $i(F)$, Please - $i(P)$ and Territory - $i(T)$. The facts that might give rise to those intentions are how friendly the Fox is (*Friendly_fox*) and how hungry he is (*Hungry_fox*). Currently, there is only one observation which is: Fox praised Crow (*Praised*). Using formula (1) we can compute the probability of each intention conditional on this observation. More details and examples can be found in [22].

4   Evolution Prospection with Intention Recognition

There are several ways an EP agent can benefit from the ability to recognize intentions of other agents, both in friendly and hostile settings. Knowing the

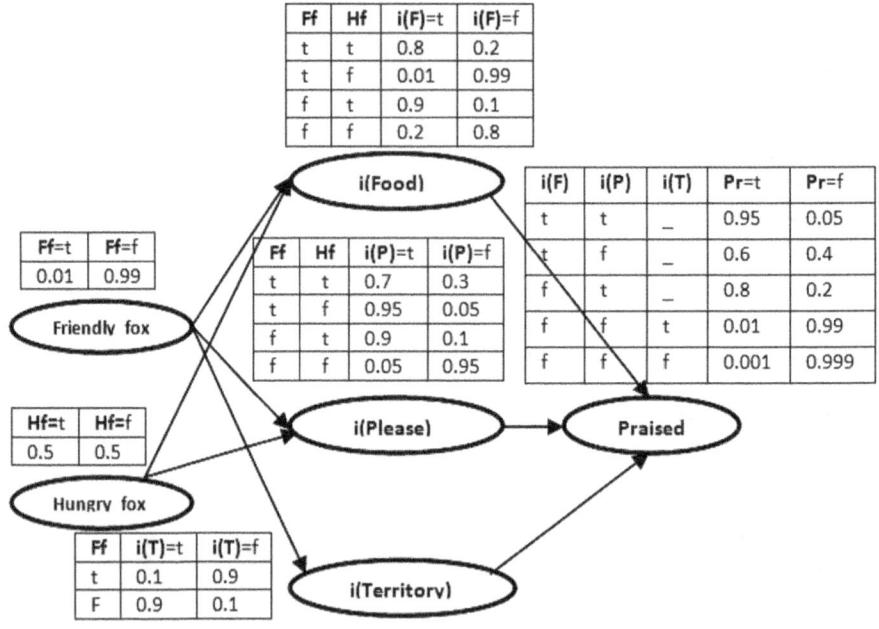

Fig. 1. Fox's Intentions CBN

intention of an agent is a means to predict what he will do next or might have done before. The recognizing agent can then plan in advance to take the best advantage of the prediction, or act to take remedial action. Technically, in EP system, this new kind of knowledge may impinge on the body of several EP constructs, such as active goals, expectation and counter-expectation rules, preference rules, context-sensitive integrity constraints, etc., providing a new kind of trigger. In the sequel we draw closer attention to some of those constructs.

4.1   Intention Triggering Active Goals

Recall that an active goal has the form

$$on\_observe(AG) \leftarrow L_1, ..., L_t \ (t \geq 0)$$

where $L_1, ..., L_t$ are domain literals. At the beginning of each cycle of evolution, those literals are checked with respect to the current evolving knowledge base and trigger the active goal if they all hold. Now, for intention triggering active goals, the domain literals in the body can be some predicate, either directly or indirectly, affected by intentions of other agents.

It is easily seen that intention triggering active goals are ubiquitous. New goals often appear when we recognize some intentions in others. In a friendly setting, we might want to help others achieve their intention, which is generally represented as follows

```
on_observe(help_achieve_goal(G)) <-
                friend(P), has_intention(P,G).
```

while in a hostile setting, we probably want to prevent the opponents to achieve their goal

```
on_observe(prevent_achieve_goal(G)) <-
                opponent(P), has_intention(P,G).
```

or, perhaps we simply want to plan in advance to take advantage of the hypothetical future obtained when the intending agent employs the plan that achieves his intention

```
on_observe(take_advantage(F)) <-  agent(P),
                has_intention(P,G), future(employ(G),F).
```

Note that the reserved Prolog predicate $has\_intention(P,G)$ holds if person P has an intention or goal G—which is validated by the presented intention recognition system (Section 3). Once a predicate $has\_intention(P,G)$ is called, the (integrated) system triggers its intention recognition component to evaluate if G is the most likely intention of the observed agent (i.e., P). One can also extend to consider N-best intention approach, i.e., assess whether the intention of the agent is amongst the most N likely intentions (see, e.g., our recent work in [9]). Sometimes one needs to be more cautious about the intention of others. Furthermore, it has been shown that by increasing N, the recognition accuracy in significantly improved [3,9]. In general, any intention recognition method can be considered, but to facilitate the integration, LP-based intention recognition method as we adopt here, is most favorable.

Let us look a little closer at each setting, providing some ideas how they can be enacted. When helping someone to achieve an intention, what we need to do is to help him/her with executing a plan achieving that intention successfully, i.e., all the actions involved in that plan can be executed. In contrast, in order to prevent an intention from being achieved, we need to guarantee that all possible plans achieving the intention cannot be executed successfully. For that, at least one action in each plan must be prevented, if the plan is conformant, i.e., a sequence of actions; in case of a conditional plan (see for example [20]), each branch is considered as a conformant plan and must be prevented.

4.2   Intention Triggering Preferences

Having recognized an intention of another agent, the recognizing agent may either favor or disfavor an abducible (*a priori* preferences), an abductive solution (*a posteriori* preferences) or an evolution (*evolution result a posteriori* preferences) with respect to another, respectively, depending on the setting they are in. If they are in a friendly setting, the one which provides more support to

achieve the intention is more favored; in contrast, in a hostile setting, the one
providing more support is disfavored. The recognizing agent may also favor the
one which takes better advantage of the recognized intention.

To illustrate the usage of intention triggering *a priori* preferences, in the sequel
we revise the Tea-Coffee example (see [19]).

Example 2 (Tea-Coffee with Intention Recognition). Being thirsty, I consider
making tea or coffee. I realize that my roommate, John, also wants to have
a drink. To be friendly, I want to take into account his intention when making
my choice. This scenario is represented with the following EP program.

```
1. abds([coffee/0, tea/0]).
2. expect(coffee).    expect(tea).
3. on_observed(drink) <- thirsty.
   drink <- tea.        drink <- coffee.
4. expect_not(coffee) <- blood_high_pressure.
5. coffee <| tea <- has_intention(john,coffee).
   tea    <| coffee <- has_intention(john,tea).
```

Fig. 2. Tea-Coffee Considering Intentions

There are two abducibles, *coffee* and *tea*, declared in line 1. Both abducibles are
expected (line 2). The only active goal is to *drink*, which is triggered when being
thirsty (line 3). The rule in line 4 states that *coffee* is not expected if the blood
pressure is high.

In line 5, the first preference says that *tea* is preferable, a priori, to *coffee* if
John intends to drink *tea*; and vice versa, if John intends to drink *coffee*, *coffee* is
preferable. The recognition of what John intends is done by the intention recognition system described above (Section 3)—which is triggered when a reserved
predicate *has_intention/2* is called.

Next, to illustrate other kind of preferences, consider the following revised
extended version of the saving city example, presented in [19].

Example 3 (Saving cities by means of intention recognition). During war time,
agent David, a general, needs to decide to save a city from his enemy's attack
or leave it to keep the military resource, which might be important for some
future purpose. David has recognized that a third party is intending to make an
attack to the enemy on the next day. David will have a good chance to defeat
the enemy if he has enough military resource to coordinate with the third party.
The described scenario is coded with the program in Figure 3.

In the first cycle of evolution, there are two abducibles, *save* and *leave*, declared
in line 1, to solve the active goal *choose*—which is triggered when *being attacked*
(line 3). Similar to the original version in [19], in the case of being a bad general
who just sees the situation at hand, David would choose to *save the city* since
it would save more people (5000 vs. 0, line 4), i.e. the *a posteriori* preference in

```
1. abds([save/0, leave/0]).
2. expect(save).           expect(leave).
3. on_observe(choose) <- be_attacked.
   choose <- save.                        choose <- leave.
4. save_men(5000) <- save.                save_men(0) <- leave.
   lose_resource  <- save.                save_resource <- leave.
5. Ai << Aj <- holds_given(save_men(Ni), Ai),
               holds_given(save_men(Nj), Aj), Ni > Nj.

6. on_observe(decide)  <-    decide_strategy.
   decide <- stay_still.
   decide <- counter_attack.
7. good_opportunity <-    has_intention(third_party,attack).
   expect(counter_attack) <- good_opportunity, save_resource.
   expect(stay_still).
8. pr(win,0.9)  <- counter_attack.
   pr(win,0.01) <- stay_still.
9. Ei <<< Ej <- holds_in_evol(pr(win,Pi), Ei),
                holds_in_evol(pr(win,Pj), Ej), Pi > Pj.
```

Fig. 3. Saving or Leaving

line 5 is taken into account immediately, to rule out the case of leaving the city since it would save less people. Then, next day, he would not be able to attack since the military resource is not saved (line 7), and that leads to the outcome with very small probability of winning the whole war (line 8).

But, fortunately, being able to look ahead plus to do intention recognition, David can see that on the next day, if he has enough military resources, he has a good opportunity to make a counter-attack on his enemy (line 7), by coordinating with a third party who exhibits the intention to attack the enemy on that day as well; and a successful counter-attack would lead to a very much higher probability of winning the conflict as a whole (line 8). The *evolution result a posteriori* preference is employed in line 9 to prefer the evolution with higher probability of winning the whole conflict.

In this example we can see, in line 7, how a detected intention of another agent can be used to enhance the decision making process. It is achieved by providing an (indirect) trigger for an abducible expectation which affects the *evolution result a posteriori* preference in line 9.

4.3   Hostile Setting

In this hostile setting, having confirmed the intention, and the plans achieving that intention being followed by the intending agent, the recognizing agent must act to prevent those plans from happening, i.e., prevent at least one action of each plan to be successfully executed; and in case of impossibility to do so, act to reduce as much as possible the losses.

Example 4 (Fox-Crow, cont'd). Suppose in Example 1, the final confirmed Fox's intention is that of getting food (to see how it is actually recognized look at ref. [20]). Having recognized Fox's intention, what should Crow do to prevent Fox from achieving it? The following EP program helps Crow with that.

```
1. abds([decline/0, sing/0, hide/2, eat/2, has_food/0, find_new_food/0]).
2. expect(decline). expect(sing). expect(hide(_,_)). expect(eat(_,_)).
3. on_observe(not_losing_cheese) <- has_intention(fox, food).
   not_losing_cheese <- decline.
   not_losing_cheese <- hide(crow,cheese), sing.
   not_losing_cheese <- eat(crow,cheese), sing.
4. expect_not(eat(A,cheese)) <- animal(A), full(A).
   animal(crow).
5. <- decline, sing.     <- hide(crow,cheese), eat(crow,cheese).
6. eat(crow,cheese) <| hide(crow,cheese).
7. no_pleasure <- decline.    has_pleasure <- sing.
8. Ai << Aj <- holds_given(has_pleasure,Ai), holds_given(no_pleasure,Aj).

9. on_observe(feed_children) <- hungry(children).
   feed_children <- has_food.    feed_children <- find_new_food.
   <- has_food, find_new_food.
10.expect(has_food) <- decline, not eat(crow,cheese).
   expect(has_food) <- hide(crow,cheese), not stolen(cheese).
   expect(find_new_food).
11.Ei <<< Ej <- hungry(children), holds_in_evol(had_food,Ei),
                holds_in_evol(find_new_food,Ej).
12.Ei <<< Ej <- holds_in_evol(has_pleasure,Ei),
                holds_in_evol(no_pleasure,Ej).
```

There are two possible ways so as not to lose the Food to Fox, either simply decline to sing (but thereby missing the pleasure of singing) or hide or eat the cheese before singing.

Line 1 is the declaration of program abducibles (the last two abducibles are for the usage in the second phase, starting from line 9). All of them are always expected (line 2). The counter-expectation rule in line 4 states that an animal is not expected to eat if he is full. The integrity constraints in line 5 say that Crow cannot decline to sing and sing, hide and eat the cheese, at the same time. The *a priori* preference in line 6 states that eating the cheese is always preferred to hiding it (since it may be stolen), of course, just in case eating is a possible solution (this is assured in our semantics of *a priori* preferences [19]).

Suppose Crow is not full. Then, the counter expectation in line 4 does not hold. Thus, there are two possible abductive solutions: *[decline]* and *[eat(crow,cheese), sing]* (since the *a priori* preference prevents the choice containing *hiding*).

Next, the *a posteriori* preference in line 8 is taken into account and rules out the abductive solution containing *decline* since it leads to having *no pleasure* which is less preferred to *has pleasure*—the consequence of the second solution that contains *sing* (line 7). In short, the final solution is that Crow eats the cheese then sings, without losing the cheese to Fox and having the pleasure of singing.

Now, let us consider a smarter Crow who is capable of looking further ahead into the future in order to solve longer term goals. Suppose that Crow knows that her children will be hungry later on, in the next stage of evolution (line 9); eating the cheese right now would make her have to find new food for the hungry children. Finding new food may take long, and is always less favourable than having food ready to feed them right away (*evolution result a posteriori* preference in line 11). Crow can see three possible evolutions: [[*decline*], [*had_food*]]; [[*hide*(*crow*, *cheese*), *sing*], [*had_food*]] and [[*eat*(*crow*, *cheese*), *sing*], [*find_new_food*]]. Note that in looking ahead at least two steps into the future, local preferences are not taken into account only after all evolution one were applied (full discussion can be found in [19]).

Now the two *evolution result a posterirori* preferences in lines 11-12 are taken into account. The first one rules out the evolution including *finding new food* since it is less preferred than the other two which includes *had food*. The second one rules out the one including *decline*. In short, Crow will hide the food to keep for her hungry children, and still take pleasure from singing.

5 Related Work

Many issues concerning intentions have been widely discussed in the literature of agent research. Some philosophers, e.g., Bratman [4,5] have been concerned with the role that intention plays in directing rational decision making and guiding future actions. Many agent researchers have recognized the importance of intentions in developing useful agent theories, architectures, and languages, such as Rao and Georgeff with their BDI model [23], which has led to the commercialization of several high-level agent languages (e.g., see [6,30]).

However, to the best of our knowledge, there has been no real attempt to model and implement the role of intentions in decision making, within a rational choice framework. Intentions of other relevant agents are always assumed to be given as the input of a decision making process; no system that integrates a real intention recognition system into a decision making system has been implemented so far.

The existent work of Pereira and Han [8,21,22] also attempts to combine the two systems, Evolution Prospection and Intention Recognition, but in a completely different manner. They use an intention recognition system to recognize the goal of the observed agent (e.g., an elder [21]), which the evolution prospection system uses to derive appropriate courses of actions to help achieve. Our approach is more general and genuinely integrated: the intention recognition system is employed also to evaluate other different kinds of information being utilized within an EP program here.

6   Conclusions and Future Work

We have summarized the existent work on Evolution Prospection and Intention Recognition and shown a coherent combination of them for decision making. The Evolution Prospection system has been proven to be a useful one for decision making, and now it has been empowered to take into account intentions of other agents—an important aspect that had not been explored so far. The fact that both systems are LP-based enabled their easy integration. We have described and exemplified several ways in which an Evolution Prospection agent can benefit from having an ability to recognize intentions of other (relevant) agents.

As a future direction, we will apply our combined system to tackle different real application domains, e.g., Ambient Intelligence [8,25] and Elder Care [21], where decision making techniques as well as intention recognition abilities are of increasing importance [26,25,8].

We also plan to provide a formal semantics for our new combined system on top of the one of Evolution Prospection—given in [19]—and the theoretical modelling of intention within a rational choice framework [24].

Acknowledgments. We thank the reviewers for useful comments. HTA acknowledges the support from FCT-Portugal, grant SFRH/BD/62373/2009.

References

1. Baral, C., Gelfond, M., Rushton, N.: Probabilistic reasoning with answer sets. Theory and Practice of Logic Programming 9(1), 57–144 (2009)
2. Blaylock, N., Allen, J.: Corpus-based, statistical goal recognition. In: Proceedings of the 18th International Joint Conference on Artificial intelligence (IJCAI 2003), pp. 1303–1308 (2003)
3. Blaylock, N., Allen, J.: Statistical goal parameter recognition. In: Zilberstein, S., Koehler, J., Koenig, S. (eds.) Proceedings of the 14th International Conference on Automated Planning and Scheduling (ICAPS 2004), pp. 297–304. AAAI (2004)
4. Bratman, M.E.: Intention, Plans, and Practical Reason. The David Hume Series. CSLI (1987)
5. Bratman, M.E.: Faces of Intention: Selected Essays on Intention and Agency. Cambridge University Press (1999)
6. Burmeister, B., Arnold, M., Copaciu, F., Rimassa, G.: BDI-agents for agile goal-oriented business processes. In: Proceedings of the 7th International Joint Conference on Autonomous Agents and Multiagent Systems: Industrial Track, AAMAS 2008, pp. 37–44 (2008)
7. Han, T.A., Pereira, L.M.: Collective intention recognition and elder care. In: AAAI 2010 Fall Symposium on Proactive Assistant Agents (PAA 2010). AAAI (2010), http://www.aaai.org/ocs/index.php/FSS/FSS10/paper/view/2178/2697
8. Han, T.A., Pereira, L.M.: Proactive intention recognition for home ambient intelligence. In: IE Workshop on AI Techniques for Ambient Intelligence, Ambient Intelligence and Smart Environments, vol. 8, pp. 91–100. IOS Press (2010)
9. Han, T.A., Pereira, L.M.: Context-dependent incremental intention recognition through bayesian network model construction. In: Nicholson, A. (ed.) Bayesian Modelling Applications Workshop (BMAW 2011), Conference on Uncertainty in Artificial Intelligence (UAI 2011). CEUR Workshop Proceedings (2011)

10. Han, T.A., Pereira, L.M., Santos, F.C.: The role of intention recognition in the evolution of cooperative behavior. In: IJCAI 2011, pp. 1684–1689. AAAI (2011)
11. Han, T.A., Pereira, L.M., Santos, F.C.: Intention recognition promotes the emergence of cooperation. Adaptive Behavior 9(3), 264–279 (2011)
12. Han, T.A., Kencana Ramli, C.D.P., Damásio, C.V.: An Implementation of Extended P-log using XASP. In: Garcia de la Banda, M., Pontelli, E. (eds.) ICLP 2008. LNCS, vol. 5366, pp. 739–743. Springer, Heidelberg (2008)
13. Heinze, C.: Modeling Intention Recognition for Intelligent Agent Systems. PhD thesis, The University of Melbourne, Australia (2003)
14. Nowak, M.A.: Five rules for the evolution of cooperation. Science 314(5805), 1560 (2006), doi:10.1126/science.1133755.
15. Pearl, J.: Causality: Models, Reasoning, and Inference. Cambridge U.P. (2000)
16. Pearl, J.: Probabilistic Reasoning in Intelligent Systems: Networks of Plausible Inference. Morgan Kaufmann (1988)
17. Pereira, L.M., Dell'Acqua, P., Pinto, A.M., Lopes, G.: Inspecting and preferring abductive models. In: Handbook on Reasoning-based Intelligent Systems. World Scientific (forthcoming, 2011)
18. Pereira, L.M., Han, T.A.: Evolution Prospection. In: Nakamatsu, K., Phillips-Wren, G., Jain, L.C., Howlett, R.J. (eds.) New Advances in Intelligent Decision Technologies. SCI, vol. 199, pp. 51–63. Springer, Heidelberg (2009)
19. Pereira, L.M., Han, T.A.: Evolution prospection in decision making. Intelligent Decision Technologies 3(3), 157–171 (2009)
20. Pereira, L.M., Han, T.A.: Intention Recognition via Causal Bayes Networks Plus Plan Generation. In: Lopes, L.S., Lau, N., Mariano, P., Rocha, L.M. (eds.) EPIA 2009. LNCS, vol. 5816, pp. 138–149. Springer, Heidelberg (2009)
21. Pereira, L.M., Han, T.A.: Elder Care via Intention Recognition and Evolution Prospection. In: Seipel, D. (ed.) INAP 2009. LNCS (LNAI), vol. 6547, pp. 170–187. Springer, Heidelberg (2011)
22. Pereira, L.P., Han, T.A.: Intention Recognition with Evolution Prospection and Causal Bayesian Networks. In: Madureira, A., Ferreira, J., Vale, Z. (eds.) Computational Intelligence for Engineering Systems: Emergent Applications, vol. 46, pp. 1–33. Springer, Heidelberg (2011)
23. Rao, A.S., Georgeff, M.P.: BDI-agents: from theory to practice. In: Proceeding of First International Conference on Multiagent Systems (1995)
24. Roy, O.: Thinking before Acting: Intentions, Logic, Rational Choice. PhD thesis, ILLC Dissertation Series DS-2008-03, Amsterdam (2009)
25. Sadri, F.: Ambient intelligence, a survey. ACM Computing Surveys (2010)
26. Sadri, F.: Logic-based approaches to intention recognition. In: Handbook of Research on Ambient Intelligence: Trends and Perspectives (2010)
27. Sigmund, K.: The Calculus of Selfishness. Princeton U. Press (2010)
28. Sukthankar, G., Sycara, K.: Robust and efficient plan recognition for dynamic multi-agent teams. In: Proceedings of International Conference on Autonomous Agents and Multi-Agent Systems, pp. 1383–1388 (2008)
29. van Hees, M., Roy, O.: Intentions and plans in decision and game theory. In: Verbeek, B. (ed.) Reasons and Intentions, pp. 207–226. Ashgate Publishers (2008)
30. Wooldridge, M.: Reasoning about rational agents. The Journal of Artificial Societies and Social Simulation 5 (2002)
31. XSB: XSB system version 3.2 vol. 2: Libraries, interfaces and packages (March 2009)

Unsupervised Music Genre Classification with a Model-Based Approach

Luís Barreira, Sofia Cavaco, and Joaquim Ferreira da Silva

CITI, Departamento de Informática
Faculdade de Ciências e Tecnologia
Universidade Nova de Lisboa
2829-516 Caparica, Portugal
lfbarreira@gmail.com, {sc,jfs}@di.fct.unl.pt

Abstract. New music genres emerge constantly resulting from the influence of existing genres and other factors. In this paper we propose a data-driven approach which is able to cluster and classify music samples according to their type/category. The clustering method uses no previous knowledge on the genre of the individual samples or on the number of genres present in the dataset. This way, music *tagging* is not imposed by the users' subjective knowledge about music genres, which may also be outdated. This method follows a model-based approach to group music samples into different clusters only based on their audio features, achieving a perfect clustering accuracy (100%) when tested with 4 music genres. Once the clusters are learned, the classification method can categorize new music samples according to the previously learned created groups. By using Mahalanobis distance, this method is not restricted to spherical clusters, achieving promising classification rates: 82%.

Keywords: Automatic music genre classification, audio indexing, unsupervised classification.

1 Introduction

Since today's digital content development triggered the massive use of digital music, an indexing process is very important to guarantee a correct organization of huge databases. While a music genre categorization would be a solution, this may be hard to achieve (manually): on the one hand, music can be associated to one or more musical genres, and on the other hand, cultural differences and human interpretations, make it difficult the attainment of common music genre taxonomy. For example, an expert could label the Gustav Mahler's 2nd symphony as *Erudite - late Romantic*, while a non-expert could label it as *Classical*. Alternatively, an automatic classification based on good audio features may prevent the occurrence of incoherencies related to manual labeling.

While many supervised automatic music genre classifiers have been proposed, these will always be dependent on a previous manual labeling of the data [4; 5; 6; 8; 9; 12; 14]. As a consequence, these will be unable to evolve with the data

L. Antunes and H.S. Pinto (Eds.): EPIA 2011, LNAI 7026, pp. 268–281, 2011.

and automatically build new clusters driven by new values in the features. Alternatively, an unsupervised approach would not have this dependency and would be able to determine the genre of the music samples only based on their audio features. Nonetheless, only a few unsupervised methods have been proposed. Rauber et al. [10] proposed a growing hierarchical self-organizing map (which is a popular unsupervised artificial neural network) with psycho-acoustic features (loudness and rhythm) to obtain a hierarchical structuring music tree. Shao et al. [11] proposed an unsupervised clustering method that fed rhythmic content, Mel-frequency cepstral coefficients (MFCCs), linear prediction coefficients and delta and acceleration values (improvements in feature extraction) to a hidden Markov model.

Here we propose not only a methodology for unsupervised clustering but also for automatic music genre classification. The clustering method consists of a learning process that is able to cluster music samples based only on their audio properties and uses no previous knowledge on the genre of the training music samples. In addition a Model-Based approach is followed to generate clusters as we do not provide any information about the number of genres in the data set. The features used are related with rhythm analysis, timbre, and melody, among others. As these features represent a large number of dimensions, a feature reduction technique is necessary to reduce the dimensionality of the data. This clustering method achieves 100% accuracy results with classical, fado, metal and reggae music samples. After the clustering process is complete, the classification method can associate new test music samples to the previously created clusters. For that, the classifier uses Mahalanobis distance so that it can consider clusters with different shapes, volumes and orientations. An accuracy of 82% is achieved when classifying new music samples.

In the next section (Feature Extraction) we describe the features we use. Section 3 explains the Clustering Method while section 4 explains the Classification Method. The Results and Conclusions and Future Work are discussed in sections 5 and 6.

2   Feature Extraction

Feature extraction is the first step to be achieved in both automatic music genre clustering and classification. In this section, we describe the features we used, which can be grouped into two distinct groups: computational features (which do not represent any musical meaning and only describe a mathematical analysis over a signal) and perceptual features (which mathematically represent music properties based on the human hearing system).

Since some of the features we used have a very high dimensionality and it is more efficient to describe them with less dimensions, we used a set of statistical spectrum descriptors (SSD) proposed by Lidy and Rauber [7]. This set of descriptors includes: the mean, median, variance, skewness, kurtosis, min and max-values. (Whenever this property is calculated, we mention it in the text below.)

Computational features are very popular and have been used in many automatic music genre classification studies [3; 4; 5; 6; 8; 9; 14]. To start with we use a set of *timbral texture features* proposed by Tzanetakis and Cook [14]. These include: spectral centroid (which is a measure of the centre of gravity of the magnitude spectrum), spectral roll-off (which corresponds to the frequency below which there is 85% of the energy of the magnitude spectrum), spectral flux (which accounts for the energy difference between successive frames of the spectrogram), zero-crossing rate (ZCR) (which is a measure of the number of times the audio waveform crosses the x-axis per time unit), and low energy (which is the percentage of frames that have lower energy than the average energy over the whole signal).

We also use the SSD of the MFCCs. The MFCCs are a very popular set of features based on the auditive human system that uses a Mel-frequency scale to group the frequency bins. In addition, three other features were also calculated: the root mean square of the spectrograms, which is an approximation of the volume (i.e., loudness) of the signal, the bandwidth, an energy-weighted standard deviation which measures the frequency range of the signal, and the uniformity, which measures the similarity of the energy levels in the frequency bands [3].

The spectral properties mentioned above can follow two different approaches: their values can be calculated over each window of a spectrogram or they can be calculated directly over the spectrum of the whole sound. Usually, these values are calculated over each window of the spectrogram, and that is the approach we used here. In addition, whenever we obtain a set of values with significant dimension (and we do not use their SSDs), we also use means and variances as features.

The perceptual features we used include rhythmic content, rhythm patterns and pitch content. The rhythmic content contains information such as the beat, the tempo, the regularity of the rhythm and time signature. In particular, the beat has been used in several studies on genre classification [4; 6] and it can be extracted from the beat histogram [14]. On the other hand, rhythm patterns represent the loudness sensation for several frequency bands in a time-invariant frequency representation [7]. We use both the SSDs and the rhythm histogram of the rhythm patterns. Finally, the pitch content is used to describe melody and harmony of a music signal. This feature is used quite often in genre classification leading to good accuracy results [4; 6; 13; 14; 15]. The pitch content can be extracted from the pitch histogram [15], and it includes the amplitudes and periods of the highest peaks in the histogram, pitch intervals between the two most prominent peaks and the overall sums of the histograms.

3   The Clustering Method

The clustering method aims to organize several music samples into clusters without any initial information besides the feature set values of these samples. This method consists of several steps as illustrated in Fig. 1, which we describe below.

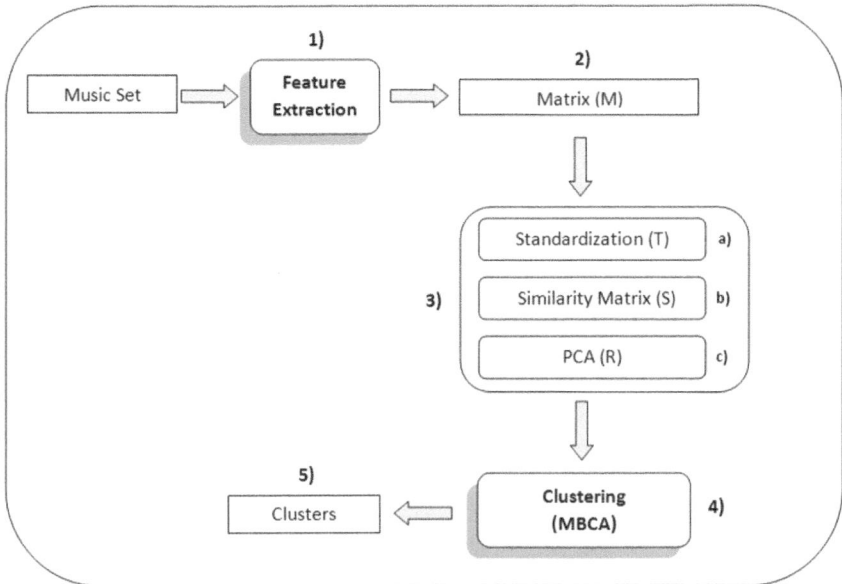

Fig. 1. The clustering method

3.1   The Feature Reduction Stage

After the audio features have been extracted, they have to be analysed to find clusters of points with similar values. For that, the method starts by representing the features by data matrix M, whose lines correspond to music samples in the training set and whose columns correspond to features. So, the $m_{s,f}$ cell of M contains the value of the fth feature for music sample s.

Once this matrix is built, the method performs some transformations as illustrated by box 3 in Fig. 1. In order to set equal importance (scale) to all columns (features) of the data set matrix, in step 3a the method performs a standardization of matrix M and creates a new matrix T with the same dimension as matrix M, that is, both matrices are $(N \times F)$, where N is the number of samples in the training set and F is the number of features. Now, the $t_{s,f}$ cell of T contains the standardized value of feature f for music s, which is given by

$$t_{s,f} = \frac{m_{s,f} - m_{.,f}}{\sqrt{var(M_f)}} \ , \tag{1}$$

where $m_{.,f}$ is the mean value of the fth column of matrix M, that is,

$$m_{.,f} = \frac{1}{N} \sum_{i=1}^{N} m_{i,f} \ , \tag{2}$$

and the variance of feature f, $var(M_f)$, is obtained from

$$var(M_f) = \frac{1}{N-1} \sum_{i=1}^{N} (m_{i,f} - m_{.,f})^2. \tag{3}$$

As we will show in Sect. 5, depending on the combination of the initial groups of features, the number of columns of M and T may be more than 800. Thus, a strong feature reduction has to be made. So, At step 3b, a sample similarity matrix S is calculated:

$$S = \begin{bmatrix} Sim(s_1,s_1) & Sim(s_1,s_2) & \dots & Sim(s_1,s_N) \\ Sim(s_2,s_1) & Sim(s_2,s_2) & \dots & Sim(s_2,s_N) \\ \vdots & \vdots & \ddots & \vdots \\ Sim(s_N,s_1) & Sim(s_N,s_2) & \dots & Sim(s_N,s_N) \end{bmatrix} \tag{4}$$

Each cell of the symmetric matrix S represents the similarity between two music samples and it is calculated by the following correlation from values of matrix T:

$$Sim(s_i,s_j) = \frac{cov(s_i,s_j)}{\sqrt{cov(s_i,s_i)} \cdot \sqrt{cov(s_j,s_j)}}, \tag{5}$$

where the covariance between music samples s_i and s_j is given by

$$cov(s_i,s_j) = \frac{1}{F-1} \sum_{f=1}^{F} (t_{s_i,f} - t_{s_i,.})(t_{s_j,f} - t_{s_j,.}), \tag{6}$$

where $t_{s_i,.}$ is the mean value of the ith line of T.

Each line of matrix S, corresponds to a music sample, now characterized by its similarity (within a range from -1 to +1) to all the other samples in the training set. On the other hand, each column of S may be seen as a new feature reflecting the similarity between a music sample and all the other samples. Clearly, there are as many columns as the number of music samples in the training set. Thus, with S, the number of features is reduced from the number of initial attributes, usually very high, to a number which is equal to the size of the training set, which may be a much smaller number. As we will show in Sect. 5 we obtained good results using a training set of 60 samples.

Since samples of the same genre tend to show high similarities and, thus, there are strong correlations between the features in S, another reduction in dimensionality can be obtained by a technique based on Principal Component Analysis (PCA) [1].

Since S is symmetric, it can be described as $S = P\Lambda P^T$, where $P = [e_1, ..., e_N]$ is the orthogonal matrix of normalized eigenvectors of S, and Λ is the diagonal matrix of its eigenvalues, $\lambda_1, ..., \lambda_N$, such that $\lambda_1 \geq ... \geq \lambda_N \geq 0$. Since Λ is symmetric, $\Lambda = \Lambda^{\frac{1}{2}}\Lambda^{\frac{1}{2}}$ and $\Lambda^{\frac{1}{2}} = (\Lambda^{\frac{1}{2}})^T$. Thus,

$$S = P\Lambda^{\frac{1}{2}}\Lambda^{\frac{1}{2}}P^T = P\Lambda^{\frac{1}{2}}(\Lambda^{\frac{1}{2}})^T P^T = P\Lambda^{\frac{1}{2}}(P\Lambda^{\frac{1}{2}})^T = QQ^T, \tag{7}$$

with
$$Q = P\Lambda^{\frac{1}{2}} \ . \tag{8}$$

The lines in matrix Q represent the music samples while the columns represent new uncorrelated features; see [1] for more details about this PCA-based technique. The leftmost columns of Q correspond to the most informative features. Thus, in order to reduce the number of features, we can discard the least informative ones, by ignoring the columns of Q having a variance (given by cells in Λ) lower that a threshold which we set to 1. We call R to this new reduced matrix (step 3c) of Fig. 1, which is a copy of the k leftmost columns of Q. This way we build a k-dimensional space in which music samples are represented: the k features correspond to the k axis in this new space, and matrix R contains the values for these features for each music in the training set. We tried other criteria associated with other threshold values, but this one provided a more reduced number of columns keeping good results. With this technique, we were able to drastically reduce the number of initial dimensions, that is, features (in S) from 60 to 7 final dimensions (in R) when we used the training set described in Sect. 5. Now, we are able to submit the resulting matrix R to the clustering stage.

3.2   The Clustering Stage

In the clustering stage (box 4 in Fig. 1) we use the Model-Based Clustering Analysis (MBCA) as proposed by Fraley and Raftery [2]. This approach uses no initial information about the number of clusters nor their shape or orientation. It represents the data by several possible models, which are characterized by different geometric properties. With this approach, data is represented by a mixture model where each element corresponds to a different cluster. Models with varying geometric properties are obtained through different Gaussian parameterizations and cross-cluster constraints. Partitions (clusters) are determined by the EM (expectation-maximization) algorithm for maximum likelihood, with initial agglomerative hierarchical clustering (see [2] for details). This clustering methodology is based on multivariate normal mixtures. So, the density function associated to cluster c has the form:

$$f_c(x_i|\mu_c, \Sigma_c) = \frac{e^{(-\frac{1}{2}(x_i-\mu_c)^T \Sigma_c^{-1}(x_i-\mu_c))}}{(2\pi)^{\frac{p}{2}}|\Sigma_c|^{\frac{1}{2}}} \ , \tag{9}$$

where vector x_i represents an element that belongs to cluster c. Clusters are ellipsoidal and centered at the means μ_c. The covariance matrix Σ_c determines the geometric characteristics of the cluster. This clustering methodology is based on the parameterization of the covariance matrix in terms of the eigenvalue decomposition in the form $\Sigma_c = \lambda_c D_c A_c D_c^T$, where D_c is the orthogonal matrix of eigenvalues, which determines the orientation of the axes. A_c is the diagonal matrix whose elements are proportional to the eigenvalues of Σ_c and which determines the shape of the ellipsoid. The volume of the ellipsoid is specified by scalar λ_c. Characteristics (orientation, shape and volume) of distributions are estimated from the input data, and can be allowed to vary between clusters, or constrained to be the same for all clusters. Once all models are created, MBCA uses the Bayesian Information Criterion (BIC) to measure the evidence of clustering for each pair (*model, number of clusters*), and the larger the value of BIC, the stronger the evidence for the pair. So, by choosing the pair having the larger BIC, the most reliable model is automatically obtained, and then a vector indicating which music samples belong to which cluster is returned by this clustering approach. In other words, clusters are automatically formed in a k-dimensional space, according to data in matrix $\boldsymbol{R}$.

4   The Classification Method

Once the clusters are learned, the classification method can be used to classify new music samples (not included in the training set). Fig. 2 shows the steps of this method, which we describe below.

4.1   Representing New Music Samples in the k-Dimensional Space Built in the Clustering Phase

Given a new (test) music sample s_t, the classification method starts by representing it with the same initial feature set as that used in the clustering method. As a result, the music sample is represented by an F-dimensional vector $\boldsymbol{m}_{s_t}$ that contains the feature values for music s_t, that is $\boldsymbol{m}_{s_t}^T = [m_{s_t,f_1}, ..., m_{s_t,f_F}]$ (step 2 in Fig. 2). Recall that matrix $\boldsymbol{M}$ (from Sect. 3.1) is a matrix whose lines are vectors of this form for the music samples in the training set.

Afterwards, vector $\boldsymbol{m}_{s_t}$ needs to be transformed into a new vector that represents music s_t in the k-dimensional space built in the clustering process. Firstly, $\boldsymbol{m}_{s_t}$ needs to be standardized (box 3a in Fig. 2). This transformation aims to set equal importance (scale) to each feature in vector $\boldsymbol{m}_{s_t}$. Despite $\boldsymbol{m}_{s_t}$ has only one value for each feature, this standardization will take into account the feature values of the music samples in the training set. Thus, the means and standard deviations calculated by the clustering method are used such that each cell of the new vector $\boldsymbol{t}_{s_t}^T = [t_{s_t,f_1}, ..., t_{s_t,f_F}]$ is given by an equation similar to (1):

$$t_{s_t,f_i} = \frac{m_{s_t,f_i} - m_{.,f_i}}{\sqrt{var(M_{f_i})}}, \tag{10}$$

where $m_{.,f_i}$ and $var(M_{f_i})$ result from (2) and (3) respectively.

Now, the method aims to calculate a similarity vector between music s_t and all the samples in the training set. For that, we could use the correlation given by (5), used to calculate the similarity between the training set samples. However, since a correlation between non-standardized variables is equivalent to a covariance between the standardization of those variables, for reasons of computational weight, we followed this last option to get the same results. So, once we have vector $\boldsymbol{t}_{s_t}$, the referred standardization corresponds to a new vector

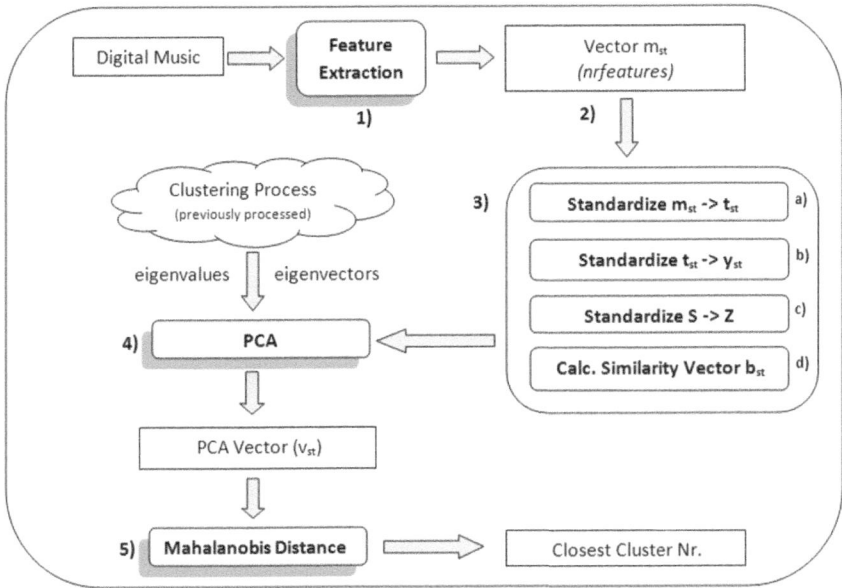

Fig. 2. The classification method

$\boldsymbol{y_{s_t}^T} = [y_{s_t,f_1}, ..., y_{s_t,f_F}]$ (box 3b in Fig. 2), where y_{s_t,f_i} is the standardized value of feature f_i for music s_t, which is given by

$$y_{s_t,f_i} = \frac{t_{s_t,f_i} - t_{s_t,.}}{\sqrt{var(T_{s_t})}} .\qquad(11)$$

$var(T_{s_t})$ stands for the variance associated to the t_{s_t,f_i} values for sound s_t along all features. So, $var(T_{s_t}) = \frac{1}{F-1} \sum_{f_i=1}^{F} (t_{s_t,f_i} - t_{s_t,.})^2$ and $t_{s_t,.} = \frac{1}{F} \sum_{f_i=1}^{F} t_{s_t,f_i}$.

At this step we need to relate our music s_t, now represented by $\boldsymbol{y_{s_t}}$, with the samples used in the learning process. In order to obtain a similarity vector $\boldsymbol{b_{s_t}}$ (box 3d in Fig.2), we need the information given by the similarity matrix $\boldsymbol{S}$, which may also be given by another matrix $\boldsymbol{Z}$ – see Appendix for details concerning matrices $\boldsymbol{S}$ and $\boldsymbol{Z}$ – such that each column of $\boldsymbol{Z}$ is a vector $\boldsymbol{z_s^T} = [z_{s,f_1}, ..., z_{s,f_F}]$ that represents the training set sample s using standardized values. In other words, each of these standardized values z_{s,f_i} is calculated by

$$z_{s,f_i} = \frac{t_{s,f_i} - t_{s,.}}{\sqrt{var(T_s)}} .\qquad(12)$$

Thus, vector $\boldsymbol{b_{s_t}}$ represents the similarity vector between $\boldsymbol{y_{s_t}}$ and each sample of the training set:

$$\boldsymbol{b_{s_t}^T} = \frac{1}{F-1} \boldsymbol{y_{s_t}^T Z} .\qquad(13)$$

Now, by using the information obtained by the PCA-based technique from Sect. 3.1, that is, with $\boldsymbol{\Lambda}$ and $\boldsymbol{P}$, we can transform $\boldsymbol{b}_{s_t}$ into a vector $\boldsymbol{u}_{s_t}$, such that

$$\boldsymbol{u}_{s_t}^T = [u_{s_t,1}, ..., u_{s_t,N}] = \boldsymbol{b}_{s_t}^T \boldsymbol{P} \boldsymbol{\Lambda}^{-\frac{1}{2}} \ , \tag{14}$$

where N is still the number of samples of the training set.

Similarly to what was mentioned in Sect. 3.1 about the most informative columns of matrix $\boldsymbol{Q}$, only the k leftmost cells of $\boldsymbol{u}_{s_t}^T$ are used to obtain a final vector $\boldsymbol{v}_{s_t}$ that represents the music sample s_t in the k-dimensional space learned by the clustering method. In other words, $\boldsymbol{v}_{s_t} = [u_{s_t,1}, ..., u_{s_t,k}]$. In Appendix, the reader may see a detailed proof that $\boldsymbol{v}_{s_t}$ is the representation of music s_t in the k-dimensional space learned by the clustering method.

4.2   The Classification Stage

Now that music s_t is represented in the k-dimensional space learned by the clustering method, we need to relate $\boldsymbol{v}_{s_t}$ to the learned clusters (box 5 in Fig. 2). Mahalanobis distance was adopted for this purpose since it takes into account the geometric properties of each cluster, which is important since distances take different impact depending on the data dispersion along each axis. (This characteristic is not achieved when using other metrics such as Euclidean or Manhattan distances.)

The method calculates the Mahalanobis distance between each cluster centroid and $\boldsymbol{v}_{s_t}$, and proposes the class represented by the cluster having a smaller distance as the most likely class for music s_t. In other words, class c will be associated to $\boldsymbol{v}_{s_t}$ if $d(\boldsymbol{v}_{s_t}, \boldsymbol{\mu}_c, \boldsymbol{\Sigma}_c^{-1}) = \min_i d(\boldsymbol{v}_{s_t}, \boldsymbol{\mu}_i, \boldsymbol{\Sigma}_i^{-1})$ where:

$$d(\boldsymbol{v}_{s_t}, \boldsymbol{\mu}_i, \boldsymbol{\Sigma}_i^{-1}) = (\boldsymbol{v}_{s_t} - \boldsymbol{\mu}_i)^T \boldsymbol{\Sigma}_i^{-1} (\boldsymbol{v}_{s_t} - \boldsymbol{\mu}_i) \ , \tag{15}$$

where $\boldsymbol{\mu}_i = [\mu_{i.,1}, ..., \mu_{i.,k}]$ is the centroid of cluster i, with $\mu_{i.,f} = \frac{1}{\|\mathcal{C}_i\|} \sum_{s \in \mathcal{C}_i} r_{s,f}$. $\mathcal{C}_i$ is cluster i, that is, the set containing all the samples in this cluster, $\|\mathcal{C}_i\|$ is its size, and $r_{s,f}$ is the value of the fth axis (i.e., final feature) for music s (this is the value corresponding to the line associated to sample s and fth column of matrix $\boldsymbol{R}$, see Sect. 3.1). So, $\boldsymbol{\mu}_i$ represents an *average* music sample of cluster i. Finally, $\boldsymbol{\Sigma}_i$ reflects the geometric properties of cluster i in the k-dimensional space:

$$\boldsymbol{\Sigma}_i = \begin{bmatrix} E_{i_{1,1}} & E_{i_{1,2}} & \cdots & E_{i_{1,k}} \\ E_{i_{2,1}} & E_{i_{2,2}} & \cdots & E_{i_{2,k}} \\ \vdots & \vdots & \ddots & \vdots \\ E_{i_{k,1}} & E_{i_{k,2}} & \cdots & E_{i_{k,k}} \end{bmatrix} \tag{16}$$

and

$$E_{i_{l,p}} = \frac{1}{\|\mathcal{C}_i\| - 1} \sum_{s \in \mathcal{C}_i} (r_{s,l} - \mu_{i.,l})(r_{s,p} - \mu_{i.,p}) \ . \tag{17}$$

The most heavy calculations needed in the classification phase, such as the matrix $\boldsymbol{\Sigma}_i^{-1}$ for each cluster, can actually be made at the end of the clustering phase, as all needed data is available for that. This way, the classification of new music samples is a fast computation.

5   Results

In order to validate our approach, we used classical, metal, and reggae music samples from Tzanetakis' GTZAN [1] data collection [14], which has 100 music samples from several different genres. In addition, we added samples from a new genre, *fado*, and therefore, our data collection has 400 music samples (all represented with a sampling frequency of 22050 Hz, 16 bits, and single channel) representing 4 different music genres: classical, fado, metal, and reggae.

Table 1. Clustering error percentages for several feature combinations. The features are: timbral texture features (ttf), rhythm patterns (rssd_rh), beat, root mean square of the spectrogram frames (rmsFrame), MFCCs, spectral centroid + bandwidth + uniformity (centBandUnif), SSD over spectrogram (specStat), low-energy over sample spectrum (lener), spectral centroid (scentroid), and ZCR

	Feature Combination	Error (%)
1	'ttf','rssd_rh','beat','rmsFrame','mfccs','centBandUnif'	0
2	'ttf','rssd_rh','beat','rmsFrame','mfccs','specStat'	0
3	'ttf', 'rssd_rh', 'centBandUnif'	0
4	'ttf', 'rssd_rh', 'specStat'	2
5	'ttf','rssd_rh','beat','rmsFrame','mfccs','lener','scentroid', 'specStat'	3
6	'ttf','rssd_rh','beat','rmsFrame','mfccs','lener','scentroid','zcr', 'specStat'	3

Even though our clustering methodology does not use any information about the number of genres nor the genre of the samples, we used this labelling information to validate the results. Thus, once the clustering process is complete, we assume that each learned cluster c corresponds to the mostly represented genre in the cluster, and count the number of samples, o_c, in the cluster that have a different labeling. The overall error percentage is given by $e = (100 \sum_c o_c)/N$, where N is the number of samples.

Table 2. Accuracy of the classification results for 3 different feature combinations

Feature Combination	Accuracy rate (%)
1	76.5
2	**81.8**
3	73.8

---

[1] http://marsyas.info/download/

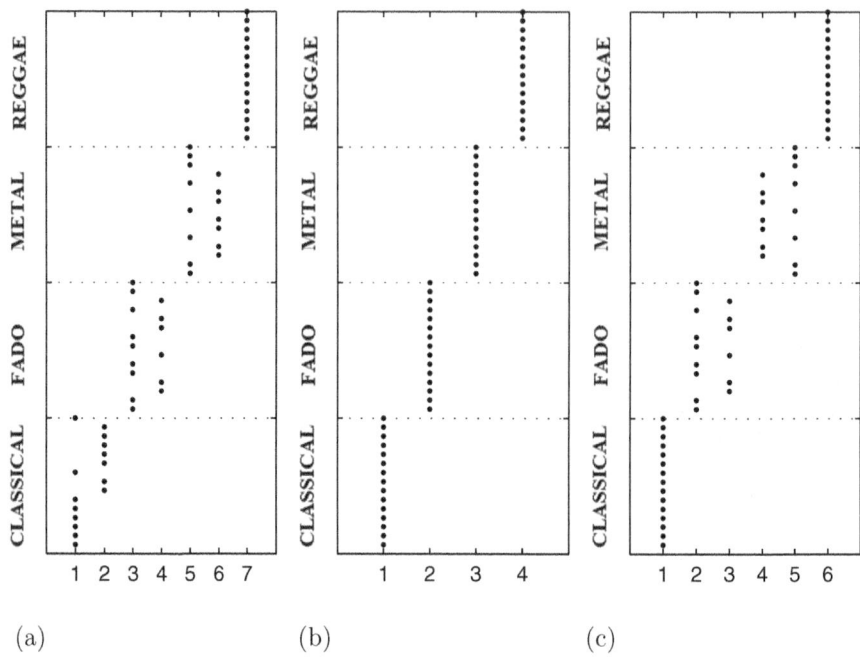

Fig. 3. Clustering results for combinations (a) 1, (b) 2 and (c) 3 from table 1. The x-axis shows the clusters learned, while the y-axis shows the initial labeling of the data. For instance, figure (a) shows that 8 classical music samples fell in cluster number 1 while 7 classical samples fell in cluster number 2

In order to validate the clustering methodology (described in Sect. 3), we used a training set composed of 60 elements (15 music samples from each of the four music genres mentioned above) and we tested many combinations of features (from Sect. 2). Table 1 shows the clustering results for the best feature combinations. The first three combinations have a 0% error rate, which shows that this approach is able to achieve perfect clustering results (assuming the initial labelling is correct).

Based on Table 1, it is clear that the third combination uses less features than the top two combinations. On the other hand, if we look at the clusters created (Fig. 3), the second combination achieves clustering results that perfectly match the initial labelling of the data. Nonetheless, this does not mean that combinations 1 and 3 achieve worse or incorrect results. It may actually be the case that these two combinations are learning sub-genres within classical, fado and metal. Each feature in the second combination actually represents a group of features as the whole number of real audio sub-features this combination represents is equal to 873.

Once the clusters were learned, we proceeded and classified new music samples. In order to evaluate the performance of the classification method (described in Sect. 4), we used a test set with the remaining 85 music samples (not used for clustering) for each of the four genres, making a total of 340 samples. Tests were made for clusters learned from each of the first three combinations from Table 1. As can be seen in Table 2, combination 2 achieves the best accuracy (precision) results with 81.8% correctly classified samples, which is a very satisfactory result, given this is an unsupervised approach.

6   Conclusions and Future Work

We proposed an unsupervised clustering and classification methodology for automatic genre classification. This kind of approach has the advantage of being totally independent of any influence from a human taxonomy. Since music genres do not present clear boundaries between them, and human genre taxonomy is hard to be achieved, we believe that an unsupervised approach is more suitable for music genre classification, while a supervised approach based on previously labelled data tends to be subjective. Besides, by learning directly from data in features, an unsupervised approach may automatically detect new genres, which is not possible for the more static nature of the supervised approaches.

In order to learn the clusters, the clustering method uses only the audio features of the training samples, and no previous knowledge on the genre of the individual samples. In addition, no information on the number of clusters is given a priori. This method achieves a perfect clustering accuracy (100%) when tested with four music genres (even though the genre labeling was not used in the training process, the results agreed with the manual labelling of the data), which shows that it is possible to achieve good accuracy results using an unsupervised method. In addition, as discussed in section 5, depending on the audio features used, the method is also able to find sub-genres within the data.

Once the clusters are learned, the classification method can categorize new music samples according to the previously learned clusters. This method uses Mahalanobis distance so that it is able to deal with clusters of different shapes, volumes and orientations. An 82% classification rate was obtained with four music genres.

We noticed that some misclassified samples were almost equidistant (in terms of Mahalanobis distance) to the chosen cluster and their actual cluster. This suggests that, as future work, a further analysis must be done to detect the existence of possible patterns of the Mahalanobis distances to every cluster, for both cases (correct classifications and incorrect ones). Such analysis may be important, for instance, to decide if a sample that is too distant from all clusters must be rejected; or to suggest that new clusters ought to be learned (which should be done by running the clustering mehtod again) because several samples are approximately equidistant to two given clusters.

Even though we only reported the clustering and classification results for four music genres, we are currently investigating how the system behaves with more genres. Although this work is still not finished, we were already able to confirm that this clustering method can achieve good results with five and six music genres (at least around 90% clustering accuracy). Working with more music genres may require the use of more (or different) audio features. There are other audio features that we did not explore yet but could be important to discriminate other genres.

Finally, in order to test different feature combinations we simply used a brute force method, that is, with no prior selection. Instead, a possible *filtering* over the extracted features should also be explored in future work, as to only process those features that present higher variances between music samples.

Acknowledgments. This work was part of the Videoflow project and partially funded by *Quadro de Referência Estratégica Nacional* (QREN) and *Fundo Europeu para o Desenvolvimento Regional* and *Programa POR Lisboa*.

Appendix

Here we prove that the test samples are represented in the k-dimensional space learned by the clustering method, that is, that $\boldsymbol{v_{s_t}}$ (see Sect. 4.1) is the translation of test music sample s_t in this k-dimensional space.

Proof. Let us suppose we want to classify a sound, say the first music of the training set, which is available in $\boldsymbol{z_1}$, the first column of matrix $\boldsymbol{Z}$. So, by (13) $\boldsymbol{b_1}^T = \frac{1}{F-1}\boldsymbol{z_1}^T\boldsymbol{Z}$ since now $\boldsymbol{y_{s_t}}$ is substituted by $\boldsymbol{z_1}$; $\boldsymbol{z_1}^T = [z_{1,1}, \ldots, z_{1,F}]$. Then $\boldsymbol{b_1}^T = [b_{1,1}, \ldots, b_{1,N}]$ where

$$b_{1,j} = \frac{1}{F-1}\sum_{i=1}^{F} z_{1,i} \cdot z_{j,i} \ . \tag{18}$$

Notice that, by statistics theory, (18) and (5) give the same result since $Sim(s_i, s_j)$ in (5) is a correlation using non-standardized values, and $b_{i,j}$ in (18) (generalizing from 1 to i) is a covariance using the standardization of those values. Then, in order to simplify this proof, let us suppose that we want to classify not just one music from the training set, but the whole training set. Then it is easy to conclude that

$$\boldsymbol{B} = \frac{1}{F-1}\boldsymbol{Z}^T\boldsymbol{Z} \ . \tag{19}$$

$\boldsymbol{B}$ would be obtained instead of $\boldsymbol{b_1}$. Note that $\boldsymbol{B} = \boldsymbol{S}$, being $\boldsymbol{S}$ the similarity matrix given by (4), because it contains the similarity vectors between each training set music and all other music samples.

Now let us work with the entire $\boldsymbol{S}$ as if we wanted to translate all training sounds into vectors in the k-dimensional space. Then, from (14) we would obtain $\boldsymbol{G} = \boldsymbol{SP\Lambda}^{-\frac{1}{2}} = \boldsymbol{SP\Lambda}^{-1}\boldsymbol{\Lambda}^{\frac{1}{2}}$, but since $\boldsymbol{P}^T\boldsymbol{P} = \boldsymbol{I}$ (where $\boldsymbol{I}$ is the identity matrix) and $\boldsymbol{S} = \boldsymbol{P\Lambda P}^T$, then $\boldsymbol{G} = \boldsymbol{P\Lambda P}^T\boldsymbol{P\Lambda}^{-1}\boldsymbol{\Lambda}^{\frac{1}{2}} = \boldsymbol{P\Lambda\Lambda}^{-1}\boldsymbol{\Lambda}^{\frac{1}{2}} = \boldsymbol{P\Lambda}^{\frac{1}{2}}$.

But $\boldsymbol{P\Lambda}^{\frac{1}{2}} = \boldsymbol{Q}$, the matrix characterizing all sounds by the PCA-based method presented before (see (8)). Since we used a *copy* of the whole training set for classification instead of just one music, we obtained a matrix ($\boldsymbol{Q}$) instead of a vector $\boldsymbol{u_1}$. Then, choosing the k leftmost columns of this matrix we would obtain matrix $\boldsymbol{R}$ referred in Sect. 3.1, which contains the representation of the whole training set in the k-dimensional space. With this, we proved that $\boldsymbol{v}_{s_t}$ is the representation of the test music sample in the k-dimensional space learned by the clustering method.

References

1. Escoufier, Y., L'Hermier, H.: A propos de la comparaison graphique des matrices de variance. Biometrical Journal 20(5), 477–483 (1978)
2. Fraley, C., Raftery, A.E.: How many clusters? which clustering method? answers via Model-Based cluster analysis. The Computer Journal 41(8), 578–588 (1998)
3. Golub, S.: Classifying recorded music. Master's thesis, University of Edinburgh - Division of Informatics (2000)
4. Koerich, A.L., Poitevin, C.: Combination of homogeneous classifiers for musical genre classification. In: 2005 IEEE International Conference on Systems, Man and Cybernetics, vol. 1, pp. 554–559 (2005)
5. Lee, C., Shih, J., Yu, K., Lin, H.: Automatic music genre classification based on modulation spectral analysis of spectral and cepstral features. IEEE Transactions on Multimedia 11(4), 670–682 (2009)
6. Li, T., Ogihara, M.: Toward intelligent music information retrieval. IEEE Transactions on Multimedia 8(3), 564–574 (2006)
7. Lidy, T., Rauber, A.: Evaluation of feature extractors and psycho-acoustic transformations for music genre classification. In: ISMIR, pp. 34–41 (2005)
8. Mckinney, M.F., Breebaart, J., Holstlaan, P.: Features for audio and music classification. In: ISMIR (2003)
9. Pye, D.: Content-based methods for the management of digital music. In: Proceedings of the IEEE International Conference on Acoustics, Speech, and Signal Processing (ICASSP), vol. 6, pp. 2437–2440, vol. 4 (2000)
10. Rauber, A., Pampalk, E., Merkl, D.: Using Psycho-Acoustic models and Self-Organizing maps to create hierarchical structuring of music by sound similarity. In: ISMIR (2002)
11. Shao, X., Xu, C., Kankanhalli, M.S.: Unsupervised classification of music genre using hidden markov model. In: IEEE International Conference on Multimedia and Expo (ICME), vol. 3, pp. 2023–2026 (2004)
12. Soltau, H., Schultz, T., Westphal, M., Waibel, A.: Recognition of music types. In: Proceedings of the 1998 IEEE International Conference on Acoustics, Speech and Signal Processing, vol. 2, pp. 1137–1140 (1998)
13. Tolonen, T., Karjalainen, M.: A computationally efficient multipitch analysis model. IEEE Transactions on Speech and Audio Processing 8(6), 708–716 (2000)
14. Tzanetakis, G., Cook, P.: Musical genre classification of audio signals. IEEE Transactions on Speech and Audio Processing 10(5), 293–302 (2002)
15. Tzanetakis, G., Ermolinskyi, A., Cook, P.: Pitch histograms in audio and symbolic music information retrieval. In: Proceedings of the Third International Conference on Music Information Retrieval (ISMIR), pp. 31–38 (2002)

Constrained Sequential Pattern Knowledge in Multi-relational Learning

Carlos Abreu Ferreira[1], João Gama[2], and Vítor Santos Costa[3]

[1] LIAAD-INESC and ISEP - Polytechnic Institute of Porto, Porto, Portugal
[2] LIAAD-INESC and Faculty of Economics - University of Porto, Porto, Portugal
[3] CRACS-INESC and Faculty of Sciences - University of Porto, Porto, Portugal

Abstract. In this work we present XMuSer, a multi-relational framework suitable to explore temporal patterns available in multi-relational databases. XMuSer's main idea consists of exploiting frequent sequence mining, using an efficient and direct method to learn temporal patterns in the form of sequences. Grounded on a coding methodology and on the efficiency of sequential miners, we find the most interesting sequential patterns available and then map these findings into a new table, which encodes the multi-relational timed data using sequential patterns. In the last step of our framework, we use an ILP algorithm to learn a theory on the enlarged relational database that consists on the original multi-relational database and the new sequence relation.

We evaluate our framework by addressing three classification problems. Moreover, we map each one of three different types of sequential patterns: frequent sequences, closed sequences or maximal sequences.

1 Introduction

Multi-relational databases are widely used to represent and store data. A multi-relational database is often composed by a *target* table and by a number of *fact* tables. The target table will represent the main objects of interest (say, patients in a medical domain); fact tables will represent the information being accumulated about the entities in the target table (say, medical visits or drug usage in the medical domain). We expect target tables to be relatively stable or to grow slowly over time; in contrast, fact tables may grow quickly. Moreover, quite often the information stored in fact tables is time-based and consists of *sequences* that reflect the evolution of a phenomenon of interest. Referring back to the medical domain, a patient is subject to a sequence of examinations, where a set of measurements, corresponding to results of different analysis, are taken.

In this work, we start from the hypothesis that the evolution of these measurements, as encoded in the fact tables, may hold relevant information for the diagnosis. The problem we address here is *how best to explore such information?* More precisely, we focus on how to learn highly descriptive and accurate decision models given multi-relational data with sequences.

The main goal of this work thus consists of exploiting heterogeneous temporal information stored in the multi-relational sequences of events. To do so, we

L. Antunes and H.S. Pinto (Eds.): EPIA 2011, LNAI 7026, pp. 282–296, 2011.
© Springer-Verlag Berlin Heidelberg 2011

propose a framework that encodes *timed data*, stored in one or several fact tables, into a separate sequence relation, uses an optimized sequence learner to find the most interesting such sequences, maps back the sequences to the relational database, and then learns a theory on the extended multi-relational database. The extended database thus contains all primitive relations and a sequential relation that stores time events for each example.

We name our framework XMuSer(eXtended MUlti-relational SEquential patteRn knowledge learning). It executes in five steps. First, we encode the multi-relational timed data into a sequence database. In this new database each example is a heterogeneous sequence that was built regarding both intra-table and inter-table relations within the temporal data. In a second phase, we use a sequence miner to find frequent sequences in the sequence database. In a third phase, we select interesting frequent sequences. By interesting we mean discriminative ones, that is, those that appear in only one class and do not appear in the others. We further use a filter to select the top-k class related sequential patterns. The fourth phase maps back the newly found sequential patterns by building a new relation where each target example is characterized by the presence or absence of one of the top-k patterns. In a last phase, we apply an Inductive Logic Programming (ILP) algorithm to learn a theory from the enlarged database, that is from the union of the original multi-relational database with the sequence relation.

This methodology allows us to explore multi-relational datasets that have different types of timed data, either sequence data or time-series data. On the one hand, we can benefit from computationally efficient sequential miners such as *PrefixSpan* [10] or *CloSpan* [14] to find the most remarkable sequential patterns. On the other hand, we still have access to the original data and can take advantage of the flexibility of ILP to learn in the extended multi-relational dataset. Indeed, we argue that the first step provides a good insight into the search space, and may enable XMuSer to perform better than classical ILP based algorithms in large search spaces. We should observe that the sequence miner and ILP learning algorithm are decoupled: we can use other sequential miners such as *SPIRIT* [9], that constrains the search space using regular expressions.

We experimentally evaluate our methodology in two datasets, addressing three prediction problems: discriminating between two hepatitis subtypes and discriminating between two sets of hepatitis fibrosis degrees, in the Hepatitis dataset, and discriminating between successful and unsuccessful loans, in the Financial dataset. The two datasets, Hepatitis and Financial, were originally introduced at PKDD Discovery Challenges.

The contributions of our work are therefore: **(i)** We introduce a framework to explore heterogeneous sources of time data, either sequence data or time-series data, stored in multi-relational database using propositional sequence miners. Our ILP based framework is highly efficient and gains both from the descriptive power of the ILP algorithms and the efficiency of the sequential miners. Moreover, we do not use classical aggregation strategies, like time windows, neither neglect valuable logic-relational information. **(ii)** We develop a new methodology to translate any multi-relational timed database into a sequence database.

Next we present the main ideas of our work and related work. In Section 3 we define concepts that will be useful to Section 4 where we discuss in detail our framework. Next, Section 5 describes the experimental setup and presents and discusses the results. We then present our conclusions and future research directions.

2 Methods and Related Work

In this section we present an overview of related work that inspired us and contributed to the overall XMuSer framework. First, we present sequential miners, a major component of our architecture. We then present ILP based algorithms that inspired us to design a framework that can benefit from using the logic-relational information without using ILP search mechanism to find sequential patterns. Last, we present different approaches to explore temporal patterns occurring in multi-relational datasets.

There exists a wide range of algorithms that can explore sequential data in an efficient way. To the best of our knowledge, Agrawal and Srikant introduced the sequential data mining problem in [2]. In [13] the same authors introduce the *GSP* algorithm, an algorithm that generalizes the original sequential pattern mining problem and that is also inspired in search procedure of the well known *APRIORI* algorithm [1]. *GSP* uses a candidate-generation strategy to find all frequent sequences, and uses a lattice to generate all candidate sequences. A more efficient approach to find the set of all frequent sequences is *PrefixSpan* [10]. This algorithm is inspired by pattern-growth strategies to efficiently find the complete set of frequent sequences. In real problems these algorithms usually find a huge number of uninteresting sequence patterns. To solve this issue of sequential miners, algorithms such as *CloSpan* [14], that returns a set of closed sequential patterns, and *SPIRIT* [9] that constrains the search space using regular expressions were developed.

A different approach, that is known to be successful, is to use post-processors, *filters*, to select interesting patterns. Some use sequential ad-hoc selection, that are model unrelated, whereas others use wrapper filters, that select features based on the induced models.

Algorithms that use Inductive Logic Programming (ILP) were the first ones to explore successfully the richness of multi-relational data. ILP approaches have an enormous representational power but are often criticized for lacking scalability [4]: ILP algorithms may not be very effective for the large search spaces induced by sequence databases.

One approach to solve the above mentioned issue is to use propositionalization with ILP [16]. The idea is to augment the descriptive power of the target table by projecting clauses (new attributes) on the target table.

Even with recent progress on scalability there exists multi-relational data which remains almost impossible to explore effectively by using only ILP based approaches. As an example, intra-table and inter-table temporal patterns remain hard to explore. One approach, followed by *WARMR* [6] is to use aggregation methodologies, unfortunately losing relevant time information. In View Learning [5] we can define and use alternative *views* of the database, i.e., we can define new fields or tables. Such new fields or tables can also be highly useful in learning, but still require searching a very large search space.

Other ILP based approaches develop specific techniques aimed at exploring the space and time information available in multi-relational datasets. The works of [11,3,7] introduce special purpose formalisms to find first-order sequential patterns in multi-relational datasets but, by using ILP based search, they suffer from traditional ILP limitations. To explore large spaces they must constrain the search space or use heuristics, otherwise the problems will be intractable. Thus, they may fail to find interesting patterns.

3 Preliminaries

We will start by remembering some key concepts of sequence mining and define the sequence mining problem.

Let $I = \{i_1, i_2, \ldots, i_n\}$ be a set of items and e an *event* such that $e \subseteq I$. A *sequence* is an ordered list of events $e_1 e_2 \ldots e_m$ where each $e_i \subseteq I$. Given two sequences $\alpha = a_1 a_2 \ldots a_k$ and $\beta = b_1 b_2 \ldots b_t$, a sequence α is called a *subsequence* of β if there exists integers $1 \leq j_1 < j_2 < \ldots < j_k \leq t$ such that $a_1 \subseteq b_{j_1}, a_2 \subseteq b_{j_2}, \ldots, a_k \subseteq b_{j_k}$. A *sequence database* is a set of tuples (sid, α) where sid is the sequence identification and α is a sequence. The *count* of a sequence α in a sequence database D, denoted $count(\alpha, D)$, is the number of sequences in D containing the α subsequence. The *support* of a sequence α is the ratio between $count(\alpha, D)$ and the number of sequences in D. We denote support of a sequence as $support(\alpha, D)$. Given a sequence database D and a minimum support value λ, the problem of sequence mining is to find all subsequences in D having at least a support value equal to λ. Each one of the obtained sequences is also known as a *frequent sequence*.

When searching for all sequential patterns in a sequence database we face one of the major problems of sequential miners, the huge number of redundant and non-interesting findings. To alleviate this we can use other sequential miners that find a constrained set of sequential patterns. A frequent sequence α is a *closed sequential pattern* if there exists no proper supersequence β having the same support as α. A frequent sequence α is *maximal sequential pattern* if it is not a subsequence of any other frequent sequence. The set of maximal sequential patterns is a subset of all closed sequential patterns.

Moreover, in a classification problem, some sequential patterns have low discriminative power and could be eliminated. Thus, we introduce the concept of Discriminative Sequential Patterns to select inter-class discriminative sequential patterns.

Definition 1. (Discriminative Sequential Patterns) *Consider two sequence database partitions D_1 and D_2, and a support threshold λ. We define the set of*

discriminative   sequential   patterns   using   the   xor   ($\dot\vee$)   operator,
$S_{disc} = \{\alpha \mid support(\alpha, D_1) \geq \lambda \, \dot\vee \, support(\alpha, D_2) \geq \lambda\}$.

Such discriminative patterns and other interesting sequential patterns represent valuable information that can be extracted from a sequence database and that could be useful at theory learning time. To do so, we introduce the concept of Sequence Relation and Enlarged Database.

Definition 2. (Sequence Relation and Enlarged Database) *Consider a multi-relational database* $\mathbf{r}$, *a sequence database* D *coded from* $\mathbf{r}$, *where each sequence id equals the primary key of* $\mathbf{r}$ *target table. Also consider the set of sequential patterns* S *obtained from solving the sequential mining problem. We define the sequence relation,* $\mathbf{r}_{sr}$, *to be the set of tuples* $(sid, \alpha_1^B, \alpha_2^B, \ldots, \alpha_n^B)$ *where sid is the sequence id in* D *and* α_i^B *is a binary attribute whose value is obtained according to the projection of the sequential pattern* α_i *in the sequence database. The enlarged database is the database resulting from the union of the multi-relational database and the sequence relation. Formally we define the enlarged database as being* $\mathbf{E_r} = \mathbf{r} \cup \mathbf{r_{sr}}$.

Our algorithm is suitable to explore timed data present in multi-relational databases, mainly sequence data. Thus, we introduce a strategy to code time-series data into a sequence database.

Definition 3. (Sequence coding) *Given a multi-relational database* $\mathbf{r}$, *having fact tables that register time events. We define sequence coding as a procedure that for all, or a subset, of database fact tables,* $\mathbf{r_i} \in \mathbf{r}$, *translates all, or a subset, of attributes* $\mathbf{r_i}.\mathbf{A_j}$ *into a sequence database.*

4   The XMuSer Algorithm

In this section we present XMuSer. This new methodology explores multi-relational information, mainly heterogeneous sequential data widespread across a multi-relational dataset, to learn a theory. The main idea of the algorithm is to explore work developed in the sequential pattern mining field to include time information in the ILP learning process.

The framework has five main steps. In the first phase we code the temporal data into a sequence database. In a second phase, we run a sequence miner to find all sequential patterns in each class partition. In a third phase, we apply two filters to select the most discriminative or class related patterns. In a fourth phase, for each example in the target table, we built a relation where the example is characterized by presence or absence of the most interesting sequential patterns. Last, we learn a theory on the enlarged database, where enlarged database is the union of the original database with the new sequence relation.

Algorithm 1 presents the pseudo-code for XMuSer. Next, we explain each one of the major components in self-contained subsections. Throughout, we follow an illustrative example, a classification problem, at Figure 1. This example is

Algorithm 1. XMuSer pseudo-code

input a multi-relational database **r**; two thresholds λ, the sequence miner support
value, and k, the number of most interesting patterns to retain
output a classifier model

Sequence Coding
$\mathbf{s} \leftarrow$ SequenceCoding($\mathbf{r}$)

Frequent Sequential Patterns
$\mathbf{s}_1, \ldots, \mathbf{s}_l \leftarrow$ partition($\mathbf{s}$)
for $i = 1$ **to** l **do**
  $sf_i \leftarrow$ SequenceMiner($\mathbf{s}_i, \lambda$)
end for

Filtering
$S_{disc} \leftarrow$ discriminate($sf_1, \ldots, sf_l$)
$S_{interesting} \leftarrow \mathbf{SU}(S_{disc}, k)$

Mapping
$\mathbf{r}_{sr} \leftarrow$ Mapping($S_{interesting}, \mathbf{r}_{targetExamples}$)
$\mathbf{E_r} \leftarrow \mathbf{r} \cup \mathbf{r}_{sr}$

Learn a Theory with an ILP algorithm
ILP Algorithm($\mathbf{E_r}$)

Blood Analysis

ID	Date	RBC	WBC
1	19750102	high	normal
1	19780203	high	high
2	19770107	high	low

Target table
Patient

ID	Sex	BornDate	Class
1	M	19520109	a
2	F	19750123	b

Urinalysis

ID	Date	Exam	Result
1	19741201	plt	high
1	19750102	alb	normal
1	19750102	ttp	normal
1	19760204	alb	normal
2	19800403	alb	normal

Fig. 1. Database Relations. ID is the patient ID, Date is in numeric format, RBC and WBC are blood parameters, and we show alb, plt and ttp urine exams.

inspired on the relational Hepatitis dataset. The example has three tables registering the follow-up of two patients. One of the tables is the target table, named *Patient*, where each record describes each patient, identified by a masked ID, and registers the class of each patient. The other two tables are fact tables registering timed blood analysis and urinalysis examinations.

Data Coding. The algorithm that we present next takes a multi-relational dataset as input, usually represented as a database of Prolog facts. To explore the richness of this representation, and namely temporal patterns, we introduce a strategy that converts multi-relational timed data into an amenable sequence database. First, we find all relations that have temporal records. Second, we sort the records in these relations by time order. We thus obtain a chronological sequence of multiple events for each example. Figure 2 shows an example event sequence for patient one. The sequence includes a sequence of blood and

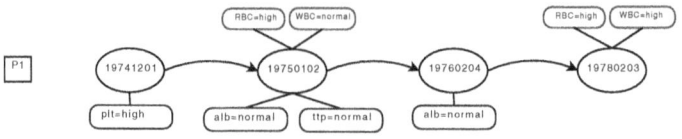

Fig. 2. Temporal Patient Events for Patient One

Table 1. Transformed Event Database. We introduce three new patients that will be useful in the following steps.

ID	Sequence	Class
1	(9) (3 5 7 8) (7) (3 6)	a
2	(3 4) (7)	b
3	(1 5) (3 4) (7)	b
4	(3 5) (7)	a
5	(8) (7)	a

urine analysis. Third, we built a temporal attribute-value sequence for each example. In this new sequence each item corresponds to all records registered at a given date/time. We have *(plt=high) (RBC=high, WBC=normal, alb=normal, ttp=normal) (alb=normal) (RBC=high, WBC=high)* for patient one. We then define a one-to-one coding map $f : Attributes \times Values \longrightarrow \mathbb{N}$. This mapping associates an *unique* number to each attribute-value pair. In the example, we use the map to code the attribute-value sequence into an integer number sequence. The definition of this map is done according to the type of attributes in each database relation. The mapping assumes discrete attribute (continuous attributes will be discretized beforehand).

Table 1 shows the transformed sequence database: each sequence tuple corresponds to an example in the target table and each subsequence corresponds to all one-time events. In the example, if we define the one-to-one map to be *f(RBC, low)=1, f(RBC, normal)=2, f(RBC, high)=3, f(WBC, low)=4, f(WBC, normal)=5, f(WBC, high)=6, f(alb, normal)=7, f(ttp, normal)=8, f(plt, high)=9*, then, patient one sequence of events is thus coded as (9) (3 5 7 8) (7) (3 6).

Table 2. Sequence Database Partition with Classes

ID	Sequence	Class
1	(9) (3 5 7 8) (7) (3 6)	a
4	(3 5) (7)	a
5	(8) (7)	a
2	(3 4) (7)	b
3	(1 5) (3 4) (7)	b

Table 3. Sequential patterns found in each class partition

Class **a** Sequential Patterns	Class **b** Sequential Patterns
(3): 2	**(3)**: 2
(3) (7): 2	**(3) (7)**: 2
(3 5): 2	(3 4) (7): 2
(3 5) (7): 2	(4): 2
(5): 2	(4) (7): 2
(5) (7): 2	**(7)** :2
(7): 3	
(8): 2	
(8) (7): 2	

This stage obtains a sequence suitable to be processed by a flat sequential miner. In Table 1 we present a sequence database registering the coded sequence of patients one and two, and three new patients that will be useful in the following steps.

Finding Frequent Sequential Patterns. We run a sequence miner in each partition of the sequence database (Table 2) in order to find *frequent* sequences, that is, having a support equal or higher than a user defined threshold. Thus, for each partition and each class we obtain all frequent patterns sf_i. In Table 3 we present the patterns that we found by running *PrefixSpan* algorithm (setting support value equal to 60%) in the two class partitions, the class **a** and class **b** partitions. Notice that several patterns occur in both classes.

Filtering. The previous step usually obtains a large number of findings. We would like to retain highly discriminative, class correlated, patterns and drop uninteresting and/or redundant patterns. To do so, we introduce a *Discriminative Filter* that selects the set S_{disc} of sequences that have support above some parameter λ in one and only one partition. The filter is implemented by using matching. The set of discriminative sequential patterns, S_{disc}, is formally defined in Section 3. In the example above, Table 4 presents the nine discriminative patterns obtained by applying this filter.

Table 4. Inter-Class Discriminative Sequential Patterns

	Discriminative Sequential Patterns
1	(3 4) (7)
2	(3 5)
3	(3 5) (7)
4	(4)
5	(4) (7)
6	(5)
7	(5) (7)
8	(8)
9	(8) (7)

Table 5. Sequence Relation

ID	S_1	S_6
1	0	1
2	1	0

Note that even after eliminating non-discriminative patterns, non-interesting patterns may remain. We use *Symmetrical Uncertainty(SU)* [15] to sort and select the top k class-related features. The SU evaluates the worth of an attribute by measuring the symmetrical uncertainty with respect to the class. For instance, if we set $k = 2$ the algorithm would select S_1, or (3 4) (7), and S_6, or (5), patterns. The sequential composition of the two filters thus results in two interesting features/sequences that we will use to build the final classification model.

Mapping Back Interesting Sequences. We have coded the timed data as a sequence database. We now want to build a relation where each example is characterized by the most interesting sequences. Table 5 shows the key idea in our approach. We construct a new table, the *sequential relation*, with an entry per example and with $k + 1$ attributes: the top-k most interesting features and the example ID. If the sequence associated with the new attribute is a subsequence of the example sequence at the sequence database, the new attribute takes value one. Otherwise, it takes the value zero.

Learning a Theory. Last, we add the new sequential relation to the initial tables and use an ILP algorithm, such as *Aleph*, to learn a set of clauses. One illustrative example of a found clause is:

```
patient_info(A,B,C,a):-
  blood_analysis(A,D,high,normal),
  urinalysis(A,E,plt,high),
  sequence_relation(A,0,1).
```

This clause calls the predicate *blood_analysis*, the predicate *urinalysis* and the predicate associated with the sequence relation, predicate *sequence_relation*. The clause explains (or covers) patient number one, a patient from class **a**, the majority class.

5   Experimental Evaluation

In this section we describe the configurations and results of our experimental evaluation.

5.1   Experimental Configuration

We evaluated our algorithm in two real-life datasets obtained from the PKDD data-mining competitions. We experimented three prediction problems: discriminating between two hepatitis subtypes and discriminating between two sets of hepatitis fibrosis degrees, in the Hepatitis dataset, and discriminating between successful and unsuccessful loans, in the Financial dataset.

Throughout, and as our framework can use any sequence miner, we tested: *PrefixSpan* algorithm to find all frequent sequences; *CloSpan* algorithm to find all closed sequential patterns; and a post-processing procedure, that works over the *CloSpan* findings, to select maximal sequential patterns, a subset of closed sequential patterns. This post-processing technique was named *MaxSpan* to easily present the results. We also use *Aleph* to learn a logical theory. We applied the following predefined parameter configuration: $k = 30$ and test two different λ values, 90% and 80%. For comparison purposes we also run the stand-alone *Aleph* algorithm to solve each one of the four problems.

We evaluate our framework using a ten-fold cross-validation procedure and compute: the mean number of patterns found after each step of XMuSer, the mean generalization accuracy and standard deviation of XMuSer and the mean time spent to complete each step of XMuSer. Using this same procedure, the 10-CV, we also compute: the mean number of rules learned by the *Aleph* algorithm, both as a component of XMuSer and as a stand-alone algorithm. We also compute the generalization accuracy and the time spent by the stand-alone *Aleph* algorithm. Concerning the generalization accuracy, we also compute the *Wilcoxon* hypothesis test p-value to measure how significantly our algorithm differs from the reference algorithm, the stand-alone *Aleph* algorithm. We set the level of significance equal to 0.10.

We use the same background knowledge and bias when running *Aleph* algorithm, either as the last component of XMuSer or as stand-alone. The major difference is that in XMuSer we introduce an extra sequence relation. The bias is relatively simple, relying on the predefined tables in the database. We evaluate our framework performance on unseen data. We further analyze the contribution of each step of XMuSer. We analyze the number of patterns found after each step, including the compactness of the theory learned by the *Aleph* algorithm. We analyze the generalization accuracy and run-time needed to complete each step of XMuSer.

5.2   Datasets and Tasks

We present below the two datasets that we used to evaluate our ILP based classifier: Hepatitis and Financial datasets. Both these datasets are available to download at the PKDD challenge web page (http://lisp.vse.cz/challenge/CURRENT/).

The Hepatitis dataset consists of seven tables registering a long term, from 1982 to 1990, monitoring of 771 patients having hepatitis B or C. One table provides personal data about patients. The other tables record blood and urinalysis examinations. We address two classification problems. Our first task is to discriminate between patients having B and C hepatitis. The second task is to determine the degree of liver fibrosis. Following previous work [12], we study fibrosis degree 2 and 3 against fibrosis degree 4. We perform limited feature selection, based on the dataset description [12]. We select GOT, GPT, TTT,

Table 6. Mean Number of Patterns in each step of XMuSer and of the stand-alone *Aleph* algorithm

		Seq. Miner					Disc.			SU-rank			Aleph(Rules)			Stand-alone Aleph (Rules)
		PrefixSpan		CloSpan		MaxSpan											
	λ	Pos.	Neg.	Pos.	Neg.	Pos.	Neg.	PrefixSpan	CloSpan	MaxSpan	PrefixSpan	CloSpan	MaxSpan	PrefixSpan	CloSpan	MaxSpan	
Hepatitis Subtype	0.9	68	0	51	0	26	2	68	51	24	30	30	30	124	124	124	129
	0.8	56011	122	21058	93	4260	38	55889	21151	4294	30	30	30	115	114	117	
Hepatitis Fibrosis	0.9	166	3	64	3	40	4	164	63	40	30	30	30	47	49	50	67
	0.8	22809	877	2773	610	1441	228	22134	3049	1648	30	30	30	50	45	47	
Financial	0.9	137	30	136	12	63	11	109	125	64	30	30	30	166	160	161	169
	0.8	7622	818	7605	313	2682	238	6900	7341	2865	30	30	30	149	150	142	

Table 7. Mean Generalization Accuracy: XMuSer against stand-alone *Aleph*

	λ	XMuSer (With Aleph)			Stand-alone Aleph	Wilcoxon p-value		
	λ	PrefixSpan	CloSpan	MaxSpan		PrefixSpan	CloSpan	MaxSpan
Hepatitis Subtype	0.9	0.79(0.10)	0.79(0.10)	0.79(0.10)	0.78(0.11)	0.269	0.201	0.524
	0.8	0.80(0.10)	0.80(0.10)	0.80(0.09)		0.073	0.093	0.308
Hepatitis Fibrosis	0.9	0.64(0.06)	0.65(0.05)	0.65(0.05)	0.58(0.09)	0.097	0.020	0.020
	0.8	0.61(0.10)	0.66(0.06)	0.64(0.13)		0.407	0.029	0.192
Financial	0.9	0.73(0.06)	0.75(0.04)	0.74(0.05)	0.71(0.07)	0.854	0.021	0.027
	0.8	0.74(0.04)	0.74(0.05)	0.73(0.07)		0.138	0.096	0.029

ZTT, T-CHO, CHE, ALB, TP, T-BIL, D-BIL, I-BIL, ICG-15, PLT, WBC and HGB features. As these features are numerical, and in fact take a wide range of values, we discretized each feature according to medical knowledge. We use three bin values, low, normal and high. For the sub-type problem we end with 206 hepatitis-B patients and 297 hepatitis-C patients. For the Fibrosis problem, we have 209 patients of {2,3}-class and 67 of the {4}-class.

The Financial dataset includes eight tables with data about clients and their accounts. A number of tables store static information on accounts, clients and regional demographics. The remaining tables register information concerning credit card types, payments, transactions, and loans for each account. The sequence database relies on the *Balance* attribute only. This attribute was then discretized into three levels, low, normal and high, based on the Chebychev inequality. Our target is predicting successful loans: 606 loans were classified as successful and 76 unsuccessful.

5.3   Results

We first present results that show the contribution of the filtering methodology and the number of rules learned by the *Aleph* algorithm (Table 6), both as a component of the XMuSer framework or a stand-alone algorithm. Then, we present results concerning the generalization accuracy of XMuSer and the stand-alone *Aleph* algorithm (Table 7). Finally, we present the running time of each component of the XMuSer and the running time of the stand-alone *Aleph* algorithm (Table 8). Moreover, we present the results that we obtain by using each one of the three sequence miners: *PrefixSpan*, *CloSpan* and *MaxSpan* (*CloSpan* + post-processing procedure).

In Table 6 we present, for each one of the three problems that we address and the two values of the λ parameter (90% and 80%), the mean number of sequential patterns found in each phase of **XMuSer** and the number of rules that were learned by *Aleph*. As we run each one of the sequential miners in each class partition, and for each one we report the number of sequential patterns found in both the positive (the Pos. column) and negative (the Neg. Column) class partitions. The positive partition is the majority class of each problem. We also present the number of discriminative patterns (the Disc. column), the patterns selected after applying the Discriminative filter, and the number of patterns that were selected using the SU filter (the SU-rank column). The maximum number of patterns to select, using SU metric, is an input parameter, the k in the framework pseudo-code above, that we set before the experiments to be 30. We obtain 30 patterns to build each sequence relation.

In Table 7 we present the mean generalization accuracy and standard deviation, inside brackets, of both the **XMuSer** framework and the stand-alone *Aleph* algorithm, the reference algorithm. This table reports results that were obtained by setting $\lambda = 90\%$, $\lambda = 80\%$ and $k = 30$, two input parameters of our framework. We also present the p-value obtained by running the *Wilcoxon* signed-rank test. We present the p-value for each sequence miner and using the stand-alone *Aleph* algorithm as the baseline.

In Table 8 we present the mean run-time of each component of our framework and the mean run-time of the stand-alone *Aleph* algorithm. The run-time for *PrefixSpan* adds the time needed to find sequential patterns in the positive and negative class partitions. These results were obtained by running **XMuSer** with the same settings that were used to measure the mean generalization error.

Table 8. Mean Run-Time, in seconds, of each **XMuSer** phase and the stand-alone *Aleph* algorithm

		XMuSer (With Aleph)												Stand-alone Aleph (Rules)
		Seq. Miner			Disc.			SU-rank			Aleph(Rules)			
	λ	PrefixSpan	CloSpan	MaxSpan	PrefixSpan	CloSpan	MaxSpan	PrefixSpan	CloSpan	MaxSpan	PrefixSpan	CloSpan	MaxSpan	
Hepatitis Subtype	0.9	3	3	13	0	0	0	3	3	3	8	8	6	129
	0.8	18	16	39	0	0	0	581	258	58	6	7	7	
Hepatitis Fibrosis	0.9	1	1	8	0	0	0	3	2	1	5	5	5	67
	0.8	3	3	17	0	0	0	138	20	11	6	4	5	
Financial	0.9	1	1	12	0	0	0	3	5	2	1	32	28	169
	0.8	3	3	20	0	0	0	151	215	75	2	22	23	

5.4   Analysis

Our goal in this study is to prove that our framework can effectively explore the timed data present in multi-relational databases.

Thus, we must be able to address multi-relational datasets, usually recording a large amount of data, and where each table has a wide number of attributes, discrete or continuous ones. Quite often, datasets are unbalanced. Unfortunately, most previous works that addresses these same problems, performs preprocessing steps, such as balancing the datasets, thus making it impossible to make a fair comparison against other works.

We present experimental results for $\lambda = 90\%$ and $\lambda = 80\%$. Lower values of λ cause a memory explosion. Further experiments with other values of k we do not show significant changes in performance, although larger k might provide more interesting patterns.

Given these constraints, we believe that we obtained very good results. In all three classification problems, XMuSer obtained significant wins against the stand-alone *Aleph* algorithm. This is especially clear when we use *CloSpan* algorithm to find sequential patterns. The best accuracy gains were obtained in the Hepatitis Fibrosis task. In this problem our mean accuracy gains range from 3% to 8%. In Subtype we obtained a significant win for a support value $\lambda = 80\%$. These results are among the best that we could find in related work that addresses this problem, and indeed outperform a previous work where the final step is a propositional classifier [8].

We further observed that the theories that we learn in all four tasks use sequences, and that XMuSer generates more compact theories. Indeed, one important point suggesting that we are in the right direction is a clear relation between improvements in the obtained classification models and the number of timed data attributes that we use to build the sequence database. In both Hepatitis tasks we eventually use a larger number of time data attributes to construct theories. In the financial task we benefit much less of time attributes: we could only use information from two time attributes in Thrombosis and one time attribute in Financial.

As expected, XMuSer execution time takes longer to learn than stand-alone *Aleph* algorithm, with execution time heavily depending on λ. To explain this behavior, consider the information of both Table 6 and Table 8. If we run XMuSer by setting a low λ we get a huge increase in the number of sequential patterns found and thus we get a run-time overhead in the initial phases of XMuSer. This is especially noticeable in two steps for the *PrefixSpan* and the *SU* filter. This problem is typical of sequential miners, the lower the support the higher the execution time and the number of patterns found. This is clear when we run *CloSpan* and *MaxSpan* technique to find a subset of sequential patterns. We get a small number of patterns and, despite the overhead of the post-processing of *MaxSpan*, we get an overall short run-time. This is clear if we observe the run-time of the *SU* filter for each sequence miner. Moreover, the last component of XMuSer uses more time when coupled with either *CloSpan* and *MaxSpan*. This is the result of getting a low number of redundant sequential patterns, sequential patterns that have the same support. The best results where obtained when we used *CloSpan* algorithm. This indicates that some frequent subsequences patterns (with the same support as a supersequence) that are not found by *MaxSpan* can contain valuable information.

The overhead observed in the *SU* filter does not result from computing the well known metric but from the preprocessing needed to build the contingency table. In order to compute the *SU* metric, for each pattern we need to know if that pattern is a subsequence of each example, which can be expensive.

Notice that the ILP search space of XMuSer is larger by a fact of 2^k, where k is the number of patterns/attributes in the sequential relation. On the other hand, our experiments do not show a significant variation in running times between *Aleph* in XMuSer and stand-alone. Observation suggests that this is because sequence-based rules can often cover a large number of positive examples, and prune well the search space.

These developments argues for the development of a sequential miner that can find class closed sequential patterns, i.e., that can find sequential patterns using class information to prune the search space and find discriminative patterns, not relying on post-processing strategies and filters. By using it we can go further ahead and explore even more timed data available in relational datasets.

6   Conclusions and Future Work

In this work, we presented XMuSer, a multi-relational framework that explores temporal information stored in a database. The methodology is an effective alternative to previous ILP-based approaches that incorporate timed data in the final first-order theory by either using time aggregation strategies, like time window aggregation, or by refining ILP clauses or predicates that explicit explore time. Differently from these approaches, our methodology, can take advantage of the strengths of sequential miners to explore efficiently any kind of timed data. Our approach can be used in conjunction with any classical ILP algorithm.

Moreover, we developed sequence coding, a technique to code timed data into a sequence database. We introduce the sequential relation to encode the patterns found by the sequential miner. Last, we define the enlarged database, a database that is the union of the original database and the sequence relation, and induce a theory over this relation. This methodology allows us to explore the valuable logic-relational information available in multi-relational datasets, especially temporal patterns, with good results as shown by empirical evaluation.

In order to improve our framework we will explore other ILP algorithms and develop a sequence miner that can use class information to prune the search space and find more interesting patterns. One way is to update *CloSpan* algorithm. New sequence miners, say *SPIRIT*, should be considered as they apply well to our algorithm.

Acknowledgments. This work was supported by the Portuguese Foundation for Science and Technology (FCT) under the projects KDUS (PTDC/EIA-EIA/098355/2008) and HORUS (PTDC/EIA-EIA/100897/2008). Carlos Abreu Ferreira was financially supported by the Portuguese Polytechnic Institute of Porto (ISEP/IPP).

References

1. Agrawal, R., Srikant, R.: Fast algorithms for mining association rules. In: Proceedings of the 20th International Conference on Very Large Data Bases, pp. 487–499. Morgan Kaufmann, Santiago de Chile (1994)

2. Agrawal, R., Srikant, R.: Mining sequential patterns. In: Eleventh International Conference on Data Engineering, Taipei, Taiwan, pp. 3–14 (1995)
3. de Amo, S., Furtado, D.A.: First-order temporal pattern mining with regular expression constraints. Data & Knowledge Engineering 62(3), 401–420 (2007); including special issue: 20th Brazilian Symposium on Databases (SBBD 2005)
4. Blockeel, H., Sebag, M.: Scalability and efficiency in multi-relational data mining. SIGKDD Explorations 5(1), 17–30 (2003)
5. Davis, J., Burnside, E., Ramakrishnan, R., Costa, V.S., Shavlik, J.: View learning for statistical relational learning: With an application to mammography. In: Proc. of the 19th International Joint Conference on Artificial Intelligence, Professional Book Center, Edinburgh, Scotland, UK, pp. 677–683 (2005)
6. Dehaspe, L., Toivonen, H.: Discovery of frequent DATALOG patterns. Data Mining and Knowledge Discovery 3(1), 7–36 (1999)
7. Esposito, F., Di Mauro, N., Basile, T.M.A., Ferilli, S.: Multi-dimensional relational sequence mining. Fundamenta Informaticae 89(1), 23–43 (2009)
8. Ferreira, C.A., Gama, J., Costa, V.S.: Sequential Pattern Mining in Multi-Relational Datasets. In: Meseguer, P., Mandow, L., Gasca, R.M. (eds.) CAEPIA 2009. LNCS, vol. 5988, pp. 121–130. Springer, Heidelberg (2010)
9. Garofalakis, M., Rastogi, R., Shim, K.: Mining sequential patterns with regular expression constraints. IEEE Transactions on Knowledge and Data Engineering 14(3), 530–552 (2002)
10. Jian, P., Han, J., Mortazavi-asl, B., Pinto, H., Chen, Q., Dayal, U., Hsu, M.: Prefixspan: Mining sequential patterns efficiently by prefix-projected pattern growth. In: Proc. of the 17th International Conference on Data Engineering, pp. 215–224. IEEE Computer Society, Heidelberg (2001)
11. Dan Lee, S., De Raedt, L.: Constraint Based Mining of First Order Sequences in SeqLog. In: Meo, R., Lanzi, P.L., Klemettinen, M. (eds.) Database Support for Data Mining Applications. LNCS (LNAI), vol. 2682, pp. 154–173. Springer, Heidelberg (2004)
12. Ohara, K., Yoshida, T., Geamsakul, W., Motoda, H., Washio, T., Yokoi, H., Takabayashi, K.: Analysis of Hepatitis Dataset by Decision Tree Graph-Based Induction. In: Proceedings of Discovery Challenge, pp. 173–184 (2004)
13. Srikant, R., Agrawal, R.: Mining Sequential Patterns: Generalizations and Performance Improvements. In: Apers, P.M.G., Bouzeghoub, M., Gardarin, G. (eds.) EDBT 1996. LNCS, vol. 1057, pp. 3–17. Springer, Heidelberg (1996)
14. Yan, X., Han, J., Afshar, R.: Clospan: Mining closed sequential patterns in large datasets. In: Proc. of the Third SIAM International Conference on Data Mining, pp. 166–177. SIAM, San Francisco (2003)
15. Yu, L., Liu, H.: Feature selection for high-dimensional data: A fast correlation-based filter solution. In: 20th Int. Conf. on Machine Learning, pp. 856–863 (2003)
16. Zelezny, F., Lavrac, N.: Propositionalization-Based Relational Subgroup Discovery with RSD. Machine Learning 62(1-2), 33–63 (2006)

Summarizing Frequent Itemsets via Pignistic Transformation

Francisco Guil-Reyes[1] and María Teresa Daza-Gonzalez[2]

[1] Dept. Languages and Computer Science
[2] Dept. Neuroscience and Health Care
University of Almería
04120 Almería
{fguil,tdaza}@ual.es

Abstract. Since the proposal of the well-known *Apriori* algorithm and the subsequent establishment of the area known as Frequent Itemset Mining, most of the scientific contribution of the data mining area have been focused on the study of methods that improve its efficiency and its applicability in new domains. The interest in the extraction of this sort of patterns lies in its expressiveness and syntactic simplicity. However, due to the large quantity of frequent patterns that are generally obtained, the evaluation process, necessary for obtaining useful knowledge, it is difficult to be achieved in practice. In this paper we present a formal method to summarize the whole set of mined frequent patterns into a single probability distribution in the framework of the Transferable Belief Model (TBM). The probability function is obtained applying the Pignistic Transformation on the patterns, obtaining a compact model that synthesizes the regularities present in the dataset and serves as a basis for the knowledge discovery and decision making processes.

In this work, we also present a real case study by describing an application of our proposal in the field of Neuroscience. In particular, our main goal is focused on the behavioral characterization, via pignistic distribution on attentional cognitive variables, of group of children pre-diagnosed with one of the three types of $ADHD$ (Attention Deficit Hyperactivity Disorder).

1 Introduction

Due to its wide range of applications, frequent patterns mining, in particular frequent itemset mining, has become one of the main data mining areas. The expressive power of frequent associations and its syntactical simplicity have been the main characteristics that have substantiated its validity as a technique for discovering useful knowledge in a growing number of application domains. Data mining is a user-centered process and can be contextualized as an essential step in the overall process of knowledge discovery from databases (KDD). The user (domain expert) has the responsibility to interpret, analyze and evaluate the whole set of patterns with the ultimate goal of obtaining useful knowledge. However, due to the huge number of patterns that are often learned in the data mining

L. Antunes and H.S. Pinto (Eds.): EPIA 2011, LNAI 7026, pp. 297–310, 2011.
© Springer-Verlag Berlin Heidelberg 2011

process, these tasks are very difficult to be carried out. One solution to this problem is the application of second-order data mining techniques, whose aim is to obtain a global model or a set of patterns that show, in a compact way and with a higher level of abstraction, the regularities presented in the dataset. Another solution is to reduce the number of discovered patterns, either during the mining process or as post-mining technique.

Reducing the number of frequent associations has been one of the main issues in the frequent pattern mining research. Traditionally, a general approach has been to obtain patterns that satisfy certain constraints. In this regard, we emphasize the well-known techniques of mining maximal frequent patterns [2,9], closed patterns [12], and non-derivable patterns [3]. Closed itemsets and non-derivable itemsets are lossless forms of compressing frequent patterns, because it is possible to recover the exact frequency values of any frequent pattern. By contrast, maximal patterns mining is a lossy compression technique, although the compression ratio is higher on average. The choice of a particular method will depend on each particular problem.

However, regardless of the compression method selected, the number of extracted frequent patterns is still too huge to be managed by the end user, being necessary to further compress them. Recently, several approaches have been proposed to tackle this issue. The idea behind these methods consist of obtaining a concise representation of the entire set of frequent patterns according to a coverage criteria, related to the choice of a small number of patterns that represent the complete set. In [1], the authors proposed to use K itemsets that maximally cover the entire set of patterns. In [20], the authors proposed a method to obtain a concise representation consisting of K clusters covered and estimated by they called *pattern profile*. In [19], the authors perform better in estimating the frequency patterns through the use of a global model known as Markov Random Field (MRF). However, as shown in [8], both approaches are heuristic in nature and do not provide any theoretical justification about the proposed methods. In this case, the authors proposed a set of regression-based approaches to summarize frequent patterns, partitioning the set of frequent patterns into K partitions. And finally, with the objective of minimizing both the runtime and the restoration error, in [21] the authors proposed a neural network and a cluster-based method to summarize the set of frequent itemsets.

In this paper we propose a very different solution to obtain a concise representation of the frequent patterns. We are particularly interested in obtaining a model of representation of uncertainty that compactly summarizes the regularities presented in the dataset. The compact representation is a probability function defined on the frequent elements (literals) of the input dataset and behaves as a covering function of the set of frequent patterns. In this proposal, instead of obtaining the K-most representative patterns (we think that each and every one of them are representative enough not to be discarded), the method examines the whole set of patterns, obtaining the probability function from the structure and frequency of each one. Intuitively, while the frequency of a pattern is just a descriptive value, the pignistic probability is a valid and informative value for decision making, as is calculated taking into account the values of frequency and structure of each pattern and all the patterns that have elements in common. The underlying idea is that if we want to make a decision based on the information provided by the set of patterns, it will be more consistent if the related uncertainty can be described by a probability distribution. Once the distribution is obtained, we can use well-known data analysis techniques in order to obtain objective measures to assist in the process of data description and inference of new knowledge.

To obtain the covering probability function, we propose the use of the *pignistic transformation* defined in the *Transfer Belief Model* (*TBM*). The key idea is to consider the set of frequent patterns as a body of evidence. Basically, *TBM* is a model for the representation of quantified beliefs held by an agent. It is a two-level model: the credal level where beliefs are held and represented by belief functions, and the pignistic level, where decisions are made by maximizing expected utilities. In order to compute these expectations, it is necessary to build a probability measure in the pignistic level. This probability measure is called *pignistic probability* and denoted as *BetP* [18]. In our approach, the agent is the data mining algorithm and the beliefs are represented by, precisely, the set of frequent patterns characterized by its frequency distribution.

As a case study, we propose the use of our approach to extract meaningful patterns based on probability distributions that characterize the functioning of attentional networks in subjects pre-diagnosed with a subtype of *ADHD* (Attention Deficit Hyperactivity Disorder). Obtaining differential patterns on each of the types of this syndrome is something of great interest because the therapeutic implications differ significantly from one type to another.

The rest of this paper is organized as follows. Section 2 introduces the notation and basic definitions necessary to define the problem. Section 3 introduces the theoretical foundations of the proposed method for summarizing frequent itemsets. Section 4 describes an empirical evaluation with a real dataset belonging to the neuroscience domain. Conclusions and future works as finally drawn in Section 5.

2 Problem Definition

Let $\mathcal{I} = \{e_1, \ldots, e_d\}$ be a set of items. A subset of $\mathcal{I}$, denoted as α, is called an itemset (or pattern). The size of the pattern is the number of items it contains. A *transactional database* D is a collection of associations, $D = \{t_1, \ldots, t_n\}$, where $t_i \subseteq \mathcal{I}$. For any pattern α, we write the transactions that contain α as $D_\alpha = \{t_i | \alpha \subseteq t_i, t_i \in D\}$.

Definition 1. *(Frequent pattern). For a transactional dataset D, a pattern α is frequent iif $|D_\alpha| \geq \sigma$, where $|D_\alpha|$ is the frequency of the pattern (the number of its occurrences in the dataset), denoted as $fr(\alpha)$, and σ is a user-defined parameter called minimum frequency (or support).*

Given a dataset D and a value for σ, the goal of frequent pattern mining is to determine in the dataset all the frequent patterns (itemsets) whose frequency are greater than or equal to $minsup$. In order to achieve this goal, the algorithm follows a level-wise pattern generation satisfying the well-known *Apriori* property (downward closure property), which states that any subset of a frequent itemset is also frequent. Although this property is used to prune the search space, it leads to an explosive number of frequent patterns. For example, a frequent itemset with n items may generate 2^n sub-associations, all of which are frequent too.

The result of the mining process consist of a set of frequent patterns denoted as $\mathcal{B}^{\sigma} = \{\alpha_1, \ldots, \alpha_m\}$, where $f_r(\alpha_i) \geq \sigma$. $\mathcal{B}^{\sigma}$ is characterized by a frequency distribution $f = \{f_1, \ldots, f_m\}$ such that $f_i = fr(\alpha_i)$. From this function f, we obtain the relative frequency distribution $\overline{f} = \{\overline{f}_1, \ldots, \overline{f}_m\}$, such that $\overline{f}_i = \frac{f_i}{\sum_i f_i}$.

Once the set of frequent patterns is extracted, the next step is its evaluation and interpretation. These tasks are designed with the aim to discover useful knowledge, and usually they are performed by the expert. However, this is a very hard process and rarely can be done due to the huge amount of frequent patterns that are often discovered. The problem is compounded if, instead of evaluating one set of patterns, the expert must evaluate different sets obtained by varying the minimum support parameter or any of the parameters of the algorithms involved in the knowledge discovery process (data cleaning or discretization, among others). In this case, the evaluation process must be carried out in parallel for each of the mined sets, thus having a much more complicated problem. If, moreover, the expert must compare two (or more) frequent pattern sets to find patterns that characterize different groups, the problem can be hardly treatable. On an experimental basis, there is no single method for assessing and interpreting the mining results, but these methods are time-consuming and they must be done using a trial-error method of problem solving. However, as is well known, decisions will be coherent if the underlying uncertainties can be described by a probability distribution [4].

So, it seems interesting to study a formal method to summarize the mined itemset-based patterns in a compact model of knowledge representation. In particular, the compact model consist of a probability distribution over the frequent items that summarizes the information contained in the patterns and aims to provide a basis for the decision-making process. Once the function is obtained, and with the aim to use quantitative objective measures to guide the analysis phase, we can use classical statistical method for studying the obtained results. This issue is of great importance if we want to compare the probability functions obtained under different experimental conditions (for example, varying the minimum support) or if we want to compare the patterns obtained from different groups of subjects in order to obtain significant differences among them.

3   Summarizing Frequent Patterns

In this section, we first introduce a brief summary of the Transferable Belief Model. Next we present the proposal method for summarizing frequent associations via pignistic transformation.

3.1   The Transferable Belief Model

The *Transferable Belief Model* [17] is an interpretation of the *Dempster-Shafer model*, a mathematical Theory of Evidence for modeling complex systems, where uncertainty is represented by quantified beliefs. The authors proposed the theory based on the fact that there is a two-level mental model: the *credal level*, where beliefs are held and represented by belief functions, and the *pignistic level*, where decisions are made by maximizing expected utilities.

   The credal level (*credal* from "credo", a Latin word that means "I believe").

   Belief measures are fuzzy measures assigned to propositions to express the uncertainty associated to them. Let us consider a k-elements frame of discernment $\Omega = \{\omega_1, \ldots, \omega_k\}$. A *basic belief assignment* (*bba*) is a mapping $m^{\Omega} : 2^{\Omega} \to [0,1]$ that satisfies $\sum_{A \subseteq \Omega} m(A) = 1$. The values of the *bba* are called *basic belief masses* (*bbm*). If $\overline{m}(\emptyset) = 0$ we say that our *bba* is normalized. The TBM postulates that the impact of a piece of evidence on an agent is translated by an allocation of parts of an initial unitary amount of belief among the propositions of Ω. For $A \in \Omega$, $m(A)$ is a part of the agent's belief that supports A.

   Every $A \subseteq \Omega$ such that $m(A) > 0$ is called a *focal element*. The main difference with probability models is that masses can be given to any proposition p of Ω, with $p \in 2^{\Omega}$, instead of only to its atoms.

   The Pignistic Level (pignistic from "pignus", a Latin word that means "a bet").

   Given the evidence available to an agent, the TBM model allows to obtain a belief function that describes the credal state on the frame of discernment. If we want to make a decision based on this credal state, will first be necessary to find a rule that allows for the construction of a probability distribution from a belief function. The authors firstly proposed a method based on the *Insufficient Reason Principle*: if we must build a probability distribution on n elements, given a lack of information, give a probability $1/n$ to each element. The probability distribution will be obtained distributing $m(A)$ among the atoms of A, so that, for each atom $\omega \in \Omega$,

$$BetP(\omega) = \sum_{A \subseteq \Omega, \omega \in A} \frac{m(A)}{|A|},$$

where $|A|$ is the number of atoms of Ω in A.

   $BetP$ is, mathematically, a probability function, but the authors called it the *Pignistic Probability Function* to stress the fact that it is the probability function in a decision context. The name of $BetP$ is selected to enhance its real nature, a probability measure for decision-making, for betting.

Once the $BetP$ is calculated for the set of atoms of Ω, for each $A \subseteq \Omega$,

$$BetP(A) = \sum_{\omega \in A} BetP(\omega).$$

Obtaining of $BetP$ from the *basic belief masses* is called as *Pignistic Transformation*. In [18], the author formalized the justification of the pignistic transformation based on linearity requirement.

3.2 Pignistic Transformation

Let $\mathcal{B}^\sigma$ be a set of frequent patterns characterized by a normalized frequency distribution $\overline{f}$. Let $\mathcal{I}^\sigma = \{e_1^\sigma, \dots, e_k^\sigma\}$ the set of frequent items presented in the set. Obviously, $\mathcal{I}^\sigma \subseteq \mathcal{I}$. Making a syntactic correspondence wit the Transferable Belief Model, we obtain that $\mathcal{I}^\sigma = \Omega$, the frame of discernment. The frequent items are called the atoms of Ω. Through this correspondence, we are interpreting the set of frequent patterns as a *body of evidence*, where $\overline{f}$ is equivalent to m, the *basic belief assignment*.

Once the correspondence is established, the summarization of the frequent patterns would be carried out by the pignistic transformation described above. From $\mathcal{B}^\sigma$ we obtain the cover probability function $BetP_{\mathcal{B}^\sigma} = \{p_1, \dots, p_k\}$, $1 \leq k \leq |\mathcal{I}^\sigma|$, where $p_i = BetP(e_i^\sigma)$ (the pignistic probability), $e_i^\sigma \in \mathcal{I}^\sigma$, and

$$p_i = \sum_{e_i^\sigma \in \alpha_i} \frac{\overline{f}(\alpha_i)}{|\alpha_i|}, \quad \forall \alpha_i \in \mathcal{B}^\sigma.$$

4   A Case Study: Attention Deficit Hyperactivity Disorder

The clinical syndrome of Attention Deficit Hyperactivity Disorder ($ADHD$) has been thought to involve specific deficits in attention. This topic has been the subject of intensive investigation over the past decade. The general accepted clinical definition of $ADHD$ is provided by $DSM - IV$ (American Psychiatric Association, 1994) according to the 4th edition of the *Diagnostic and Statistical Manual of Mental Disorders*. The diagnostic algorithm results in three possible diagnostic subtypes: (1) inattentive ($ADHD - I$); (2) hyperactive/impulsive ($ADHD - H/I$); and (3) combined type ($ADHD - C$), depending on whether individuals meet the criteria in several sets of symptoms.

The main behavioral assessment techniques used to determine diagnosis of this disorder include parent and teacher rating scales [11]. These behavioral ratings from parents and teachers reliably differentiate two symptom domains labeled as "inattention" and "hyperactivity-impulsivity" [10]. However, attention is a complex cognitive function. Rather than involving a single unified system, attentional functioning is executed by separate independent systems [14]. According to the *Attention Networks Theory* the systems can be subdivided into an *alerting or vigilance* network, a network of *orientation or selection*, and an *executive or conflict* network. A range of experimental, neuroimaging and clinical studies have supported this theory.

The aim for this study was to assess the functioning of attention networks in a group of 23 children, aged 7 to 10 years, who appear at significantly risk to receive a diagnosis of *ADHD*, compared to age-matched control children (n=31). We use a version of the revised *Conners' Rating Scales* for the assessment of *ADHD*. The associated scales has been standardized for school-aged children aged 6 to 12 years. To assess the functioning of the attentional networks we have used a version of the *Attentional Network Test (ANT)* [16]. The test is a child-friendly version of a neuropsychological task with alerting and orienting cues. The children have to respond based on whether a central stimulus (in this case is a goldfish) was pointing to the left or right by pressing the corresponding left or right key on the mouse (see Figure 1). On congruent trials, the central fish is pointing in the same direction; on incongruent trials, the marginal fish point in the opposite direction from the central fish; and on neutral trials, the central fish appears alone. Each target is preceded by one of four warning cue conditions: a center cue, a double cue, a spatial cue, or no cue. In the center cue condition, an asterisk is presented at the location of the fixation cross. In the double cue condition, an asterisk appears at the locations of the target above and below the fixation cross. Spatial cues involve a single asterisk presented in the position of the upcoming target.

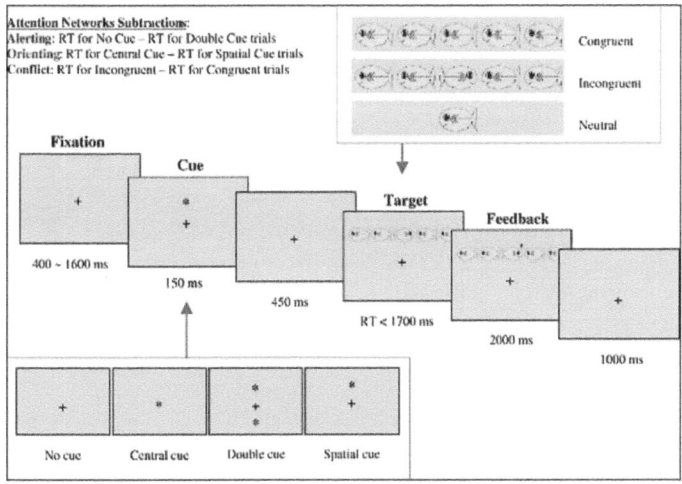

Fig. 1. Schema of the child version of ANT task

A session of the ANT task consist of a total of 24 practice trials and one experimental block of 48 trials. Each trial represents one of the 12 conditions in equal proportions: three target types (congruent, incongruent and neutral) four cues (no cue, central cue, double cue and spatial cue). Participants indicate their responses via right or left button-press on a mouse. Accuracy and reaction time are recorded. From the values of reaction time and accuracy obtained after the execution of the task, we obtain the values of the cognitive variables that shape the functioning of attentional networks. This is performed applying the *subtractive method* [13] which is based on the assumption that mental operations can be measured by decomposing complex cognitive tasks in sequences of simpler tasks. This method assumes that the effect of each mental operation is additive and that it is possible to isolate the effect of a single mental operation by comparing two tasks that differ only by the presence or absence of that mental operation

So, as we can see in Figure 4, six subtractions are computed to obtain the alerting, orienting and conflict scores for each participant (three subtractions with reaction time -*RT*- and three with error rate -*ER*-). Through this six cognitive measures we can characterize the three neural networks associated with the attentional cognitive function. The goal now is to find patterns that characterize each of the populations described above.

4.1   The Mining Process

As a result, we obtain a dataset composed of 53 rows and 7 columns (6 cognitive attributes and 1 class attribute). The cognitive attributes are numeric, while the attribute class (type of *ADHD*) is nominal. Table 1 shows a statistical description of the numerical attributes.

Table 1. Description of the cognitive attributes

Cognitive Attribute	Minimum	Maximum	Mean	StdDev	#$Interval$
Alerting (TR)	-187.000	444.500	61.000	102.810	7
Orienting (TR)	-133.750	268.750	48.500	81.911	5
Executive (TR)	-169.000	445.000	107.943	94.060	7
Alerting (ER)	-0.250	0.170	0.001	0.089	5
Orienting (ER)	-0.170	0.250	-0.006	0.087	5
Executive (ER)	-0.120	0.500	0.038	0.107	6

In Table 2 we can observe the distribution of the 53 subjects according to the class attribute, which can be *Control* (subjects who present no risk of *ADHD*) or one of the three subtypes of *ADHD*. Although the proportion of subjects with and without risk of *ADHD* is more or less balanced, in our study we want to characterize the functioning of the attentional networks of each group independently. This is an important point as the therapeutic implications of each risk group are quite different.

Perhaps, due to the number of subjects per group is rather unbalanced (it varies significantly) or by the nature of the dataset, it is difficult to find meaningful patterns that characterize each subgroup. In order to prove this, we have

Table 2. Description of the class attribute

Class Attribute	#Subjects	%
Control	31	58.49
ADHD − I	9	16.98
ADHD − H/I	7	13,21
ADHD − C	6	11,32

conducted a previous experimental process using classical machine learning techniques like those implemented in *WEKA* [7]. In particular, we have use the implemented *ZeroR* (majority class classifier), *Logistic Regression, J*48 (a clone of the well-known decision tree C4.5 [15]), and the *Naive-Bayes* classifiers in order to obtain different classification models with no cross-validation and two different values for the cross-validation process. In particular, for the cross-validation parameter we have used two difference values, 10-fold cross-validation (commonly used) and 53-fold cross-validation (equivalent to the *leave-one-out* method, very appropriated for small datasets). In Table 3 we can see the classification accuracy of each classifier vs. the number of folds used in the model evaluation.

Table 3. Classification Accuracy (%)

	Training set	*10 − fold*	*53 − fold*
ZeroR	58.49	58.49	58.49
Logistic Regression	64.15	52.83	58.49
Neural Network	92.45	54.72	52.83
Naive-Bayes	75.47	56.60	58.49
C4.5 decision tree	71.70	50.94	58.49
Decision Table	66.04	**60.38**	58.49

At this point it is noteworthy that classical machine learning techniques cannot find an accurate model. In the case of the common 10-fold cross-validation, the results indicate, practically, that all subjects are classified as "Control", which provides very little information. Some classifiers get some additional rules for classifying several subjects belonging to one of the four groups. For example, the *Decision Table* classifier (for 10-fold cross-validation) obtains the rule R_1 associated with the contingency matrix shown in Table 4. As we can see, only the *Control* subjects and one of the *combined* group are correctly classified. In conclusion, we can establish that the information provided by intelligent classifiers is not significant enough to obtain useful knowledge.

$$R_1 : if(\text{Executive (ER)} \in (0.095, \infty)) \Rightarrow \text{ADHD-C else Control}$$

However, previous analysis of variance between groups (ANOVA) showed significant differences in the Orienting and Executive networks of the $ADHD − I$ and

Table 4. Contingency Matrix

	Control	ADHD-I	ADHD-H/I	ADHD-C
Control	31	0	0	0
ADHD-I	8	0	0	1
ADHD-H/I	7	0	0	0
ADHD-C	5	0	0	1

$ADHD - C$ groups versus the *Control* group. These results, in conjunction with the social interest of the domain, are a great motivation to keep studying new intelligent techniques for obtain useful knowledge. At this point, it seems interesting the application of local data mining techniques to obtain useful patterns that help us in our goal. Regardless of the size of the dataset, its nature makes it relevant in this study because it is difficult to find patterns that characterize computationally the four groups. Thus, it would be interesting to divide the dataset into 4 partitions according to the class label (as done in [5] for mining emerging patterns). And then, we can use the *Apriori* algorithm on each of the groups, obtaining four sets of patterns.

An important aspect to consider before proceeding with the data mining process is that *Apriori*-based algorithms cannot handle numeric attributes. Since the dataset contains continuous attributes, very common in real datasets, it will require a discretization method to obtain a dataset with only nominal attributes. Basically, the discretization techniques consist of the partition of the numerical domain into a set of intervals, treating each interval as a category. The choice of a discretization method is given a task of great importance since it affects significantly on the set of patterns obtained. In [6] the authors proposed a formal method for evaluating the data mining results in terms of quality in the information presented in the set of patterns. In this case, we have selected a statistical discretization method implemented in the $WEKA$ workbench. In Table 1, the $\#Interval$ column shows the number of categories (intervals) associated with each cognitive variable. In total, there are 35 intervals, which are those that form the input set of literals (items) for the mining algorithm. Each literal is denoted as the pair $(attribute - value)$, where $attribute$ indicates one of the six cognitive variables, and $value$ will indicate one of the intervals of categories associated. Usually, each interval is associated with a given linguistic label, for example, 'extremely low", "very low", "low", and so on.

For the minimum support parameter, we have used two different values: $\sigma = .3$, and $\sigma = .1$. The choice of the values of this parameter was a decision guided by the expert of the domain. Usually, the most appropriate value of the parameters is obtained after an incremental process of trial and error, trying to find in a reasonable time the best set of patterns based on the number and length of them. In this case, .1 and .3 are two good alternatives as values for the minimum support parameter. The obtained results are shown in Table 5. The N_i denotes the number of frequent items associated with each group (pairs $attribute - value$) and minimum support value. The N_p parameter indicates the total number of extracted patterns, and L_{max} the maximum length of them.

Table 5. Mined patterns summary

Group	Parameters	$\sigma = .3$	$\sigma = .1$
	N_i	7	18
Control	N_p	31	173
	L_{max}	4	5
	N_i	7	21
ADHD − I	N_p	29	454
	L_{max}	4	6
	N_i	10	17
ADHD − H/I	N_p	58	353
	L_{max}	5	6
	N_i	8	21
ADHD − C	N_p	19	352
	L_{max}	3	6

To find the main differences between groups and to find meaningful patterns, we must first analyse each of the patterns and compare them with the rest of the patterns belonging to other groups. Although in this case the sets of patterns obtained are small, the task of analysis is quite complicated if we want to perform precise work. Also, what is the most important parameter in a pattern: its length or frequency? In the next section we present the results obtained applying the proposed summarization technique.

4.2   The Summarization Process

In this section, we will focus on the results obtained with the mining algorithm, establishing $\sigma = .3$, because this is a representative case. For corresponding sets of frequent itemsets we have obtained the pignistic probability distribution that summarizes the regularities present in each one of them. The result is 4 probability functions that could be considered as 4 patterns that describe the regularities presented in each of the groups of subjects studied. Instead of displaying the numerical values of the 4 probability distributions, in Table 6 (Figure a) , we show, graphically, the obtained probability distributions. The X-axis shows the different discretized values associated with the 6 cognitive variables. In total there are 16 values that correspond to the union of the items ($attribute - value$ pair, for example $AlertingTR - low$, $ExecutiveER - high$) common for the 4 groups. As we can see, each group shows a different behaviour pattern, reinforcing the initial hypothesis about differences in the functioning of attentional networks in subjects without risk and at risk in any of the three types of $ADHD$. In order to study what are the significant differences in each risk group with respect to the control group, in Table 6 (Figures b, c, and d), we show separately the obtained probability distributions. Items marked in bold indicate a statistically significant difference ($p < 0.01$ by t-test) and marked in grey show a marginally significant difference ($p < 0.05$). As can be seen, each risk group shows a different set of

Table 6. Experimental results

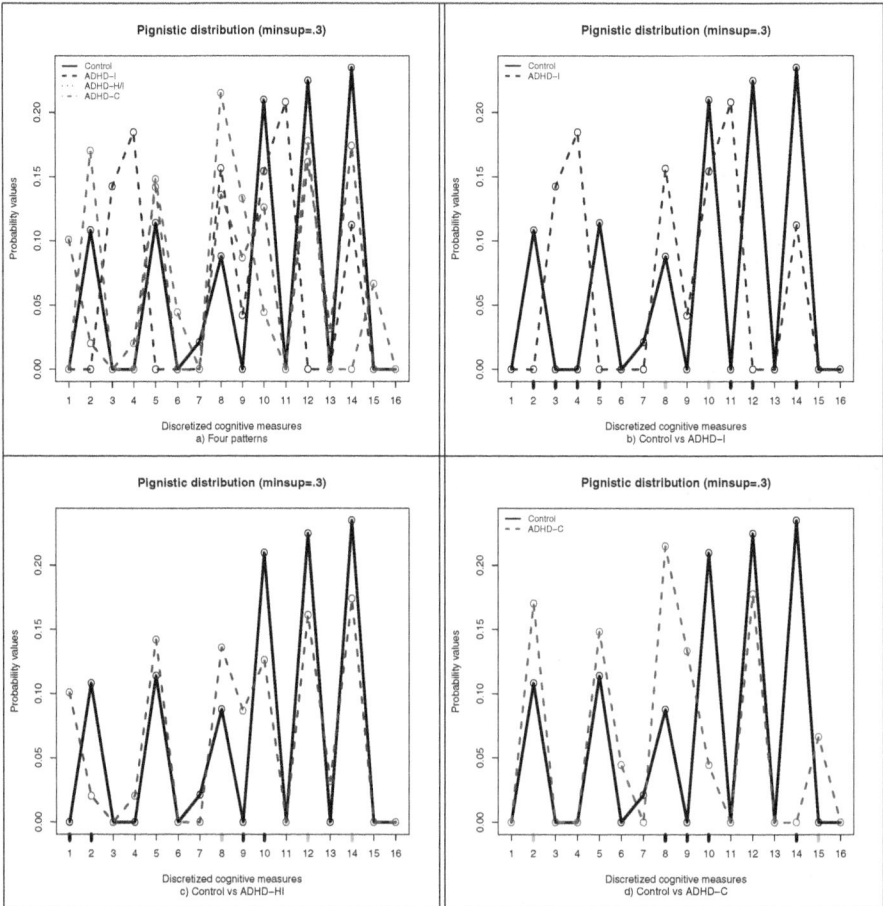

items which are differences with the control group. For more information, the relevant data are shown in Table 7.

On the one hand, this method provides a quantitative mechanism and a easy visual interpretation for assessing and interpreting the results obtained from the mining process. On the other hand, the computational nature of the result would be interesting to establish an automatic mechanism for the interpretation and comparison of different probability distributions and variable selection method to determine where there are more significant differences between groups. This could lead, for example, to the implementation of a reduced version of the *ANT* task, focused only on the trials related to this type of attentional network, thus obtaining a quick and effective pre-diagnostic tool.

Table 7. Mined patterns summary

Group	p-value	Items
Control vs ADHD-I	$p < 0.01$	{2,3,4,5,11,12,14}
	$p < 0.05$	{8,10}
Control vs ADHD-H/I	$p < 0.01$	{1,2,9,10}
	$p < 0.05$	{8,12,14}
Control vs ADHD-C	$p < 0.01$	{8,9,10,14}
	$p < 0.05$	{2,15}

One important aspect that we wish to emphasize is that from a computational point of view, the summarization method is quite efficient because the method is linear in size.

5 Conclusions and Future Work

In this work we have presented a formal method to summarize, efficiently, frequent patterns into a compact model that synthesizes the regularities present in the dataset and serves as a basis for the knowledge discovery and decision making processes. Treating the whole mined patterns as a body of evidence, the method is based on the pignistic transformation proposed in the Transfer Belief Model. As an application example we have presented a case study in order to find patterns that characterize the functioning of attentional networks in patients without risk and at risk for a particular type of *ADHD*. The results show different behaviour patterns for each of the different groups of subjects, which indicates that the attentional networks operate differently depending of the subtype of *ADHD*.

As future work, we propose to extend the work incorporating new descriptive techniques that help in the interpretation of the obtained results. We also propose the use of the mined pignistic distributions as a basis for obtaining an automatic classifier.

Acknowledgments. This work has been partially supported by contributions from the Spanish MEC under the National Project TIN2006-15460-C04-01 and the Excellence Project of the Junta of Andalusia with number P07-SEJ-03214.

References

1. Afrati, F., Gionis, A., Mannila, H.: Approximating a collection of frequent sets. In: Proc. of the 10th ACM SIGKDD Int. Conf. on Knowledge Discovery and Data Mining, pp. 12–19 (2004)
2. Bayardo, R.J.: Efficiently mining long patterns from databases. In: Proc. of the ACM SIGMOD Int. Conf. on Management of Data (SIGMOD 1998), pp. 85–93 (1998)

3. Calders, T., Goethals, B.: Non-derivable itemset mining. Data Mining and Knowledge Discovery 14(1), 171–206 (2007)
4. DeGroot, M.H.: Optimal Statistical Decisions. McGraw-Hill, New York (1970)
5. Dong, G., Li, J.: Efficient mining of emerging patterns: Discovering trends and differences. In: Proc. of the 5th ACM SIGKDD Int. Conf. on Knowledge Discovery and Data Mining, pp. 43–52 (1999)
6. Guil, F., Palacios, F., Campos, M., Marín, R.: On the evaluation of mined frequent sequences. an evidence theory-based method. In: Proc. of the 3rd Int. Conf. on Health Informatics (HEALTHINF 2010), pp. 263–268 (2010)
7. Hall, M., Frank, E., Holmes, G., Pfahringer, B., Reutemann, P., Witten, I.H.: The weka data mining software: An update. SIGKDD Explorations 11(1), 10–18 (2009)
8. Jin, R., Abu-Ata, M., Xiang, Y., Ruan, N.: Effective and efficient itemset pattern summarization: regression-based approaches. In: Proc. of the 14th ACM SIGKDD Int. Conf. on Knowledge Discovery and Data Mining (KDD 2008), pp. 399–407 (2008)
9. Lin, D.-I., Kedem, Z.M.: Pincer Search: A New Algorithm for Discovering the Maximum Frequent Set. In: Schek, H.-J., Saltor, F., Ramos, I., Alonso, G. (eds.) EDBT 1998. LNCS, vol. 1377, pp. 105–119. Springer, Heidelberg (1998)
10. McBurnett, K., Pfiffner, L., Willcutt, E., Tamm, L., Lerner, M., Ottolini, Y., Furman, M.: Experimental cross-validation of dsm-iv types of adhd. Journal of the American Academy of Child and Adolescent Psychiatry 38, 17–24 (1999)
11. Mitchell, W.G., Chavez, J.M., Baker, S.A., Guzman, B.L., Azen, S.P.: Reaction time, impulsivity, and attention in hyperactive children and controls: a video game technique. Journal of Child Neurology 5, 195–204 (1990)
12. Pasquier, N., Bastide, Y., Taouil, R., Lakhal, L.: Discovering frequent closed itemsets for association rules. In: Proc. of the 7th Int. Conf. on Database Theory, pp. 398–416 (1999)
13. Posner, M.I.: Chronometric Explorations of the Mind. Lawrence Erlbaum Associates (1976)
14. Posner, M.I., Petersen, S.E.: The attention system of the human brain. Annual Review of Neuroscience 13, 25–42 (1990)
15. Quinlan, J.R.: C4.5: Programs for Machine Learning. Morgan Kaufman Publishers (1993)
16. Rueda, M.R., Fan, J., McCandlis, B.D., Halparin, J.D., Gruber, D.B., Lercar, L.P., Posner, M.I.: Development of attention networks in chilhood. Neuropsychologia 42, 1029–1040 (2004)
17. Smets, P., Kennes, R.: The transferable belief model. Artificial Intelligence 66, 191–234 (1994)
18. Smets, P.: Decision making in the tbm: the neccessity of the pignistic trasformation. Int. Journal of Approximate Reasoning 38, 133–147 (2005)
19. Wang, C., Parthasarathy, S.: Summarizing itemset patterns using probabilistic models. In: Proc. of the 12th ACM SIGKDD Int. Conf. on Knowledge Discovery and Data Mining (KDD 2006), pp. 730–735 (2006)
20. Yan, X., Cheng, H., Han, J., Xin, D.: Summarizing itemset patterns: a profile-based approach. In: Proc. of the 11th ACM SIGKDD Int. Conf. on Knowledge Discovery and Data Mining (KDD 2005), pp. 314–323 (2005)
21. Zhao, Z., Qian, J., Cheng, J., Lu, N.: Frequent itemsets summarization based on neural network. In: Proc. of the 2nd IEEE Int. Conf. on Computer Science and Information Technology, pp. 496–499 (2009)

A Simulated Annealing Algorithm
for the Problem of Minimal Addition Chains

Adan Jose-Garcia, Hillel Romero-Monsivais, Cindy G. Hernandez-Morales,
Arturo Rodriguez-Cristerna, Ivan Rivera-Islas, and Jose Torres-Jimenez

Information Technology Laboratory, CINVESTAV-Tamaulipas Km. 5.5 Carretera Cd.
Victoria-Soto la Marina, 87130, Cd. Victoria Tamps., Mexico
{ajose,hromero,chernandez,arodriguez,rrivera}@tamps.cinvestav.mx,
jtj@cinvestav.mx

Abstract. Cryptosystems require the computation of modular exponentiation, this operation is related to the problem of finding a minimal
addition chain. However, obtaining the shortest addition chain of length
n is an NP-Complete problem whose search space size is proportional to
$n!$. This paper introduces a novel idea to compute the minimal addition
chain problem, through an implementation of a Simulated Annealing
(SA) algorithm. The representation used in our SA is based on Factorial Number System (FNS). We use a fine-tuning process to get the best
performance of SA using a Covering Array (CA), Diophantine Equation
solutions (DE) and Neighborhood Functions (NF). We present a parallel implementation to execute the fine-tuning process using a Message
Passing Interface (MPI) and the Single Program Multiple Data (SPMD)
model. These features, allowed us to calculate minimal addition chains
for benchmarks considered difficult in very short time, the experimental results show that this approach is a viable alternative to solve the
solution of the minimal addition chain problem.

1 Introduction

The modular exponentiation (repetition of modular multiplications) is one of
the most important operations for data scrambling and it is used by several
public-key cryptosystems such as RSA encryption scheme [1]. Moreover, since
cryptographic applications often use very large fields or rings, the cost of one
multiplication is rather high and reducing the number of multiplications needed
in exponentiation can significantly improve the speed of a cryptosystem. Therefore, the performance of a cryptosystem is determined by the efficiency of the
implementation of modular multiplications [9].

The problem of determining the optimal sequence of multiplications required
for performing a modular exponentiation can be formulated using the concept
of *addition chains*.

An addition chain for a positive integer α is a sequence of positive integers
$C = (\alpha_0, \alpha_1, \ldots, \alpha_r)$ s.t [10]:

L. Antunes and H.S. Pinto (Eds.): EPIA 2011, LNAI 7026, pp. 311–325, 2011.

$$\alpha_i = \begin{cases} 1 & \text{if } i = 0 \text{ and } \alpha_r = \alpha \\ \alpha_j + \alpha_k & \text{if } i > 0 \text{ for some } j, k < i \end{cases}$$

The integer r is the length of the chain C and is denoted by $\mid C \mid$. The chain length $l(\alpha)$ of α is the minimal length of all possible chains for α. Addition chains for α generate multiplications for the computation x^α. For instance, the chain $(1, 2, 4, 5, 9, 18, 23)$ leads to the following scheme for the computation of x^{23},

$$xx = x^2, x^2x^2 = x^4, xx^4 = x^5, x^4x^5 = x^9, x^9x^9 = x^{18}, x^5x^{18} = x^{23}$$

The addition chain-based methods use a sequence of positive integers s.t. the first number of the chain is 1 and the last is the exponent α, and in which each member is the sum of two previous members of the chain. Clearly this establishes a correspondence between addition chains and such schemes. Therefore, the length of an addition chain α is equal to the corresponding number of multiplications required for the computation of x^α. Thus, the smallest of such multiplications is given by the chain length $l(\alpha)$ of α.

Addition chains have been studied extensively, Downey *et al.* have shown that the problem of determining an addition chain for α with the shortest length $l(\alpha)$ belongs to the family of NP-complete [4] problems. Therefore, it is recommended to use some kind of heuristic strategy in order to find an optimal addition chain when dealing with large exponents α.

In this paper, we present a Simulated Annealing (SA) to solve the problem of minimal addition chains. We subdivide the remaining of the paper into six sections. Section 2 describes the relevant related work to our research. Section 3 contains the description and the implementation details of our SA. Section 4 presents the experimental design using Covering Array and Diophantine Equation solutions, and describes a parallel approach to execute the SA. Section 5 reports the tuning parameters, the benchmark used to measure the performance of the SA, and the experimental results obtained. Finally, in Section 6 some conclusions are drawn.

2   Relevant Related Work

Let x be an arbitrary integer in the range $[1, n - 1]$, and α an arbitrary positive integer. Then, finding the modular exponentiation is defined as the problem of finding a unique integer $\beta \in [1, n - 1]$ that satisfies

$$\beta = x^\alpha \bmod n$$

A naive method to compute β requires $\alpha - 1$ modular multiplications. The sequence of powers is described as:

$$x \rightarrow x^2 \rightarrow x^3 \rightarrow \ldots \rightarrow x^{\alpha-1} \rightarrow x^\alpha$$

For instance, to compute x^8, this method performs seven multiplications: $x \rightarrow x^2 \rightarrow x^3 \rightarrow x^4 \rightarrow x^5 \rightarrow x^6 \rightarrow x^7 \rightarrow x^8$. However, x^8 can be computed using only three multiplications $x \rightarrow x^2 \rightarrow x^4 \rightarrow x^8$. The correct sequence of multiplications required for performing a modular exponentiation can be formulated using the concept of *addition chains*.

One possible way to compute x^α for $\alpha \in \mathbb{N}$ is by the *binary method* [3]. The **Equation** 1 presents a recursive description of this method.

$$x^\alpha = \begin{cases} x & \text{if } \alpha = 1 \\ x^{\frac{\alpha}{2}} \cdot x^{\frac{\alpha}{2}} & \text{if } \alpha \text{ is even} \\ x^{\alpha-1} \cdot x & \text{otherwise} \end{cases} \quad (1)$$

This method is faster than the naive method that computes $\alpha - 1$ multiplications. For example, 7 multiplication are necessary to compute x^{23}, the powers x^2, x^4, x^5, x^{10}, x^{11}, x^{22} and x^{23} are computed from previous ones using one multiplication. However, the shortest way to compute x^{23} requires only 6 multiplications for $\alpha = 23$, the powers involved are: $x^2, x^4, x^5, x^9, x^{18}$ and x^{23}.

For some $\alpha \in \mathbb{N}$, let $B(\alpha)$ the addition chain corresponding to the *binary method*, it is defined in **Equation** 2, i.e. $B(\alpha)$ is the list of integers s_i where x^{s_i} is an intermediary result while computing x^α. Therefore, we can describe $B(\alpha)$ as follows:

$$B(\alpha) = \begin{cases} 1 & \text{if } \alpha = 1 \\ B(\frac{\alpha}{2}), \alpha & \text{if } \alpha \text{ is even} \\ B(\alpha - 1), \alpha & \text{otherwise} \end{cases} \quad (2)$$

According to the *binary method* the length of $B(\alpha)$ is $\lfloor \log_2(\alpha) \rfloor + 1$ plus the number of bits equal to 1 in the binary representation of α. If we denote this value by $v(\alpha)$, we get the immediate consequence:

$$l(\alpha) \leq \lfloor \log_2(\alpha) \rfloor + v(\alpha) - 1 \leq 2\lfloor \log_2(\alpha) \rfloor$$

There are a large number of reported algorithms for finding optimal addition chains. Reported strategies include: *binary method, m-ary method, window-based method* [3,4]. Those algorithms have in common that they strive to keep the number of required field multiplications as low as possible through the usage of a particular heuristic. Different heuristics for finding short, but not necessarily minimal, addition chains for α large have been proposed [4,5,7]. In the reviewed bibliography we do not find reported a SA approach for computing minimal addition chains. Simulated Annealing algorithms have been successfully applied to solve a broad variety of optimization problems [11,13]. In this paper, we present a SA applied to optimize addition chains. The results obtained using this algorithm suggest that this approach is a competitive alternative to solve the problem of minimal addition chain.

3   Proposed Approach

In the proposed approach, the search space of the minimal addition chain problem has a lower bound and upper bound, where the *lower bound* value in the binary representation for an addition chain α is denoted by φ,

$$\varphi = \lfloor \log_2(\alpha) \rfloor$$

while the *upper bound* is denoted by ψ,

$$\psi = 2\lfloor \log_2(\alpha) \rfloor$$

The initialization for an addition chain α assures that it has an initial length equal to $\varphi + 1$, then the minimal length for α is,

$$\varphi \leq \alpha < \psi$$

We handled an addition chain C by using another one Cr is based on Factorial Number System [16]. This representation gives a better operation flexibility than the values of original addition chain in agreement to neighborhood functions are depicted in **Subsection 3.3**. For each C corresponds one and only one Cr; these are defined by **Equations 3** and **4**, respectively.

$$Cr(i) = \begin{cases} \{0, 1, \ldots, n-1\} & \textbf{if } i > 0 \\ indefinite & \textbf{if } i = 0 \end{cases} \tag{3}$$

and

$$C'(i) = \begin{cases} C(i-1) + C(Cr(i)) & \textbf{if } i > 0 \\ 1 & \textbf{if } i = 0 \end{cases} \tag{4}$$

where $C'(i)$ is the addition chain generated with our approach.

3.1   The Proposed Simulated Annealing Algorithm

To evaluate the practical usefulness of the evaluation function Ψ, a SA was developed. Next some details of the implementation proposed are presented:

Evaluation function. In this implementation we have included the evaluation function Ψ whose formal definition is presented in the subsection 3.2.

Neighborhood function. The neighborhood of a solution $\mathcal{N}(\phi)$ in our implementation contains all the arrangements ϕ' obtained by changing one or two random selected places of the current arrangement ϕ.

Initial solution. The initial solution is the starting configuration used as input for the algorithm. This SA implementation generates random the initial arrangement.

Termination condition. The SA finishes with any of the following stop conditions: a) when the temperature reaches the value of 1×10^{-4}; b)
when the maximum number of iterations for the SA is attained; or c)
if a number of calls to the evaluation function equals an established
limit.

3.2 Evaluation Function

A neighbor of the current possible solution is evaluated by two factors. The first
factor $\mathcal{L}$ is the length of the addition chain α'. The second factor α' represents
the final value of the addition chain. A solution whose α' is near the α value has
an evaluation determined by the length of the addition chain, on the other hand
a solution whose α' is far away the α value has an evaluation determined by the
absolute difference between α' and α multiplied by the addition chain length.
Equation 5 describe the evaluation function denoted by Ψ.

$$\Psi = \mathcal{L} \mid \alpha' - \alpha \mid + \mathcal{L} \tag{5}$$

3.3 Neighborhood Functions

The *Neighborhood Functions* (NFs) are used to manipulate the Cr chain modifying their positions bounded by the allowable range. In order to generate C',
we use two main neighborhood functions named $\mathcal{N}_1$ and $\mathcal{N}_2$, which are used to
derivate other two functions $\mathcal{N}_3$ and $\mathcal{N}_4$. All these functions are explained below:

$$\mathcal{N}_1(s) \rightarrow (s' \mid s' = s) \wedge (s'_i = \phi(l)) \wedge (s'_i \neq s_i) \wedge (2 \leq \phi(l) \leq l) \wedge (2 \leq i \leq l)$$
$$\mathcal{N}_2(s) \rightarrow (s' \mid s' = s) \wedge (s'_i = \phi(l)) \wedge (s'_j = \phi(l)) \wedge (i \neq j) \wedge (s'_i \neq s_i) \wedge$$
$$(s'_j \neq s_j) \wedge (2 \leq \phi(l) \leq l) \wedge (2 \leq i, j \leq l)$$

$\mathcal{N}_1(s)$	choose a random i position from Cr and modify it value with another value choosed randomly.
$\mathcal{N}_2(s)$	choose two different random i positions from Cr and modify its value with another values choosed randomly.
$\mathcal{N}_3(s)$	it is derived from $\mathcal{N}_1$, but here it is picked up the best value for the i position according to the evaluation function.
$\mathcal{N}_4(s)$	it is derived from $\mathcal{N}_2$, but here it is picked up the best values for the two i positions according to the evaluation function.

3.4 Distribution Functions

The NFs allow the construction of solutions where an addition chain of length l
for an integer α is a sequence S of positive integers, $s_0 = 1, s_1, s_2, \ldots, s_l = \alpha$ s.t.
for each $i > 1$, $s_i = s_j + s_k$ for some j and k with $0 \leq j \leq k < i$. Therefore, the
neighborhood function selects in a randomly way a position i s.t. $2 < i \leq l - 1$,
the distribution for i is determinated by:

$$\tau = \sum_{i=2}^{l-1} i = \frac{(l-1)\times |\,l\,|}{2} - 1$$

The **Equation 6** calculate a random value for some x with $0 \le x \le \tau$, where $\mathcal{F}_1$ and $\mathcal{F}_2$ are *Distribution Functions (DF)* and represents the selection for one point i for neighborhood function $\mathcal{N}_1$, $\mathcal{N}_2$, $\mathcal{N}_3$ or $\mathcal{N}_4$.

$$\phi(l) = \begin{cases} \mathcal{F}_1 = \left\lfloor \dfrac{1+\sqrt{1+8(x+1)}}{2} \right\rfloor \\[2ex] \mathcal{F}_2 = l - \left\lfloor \dfrac{1+\sqrt{1+8(x+1)}}{2} \right\rfloor \end{cases} \tag{6}$$

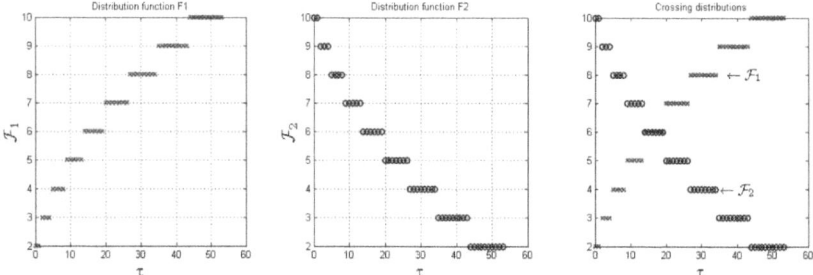

Fig. 1. Distribution functions $\mathcal{F}_1$ and $\mathcal{F}_2$ to select a position s_i in an addition chain

It is easy to deduce that a change in the cost of the Cr implies a dramatic change in the required α value, and that a change in the right side of the Cr implies a small change in the α value. Then, we can conclude that the change in the left side of the Cr are more exploitatory and the changes in the right side of the Cr are more exploitatory.

We can deduce that a change near the leftmost position of the Cr have more impact in the required since of the chain length, then this kind of changes are more exploratory. The changes made on the right part of the chain are more exploitatory given that this kind of changes does not have a significant effect on the required length of the chain. We expect that in the beginning the algorithm behavior must be more exploratory and at the end of the algorithm the behavior will be more exploitatory.

4   Experimental Design

The process of fine-tune the parameters of an algorithm is generally done through an experimental design. Sometimes, the parameters values are taken from the literature without validating its effectiveness. The methodology followed to tune the values of the parameters for the SA algorithm is based on the study of the effect over the quality of the solution by the interaction between parameters.

In our approach, the tuning of the parameters is based on mathematical structures called Covering Arrays (CA). A Covering Array [11] $CA(N; t, k, v)$ is a matrix of size $N \times k$ where every $N \times t$ subarray contains all ordered subsets from v symbols of size t at least once. The value t is called the strength of the CA, the value k is the number of columns or parameters and v is called the alphabet. The value t is called the strength of the CA, the value k is the number of columns or parameters and v is called the alphabet.

The tuning process was done using a CA of strength $t = 2$ with seven parameters $k = 7$ (Table 1b), four of them of alphabet five and two of alphabet three, $v = 5^4 2^3$. Then, the resulting $CA(25; 2, 7, 5^4 2^3)^1$ consists of 25 rows, each row corresponds to a combination of values for the parameters. Together, all the rows contain all the interactions between pair of parameter values, used during the experiment. Table 1a shows the CA values. The first column indicates the number of configuration, the second to fifth are the probability of use certain distribution function, the sixth is the initial temperature, the seventh indicates the cooling rate and the eighth column is the Markov chain.

It is well-known that the performance of a SA algorithm is sensitive to parameter tuning. In this sense, we follow a methodology for a fine tuning of the four NFs used in our SA algorithm. The probabilities to use the neighborhood functions was based in the linear Diophantine Equation,

$$p_1 + p_2 + p_3 + p_4 = q \tag{7}$$

In the previous **Equation 7**, every time that a new neighborhood should be created, the used neighborhood function will be $\mathcal{N}_1$ with probability p_1, $\mathcal{N}_2$ with probability p_2, $\mathcal{N}_3$ with probability p_3 and $\mathcal{N}_4$ with probability p_4, where, p_i is a value in $\{0.1, 0.2, \ldots, 1.0\}$ that represents the probability of executing $\mathcal{N}_i$, and q is set to 1.0 which is the maximum probability of executing any $\mathcal{N}_i$. Then, the Diophantine Equation has 286 solutions that correspond to $\binom{10+4-1}{4-1}$.

4.1   Parallel Implementation

In this subsection we will define the relevant features of the parallel implementation used to run the experimentation of the implemented SA. Problems solved with a SA algorithm are considered a sequential search process, by the nature of the algorithm each new state involved the selection of a next state. Considering that is necessary to run a finite set of tests with different variables and that the communication between test is null, it is possible implement a parallelization experiment, which will distribute the search for solutions among different processors converging in less time and showing a better performance [13].

The Message Passing Interface (MPI) standard is the most important message-passing specification supporting parallel programming for distributed memory.

---

[1] Available under request in http://www.tamps.cinvestav.mx/~jtj

Table 1. $CA(25; 2, 7, 5^4 2^3)$: c (*initial temperature*), V (*cooling rate*), L (*Markov chain*), and $\mathcal{P}_1$, $\mathcal{P}_2$, $\mathcal{P}_3$ and $\mathcal{P}_4$ indicates the probability of using certain distribution function $\mathcal{F}_1$ and $\mathcal{F}_2$ in neighborhood functions $\mathcal{N}_1$, $\mathcal{N}_2$, $\mathcal{N}_3$ and $\mathcal{N}_4$.

(a) CA values for the parameters of the SA contained in $CA(25; 2, 7, 5^4 2^3)$

Ind	$\mathcal{P}_1$	$\mathcal{P}_2$	$\mathcal{P}_3$	$\mathcal{P}_4$	c	V	L
1	1	1	3	3	0	1	1
2	1	4	4	4	1	0	0
3	4	3	3	4	2	0	2
4	4	4	0	1	0	2	1
5	1	2	2	0	1	0	0
6	4	1	2	2	1	2	1
7	2	0	2	4	0	1	0
8	0	3	1	0	0	0	1
9	3	4	3	0	2	2	2
10	2	2	3	1	1	2	1
11	0	1	4	1	2	0	2
12	0	2	0	4	1	1	2
13	1	3	0	2	2	1	1
14	3	2	4	2	0	0	0
15	1	0	1	1	0	2	2
16	2	1	0	0	2	0	0
17	0	4	2	3	2	1	2
18	3	1	1	4	2	2	1
19	4	0	4	0	2	1	1
20	3	0	0	3	1	0	0
21	0	0	3	2	2	2	0
22	3	3	2	1	1	1	0
23	4	2	1	3	2	2	0
24	2	3	4	3	2	2	0
25	2	4	1	2	1	1	2

(b) Values for each instance of CA

Values	0	1	2	3	4
c	1	4	10	-	-
V	0.85	0.9	0.99	-	-
L	$\log_2 v$	$5 \cdot \log_2 v$	$10 \cdot \log_2 v$	-	-
$\mathcal{P}_1$	$0,1$	$\frac{1}{4},\frac{3}{4}$	$\frac{2}{4},\frac{2}{4}$	$\frac{1}{4},\frac{3}{4}$	$1,0$
$\mathcal{P}_2$	$0,1$	$\frac{1}{4},\frac{3}{4}$	$\frac{2}{4},\frac{2}{4}$	$\frac{1}{4},\frac{3}{4}$	$1,0$
$\mathcal{P}_3$	$0,1$	$\frac{1}{4},\frac{3}{4}$	$\frac{2}{4},\frac{2}{4}$	$\frac{1}{4},\frac{3}{4}$	$1,0$
$\mathcal{P}_4$	$0,1$	$\frac{1}{4},\frac{3}{4}$	$\frac{2}{4},\frac{2}{4}$	$\frac{1}{4},\frac{3}{4}$	$1,0$

Virtually every commercial parallel computer support MPI, and free libraries meeting the MPI standard are available for "homemade" commodity clusters. The underlying hardware is assumed to be a collection of processors, each with its own local memory. A processor has direct access only to the instruction and data stored in its local memory. However, an interconnection network support message-passing between processors. Processor A may send a message containing some of its local data values to processor B, giving processor B indirect access to these values [12].

The model Single Program Multiple Data (SPMD) is a technique employed to achieve parallelism, tasks are split up and run simultaneously on multiple processors with different input in order to obtain results faster [14].

A SPMD using MPI are used in the development of this algorithm maximizing the potential of a parallel program. Our SPMD has three sections to tackle the problem:

- **Data distribution.** This point considers the reading of CAs and Diophantine Equations Solutions. Each processor read their own environment variables independent of the others (no partitioning).
- **Computing.** In this point each processor calculates the number of threads launched to cover the entire tasks after all threads finish the results will come out in a temporary file with all the information computed.
- **Collection of results.** When each one thread finish alert to main subprocess, and when all subprocess finished alert to main processor and that charge the union of all temporary files generating a SQL (Structured Query Language) file. Once generated SQL file, temporary files generated for threads are removed.

Partitioning Model. The election of a correct partition scheme has a significant impact on experimentation about the balance between the weight of each task assignment to each processor and to exploit the available resources efficiently [12]. For those reasons, we decide to use a Static/Cyclic 3D block partition model as show in Figure 2, were the sides of our cube are defined by: CA 1b rows, **Diophantine Equation 7** solutions and the required number of iterations.

The Scheme Static/Cyclic 3D block is defined by:

- Static Variables, CA and Diophantine Equation solutions are static while Algorithm life cycle.
- Cyclic. All the task are divided by the total of processors.
- Block. The balance of work need be divided the rows by blocks.

The SPMD used for the collection stage is *master-slave*. Within the stages of this model the data loading was suppressed, since each process has the necessary information and calculates its range of blocks to make its computing operations. The stages of this model that are not deleted are independent computing stage and data collection for the master process.

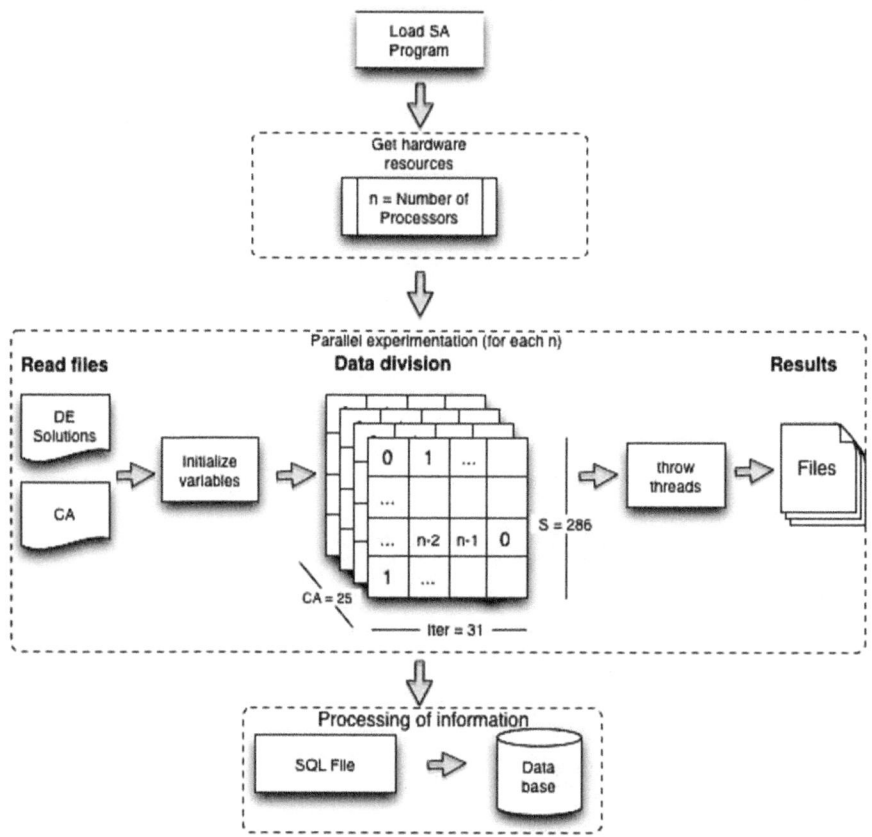

Fig. 2. Diagram for the parallelization of our SA implementation

4.2   Running Time Analysis: Sequential Program vs Parallel Program

The performance analysis for SA program with running time of $O(n)$ is divided as following: L is the total number of experiments, p indicates the numbers of processors, ts is the latency of the communication channel used, tw is the bandwidth of the communication channel, Tp indicates the parallel time, Ts indicates the sequential time and S represented the speedup (time savings over the sequential algorithm).

The total running time for *data loading* is determinate by the number of experiments L, while that the *computation time* is determined by **Equation 8**. Finally, the total running time for *collection results* is estimate by **Equation 9**. The parallel time is obtained from **Equation 10**.

$$Tp_2 = \frac{L \cdot n}{p} \tag{8}$$

$$Tp_3 = (p-1)(ts+tw) \tag{9}$$

$$Tp \approx \frac{L \cdot n}{p} + L \tag{10}$$

The sequential SA algorithm has a running time of $O_s(L \cdot n)$ while running time for parallel algorithm is $O_s(\frac{L \cdot n}{p} + L)$. We can see that the analysis of the sequential and parallel SA program is a difference in performance and efficiency. The efficiency of a parallel program is a measure of processor utilization. The analysis of efficiency of parallel SA program is given by the **Equation 11**.

$$E = \frac{L \cdot n}{p(\frac{L \cdot n}{p} + L)} = \frac{L \cdot n}{L \cdot n + L \cdot p} \tag{11}$$

Finally, the analysis of SA algorithm can be summarized in **Equation 12**, where the *speed-up* is the ratio between sequential running time and parallel running time.

$$S = \frac{L \cdot n}{\frac{L \cdot n}{p} + L} = \frac{L \cdot n \cdot p}{L \cdot n + L \cdot p} \tag{12}$$

5   Experimental Results

The experiments were run on a cluster with 10 nodes (8 cores per node), where each node has 2 Intel quad core 5550 to 2.67 GHz, 16 GB of RAM and a communication channel *infiniband*. The SA algorithm and the experimental method were implemented using C language and compiled with mpiicc intel compiler version 11.

5.1   Fine-Tuning the Experiment

The algorithm was tuned using a CA of size 25×7 and the 286 solutions obtained from Diophantine **Equation 7**. Each combination was executed 31 times (for provide statistical validity to experiment) for $\alpha = 14143037$ and $\alpha = 379$, this values is difficult to obtain the minimal addition chain as stated in [5]. The total number of experiments is determinate by **Equation 13**.

$$\mathcal{T} = \mathcal{C} \times \mathcal{D} \times \mathcal{B} \tag{13}$$

where $\mathcal{C}$ represents the size of CA, $\mathcal{D}$ represents the number of Diophantine Equation Solution and $\mathcal{B}$ is the number of times that the experiment was execute. Since $\mathcal{C} = 25$, $\mathcal{D} = 286$, $\mathcal{B} = 31$, then the total number of experiments is $25 \times 286 \times 31$.

According to the results of this experiment, the neighborhood functions $\mathcal{N}_2$ and $\mathcal{N}_4$ that modify two elements at the same time, produce better results than $\mathcal{N}_1$ and $\mathcal{N}_3$, while the neighborhood functions considered exploratory $\mathcal{N}_1$ and $\mathcal{N}_2$ generate best results when this modify the end of addition chains, and the exploitatory functions $\mathcal{N}_3$ and $\mathcal{N}_4$ generate better results when this modify the begin of addition chains. Therefore, the final CA row must be $\boxed{3}\boxed{3}\boxed{1}\boxed{1}\boxed{2}\boxed{0}\boxed{2}$ and the solution for Diophantine Equation is $\boxed{0}\boxed{9}\boxed{1}\boxed{0}$, this configuration get the best results.

5.2   Benchmark Description

We calculate the minimal addition chain for $\alpha \in [1, 1000]$ using the setup described in the fine-tuning experiment, the results obtained are presented in Table 2.

Table 2. Measure for all $\alpha \in [1, 1000]$ computed

Measure	Results
Optimal value	10808
Best case	10823
Median	10834
Average	10829.53
Standard deviation	3.11

The results obtained with the proposed approach are better compared with other approaches reported [5,9]. A spacial class of exponents difficult to optimize are presented by Edward Thurber and a compilation of them can be found in [15]. These values considered as difficult to obtain were processed, Table 3b presents the results obtained from the experiment described, show the α value, the minimal length (optimal), and the number of times that it was found.

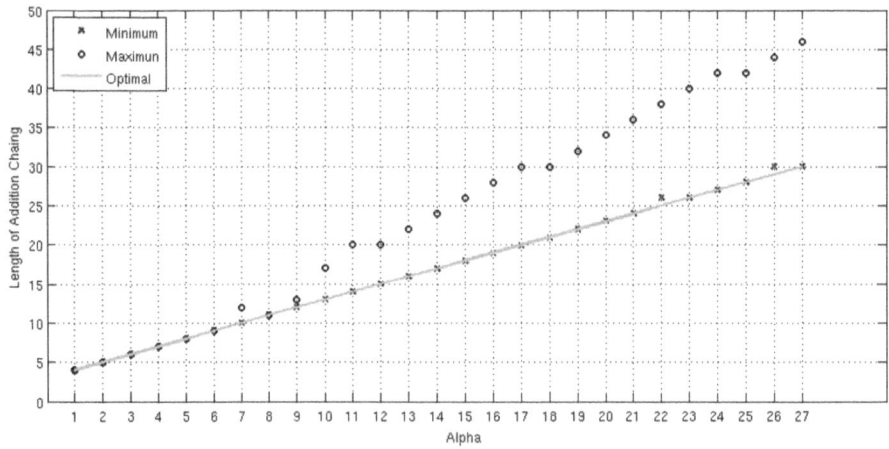

Fig. 3. Comparison of minimum, maximum and optimal values for different α values

Table 3. Best results obtained from computational experiment

(a) Addition chains considered difficult to compute, the minimal length, and the number of hits from 32 experiments

Id	α	Minimal length	Hits
1	7	**4**	32
2	11	**5**	32
3	19	**6**	32
4	29	**7**	32
5	47	**8**	32
6	71	**9**	32
7	127	**10**	30
8	191	**11**	30
9	379	**12**	30
10	607	**13**	28
11	1087	**14**	13
12	1903	**15**	6
13	3583	**16**	1
14	6271	**17**	1
15	11231	**18**	1
16	18287	**19**	2
17	34303	**20**	1
18	65131	**21**	2
19	110591	**22**	2
20	196591	**23**	2
21	357887	**24**	1
22	685951	**26**	1
23	1176431	**26**	1
24	2211837	**27**	2
25	4169527	**28**	1
26	7624319	**30**	1
27	14143037	**30**	2

(b) Minimal addition chains obtained using our proposed approach

α	Minimal addition chain	Minimal length
2211837	1 - 2 - 3 - 5 - 7 - 14 - 21 - 42 - 63 - 64 - 127 - 254 - 508 - 1016 - 2032 - 4064 - 8128 - 8636 - 17272 - 34544 - 69088 - 138176 - 276352 - 552704 - 552958 - 1105916 - 1105921 - 2211837	27
4169527	1 - 2 - 4 - 8 - 12 - 13 - 21 - 42 - 84 - 168 - 336 - 672 - 1344 - 1357 - 2714 - 5428 - 10856 - 21712 - 21724 - 43436 - 65148 - 130296 - 260592 - 521184 - 1042368 - 2084736 - 2084757 - 2084770 - 4169527	28
14143037	1- 2 - 3 - 5 - 10 - 20 - 30 - 31 - 61 - 92 - 184 - 368 - 736 - 1472 - 2944 - 5888 - 8832 - 9568 - 18400 - 36800 - 73600 - 147200 - 220800 - 441600 - 883200 - 1766400 - 3532800 - 7065600 - 7071488 - 14142976 - 14143037	30

Figure 3 represent the α values computed, the maximum and minimum values of our experiment compared to the optimal value presented in Table 3a.

6   Conclusions

The quality of our experimental results using a Simulated Annealing (SA) algorithm demonstrates the strength of the proposed approach, which is based on: a) a novel representation based on Factorial Number System; b) the use of Diophantine Equation solutions for fine-tuning the probabilities functions $\mathcal{F}_1$ and $\mathcal{F}_2$ in Neighborhood Functions; c) the use of a Covering Array to finetune the parameters of SA; and d) the use of a parallel implementation based on the Message Passing Interface (MPI) using Single Program Multiple Data model.

The results obtained evidence the suitability of our approach to compute minimal addition chains for cases considered difficult.

Based on the proposed parallelization implementation, we suggest to follow a parallel mechanism to reduce the time needed in the tuning process of algorithms, using CAs and Diophantine Equations.

Acknowledgments. This research was partially funded by the following projects: CONACyT 58554-Cï¿œlculo de Covering Arrays, 51623- Fondo Mixto CONACyT y Gobierno del Estado de Tamaulipas.

References

1. Michalewicz, Z.: Genetic Algorithms + Data Structures = Evolution Program, 3rd edn. Springer, Heidelberg (1996)
2. Nedjah, N., de Macedo Mourelle, L.: Minimal Addition Chain for Efficient Modular Exponentiation Using Genetic Algorithms. In: Hendtlass, T., Ali, M. (eds.) IEA/AIE 2002. LNCS (LNAI), vol. 2358, pp. 88–98. Springer, Heidelberg (2002)
3. Gordon, D.M.: A Survey of Fast Exponentiation Methods. Journal of Algorithms, 129–146 (1998)
4. Downey, P., Leong, B., Sethi, R.: Computing Sequences with Addition Chains". SIAM Journal on Computing, 638–646 (1981)
5. Cruz-Cortés, N., Rodríguez-Henríquez, F., Juárez-Morales, R., Coello Coello, C.A.: Finding Optimal Addition Chains Using a Genetic Algorithm Approach. In: Hao, Y., Liu, J., Wang, Y.-P., Cheung, Y.-m., Yin, H., Jiao, L., Ma, J., Jiao, Y.-C. (eds.) CIS 2005. LNCS (LNAI), vol. 3801, pp. 208–215. Springer, Heidelberg (2005)
6. de Castro, L.N., Timmis, J.: An Introduction to Artificial, Immune Systems: A New Computational Intelligence Paradigm. Springer, Heidelberg (2002) ISBN: 978-1-85233-594-6
7. Nedjah, N., Mourelle, L.M.: Efficient Pre-processing for Large Window-Based Modular Exponentiation Using Genetic Algorithms. In: Chung, P.W.H., Hinde, C.J., Ali, M. (eds.) IEA/AIE 2003. LNCS, vol. 2718, pp. 165–194. Springer, Heidelberg (2003)
8. Bos, J.N.E., Coster, M.J.: Addition Chain Heuristics. In: Brassard, G. (ed.) CRYPTO 1989. LNCS, vol. 435, pp. 400–407. Springer, Heidelberg (1990)

9. Bleichenbacher, D., Flammenkamp, A.: An efficient Algorithm for Computing Shortest Additions Chains, Bell Labs (1997),
   http://www.homes.uni-bielefeld.de/achim/ac.ps.gz
10. Bergeron, F., Berstel, J., Brlek, S.: Efficient Computation of Addition Chains. Journal de Théorie des Nombres de Bordeaux tome 6, 21–38 (1994)
11. Lopez-Escogido, D., Torres-Jimenez, J., Rodriguez-Tello, E., Rangel-Valdez, N.: Strength Two Covering Arrays Construction Using a SAT Representation. In: Gelbukh, A., Morales, E.F. (eds.) MICAI 2008. LNCS (LNAI), vol. 5317, pp. 44–53. Springer, Heidelberg (2008)
12. Quinn Reference, M.J.: Parallel Programming in C with MPI and OpenMP. Mc Graw Hill Higher Education (2004) ISBN 0-07-282256-2
13. Grabysz, A.D., Rabenseifner, R.: Nesting OpenMP in MPI to Implement a Hybrid Communication Method of Parallel Simulated Annealing on a Cluster of SMP Nodes. In: Nagel, W.E., Walter, W.V., Lehner, W. (eds.) Euro-Par 2006. LNCS, vol. 4128, pp. 1075–1084. Springer, Heidelberg (2006)
14. Strout, M.M., Kreaseck, B., Hovland, P.D.: Data-Flow Analysis for MPI Programs. In: IEEE Computer Society, International Conference on Parallel Processing, pp. 175–184. IEEE Compurer Society (2006) ISBN 0-7695-2636-5
15. Thurber, E.W.: Shortest Addition Chains,
   http://wwwhomes.uni-bielefeld.de/achim/addition_chain.html
16. Laisan, C.A.: Sur la Numeration Factorielle, Application Aux Permutations. Bulletin de la Societe Mathematique de France, 176–183 (1888)

Novelty Detection Using Graphical Models for Semantic Room Classification

André Susano Pinto[1,2], Andrzej Pronobis[2], and Luis Paulo Reis[1,3]

[1] Dep. Informatics Engineering, Faculty of Engineering of the University of Porto
[2] Centre for Autonomous Systems, The Royal Institute of Technology (KTH)
SE100-44 Stockholm, Sweden
[3] LIACC - Artificial Intelligence And Computer Science Lab. - University of Porto
Rua Dr. Roberto Frias, s/n 4200-465 Porto, Portugal

Abstract. This paper presents an approach to the problem of novelty detection in the context of semantic room categorization. The ability to assign semantic labels to areas in the environment is crucial for autonomous agents aiming to perform complex human-like tasks and human interaction. However, in order to be robust and naturally learn the semantics from the human user, the agent must be able to identify gaps in its own knowledge. To this end, we propose a method based on graphical models to identify novel input which does not match any of the previously learnt semantic descriptions. The method employs a novelty threshold defined in terms of conditional and unconditional probabilities. The novelty threshold is then optimized using an unconditional probability density model trained from unlabelled data.

Keywords: Novelty detection, semantic data, probabilistic graphical models, room classification, indoor environments, robotics, multi-modal classification.

1 Introduction

There has been several efforts in the areas of artificial intelligence and robotics in creating robots that are able to interact with humans and their environments. One of the important aspects is to endow those robots with a deeper understanding of human environments, not just for the purpose of navigation and obstacle avoidance, but also in terms of human semantics and functionality. An important problem in creating reliable representations of space for robots that are to be deployed in new and unknown realistic environments is to be able to automatically identify gaps in robot's knowledge and act in order to fill those gaps.

This article addresses the problem of *novelty detection* within the context of semantic mapping i.e. generating maps containing *semantic information* about indoor environments, such as homes or offices. In that context, the ability to detect that the observations result from a semantic concept unknown to the robot, and cannot be explained by one of its models, is crucial for generating fully

L. Antunes and H.S. Pinto (Eds.): EPIA 2011, LNAI 7026, pp. 326–339, 2011.

autonomous and reliable behavior. In order to be robust, the robot must identify novel concepts and instead of making a costly error, refrain from the decision and initiate learning. In particular, we address the problem of novelty detection for room categorization i.e. detecting whether the area in the environment identified as a separate room can be assigned to one of the semantic labels that the robot knows (e.g. *a kitchen* or *an office*) or belongs to an unknown semantic category.

The novelty detection algorithm is implemented on a cognitive robot Dora the Explorer, which already uses a developed architecture based on probabilistic graphical models oriented towards dealing with and reasoning about uncertain semantic information [1]. One of the major problems with using the probabilistic models of the environments for novelty detection is selecting the optimal threshold above which the test sample is considered novel. This problem becomes more difficult when, as in case of Dora, the representation grows as the robot explores the environment. Methods are required to find the right threshold given the current structure of the model which constantly changes. To this end, this work studies methods for novelty threshold selection using probabilistic graphical models.

The rest of this paper is structured as follows. First, we briefly review the related work related to room categorization and novelty detection (Section 2). Then, we give an outline of the architecture of the Dora system and the structure of the conceptual map representing the semantic information (Section 3). Next, we discuss the methods for novelty detection (Section 4) and present results of our preliminary experiments (Section 5). This paper concludes with a summary in Section 6.

2   Related Work

The problem of room categorization based on visual information was first addressed in the computer vision community. The research focused mainly on the problem of classifying single images captured in indoor or outdoor environments (scene classification) [2,3]. At the same time, robotics researchers initially employed the 2D laser range sensor being much more robust to variations occurring in the environment and much easier to handle computationally in real time [4].

Multi-modal approaches, such as combining semantic data extracted from several sources or classifiers are expected to have better performance on scene recognition than single-cue approaches. Quattoni and Torralba [5] showed that most scene recognition models work poorly in indoor scenes when compared to outdoor scenes since the properties that characterize rooms changes depending on the category. For instance corridors are well described by global properties and bookstores are well described by the presence of specific objects (books). Galindo et al. [6] also exploits this by defining a bidirectional relation between object and room category, where object defines a room category and a room category provides information on where objects may be found.

Probabilistic representations have been frequently used for spatial modelling in robots operating in the real-world [7,8]. Boutell et al. [9] have studied outdoor scene classification using *factor graphs* and modelling spatial relations between objects in the scene to extract better knowledge from semantic (high-level) features.

Although our approach is presented in the context of mobile robotics it relies on standard concepts and techniques such as semantic data and graphical models. Those are often used in the area of information retrieval. An interesting example is the usage of an hidden concept layer between visual features and text information to provide automatic image annotation [10].

Novelty detection has been studied for many years and there are several approaches based on statistical analysis [11]. Graphical models have been used to learn distributions of variables, both in supervised and unsupervised ways and by using thresholds on those distributions based solely on the conditional probability, as seen on Bishop [12], a novelty system can be trivially implemented.

However to the knowledge of the authors there is no reference on how to perform novelty detection using graphs that are dynamically generated.

3   Dora Architecture Overview

The Dora system [1] consists of several co-operating sub-systems, all of which actively use or maintain the spatial knowledge representation (see Figure 1). Only the *conceptual layer* of the representation is of interest to this article. Its role is to aggregate the following semantic information coming from other sub-systems:

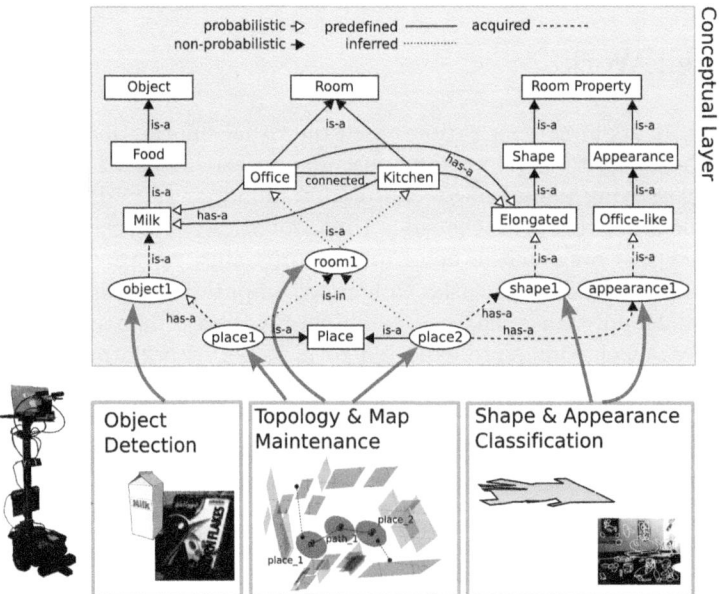

Fig. 1. Interaction between the sub-systems of Dora with special focus on the conceptual layer

Doorway detection is used to segment the continuous space into rooms and map connectivity between them.

Room size and shape are obtained by classifying 2D laser scans from laser range finder mounted on the robot and are used as properties of a room. The system utilizes pre-trained set of classifiers to extract rooms sizes (either large, medium or small) and shapes (rectangular or elongated).

Object detection is performed in images acquired by the robot through its camera. The system keeps track of the number of objects of each type in each room. Objects are detected by running a pre-trained set of detectors for the following object types: book, cereal box, computer, robot, stapler, toilet paper.

Room appearance is categorized from the visual input by using global visual features and a pre-trained set of 7 different models.

As Dora moves through the environment its *conceptual layer* builds a structural and probabilistic representation of space instantiated as a *graphical model*. It includes taxonomy of human-compatible spatial concepts which are linked to the sensed instances of these concepts drawn from lower layers. It is the conceptual layer which contains the information that kitchens commonly contain cereal boxes and have certain general appearance and allows the robot to infer that the cornflakes box in front of the robot makes it more likely that the current room is a kitchen. The conceptual layer is described in terms of a probabilistic ontology defining spatial concepts and linking those concepts to instances of spatial entities (see the example of the ontology in Figure 1).

3.1 Conceptual Map

Based on this design, a *chain graph* [13] model is proposed as a representation for performing inferences on the knowledge represented in the conceptual layer. Chain graphs are probabilistic graphical models that combine the properties of both Bayesian Networks and Random Markov Fields. This results in an efficient approach to probabilistic modeling and reasoning about conceptual knowledge.

An exemplary chain graph corresponding to the conceptual map ontology is presented in Figure 2. Each discrete place identified in the environment is represented by a set of random variables, one for each class of relation linked to that place. These are each connected to a random variable over the categories of rooms, representing the "is-a" relation between rooms and their categories. Moreover, the room category variables are connected by undirected links to one another according to the topological map. The remaining variables represent: shape and appearance properties of space as observed from each place, and the presence of objects. These are connected to observations of features extracted directly from the sensory input. Finally, the distributions $p_s(\cdot|\cdot)$, $p_a(\cdot|\cdot)$, $p_{o_i}(\cdot|\cdot)$ represent the common sense knowledge about shape, appearance, and object co-occurrence, respectively. They allow for inference about other properties and room categories e.g. that the room is likely to be a kitchen, because you are likely to have observed cornflakes in it.

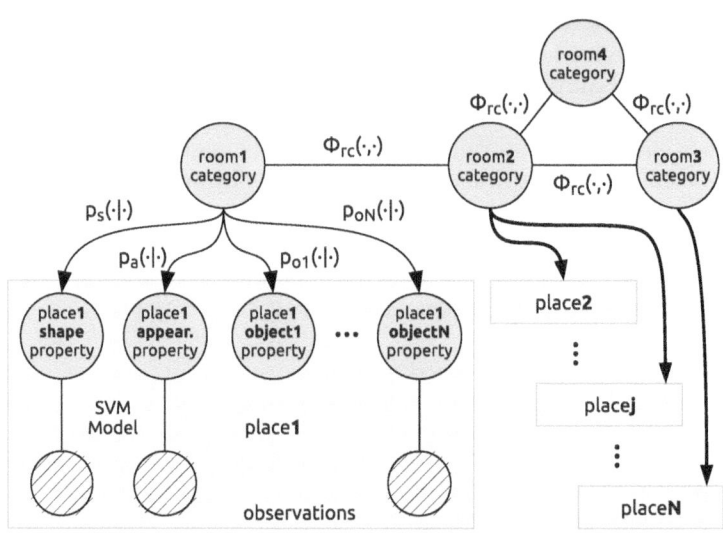

Fig. 2. Example of a chain-graph produced by the *conceptual layer*

The use of graphical models to describe distributions of variables has useful properties. First, they permit inference about uncertain conceptual knowledge. At the same time, they are generative models and therefore allow to calculate the probability on any given subset of variables of the graph, allowing the system to work even when some information is missing.

3.2   Factor Graphs

Although the conceptual layer works with *chain graphs*, those can be converted into *factor graphs* [14]. Factor graphs are used throughout this paper as they provide an easier manipulation due to factorization. Moreover, there exist efficient implementations of inference engines operating on factor graph representations [15]. Describing the distribution function in terms of graphs allows to use those engines to efficiently calculate marginals over any given subset of variables by exploiting conditional independence between variables.

A *factor graph* is a bipartite graph connecting two sets of nodes X_G and F_G representing random variables and factors. Each factor is described by a function ϕ dependent only on the variables x_ϕ to which the factor is connected. Thus, a factor graph can be seen as a description of probability density function obtained by a product of all the factors. In order to represent the probability, a normalization factor needs to be introduced, resulting in the following equation:

$$P_G(x) = \frac{1}{Z} \prod_{\phi \in F_G} \phi(x_\phi), \qquad Z = \sum_{X_G} \prod_{\phi \in F_G} \phi(x_\phi) \qquad (1)$$

4   Novelty Detection

The aim of novelty detection is to identify data samples originating from a distribution different than one of those the system knows about [11]. It is harder than classification as only positive samples of the class are available rendering normal classification methods unusable. Adding novelty detection capabilities allows to increase reliability of the system. Novelty signal can be used to inform the system that it should proceed with caution as its knowledge does not correctly describe the environment.

Due to the nature of the sensed data which are noisy and uncertain, novelty ought to be treated in a probabilistic fashion. In such case, each sample is associated with certain probability of being generated by a class not known to the agent and a complementary probability $P(\overline{novel}|x)$ of being generated by a known class. The true positive and false positive rate of a novelty detection system which classifies the set N of samples as novel is given by:

$$P(\text{true positive}) = \sum_{x \in N} P(novel|x)P(x) \tag{2}$$

$$P(\text{false positive}) = \sum_{x \in N} P(\overline{novel}|x)P(x) \tag{3}$$

It follows that by extending the set N with new samples, the true positive and false positive rate can never be decreased, leading to the problem where in order to increase detection, the system needs to increase its error. This describes the base of the *error and rejection tradeoff* [16], which states that a system aiming at increasing the true-positive probability will eventually increase its false-positive error.

This way an optimal detector can be formulated by achieving the maximum true-positive probability without its false-positive probability increasing beyond a given limit. This is equivalent to the *continuous knapsack problem* which allows for a greedy solution by sorting the items with a value per weight function. In the case of a detection system, this can be defined as:

$$value(x) = P(novel|x) \tag{4}$$

$$weight(x) = P(\overline{novel}|x) \tag{5}$$

$$cost(x) = value(x)/weight(x) \tag{6}$$

$$= \frac{P(novel|x)P(x)}{P(\overline{novel}|x)P(x)} \tag{7}$$

Therefore, a novelty detection system before classifying a sample a as novel, should (greedily) classify any sample b with a smaller cost as that would achieve a higher true positive probability given a fixed false positive one.

$$\frac{P(novel|b)}{P(\overline{novel}|b)} < \frac{P(novel|a)}{P(\overline{novel}|a)} \tag{8}$$

This relation between a and b can further be simplified into:

$$P(\overline{novel}|b) < P(\overline{novel}|a) \tag{9}$$

Based on this, it can be said that a novelty detection system aims at defining an order relation on all the possible inputs equivalent to the order defined by the function: $P(\overline{novel}|x)$. Then, the detector can be described by the largest $P(\overline{novel}|x)$ accepted which is the principle of thresholding.

Using Bayes rule and assuming a constant $P(\overline{novel})$, a ratio between conditional and unconditional probabilities of the input x is obtained. Such a ratio is a suitable function for implementing a novelty detector system with optimal thresholding.

$$P(\overline{novel}|x) = \frac{P(x|\overline{novel})P(\overline{novel})}{P(x)} \propto \frac{P(x|\overline{novel})}{P(x)} \tag{10}$$

Note, however that in the case of dynamic graph structures, an assumption on a constant $P(novel)$ is quite strong. Although in our first approach, it is assumed to be constant, the authors acknowledge that structure plays an important role and should be used as prior information when calculating $P(novel)$.

4.1   Conditional Probability

The conditional probability models the distribution of variables given that the agent knowledge holds true and can be used to describe the generating classes of the sensed sample. Under that its natural to model it with a graphical model that combines the learned variables and categories as well with the relations between them.

In Dora case, this is equivalent to use the graphical model used by it to describe its current believes on the variables modelled by the system. That graph is built, by the conceptual layer, by instancing the information extracted from other layers together with the conceptual knowledge such as: objects are properties of rooms, rooms are connected between each other.

Figure 3 illustrates a graph G built from the conceptual layer to represent the conditional probability on the sensed variables x. A set of hidden variables is added to represent the conceptual knowledge the system is aware of. In the presented graph variables R_i were added to model the room categories that influence the directly sensed features on each physical room, as well connectivity factors between each room.

The factors connecting the variables are trained by the system by searching databases of common knowledge to build potentials describing how likely it is that a specific set of values of certain variable types is likely to occur. For instance: it is very likely to find a cereal box in a kitchen; and it is unlikely to find bathroom connected to another bathroom. We propose using such a graph G built by the conceptual layer for modelling $P_G(x)$ as an approximation for $P(x|\overline{novel})$.

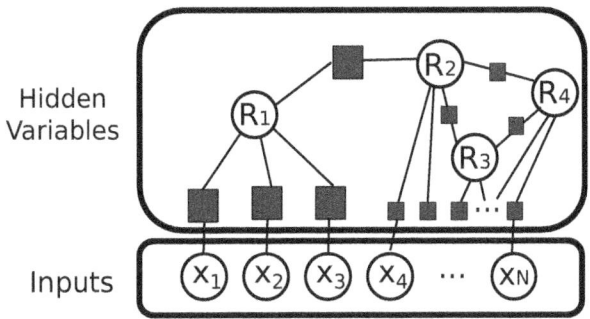

Fig. 3. Illustration of a factor graph modelling a distribution of a set x of sensed variables.

4.2   Unconditional Probability

With only access to labelled data a common approach is to define a threshold assuming that $P(x)$ is constant through all the samples.

Its important to notice that in several cases assuming it to be constant leads to discarding the factor. Nonetheless, here the distributions are dynamically growing as the system learns more on the environment. So the normalizing argument $P(x)$ has to be evaluated for each new subset of x.

Assuming that the unconditional distributions generates all possible outcome with the same probability we can model it with $\prod 1/\#x_i$, where $\#x_i$ denotes the cardinality of the state space of variable x_i. In graphical model terms this is represented to a factor graph U with the variables but without any factors as illustrated on Figure 4.

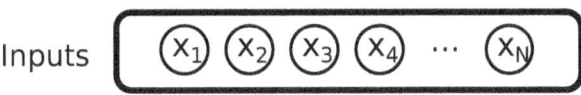

Fig. 4. Without any existing factors, this graph U represents a uniform distribution over any set of its variables

Having a graphical model G built to model the known data distribution and a model U for the unconditional probability a novelty threshold would be given by: $P_G(x)/P_U(x)$. Here $P_U(x)$ can be seen as a normalizing factor to lever all the $P_G(x)$ on any set of variables x into the same measure units (error rate), such that a static threshold can be implemented. For example the conditional probability would yield very small values on large sets of variables x than in small sets due to the spreading over the dimensions of the sample space. A novelty measure is seen as a ratio on how much introducing the known concepts helps to understand the observed result.

4.3   Semi Supervised: Using Unlabelled Data

Nonetheless, its often the case that there is access to extra data that allows to obtain a better approximation to the unconditional probability than the uniform one. In specific, all the knowledge of the agent can be considered to hold true apart from complete knowledge on the categories of a room. In that case a single big factor can be used to model all the variables directly dependent on the possibly novel room as illustrated in Figure 5.

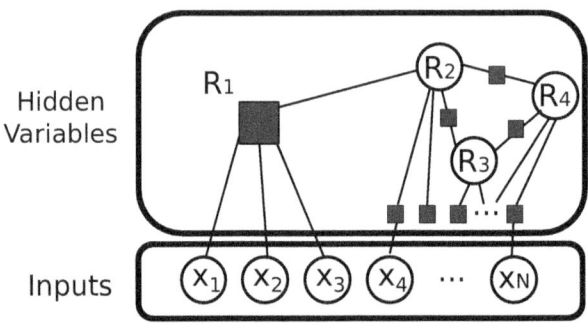

Fig. 5. Without being able to model variable R_1 all the variables directly dependent on it become dependent between each other introducing a single big factor

For practical reasons, it is impossible to train such a factor, and simplifications need to be performed. Here, it was assumed that it can be approximated by factorizing it in several single factors such that all variables become independent. Additionally those single factors can easily be trained by using unlabelled data. Obtaining this way a graphical model I, illustrated in Figure 6, to be used as approximation for the unconditional probability

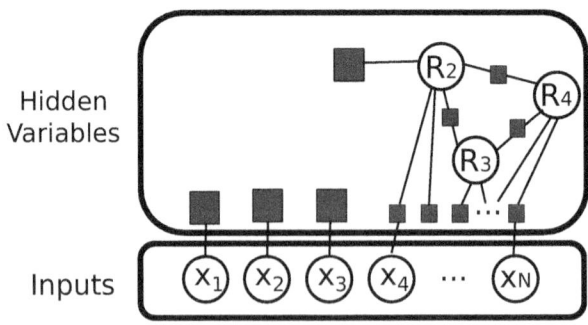

Fig. 6. By factorizing the single factor introduced by room 1 not being necessarily known, several single factors are obtained that can be trained from unlabelled data

Once again the novelty threshold would be given by $P_G(x)/P_I(x)$. This time the addition of the unconditional factors can be understood as an attempt to compensate for an existing bias on the unconditional distribution. This is an important step to achieve a correct order-relation of the inputs sample for implementing a novelty threshold.

5 Results

In order to verify the performance of the proposed threshold functions, a synthetic dataset was generated. As the point on this initial work was only to test the correctness of the presented threshold function and approximation capabilities by using a uniform or an independent model, only information regarding direct features of a room were modelled and no structured knowledge such as room connectivity was taken in account.

The synthetic distribution assumes that an independent and variable number of features x is generated by a given room category. In whole there was 11 different room categories and 9 different measured feature types. The number of sensed features is dynamic and mimic the type of information extracted when running on the robot. Due to that it is possible that on a given sample a certain feature type can be present more than once or not be present at all (e.g.: room shape is extracted from 2D laser scans in more than one position in the room, and information about detected objects is only present when the robot previously tried to detect objects on a given room).

The sensed properties and room categories were chosen to mimic as close as possible the reality and they are based on a previously built ontology from web data. There is in total 11 different room categories ranging from: corridor, hallway, 1 person office, 2 person office, bathroom, conference hall, etc..., and there is 9 different extracted features: room size, room shape, room appearance and 6 different objects (e.g. book, cereal box, computer).

From the distribution, 100 labelled samples for 5 of the 11 room categories were drawn to represent the known categories and 1000 unlabelled samples were drawn from all the room categories for learning the unconditional probability distribution. Using those samples, factors were learnt for the graphs used to model the conditional distribution and the independent unconditional distribution. Figure 7 shows the graph structure used for approximate the trained conditional and unconditional distributions. Its important to notice that graph G used to model the known classes when given enough labelled data is able to exactly learn the conditional distribution as it uses the same structure as the created synthetic distribution.

Using the learned models G, U and I, two thresholds were trained: $P_G(x)/P_U(x)$ assuming a uniform unconditional distribution and $P_G(x)/P_I(x)$ assuming an independent unconditional distribution. Since the distribution is synthetic there is access to $P(x)$ and $P(x|\overline{novel})$ and a perfect threshold function could also be created to test how far the presented thresholds are from optimal.

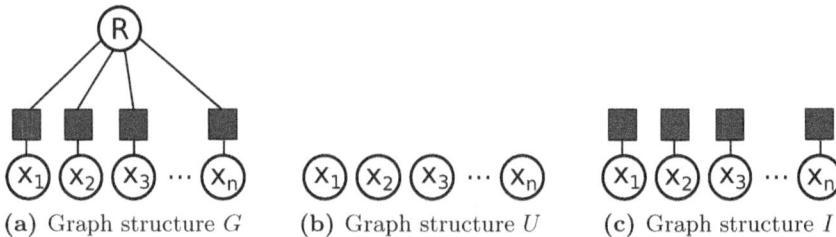

(a) Graph structure G (b) Graph structure U (c) Graph structure I

Fig. 7. The graph structures used to model the conditional and unconditional probability for implementing the novelty thresholds $P_G(x)/P_U(x)$ and $P_G(x)/P_I(x)$

5.1   Probability Ratio Comparison

First, the performance of the novelty threshold selection was plotted for a set of 1000 samples taken from the whole distribution (Figure 8). The samples where uniformly generated by graphs with 5, 10, 15, 20, 35, 50 features. Additionally the feature types were also uniformly sampled, for that it is possible that in certain samples some feature types were sensed more than once and other were not sensed at all. This was chosen to mimic the dynamic properties expected to see when implemented on a robot.

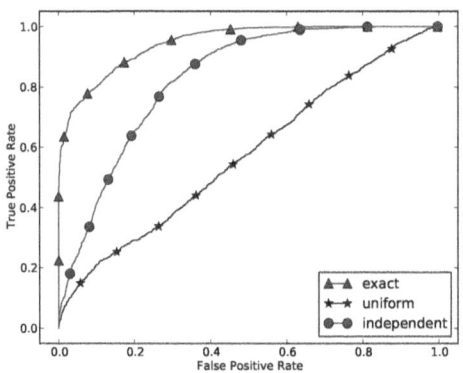

Fig. 8. ROC curve comparing novelty detection performance under samples with variable size of sensed properties

The convex shape of the optimal threshold shows that the ratio between conditional and unconditional probability is indeed a suitable detector for implementing a threshold when the samples are taken from dynamic distributions when $P(novel)$ is constant (e.g. some samples where there is only access to room size versus samples where there is a lot of information about the room properties).

Its also possible to see how important it is to estimate a correct unconditional probability in order to obtain a correct novelty measure on the inputs. The assumption of a uniform unconditional probability has led to very poor results.

That is probably explained by the semantic properties being highly biased towards some values. This shows that bias plays an important role in detecting whether a given sensed value is a valuable cue about the room category.

5.2   Performance Changes with Amount of Available Information

In order to measure the performance impact as more semantic information becomes available, ROC curves were plotted for samples grouped by the number of sensed semantic features.

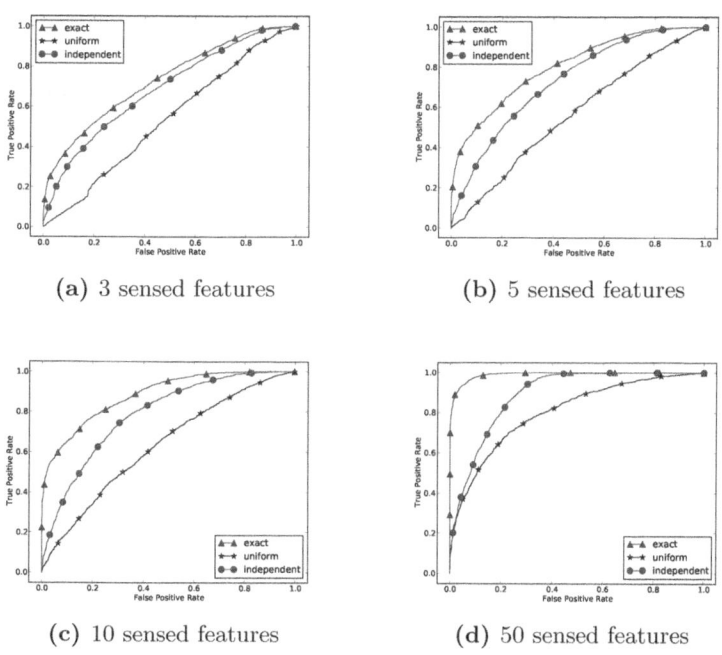

(a) 3 sensed features   (b) 5 sensed features

(c) 10 sensed features   (d) 50 sensed features

Fig. 9. ROC curves plotted showing performance of the presented novelty detection method for graphs generated for different amount of sensed features

It is possible to see that as the system gains more semantic information, it becomes easier to detect novelty. The size of the input space increases and allows the existing classes to become more easily distinguished.

The performance of the independent threshold decreases as the number of sensed features increases. This is easily explained by the fact that the graph I is not able to model the existent dependence between the features. This becomes obvious as the number of features increases (e.g. graph I perfectly models $P(x)$ in the case where only 1 feature is sensed).

The uniform threshold shows poor performance especially for samples with small amount of features where it performs almost no better than random. The performance increases as the size of sensed features increases but nonetheless is very small when compared to the optimal threshold.

6 Conclusions and Future Work

In this paper we presented how to define a stable novelty threshold function on top of *probabilistic graphical models* instantiated dynamically from sensed semantic data. The presented technique is based on the ratio between a conditional and unconditional probability and when perfect information exists it performs an optimal novelty detection under the assumption that $P(novel)$ is constant across all the graph structures.

It was also shown that a correct estimation of unconditional probability plays an important role specially on small input spaces. Moreover semi-supervised techniques, implemented with the access to unlabelled data, can be used to significantly improve novelty detection performance.

Given the synthetic distribution, an assumption on an uniform distribution has led to very poor results. The same behaviour is expected to in real world distributions based on semantic data. For that reason, and due to easy access to unlabelled data, special attention will be given to using semi-supervised techniques for novelty detection.

After this initial study on how to detect new semantic classes based on graphical models, future work will focus on how to use the structured information available from the conceptual layer to be able to detect which variable of the graph is novel and what makes it different from other previously learned classes. That will lead to generation of useful information that can be used for communication with the user and performing active learning of new room categories.

References

1. Hanheide, M., Gretton, C., Dearden, R.W., Hawes, N.A., Wyatt, J.L., Pronobis, A., Aydemir, A., Göbelbecker, M., Zender, H.: Exploiting probabilistic knowledge under uncertain sensing for efficient robot behaviour. In: Proceedings of the 22nd International Joint Conference on Artificial Intelligence (IJCAI 2011), Barcelona, Spain (2011)
2. Oliva, A., Torralba, A.: Building the gist of a scene: The role of global image features in recognition. Progress in Brain Research 155, 23–36 (2006)
3. Torralba, A.: Contextual priming for object detection. International Journal of Computer Vision 53(2), 169–191 (2003)
4. Mozos, O.M., Stachniss, C., Burgard, W.: Supervised learning of places from range data using adaboost. In: Proceedings of the 2005 IEEE International Conference on Robotics and Automation, ICRA 2005, pp. 1730–1735. IEEE (2005)
5. Quattoni, A., Torralba, A.: Recognizing indoor scenes (2009)
6. Galindo, C., Saffiotti, A., Coradeschi, S., Buschka, P., Fernández-Madrigal, J.A., González, J.: Multi-hierarchical semantic maps for mobile robotics. In: 2005 IEEE/RSJ International Conference on Intelligent Robots and Systems, (IROS 2005), pp. 2278–2283. IEEE (2005)
7. Gross, H.M., Boehme, H., Schroeter, C., Mueller, S., Koenig, A., Einhorn, E., Martin, C., Merten, M., Bley, A.: TOOMAS: Interactive Shopping Guide robots in everyday use-final implementation and experiences from long-term field trials. In: IEEE/RSJ International Conference on Intelligent Robots and Systems, IROS 2009, pp. 2005–2012. IEEE (2009)

8. Maier, W., Steinbach, E.: A probabilistic appearance representation and its application to surprise detection in cognitive robots. IEEE Transactions on Autonomous Mental Development 2(4), 267–281 (2010)
9. Boutell, M.R., Luo, J., Brown, C.M.: Factor Graphs for Region-based Whole-scene Classification. In: Proceedings of the 2006 Conference on Computer Vision and Pattern Recognition Workshop, page 104. IEEE Computer Society (2006)
10. Zhang, R., Zhang, Z., Li, M., Ma, W.Y., Zhang, H.J.: A probabilistic semantic model for image annotation and multi-modal image retrieval. Multimedia Systems 12(1), 27–33 (2006)
11. Markou, M., Singh, S.: Novelty detection: a review–part 1: statistical approaches. Signal Processing 83(12), 2481–2497 (2003)
12. Bishop, C.M.: Novelty detection and neural network validation. IEE Proc.-Vls. Image Signal Process 141(4), 217 (1994)
13. Lauritzen, S.L., Richardson, T.S.: Chain graph models and their causal interpretations. Journal of the Royal Statistical Society: Series B (Statistical Methodology) 64(3), 321–348 (2002)
14. Kschischang, F.R., Frey, B.J., Loeliger, H.A.: Factor graphs and the sum-product algorithm. IEEE Transactions on Information Theory 47(2), 498–519 (2001)
15. Mooij, J.M.: libDAI: A free and open source C++ library for discrete approximate inference in graphical models. Journal of Machine Learning Research 11, 2169–2173 (2010)
16. Chow, C.: On optimum recognition error and reject tradeoff. IEEE Transactions on Information Theory 16(1), 41–46 (1970)

A Reinforcement Learning Based Method for Optimizing the Process of Decision Making in Fire Brigade Agents

Abbas Abdolmaleki[1,2], Mostafa Movahedi[5], Sajjad Salehi[6],
Nuno Lau[1,4], and Luís Paulo Reis[2,3]

[1] IEETA – Institute of Electronics and Telematics Engineering of Aveiro, Portugal
[2] LIACC – Artificial Intelligence and Computer Science Lab., Porto, Portugal
[3] DEI/FEUP – Informatics Engineering Dep., Faculty of Engineering,
Univ. of Porto, Portugal
[4] DETI/UA – Electronics, Telecommunications and Informatics Dep.,
Univ. of Aveiro, Portugal
[5] Sheikh Bahaee University, Department of Computer Engineering, Isfahan, Iran
[6] Young researchers club, Qazvin branch, Islamic azad university, Qazvin, Iran
{Abbas.Abdolmaleky,mr.mos.movahedi,salehi.sajjad.ai}@gmail.com,
nunolau@ua.pt, lpreis@fe.up.pt

Abstract. Decision making in complex, multi agent and dynamic environments such as disaster spaces is a challenging problem in Artificial Intelligence. Uncertainty, noisy input data and stochastic behavior which are common characteristics of such environment makes real time decision making more complicated. In this paper an approach to solve the bottleneck of dynamicity and variety of conditions in such situations based on reinforcement learning is presented. This method is applied to RoboCup Rescue Simulation Fire brigade agent's decision making process and it learned a good strategy to save civilians and city from fire. The utilized method increases the speed of learning and it has very low memory usage. The effectiveness of the proposed method is shown through simulation results.

Keywords: RoboCup Rescue Simulation, Fire Brigade, Decision Making, Reinforcement Learning.

1 Introduction

Disaster rescue is one of the most serious social issues which involve very large numbers of heterogeneous agents in an hostile environment. The intention of the RoboCupRescue project is to promote research and development in this socially significant domain at various levels which involve multi-agent team work coordination. In the RoboCup Rescue Simulation League, a generic urban disaster simulation environment was constructed. Heterogeneous intelligent agents such as fire brigades, police forces, ambulances and civilians conduct search and rescue activities in this virtual disaster world [1, 2].

L. Antunes and H.S. Pinto (Eds.): EPIA 2011, LNAI 7026, pp. 340–351, 2011.
© Springer-Verlag Berlin Heidelberg 2011

In rescue simulation, agents must perform a sequence of actions to fulfill their tasks efficiently, like extinguishing fires or rescuing injury civilians. Each action alters the environment and also influences on choosing the next action. The goal of the agents is to achieve the best score at the end of the simulation. So agents should do a sequence of actions by which best score is achieved.

Fire brigades are responsible to control the fire. The spread of fire depends upon wind speed and wind direction. Due to fire, buildings get damaged and collapse which resulting in blocked roads. Additionally, fire can result in civilian death and agents getting hurt. The most important issue is to select and extinguish the best fiery point to reduce damage of the civilians and city. In this paper we are going to present an approach to select the best fiery building for extinguishing in each cycle which finally leads to the best score.

So far many methods are proposed for decision making of fire brigade agents. As in [3] is claimed, since fires start in separate locations and spread outwards, the buildings on fire tend to form clusters around their respective points of origin. And because smaller clusters of less intense fires are much easier and vital to extinguish, it's proposed to put out the smallest fiery points first then start to control larger fiery areas. Although this idea may provide good performance to save a city from fire, it doesn't consider civilian positions; hence it does not give any guarantee to save civilians from fire.

Another method is to prioritize fire sites from a fire brigade's perspective [4]. In this method each fire brigade gives a priority to each fire site based on its properties like volume of fire and civilians around it and selects the critical building in highest priority fire site to extinguish. Because this process has been done in each time step, it consumes a lot of memory and CPU usage.

In [5] a decision tree for priority extraction has been proposed. Decision tree for decision making has been widely used in multi agent systems. But in this paper the extracted features to describe environment are not enough.

In [6] an evaluative fuzzy neural network is proposed for fire prediction and fire selection. Fuzzy logic has been used to solve many decision-making problems. Because of intricacy of fire prediction and selection in Rescue Fire brigade agents; this method seems to be good for solving this problem.

To make coordination between agent three options are proposed in [7]: environment partitioning, centralized direct supervision and decentralized mutual adjustment. Among these three approaches decentralized approach is more flexible but it does not mean it is always better. In [8] a hybrid approach is proposed to use advantages of both centralized and decentralized approach. In order to make a decision about the number of ambulances which should cooperate to rescue civilian, evolutionary reinforcement learning is utilized by [9]. In [10] fire brigade learns to do task allocation and learns how to choose the best building for extinguishing to maximize the score. The learning features include number of civilians in the building, area of the building, level of fieriness of building and building material. In [11] fire brigades learn how to distribute in the city using neural reinforcement learning.

The major contribution made in this paper is to introduce a novel algorithm based on reinforcement learning for fire brigade agents to save city and civilians from fire.

With regard to the limits on the environment cognition of agents and the fact that agents don't know the effect of their action on environment, using reinforcement learning seems a good method to solve this problem. Reinforcement learning is learning what to do (how to map situations to actions) so as to maximize a numerical reward signal. The learner is not told which actions to take, as in most forms of supervised machine learning, but instead must discover which actions yield the most reward by trying them [12].

The mechanism which we have used to learn the agents is based on Temporal Difference method of reinforcement learning [12]. Because TD methods update estimates based in part on other learned estimates, without waiting for a final outcome (they bootstrap), So this method is applicable for complex environment like rescue simulation which an action should be selected in each cycle.

The rest of this paper is organized as follow. In Section 2 we explain the test bed which is used to implement and test our approach. In section 3 a brief explanation of reinforcement learning is presented. Section 4, discusses how to design and model the problem for applying reinforcement learning, in detail. In Section 5 details of implementation are presented and achieved results are shown. And Section 6 concludes.

2   Test Bed

RoboCup rescue simulation is a simulation of an earthquake effect in a city. The aim of creating this environment is to learn the best rescue strategies for humans and also extinguishing fires tactics caused by the earthquakes in real situations. It is currently a major league in RoboCup simulation competitions. The simulation consists of a kernel as the main part of the simulation, and some simulators which simulate the earthquake, fire, blockades and civilians of the city and also a viewer which shows the city changes during the time.

There are three kinds of agent including Police Force Agent, Ambulance Agent and Fire Brigade Agent in the simulation that should perform the job of rescue. Police Force Agents control the traffic in the city and open the blockades of the closed roads. Ambulance agent's duty is to find the injured civilians and give first-aid to them until they are carried out of buried building and they should also carry them to refuges. The main job of the Fire Brigade Agents is to extinguish the fire caused by the earthquake. The performance score of a simulation is usually computed from the quantity of alive civilians and the safe areas of the city. When the fire brigade agents want to extinguish fire they have to sort the set of the fired buildings to extinguish them one after another. We have applied our method to this environment to find an optimum policy for doing best action in each state of environment.

Fig. 1. Rescue Simulation Environment

Fig.1 shows a screenshot of the simulation environment. It displays map of Kobe city after an earthquake. The blue, white and red circles demonstrate the police force agents, Ambulance agents and fire brigade agents respectively. The bright and dull green circles display healthy and hurt civilians and black circle represent the dead civilian. The yellow, orange, dark red and black building represents level of fieriness of a building. Yellow shows least level of fieriness of a building and black shows a burned out building. There is a special type of buildings: the buildings that are marked with home icon are refuges where saved civilians are taken. Black areas on the roads represent blockades. The simulation runs for 300 time steps.

In order to reduce the complexity of simulation process we designed and implemented a simple rescue simulation environment that has all features we need such as fire simulator. So our simple rescue simulation is much faster than the official rescue simulation while keeping the necessary capabilities. In Fig.2 a screen shot of simple rescue simulation is displayed.

This environment has a fire simulator with the same algorithms of Rescue Simulation. So burning of buildings and fire spread is the same as original Rescue Simulation. Also the algorithm to calculate the health of civilians is the same as the one in misc simulator of Rescue Simulation.

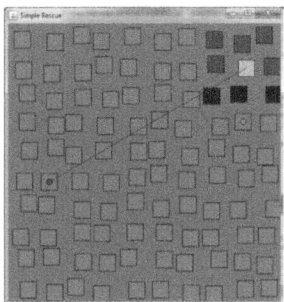

Fig. 2. Simple rescue simulation

The learning of agent and test phases are performed in this simulator and the result with a little modification is used in original rescue simulation.

3   Reinforcement Learning

Reinforcement learning is a branch of machine learning, in which an agent tries to find optimum actions to achieve the goal, without having full knowledge about environment or about the impact of its actions on the environment. In this type of learning the agent is not told how to do, but the agent is told what to do instead. The agent tries to achieve the goal by optimizing the received rewards.

In [12] three fundamental classes of methods for solving the reinforcement learning problems are described. Dynamic programming (DP), Monte Carlo methods, and Temporal Difference (TD) learning are these three classes.

Each class of methods has its strengths and weaknesses. Dynamic programming methods are well developed mathematically, but require a complete and accurate model of the environment. In rescue simulation, agents don't have a complete view of the world model, so this class of methods cannot be applied on rescue simulation.

Monte Carlo methods do not require a model and are conceptually simple, but are not suited for step-by-step incremental computation as is the case in rescue simulation. They don't update their value estimates on the basis of the value estimates of successor states. In other words, it is because they do not bootstrap.

Finally, temporal-difference methods require no model. Like DP(dynamic programming), TD(temporal-difference) methods update estimates based in part on other learned estimates, without waiting for a final outcome (they bootstrap). So we chose this method for applying on fire brigade agent.

4 Design

This section presents necessary issues including description of environment, the design of the reward function and the learning algorithm used in our approach. Finally a proposed learning process for the problem which is called lesson-by-lesson learning is presented.

4.1 Description of Environment

Although in Temporal Differential methods the model of environment is not required, it needs a clear description of the states of the model. So in this part we are going to describe rescue simulation environment with some discrete and finite states.

The states of environments are modeled by the following features.

1. freeEdges: The number of edges that fire can spread from represented by 0,1,2,3 or 4. For example in Figure2 the fire can spread only from left side, so number of freeEdges is 1.
2. distanceFromCenter: The distance of nearest fiery building to center of city, that is low, medium or high.
3. distanceFromCivilian: The distance of nearest fiery building to nearest civilian, that is low, medium or high
4. volumeOfFieryBuildings: The total volume of fiery buildings, that is veryLow, low, medium, high, veryHigh or huge.

So there are 5×3×3×7=270 different states. These states can describe all conditions in all maps.

After defining the potential states for environment, the possible actions for agents should be determined as well. So agents in each state can perform one of following actions:

1. extinguishEasyestBuilding: Extinguish the building that has lowest temperature.
2. extinguishNearestToCivilian: Extinguish the nearest fiery building to civilian.
3. extinguishNearestToCenter: Extinguish the nearest building to center of city.
4. extinguishNearestToMe: Extinguish the nearest building to fire brigade.

These four actions are the most reasonable actions that a fire brigade in different situations can do. The problem is to find the best action in each state of environment.

4.2 Reward Function

Reward function plays an important role in RL as it directs learning process to a solution.

In a burning city the world situation is getting worse over time, hence the agent always gets penalty. So the Fire brigade must reduce taken penalties.

Reward function is defined as follows:

- For each *waterDamaged* building (the building that is extinguished and is damaged because of too much water) in each time step the agent takes a reward -1.

- For each fiery building in each time step the agent takes reward -2.

- For each burned out building in each time step the agent takes reward -3.

- For each damaged civilian in each time step the agent takes reward *HP-1000*. The *HP* (Health Points) of each healthy civilian is 1000 and if the building that contains a civilian ignited, the civilians' *HP* reduces over the time.

- The first reward causes the fire brigade to putout fire without wasting water. The second and third rewards forces the fire brigade to putout fire as soon as possible and the last reward causes the fire brigade to save civilians from fire.

4.3 Learning Algorithm

The meaning of learning here is not to learn the way of doing an action by agents, but the agents learn to do which behavior in which time or situation.

Fire brigade agent with regards to environment state learns which one of the mentioned actions is most valuable (leads to best score) in a specific state. To do so, the agents are taught by an on-policy TD control method called SARSA [12]. Q values of each action in each state were updated by formula (1) considering their rewards or penalties.

$$Q(s_t, a_t) \leftarrow Q(s_t, a_t) + \alpha[r_{t+1} + \gamma Q(s_{t+1}, a_{t+1}) - Q(s_t, a_t)] \qquad (1)$$

In above equation s_t is previous state before taking action a, s_{t+1} is current state after action a, r is reward received from the environment after taking action a in previous state and reaching the current state, a is the last executed action, Q is the Q-Table, *alpha* is learning rate and *gamma* is discount factor.

The learning rate specifies to what extent the newly learned information affect the learned old information. The learning rate, set between 0 and 1. Setting it to 0 will make the agent learn nothing. Setting a high value will make the agent learn quickly.

The discount factor determines to what extent future rewards are important. The discount factor, set between 0 and 1. It implies that immediate rewards are more worth than future rewards.

The learning algorithm is as follows:

Initialise $Q(a, s)$ arbitrarily
Repeat (for each episode):
    Initialise s
    Choose a from s using policy derived from Q $(e − greedy)$

Repeat (for each step of episode):
   Take action a, observe r, s'
   Choose a' from s' using policy derived from Q (e − greedy)
   $Q(s_t, a_t) \leftarrow Q(s_t, a_t) + \alpha[r_{t+1} + \gamma Q(s_{t+1}, a_{t+1}) - Q(s_t, a_t)]$
   $s \leftarrow s'; a \leftarrow a'$
until s is terminal

This algorithm uses the e-greedy action selection algorithm. It means that with probability of e the agent takes random action (exploration) and in all other times takes the max valued action from previous experiences of agent (exploitation).

4.4 Lesson-by-Lesson Learning

In reinforcement learning methods the rewards provide guidance to the agent. If agent gets good reward it understands that it did right action sequence, and if it gets bad reward it understand that it did the wrong action. Now if agent gets bad rewards most of the time, it can't find the right action. In a very complicated environment such as rescue simulation, the problem search space is huge. So the agent may get bad rewards most of the time and can't learn the optimum policy.

To overcome this problem we started our learning phase in a Lesson-by-Lesson mode. In this mode we have these five steps:

1. First step is to teach the agent how to extinguish some easy scenarios, which have just one fiery point in the city. After a while the agent finds out the optimum policy to extinguish that fire.
2. Second step is to find the optimum policy to extinguish some scenarios, which have some fiery points in different parts of a city with different fieriness levels.
3. Third step is to find the optimum policy to extinguish some scenarios, which have some fiery points in different parts of a city and one civilian.
4. Forth step is to find the optimum policy to extinguish some scenarios, which have some fiery points in different parts of the city and some civilians in different parts of the city
5. In the fifth step extinguish distance has been restricted as it is in rescue simulation. So the fire brigade has to move to near the fiery points to extinguish them.
6. In the last step we limit the tanker capacity of the fire brigade as is case in rescue simulation. It means that after a while the water of fire brigade will be finished and it will has to refill it. This process takes usually between 10 and 20 cycle based on distance from refuge.

In each step the fire brigade uses obtained Q-Table in its previous step as the initial knowledge and the output is a better Q-Table.

5   Implementation and Results

In this section the achieved results of using explained method in simplified and original rescue simulation are presented and the advantages of using lesson-by-lesson learning are discussed.

5.1   Implementation in Simplified Rescue Simulation

In order to train the fire brigade, the parameters γ and α of Q-Table are considered 0.5 and 0.7, respectively based on trial and error. The value of e factor in e-greedy selection algorithm is set in each episode based on $e = e * 0.9^{episode}$ formula. This formula causes agent in primary episodes more explore the search space and in end episodes more uses its obtained experience in previous episodes and try to converge to a good solution.

In order to train agents, five scenario sets were chosen which in each set there are 4 maps as was explained in lesson-by-lesson learning section. In the first episode the initial Q-Table is set to zero. And in next episodes the previous obtained Q-Table is used for initializing. The end condition of each training phase is when, agent converge to an optimum policy for a scenario. Agent learns appropriate action in each state after training in all scenario sets using lesson-by-lesson learning approach.

To test the trained agent, 2 sample scenarios were used. One of them has 3 fire point (2 point near a civilian and 1 far from civilian) and another one has 4 fire point (3 point near the civilian and 1 far from civilian). In these scenarios five other extinguishing strategies including random, extinguish nearest fire to civilian, extinguish nearest fire to center, extinguish easiest fire and extinguish nearest fire to me to compare with trained agent strategy are tested. The result is shown in Table 1. It can be noticed that our trained agent shows better results than other agents. In Figures 3, 4,5,6,7 and 8 a snapshot of end of simulation in scenario 1 using different strategies for fire extinguishing is displayed. In Fig. 3 which shows the final result of using RL based agent, it is obvious that fire is extinguished and civilian is alive. This strategy seems has a good performance and considering chosen actions, states and our method, it is optimum policy among possible policies. Fig 4 represents the result of using extinguishing nearest fire to civilian strategy. Although fire has been extinguished before it could damage the civilian, about half of the city burnt out. So it seems not to be a good strategy.

Table 1. Map score based on strategy performed

	trained Firebrigade	Nearest fire to civilian	Nearest fire to center	Easiest fire	Nearest fire to me	Random action selection
Score(scenario1)	1.953	1.725	0.889	0.306	0.695	0.685
Score(scenario2)	1.509	1.215	0.367	0.400	0.541	0.567

In Fig 5 which shows the performance of random strategy, about half of the city is burnt out and fire has reached to civilian. So it is not a good strategy at all. In Fig 6 which shows the performance of extinguishing the nearest fire to center strategy, the fire is extinguished well but there is a civilian that has gotten stuck in one of the fiery buildings. Therefore it does not have any advantage to our explained method. Fig 7 and 8 display results of using extinguishing easiest fire strategy and extinguishing nearest fire to me strategy. It is obvious that the fire brigade couldn't save the city and civilian well. So they are not acceptable strategies.

Fig. 3. Learned Fire Brigade

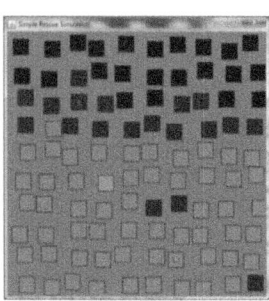

Fig. 4. Extinguishing nearest fire to civilian

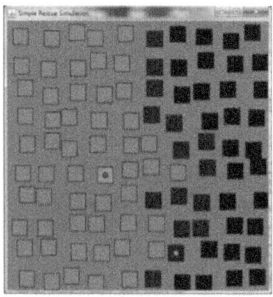

Fig. 5. Random Strategy

Fig. 6. Extinguishing the nearest fire to center

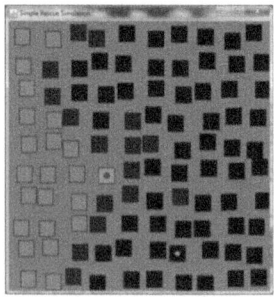

Fig. 7. Extinguishing easiest fire

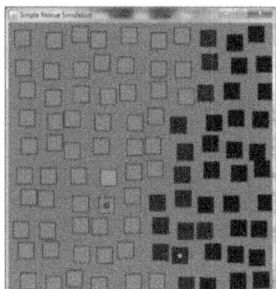

Fig. 8. Extinguishing nearest to me

5.2 Implementation in Original Rescue Simulation

In order to observe the performance of our approach in official rescue simulation, we selected two standard maps and compared the extinguishing fire task of our agent using trained Q-table with the strategy of the first three teams of 2010 worldwide rescue simulation competitions in Singapore and also with previous sterategy of our team (which is championship of IranOpen 2011). The scores were extracted by using the released codes of RoboAKUT, IAMRescue and ZJUBase rescue teams.

To use the obtained Q-table of simple rescue simulation in official rescue simulation we should first adapt the Q-table with this environment so we first trained the agent 50 episodes by using obtained Q-table in simple rescue simulation as initial table in original rescue simulation. The parameters are set same as which were set in simplified rescue simulation.

Table 2 shows the results of the experiments on all maps. It can be noticed that on all test cases our agent show better results when using the trained Q-Table. This demonstrates that the reinforcement learning system tends to refine and keep better strategy for the fire brigade agent.

Table 2. Results of the experiments

Team	Kobe1	Kobe2
BraveCircles(Old)	37.116	11.892
BraveCircles(New)	43.698	14.972
RoboAKUT	24.692	9.5
IAMRescue	38.862	8.93
ZJUBase	27.475	9.284

5.3 Discussion on Lesson by Lesson Learning

Now to show the advantage of lesson-by-lesson learning, we compare the speed learning of two different agents. One of them is experienced agent which has passed first five mentioned steps in lesson by lesson learning process and other one is newbie agent which has not trained at all. After limiting the tanker capacity of agent(step 6), and starting the train process for both experienced and newbie agents, we observed that the learning speed of experienced agent was much more than newbie agent. Table 3 compares the speed of learning in two experienced and newbie agents in 3 different scenarios. Results show an experienced agent can adapt itself to changes, faster than a newbie agent. Fig 9,10 show the convergence of experienced and newbie agents in a same scenario. In Fig 9,10 horizontal axis is episode and vertical axis is score.

Table 3. Expert vs. newbie fire brigade

Map	Score	Expert Firebrigade	Newbie Firebrigade
Map1 (Easy)	Score	2.853	2.853
	Episode of convergence	83	95
Map2 (medium)	Score	1.888	1.888
	Episode of convergence	56	220
Map3 (hard)	Score	2.907	2.907
	Episode of convergence	188	4760

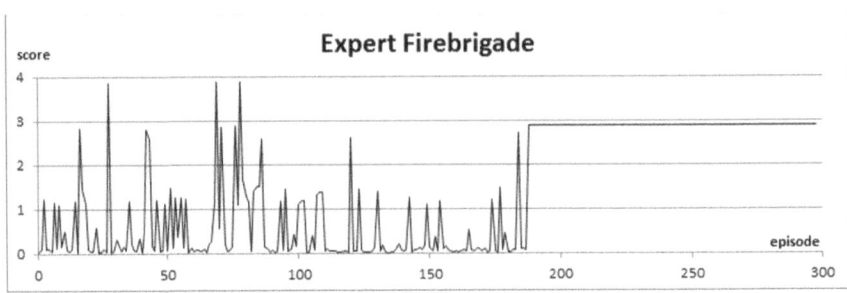

Fig. 9. Expert Firebrigade (Map3)

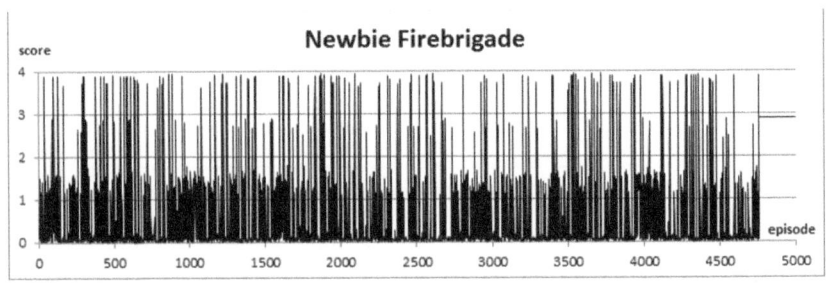

Fig. 10. Newbie fire brigade (Map3)

6   Conclusions and Perspectives for Future Research

In this paper we discussed about using temporal difference learning to find the optimum policy for fire extinguishing task of fire brigades, and results show that the trained agent learned by TD has a good performance to extinguishing fires. This method increases the speed of learning and it has very low memory usage as well. Also we used a lesson-by-lesson learning process and according to results the experienced agent can learn the optimum policy much faster than the newbie agent.

For next steps we are going to find the best parameters for RL by using optimization algorithms like Genetic and PSO. Also as a future work we want to use this method for other kind of agents like police and ambulance agent to find the optimum policy for their task.

Because the rescue simulation environment is a multi-agent system, the coordination between agents is an important issue. So in future we are going to design and implement a method based on RL to achieve an optimum policy for coordination of different agents.

Acknowledgment. We would like to acknowledge the support of FCT – Fundação para a Ciência e Tecnologia, through the project "Intelligent wheelchair with flexible multi modal interface" under grant FCT/RIPD/ADA/109636/2009.

References

1. Kitano, H., Tadokoro, S.: RoboCup rescue: A grand challenge for multiagent and intelligent systems. AI Magazine 22(1), 39–52 (2001)
2. Takeshi, M.: How to develop a RoboCupRescue agent (2000)
3. Nanjanath, M., Erlandson, A.J., Andrist, S., Ragipindi, A., Mohammed, A.A., Sharma, A.S., Gini, M.: Decision and Coordination Strategies for RoboCup Rescue Agents. In: Ando, N., Balakirsky, S., Hemker, T., Reggiani, M., von Stryk, O. (eds.) SIMPAR 2010. LNCS, vol. 6472, pp. 473–484. Springer, Heidelberg (2010)
4. Fave, F.M.D., Packer, H., Pryymak, O., Stein, S., Stranders, u., Tran-Thanh, L., Vytelingum, P., Williamson, S.A., Jennings, N.R.: RoboCupRescue 2010 Rescue Simulation League Team Description IAMRescue, United Kingdom) (2010)
5. Shahbazi, H., Zafarani, R.: Priority Extraction Using Delayed Rewards in Multi Agents Systems: A Case Study in RoboCup. In: CSICC 2006, Iran, pp. 571–574 (2006)
6. Shahgholi Ghahfarokhi, B., Shahbazi, H., Kazemifard, M., Zamanifar, K.: Evolving Fuzzy Neural Network Based Fire Planning in Rescue Firebrigade Agents. In: SCSC 2006, Canada (2006)
7. Paquet, S., Bernier, N., Chaib-draa, B.: Comparison of Different Coordination Strategies for the RoboCup Rescue Simulation. In: Orchard, B., Yang, C., Ali, M. (eds.) IEA/AIE 2004. LNCS (LNAI), vol. 3029, pp. 987–996. Springer, Heidelberg (2004)
8. Mohammadi, Y.B., Tazari, A., Mehrandezh, M.: A new hybrid task sharing method for cooperative multi agent systems. In: Canadian Conf. on Electrical and Computer Engineering (May 2005)
9. Martínez, I.C., Ojeda, D., Zamora, E.A.: Ambulance Decision Support using Evolutionary Reinforcement Learning in RoboCup Rescue Simulation League. In: Lakemeyer, G., Sklar, E., Sorrenti, D.G., Takahashi, T. (eds.) RoboCup 2006: Robot Soccer World Cup X. LNCS (LNAI), vol. 4434, pp. 556–563. Springer, Heidelberg (2007)
10. Paquet, S., Bernier, N., Chaib-draa, B.: From global selective perception to local selective perception. In: AAMAS, pp. 1352–1353 (2004)
11. Amraii, S.A., Behsaz, B., Izadi, M.: S.o.s 2004: An attempt towards a multi-agent rescue team. In: Proc. 8th RoboCup Int'l Symposium (2004)
12. Sutton, R.S., Barto, A.G.: Reinforcement Learning: An Introduction. MIT Press, Cambridge (1998)

Humanoid Behaviors: From Simulation to a Real Robot

Edgar Domingues[1], Nuno Lau[1], Bruno Pimentel[1,2], Nima Shafii[2],
Luís Paulo Reis[2], and António J.R. Neves[1]

[1] DETI/UA - Dep. of Electronics, Telecommunications and Informatics,
IEETA - Inst. of Electronics and Telematics Engineering of Un. of Aveiro,
Campus Universitário de Santiago, 3810-193 Aveiro, Portugal
{edgar.domingues,nunolau,brunopimentel,an}@ua.pt
[2] DEI/FEUP - Dep. Informatics Engineering, Faculfy of Engineering of the
University of Porto,
LIACC - Artificial Intelligence And Computer Science Lab. of the University of Porto,
Rua Dr. Roberto Frias, s/n 4200-465 Porto, Portugal
nima.shafii@gmail.com, lpreis@fe.up.pt

Abstract. This paper presents the modifications needed to adapt a humanoid agent architecture and behaviors from simulation to a real robot. The experiments were conducted using the Aldebaran Nao robot model. The agent architecture was adapted from the RoboCup 3D Simulation League to the Standard Platform League with as few changes as possible. The reasons for the modifications include small differences in the dimensions and dynamics of the simulated and the real robot and the fact that the simulator does not create an exact copy of a real environment. In addition, the real robot API is different from the simulated robot API and there are a few more restrictions on the allowed joint configurations. The general approach for using behaviors developed for simulation in the real robot was to: first, (if necessary) make the simulated behavior compliant with the real robot restrictions, second, apply the simulated behavior to the real robot reducing its velocity, and finally, increase the velocity, while adapting the behavior parameters, until the behavior gets unstable or inefficient. This paper also presents an algorithm to calculate the three angles of the hip that produce the desired vertical hip rotation, since the Nao robot does not have a vertical hip joint. All simulation behaviors described in this paper were successfully adapted to the real robot.

Keywords: humanoid robot, behaviors, simulation, real robot, Aldebaran Nao.

1 Introduction

Since the year 2004 the FC Portugal team [8,13] participates in the RoboCup 3D Simulation League. This year, for the first time, following this collaboration and also including members from CAMBADA [10] and 5DPO [3] teams, the Portuguese Team will participate in the Standard Platform League (SPL).

L. Antunes and H.S. Pinto (Eds.): EPIA 2011, LNAI 7026, pp. 352–364, 2011.

In the RoboCup 3D Soccer Simulation League robotic soccer games are run on a simulator. Each participating team programs an agent to control a humanoid robot that models the Nao robot from Aldebaran. In the SPL all robots are the same, being nowadays the Nao robot, the only difference being the software which controls them. Since 2008 both the Simulation League and the SPL use the same robot model: the Aldebaran Nao (Fig. 1 and 2). This would allow to develop agents that can compete on both leagues, test the agent on a simulated robot before running it on the real one, and use on the latter the results of machine learning techniques ran in the simulation.

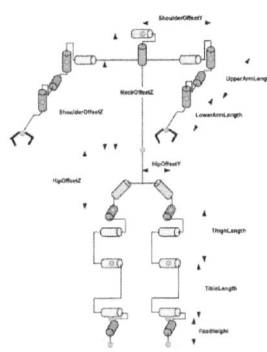

Fig. 1. The Nao robot   **Fig. 2.** Nao joints (adapted from [5])

However, the simulator cannot reproduce reality with precision. The simulator does not detect collisions between body parts of the same robot. The ground friction is low compared to the official SPL field. The simulated robot model is not an exact copy of the real one, since it has a few differences in the dimensions and one more degree of freedom. Hence, some modifications must be made to adapt the low level behaviors from simulation to a real robot. Once we have stable low level behaviors on both the simulation and the real robot, the same high level behaviors can be used on both, with little or no modifications.

This paper presents the modifications needed on the FC Portugal simulation agent to run on the real Naos. All behaviors needed for the real robot are present in the simulation agent. However, the get up behaviors were developed from scratch, since those used in the simulation execute motions that are not possible on a real robot. A kick to the side that did not exist in the simulation was also created. The next two sections describe the simulation environment and the real robot. Section 4 describes the architecture of our agent. Section 5 and Section 6 describes the behaviors we use in simulation and their adaptation to the SPL. Section 7 shows the results achieved with real Naos. The last section presents the conclusions.

2   RoboCup 3D Simulator

RoboCup 3D Soccer Simulation League uses SimSpark simulator [2] to simulate the games. This simulator uses the Open Dynamics Engine (ODE) [16] to simulate the physical environment. The robot model used is based on the Aldebaran Nao robot. However, it is not a precise replica of the real one, since it has a few differences in the dimensions and 22 degrees of freedom (one more than the real version). The joints are controlled specifying the desired angular speed for each joint. The simulated robot, like the real robot, has a gyroscope and an accelerometer on the torso and foot force sensors. Instead of cameras, the vision information is given as spherical coordinates of the objects, but this is not important for this paper, since we talk only about low level behaviors.

3   Nao Hardware and Software

The real Nao robot has been developed by the French company Aldebaran Robotics [4]. It is a humanoid robot with 21 degrees of freedom, one less than the simulation because the two hip yaw pitch joints are controlled by the same motor. Joint control is based on providing angle targets. Each joint has an internal PID controller which then acts autonomously to achieve its target. This means that the joint control on the real robot is higher level than the used on simulation. Also unlike in simulation, the real robot provides stiffness control for each joint, which can be useful to save energy and improve the behavior robustness and speed [7].

Like the simulated robot, the real robot has a gyroscope and an accelerometer in the torso. Aldebaran provides a programming interface that delivers the filtered angles of the torso position relative to the vertical. Each foot has 4 force sensitive resistors and there are two cameras in the head.

The Nao robot runs NaoQi [4] which is a framework that manages the execution of modules. These modules are programming libraries that can be created and executed on the robot. Aldebaran made available some modules with the robot: modules to control motions, LEDs, text-to-speech, etc. One of these modules is Device Communication Manager (DCM), that allows low level access to the actuators and sensors of the robot. Modules can only communicate between each other, so it is necessary to create our own module in order to use any of the modules provided.

4   Portuguese Team Agent Architecture

Our approach to allow the agent architecture access the robot hardware, like many others SPL teams [14,12,9], was to create a simple NaoQi module to communicate with the DCM. This module reads the actuators values from the shared memory, sends them to be executed by the DCM, and copies the sensors values from the DCM to the shared memory. By accessing the shared memory, our agent becomes independent from the Aldebaran software.

The architecture of the agent that uses the shared memory is identical to our agent in the simulation league. The only modifications made were in the low level communications. In simulation, the agent communicates with the server to send the actuator values and receive the sensor values. In the real robot, instead of communicating with a server, the agent sends the actuator values and reads the sensor values from the shared memory. As presented above, the real robot has a PID controller for each joint as opposed to the simulated robot, whose PID controllers must be implemented by the user. Therefore, it is unnecessary to copy this software PID controller from simulation to the real robot.

The high level behaviors can be ported from the simulated robot to the real robot with no modification, as long as the low level behaviors are developed for both the simulated and the real robot.

5   Behavior Models

5.1   Slot Behaviors

A Slot Behavior is defined by a sequence of slots. Each slot has a time duration and a target angle for each joint to be controlled. When a slot is executed the joints are moved with a sinusoidal trajectory, from the angle they have at the beginning of the slot, to the target angle that is achieved at the end of the slot. The sinusoidal trajectory is used since the initial and final speed are zero, and it assures the lowest second derivative maximum, hence the acceleration will be minimized [1]. This produces a smooth motion for the joints. Slot behaviors are defined in Extensible Markup Language (XML) files, so they can be easily manipulated without need to recompile the agent.

This was used to develop some behaviors, namely: kick the ball, get up, among others.

5.2   CPG Behaviors

Central Pattern Generator (CPG) Behaviors execute, on each targeted joint, a trajectory defined by a sum of sine waves [11]. Each sine has four parameters: amplitude, period, phase, and offset. Like Slot Behaviors, CPG Behaviors are defined using XML files, to be easily manipulated without the need to recompile the agent.

This was used to create periodic behaviors: walk, turn in place, rotate around the ball, etc.

5.3   OmnidirectionalWalk Behavior

This behavior was based on Sven Behnke's work [1]. It is an open-loop omnidirectional walk developed for the humanoid robot Jupp. It had to be adapted to the simulated Nao, since the two robots are different.

This behavior uses a Leg Interface to control the leg movements. The Leg Interface allows to specify leg positions using three parameters: leg angle, roll, pitch, and yaw angles between the torso and the line connecting hip and ankle; leg extension, the distance between the hip and the ankle; and foot angle, roll, and pitch angles between the foot and the perpendicular of the torso. This behavior uses this interface to generate a stable omnidirectional walk.

The leg angle, $\theta_{\text{Leg}} = \left(\theta_{\text{Leg}}^{\text{r}}, \theta_{\text{Leg}}^{\text{p}}, \theta_{\text{Leg}}^{\text{y}}\right)$, is the angle between the torso and the line connecting hip and ankle. $\theta_{\text{Leg}}^{\text{r}} = 0$ when the leg is parallel to the trunk, on the coronal plane, and $\theta_{\text{Leg}}^{\text{r}} > 0$ when the leg is moved outwards to the side. $\theta_{\text{Leg}}^{\text{p}} = 0$ when the leg is parallel to the trunk, in the sagittal plane, and $\theta_{\text{Leg}}^{\text{p}} > 0$ when the leg is moved to the front. $\theta_{\text{Leg}}^{\text{y}} = 0$ when the foot points to the front and $\theta_{\text{Leg}}^{\text{y}} > 0$ when the leg is rotated vertically pointing the foot outward. The foot angle, $\theta_{\text{Foot}} = (\theta_{\text{Foot}}^{\text{r}}, \theta_{\text{Foot}}^{\text{p}})$, controls the foot angle relative to the torso. If $\theta_{\text{Foot}} = (0,0)$ then the foot base is perpendicular to the torso. Hence, if the torso is perpendicular to the ground, the foot base is parallel to the ground. The leg extension, $-1 \le \gamma \le 0$, denotes that the leg is fully extended when $\gamma = 0$ and is shortened when $\gamma = -1$. When shortened, the leg is η_{min} of its fully extended length. The relative leg length can be calculated as $\eta = 1 + (1 - \eta_{\text{min}})\gamma$.

5.4   TFSWalk Behavior

The TFSWalk gait combines two approaches: a gait where the joints trajectory is generated using Partial Fourier Series optimized with Genetic Algorithms [11]; and a gait using Truncated Fourier Series, which are imitated from human walk, to generate the joints angles optimized with Particle Swarm Optimization [15]. This gait allows the robot to walk forward and backward (either straight or turning) and to turn in place.

6   Adaptation to the Real Robot

When adapting the behaviors from simulation to a real robot we tried to keep the architecture of the agent similar on both. Since all behaviors on simulation produce joints angles which are then passed to a software PID controller, this was replaced with some software to send these angles to the shared memory (as explained on Section 4), and convert them to the angle range of the real robot joints if needed.

6.1   Leg Interface

The Leg Interface calculates the angles of the 6 leg joints using: leg angle, $\theta_{\text{Leg}} = \left(\theta_{\text{Leg}}^{\text{r}}, \theta_{\text{Leg}}^{\text{p}}, \theta_{\text{Leg}}^{\text{y}}\right)$; foot angle, $\theta_{\text{Foot}} = (\theta_{\text{Foot}}^{\text{r}}, \theta_{\text{Foot}}^{\text{p}})$; and leg extension, γ. In the Jupp robot the hip joints are in the following order, from up to down: roll, pitch and yaw. But in the Nao robot the hip joints are in a different order: yaw pitch, roll, pitch. This means that we cannot use all the formulas presented in Sven Behnke's paper [1]. So we have developed our own formulas. Given the leg extension γ we calculate the relative leg length as $\eta = 1 + (1 - \eta_{\text{min}})\gamma$ which is then multiplied by the length of the fully extended leg to get the desired leg length, $l = \eta(l_{\text{upperLeg}} + l_{\text{lowerLeg}})$. Where l_{upperLeg} and l_{lowerLeg} are the lengths of the thigh and shank, respectively. Using the law of cosines (1), the knee joint angle can be calculated by (2). It is subtracted by π because it is the outside angle, contrary to the one used in the law of cosines which is the inside angle. θ_{Knee} is 0 when the leg is stretched and negative when the leg is retracted. In the same way, the law of cosines is used to calculate the angles of the hip (3) and ankle (4) to compensate the knee angle. Therefore, when the leg extension changes, θ_{Leg} and θ_{Foot} are not changed.

$$c^2 = a^2 + b^2 - 2 \cdot a \cdot b \cdot \cos(\theta) \tag{1}$$

$$\theta_{\text{Knee}} = \arccos\left(\frac{l_{\text{upperLeg}}^2 + l_{\text{lowerLeg}}^2 - l^2}{2 \cdot l_{\text{upperLeg}} \cdot l_{\text{lowerLeg}}}\right) - \pi \tag{2}$$

$$\Delta\theta_{\text{Hip}}^{\text{p}} = \arccos\left(\frac{l_{\text{upperLeg}}^2 + l^2 - l_{\text{lowerLeg}}^2}{2 \cdot l_{\text{upperLeg}} \cdot l}\right) \tag{3}$$

$$\Delta\theta_{\text{Ankle}}^{\text{p}} = \arccos\left(\frac{l_{\text{lowerLeg}}^2 + l^2 - l_{\text{upperLeg}}^2}{2 \cdot l_{\text{lowerLeg}} \cdot l}\right) \tag{4}$$

The remaining leg joint angles can be calculated by equations (5), where ls (leg side) is -1 for the left leg and 1 for the right leg. The hip angles are the θ_{Leg} angles with the compensation $\Delta\theta_{\text{Hip}}^{\text{p}}$. The ankle angles are the differences between θ_{Foot} angles and the θ_{Leg} angles with the compensation $\Delta\theta_{\text{Ankle}}^{\text{p}}$.

$$\theta_{\text{HipYawPitch}} = \theta_{\text{Leg}}^{\text{y}} \tag{5}$$
$$\theta_{\text{HipRoll}} = -ls \cdot \theta_{\text{Leg}}^{\text{r}}$$
$$\theta_{\text{HipPitch}} = \theta_{\text{Leg}}^{\text{p}} + \Delta\theta_{\text{Hip}}^{\text{p}}$$
$$\theta_{\text{AnklePitch}} = \theta_{\text{Foot}}^{\text{p}} - \theta_{\text{Leg}}^{\text{p}} + \Delta\theta_{\text{Ankle}}^{\text{p}}$$
$$\theta_{\text{AnkleRoll}} = -ls \cdot \left(\theta_{\text{Foot}}^{\text{r}} - \theta_{\text{Leg}}^{\text{r}}\right)$$

Note that in (5) the leg vertical rotation is applied directly in the hip yaw pitch joint. The Nao robot, unlike Jupp, has no vertical hip joint. It has only one joint rotated 45 degrees that can be used to obtain vertical rotation of the leg. When we applied the leg yaw rotation directly to the hip yaw pitch joint, the robot was stable but the torso oscillated forwards and backwards.

To correct this oscillation we used a formula that, given the desired yaw hip rotation, calculates the correct angles to apply on the three hip joints of the Nao robot. We start by calculating the rotation matrix of the thigh in relation to the hip, aligned with the hip yaw pitch joint axis, using (6).

$$RotHip = Rot_x\left(ls \cdot \frac{\pi}{4}\right) \cdot Rot_z(\theta_{\text{Leg}}^y) \tag{6}$$

$$= \begin{pmatrix} 1 & 0 & 0 \\ 0 & \frac{\sqrt{2}}{2} & -ls \cdot \frac{\sqrt{2}}{2} \\ 0 & ls \cdot \frac{\sqrt{2}}{2} & \frac{\sqrt{2}}{2} \end{pmatrix} \cdot \begin{pmatrix} \cos(\theta_{\text{Leg}}^y) & -\sin(\theta_{\text{Leg}}^y) & 0 \\ \sin(\theta_{\text{Leg}}^y) & \cos(\theta_{\text{Leg}}^y) & 0 \\ 0 & 0 & 1 \end{pmatrix}$$

$$= \begin{pmatrix} \cos(\theta_{\text{Leg}}^y) & -\sin(\theta_{\text{Leg}}^y) & 0 \\ \frac{\sqrt{2}}{2} \cdot \sin(\theta_{\text{Leg}}^y) & \frac{\sqrt{2}}{2} \cdot \cos(\theta_{\text{Leg}}^y) & -ls \cdot \frac{\sqrt{2}}{2} \\ ls \cdot \frac{\sqrt{2}}{2} \cdot \sin(\theta_{\text{Leg}}^y) & ls \cdot \frac{\sqrt{2}}{2} \cdot \cos(\theta_{\text{Leg}}^y) & \frac{\sqrt{2}}{2} \end{pmatrix}$$

Then, part of the B-Human inverse kinematics [6] can be used to calculate the angles of the three hip joints that produce the desired rotation. First, the rotation matrix produced by the three hip joints is constructed (the matrix is abbreviated, e.g. c_x means $\cos(\delta_x)$) (7).

$$RotHip = Rot_z(\delta_z) \cdot Rot_x(\delta_x) \cdot Rot_y(\delta_y) \tag{7}$$

$$= \begin{pmatrix} c_x c_z - s_x s_y s_z & -c_x s_z & c_z s_y + c_y s_x s_z \\ c_z s_x s_y + c_y s_z & c_x c_z & -c_y c_z s_x + s_y s_z \\ -c_x s_y & s_x & c_x c_y \end{pmatrix}$$

It is now clear that the angle of the hip roll (x axis) can be calculated using s_x, as shown in (8). This angle has to be rotated 45 degrees, because the hip roll joint space is rotated according to the hip yaw pitch axis.

$$\delta_x = \arcsin(s_x) - ls \cdot \frac{\pi}{4} = \arcsin(ls \cdot \frac{\sqrt{2}}{2} \cdot \cos(\theta_{\text{Leg}}^y)) - ls \cdot \frac{\pi}{4} \tag{8}$$

Calculating each of the other two hip joints requires 2 matrix entries. Equation (9) shows how we can obtain the rotation along the z axis by combining 2 entries of the rotation matrix. With this, the remaining two hip joints can be calculated using (10) and (11).

$$\frac{c_x \cdot s_z}{c_x \cdot c_z} = \frac{\cos(\delta_x) \cdot \sin(\delta_z)}{\cos(\delta_x) \cdot \cos(\delta_z)} = \frac{\sin(\delta_z)}{\cos(\delta_z)} = \tan(\delta_z) \tag{9}$$

$$\delta_z = \text{atan2}(c_x \cdot s_z, c_x \cdot c_z) = \text{atan2}(\sin(\theta_{\text{Leg}}^y), \frac{\sqrt{2}}{2} \cdot \cos(\theta_{\text{Leg}}^y)) \tag{10}$$

$$\delta_y = \text{atan2}(c_x \cdot s_y, c_x \cdot c_y) = \text{atan2}(-ls \cdot \frac{\sqrt{2}}{2} \cdot \sin(\theta_{\text{Leg}}^y), \frac{\sqrt{2}}{2}) \tag{11}$$

These three angles (δ_x, δ_y and δ_z) are added to the calculated joints angles of (5) producing the final angles of the three hip joints:

$$\theta_{\text{HipYawPitch}} = \delta_z \tag{12}$$
$$\theta_{\text{HipRoll}} = -ls \cdot \theta^{\text{r}}_{\text{Leg}} + \delta_x$$
$$\theta_{\text{HipPitch}} = \theta^{\text{p}}_{\text{Leg}} + \Delta\theta^{\text{p}}_{\text{Hip}} - \delta_y$$

This method can be used whenever a Nao robot's leg must rotate around its vertical axis. After this the differences between the Nao robot and other humanoid robots are less significant. Most humanoid robots have a joint to rotate each leg vertically, but the Nao robot does not. This means that code developed for a generic humanoid robot could not be applied to the Nao robot. With this algorithm, we obtain a virtual joint on the hip that rotates around the vertical axis, permitting to easily adapt code from other humanoid robots to this one.

6.2   Slot Behaviors

Adapting the Slot Behavior algorithm was easy, since it only generates joint trajectories based on XML files. Only the behaviors themselves needed to be adapted.

The get up behaviors (after falling forward and backward) were developed from scratch because the get up behaviors used on the simulation execute motions that are not possible on a real robot. The sequence of poses of our get up behaviors were based on Aaron Tay's thesis [17] and on the B-Human 2010 Code Release [14].

The kick forward behavior was adapted from simulation and had to be changed to be stable on the real robot. Some slot durations were increased to make the behavior slower. Compared to the simulation, the slot durations were increased an average of 23%. The angles of the joints that move on the coronal plane (hip roll and ankle roll) were also increased from 11 to 20 degrees, so the center of mass is better shifted to the support foot.

Was also created a kick to the side that did not exist on the simulation.

6.3   CPG Behaviors

In the CPG Behaviors, like the Slot Behaviors, the algorithm itself was easily adapted but the behaviors needed to be changed. The common modification was to slowdown the behaviors and reduce the joint amplitudes. However, each behavior needed to be independently adapted to the real robot. The forward walk created as a CPG Behavior was adapted from the simulation to the real robot by reducing the joint angle amplitudes 50% and slowing it down 67%. The rotate around the ball was slowdown in 58% and the joint angle amplitudes had an average reduction of 74%.

6.4   OmnidirectionalWalk Behavior

As said above, this behavior produces the desired motion calculating the trajectories of three parameters for each leg, and then uses the Leg Interface to convert these parameters into joint angles. To adapt this behavior from the simulation to the real robot we modified the Leg Interface according to the dimensions of the real robot.

A difference from the Nao robot to most humanoid robots (like the Jupp robot for which Sven Behnke's walk [1] was developed) is that the hip yaw pitch joints in both legs are controlled by a single motor. This means that any rotation is applied to both joints at the same time. In the simulation the two joints can be controlled independently. So, when adapting this walk from simulation to the real robot, we need to combine the two motions into one that produces the desired stable gait. When, in the simulation, the two motions are equal, we can apply them directly to the real robot. When these two motions are different, the task is more difficult and each specific case must be analyzed.

For this walk the desired yaw motion of the legs is shown in Fig. 3. Each leg is composed by two kinds of movements. A sinusoidal movement, when the leg is in the air, and a linear movement, when it is on the ground. As presented in Fig. 3, the two motions are significantly different and cannot be directly executed by the common joint of the real Nao robot. We needed to combine the two motions creating one that is a continuous function (so the joint can follow it smoothly) and that produces a stable walk. Our choice was to follow a linear-like motion which results from selecting the motion of the leg that is on the ground (Fig. 4). To do this we switch between left and right leg yaw trajectory followed in the intersection of the two motions when both legs follow linear movements.

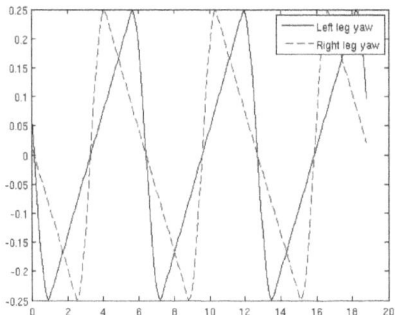

Fig. 3. Yaw motions of the legs in the OmnidirectionalWalk

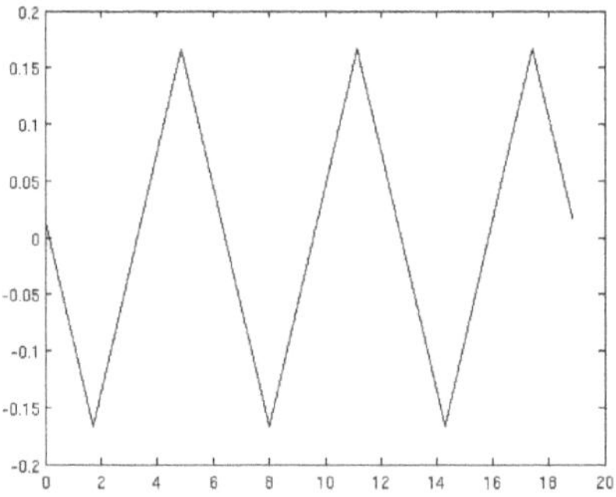

Fig. 4. Common joint motion, result of the combination of the two motions shown in Fig. 3

6.5 TFSWalk Behavior

This gait was optimized on the simulator, resulting in the fastest gait we have. However, the ground on the simulator has low friction and so the gait learned to use this as an advantage, not lifting the feet too much and almost not using coronal movement. When adapting this behavior to the real robot it only worked on slippery ground. On the official SPL field, the carpet has more friction, making the robot stumble and fall.

The first approach we took to solve this problem was to add coronal movement, allowing the robot to better shift its Center of Mass (CoM), resulting in a more stable gait. The coronal movement used is shown in Fig. 5 and was based on [15]. It consists of rotating the hip roll joint of the support leg (the one on the ground). This lifts the other leg (the swinging leg) from the ground. The ankle roll joint of the swinging leg is rotated with the same angle as the support leg hip, so the foot is always parallel to the ground.

The hip motion is defined using (13) and produces the trajectories shown in Fig. 6. The left and right hips have a phase shift of π. θ is the amplitude and T the period. The latter is the same for the sagittal and the coronal movements. On the other hand, the coronal movement has a phase shift of $\pi/2$ relative to the sagittal movement. The amplitude is empirically defined and depends on the walking speed. When the latter increases, the coronal movement amplitude decreases.

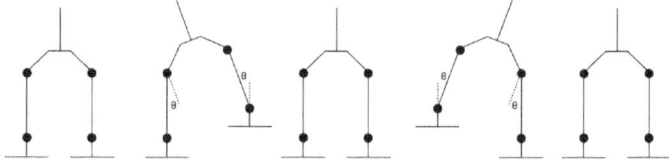

Fig. 5. Coronal movement

Fig. 6. Left and right hip roll joints trajectories

$$f(t) = \begin{cases} \theta \sin\left(\frac{2\pi}{T}t\right) , & \text{if } t < \frac{T}{2} \\ 0 & , \text{ otherwise} \end{cases} \tag{13}$$

The other approach we took, instead of using coronal movement to shift the CoM, lifts the feet higher and rapidly making use of the robot dynamics to keep the robot balanced. In order to increase the height of the feet trajectories, the knee, hip and ankle joints trajectories should be changed in a coordinated way. However, the specification of TFSWalk, by defining each joint trajectory independently of the others, does not provide a good model for controlling the foot trajectory. To achieve a more controllable model, the TFSWalk specification was converted to use the Leg Interface (Section 6.1). With this new model it is easy to control foot height trajectory just changing the leg extension parameter.

In addition, the velocity of the robot, controlled by the leg angle amplitude, becomes independent on the foot trajectory height. As an example, the feet may be kept moving up and down in the same place by setting the leg angle amplitude to zero while keeping the normal value of the leg extension. This up and down movement is very useful to initiate and finish the walking behavior. The increase in the foot height trajectory diminished foot collisions with the ground and resulted in a stable walk without needing coronal movement.

Feedback was also added to make the behavior more robust against external disturbances, such as uneven ground and collisions with obstacles. The feedback is calculated using a filtered value of the accelerometer in the x direction (front), as shown in (14), and it is based on the rUNSWift Team Report 2010 [12]. This value is summed to the two hip pitch joints, to balance the torso. This way, when the robot is falling to the front, the accelerometer will have a positive value that is added to the hip pitch joints, making the legs to move forward, to compensate. The inverse is also true, when the robot is falling backwards.

$$filAccelX = 0.5 * filAccelX + 0.5 * accelX * 0.2; \qquad (14)$$

7 Results

All the described behaviors were tested and are stable.

The get up behaviors developed as Slot Behaviors put the robot in the stand position when the robot is lying down on the ground in 16.5 seconds, when lying on the front, and 8.1 seconds, when lying on the back. The kick forward behavior takes 3.2 seconds to hit the ball and then 2.2 seconds to return to the stand up position. The distance traveled by the ball was measured for five kicks and the average distance was 2.87 ± 0.46 meters. The kick to the side behavior hits the ball in 3 seconds and returns to the stand position in 3.2 seconds. This kick was executed five times and the ball traveled an average distance of 1.05 ± 0.17 meters.

The forward walk created using a CPG Behavior was executed five times in the real robot and had an average velocity 6 ± 0.4 cm/s. The turn in place and the rotate around the ball had an average velocity of 27 °/s. In the simulation, the forward walk had a velocity of 7.33 cm/s, the turn in place had an average velocity of 49.47 ± 0.42 °/s, and the rotate around the ball had an average velocity of 49.98 ± 1.24 °/s

The OmnidirectionalWalk achieved in the real robot a forward velocity of 5 cm/s, a side velocity of 1 cm/s, and rotates at 35 °/s. In the simulation this behavior achieved a forward velocity of 10 cm/s, a side velocity of 3 cm/s, and rotates at 12 °/s.

The TFSWalk was executed 10 times in the real robot and has an average forward velocity of 22 ± 1 cm/s and rotates at 25 °/s. In the simulation the TFSWalk behavior achieves a forward velocity of 51 cm/s and rotates at 42 °/s.

8   Conclusion

This paper described the adaption of behaviors from the simulation to a real robot. All presented behaviors were successfully adapted and tested on a real robot. It is harder to develop behaviors for a robot in a real environment instead of simulated environment because of factors that make behaviors unstable, namely: external disturbances; uneven ground; or motor strength that affects the precision the joints follow the trajectories. We proved that it is possible to use the behaviors of simulation in a real robot. The simulation environment can be used to test the behaviors before executing them on a real robot, or to use machine learning techniques to optimize them. Some modification were needed since the simulated and the real robot are different. In the TFSWalk behavior, feedback was added to increase stability and robustness, making the behavior adapt to changes in the environment. The process of adapting behaviors from simulation to a real robot allowed us to find and correct some bugs on the existing behaviors.

Was also presented in this paper an algorithm to calculate the angles of the three hip joints that produce the desired vertical rotation of the hip, since the Nao robot does not have a vertical hip joint. This algorithm was used with success on the presented behaviors and can be used in the future for any behavior, either in simulation or in a real robot, that needs to rotate the hip around the vertical axis.

The agent architecture and the high level behaviors are the same on both simulation and real robot. This allows to use the same agent on both the RoboCup 3D Simulation League and the Standard Platform League.

Future work include the optimization of the adapted behaviors in the real robot, making them more stable and faster.

References

1. Behnke, S.: Online trajectory generation for omnidirectional biped walking. In: Proceedings 2006 IEEE International Conference on Robotics and Automation, ICRA 2006, Orlando, Florida, USA, pp. 1597–1603 (May 2006)
2. Boedecker, J., Asada, M.: Simspark concepts and application in the Robocup 3D Soccer Simulation League. In: Proceedings of SIMPAR-2008 Workshop on The Universe of RoboCup simulators, Venice, Italy, pp. 174–181 (November 2008)
3. Conceição, A., Moreira, A., Costa, P.: Design of a mobile robot for Robocup Middle Size League. In: 6th Latin American Robotics Symposium (LARS 2009), Chile, pp. 1–6 (October 2009)
4. Gouaillier, D., Hugel, V., Blazevic, P., Kilner, C., Monceaux, J., Lafourcade, P., Marnier, B., Serre, J., Maisonnier, B.: The NAO humanoid: a combination of performance and affordability, ArXiv e-prints (July 2008)
5. Gouaillier, D., Hugel, V., Blazevic, P., Kilner, C., Monceaux, J., Lafourcade, P., Marnier, B., Serre, J., Maisonnier, B.: Mechatronic design of NAO humanoid. In: IEEE International Conference on Robotics and Automation, ICRA 2009, Kobe, Japan, pp. 769–774 (May 2009)

6. Graf, C., Härtl, A., Röfer, T., Laue, T.: A robust closed-loop gait for the Standard Platform League humanoid. In: Zhou, C., Pagello, E., Menegatti, E., Behnke, S., Röfer, T. (eds.) Proceedings of the Fourth Workshop on Humanoid Soccer Robots in Conjunction with the 2009 IEEE-RAS International Conference on Humanoid Robots, Paris, France, December 2009, pp. 30–37 (May 2009)
7. Kulk, J.A., Welsh, J.S.: A low power walk for the NAO robot. In: Australasian Conference on Robotics and Automation, ACRA (December 2008)
8. Lau, N., Reis, L.P.: FC Portugal - high-level coordination methodologies in soccer robotics. In: Lima, P. (ed.) Robotic Soccer, pp. 167–192. InTech (December 2007)
9. Liemhetcharat, S., Coltin, B., Meriçli, Ç., Tay, J., Veloso, M.: Cmurfs: Carnegie mellon united robots for soccer (2010)
10. Neves, A.J.R., Azevedo, J.L., Cunha, B., Lau, N., Silva, J., Santos, F., Corrente, G., Martins, D.A., Figueiredo, N., Pereira, A., Almeida, L., Lopes, L.S., Pinho, A.J., Rodrigues, J., Pedreiras, P.: CAMBADA soccer team: from robot architecture to multiagent coordination. In: Papić, V. (ed.) Robot Soccer, pp. 19–45. InTech (January 2010)
11. Picado, H., Gestal, M., Lau, N., Reis, L.P., Tomé, A.M.: Automatic Generation of Biped Walk Behavior Using Genetic Algorithms. In: Cabestany, J., Sandoval, F., Prieto, A., Corchado, J.M. (eds.) IWANN 2009. LNCS, vol. 5517, pp. 805–812. Springer, Heidelberg (2009)
12. Ratter, A., Hengst, B., Hall, B., White, B., Vance, B., Sammut, C., Claridge, D., Nguyen, H., Ashar, J., Pagnucco, M., Robinson, S., Zhu, Y.: rUNSWift team report 2010 Robocup Standard Platform League (October 2010)
13. Reis, L.P., Lau, N.: FC Portugal Team Description: RoboCup 2000 Simulation League Champion. In: Stone, P., Balch, T., Kraetzschmar, G.K. (eds.) RoboCup 2000. LNCS (LNAI), vol. 2019, pp. 29–40. Springer, Heidelberg (2001)
14. Röfer, T., Laue, T., Müller, J., Burchardt, A., Damrose, E., Fabisch, A., Feldpausch, F., Gillmann, K., Graf, C., de Haas, T.J., Härtl, A., Honsel, D., Kastner, P., Kastner, T., Markowsky, B., Mester, M., Peter, J., Riemann, O.J.L., Ring, M., Sauerland, W., Schreck, A., Sieverdingbeck, I., Wenk, F., Worch, J.H.: B-Human team report and code release (October 2010), http://www.b-human.de/file_download/33/bhuman10_coderelease.pdf
15. Shafii, N., Reis, L.P., Lau, N.: Biped Walking Using Coronal and Sagittal Movements Based on Truncated Fourier Series. In: Ruiz-del Solar, J., Chown, E., Plöger, P. (eds.) RoboCup 2010. LNCS, vol. 6556, pp. 324–335. Springer, Heidelberg (2010)
16. Smith, R.: Open dynamics engine (July 2008), http://www.ode.org
17. Tay, A.J.S.B.: Walking Nao Omnidirectional Bipedal Locomotion (August 2009)

Market-Based Dynamic Task Allocation Using Heuristically Accelerated Reinforcement Learning

José Angelo Gurzoni Jr., Flavio Tonidandel, and Reinaldo A.C. Bianchi

Department of Electrical Engineering
Centro Universitário da FEI, São Bernardo do Campo, Brazil
jgurzoni@ieee.org, {flaviot,rbianchi}@fei.edu.br

Abstract. This paper presents a Multi-Robot Task Allocation (MRTA) system, implemented on a RoboCup Small Size League team, where robots participate of auctions for the available roles, such as attacker or defender, and use Heuristically Accelerated Reinforcement Learning to evaluate their aptitude to perform these roles, given the situation of the team, in real-time.

The performance of the task allocation mechanism is evaluated and compared in different implementation variants, and results show that the proposed MRTA system significantly increases the team performance, when compared to pre-programmed team behavior algorithms.

Keywords: Multi-Robot Task Allocation, Reinforcement Learning, RoboCup Robot Soccer.

1 Introduction

Effective cooperation in teams of autonomous robots, operating in highly dynamic environments, poses a significant challenge. The robotic team members have to adapt or change their assigned tasks in real-time, in response to new and possibly unforeseen situations, while ensuring the team's long-term goals progression. Under these circumstances, efficient dynamic task allocation systems are desirable even in small and homogeneous teams, if only to minimize physical interference. However, on larger, often heterogeneous teams, it becomes essential.

Designing these allocation systems, often referred to as Multi-Robot Task Allocation (MRTA) systems, involves dealing with a straight-forward question, but of difficult answer ([8]): "which robot should execute which task?". To answer this question, robots need to perceive their environment, evaluate their aptitudes and communicate with their teammates, to avoid interferences, effort duplication, and deficient task resolution. Such challenges motivated researchers to develop a number of solutions to solve robotic task allocation problems. In-depth surveys of the MRTA field can be found on [13], [8], and [9].

L. Antunes and H.S. Pinto (Eds.): EPIA 2011, LNAI 7026, pp. 365–376, 2011.
© Springer-Verlag Berlin Heidelberg 2011

Achieving fully cooperative behavior into a team of the RoboCup Small Size League (SSL) is specially challenging because, besides being also an adversarial domain, in an SSL game, the number of robots involved is relatively large and these robots are highly dynamical, able to reach speeds above five meters per second. Although many works addressed team play behavior and cooperation in robot soccer, most of them dealt with specific actions or situations, such as passing the ball [11]. Few addressed the creation of fully cooperative team architectures, such as the STP [2], [3]. The paper presents a task allocation system that can contribute towards the creation of a fully cooperative architecture, and that is flexible enough to be attached to an existing strategy system.

The proposed MRTA system uses auctions to allow an autonomous strategy expert, the Coach module, to offer roles for the robots to perform during the game. The robots use a reinforcement learning accelerated by heuristics [1] [4] technique, the Heuristically Accelerated Q(λ) (HAQ(λ)), to learn their fitness for each of the roles, given the present game situation.

The auctioning forms a computationally cheap and efficient method for achieving team behavior, while the reinforcement learning allows the robots to reason by themselves about their functions on the team.

The rest of this paper is organized as follows: Sec. 2 discusses the market-driven methods for task allocation in robotic teams, while Sec. 3 describes the heuristically accelerated reinforcement learning algorithm used. The proposed MRTA system is described in Sec. 4, and experiments and results are shown in Sec. 5. The paper ends with conclusions and future works, in Sec. 6.

2   Market-Based Methods for MRTA

Market-based methods take inspiration from the theory of market economies, where self-interested agents and groups trade goods and services, seeking to maximize their own profits, and while doing that inherently improve their economy as a whole, making it more efficient. These market-based methods are centered on the concept of *utility* functions (sometimes called cost or profit functions), which represent the ability of the agent to measure its own interest in a particular item available for trading. In MRTA systems, utility functions commonly express some measure of the robot's fitness towards performing a certain task, a function of the cost estimated to perform it or a junction of both.

Several market-based approaches were developed for multi-robot coordination, with different characteristics. A good survey on these approaches is provided on [6] and [9].

Many market-based MRTA systems operate, in one way or another, through auction mechanisms. In general, auctions are, scalable, computationally cheap, and have reduced communication requirements. They can be performed centrally, by an auctioneer, or by the robots themselves, in a distributed way. The MURDOCH [7] architecture, for example, uses first-price auctions, where robots submit their bids for the tasks being offered, and the highest value wins. This architecture also allows robots to renegotiate, selling their tasks to other robots, and contracts are time-limited (the auctioneer can reclaim a task after some time), two features that give fault-tolerance capabilities to the architecture.

Another interesting architecture is the TraderBots [5], which employs a fully distributed, fault-tolerant, trading system and applies it into spatial exploration robotic teams. In the TraderBots, a more sophisticated economy is created, where robots seek to accumulate profits on the long-term, and can also subcontract as a mean to take profits for giving guidance to other robots. The MRTA system proposed in this paper takes a simpler approach, primarily because of the different goals in the robot soccer domain. Nevertheless, these architectures are rich references for those who want to understand market-based task allocation systems.

In a work similar to the proposed in this paper, Kose et al. [10] applied an market-based MRTA system with Q(λ) to simulated robot soccer, but in a different abstraction level. In that work, the MRTA system is applied directly to the control of the robot actions, such as to kick or to defend the goal, while in this paper the MRTA system operates on a higher level abstraction. This paper also proposes the use of heuristic functions into the reinforcement learning algorithm.

Finally, one interesting direction, not taken in this work, would be to explore combinatorial auctions and other dynamic programming techniques, as described in [14]. These methods may improve performance, as they do not act similarly to greedy task schedulers like first-price auctions do.

3   Heuristically Accelerated Q(λ)

In a Reinforcement Learning problem, the agent learns through iterations with the environment, by experience, without the need for an environment's model. On each iteration step, the agent senses the current state s, takes an action a, altering the state s, and receives a reinforcement signal r. The agent's goal is to learn a policy π that maximizes its returns. The Q(λ) algorithm is an extension of the popular Q-Learning [21], where reinforcements are not given only for the terminal states, but also to the states recently visited, making the convergence to the optimal policy faster. The Q(λ) update rule is as follows:

$$Q(s,a) \leftarrow Q(s,a) + \alpha \delta e(s,a) \tag{1}$$

$$\delta \leftarrow r + \gamma Q(s',a^*) - Q(s,a) \tag{2}$$

$$e(s,a) \leftarrow \gamma \lambda e(s,a) \tag{3}$$

Where s is the current state; a is the action performed in s; r is the reward received; s' is the new state; γ is the discount factor $(0 < \gamma < 1)$; α is the learning rate; and λ controls the decay in the reinforcement for states farther in the past.

The action selection of the algorithm is:

$$a \leftarrow arg \max_a \left(Q(s',a) \right) \tag{4}$$

A heuristically accelerated reinforcement learning algorithm [1], including the Heuristically Accelerated Q(λ), is a way to solve the RL problem with explicit use of a heuristic function $H : S \times A \rightarrow \Re$ for influencing the choice of actions taken by the learning agent. The heuristic serves as a starting point to the agent, a prior knowledge about the domain that biases the decision towards one of the options. It helps the algorithm to converge faster. The heuristic $H_t(s, a)$ indicates the importance of performing action a when visiting state s, at time t. The use of heuristics does not alter the convergence proofs of the Q(λ) ([21], [17]), because the only change introduced is in the action selection, that is modified from (4) to:

$$a \leftarrow arg \max_a [Q(s', a) + \xi H(s, a)] \tag{5}$$

Heuristic functions are flexible, they can be adapted or modified on-line, as learning evolves and new information becomes available, and either prior domain information or knowledge acquired during initial stages of the learning can be used to define heuristics.

The next section shows the implemented task allocation system.

4 The Implemented Task Allocation System

The proposed MRTA system, shown in Fig. 1, has essentially two parts: (i) the auctioning module, that uses first-price auctions as task allocation mechanism, as shown in Sec. 2, and (ii) the RL module described in Sec. 3, which is present in each robot and uses the HAQ(λ) algorithm to learn the robot's utility functions, or interests, towards bidding for each of the roles offered on the auction. These modules are described in details along this section, after a brief outline of the strategy system where the MRTA is inserted in.

Market-based methods use the sound theory of market equilibrium, from economy. They are computationally efficient, and can be enhanced with machine learning algorithms in a simple manner, in the form of utility functions. However, the efficiency of market-based methods is only as good as these utility functions, used by the agent to reason about its aptitudes and interests. As the creation of high-level reasoning algorithms that could serve as utility functions in robot soccer is a difficult task, due to the complexity and dynamism of the environment, using reinforcement learning as the machine learning algorithm is attractive, as RL can learn by experiences, without environment models. Also, as shown in Sec. 3, the encoding of the domain's knowledge in the form of heuristics allows faster convergence of the RL algorithm.

The proposed MRTA system is not inspired only by market-based methods, though. As many of the more recent task allocation architectures [18], [12], it has characteristics from other MRTA paradigms, such as the use of roles, commonly found in socially inspired MRTA systems.

The remainder of this section briefly describes the strategy module where the task allocation is performed, to help on its understanding, and then presents a description of the actual MRTA system proposed.

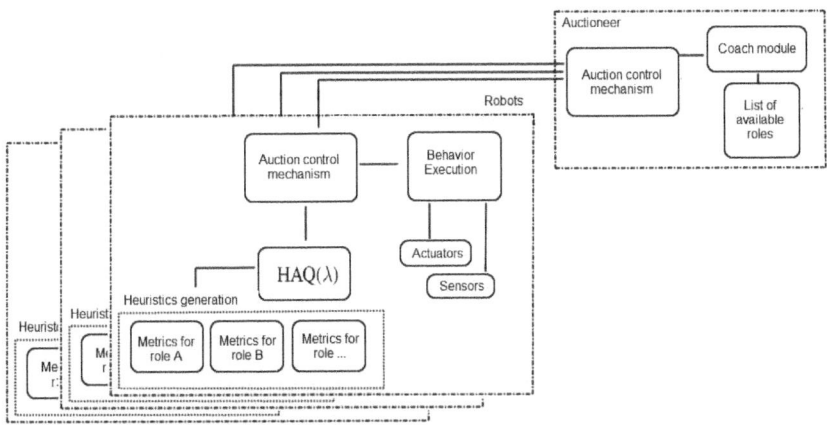

Fig. 1. Block diagram of the implemented MRTA system, showing the modules of the participating robots and the auctioneer

4.1   Strategy System

This MRTA system is implemented in a strategy module formed by three abstraction layers: *primitives, skills* and *roles*. The lowest layer has the *primitives*, that are simple actions like activating the kicking or dribbling devices, or the ball presence sensor. On top of this layer is the *skills* layer, that contains short duration actions which involve the use of one or more primitives and additional computation, such as speed estimation, forecasting of the positions of objects and measurement of the completion of primitive tasks. Passing the ball or aiming and shooting to the goal are examples of skills.

The top layer is the *roles* layer, which are created using combinations of skills and the logic required to coordinate their execution, and are intended to be executed for longer periods. The existent roles are fullback, defender, mid-fielder, striker, forward and attacker.

4.2   Auctioning Module

The Auctioning module is executed by the Coach agent, responsible for analyzing the game situation and defining, according to its reasoning of the conditions, what roles, and in how many instances, will be available for the robots to bid. At a certain stage of the match, for instance, the Coach can decide that the team should be more offensive, and then auction more offensive instances, like Strikers and Attackers, and only one instance of the Defender role. A note: the number of roles offered can be larger than the number of robots, so as to give to the robots more selection choices. The Coach also prioritizes the order in which the roles will be offered, starting by the offensive roles (Attacker, Striker) when the team is attacking and by the defensive roles otherwise.

The first-price auction algorithm works as follows:

1. **Auction announcement.** The coach agent starts an auction, offering the role of highest priority available. A message is sent to the robots, informing about the open auction and the role being offered.
2. **Biddings formulation.** Each robot evaluates its utility function and submits a bidding towards that role.
3. **Auction result.** The coach defines the winner of the auctioned role and sends a message to the robots, informing them.
4. **Repetition.** The process is repeated, without the winner robot and the previously auctioned role, until there are no more robots without tasks.

An important parameter for the auctioning module is the interval between auctions. If the auctioneer could monitor the robot's progress regarding a given task, to decide whether that task should be auctioned again or not, the adjustment of the interval would not be an issue, but, in the implemented MRTA system, the tasks (roles), have no defined duration or terminal states. To overcome this issue, the implemented system expresses the allocation problem as an instantaneous iterated allocation problem, like in [23] and [22], and the adjust of the most suitable interval must be made empirically.

4.3   Reinforcement Learning Algorithm - HAQ(λ)

Each robot on the team runs its own HAQ(λ) algorithm (seen in Sec. 3), which is used to formulate the bidding towards each of the roles being offered by the auctioneer. The robot's experience is not shared with teammates. To create the notion of team work, though, the reinforcements received by all agents are equal, and given only when a goal is scored or suffered.

The design of the sate space for the RL algorithm is key to effective learning. In the role selection abstraction level, the RL algorithm's state space needs to represent aspects of the dynamics of the environment, such as robot speeds and positioning over time, statistical distributions about passing skills, as well as ball positions over time. This concept of capturing attributes with broader temporal significance appears on other works that operate in similar levels of abstraction, like [20] and [16].

The state space implemented has 27 dimensions. It uses features obtained by an algorithm that keeps a histogram of the last 10 x, y positions of the robots and ball on the field, sampled once per second. At each iteration cycle, this algorithm extracts dynamical characteristics of the robots using the histogram data, such as distance traveled and region of the field where the robot stayed the most, recently. The state space also has features measuring the number of recent kicks to the goal from both teams, and the amount of time the ball stayed in the offensive field.

Even using higher level features, the resulting state is still too large and time to convergence would be prohibitive. Also, the robot cannot be expected to experience all possible states, is has to learn with limited experience and having visited only a sparse sample of the state space. Therefore, the Q-value tables must be approximated using some representation with fewer variables, a technique known as *function approximation*. In this work, CMACs with tile coding and hashing, implemented similarly to the proposed in [21], are used for function approximation. The CMAC and tile coding detailed description can be found in details in [15].

For each of the available roles, a set of metrics was created to serve as heuristics for the RL algorithm, using programmer's domain knowledge. Mostly, these metrics were extracted from the hand-coded role selection system that existed previously in the team's strategy software, such as logic to evaluate the opponent's positions in the field and determine if there's need for defending the goal. The advantage of the heuristic functions proposed by [1] is that any function capable of producing a scalar output can be used. Also, if the heuristic is incorrect, as the RL algorithm operates and gains experience this heuristic will be superseded.

Reinforcements are given to all robots on the team when a terminal state, defined as a goal in favor or against, is reached. For goals in favor the value of the scalar reinforcement is +100, and -100 for goals suffered. A small negative reinforcement, -1, is given for every transition that does not reach a terminal state (goal). This small reinforcement serves to prevent the robots from learning to do nothing. The heuristic functions are normalized to one order of magnitude less than the reinforcements, between -10 and +10.

The parameters employed on the HAQ(λ) algorithm are: ϵ-*greedy* algorithm, with $\epsilon = 0.1$, $\gamma = 0.9$, $\lambda = 0.3$, and eligibility trace by substitution.

The next section details the experiments performed with the MRTA system and results obtained.

5   Experiments and Results

To validate the proposed MRTA system, four experiments were performed in simulation. Three partial implementations of the system, to be serve as benchmarks, and the actual MRTA system. One short experiment using real robots was also performed, during the Latin American Robotics Competition 2010.

The experiments, both simulated and with real robots, used two instances of the RoboFEI 2010 software, one using the MRTA system and the other without it. This software is the same used by the RoboFEI team during the RoboCup 2010 in Singapore.

The experiments used a computer with two Intel Xeon Quad Core 2.26GHz processors and 32GB RAM memory. The simulated results were performed using the simulator's time acceleration of 30 times, meaning that 1 second in real time corresponded to 30 seconds in simulation, thus allowing each trial to be performed in around 5 days.

5.1   Simulated Experiments

Four simulated experiments, described ahead, were executed. All of them were executed in 5 independent trials, each containing 20,000 games. The opponent

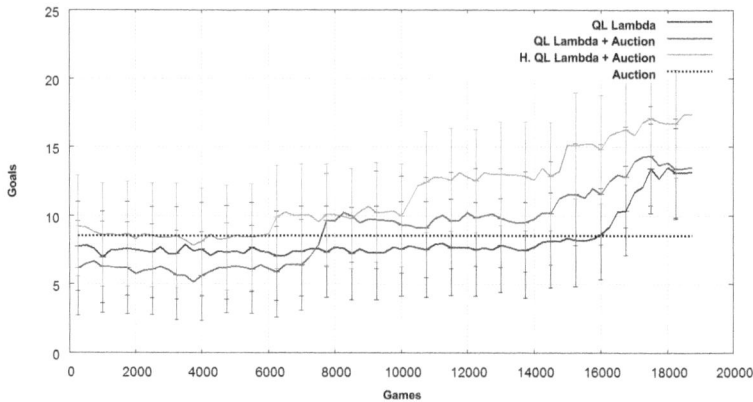

Fig. 2. Comparative of the different MRTA experiment results. The graph shows the average and standard deviation of the goals scored per match, in groups of 10 games, for 5 independent trials. Each trial had 20,000 games, except the Auction-only experiment.

team in all the experiments was an instance of the same software, but using pre-programmed role selection instead of the MRTA algorithm.

Auction-only Experiment - On this experiment, the auctioning module was used and the robots participated without using reinforcement learning. The robot's utility functions towards each role were the result of the metrics developed to be heuristics. The goal was to find the average goals per match at different intervals between auctions. A notice: as this experiment involved no learning algorithm, each of its trial had only 500 games.

Table 1 shows the results, with the average goals and respective standard deviation. These results indicate low performance with 5 seconds interval, mostly because the actions the robots perform are often cut before conclusion by a too short time between auctions. The best result occurs with 15 seconds, and further increasing the interval, to 30 or 45 seconds, results in performance degradation, due to slow response to changes in the game conditions.

Q(λ) Experiment - This experiment consisted in using the Q(λ) algorithm instead of the auctioning module on the Coach agent, thus allowing it to select the roles of all the robots centrally. The Q(λ) was used with the same parameters, $\epsilon = 0.1$, $\gamma = 0.9$ e $\lambda = 0.3$. The interval between iterations of the algorithm

Table 1. Influence of the different interval between auctions (in seconds) on the average goals per match. Table shows average and standard deviation for each case.

Interval (s)	Average goals per match
5	2.06 ± 1.43
15	8.62 ± 3.47
30	6.31 ± 2.92
45	3.35 ± 1.76

was set to 15 seconds or when a terminal state was reached, as in all other experiments. This interval was used for it was the best result of the experiment with different intervals described earlier.

This experiment shows how the original $Q(\lambda)$ algorithm, operating centrally, performs. The result shows that learning occurs, but the time taken for outperforming the hand-coded algorithm is high, as expected.

Auction-$Q(\lambda)$ Experiment - On this experiment the MRTA system is employed with its two modules, the auctioning on the Coach and the RL module on the robots. Only the heuristics were absent from the algorithm, thus leaving all the computation of utility functions for the $Q(\lambda)$ algorithm. The initial performance is low, but after approximately 7,000 games the performance is already above the performance of the hand-coded and RL-only algorithms.

Auction-HAQ(λ) Experiment - This experiment uses the proposed MRTA system in full, with the auctioning module being performed by the Coach and the robots executing the Heuristically Accelerated $Q(\lambda)$ (HAQ(λ)) algorithm. Equation (6) was used to define the value of the function $H(p, s)$, used on (5). This function is simple: the heuristics of all roles p' were calculated for the state s and, for the heuristic with higher value, the result $H(p)$ was used. For all the other roles, the $H(p)$ value was zero.

$$H(p, s) = \begin{cases} H(p, s) & \text{in case } arg \max_{p'}(metric(p', s)) \\ 0 & \text{otherwise} \end{cases} \tag{6}$$

The results of all the experiments are shown in Fig. 2. Observing these results, some conclusions can be drawn: the first is that using the $Q(\lambda)$ only leads to a considerably longer learning curve, what was expected. Another conclusion is that the Auction-$Q(\lambda)$ experiment has an initial average of 6.2 goals per match, inferior to both the $Q(\lambda)$ only and Auction only experiments (8.62 and 7.50, respectively). This happened because the RL algorithm has no initial domain knowledge, so it bids random values on the auctions, what results in deficient allocations. However, after some time, the performance becomes superior to the experiment without RL. On the case of the Auctions-$Q(\lambda)$, after 7,000 games the result is above the Auction only, and after 17,000 games, the average was already above 13 goals, an increase of more than 50% in comparison Auction only experiment.

5.2   Experiments with Real Robots

The experiment with real robots is an attempt to transfer what was learned during simulation, as it is not feasible to execute the number of games for the RL algorithm to converge, using real robots. The experiment consisted of copying the Q-values contained in the CMACs and other learning parameters of the HAQ(λ) algorithm, after the execution of different amounts of simulated games, and then using this simulated memory blocks in real robots. Table 2 shows the results of 3 independent trials of 50 games each, averaged. These results have trends similar to the simulated ones, indicating that experience learned from the simulated environment can be used on the real robots.

Table 2. Result of the copying of RL knowledge gained during simulation to real robots, compared to the performance of the Base team, without any MRTA system.

Algorithm	Sim. games transferred	Average goals (real robots)
Base team (without MRTA)	-	3.12 ± 1.52
Auction-HAQ(λ)	0	3.01 ± 2.34
	5000	5.20 ± 3.11
	10000	7.69 ± 3.65
	15000	7.44 ± 3.04

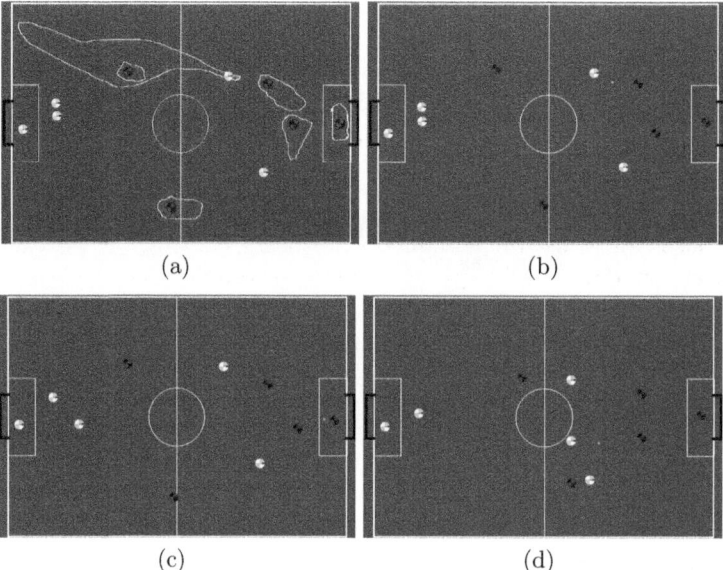

(a)   (b)

(c)   (d)

Fig. 3. Logs of the match (between real robots) on the 2010 competition. The RoboFEI team is shown in yellow and white.

One empirical test was also made during an official RoboCup game in 2010, when the team executed in the real robots the MRTA system with the knowledge gained during 20,000 simulated games. Fig. 3 shows extracted logs of this game, where the RoboFEI team is the yellow. In Fig. 3(a), the delineated areas in white represent where ball and adversaries have been in the last few seconds, according to the histograms described in Sec. 4.3.

A full description of the logs shown in Fig. 3 is the following: (a) the moment an auction occurs and one of the defense robots takes an Attacker role, while its teammate heads to the goal, in possession of the ball. (b) A kick to the goal occurs. (c) Blue team's goalkeeper defends, bouncing the ball back. Meanwhile, the yellow robot who just changed roles heads to the ball from the defensive field. (d) Ball rolls towards the middle of the field, and the mentioned yellow robot approaches the ball. The robot takes the ball and kicks again to the goal.

Although not resulting in a scored goal, the logs show an example situation where the MRTA system gave extra offensive strength to the team. The authors acknowledge this is an evidence rather than a proof. Nevertheless, it is an evidence in line with the results of the experiments performed in simulation and laboratory.

6 Conclusions and Future Work

The results of the experiments indicate that the proposed MRTA system was capable of improving the performance of the team, and that the union of market-based methods and reinforcement learning resulted in superior performance than their separated usage, what can be seen on the Fig. 2 graph.

The results also show that the heuristic acceleration could overcome the initial stage deficiency of RL algorithm, as the heuristics provide the initial domain knowledge the RL lacks. The heuristically accelerated algorithm also demonstrated the best results, another evidence in favor of the use of heuristics in the RL algorithm.

Nevertheless, even with the heuristic acceleration, the reinforcement learning algorithm still has convergence times too high for use directly. However, it is very useful for training the task allocation system off-line, outperforming the hand-coded approaches. This prior training can become particularly powerful if a solution for modeling the opponent's behavior is used. There are works on this regard, although the authors have not researched into it as of the writing of this paper.

The authors also believe that more research into the topic of transfer learning for RL domains [19] could considerably improve the capabilities to apply what was learned in simulation to real robots.

References

1. Bianchi, R.A.C., Ribeiro, C., Costa, A.: Accelerating autonomous learning by using heuristic selection of actions. Journal of Heuristics 14, 135–168 (2008)
2. Browning, B., Bruce, J., Bowling, M., Veloso, M.: STP: Skills, tactics and plays for multi-robot control. IEEE Journal of Control and Systems Engineering 219, 33–52 (2005)
3. Bruce, J., Zickler, S., Licitra, M., Veloso, M.: Cmdragons: Dynamic passing and strategy on a champion robot soccer team. In: Proceedings of the IEEE Int. Conf. on Robotics and Automation (ICRA), Pasadena, CA (2008)
4. Celiberto Jr., L.A., Ribeiro, C.H.C., Costa, A.H.R., Bianchi, R.A.C.: Heuristic Reinforcement Learning Applied to RoboCup Simulation Agents. In: Visser, U., Ribeiro, F., Ohashi, T., Dellaert, F. (eds.) RoboCup 2007: Robot Soccer World Cup XI. LNCS (LNAI), vol. 5001, pp. 220–227. Springer, Heidelberg (2008)
5. Dias, M.B., Zlot, R.M., Zinck, M.B., Gonzalez, J.P., Stentz, A.T.: A versatile implementation of the traderbots approach for multirobot coordination. In: Int. Conf. on Intelligent Autonomous Systems (2004)

6. Dias, M., Zlot, R., Kalra, N., Stentz, A.: Market-based multirobot coordination: A survey and analysis. Proceedings of the IEEE 94(7), 1257–1270 (2006)
7. Gerkey, B., Matarić, M.: Sold!: auction methods for multirobot coordination. IEEE Transactions on Robotics and Automation 18(5), 758–768 (2002)
8. Gerkey, B.P., Matarić, M.J.: Multi-robot task allocation: analyzing the complexity and optimality of key architectures. In: Proceedings of IEEE Int. Conf. on Robotics and Automation, ICRA 2003, vol. 3, pp. 3862–3868 (September 2003)
9. Gerkey, B.P., Matarić, M.J.: A formal analysis and taxonomy of task allocation in multi-robot systems. Int. Journal of Robotics Research 23(9), 939–954 (2004)
10. Kose, H., Tatlidede, U., Mericli, C., Kaplan, K., Akin, H.L.: Q-learning based market-driven multi-agent collaboration in robot soccer. In: Proceedings of the Turkish Symposium on Artificial Intelligence and Neural Networks, pp. 219–2228 (2004)
11. Kyrylov, V.: Balancing Gains, Risks, Costs, and Real-Time Constraints in the Ball Passing Algorithm for the Robotic Soccer. In: Lakemeyer, G., Sklar, E., Sorrenti, D.G., Takahashi, T. (eds.) RoboCup 2006: Robot Soccer World Cup X. LNCS (LNAI), vol. 4434, pp. 304–313. Springer, Heidelberg (2007)
12. Parker, L.E., Tang, F.: Building multirobot coalitions through automated task solution synthesis. Proceedings of the IEEE 94(7), 1289–1305 (2006)
13. Parker, L.E.: Distributed intelligence: Overview of the field and its application in multi-robot systems. Journal of Physical Agents 2(1), 5–14 (2008); special issue on Multi-Robot Systems
14. Sandholm, T., Suri, S.: Improved algorithms for optimal winner determination in combinatorial auctions and generalizations. In: Proceedings of the Seventeenth National Conf. on Artificial Intelligence, pp. 90–97 (2000)
15. Stone, P., Sutton, R.S., Kuhlmann, G.: Reinforcement learning for RoboCup-soccer keepaway. Adaptive Behavior 13(3), 165–188 (2005)
16. Sukthankar, G., Sycara, K.: Robust recognition of physical team behaviors using spatio-temporal models. In: AAMAS 2006: Proceedings of the Fifth Int. Joint Conf. on Autonomous Agents and Multiagent Systems, pp. 638–645. ACM (2006)
17. Sutton, R.S., Barto, A.G.: Reinforcement Learning: An Introduction. MIT Press, Cambridge (1998)
18. Tang, F., Parker, L.E.: A complete methodology for generating multi-robot task solutions using asymtre-d and market-based task allocation. In: 2007 IEEE Int. Conf. on Robotics and Automation, pp. 3351–3358 (April 2007)
19. Taylor, M.E., Stone, P.: Transfer learning for reinforcement learning domains: A survey. Journal of Machine Learning Research 10(1), 1633–1685 (2009)
20. Vail, D., Veloso, M.: Feature selection for activity recognition in multi-robot domains. In: AAAI 2008, Twenty-third Conf. on Artificial Intelligence (2008)
21. Watkins, C.J.C.H.: Learning from Delayed Rewards. Ph.D. thesis, University of Cambridge (1989)
22. Weigel, T., Auerbach, W., Dietl, M., Dümler, B., Gutmann, J.-S., Marko, K., Müller, K., Nebel, B., Szerbakowski, B., Thiel, M.: CS Freiburg: Doing the Right Thing in a Group. In: Stone, P., Balch, T., Kraetzschmar, G.K. (eds.) RoboCup 2000. LNCS (LNAI), vol. 2019, p. 52. Springer, Heidelberg (2001)
23. Werger, B., Mataric, M.J.: Broadcast of local eligibility for multi-target observation. In: 5th Int. Symposium on Distributed Autonomous Robotic Systems (DARS), pp. 347–356 (2000)

Shop Floor Scheduling in a Mobile Robotic Environment

Andry Maykol Pinto, Luís F. Rocha,
António Paulo Moreira, and Paulo G. Costa

INESC Porto – Institute for Systems and Computer Engineering of Porto, and
Department of Electrical and Computer Engineering, University of Porto, Portugal
{amgp,lfr,amoreira,paco}@fe.up.pt
http://www.inescporto.pt/
http://www.fe.up.pt/

Abstract. Nowadays,it is far more common to see mobile robotics working in the industrial sphere due to the mandatory need to achieve a new level of productivity and increase profits by reducing production costs. Management scheduling and task scheduling are crucial for companies that incessantly seek to improve their processes, increase their efficiency, reduce their production time and capitalize on their infrastructure by increasing and improving production.

However, when faced with the constant decrease in production cycles, management algorithms can no longer solely focus on the mere management of the resources available, they must attempt to optimize every interaction between them, to achieve maximum efficiency for each production resource.

In this paper we focus on the presentation of the new competition called Robot@Factory, its environment and its main objectives, paying special attention to the scheduling algorithm developed for this specific case study. The findings from the simulation approach have allowed us to conclude that mobile robotic path planning and the scheduling of the associated tasks represent a complex problem that has a strong impact on the efficiency of the entire production process.

Keywords: Robot Scheduling, Mobile Navigation, Tasks Architecture, AGV (Autonomous Guided Vehicle).

1 Introduction

Efficient planning and supervision of the different resources in a productive system is a highly complex problem, particularly in the current dynamic and competitive environment in which companies work[2]. This environment is strongly affected by changes in the market, and this create a challenge for production management systems to guarantee the delivery of a wide variety of the product, in their operative time.

Both centralized and decentralized approaches have been studied in order in order to meet these requirements.

L. Antunes and H.S. Pinto (Eds.): EPIA 2011, LNAI 7026, pp. 377–391, 2011.

A decentralized structure is presented in [3], where the task scheduling is performed by dividing time into different slots. However, in some cases the possibility of unexpected changes occurring in the production orders is not considered.

André Thomas, in [4] proposed a new structure, called the Master Production Scheduling, that attempts to identify and eliminate bottlenecks. This is not exactly what we aim to achieve in our work.

Caterina Genua et al. [5] has developed a framework to analyze the production flow based on statistical methods for equipment, operators and flow linearity parameters. In this work, the impact of shop floor policies on actual production rates was measured by comparing data collected in the real application with the results retrieved by the simulator.

Tthe information system offered by MES (Manufacturing Execution System) is often fundamental for companies in order to retrieve all of the information related to the production process. In this level MES is responsible for production management and control, the planning of the manufacturing resources and aims to manage and optimize production processes. In another words, this level deals with:

- Information on Production Capacity, which is available for use.
- Information on Product Definition, how to make a product.
- Production Scheduling: what to make and use.
- Production Performance: what was made and used.

To optimize scheduling criteria such as time and the type of resources that are available must be considered. This integration is a difficult task to accomplish. Therefore, the strategy must define local and global rules in order to optimize different performance objectives [9].

This approach focuses on the development of a reactive decentralized architecture for system scheduling, which reduces the complexity of production planning and control, and is able to deal with a dynamic environment, based on a set of local rules.

1.1 Scheduling

The theory of scheduling is characterized by a virtually unlimited number of types of problems. However, in this paper we develop a new heuristic for scheduling, that is based on an automation architecture and optimality criterion. This work focuses on the Robot@Factory scheduling problem as a case study.

Schedule Concepts. It is important to describe the concepts used in this paper further before moving forward.

The Job, J_i, is based on a n-number of operations ($Op_{ij}, j \epsilon \{0, ..., n\}$). Each operation has a processing requirement linked to it, p_{ij}, and a set of machines $M_{ij} \subseteq \{M_i, M_2, ..., M_m\}$. Throughout the environment, m machines are available and the operation Op_{ij} could be processed by any machine belonging to M_{ij}.

When there are no two time intervals overlapping on the same machine/ resource, doing the same job, and if schedule meets the problem-specific characteristics, the schedule is feasible.

Scheduling problems can be specified in terms of a three-field classification system: the machine environment, the job characteristics, and the optimality criterion [1]. This classification scheme was introduced by Graham et al. [6].

The machine environment defines the type of processing machinery that the problem should have. Each job must be processed on a specifically allocated machine, or using parallel machines. This means that each job can be processed on a set of machines (identical, uniform or unrelated).

The job characteristics define attributes such as:

Preemption (or job splitting), this means that the machine processing can be interrupted and can the resume at later time, even on another machine.

Precedence describes relationship between jobs. A precedence relation is common in industrial scheduling problems;

Release dates for the jobs may also be specified, depending on the context of the problem context;

Processing requirements defines the restrictions on the processing times, or on the number of operations for a Job.

To analyze the performance for the feasible solution of the scheduling problem, cost functions $f_i(t)$ are commonly used to measure the cost for completing the job J_i, in time t. Different jobs can be compared with respect to their priorities, in order to obtain the cost for a whole set of jobs.

Problem Description. Following this explanation, the problem can characterized as a scheduling problem without preemption and released dates, but with precedence and processing requirements for the jobs.

Initially the problem may appear simple. However when studied more closely it can be seen that many aspects must be taken into account, and hence, from a mathematical pint of view, there is no optimal solution that could be implemented in a computer, because this type of problem is a NP-hard problem.

One possible method used to solve NP-hard problems is to apply approximation algorithms, called heuristics. Some recent works have addressed the issue of solving scheduling problems using genetic algorithms for solving scheduling problem has also been addressed, and thus reducing computational cos involved in solving these kind of problems [10].

However, we have developed a new heuristic to plan shop floor scheduling, based on a well know navigation architecture.

This architecture is relatively flexible with regard to changes in the number of AGVs transport the product between the machines and the warehouses.

2   Robot@Factory

Robot@Factory is a new competition introduced for the 11th Portuguese Robotics Open and it focuses on the manufacturing management of AGV's (Automated

Guided Vehicles) in a small automated factory. The main goals that are proposed to produce a specific number of parts that differ in type, in operation sequences and quantities, in the least time possible.

On the shop floor level, the factory consists of an input warehouse, with a maximum capacity of five parts, where the parts that have to be produced are located, one output warehouse with a maximum capacity of five parts, where the parts that are produced have to be stored, eight production machines where the parts have to be processed, according to their operation sequence, before being transported to the output warehouse, and one or more AGV's responsible for transporting the parts between each of the resources previously described.

In the competition format, there are three different types of parts, each one with different operation sequences in the machines. It is important to note that each machine can only process one part at a time, and it has different operation times according to the type of part being processed.

To help the participants prepare for the competition, an oficial Robot@Factory simulator, figure 1 was available that faithfully represents the real environment that the participants will confront in the conference.

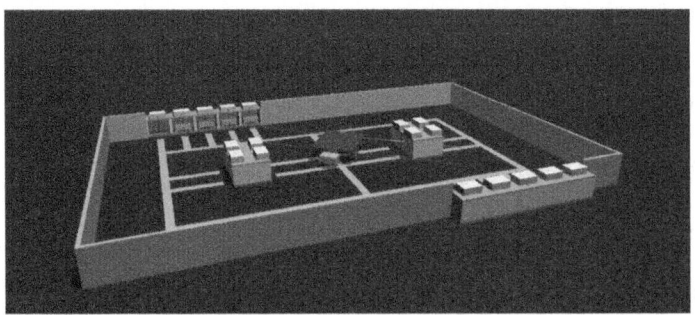

Fig. 1. Official Simulation Environment for the Robot@Factory competition

In summary, the competition presents some interesting challenges in the fields of navigation, control, cooperation and task scheduling for AGV's and managing the availability of the machines and warehouses. In another words it simulates the entire Manufacturing Execution System (MES).

Therefore, this paper will only focus on the path planning and scheduling of the task for the vehicles which will be tested later using the previously mentioned simulator.

3   System Architecture

One of the first steps was to develop a modular architecture that would allow us to easily increment or decrement the number of vehicles used. This modular structure function well when a quick adjustment on production capacity and functionality is required.

Each AGV has an architecture for job scheduling that makes it possible to plan and execute the most appropriate job (or task), in an autonomous manner. The architecture developed, figure 2, makes the AGV smarter and less dependent on centralized control. The advantage of decentralized control, is that it increases reactivity on the "shop floor", making the process more dynamic and minimizing the probability of having to stop the entire production process due to a problem in the centralized scheduler.

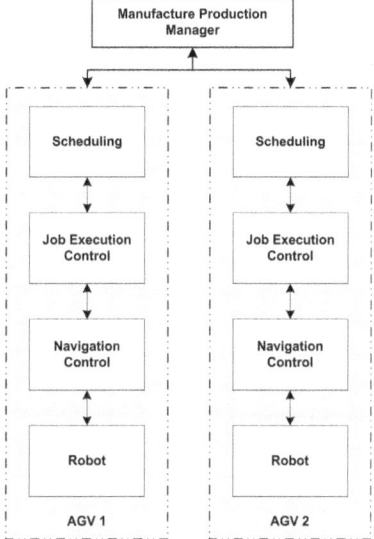

Fig. 2. System architecture

Consequently, each AGV has built-in scheduling that will create and allocate the job and the respective path for the corresponding AGV based on the production order that still needs to be executed and the resources available. In addiction to job planning, in this architecture there are a set of subsystems that enable the robot to perform the jobs that were assigned to it properly, such as, environment navigation without losing it self or colliding with obstacles. However, AGV navigation control does not come within the scope of this article, thus it will not be presented in detail.

This paper will focus on the management modules, job planning and execution. One single AGV will be considered to streamline the heuristic planning algorithm, although a larger number of robots could be used simultaneously.

When the AGV is task free, it analyzes the specifications of the production order, and based on the environment characteristics, it will attempt to create a job to perform. If the job can be created, the system's job execution control and navigation ensures that the robot carries out the necessary actions properly to successfully complete the task. After the robot has completed the entire allocated task, it will request a new task from the scheduling manager.

In the next subsections some of the architecture system levels will be discussed in more detail.

3.1   Manufacture Production Manager

The Manufacture Production Manager is responsible for the overall management of production, ensuring that each AGV that is working in the environment has updated its information concerning the execution state of the production order. It also provides information about the parts that are currently being produced and those that are already ready to be sent to the output warehouse.

It makes it possible to dynamically control production using different AGVs and also illustrate the production status for management applications.

3.2   Scheduling

The scheduling that has been developed is the upper layer algorithm. It can be divided into two levels: the Job scheduler and Path planning, figure 3.

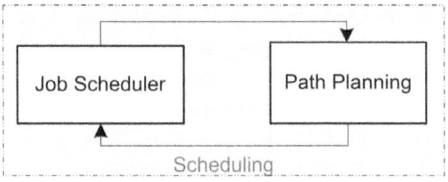

Fig. 3. Scheduling level

The Job Scheduler is responsible for transforming the sequences of complex operations into simple sequences, by using the information on the product definition, and at the same time managing the resources available (robots, machinery and warehouses).

The Path Planner is responsible for defining the whole path for the AGV, for the execution of the task.

Eventually each robot will have a task allocated to it,according to an optimization scheme that takes the type of task, the available machinery, the original position of the part, and the path to the location of the part and from there to the machine or the output warehouses all into account.

Path Planner. One of the major initial problems is related to the AGV's path planning when performing a determined task. With just one AGV this problem is simpler and the only concern is getting to the destination as quickly as possible. With the increase in the number of vehicles, and considering the dimensions of the track of the Robot@Factory competition, we can only consider a maximum of two AGVs, and cooperation and collision avoidance strategies must be taken into account.

A graph as been developed for just one AGV showing the mapping of the factory shop floor end the routes that the AGV can use during the transportation of the parts between different stations, see figure 4 (all the routes are bi-directional). The nodes SA1 to SA4 and SB1 to SB4 represent the entrance nodes of the machines, while nodes WI1 to WI5 and WO1 to WO5 represent the entrance of the input and output warehouse respectively.

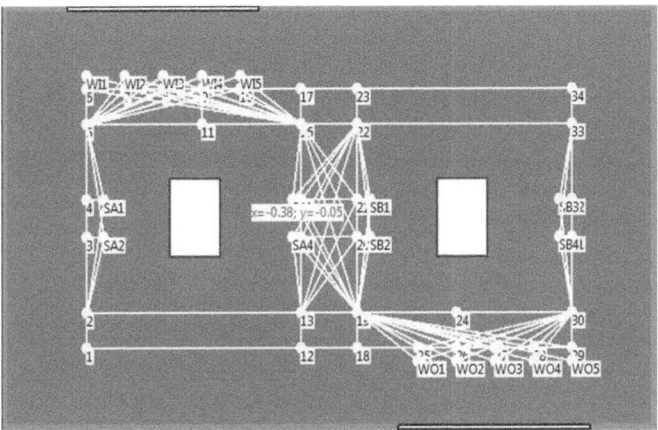

Fig. 4. Shop Floor Graph representing the routes available(white lines represent the possible AGV routes and white dots represent the nodes of the graph)

Having defined the routes available, an algorithm is now required that will compute the optimal route from the initial position to the final position (from one node to another node of the graph) for each attributed task.

To accomplish this objective we used the A* (AStar) algorithm, that has already proven its value in this field [7,8]. This algorithm is different from the others due to the fact that it uses a heuristic function, usually denominated F(n), which measures the proximity to the final desired position. This function F(n) is composed of the sum of two other functions g(n) and h(n).

$$F(n) = g(n) + h(n) \tag{1}$$

A more detailed description of each function is shown bellow.

- g(n) - This function is deterministic, and indicates the cost of traveling from the initial node to the current node being searched.
- h(n) - This function returns an estimate of the cost of traveling from the actual searched node to the objective node, in this case this means the euclidean distance between the actual node and the final node. This feature make it possible to embed knowledge about the domain of the problem when searching for solutions.

– F(n) - This function, as previously mentioned, represents the sum of the function g(n) and h(n), thus indicating the minimum estimated cost of traveling from the initial node to the terminal node. The A * algorithm begins its search with the starting node, keeping a line open with the neighbor of the current node. The smaller F(n) is the highest priority of the neighbor.

Job Scheduler. It is important to note that, with refard to the AGV, the problem could be summarized as collecting and placing the parts. Following a careful analysis of the scheduling problem characterized above, we can define three different types of jobs:

ToDoJobs. Picking up the parts in the input warehouse or from a machine and transporting them to another machine.

OutJobs. Picking up the parts in the input warehouse or from a machine and transporting them to the output warehouse for further expedition.

ProcessingJobs. This depicts the processing of the parts by the production machines. This is not executed directly by the AGV, but information on the parts that are currently being processed is very important for the autonomy of the robot.

Our algorithm is based on three lists of Jobs that are currently being executed. Whenever a new task is created, it is added to one of these lists (a proper one). In fact, using this logic we can easily synchronize the Jobs between all of the AGVs, making it possible to update the information on the resources available (communicating jobs that are created and executed, each AGV can "'know"' what the others are doing).

Initially the characteristics of the "'shop floor"', such as the production type, duration and location of each machine, the initial position and types of the parts are all imported.

The status of the production order is constantly being updated, along with the availability of the resources (machines, warehouses and parts). This periodic updating is essential to avoid deadlocks on shared resources(when two jobs are performed by the same machine during the same period of time, by different AGVs).

One job is fully defined by the initial path between the robot and the part (that could be in the input warehouse or in a machine), and the final path between the part and the destination for that part (in the output warehouse or in a machine). Each job has an identity, such as the machine, warehouse, and the part that is transported or being transported or processed by the machine, making it possible to the reserve resources.

The structure for the job planning can be seen in figure 5.

If the robot is waiting for a job and full production has not been reached, their scheduling algorithm will try to create a new job and the type depends on decision criteria. The AGV aims to transport the part to the exit warehouse, immediately, whenever one specific type of part is available on the shop floor and production of this type has not yet reached the desired value. The reason for this criterion is to allow other AGVs to reuse the machine as soon as possible.

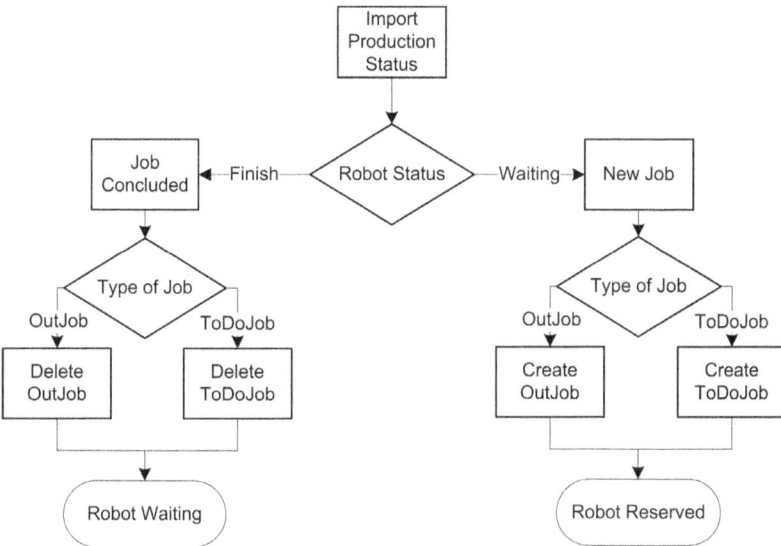

Fig. 5. Job Scheduler structure

Assuming that there are many parts that need to be dispatched, the algorithm identifies the one with the lower cost for the AGV. This cost is calculated in terms of the difficulty of performing the transportation task, using the path planning algorithm (Astar) based on the graph that was developed for the map. This means that the Job path for all parts is determined and compared, in order to find one with lower operation costs.

If the robot does not have a part that is ready for dispatch, it cannot create an OutJob. In this cases, the scheduling analyzes the production needs, creating a new ToDoJob. In order to evaluate if a ToDoJob is necessary, the number of parts already produced, plus the ones being produced, plus the parts available in the shop floor must be less than the number of parts required and this must be analyze for each type of part. If there are n different types of parts that are needed, or different feasible combinations of machines for the same type of part, then the algorithm analyzes the part-machine relationship with the lowest cost function. This cost is calculated differently to the OutJob, because in has added transportation and operating time machine costs. The ToDoJob should then be a minimization of the trajectory that will be followed by the robot along with the time taken by the machine to process the part.

Once the AGV begins to perform the Job, it is converted into ProcessingJob, and its execution is dependent on the processing time. At the end of this job, the part is available for dispatch or to be processes by another machine. It should be emphasized that the machine operating times, in the current terms, are considered deterministic.

Finally, assuming that the robot concludes a Job, it must be eliminated from the corresponding list, sending this information to the manufacturing production manager, making it possible for the other AGVs, to update their lists.

3.3   Job Execution Control

After one job (ToDoJob or OutJob) is created, the AGV will perform all of the necessary procedures to control the tasks execution. Figure 6 represents a simple sequence of tasks that are described by the specific job. Further details about the behavior of this system do not come within the scope of this paper. However must be noted that the process of collecting the part is something that needs to be very robust and this approach is based on sensor information (infra-red sensors).

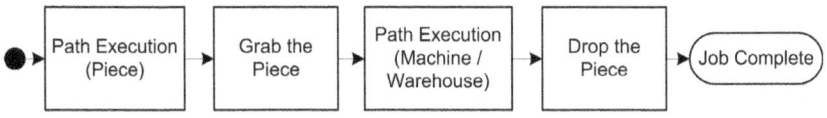

Fig. 6. Job Execution Control

3.4   Navigation Control and the Robot

A differential robot was used to validate this heuristic scheduling algorithm and it was modeled in a 3D simulation program called SimTwo that can be seen in 1. It was also provided with a navigation control system that ensures that the robot moves without colliding with other obstacles, and without the problem of localization inside the shop floor environment. These aspects will not be specified in more detail in this paper.

4   Results

As explained at the beginning of this paper, three types of parts are considered for production for the Robot@Factory competition. Therefore, having taken the production orders into account, the parts have to be transferred from the input warehouse to each machine or directly to the output (if the type of part matches the production order). This production process must take into consideration the specificity of each machine presented in Table 1. In this Table A1 to A4 represents the first group of machines and B1 to B4 represents the second group.

The next subsection outlies an example of the following production order: 2 parts of type A and 2 parts of type B (it is possible to see a video about this experience[1]) to demonstrate the results of this architecture, Scheduling, Path

---

[1] Video about the production scheduling,
http://feupload.fe.up.pt/get/XJ2OcPQ28gPLKpU

Table 1. Machines Criteria and Production Times

Name	From Type	To Type	Process Time(sec.)
A1	A	B	30
A2	B	A	50
A3	B	C	5
A4	B	C	5
B1	A	B	10
B2	B	C	0.5
B3	C	B	50
B4	A	B	500

Planning and Navigation Control. It must be noted that five parts are already in the input warehouse (all parts available are type A), and because of this, the production of the type A part just consists of transporting the part between the two warehouses. On the contrary, for the production of type B, type A parts need to be processed by a specific machine (see table 1).

4.1 Production Example: 2-A, 2-B

With this production order, the scheduling algorithm will give priority to the jobs that only require the transportation of parts directly to the output, these are all of the OutJobs.

As shown in Figure 7, the robot begins its movement in the lower-left-hand corner of the image and selects the best type A part (the one with the lowest transportation cost in the all path). It then moves to the entrance of the input warehouse, collects the part and transports it to the output.

Figures 8 and 9 show the path that was taken by the robot to transport the second type A part. The first figure illustrate the movement of the robot from the output to the input warehouse, collecting another type A part, and the second illustrates how it is transported to the output warehouse.

After producing the first two parts (type A), the scheduling algorithm will create ToDoJobs in order to select a part and a machine combination with the lowest results of production costs. Figures 10 and 11 show the robot moving from its current position to the input warehouse, collecting the respective type A part, and moving it to the machine, in order to create type B parts.

After the type A parts have been transformed into type B parts by the respective machine, the robot will collect them and transport them to the output warehouse, concluding the production orders (figure 12 and 13).

This is a simple example of the results that were obtained using our scheme. Although the production of type C parts was not illustrated, since its production is very similar to the type B part but with more ToDoJobs, we simulated its production successfully in the Robot@Factory 2011 competition.

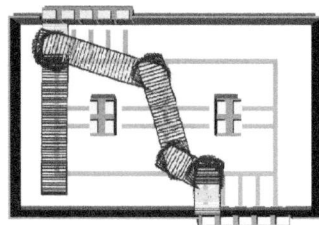

Fig. 7. Production of the first type A part - transportation from the input to the output warehouses

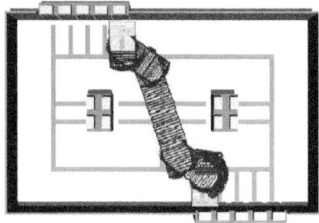

Fig. 8. Production of the second type A part - movement of the robot from the output to the input warehouses

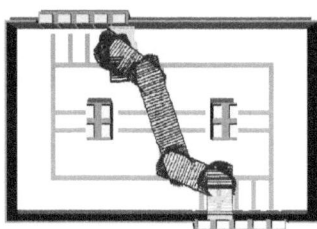

Fig. 9. Production of the second type A part - transportation from the input to the output warehouses

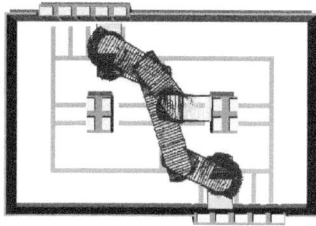

Fig. 10. Processing of the first type B part

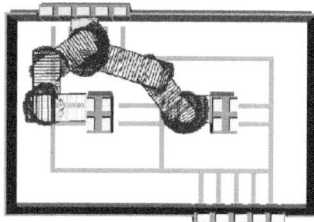

Fig. 11. Processing of the second type B part

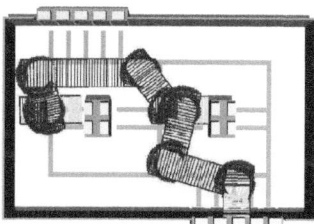

Fig. 12. Transportation of the first type B part to the output

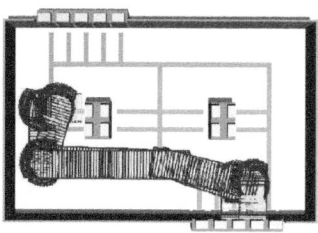

Fig. 13. Transportation of the second type B part to the output

5   Conclusions

The first edition of the Robot@Factory, presented in the 2011 Robotic meeting in Lisbon, was a success. This competition tries to portray the problem of scheduling several automated resources in an industrial environment. The scheduling has several important constraints that have a major impact on production efficiency, and consequently on the enterprise's profits. Although these problems are carefully studied, the approaches found in the literature do not cover the environment presented in the competition, which was characterized by challenges without valid and well defined mathematical model.

For that competition we developed a new architecture that makes it possible to transport the parts between the warehouses and machines using AGVs. A decentralized scheduling algorithm, a job controller, a path planner, and a navigation control system were implemented in order to improve the performance in terms of time spent on producing the entire production order.

With our work we took a new approach to scheduling the resources available on the shop floor, for multiple AGVs. The allocation of tasks is properly optimized, under the most relevant aspects such as transportation and processing time. Based on these, we defined a cost function that is used to define which tasks are the most appropriate to be executed in each moment. Beyond this function, our scheduling architecture is quite flexible, making on-line task planning possible, which is critical when AGV are interacting.

This architecture enables each AGV to identify the necessary procedures, creating jobs that improve the state of the producing order. There are two different types of jobs according to the final destination that is desired: if the final location is one machine, then it is necessary to create a ToDoJob; if the final location is the output warehouse, then a OutJob is created. The jobs are created based on a set of optimization criteria, which allow the Job Scheduler to compare different Jobs, choosing the one with the lowest cost function.

In this paper we demonstrate the result of one production order, and by analyzing all of the trajectories, it is possible to see that the best part and machine with respect to the actual position of the AGV is selected. The job execution controller and the navigation system enabled the AGV to reach all of the desired locations in order to successfully complete the desired job.

With these decentralized structure, there is no overlapping in the usage of the resources, avoiding deadlocks between shared ones. Finally it can be said, that with this innovative scheduling approach we can contribute to expanding the fields covered in the Portuguese robotics open.

6   Future Work

Considering the fact that our scheduling architecture already allows for the incorporation of more than one AGV, we need to upgrade our path planning algorithm with the aim of trying to avoid collisions with the vehicles. Our main idea is to simplify the graph of the routes optimized for one vehicle and reserve a route for each vehicle, meaning that the routes cannot be the same. In addition, we will define the priorities between them taking the task that they are assigned into consideration. This needs to be considered for the cases of routes that intersect with others, where a possible collision may occur. For this we need to enable communication between them, so each robot knows the task and position of the other robots.

We will perform tests with the updated approach in a real environment in order to compare the expected (simulation) and the achieved (real) system behavior.

References

1. Brucker, P.: Scheduling algorithms, 5th edn. Springer, Heidelberg (2007)
2. Benito Martin, J.J., Angulo, P.S.: VSFC - virtual shop floor control: Approach to the problem of programming and control of production through intelligent agents and Breeding Environments. In: 5th Iberian Conference on Information Systems and Technologies (CISTI), pp. 1–6. IEEE Press (2010)
3. Spath, D., Barrho, T., Klinkel, S.: Flexible shop floor control in a decentralized production environment. In: IEEE Conference on Systems, Man, and Cybernetics, vol. 5, pp. 3291–3296. IEEE Press (2001)
4. Thomas, A.: Model reducing method for rescheduling manufacturing shop floor. In: Proceedings of the 35th Southeastern System Theory, pp. 186–190. IEEE Press (2003)
5. Genua, C., Giuffrida, S., Rinaudo, S.: Gantt charts for production flow: a performance analysis framework for shop floor control effectiveness evaluation and monitoring. In: 10th IEEE Conference on Emerging Technologies and Factory Automation, pp. 921–928. IEEE Press (2005)
6. Graham, R.L., Lawler, E.L., Lenstra, J.K., Rinnooy Kan, A.H.G.: Optimization and approximation in deterministic sequencing and scheduling: A survey. Annals of Discrete Mathematics 5, 287–326 (1979)
7. Nascimento, T., Conceição, A.G.S., Moreira, A.P.: A Modified A* Application to a Highly Dynamic Unstructured Environment. In: 11th Portuguese Robotics Open, Lisbon (2011)
8. Khantanapoka, K., Chinnasarn, K.: Pathfinding of 2D & 3D game real-time strategy with depth direction A* algorithm for multi-layer. In: Eighth International Symposium on Natural Language Processing, SNLP 2009, October 20–22, pp. 184–188 (2009)
9. Pu, P., Hughes, J.: Integrating AGV schedules in a scheduling system for a flexible manufacturing environment. In: Proceedings IEEE International Conference on Robotics and Automation, pp. 31-49 (1994)
10. Morandin Jr., O., Deriz, A.C., Sanches, D.S., Kato, E.R.R.: A Search Method using Genetic Algorithm for Production Reactive Scheduling Manufacturing Systems. In: IEEE International Symposium on Industrial Electronics, pp. 1843–1848 (2008)

Humanized Robot Dancing:
Humanoid Motion Retargeting Based in a
Metrical Representation of Human Dance Styles

Paulo Sousa[1,3], João L. Oliveira[1,2,3], Luis Paulo Reis[1,3], and Fabien Gouyon[2]

[1] DEI/FEUP - Informatics Engineering Dep.,
Faculty of Engineering of the University of Porto, Portugal
[2] INESC Porto - Systems and Computers Engineering National Inst., Porto, Portugal
[3] LIACC - Artificial Intelligence and Computer Science Lab.,
Univ. of Porto, Portugal
{ei06047,joao.lobato.oliveira,lpreis}@fe.up.pt, fgouyon@ienscporto.pt

Abstract. Expressiveness and naturalness in robotic motions and behaviors can be replicated with the usage of captured human movements.
Considering dance as a complex and expressive type of motion, in this paper we propose a method for generating humanoid dance motions transferred from human motion capture (MoCap) data. Motion data of samba
dance was synchronized to samba music, manually annotated by experts,
in order to build a spatiotemporal representation of the dance movement
with variability, in relation to the respective musical temporal structure
(musical meter). This enabled the determination and generation of variable dance key-poses according to the captured human body model. In
order to retarget these key-poses from the original human model into
the considered humanoid morphology, we propose methods for resizing
and adapting the original trajectories to the robot joints, overcoming its
varied kinematic constraints. Finally, a method for generating the angles for each robot joint is presented, enabling the reproduction of the
desired poses in a simulated humanoid robot NAO. The achieved results
validated our approach, suggesting that our method can generate poses
from motion capture and reproduce them on a humanoid robot with a
good degree of similarity.

Keywords: Humanoid Robot Motion Generation, Motion Retargeting,
Robot Dance.

1 Introduction

Robotics applications grow daily, and the creation of realistic motion for humanoid robots increasingly plays a key role. Since motion can be regarded as
a form of interaction and expression, that allows to enrich communication and
interaction, improving humanoid robot motion expressiveness and realism, as a
form to accomplish better and richer human-robot interaction. A form of achieving more humanized robotic motion is to feasibly reproduce, and imitate, the
motions performed by humans. This would allow not only more expressiveness,

L. Antunes and H.S. Pinto (Eds.): EPIA 2011, LNAI 7026, pp. 392–406, 2011.
© Springer-Verlag Berlin Heidelberg 2011

diversity and realism in the humanoid robot motion, but also a simple, less time-consuming and automatic form of creating and converting diverse human motion to robotics. Considering dance as a rich and expressive type of motion, constituting a form of non-verbal communication in social interactions, also transmitting emotion, it imposes a good case study of clear humanized motion.

This paper presents methods for generating humanoid robot dancing movement from human motion capture data. The presented methods aim to generate robot dance motion with a good degree of similarity and musical expressiveness according to the original dance. All the described processes are built upon [11]'s method of dance motion analysis. This method was applied to motion data of samba dance style synchronized to samba music (manually annotated by experts), for building a spatiotemporal representation of the original dance movement, with variability, in relation to the respective music temporal structure (musical meter), allowing the determination of the fundamental key-poses of the dance style. Using this representation as starting point, we present methods for resizing the body segments and retargeting the joint trajectories towards different humanoid body morphologies. As such, we firstly synthesize stochastic variations of the determined key-poses, then resize these according to the targeted segment lengths and finally process the necessary adjustments to retarget the generated joint trajectories onto the considered humanoid morphology. The method was tested on a humanoid robot NAO [2], simulated on the SimSpark simulation environment [1] and using the FC Portugal agent [13] [5], in order to generate and reproduce the original samba dance.

The remainder of this paper is structured as follows: Section 2 presents a review of related work on human dance motion analysis and humanoid dance motion generation. Section 3 presents the proposed methods and the developed work. In Section 4, the main results are presented and discussed and a evaluation of the similarity between the original human motion and the generated humanoid motion is performed. Finally, in Section 5 the conclusion and future work are presented.

2   Related Work

Nowadays many attempts have been made to achieve realistic humanoid motion based on human motion. The techniques referred below use dance motion capture data and synthesize new motion from it. The first step in this process typically consists on determining the most important key-poses from the motion capture data. This choice impacts the overall aspect of the final motion, and must be accurate at determining the key-poses that best represent the original dance style. From these key-poses, the motion is then transfered to the robot, by trying to achieve the greater similarity possible with the original motion capture data.

2.1   Motion Analysis

One of the traditional methods to generate dance motion in computer animation and in robotics, is to interpolate transition motion between defined key-poses.

As such, key-poses most show representative instances of the motion. The techniques to analyze and determine the appropriate key-poses, in dance motion, are based on a spatial analysis of the performed body motion [6] [4] or by analyzing the dance music [11], and, in some cases, a simultaneous analysis of both aspects [18] [19] [17] [10]. Working on Latin dances [6], uses information about the main characteristics of Merengue dance style, focusing only on components of the rotation of shoulders and hips. By analyzing Japanese folk dance, [8] and [7] segmented the dance MoCap data in key-poses, in terms of minimum velocity of the end-effectors' (hands and feet). Then these key-poses were clustered and interpolated to generate the original dance. [4] extracts rapid directional changes in motion.

On the other hand, a combination of music and motion analysis is applied in [18] [19] [17] [10]. [18]. These authors identify key-points of the motion rhythm, as local minimums of the 'weight effort' (linear sum of rotation of each body joint), which indicate stop-motions and are recognized as key-poses; and motion intensity points as the average of instant motion from the previous key-pose. Their music analysis focus on the music intensity (sound chunk whose spectral power is strongest between the neighboring frequency sounds) and music rhythm, that is found by analyzing the repetition of several phrases and patterns presented in the music structure. In [19], stop frames of the hand motion are considered as key-frames, and motion intensity is determined as the difference in the velocity of the hands between frames. The performed musical analysis is similar to [18], extracting the music beat and degree of chord change for the beat structure analysis, and music intensity for mood analysis. There are also works based on temporal scaling techniques, for upper-body [17] and leg [10] motion. In [17] the dance motion is captured at different speeds and by comparison of the variance in the motion, the authors observed that some poses stay preserved. These poses are considered key-poses since they tend to represent important moments to the music. The analysis in [10] is similar to [17], but focusing on the analysis of the step motion. The determination of the key-poses is made by using the indication that the original timings for tasks around key-poses are maintained and that stride length is also close to the original, even at different speeds. In [11] a spatiotemporal dance analysis model is presented, based on the Topological Gesture Analysis (TGA) [9], that conveys a discrete point-cloud representation of the dance. This method describes the space that the dancer occupies at each musical class (1 beat, half-beat, 2 beats, etc) in terms of point-clouds, and generates a spatiotemporal representation of the occupied positions, by projecting musical cues onto spatial trajectories. We followed [11]'s method of dance motion analysis since it conveys a parameterizable representation of the original dance, incorporating its intrinsic variability and musical rhythmic qualities.

2.2   Dance Motion Generation

Dance motion generation techniques typically aim at generating motion from the key-poses extracted in a prior motion analysis phase. [14], [15], [3], [10] and [8] apply Inverse Kinematics (IK) to transform the marker positions from the motion capture data into robot joint angles. In [3], the Inverse Kinematics is only applied to the upper-body, while the pelvis, leg and feet motion is generated by optimization based on the Zero Motion Point (ZMP) trajectory and dynamic mapping. [10] checks the intervals between steps to keep a stable ZMP and then also applies Inverse Kinematics to map the leg joint positions to robot joint angles. [14] applies optimization to ensure the physical restrictions of the robot, ensuring that the angle, velocity and acceleration limits are met. [15] applies sequential motion restrictions by optimization by firstly limiting the joint angles, then solving self-collision avoidance and, finally, overcoming velocity and dynamic force constraints. For assuring self-collision avoidance the authors increased the critical distances for the periods that present collisions [20], by applying kinematics mapping to translate the motion capture data to the humanoid, using similarity functions based on the value of the angles of the original data. For improving the balance, an algorithm based on the number of feet in contact with the ground is used, modifying the hip trajectory to satisfy a constraint based on a ZMP criterion. [8] also limits angles and the angular velocity and finally modifies the ZMP trajectory in order to keep balance. [4] and [18] use motion graphs to represent the motion. [4] aimed to generate motion on-the-fly, using a library of motion graphs, matching then the motion represented by the graph with the input sound signal. On its hand, [18] traces the motion graph based on the correlation between the music and motion features. Finally, [18] determines the better graph path by choosing the highest value from the correlation between the music and motion intensity and a correlation between music beats and motion key-frames. In terms of motion retargeting, [4] uses a real-time algorithm to adapt the motion to the target character. [17] applies optimization to overcome the joint angles limitations of the target character. [16] presents a way to extract the joint angles from the three-dimensional point representation of a pose. This method can be applied not only for computer animation but also for robotics. After generating the motion, [14] applies a phase of motion refinement to detect trajectory errors and correct them. On its hand, [10] makes a final refinement of the generated movement in order to keep the robot's balance and avoid self-collisions. To generate the robot joint angles from the pose point representation obtained from the previous dance movement representation [11] we based our approach in [16], for extracting Euler angles in the 3 dimensions based on a body-centered axis system.

3   Methodology

3.1   Dance Movement Analysis

Our motion analysis stage is based on the approach presented in [11]. As such, we recurred to the same dance sequences of Afro-Brazilian samba, which were captured with a MoCap system, and synchronized to the same genre of samba music (manually annotated by experts). Upon these, we also applied the TGA

(Topological Gesture Analysis) method [9] for building a spatiotemporal representation of the original dance movement in relation to the respective music temporal structure (musical meter). This method relies in the projection of musical metric classes onto the motion joint trajectories, and generates a spherical distribution of the three-dimensional space occupied by each body joint according to every represented metric class. In such way, this representation model offers a parameterizable spatiotemporal description of the original dance, which translates both musical qualities and variability of the considered movement.

3.2   Dance Movement Generation

The actual dance movement generation is based on three steps: (a) key-poses synthesis from the given representation model, (b) morphological adaption of the key-poses to the used body model, in terms of segments length and number of joints, and (c) the actual key-poses' retargeting from the used character to the simulated robot NAO.

Key-Poses Synthesis. By following [12], the synthesis of key-poses consisted on calculating a set of full-body joint positions (one for each considered metric class). In order to translate the variability imposed in the original dance, for every key-pose the joint positions were calculated by randomly choosing rotations circumscribed by every joints' TGA distributions without violating the fixed geometry of the human body.

As described in [12], each key-pose is split into 5 kinematic chains. From the anchor to the extremity of each kinematic chain, each joint position p_j^m is calculated based on a random rotation circumscribed by the possible variations of its rotation quaternion $qv_{s_j^m}$ (*i.e.* the 3d rotation of a target unity vector $\vec{v}'_{s_j^m}$ around its base unity vector $\vec{v}_{s_j^m}$) between every two body segments:

$$p_j^m = p_{j-1}^m + l_{j-1,j} * \vec{v}'_{s_j^m} : p_j^m \in T_j^m, \tag{1}$$

where m is the considered musical metrical class, p_j^m is the current joint position under calculation, p_{j-1}^m is the former calculated joint position, $l_{j-1,j}$ is the current segment length (linking p_{j-1}^m to p_j^m), and T_j^m is the current TGA spherical distribution.

Morphological Adaption. In order to get a representation of the key-poses in the target humanoid morphology the prior spatiotemporal representation of the dance must be adapted, maintaining the spatial relationship and expression of the poses across all the represented metrical classes. To achieve this, we must look at the target morphology in terms of segment lengths, number of joints, joints' degree of freedom, and other target physical kinematic constraints.

(a) Different Segment Lengths: Prior to the actual humanoid key-poses' generation, the segment lengths of each body part length must be changed to those of the target body model (figure 1). For each joint j, $l_{j-1,j}$ is the length of the segment that connects $j - 1$ to j, and D_j^m is the spherical distribution, with radius r_j^m and center o_j^m. The distance from o_{j-1}^m to o_j^m is considered as $d_{j-1,j}^m$ and direction vector from o_{j-1}^m to o_j^m is $\vec{vo}_{j-1,j}^m$. In order to change the segment length from $l_{j-1,j}$ to $l'_{j-1,j}$ we proceeded as follows:

$$
\begin{cases}
redim = l'_{j-1,j}/l_{j-1,j} \\
d'^m_{j-1,j} = d^m_{j-1,j} * redim \\
r'^m_j = r'^m_j * redim \\
o'^m_j = o^m_{j-1} + d'^m_{j-1,j} * \vec{vo}^m_{j-1,j}
\end{cases}
\quad , \qquad (2)
$$

where $bd'^m_{j-1,j}$ is the new distance from o_{j-1}^m to o_j^m, and r'^m_j the adapted radius of the spherical distribution D_j^m, with o'^m_j as its new center point. The translation that was applied from o_j^m to o'^m_j is then applied to all the following joint centers in the considered kinematic chain (noted by $trans$ in figure 1). This method allows to resize any body segment by manipulating the spherical distributions of the movement representation according to the target segment lengths. Only the anchor sphere of the body model isn't resized or moved. The relation between the segment length and the radius was considered linear, as pointed by $redim$ in eq. (2). The method only performs a translation of the spherical distributions centers maintaining the relation between the spherical distributions centers. The change in the spherical distribution radius, is regarded a linear an adaption of the segment reach.

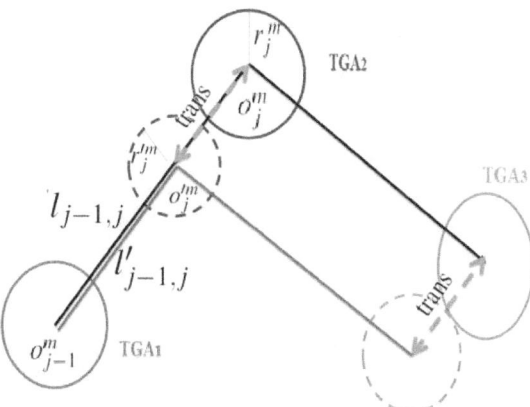

Fig. 1. Resize of a segment (from TGA1 to TGA2) and application of the translation over the rest of the kinematic chain (TAG3)

(b) Different Number of Joints:

In order to have the correct body morphology, it may be necessary to remove joints from the MoCap body model that the target humanoid body doesn't have. Considering three adjacent joints j, $j + 1$ and $j + 2$, the goal is to determine a segment from j to $j + 2$ that is closest to the spherical distribution of $j + 1$, and discard $j + 1$ as required (figure 2a). So, in order to choose points inside D_j^m

and D^m_{j+2}, we calculate the interception, in the form of the spherical cap C^m_j, of the spherical distribution D^m_{j+2} for the target joint, with a sphere S^m_j centered in the position of the first joint, p^m_j, and with radius equal to $l_{j,j+1} + l_{j+1,j+2}$. Then, if C^m_j is an empty set, the center of D^m_{j+2} is translated in the direction of the vector from p^m_j to o^m_{j+2} (figure 2b)), increasing or decreasing the distance, from p^m_j to o^m_{j+2}, this way assuring to obtain interception between D^m_{j+2} and S^m_j. Finally, a search for a point in C^m_j is employed. The segment that connects this calculated point to p^m_j must be closest to the sphere center of the eliminated joint, o^m_{j+1}.

In the special case where p^m_j is the anchor point of the model (and first point to be determined in the model), we can also move p^m_j, keeping it inside D^m_j, in the direction of the vector from p^m_j to o^m_{j+2} to enable the interception, and then move p^m_j, in order to approach it from p^m_{j+1}. This enables a better fit of the segment that will be traced. This method is applied to erase the middle joint of the spine (joint MSP in figure 4).

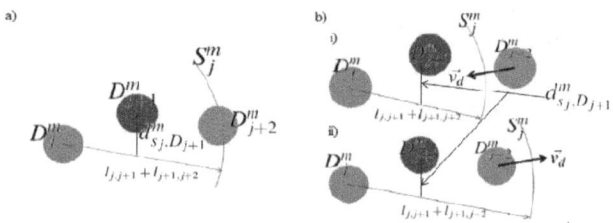

Fig. 2. Morphological adaption of a kinematic chain with three joints $(j, j+1, j+2)$ to one with two joints $(j, j+2)$, by "erasing" the middle joint

(c) Additional Physical Restrictions: Another problem that we faced in the motion retargeting was related to the necessity to ensure that certain body segments were collinear. In order to ensure that the segment from j to j' is collinear with the segment from $j2$ to $j2'$, firstly we generate, to the segment from j to j', the random quaternion and its corresponding direction vector $\overrightarrow{v}^m_{j,j'}$ and try to apply it to the other segment involved (situation exemplified in figure 3).

$$(p^m_j + \overrightarrow{v}^m_{j,j'} * l_{j,j'}) \in D^m_{j'} \tag{3}$$

$$(p^m_j + \overrightarrow{v}^m_{j,j'} * l_{j,j'}) \in D^m_{j'} \wedge (p^m_{j2} + \overrightarrow{v}^m_{j,j'} * l_{j2,j2'}) \in D^m_{j2'}. \tag{4}$$

Condition (3) is the constraint applied to accept a generated point in [12], which the algorithm will try to ensure while not reaching the maximum number of attempts. When this maximum is reached, condition (4) must be ensured instead, by using symmetric vectors in both segments. If the method can't generate a vector that satisfies condition (4), then it will generate a vector that can satisfy (3), and calculate a new point $pf^m_{j2'}$ from it:

$$pf_{j2'}^m = p_{j2}^m + \overrightarrow{v}_{j,j'}^m * l_{j2,j2'}.$$ (5)

This new point will be outside the spherical distribution $D_{j2'}^m$ of $j2'$, so the distance from $pf_{j2'}^m$ to the center of $D_{j2'}^m$ is assumed as the translation that must be applied to all the following spherical distribution center points in the same kinematic chain. This translation ensures that the centers of the remaining spherical distributions still maintain their spatial relation until the extremity joint of the chain, this way keeping the original key-pose posture without compromising the fixed geometry of the humanoid body model. This method was applied over the hip segments of the model.

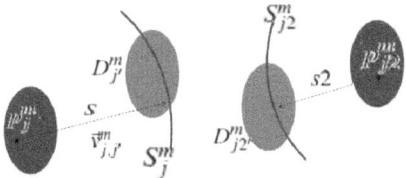

Fig. 3. Physical Restriction example. Segment s and $s2$ must be parallel.

Key-Poses Retargeting. To generate the actual robot joint angles from the representation obtained from the application of the previous adaption methods, we applied a motion retargeting technique based on [16] for extracting the Euler angles of each joint, in the 3 dimensions, based on a body-centered axis system. This method considered the robot joint's degrees-of-freedom (DoFs) and their singularities of different natures. The following procedure was applied to all synthesized key-poses adjusted to the target humanoid morphology.

At first, the local coordinate system for the robot upper-body R_{ub} is defined in the chest of the previously resized and adapted body model, as follows:

$$\begin{cases} R_{ub}^y = v_{SH} = p_{LSH} - p_{RSH} \\ R_{ub}^{z'} = v_{SP} = p_{CSH} - p_{CHIP} \\ R_{ub}^x = R_{ub}^y \times R_{ub}^{z'} \\ R_{ub}^z = R_{ub}^x \times R_{ub}^y \\ R_{ub} = [norm(R_{ub}^x), norm(R_{ub}^y), norm(R_{ub}^z)] \end{cases}$$, (6)

where $norm(X) = \frac{X}{|X|}$ and $\times$ is the cross product between two dimensions. And p_{Name} represents the position of a body joint illustrated in 4a).

Using the vectors from the global coordinate system, the correspondent vectors in the local coordinate system are calculated and the angles in all the axis determined. As R_{ub} represents a rotation matrix, the product of that rotation matrix and a global unit vector will result in the corresponding local unit vector.

The retargeting started at the left shoulder, considering a vector v_{LSE} connecting the shoulder's joint global position p_{LSH} to the elbow's p_{LELB}.

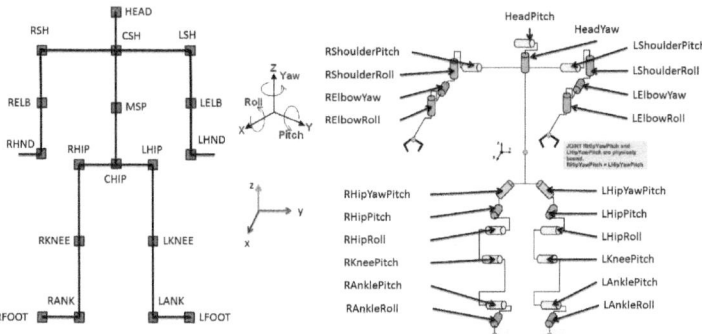

Fig. 4. a) Considered labels for the joints of the original MoCap body model (left); b) Joint rotation and Cartesian axes. c) Humanoid robot NAO body model description.

According to the specified joint's characteristics, firstly, the rotation angle in the y axis is extracted, which corresponds to the pitch rotation of the robot's left shoulder $LShoulderPitch$:

$$\begin{cases} \boldsymbol{v}_{LSE} = p_{LELB} - p_{LSH} \\ \boldsymbol{v'}_{LSE} = R_{ub}^T \times \boldsymbol{v}_{LSE} \\ LShoulderPitch = atan2(\boldsymbol{v'}_{z_{LSE}}, \boldsymbol{v'}_{x_{LSE}}) \end{cases}. \qquad (7)$$

Then, the rotation of the left shoulder in the z axis (roll rotation $LShoulderRoll$), is calculated:

$$\begin{cases} \boldsymbol{v''}_{LSE} = R_y(LShoulderPitch) \times \boldsymbol{v'}_{LSE} \\ LShoulderRoll = atan2(\boldsymbol{v''}_{y_{LSE}}, \boldsymbol{v''}_{x_{LSE}}) \end{cases}, \qquad (8)$$

where $R_y(LShoulderPitch)$ is the rotation matrix in y by $LShoulderPitch$ degrees.

To complete the left shoulder rotation, only the x rotation is missing, that represents the rotation of the shoulder over itself. This rotation is applied on the robot elbow. For such, it is now considered the vector $\boldsymbol{v}_{LEH}$ from the elbow joint p_{LELB}, to the hand's p_{LHND}. Following this, both existing elbow rotations, in the coronal (roll rotation) $LElbowRoll$ and transverse (yaw rotation) planes $LElbowYaw$, are also calculated in relation to R_{ub}:

$$\begin{cases} \boldsymbol{v}_{LEH} = p_{LHND} - p_{LELB} \\ \boldsymbol{v'}_{LEH} = R_z(LShoulderRoll) \times R_y(LShoulderPitch) \times R_{ub}^T \times \boldsymbol{v}_{LEH} \\ LElbowRoll = atan2(\boldsymbol{v'}_{z_{LEH}}, -\boldsymbol{v'}_{y_{LEH}}) \\ \boldsymbol{v''}_{LEH} = R_x(LElbowRoll) \times \boldsymbol{v'}_{LEH} \\ LElbowYaw = atan2(\boldsymbol{v''}_{y_{LEH}}, \boldsymbol{v''}_{x_{LEH}}) \end{cases}. \qquad (9)$$

For the extraction of the legs' joint rotations a new local coordinate system R_{lb} is defined. This new coordinate system R_{lb} use both the hip and the spine directions, as follows:

$$\begin{cases} R_{lb}^{y} = \boldsymbol{v}_{HIP} = p_{LHIP} - p_{RHIP} \\ R_{lb}^{z'} = \boldsymbol{v}_{SP} = p_{CSH} - p_{CHIP} \\ R_{lb}^{x} = R_{lb}^{y} \times R_{lb}^{z'} \\ R_{lb}^{z} = R_{lb}^{x} \times R_{lb}^{y} \\ R_{lb} = [norm(R_{lb}^{x}), norm(R_{lb}^{y}), norm(R_{lb}^{z})] \end{cases} \tag{10}$$

The following steps describe the calculation of all joints presented in the robot's left leg, starting with the hip joint rotations, proceeding to the knees, and finally to the feet joints.. Firstly we extracted the robot's left hip roll rotation $LHipPitch$, that controls the hip movement along the body's sagittal plane:

$$\begin{cases} \boldsymbol{v}_{LKH} = p_{LHIP} - p_{LKNEE} \\ \boldsymbol{v'}_{LKH} = R_{lb}^{T} \times \boldsymbol{v}_{LKH} \\ LHipPitch = atan2(\boldsymbol{v'}_{x_{LKH}}, -\boldsymbol{v'}_{z_{LKH}}) \end{cases} \tag{11}$$

Then the $LHipRoll$ is extracted:

$$\begin{cases} \boldsymbol{v''}_{LKH} = R_{y}(LHipPitch) \times \boldsymbol{v'}_{LKH} \\ LHipRoll = atan2(\boldsymbol{v''}_{x_{LKH}}, -\boldsymbol{v''}_{z_{LKH}}) \end{cases} \tag{12}$$

Using this new vector, the last angle for the hip section $LHipYawPitch$, over the local z axis is determined. This hip freedom corresponds to an actuator that is shared by both legs. As such, this rotation defines the hip movement, symmetrically, for both legs:

$$\begin{cases} \boldsymbol{v}_{LAK} = p_{LKNEE} - p_{LANK} \\ \boldsymbol{v'}_{LAK} = R_{x}(LHipRoll) \times R_{y}(LHipPitch) \times R_{lb}^{T} \times \boldsymbol{v}_{LAK} \\ LHipYawPitch = atan2(\boldsymbol{v'}_{y_{LAK}}, -\boldsymbol{v'}_{x_{LAK}}) \end{cases} \tag{13}$$

Having calculated the three rotations for the hip section we proceeded to the calculation of the knee and ankle (both knee and ankle only present 1 DoF), that enables rotations over the y axis of the local coordinate system. The robot body model also presents another rotation freedom at the ankles concerning its roll rotation around the z plane, $RAnkleRoll$ and $LAnkleRoll$. Yet, this rotation was not considered since it doesn't have any correspondence to our synthetic body model, being rather important for the maintenance of the robot's balance (which is outside the scope of this paper). As such, the respective pitch rotations of the left knee $LKneePitch$ and left ankle $LAnklePitch$ were calculated as follows:

$$\begin{cases} \boldsymbol{v''}_{LAK} = R_{z}(LHipYawPitch) \times \boldsymbol{v'}_{LAK} \\ LKneePitch = atan2(\boldsymbol{v''}_{x_{LAK}}, -\boldsymbol{v''}_{z_{LAK}}) \end{cases} \tag{14}$$

$$\begin{cases} \boldsymbol{v}_{LFA} = p_{LANK} - p_{LFOOT} \\ \boldsymbol{v'}_{LFA} = R_{y}(LKneePitch) \times R_{z}(LHipYawPitch) \times R_{x}(LHipRoll) \\ \qquad\qquad\qquad \times R_{y}(LHipPitch) \times R_{lb}^{T} \times \boldsymbol{v}_{LFA} \\ LAnklePitch = atan2(\boldsymbol{v'}_{z_{LFA}}, -\boldsymbol{v'}_{x_{LFA}}) \end{cases} \tag{15}$$

All the former processes are similarly applied to the right arm and leg of the robot.

3.3   Robot Motion Generation

The actual generation of the robot dance motion is accomplished by the cyclic repetition of the synthesized and adapted key-poses, generating this way the dance motion in the humanoid. The several key-poses are concatenated, and are synthesized with variability at quarter-beat resolution, according to the dance TGA representation. The obtained key-poses are ordered accordingly with the dance representation model (at the considered metrical resolution), and the poses concatenation is done by sine interpolation of the joint angles.

The beat-synchrony aspect of the dance was achieved by determining the dance motion frames where the determined key-poses occur and then calculating for each pose p the correspondent duration of the transition until the next pose $p + 1$.

4   Evaluation and Results

4.1   Key-Poses Comparison

The following figure 5 present a comparison between the synthesized-adjusted key-poses and their representation in the simulated humanoid NAO.

4.2   Key-Poses Degree of Similarity

In order to compare the similarity between poses we defined a measure based on the evaluation of distances between the same joint positions among both poses, as follows:

$$s_{pose_{i,j}, robot_{i,j}} = \sum_{j}^{n} \sum_{i=j}^{n} \left| d_{pose_{i,j}} - d_{robot_{i,j}} \right| \tag{16}$$

This similarity function aims to compare the feasibility of the humanoid reproduction of the adapted synthesized key-poses. For such, the distance between the whole joint positions of the adapted synthesized key-poses, $d_{pose_{i,j}}$ in (16), is compared with the actual humanoid joints position that resulted from the angles retargeted onto the robot, $d_{robot_{i,j}}$ in (16).

With this evaluation function, the joint positions of the adapted synthesized key-poses were extracted (noted as $synthPose$), and compared with the spatial joint positions of the same robot key-poses generated by: i) the respective joint angles extracted from the adapted synthesized poses, $s_{synthPose, poseAdapted}$; ii) setting all its joint angles to 0 (neutral pose), $s_{synthPose, poseNeutral}$; and iii) the angles extracted from synthesized key-poses without morphological adaptation, $s_{synthPose, poseNormal}$.

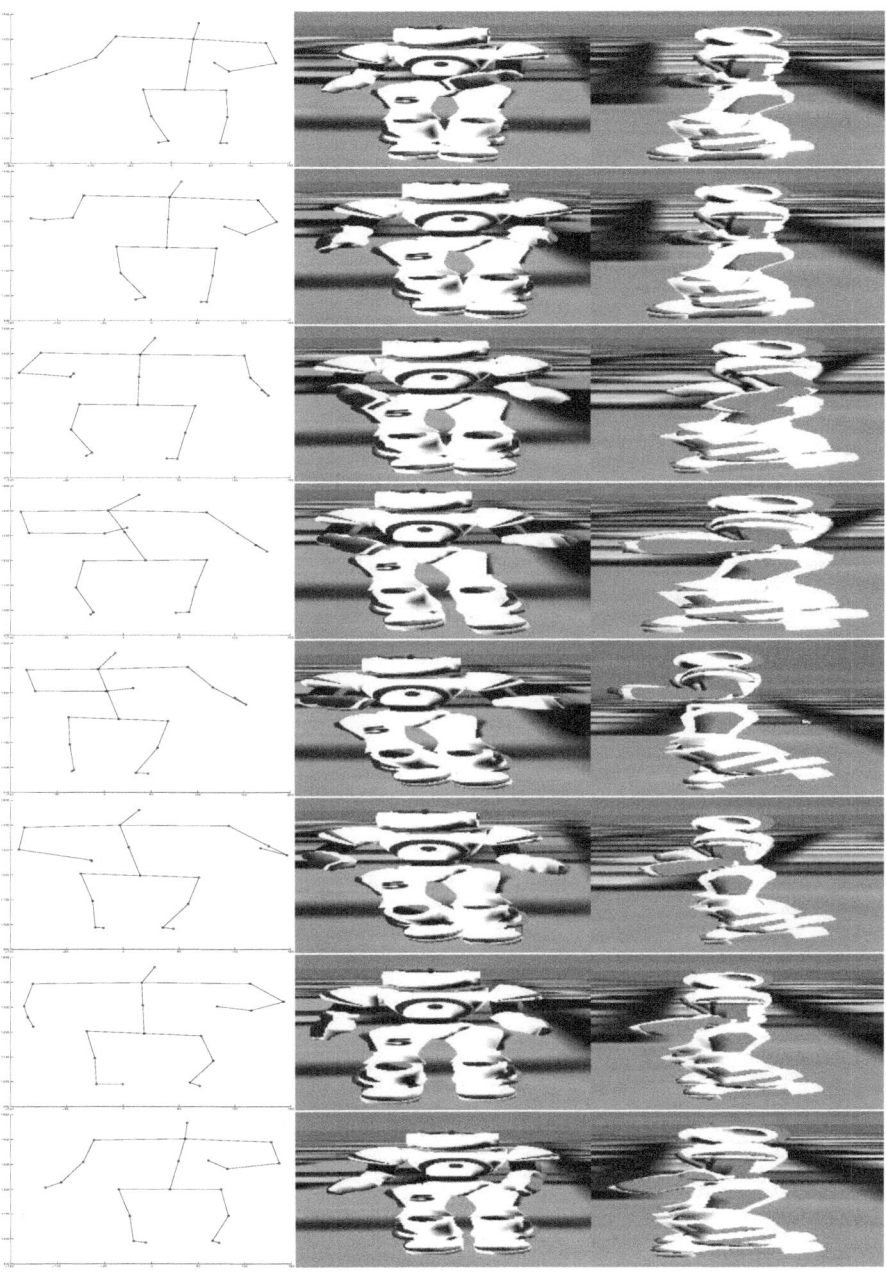

Fig. 5. Visualization of key-poses, from key-pose 1 (*top*) to key-pose 8 (*bottom*) (each row represents a new pose), synthesized at "variability-4": a) Synthesized-adjusted (*left*); b), c) Retargeted to simulated humanoid NAO, in frontal (*middle*) and lateral (*right*) views

Table 1. Similarity comparison of the generated poses, values in mm

Pose	$s_{synthPose,poseAdapted}$			$s_{synthPose,poseNeutral}$			$s_{synthPose,poseNormal}$		
	Arms	Legs	Total	Arms	Legs	Total	Arms	Legs	Total
1	571	543	2714	818	546	6300	328	637	3166
2	328	741	2911	489	934	6566	346	633	3146
3	182	1008	2847	494	826	6397	187	686	2776
4	393	739	2918	517	714	6687	434	676	3173
5	514	802	3177	516	819	6400	709	864	4041
6	368	912	3051	453	1081	7164	417	829	3199
7	322	1013	2912	424	750	6432	385	884	3065
8	220	579	1896	453	662	6762	230	579	2012

The results of these comparisons are presented in table 1.

4.3   Discussion

The method for resizing the body model by scaling and translating the joints' spherical distributions seems effective, presenting the desired result. The methods presented for changing the body morphology work separately, and aim to solve their specific problem without considering the global solution. This can signify that the achieved solution isn't the optimal solution, but rather a possible solution facing the individual constraints applied.

The key-pose angle extraction shows good overall similarity results. These results seem feasible, both numerically and visually, at the arms section, but the legs section seem to compromise it, especially in the hips. This difference can be explained by a bigger morphological similarity between the human's and the robot's arms than between the human's and robot's legs. The greater legs difference is mainly caused by the hip section due to the angular limits of the correspondent humanoid joint that makes impossible to achieve certain types of positions.

The results in table 1 shows better similarity results for the extracted angles from the adapted key-poses than from the non-adapted key-poses. In this table, the comparison with the neutral poses helps to have a sense of gain in similarity by the application of the angle extraction method, which seems to be in the order of 135%.

The proposed measure to evaluate the similarity between the synthesized and humanoid key-poses seemed to correctly qualify the degree of similarity and compare the different key-poses, but still lacks a zero value in order to get normalized similarity measures.

5   Conclusions and Future Work

In this study we proposed and evaluated methods for adjusting and retargeting human dance MoCap data, of Samba, into a target humanoid body model.

The process starts from information extracted from the analysis and representation of human dance MoCap data, and then generates random joint rotations that satisfy the desired model. Finally, the key-poses are extracted and reproduced into the target humanoid. The overall results seem to show that the methods used for the resizing and retargeting are valid, needing further evaluation with different humanoid morphologies and experimentation with MoCap data of different dance styles. In relation to the angle extraction method, it seems to be valid and obtains good results to the upper-body, but still only offers an approximation for the legs. The importance of the hip movements for the considered dance style, and the several morphological differences between the human hip and the robot hip, present obstacles that reduce the legs similarity. In a way to improve the degree of similarity of the humanoid poses with the original ones, the proposed similarity evaluation function may be also used as the fitness function in an optimization process. Finally, in order to obtain stable movements, motion refinement should be also performed, in order to assure the robot biped balance and avoiding self-collisions.

References

1. Simspark comunity, http://simspark.sourceforge.net/wiki
2. Gouaillier, D., Hugel, V., Blazevic, P., Kilner, C., Monceaux, J., Marnier, B., Serre, J., Serre, J.: The nao humanoid: a combination of performance and affordability. Computing Research Repository CoRR, abs/0807 3, 1–10 (2008)
3. Kim, S., Kim, C.H., You, B., Oh S.: Stable Whole-Body Motion Generation for Humanoid Robots to Imitate Human Motions. In: IEEE/RSJ International Conference on Intelligent Robots and Systems (IROS), St. Louis, MO, USA, pp. 2518–2524. IEEE (October 2009)
4. Kim, T.-h., Park Il, S., Shin, S.Y.: Rhythmic-Motion Synthesis based on Motion-Beat Analysis. ACM Transactions on Graphics 22(3), 392 (2003)
5. Lau, N., Reis, L.P.: High-level coordination methodologies in soccer robotics, robotic soccer, pp. 167–192. Itech Education and Publishing (2007)
6. Nagata, N., Okumoto, K., Iwai, D., Toro, F., Inokuchi, S.: Analysis and Synthesis of Latin Dance Using Motion Capture Data. Advances in Multimedia Information Processing 3333, 39–44 (2005)
7. Nakaoka, S., Nakazawa, A., Yokoi, K., Hirukawa, H., Ikeuchi, K.: Generating Whole Body Motions for a Biped Humanoid Robot from Captured Human Dances. In: IEEE International Conference on Robotics and Automation (ICRA), vol. 3, pp. 3905–3910. IEEE (2003)
8. Nakazawa, A., Nakaoka, S., Ikeuchi, K., Yokoi, K.: Imitating human dance motions through motion structure analysis. In: Proceedings of the International Conference on Intelligent Robots and Systems (IROS), pp. 2539–2544 (2002)
9. Naveda, L., Leman, M.: The spatiotemporal representation of dance and music gestures using Topological Gesture Analysis (TGA). Music Perception 28(1), 93–111 (2010)
10. Okamoto, T., Shiratori, T., Kudoh, S., Ikeuchi, K.: Temporal Scaling of Leg Motion for Music Feedback System of a Dancing Humanoid Robot. In: IEEE/RSJ International Conference on Intelligent Robots and Systems (IROS), Taipei, Taiwan, pp. 2256–2263 (2010)

11. Oliveira, J.L., Naveda, L., Gouyon, F., Leman, M., Reis, L.P.: Synthesis of Dancing Motions Based on a Compact Topological Representation of Dance Styles. In: Workshop on Robots and Musical Expressions (WRME) at IROS, Taipei, Taiwan (2010)
12. Oliveira, J.L., Naveda, L., Gouyon, F., Leman, M., Reis, L.P., Sousa, P.: A Spatiotemporal Method for Synthesizing Expressive Dance Movements of Virtual Humanoid Characters. Special Issue on Music Content Processing by and for Robots from the EURASIP Journal on Audio, Speech, and Music Processing (submitted to, 2011)
13. Lau, N., Reis, L.P.: FC Portugal 2001 Team Description: Flexible Teamwork and Configurable Strategy. In: Birk, A., Coradeschi, S., Tadokoro, S. (eds.) RoboCup 2001. LNCS (LNAI), vol. 2377, pp. 515–518. Springer, Heidelberg (2002)
14. Ruchanurucks, M., Nakaoka, S., Kudoh, S., Ikeuchi, K.: Generation of Humanoid Robot Motions with Physical Constraints using Hierarchical B-Spline. In: IEEE/RSJ International Conference on Intelligent Robots and Systems (IROS), pp. 674–679. IEEE (2005)
15. Ruchanurucks, M., Nakaoka, S., Kudoh, S., Ikeuchi, K.: Humanoid Robot Motion Generation with Sequential Physical Constraints. In: IEEE International Conference on Robotics and Automation (ICRA), Orlando, FL, USA, pp. 2649–2654. IEEE (2006)
16. Shiratori, T.: Synthesis of Dance Performance Based on Analyses of Human Motion and Music. Phd thesis, University of Tokyo (2006)
17. Shiratori, T., Kudoh, S., Nakaoka, S., Ikeuchi, K.: Temporal Scaling of Upper Body Motion for Sound Feedback System of a Dancing Humanoid Robot. In: IEEE/RSJ International Conference on Intelligent Robots and Systems (IROS), San Diego CA, USA, pp. 3251–3257. IEEE (October 2007)
18. Shiratori, T., Nakazawa, A., Ikeuchi, K.: Dancing-to-Music Character Animation. In: EUROGRAPHICS, vol. 25, pp. 449–458 (September 2006)
19. Shiratori, T., Nakazawa, A., Ikeuchi, K.: Synthesizing Dance Performance using Musical and Motion Features. In: IEEE International Conference on Robotics and Automation, pp. 3654–3659. IEEE (2006)
20. Zhao, X., Huang, Q., Du, P., Wen, D., Li, K.: Humanoid Kinematics Mapping and Similarity Evaluation Based on Human Motion Capture. In: International Conference on Information Acquisition (ICIA), pp. 426–431. IEEE (2004)

Bankruptcy Trajectory Analysis on French Companies Using Self-Organizing Map

Ning Chen[1], Bernardete Ribeiro[2], and Armando S. Vieira[1]

[1] GECAD, Instituto Superior de Engenharia do Porto, Portugal
[2] CISUC, Department of Informatics Engineering, University of Coimbra, Portugal
{cng,asv}@isep.ipp.pt, bribeiro@dei.uc.pt

Abstract. As one of the major business problems, corporate bankruptcy has been extensively studied using a large variety of statistical and machine learning approaches. However, the trajectory of bankruptcy behavior is seldom explored in the literature. In this paper, we use self-organizing map neural networks to analyze the changes of financial situation of companies in several consecutive years through a two-step clustering process. Firstly, the bankruptcy risk is characterized by a feature map, and therefore the temporal sequence is converted to the trajectory vector projected on the map. Afterwards, the trajectory map clusters the trajectory vectors to a number of evolution patterns. The approach is applied to a large database of French companies which contains the financial ratios spawning over a period of four years. Typical behaviors such as the deterioration and amelioration associated with the bankruptcy risk, as well as the influence of financial ratios can be revealed by means of visual interpretation.

Keywords: corporate bankruptcy, trajectory analysis, self-organizing map, visual clustering.

1 Introduction

In recent years, more and more companies are faced with the bankruptcy crisis under the severe challenges of worldwide economics. The increase of financial failure accelerates the economic deterioration and yields a lot of social problems. It becomes critically important to explore the potential bankruptcy behaviors and understand the implicit patterns from the perspective of early warning and decision support. To date, a large amount of research has been carried out using different methods, e.g., univariate and multivariate analysis, neural network, support vector machine, and rough set. However, most of the prior studies are focused on financial distress prediction problem based on the historical financial statements. To our knowledge, bankruptcy trajectory receives little attention and only a few work attempted to analyze the temporal sequence of financial statements [9,17].

Self-organizing map (SOM), a non-parametric neural network visualization tool, is a well-known visual clustering method in a wide range of applications due

L. Antunes and H.S. Pinto (Eds.): EPIA 2011, LNAI 7026, pp. 407–417, 2011.

to the desirable combination of data abstraction and spatialization. In this paper, we study the changes of financial situation to examine the trajectory patterns through a two-step clustering process. Initially, a feature map is constructed to characterize the bankruptcy risk of companies. Afterwards, the instantaneous observations of temporal sequence are successively projected on the map and the positions are concatenated to a trajectory vector. The trajectory patterns are then learned by a trajectory map and shown through an appropriate visual representation.

The proposed approach is applied to a data set of French companies containing financial ratios in 4 consecutive years (2003~2006) and the final state in the following year (2007). The main purpose of this study is not to deliver a reliable classification or prediction regarding the bankruptcy risk as was done in most of the previous research. Instead, we aim at a compact visualization of the complex temporal behaviors in financial statements. Taking the perspective of decision support, the described method might give experts insight into patterns that are a pretense for bankruptcy or healthy company development. The experimental results demonstrate the promising functionality of SOM for bankruptcy trajectory analysis.

The remainder of this paper is organized as follows. Section 2 reviews the related studies on bankruptcy prediction and analysis. Section 3 presents the data set under exploration and the methodology of a SOM-based trajectory analysis approach. In section 4, the experimental results are reported. Lastly, the contributions and future remarks are discussed in section 5.

2   Related Work

Self-organizing map (SOM) is an unsupervised neural network proposed by Kohonen [12] for visual cluster analysis. The neurons of the map are located on a regular grid embedded in a low (usually 2 or 3) dimensional space, and associated with the cluster prototypes. In the course of learning process, the neurons compete with each other through the best matching principle, i.e., the input is projected to the nearest neuron using a defined distance metric. The winner neuron and its neighbors on the map are adjusted towards the input in proportion with the neighborhood distance, consequently the neighboring neurons likely represent the similar patterns of the input data space. Due to the data clustering and spatialization through the topology preserving projection, SOM is widely used in the context of visual clustering applications. In a recent study, the visual capabilities of SOM are explored to predict the emergence of currency crises and understand the factors and conditions [16]. Despite the unsupervised nature, the applicability of SOM is extended to classification tasks by means of a variety of ways, such as neuron labeling, semi-supervised learning [7] and supervised learning vector quantization (LVQ) [12].

SOM is one of the frequently used models to analyze the high-dimensional financial statement and understand the bankruptcy phenomenon. A wide range of research groups concentrate on the bankruptcy prediction problem, usually solved as a classification task to separate the companies into distress and healthy category (binary) or a number of predefined credit rates (multi-class). The predictive models are constructed from the training data in terms of some criteria, such as the overall misclassification error, the Neyman-Pearson criterion [11], and the total misclassification cost [4]. The capability of SOM and its supervised variants has been demonstrated in comparison with statistical and other intelligent methods. Recent examples are as follows. SOM is used to determine the credit class through a visual exploration [13]. An enhanced version of LVQ can boost the prediction performance of multi-layer perceptron neural network [14]. It is demonstrated that LVQ outperformed other neural networks, support vector machines and multivariate statistical methods on predicting the financial failure of Turkish banks [2].

From the perspective of decision support, a more important issue than accurate prediction is the deep analysis and well understanding of financial statements, e.g., which factors contribute mostly to the difference between the bankrupt companies and healthy ones? And how to characterize the evolution of companies? Trajectory mining is a promising approach to these issues.

Trajectory data has attracted considerable attention in many applications where the object movements are routinely collected as time-dependent observation sequences, such as traffic monitoring, visual surveillance, robotic navigation, and stock prediction. Trajectory mining attempts to explore the implicit patterns from trajectory data. SOM has been introduced as a valuable tool for trajectory mining. In a robotic navigational environment, SOM clusters the motion trajectory of moving objects and predicts the next instant position [15]. A SOM model provides a powerful tool to visualize the dynamic behaviors of industrial processes for human supervision and fault detection [5]. In the field of bankruptcy analysis, trajectory data of financial statements makes it possible to detect the time evolution of companies and recognize the trajectory patterns. In [17], a SOM-based framework is designed for general-purpose enabling users to visually monitor and interactively control the clustering process of trajectory data. In [9], a hierarchical SOM model is employed to explore the year-to-year trajectory of enterprises and view the time evolution of the state. The study of this paper follows this idea, examining the trajectory of financial statements of companies and recognizing the influential factors relevant to bankruptcy risk.

3   Methodology of Bankruptcy Trajectory Clustering

The trajectory clustering procedure is basically a two-step consecutive learning process. The first SOM is trained without explicitly taking the temporal aspects of the data into account. The high-dimensional vectors of financial indicators in each time step are mapped to the respective best-matching units (BMUs) in the SOM grid. Their grid coordinates then serve as a compressed two-dimensional representation of the original data, while the temporal progression for each company can be represented as the sequence of respective BMUs in the order of the given time steps. Therefore, each company yields a trajectory in the SOM neighborhood grid. These trajectories are visualized and clustered by a second SOM to find the behavioral patterns in the set of trajectories.

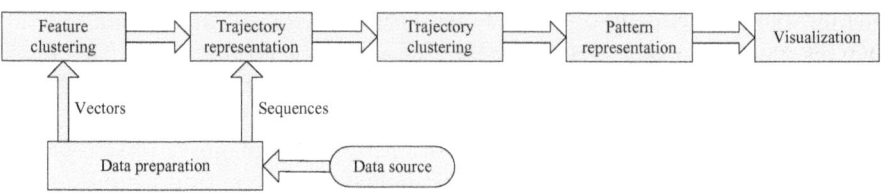

Fig. 1. Methodology of trajectory clustering

As shown in Figure 1, the trajectory clustering approach is composed of 6 phases, namely data preparation, feature clustering, trajectory representation, trajectory clustering, pattern representation, and visualization. Each of the phases is described as follows.

3.1 Data Set Description

The original French database contains the financial statements of 110,723 companies in small or middle size during the year 2003 to 2006. From the financial statements, 29 numerical financial ratios are derived shown in Table 1. The class indicator gives the financial state of companies in the year 2007 in which 2792 observations go bankrupt eventually.

3.2 Data Preparation

Data preprocessing is an integral part of clustering procedure, usually composed of sampling, logarithmization, and normalization. Since we consider the time evolution of companies during several years, only the companies which are in operation in all the years are preserved. Firstly, we chose all companies distressed in 2007 with at least 20% available values in each fiscal year, yielding a total of 718 distressed companies. To create a balanced data set, a random sampling was then employed to select a same number of healthy companies. Consequently, a balanced sample is made up of 1436 observation sequences. Afterwards, a logarithmized operation is applied to mitigate the scatter distribution of the data [4]. The financial ratios vary greatly in scales, thereby a normalization is needed before the distance calculation. We use the zero-mean normalization so that each feature has zero mean and unit variance.

3.3 Feature Clustering

As demonstrated in previous studies, one-year prior to the bankruptcy announcement is able to derive accurate bankruptcy models, therefore, we use the data composed of the financial ratios one year before bankruptcy to generate a feature SOM (FSOM) which characterizes the bankruptcy risk. Using the update rule, the network is trained by adjusting the prototypes iteratively on the foundation of winner-take-all principle. Assuming the map consists of m neurons, each associated with a prototype, the learning process is performed as follows:

Table 1. Financial ratios of French Database

Variable Description	
x_1 - Number of Employees Previous year	x_{16} - Cashflow / Turnover
x_2 - Capital Employed / Fixed Assets	x_{17} - Working Capital / Turnover days
x_3 - Financial Debt / Capital Employed	x_{18} - Net Current Assets/Turnover days
x_4 - Depreciation of Tangible Assets	x_{19} - Working Capital Needs / Turnover
x_5 - Working Capital / Current Assets	x_{20} - Added Value per Employee in k Euros
x_6 - Current ratio	x_{21} - Total Assets Turnover
x_7 - Liquidity Ratio	x_{22} - Operating Profit Margin
x_8 - Stock Turnover days	x_{23} - Net Profit Margin
x_9 - Collection Period days	x_{24} - Added Value Margin
x_{10} - Credit Period days	x_{25} - Part of Employees
x_{11} - Turnover per Employee k Euros	x_{26} - Return on Capital Employed
x_{12} - Interest / Turnover	x_{27} - Return on Total Assets
x_{13} - Debt Period days	x_{28} - EBIT Margin
x_{14} - Financial Debt / Equity	x_{29} - EBITDA Margin
x_{15} - Financial Debt / Cashflow	x_{30} - Class (bankrupt, healthy)

1. For $p = 1, ..., m$, initialize the map with prototypes m_p;
2. For each input instance x, project it to the BMU c using Euclidean distance, defined as:

$$c = \underset{1 \leq i \leq m}{\text{argmin}} \, ||x - m_i||$$

3. For $p = 1, ..., m$, update the prototypes proportional to the learning rate $\alpha(t)$ and the neighborhood function $h(t)$:

$$m_p(t + 1) = m_p(t) + \alpha(t)h_{c,p}(t)(x - m_p)$$

4. Repeat from Step 2 a few iterations until the termination condition is satisfied.

A labeling operation is applied to the learned map in a supervised fashion. By projecting the instances to the BMUs, the input space is divided into a finite number of Voronoi sets, constituted by the instances projected on the underlying neuron:

$$V_p = \{x \mid p = \underset{1 \leq i \leq m}{\text{argmin}} \, ||x - m_i||\}$$

Thereby, the neurons can be labeled by the majority class indicator, i.e., the one most frequently occurring in the corresponding Voronoi sets.

For the quantitative analysis, each neuron is associated with a percentage of default (PD) value defined as the ratio of bankrupt companies in the Voronoi set:

$$PD(m_p) = \frac{|\{x \in V_p \mid Class(x) = bankrupt\}|}{|V_p|}$$

For the zero-hit neurons, the PD value is 0. By taking the value as a real number in [0,1], PD conveys the information of bankruptcy risk. The higher the value of PD, the higher the risk of a company being distressed in the following year.

3.4   Trajectory Representation

For trajectory clustering, of primary concern is the meaningful interpretation of trajectory data. In this study, we consider a simple vector representation. As a projection-based approach, SOM preserves the topological property between the input space and the output space, thereby the feature vector can be represented by the position of projected neuron on the map. The projection operation is applied to the financial vectors yearly for each company. The concatenation of the projected coordinates along the sequence yields the trajectory vector. As a result, we have a trajectory vector formatted as $\bar{x} =< p_{x1}, p_{y1}, ..., p_{xk}, p_{yk} >$ where (p_{xi}, p_{yi}) are the x-coordinate and y-coordinate values on the map, k is the number of years.

3.5   Trajectory Clustering

In the same manner as feature clustering, a trajectory SOM (TSOM) is then constructed taking as input the produced trajectory vectors. The learning is also conducted in an unsupervised manner so that the trajectories are grouped into a number of clusters based on the similarity.

3.6   Pattern Representation

On the basis of the clusters obtained by TSOM, the trajectory patterns can be extracted, taking the medoid of the cluster as the representative trajectory. For each cluster, the pairwise distance is calculated and then the medoid associated with the cluster is the trajectory where the sum of Euclidean distance from others is minimum. The medoid of cluster C is formally defined as:

$$medoid(C) = \underset{i}{\arg\min} \sum_{i \neq j} ||\bar{x}_i - \bar{x}_j||, \ \bar{x}_i, \bar{x}_j \in C$$

Since the final status of companies is known, the trajectory cluster can be labeled on the foundation of major voting scheme.

3.7   Visualization

Compared with other clustering methods, one of the advantages of SOM is its capability of data exploration through a visual approach. Due to the topology-preserving projection, both the cluster structure and the non-linear correlation between variables can be detected from the link of neurons on the map [6]. Several kinds of visualization can be applied for the sake of visual trajectory inspection. Firstly, the PD value associated with each cluster is visualized stating the risk of bankruptcy. Secondly, the trajectory of a company projected on the map discloses the evolution of its financial statements. Thirdly, the patterns derived from the trajectory clusters represent some typical behaviors such as the deterioration or amelioration associated with the bankruptcy risk. Lastly, the important financial ratios can be recognized which have significant influence on the state of companies.

4 Experiments and Results

In this section, we show the results on the French data set, from which the bankruptcy trajectory patterns are detected and the influential factors are analyzed.

4.1 SOM Parameterization

As described in section 3, the clustering model consists of two self-organizing maps (see Figure 2): the FSOM clusters the feature vectors $x =< x_1, x_2,...,x_{29} >$ composed of financial statements in 2006, and the TSOM clusters the trajectory vectors $\bar{x} =< p_{x1}, p_{y1}, p_{x2}, p_{y2}, p_{x3}, p_{y3}, p_{x4}, p_{y4} >$ composed of the positions on the FSOM. In the presented study, we use the same [4 x 4] hexagonal topology with gaussian neighborhood kernel for the two maps (a 4 x 4 map is used for easy interpretation and visualization). As a result, the coordinate value of p_{xi} and p_{yi} is restricted to {1, 2, 3, 4}.

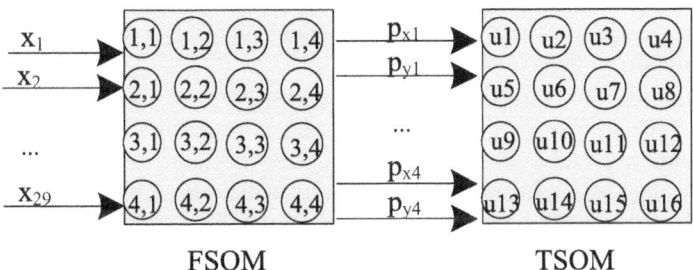

Fig. 2. The structure of SOM model

4.2 Trajectory Pattern Analysis

Each neuron of the TSOM corresponds to a trajectory cluster, represented by the medoid called trajectory pattern. In Figure 3, the trajectory patterns are shown on FSOM (one map per cluster) by connecting the projected units of yearly vector. The gray level indicates the PD value associated with the cluster (the higher risk is marked by a darker color). It can be seen that the top region has the high bankruptcy risk, and the bottom region has the low risk relatively. For each trajectory, the color indicates the final state (red: 'bankrupt', green: 'healthy'), and the arrow denotes the trajectory from start point (2003) to end point (2006).

We can summarize the patterns into 6 typical types in Table 2, in which the first three types are risky and go bankrupt eventually:

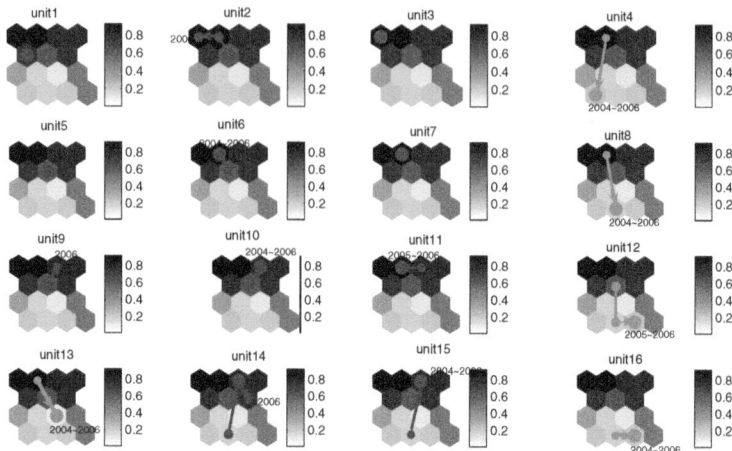

Fig. 3. Trajectory patterns visualization: 2003~2006. (Shown are the FSOM with PD value, and trajectory patterns corresponding to each cluster.)

- 'a' denotes the stably bad trajectory which always stays in the high risk region;
- 'b' denotes the slightly worse trajectory which keeps the high risk with a slight deterioration;
- 'c' denotes the significantly worse trajectory jumping from low risk to high risk, e.g., the pattern of u14 has an increasing PD value, entering the high risk region in 2004;
- 'd' implies the significantly better trajectory jumping from the high risk to the low risk, e.g., the pattern of u4 has a distinct improvement in 2004 compared with 2003;
- 'e' means the slightly better trajectory which has a moderate amelioration of PD, taking u12 as an example;
- 'f' denotes the stably good trajectory which always stays in the low risk region during four years, indicating the perfect development of companies.

Table 2. Trajectory patterns

Type	neurons	description	class	x_1	x_6	x_{14}	x_{16}	x_{28}
a	u1, u3, u5, u7	stably bad	bankrupt	-	-	-	-	-
b	u2, u6, u9, u10, u11	slightly worse	bankrupt	-	↓	↓	↓	↓
c	u14, u15	significantly worse	bankrupt	-	↓	-	↓	↓
d	u4, u8, u13	significantly better	healthy	-	↑	↑	↑	↑
e	u12	slightly better	healthy	-	↑	-	↑	↑
f	u16	stably good	healthy	-	-	-	-	-

4.3   Component Planes

Normally, financial data is described by a large number of variables, in which some have direct relation with the bankruptcy risk. Strategies commonly used for

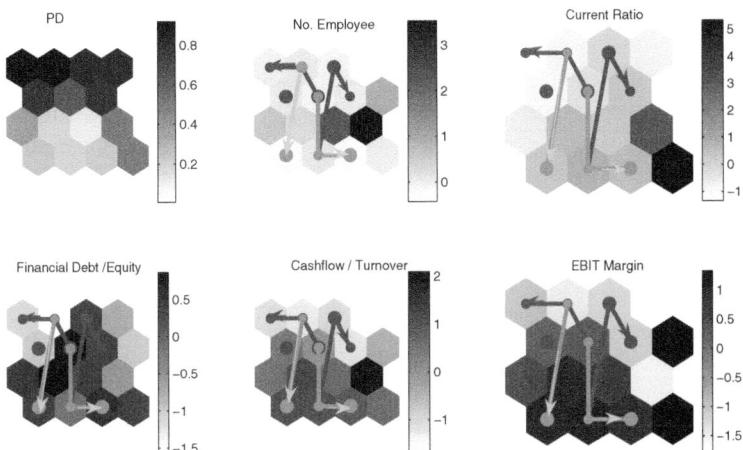

Fig. 4. Trajectory patterns on component planes: 2003∼2006. (Shown are the FSOM with PD value, component planes of x_1, x_6, x_{14}, x_{16}, and x_{28} with representative trajectory patterns.)

explanatory variable selection include suggestion by previous studies or financial experts, univariate statistical test, and automatic search using a predefined evaluation criterion [8]. According to the previous work [1], 5 ratios are pinpointed as important factors with significant influence on the financial state, which are x_1 (No. of Employees), x_6 (Current Ratio), x_{14} (Financial Debt / Equity), x_{16} (Cashflow / Turnover), and x_{28} (EBIT Margin).

Component plane representation displays the scalar values of variables across the map using a multi-spectral image. We use a grey-level image in the way that the larger value is marked by a darker color. From the component planes, not only the mutual relation between financial ratios, but also the relation between univariate ratio and the bankruptcy risk are easily recognized. In Figure 4, the first plane shows the PD value on the FSOM, and the others show the variable map of important factors (one map per factor). For simplicity of interpretation, one pattern is selected from each type as example, which are the patterns corresponding to u1, u2, u4, u12, u14 and u16. As can be seen in the planes, the variables x_6 , x_{16} , x_{28} are inversely related to PD, in other words, they take apparently bigger values in the low risk region (the bottom region on the map) than the high risk region (the top region). It means these factors discriminate effectively the healthy companies and the bankrupt ones. Particularly, x_{16} and x_{28} are positively correlated due to the fact that when the former takes high value on the map the latter is also high in most cases, and vice verse. In contrast, no direct correlation is presented between the value of x_1, x_{14} and PD from their component planes. In fact, there is a mixture of high and low values for x_1 corresponding to the low risk region. Regarding x_{14}, the companies positioned in the high risk region take a wide variety of values.

For each trajectory pattern, the changes of important factors are summarized in Table 2 (':↓': the value decreases along the trajectory, '↑': the value increases along the trajectory, and '-': no significant tendency). It therefore can be concluded that increasing the current ratio, cashflow turnover, and EBIT margin contributes to the improvement of financial state, yielding a low bankruptcy risk in successive years.

5   Conclusions

Self-organizing map is a useful visual data mining approach to explore the large amount of data. The projection defined from the high dimensional input data to a 2D visualization space makes the patterns graphically represented and easily recognizable. Despite the fact that SOM has been considerably used for financial failure prediction, its potential on trajectory analysis in finance field is not fully developed. In this paper, a SOM-based clustering approach is used to visualize and cluster the characteristics of the temporal progression of financial indicators. The idea is to assist experts with observing the development of companies over time and assessing their risk of corporate bankruptcy. The feature map clusters the financial vectors of observations and converts the temporal sequences to trajectory vectors by successive projection of the instantaneous statements to its BMU. Sequentially, the trajectory map clusters the trajectory vectors and trajectory patterns can be detected from the resulted clusters. Through the visual interpretation, the important factors are analyzed qualitatively to understand the influence on bankruptcy risk. Experimental results are presented on a data set of financial statements from French companies, where the class label indicate whether the companies ended up bankrupt or not. As indicated by the results, the proposed SOM-based model demonstrates the practicability on bankruptcy trajectory analysis.

The future study will consider the following research directors. Firstly, to verify the afore-mentioned findings presented in the paper more experiments will be conducted on the comparison to existing methods, the evaluation on different data sets, and the qualitative statement about the single step in the algorithmic chain. Since the rationale behind the trajectory representation is the topology-preservation projection of SOM, the topographic function [10] can be served as the measurement for parameter specification and model evaluation. Secondly, feature selection is an issue of interest for the visualization of component planes. At present, the significant predictors for component analysis are chosen from the previous studies, however, the quantitative analysis based on conventional feature selection methods would be helpful to measure the significance of variables and select the informative factors. Thirdly, although the present study does not aim for the prediction models, the derived trajectory model can be used in corporate bankruptcy prediction by classifying the observed sequence to the best matching cluster. The applicability of the model in prediction will be justified through performance comparison. Finally, apart from the financial data set, the proposed approach could be valuable to other trajectory-relevant data mining applications.

References

1. AIRES: Advanced Intelligent Risk Evaluation,
   `http://aires.dei.uc.pt/AiresII/development/visualization.jsp`
2. Boyacioglu, M.A., Kara, Y., Baykan, O.K.: Predicting Bank Financial Failures using Neural Networks, Support Vector Machines and Multivariate Statistical Methods: A Comparative Analysis in the Sample of Savings Deposit Insurance Fund (SDIF) Transferred Banks in Turkey. Expert Systems with Applications 36, 3355–3366 (2009)
3. Chen, N., Vieira, A.: Bankruptcy Prediction Based on Independent Component Analysis. In: Filipe, J., Fred, A., Sharp, B. (eds.) ICAART 2009. Communications in Computer and Information Science, vol. 67, pp. 150–155. Springer, Heidelberg (2010)
4. Chen, N., Vieira, A.S., Duarte, J., Ribeiro, B., Neves, J.C.: Cost-Sensitive Learning Vector Quantization for Financial Distress Prediction. In: Lopes, L.S., Lau, N., Mariano, P., Rocha, L.M., et al. (eds.) EPIA 2009. LNCS (LNAI), vol. 5816, pp. 374–385. Springer, Heidelberg (2009)
5. Fuertesa, J.J., Domingueza, M., Regueraa, P., Pradaa, M.A., Diazb, I., Cuadrado, A.A.: Visual Dynamic Model based on Self-Organizing Maps for Supervision and Fault detection in Industrial Processes. Engineering Applications of Artificial Intelligence 23(1), 8–17 (2010)
6. Hanafizadeh, P., Mirzazadeh, M.: Visualizing Market Segmentation using Self-Organizing Maps and Fuzzy Delphi Method-ADSL Market of a Telecommunication Company. Expert Systems with Applications 38, 198–205 (2011)
7. Heikkonen, J., Koikkalainen, P., Oja, E.: Self-Organizing Maps for Collision Free Navigation. World Congress on Neural Networks. In: International Neural Network Society Annual Meeting, vol. III, pp. 141–144 (1993)
8. Jardin, P.: Predicting Bankruptcy using Neural Networks and Other Classification Methods: The Influence of Variable Selection Techniques on Model Accuracy. Neurocomputing 73, 2047–2060 (2010)
9. Kiviluoto, K., Bergius, P.: Exploring Corporate Bankruptcy with Two-level Self-organizing Maps. In: Proceedings of Fifth International Conference on Computational Finance, pp. 373–383 (1998)
10. Kiviluoto, K.: Topology Preservation in Self-Organizing Maps. In: IEEE International Conference on Neural Networks, vol. 1, pp. 294–299 (1996)
11. Kiviluoto, K.: Predicting Bankruptcies with the Self-Organizing Map. Neurocomputing 21, 191–201 (1998)
12. Kohonen, T.: Self-Organizing Maps, 3rd edn. Springer, Heidelberg (2001)
13. Merkevicius, E., Garsva, G., Simutis, R.: Forecasting of Credit Classes with the Self-Organizing Maps. Informaciens Technologijos Ir Valsymas 4(33), 61–66 (2004)
14. Neves, J.C., Vieira, A.: Improving Bankruptcy Prediction with Hidden Layer Learning Vector Quantization. European Accounting Review 15(2), 253–271 (2006)
15. Rajpurohit, V.S., Manohara, P.M.: Using Self Organizing Networks for Moving Object Trajectory Prediction. ICGST-AIML Journal 9(1), 27–34 (2009)
16. Sarlin, P., Marghescu, D.: Visual Predictions of Currency Crises using Self-Organizing Maps. Systems in Accounting, Finance and Management 18(1), 15–38 (2011)
17. Schreck, T., Bernard, J., Tekusova, T., Kohlhammer, J.: Visual Cluster Analysis of Trajectory Data with Interactive Kohonen Maps. Information Visualization 8, 14–29 (2009)

Network Node Label Acquisition and Tracking

Sarvenaz Choobdar, Fernando Silva, and Pedro Ribeiro

CRACS & INESC-Porto LA
Faculty of Science, University of Porto, Portugal
{sarvenaz,fds,pribeiro}@dcc.fc.up.pt

Abstract. Complex networks are ubiquitous in real-world and represent a multitude of natural and artificial systems. Some of these networks are inherently dynamic and their structure changes over time, but only recently has the research community been trying to better characterize them. In this paper we propose a novel general methodology to characterize time evolving networks, analyzing the dynamics of their structure by labeling the nodes and tracking how these labels evolve. Node labeling is formulated as a clustering task that assigns a classification to each node according to its local properties. Association rule mining is then applied to sequences of nodes' labels to extract useful rules that best describe changes in the network. We evaluate our method using two different networks, a real-world network of the world annual trades and a synthetic scale-free network, in order to uncover evolution patterns. The results show that our approach is valid and gives insights into the dynamics of the network. As an example, the derived rules for the scale-free network capture the properties of preferential node attachment.

Keywords: Network Characterization, Node labeling, Clustering, Association Rules.

1 Introduction

Advances in information technology led the world activity to become very much centered on information data. The explosive growth in data that we are witnessing naturally opens an enormous opportunity for researchers to develop new methodologies to dynamically extract useful information and knowledge from the data. Real life data inherently contains structural information on objects and their relationships. This structure can be modeled with networks, or graphs, that are abstract representations of a set of nodes and the connections between them.

Most real world networks are complex, in the sense that they present non trivial topological features. Research on complex network data analysis has been very prolific and a large variety of characterization methodologies emerged, such as node classification [23], graph clustering [13,25], frequent subgraph mining [32,16] or network motifs discovery [24]. These approaches treat the network as a static object. However, many networks are intrinsically dynamic and change over time. Only recently has the research community started to analyze the temporal evolution of networks [22,4,21,7,6]. Most of these studies have characterized network

L. Antunes and H.S. Pinto (Eds.): EPIA 2011, LNAI 7026, pp. 418–430, 2011.

structure by directly examining the topology of the network. Nevertheless, more indirect methods that use network measurements such as degree centrality or clustering coefficient can be a rich source of information [9]. Our aim is precisely to use the evolution of these kind of metrics to study the dynamics of networks that change over time.

We propose a novel two-phase general methodology designed to characterize time evolving networks. First we group and classify nodes based on their role in the network, and then we track changes and find patterns on how this classification evolves.

The first phase involves looking at the network from a static point of view and creating a node classification. We can either use a predefined label for each node, or an automated classification of nodes based on their local properties, such as degree or betweenness, which have been shown to be very fruitful in node characterization [9]. It is however not an easy task to choose a set of these kind of metrics that best generally describes and distinguishes nodes. Costa et al. [10] presented a node label acquisition methodology based on these metrics, but they only identify outliers, that is, singular nodes that are the most different from the others. Instead, we apply these measurements to all nodes in the networks and then employ clustering techniques to group nodes and attribute them labels.

The second phase involves tracking the evolution of node labels over time. We mine association rules using the *apriori* algorithm [2], characterizing the dynamics of the networks in order to uncover emerging patterns that show the appearance, change and disappearance of groups of nodes.

We apply our methodology on a real and on a synthetic complex network in order to demonstrate the validity and usefulness of our approach. We show that our method can discover interesting insights in dynamic networks.

In the remainder of the paper, we start by presenting recent and state of the art work regarding network evolution. After, we describe the proposed methodology in section 3, with all intermediate steps and techniques used. Then, we evaluate our approach in section 4, by applying it to two network datasets. Finally, we draw some conclusions.

2 Related Work

2.1 Graph Clustering

Graph clustering is the task of grouping the nodes of the graph into clusters taking into consideration the edge structure of the graph in such a way that there should be many edges within each cluster and relatively few between the clusters [25].

2.2 Label Acquisition

Label acquisition, as most commonly defined in the literature, involves determining the label for a node in a network that is partially labeled. Normally, it is assumed that at least some of the nodes have a predefined label and only the labels for remaining nodes are predicted using relational classifiers [30].

With networked data, the label of a node may influence the label of a related node. Furthermore, nodes not directly connected may be related through chains of links. This complex dependencies thus suggest that it may be beneficial to predict the label of all nodes simultaneously. Regarding the values of an attribute or attributes for multiple connected nodes for which some attribute values are unknown, a simultaneous statistical assessment is required and this can be done by using collective inferencing [17]. Networked data allow collective inference, meaning that various interrelated values can be derived simultaneously [23]. Macskassy et al. [12] used another source of information in networks that is independent of the available node labels and improved the accuracy of node's label by adding label independent features which include nodes local measurements like degree and betweenness.

All the mentioned studies aim to find the label of a node in a partial labeled network and rely mainly on the available information about label of some of the nodes, they predict labels of nodes instead of assigning the labels. Our work in label acquisition follows the work by Costa et. al. [10], but differs in that we consider all the nodes in the network instead of just the singular node-motifs.

2.3   Node Evolution

Different approaches for explaining network evolution have been reported in the literature. Some have focused on the global evolution of networks by an exploratory point of view. Leskovec et al. [22] discovered the shrinking diameter phenomena on time-evolving networks. Backstrom et al. [4] studied the evolution of communities in social networks. Still from an exploratory perspective, Leskovec et al. [21] studied the evolution of networks but at a more local level. Using a methodology based on the maximum-likelihood principle, they investigate a wide variety of network formation strategies, and show that edge locality plays a critical role in evolution of networks[6].

Braha and Bar-Yam [7] described node's centrality changes over time and showed that hubs do not remain a hub for the all time. They use nodes degree over time to computes correlation between pairs of daily networks.

Other recent papers, present algorithmic tools for the analysis of evolving networks. Tantipathananandh et al. [29] focus on assessing the community affiliation of users and how this changes over time. Sun et al. [27], apply the minimum description length (MDL) principle to the discovery of communities in dynamic networks, developing a parameter-free framework. This is the main difference to previous work such as [1,28]. However, as in [31], the focus lies on identifying approximate clusters of users and their temporal change. No exact patterns are found, nor is time part of the results obtained with these approaches. Ferlez et al. [11] use the MDL principle for monitoring the evolution of a network.

Network motifs as small subgraphs that show the topological properties of the network have also been used in [19] to monitor temporal changes in the structure of an email communication network. They considered z-score of motif as its significance in the network and trace it over time. The dynamics of network is studied in [7] by calculating the network motifs frequency over the time.

3   Methodology

We address the characterization of network dynamics by tracking the evolution of groups of nodes over time, and deriving rules that explain that evolution. We do this in two phases: one deals with node label acquisition, and the other deals with node evolution pattern discovery. The first phase has multiple steps, as illustrated in figure 1, that include measuring local properties of nodes, determining the proper number of clusters in the network, clustering the nodes, coordinating the clusters of nodes in the network time span and assigning labels to groups of nodes. The second phase comprises defining a time granularity for network pattern detection and mining association rules.

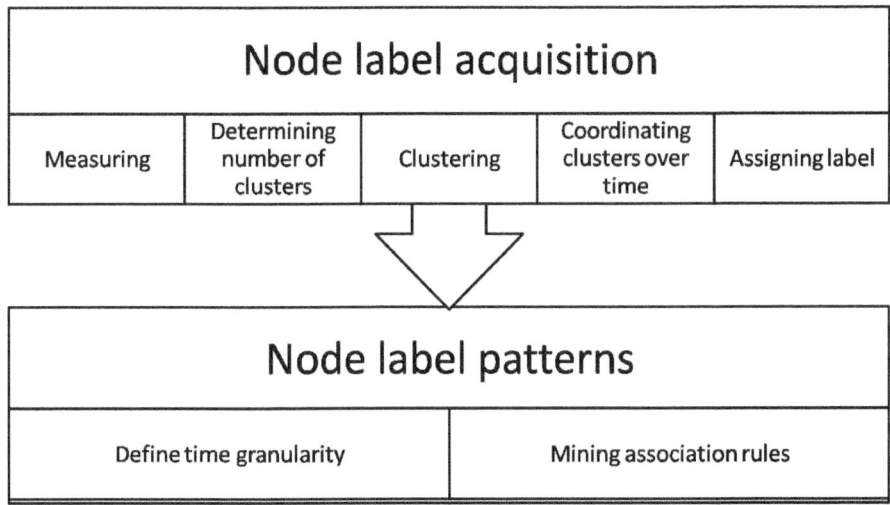

Fig. 1. Proposed methodology

3.1   Node Label Acquisition

The label of a node is determined based on its properties in the network. The same label is assigned to the nodes that have similar properties. Therefore, first we select a set of local measurements that best characterize nodes in the network structure, and then we determine the groups of nodes that have similar properties. The same metrics as introduced by Costa et al. [10] are employed as "feature vector" for nodes. This feature vector measures the connectivity of a node in the neighborhood structure. It includes:

- the normalized average degree (r),
- the coefficient variation of the degrees of the immediate neighbors of a node (cv),

- the clustering coefficient (cc),
- the locality index (loc), which is an extension of the matching index and takes into account all the immediate neighbors of each node, instead of individual edges, and
- the normalized node degree (K).

Label assignment requires the grouping of identical nodes, that is, nodes that share the same feature vector. We use multivariate statistics and pattern recognition [18] techniques to find these identical nodes. Clustering is a method widely used for finding groups of objects, called clusters, in the dataset such that the objects in the same group are more similar to each other than they are to objects of other groups. We use the well known k-means clustering algorithm [15], which bases its operation on the euclidean distance between nodes. If two nodes have similar feature vectors, they are clustered into the same group. After clustering the nodes, coherent groups of nodes are derived and, therefore, can be said to have the same role or label in the network.

The number of potential groups of nodes in the network is equal to the number of clusters in the dataset. Determining the actual number of groups in a dataset is a fundamental and largely unsolved problem in cluster analysis. We employ the method by Sugar and James [26], since it does not require parametric assumptions, is independent of the method of clustering, and was shown to achieve excellent results. This method uses a theoretic information approach that considers the transformed distortion curve $d_K^{-p/2}$ [26]. *"Distortion"* is a measure of within cluster dispersion which is a kind of average Mahalanobis distance between the data and the set of cluster centers as a function of the number of clusters, K. This method is called the *"jump method"*. First, it runs the k-means algorithm for different numbers of clusters, K, and calculates the corresponding distortions, $\hat{d}_K$. Then it transforms the distortion by power transformation of $Y = p/2$, where p is the number of dimensions in the dataset. The "jumps" in the transformed distortion are calculated by $J_k = \hat{d}_K^{-y} - \hat{d}_{K-1}^{-y}$. Finally, the appropriate number of clusters for the data is equal to $K^* = argmax_k J_k$.

We use the jump method to determine the groups of nodes in the network at each time instance. The nodes that are members of the same cluster hold the label of that cluster. The labels can be the numbers of the cluster or be defined manually by the domain expert, based on the properties of the cluster's center.

After labeling the nodes in the network over time, we are left with a series of networks whose nodes are independently labeled. For example, it may be the case that group number one at time t includes nodes with low degree, low clustering coefficient and high neighborhood degree, but at time $t + 1$ it may be that it is group number two that includes nodes with this feature vector. That is, the same feature vector might appear in different groups of nodes at different times. Therefore, a coordination is required for the labels. For the coordination, we use the centers of the detected clusters at each time. We assign the same label to the nodes belonging to two clusters whose centers are close to each other. By considering the Mahalanobis distance of the centers, we can derive an universal label for the nodes of the network for all the time instances.

At the end of this phase, coherent groups of nodes at each time instance are derived and labeled. Therefore, a sequence of labels is generated for each node over time. In the next phase, we attempt to extract from these sequences patterns that explain nodes evolution.

3.2   Node Evolution Patterns

Having a time-evolving network with labeled nodes is a requirement for determining evolution patterns. Our goal is to find rules that explain transitions of nodes between groups over time. We consider the labels of nodes as items, and thus an itemset in our case is the sequence of node's labels over the time. Frequent itemsets are the sequences or subsequences which have minimum support. Therefore, the patterns of node evolution are the extracted frequent itemsets and association rules.

The Apriori algorithm is a powerful tool for mining associations, correlations, causality and sequential patterns [2,3,8,20]. Association rules mining has two main steps [2]:

1. Finding all sets of items (itemsets) whose transaction support is above a minimum support threshold. The support for an itemset is the number of transactions that contain the itemset. Itemsets with minimum support are called large itemsets, and all others small itemsets.
2. Use the large itemsets to generate the desired rules for every large itemset l and all non-empty subsets of l. For every such subset l_a, output a rule of the form $l_a \Rightarrow (l - l_a)$ if the ratio of support (l) to support (l_a) is at least minimum confidence. We need to consider all subsets of l to generate rules with multiple consequents.

The apriori algorithm generates the frequent itemsets with different time granularity. Patterns of evolutions are generated using a sliding window method that enable us to detect changes at different stages of network lifetime. At each time window, rules with different time granularity are extracted.

4   Experiments

The proposed methodology for automatically assigning labels to nodes and track their evolution over time was implemented in R. We evaluated it on two different datasets, a network of the world countries' global trade (GDP data) [14], and a synthetic scale free network. We start by describing these datasets in some detail and then present our evaluation results.

4.1   Datasets

The first data set is created from the publicly available Expanded Trade and GDP Data [14]. The data represents the yearly imports and exports, total trade and gross domestic product (GDP) of 196 countries spanning the 52 years from 1948 till 2000. The time series for each country is the proportion of its share in the global economy according to its GDP for that year. The time series for GDP-Norm is the normalized value of each individual annual GDP, divided by the total GDP for all countries during that year. The topology for the graph was created by comparing the yearly total trade for each country and its trade with each of the other countries. If the trade between country A and country B in any given year accounts for more than 10% of either country's total trade for that year, an edge is created between the two countries.

The second dataset is a synthetic scale-free network generated based on the Barabasi-Albert model for graph generation [5]. It is a model of network growth that is based on two basic parameters: growth and preferential attachment. The basic idea is that in the network nodes with high degrees acquire new edges at higher rates than low-degree nodes. An undirected graph is constructed as follows. Starting with m_0 isolated nodes, at each time step $t = 1, 2, \ldots, N$ a new node j with $m \leq m_0$ links is added to the network. The probability that a link will connect j to an existing node i is linearly proportional to the actual degree of node i given by

$$P(k_i) = k_i / \sum_j k_j \qquad (1)$$

4.2   Results

Table 1 provides details on the networks used in our experiments, namely the number of nodes in each network, the number of time instances of network evolution, and the number of labels produced with the application of the first step of our method for node label acquisition.

Table 1. Statistics of used networks

Dataset	# nodes	# times	# labels
GDP	171	52	4
Scale-Free	200	100	7

Figures 2 and 3 show the profile of the groups found in each network. The profile depicts the values of the feature vector of each group in the network. As explained earlier, the feature vector includes the metrics normalized average degree (r), coefficient variation of the degrees of immediate neighbors (cv), the clustering coefficient (cc), the locality index (loc), and the normalized node degree (K).

GDP network: in this global trade network of countries, our method found four distinguishable groups of nodes. Each group has a different feature vector as illustrated in Figure 2. The first group includes nodes that represent countries with very high degree and many low degree nodes connected to them. Neighbors

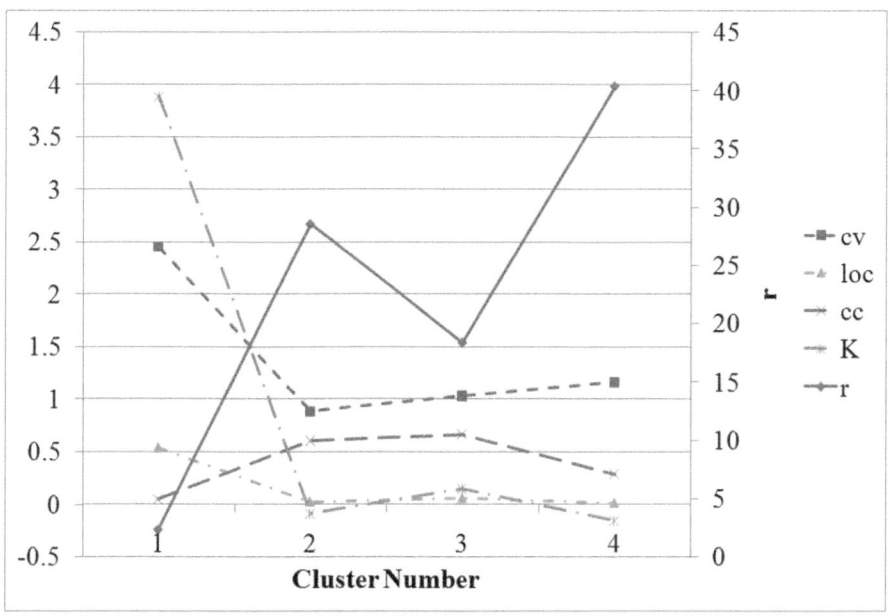

Fig. 2. Profile of feature vector in groups of nodes in the GDP network. The feature of **r** is depicted in the right side of the vertical axis.

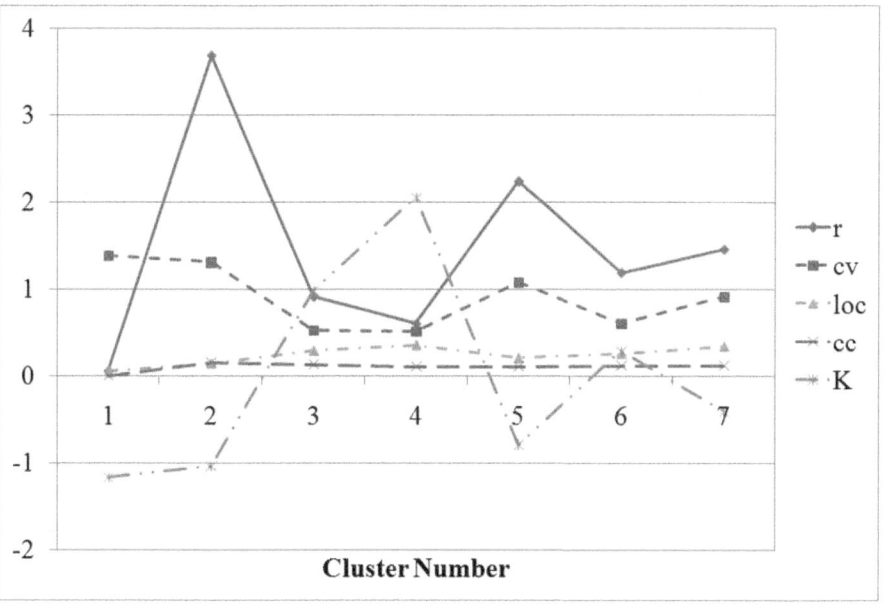

Fig. 3. Profile of feature vector in groups of nodes in the Scale-Free network

of these nodes have low degree since the normalized average degree of the immediate neighbors of a node for this group is very low. This means that nodes of group one behave as hubs in the network, that is, as hub countries in global trade, with commercial transactions with many other countries that have a high variation of degree in neighborhood (cv). According to the value of loc and cc, respectively, the locality index and the clustering coefficient, nodes of this group are highly connected in their neighborhood. United States of America, Canada and France are members of this group.

Figure 4 depicts the evolution on the frequency of each label over time in each network. At the initial stages of network evolution, groups with label 1 and 3 are rather common, but they become rare as the network evolves. These groups have different sizes (number of nodes) at each time step, but they never vanish. Over time, one can notice a transition from group 3 to group 2. After the initial stages of network lifetime, a new group emerges, in this case group number four.

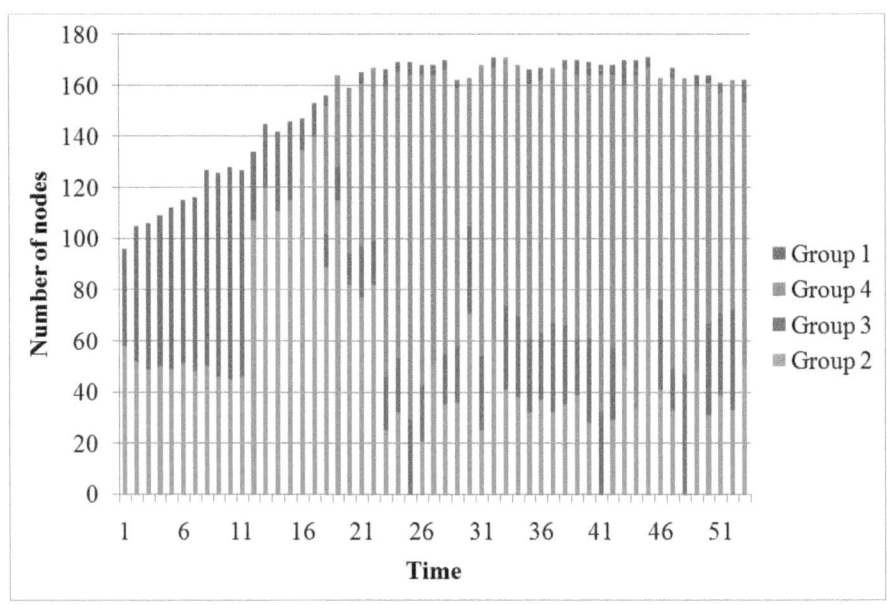

Fig. 4. Frequency of groups in the GDP network over time

The changes over time in the role of the nodes of certain groups is described by the extracted patterns of node's evolution. Table 2 shows the stronger rules, in terms of support and level of confidence, that are extracted from the label sequence of nodes in GDP network. We used a sliding window to find out the changes in the network. The sliding window (SW) parameter helps to narrow down the search interval to find more precisely rules that describe changes of node's label. The size of the SW can be determined by the Fig. 4 which shows the trend of node's membership. With different SW size, several rules could be

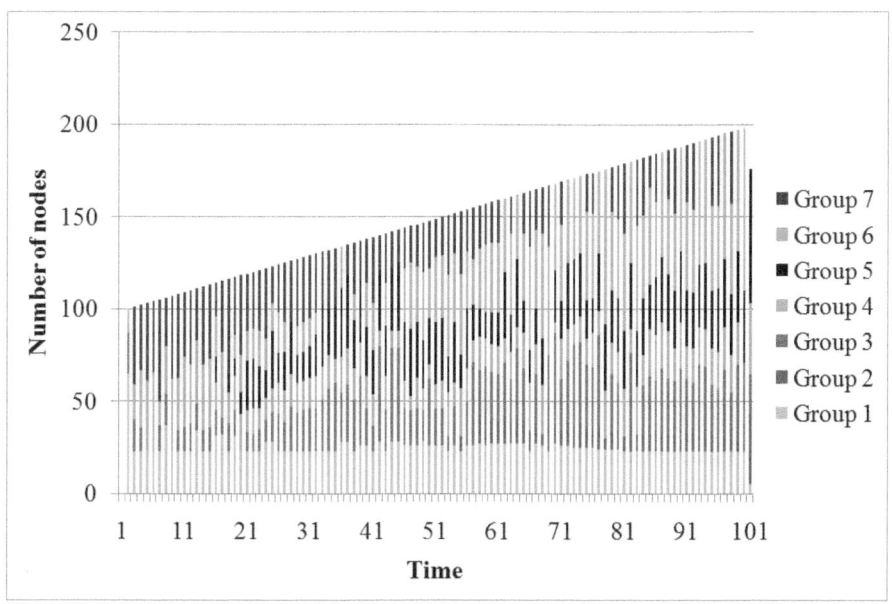

Fig. 5. Frequency of groups in the Scale-Free network over time

found. The significant ones that characterize the appearance and disappearance of the groups are listed in table 2.

For example group four does not exist in the network before time step 18. This pattern of change is detected and described by rule of $\{t14 = 2, t17 = 2\} \Rightarrow \{t18 = 4\}$. This rule says that nodes that were in the second group at time 14 and 17 are likely to change to the fourth group at time 18. The support for this rule is 11%, but its confidence is 87%.

Table 2. Derived rules for the networks

Network	Rule	Support	Confidence
GDP	$\{t1 = 2, t4 = 2, t8 = 3\} \Rightarrow \{t9 = 2\}$	18%	95%
	$\{t14 = 2, t17 = 2\} \Rightarrow \{t18 = 4\}$	11%	87%
	$\{t28 = 4, t29 = 1\} \Rightarrow \{t30 = 3\}$	16%	70%
	$\{t40 = 2, t41 = 3, t42 = 3\} \Rightarrow \{t43 = 2\}$	17%	75%
Scale Free	$\{t16 = 7, t = 18 = 7\} \Rightarrow \{t20 = 5\}$	6%	72%
	$\{t57 = 5\} \Rightarrow \{t60 = 2\}$	11%	82%
	$\{t75 = 6, t76 = 6\} \Rightarrow \{t80 = 3\}$	7%	81%

Scale-Free network: this network was generated with 200 nodes and we sampled 100 networks from its evolution time. Our method detected seven different groups of nodes with distinct feature vectors, as illustrated in Figure 3. The first group of nodes in this network are those that are weakly connected such that all of their local connectivity properties in the feature vector of this group have the lowest values between the nodes but have a very high variation of degree in their neighborhood (cv). A reason for this is that the neighbors of these nodes are mostly low degree nodes that, however, are connected to a hub in the network with very high degree. As shown in Figure 3, these groups have different numbers in each time but never cut down in the network lifetime. Second group are nodes with low degree. This group was formed almost at the middle of network evolution time span (low K). They are connected to very high degree nodes (high c and high cv). This group of nodes appears after the 50th time instance. The third and forth group of nodes are highly connected nodes (high K and cc) with neighborhood of low degree nodes (low r and cv). The other three groups of nodes, 5, 6 and 7, are low degree nodes, but the nodes in fifth group are also connected to a hub, which does not happen with the others groups. The sixth group emerges at the beginning stage of the network development and becomes more frequent as time goes on.

Extracted rules in table 2 describe the stronger trends in nodes' transitions between groups. For example $\{t57 = 5\} \Rightarrow \{t60 = 2\}$ shows that nodes in fifth group, after a while, change to the second group. This rule also shows that as time goes on, regarding the generation model of the scale-free network, although the neighborhood of the nodes get more crowded (r and cv increases), their degree remains low. Nodes in the fifth group have low degree, thus they can not absorb new connections and their degree does not increase.

5   Conclusions

Many networks are intrinsically dynamic and evolve over time. Discovering topological features in these networks is far from an easy task. In this work, we proposed a network characterization method that considers both, a static and a dynamic point of view. It is a two phase methodology that automatically assigns labels to nodes of the network based on their local properties and extracts patterns of nodes evolution. The static view provides a general description of the network through label assignment to groups of nodes. Each group in the network is well characterized by the corresponding feature vector profiling. From a dynamic point of view, the methodology discovers patterns of network evolution at node level. Extracted patterns are general rules that describe how one node's label changes from time to time.

We applied our method to two networks to demonstrate and assess its capabilities. It successfully clusters nodes in groups performing a similar role in the network, labels the groups and, through association rule mining, derives rules that explain, with high confidence, patterns of network evolution. The rules show node transitions between groups as time evolves.

Future research will be pursued to extend this methodology, so that we do not just look at individual nodes but subgraphs in the network. In particular, given our prior work on efficient methods for motifs discovery, we are specially concerned with using subgraph motifs as a metric for network characterization, and then studying network evolution based on such larger entities.

Acknowledgments. Sarvenaz Choobdar is funded by an FCT Research Grant (SFRH/BD/72697/2010). The authors are also thankful to Maria Eduarda Silva for the fruitful discussions and suggestions that she made.

References

1. Aggarwal, C.C., Yu, P.S.: Online analysis of community evolution in data streams. In: Proceedings of the 15th SIAM International Conference on Data Mining, Society for Industrial Mathematics page 56(2005)
2. Agrawal, R., Srikant, R.: Fast algorithms for mining association rules. In: Proceedings of 20th International Conference on Very Large Data Bases, vol. 1215, pp. 487–499 (1994)
3. Agrawal, R., Srikant, R.: Mining sequential patterns. In: 11th International Conference on Data Engineering, page 3 (1995)
4. Backstrom, L., Huttenlocher, D., Kleinberg, J., Lan, X.: Group formation in large social networks: membership, growth, and evolution. In: Proceedings of the 12th ACM SIGKDD International Conference on Knowledge Discovery and Data Mining, pp. 44–54 (2006)
5. Barabási, A.L., Albert, R.: Emergence of scaling in random networks. Science 286(5439), 509 (1999)
6. Berlingerio, M., Bonchi, F., Bringmann, B., Gionis, A.: Mining Graph Evolution Rules. In: Buntine, W., Grobelnik, M., Mladenić, D., Shawe-Taylor, J. (eds.) ECML PKDD 2009. LNCS, vol. 5781, pp. 115–130. Springer, Heidelberg (2009)
7. Braha, D., Bar-Yam, Y.: Time-dependent complex networks: Dynamic centrality, dynamic motifs, and cycles of social interactions. Adaptive Networks 51, 39–50 (2009)
8. Brin, S., Motwani, R., Silverstein, C.: Beyond market baskets: Generalizing association rules to correlations. ACM SIGMOD Record 26, 265–276 (1997)
9. Costa, L.F., Rodrigues, F.A., Travieso, G., Villas Boas, P.R.: Characterization of complex networks: A survey of measurements. Advances In Physics 56, 167 (2007)
10. Costa, L.F., Rodrigues, F.A., Hilgetag, C.C., Kaiser, M.: Beyond the average: detecting global singular nodes from local features in complex networks. EPL (Europhysics Letters) 87, 18008 (2009)
11. Ferlez, J., Faloutsos, C., Leskovec, J., Mladenic, D., Grobelnik, M.: Monitoring network evolution using MDL. In: IEEE 24th International Conference on Data Engineering, pp. 1328–1330 (2008)
12. Gallagher, B., Eliassi-Rad, T.: Leveraging Label-Independent Features for Classification in Sparsely Labeled Networks: An Empirical Study. In: Giles, L., Smith, M., Yen, J., Zhang, H. (eds.) SNAKDD 2008. LNCS, vol. 5498, pp. 1–19. Springer, Heidelberg (2010)
13. Girvan, M., Newman, M.E.J.: Community structure in social and biological networks. PNAS 99(12), 7821–7826 (2002)
14. Gleditsch, K.S.: Expanded trade and GDP data. Journal of Conflict Resolution 46(5), 712 (2002)
15. Hartigan, J.A., Wong, M.A.: A k-means clustering algorithm. Journal of the Royal Statistical Society C 28(1), 100–108 (1979)
16. Huan, J., Wang, W., Prins, J.: Efficient mining of frequent subgraphs in the presence of isomorphism. In: Proceedings of the Third IEEE International Conference on Data Mining, ICDM 2003, pp. 549-552 (2003)

17. Jensen, D., Neville, J., Gallagher, B.: Why collective inference improves relational classification. In: Proceedings of the 10th KDD (2004)
18. Johnson, R.A., Wichern, D.W.: Applied multivariate statistical analysis, vol. 5. Prentice Hall Upper Saddle River, NJ (2002)
19. Juszczyszyn, K., Kazienko, P., Musial, K., Gabrys, B.: Temporal Changes in Connection Patterns of an Email-Based Social Network. In: IEEE/WIC/ACM International Conference on Web Intelligence and Intelligent Agent Technology, vol. 3, pp. 9–12 (2008)
20. Klemettinen, M., Mannila, H., Ronkainen, P., Toivonen, H., Verkamo, A.I.: Finding interesting rules from large sets of discovered association rules. In: Proceedings of the 3rd International Conference on Information and Knowledge Management, pp. 401–407 (1994)
21. Leskovec, J., Backstrom, L., Kumar, R., Tomkins, A.: Microscopic evolution of social networks. In: Proceeding of the 14th ACM SIGKDD International Conference on Knowledge Discovery and Data Mining, pp. 462–470 (2008)
22. Leskovec, J., Kleinberg, J., Faloutsos, C.: Graphs over time: densification laws, shrinking diameters and possible explanations. In: Proceedings of the 11th ACM SIGKDD International Conference on Knowledge Discovery in Data Mining, pp. 177–187 (2005)
23. Macskassy, S.A., Provost, F.: Classification in networked data: A toolkit and a univariate case study. The Journal of Machine Learning Research 8, 935–983 (2007)
24. Milo, R., Shen-Orr, S., Itzkovitz, S., Kashtan, N., Chklovskii, D., Alon, U.: Network Motifs: Simple Building Blocks of Complex Networks. Science 298(5594), 824–827 (2002)
25. Schaeffer, S.E.: Graph clustering. Computer Science Review 1(1), 27–64 (2007)
26. Sugar, C.A., James, G.M.: Finding the number of clusters in a dataset. Journal of the American Statistical Association 98(463), 750–763 (2003)
27. Sun, J., Faloutsos, C., Papadimitriou, S., Yu, P.S.: Graphscope: parameter-free mining of large time-evolving graphs. In: Proceedings of the 13th ACM SIGKDD International Conference on Knowledge Discovery and Data Mining, pp. 687–696 (2007)
28. Sun, J., Tao, D., Faloutsos, C.: Beyond streams and graphs: dynamic tensor analysis. In: Proceedings of the 12th ACM SIGKDD International Conference on Knowledge Discovery and Data Mining, pp. 374–383 (2006)
29. Tantipathananandh, C., Berger-Wolf, T., Kempe, D.: A framework for community identification in dynamic social networks. In: Proceedings of the 13th ACM SIGKDD International Conference on Knowledge Discovery and Data Mining, pp. 717–726 (2007)
30. Taskar, B., Abbeel, P., Koller, D.: Discriminative probabilistic models for relational data. In: 18th Conference on Uncertainty in Artificial Intelligence, pp. 895–902 (2002)
31. Vanetik, N., Shimony, S.E., Gudes, E.: Support measures for graph data. Data Mining and Knowledge Discovery 13(2), 243–260 (2006)
32. Yan, X., Han, J.: gspan: Graph-based substructure pattern mining. In: Proceedings of the 2002 IEEE International Conference on Data Mining, ICDM 2002, pp. 721–724 (2002)

Learning to Rank for Expert Search
in Digital Libraries of Academic Publications[*]

Catarina Moreira, Pável Calado, and Bruno Martins

Instituto Superior Técnico, INESC-ID
Av. Professor Cavaco Silva, 2744-016 Porto Salvo, Portugal
{catarina.p.moreira,pavel.calado,bruno.g.martins}@ist.utl.pt

Abstract. The task of expert finding has been getting increasing attention in information retrieval literature. However, the current state-of-the-art is still lacking in principled approaches for combining different sources of evidence in an optimal way. This paper explores the usage of learning to rank methods as a principled approach for combining multiple estimators of expertise, derived from the textual contents, from the graph-structure with the citation patterns for the community of experts, and from profile information about the experts. Experiments made over a dataset of academic publications, for the area of Computer Science, attest for the adequacy of the proposed approaches.

1 Introduction

The automatic search for knowledgeable people in the scope of specific user communities, with basis on documents describing people's activities, is an information retrieval problem that has been receiving increasing attention [17]. Usually referred to as *expert finding*, the task involves taking a short user query as input, denoting a topic of expertise, and returning a list of people sorted by their level of expertise in what concerns the query topic.

Several effective approaches for finding experts have been proposed, exploring different retrieval models and different sources of evidence for estimating expertise. However, the current state-of-the-art is still lacking in principled approaches for optimally combining the multiple sources of evidence that can be used to estimate expertise. In traditional information retrieval tasks such as ad-hoc retrieval, there has been an increasing interest on the usage of machine learning methods for building retrieval formulas capable of estimating relevance for query-document pairs [13]. The general idea is to use hand-labeled data (e.g., document collections containing relevance judgments for specific sets of queries, or information regarding user-clicks aggregated over query logs) to train ranking

[*] This work was partially supported by the Fundação para a Ciência e Tecnologia (FCT), through project grant PTDC/EIA-EIA/115346/2009 (SMARTIES), and by the ICP Competitiveness and Innovation Framework Program of the European Commission, through the European Digital Mathematics Library (EuDML) project – http://www.eudml.eu/

L. Antunes and H.S. Pinto (Eds.): EPIA 2011, LNAI 7026, pp. 431–445, 2011.

models, this way leveraging on data to combine the different estimators of relevance in an optimal way. However, few previous works have specifically addressed the usage of learning to rank approaches in the task of expert finding.

This paper explores the usage of learning to rank methods in the expert finding task, specifically combining a large pool of estimators for expertise. These include estimators derived from the textual similarity between documents and queries, from the graph-structure with the citation patterns for the community of experts, and from profile information about the experts. We have built a prototype expert finding system using learning to rank techniques, and evaluated it on an academic publication dataset from the Computer Science domain.

The rest of this paper is organized as follows: Section 2 presents the main concepts and related works. Section 3 presents the learning to rank approaches used in our experiments. Section 4 introduces the multiple features upon which we leverage for estimating expertise. Section 5 presents the experimental evaluation of the proposed methods, detailing the dataset and the evaluation metrics, as well as the obtained results. Finally, Section 6 presents our conclusions and points directions for future work.

2   Concepts and Related Work

Serdyukov and Macdonald have surveyed the most important concepts and representative previous works in the expert finding task [17, 15]. Two of the most popular and well-performing types of methods are the profile-centric and the document-centric approaches [6, 21]. Profile-centric approaches build an expert profile as a pseudo document, by aggregating text segments relevant to the expert [1]. These profiles of experts are latter indexed and used to support the search for experts on a topic. Document-centric approaches are typically based on traditional document retrieval techniques, using the documents directly. In a probabilistic approach to the problem, the first step is to estimate the conditional probability $p(q|d)$ of the query topic q given a document d. Assuming that the terms co-occurring with an expert can be used to describe him, $p(q|d)$ can be used to weight the co-occurrence evidence of experts with q in documents. The conditional probability $p(c|q)$ of an expert candidate c given a query q can then be estimated by aggregating all the evidences in all the documents where c and q co-occur. Experimental results show that document-centric approaches usually outperform profile-centric approaches [21].

Many different authors have proposed sophisticated probabilistic retrieval models, specific to the expert finding task, with basis on the document-centric approach [1, 16, 17]. For instance Cao et al. proposed a two-stage language model combining document relevance and co-occurrence between experts and query terms [4]. Fang and Zhai derived a generative probabilistic model from the probabilistic ranking principle and extend it with query expansion and non-uniform candidate priors [10]. Zhu et al. proposed a multiple window based approach for integrating multiple levels of associations between experts and query topics in expert finding [27]. More recently, Zhu et al. proposed a unified language model integrating many document features for expert finding [22]. Although the above models are capable of employing different types of associations among query terms, documents and experts, they mostly ignore other important sources of evidence, such as the importance of individual documents, or the co-citation patterns between experts available from citation graphs. In this paper, we offer a principled approach for combining a much larger set of expertise estimates.

In the Scientometrics community, the evaluation of the scientific output of a scientist has also attracted significant interest due to the importance of obtaining unbiased and fair criteria. Most of the existing methods are based on metrics such as the total number of authored papers or the total number of citations. A comprehensive description of many of these metrics can be found in [19, 20]. Simple and elegant indexes, such as the Hirsch index, calculate how broad the research work of a scientist is, accounting for both productivity and impact. Graph centrality metrics inspired on PageRank, calculated over citation or co-authorship graphs, have also been extensively used [14]. In the context of academic expert search systems, these metrics can easily be used as query-independent estimators of expertise, in much the same way as PageRank is used in the case of Web information retrieval systems.

For combining the multiple sources of expertise, we propose to leverage on previous works concerning the subject of learning to rank for information retrieval (L2R4IR). Tie-Yan Liu presented a good survey on the subject [13], categorizing the previously proposed algorithms into three groups, according to their input representation and optimization objectives:

- **Pointwise approach** - L2R4IR is seen as either a regression or a classification problem. Given feature vectors of each single document from the data for the input space, the relevance degree of each of those individual documents is predicted with scoring functions which can sort all documents and produce the final ranked list.
- **Pairwise approach** - L2R4IR is seen as a binary classification problem for document pairs, since the relevance degree can be regarded as a binary value which tells which document ordering is better for a given pair of documents. Given feature vectors of pairs of documents from the data for the input space, the relevance degree of each of those documents can be predicted with scoring functions which try to minimize the average number of misclassified document pairs. Several different pairwise methods have been proposed, including SVM$rank$ [12].
- **Listwise approach** - L2R4IR is addressed in a way that takes into account an entire set of documents, associated with a query, as instances. These methods train a ranking function through the minimization of a listwise loss function defined on the predicted list and the ground truth list. Given feature vectors of a list of documents of the data for the input space, the relevance degree of each of those documents can be predicted with scoring functions which try to directly optimize the value of a particular information retrieval

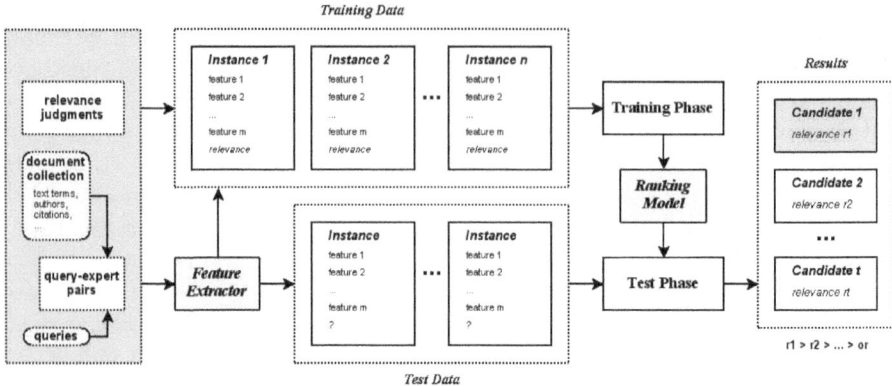

Fig. 1. The general procedure of learning to rank for expert search

evaluation metric, averaged over all queries in the training data [13]. Several different listwise methods have also been proposed, including SVM*map* [25].

In this paper, we made experiments with the application of representative learning to rank algorithms from the pairwise and the listwise approaches, namely the SVM*rank* and the SVM*map* algorithms, in a task of expert finding within digital libraries of academic publications.

3   Learning to Rank Experts

In this paper, we follow a general approach which is common to most supervised learning to rank methods, consisting of two separate steps, namely training and testing. Figure 1 provides an illustration.

Given a set of queries $Q = \{q_1, \ldots, q_{|Q|}\}$ and a collection of experts $E = \{e_1, \ldots, e_{|E|}\}$, each associated with specific documents describing the topics of expertise, a training corpus for learning to rank is created as a set of query-expert pairs, each $(q_i, e_j) \in Q \times E$, upon which a relevance judgment indicating the match between q_i and e_j is assigned by a labeler. This relevance judgment can be a binary label, e.g., relevant or non-relevant, or an ordinal rating indicating relevance, e.g., definitely relevant, possibly relevant, or non-relevant. For each instance (q_i, e_j), a feature extractor produces a vector of features that describe the match between q_i and e_j. Features can range from classical IR estimators computed from the documents associated with the experts (e.g., term frequency, inverse document frequency, BM25, etc.) to link-based features computed from networks encoding relations between the experts in E (e.g., PageRank). The inputs to the learning algorithm comprise training instances, their feature vectors, and the corresponding relevance judgments. The output is a ranking function, f, where $f(q_i, e_j)$ is supposed to either give the true relevance judgment for (q_i, e_j), or produce a ranking score for e_j so that when sorting experts according to these scores the more relevant ones appear on the top of the ranked list.

During the training process, the learning algorithm attempts to learn a ranking function capable of sorting experts in a way that optimizes a particular bound on an information retrieval performance measure (e.g., Mean Average Precision). In the test phase, the learned ranking function is applied to determine the relevance between each expert e_j in E and a new query q. In this paper, we experimented with the following learning to rank algorithms:

- SVM*rank* [12] : This pairwise method builds a ranking model in the form of a linear scoring function, i.e. $f(x) = w^T x$, through the formalism of Support Vector Machines (SVMs). The idea is to minimize the following objective function over a set of n training queries $\{q_i\}_{i=1}^n$, their associated pairs of experts $(x_u^{(i)}, x_v^{(i)})$ and the corresponding relevance judgment $y_{u,v}^{(i)}$ over each pair of experts (i.e., pairwise preferences resulting from a conversion from the ordered relevance judgments over the query-expert pairs):

$$\min \frac{1}{2}||w||^2 + C \sum_{i=1}^n \sum_{u,v:y_{u,v}^{(i)}} \xi_{u,v}^{(i)} \qquad (1)$$

s.t. $w^T(x_u^{(i)} - x_v^{(i)}) >= 1 - \xi_{u,v}^{(i)}$, if $y_{u,v}^{(i)} = 1, \xi_{u,v}^{(i)} >= 0$, $i = 1, \ldots, n$

Differently from standard SVMs, the loss function in SVM*rank* is a hinge loss defined over document pairs. The margin term $\frac{1}{2}||w||^2$ controls the complexity of the pairwise ranking model w. The method introduces slack variables, $\xi_{u,v}^{(i)}$, (i.e., a variable that is added to an optimization constraint to turn an inequality into an equality where a linear combination of variables is less than or equal to a given constant), which measure the degree of misclassification of the datum x_i. The coefficient C affects the trade-off between model complexity and the proportion of non-separable samples. If it is too large, we have a high penalty for non-separable points and we may store many support vectors and overfit. If it is too small, we may have underfitting. The objective function is increased by a function which penalizes non-zero $\xi_{u,v}^{(i)}$, and the optimization becomes a trade off between a large margin, and a small error penalty.

- SVM*map* [25] : This listwise method builds a ranking model through the formalism of structured Support Vector Machines [23], attempting to optimize the metric of Average Precision (AP). Suppose $x = \{x_j\}_{j=1}^m$ is the set of all the experts associated with a training query q, and $y_{u,v}^{(i)}$ represents the corresponding ground truth labels. Any incorrect label for x is represented as y^c. The SVM*map* approach can be formalized as follows, where AP is used in the constraints of the structured SVM optimization problem.

$$\min \frac{1}{2}||w||^2 + \frac{C}{n}\sum_{i=1}^{n}\xi^{(i)}$$

$$\text{s.t. } \forall y^{c(i)} \neq y^{(i)}, w^T\Psi(y^{(i)}, x^{(i)}) >= w^T\Psi(y^{c(i)}, x^{(i)}) + 1 - AP(y^{c(i)}) - \xi^{(i)} \tag{2}$$

In the constraints, Ψ is called the joint feature map, whose definition is:

$$\Psi(y, x) = \sum_{u,v:y_u=1,y_v=0}(x_u - x_v)$$

$$\Psi(y^c, x) = \sum_{u,v:y_u=1,y_v=0}(x_u^c, y_v^c)(x_u - x_v) \tag{3}$$

Since there are an exponential number of incorrect labels for the documents, it is a big challenge to directly solve the optimization problem involving an exponential number of constraints for each query. The formalism of structured SVMs efficiently tackles this issue by maintaining a working set with those constraints with the largest violation:

$$Violation \triangleq 1 - AP(y^c) + w^T\Psi(y^c, x) \tag{4}$$

The survey by Tie-Yan Liu discusses the above methods in more detail [13].

4   Features for Estimating Expertise

The considered set of features for estimating the expertise of a researcher towards a given query can be divided into three groups, namely textual features, profile features and graph features. The textual features are similar to those used in standard text retrieval systems and also in previous learning to rank experiments (e.g., TF-IDF and BM25 scores). The profile similarity features correspond to importance estimates for the authors, derived from their profile information (e.g., number of papers published). Finally, the graph features correspond to importance and relevance estimates computed from the author co-authorship and co-citation graphs.

4.1   Features Based on Textual Similarity

Similarly to previous expert finding proposals based on document-centric approaches, we also use textual similarity between the query and the contents of the documents to build estimates of expertise. In the domain of academic digital libraries, the associations between documents and experts can easily be obtained from the authorship information associated to the publications. For each topic-expert pair, we used the Okapi BM25 document-scoring function, to compute the textual similarity features. Okapi BM25 is a state-of-the-art IR ranking mechanism composed of several simpler scoring functions with different parameters and components (e.g., term frequency and inverse document frequency). It can be computed through the formula shown in Equation 5, where $Terms(q)$ represents the set of terms from query q, $Freq(i, d)$ is the number of occurrences of term i in document d, $|d|$ is the number of terms in document d, and $\mathcal{A}$ is the average length of the documents in the collection. The values given to the parameters k_1 and b were 1.2 and 0.75 respectively. Most previous IR experiments use these default values for the k_1 and b parameters.

$$BM25(q, d) = \sum_{i \in Terms(q)} \log \left(\frac{N - Freq(i) + 0.5}{Freq(i) + 0.5} \right) \times$$

$$\frac{(k_1 + 1) \times \frac{Freq(i,d)}{|d|}}{\frac{Freq(i,d)}{|d|} + k_1 \times (1 - b + b \times \frac{|d|}{\mathcal{A}})} \quad (5)$$

We also experimented with other textual features commonly used in ad-hoc IR systems, such as *Term Frequency* and *Inverse Document Frequency*.

Term Frequency (TF) corresponds to the number of times that each individual term in the query occurs in all the documents associated with the author. Equation 6 describes the TF formula, where $Terms(q)$ represents the set of terms from query q, $Docs(a)$ is the set of documents having a as author, $Freq(i, d_j)$ is the number of occurrences of term i in document d_j and $|d_j|$ represents the number of terms in document d_j.

$$TF_{q,a} = \sum_{j \in Docs(a)} \sum_{i \in Terms(q)} \frac{Freq(i, d_j)}{|d_j|} \quad (6)$$

The Inverse Document Frequency (IDF) is the sum of the values for the inverse document frequency of each query term and is given by Equation 7. In this formula, $|D|$ is the size of the document collection and $f_{i,D}$ corresponds to the number of documents in the collection where the i_{th} query term occurs.

$$IDF_q = \sum_{i \in Terms(q)} \log \frac{|D|}{f_{i,D}} \quad (7)$$

Other features used were the number of unique authors associated with documents containing the query topics, the range of years since the first and last publications of the author containing the query terms, and the document length, in terms of the number of words, for all the publications associated to the author.

In the computation of these textual features, we considered two different textual streams from the documents, namely (i) a stream consisting of the titles, and (ii) a stream using the abstracts of the articles.

4.2   Features Based on Profile Information

We also considered a set of profile features related to the amount of published materials associated with authors, generally taking the assumption that highly prolific authors are more likely to be considered experts. Most of the features based on profile information are query independent, meaning that they have the same value for different queries. The considered set of profile features are based on the temporal interval between the first and the last publications, the average number of papers and articles per year, and the number of publications in conferences and in journals with and without the query topics in their contents.

4.3 Features Based on Graphs Co-citation and Co-authorship

Scientific impact metrics computed over scholarly networks, encoding co-citation and co-authorship information, can offer effective approaches for estimating the importance of the contributions of particular publications, publication venues, or individual authors. Thus, we considered a set of features that estimate expertise with basis on co-citation and co-authorship information. The features considered are divided in two sets, namely (i) citation counts and (ii) academic indexes. In what regards citation counts, we used the total, the average and the maximum number of citations of papers containing the query topics, the average number of citations per year of the papers associated with an author and the total number of unique collaborators which worked with an author. On what regards academic impact indexes, we used the following features:

- **Hirsch index** of the author and of the author's institution, measuring both the scientific productivity and the apparent scientific impact [11]. An author/institution has an Hirsch index of h if h of his N_p papers have at least h citations each, and the other $(N_p - h)$ papers have at most h citations each. Authors with a high Hirsch index, or authors associated with institutions with a high Hirsch index, are more likely to be considered experts.
- The **h-b-index**, which extends the Hirsch index for evaluating the impact of scientific topics in general [2]. In our case, the scientific topic is given by the query terms and thus the query has an h-b-index of i if i of the N_p papers containing the query terms in the title or abstract have at least i citations each, and the other $(N_p - i)$ papers have at most i citations each.
- **Contemporary Hirsch index** of the author, which adds an age-related weighting to each cited article, giving less weight to older articles [18]. A researcher has a contemporary Hirsch index h^c if h^c of his N_p articles get a score of $S^c(i) >= h^c$ each, and the rest $(N_p - h^c)$ articles get a score of $S^c(i) <= h^c$. For an article i, the score $S^c(i)$ is defined as:

$$S^c(i) = \gamma * (Year(now) - Year(i) + 1)^{-\delta} * |CitationsTo(i)| \qquad (8)$$

  The γ and δ parameters are set to 4 and 1, respectively, meaning that the citations for an article published during the current year account four times, the citations for an article published 4 years ago account only one time, the citations for an article published 6 years ago account 4/6 times, and so on.
- **Trend Hirsch index** [18] for the author, which assigns to each citation an exponentially decaying weight according to the age of the citation, this way

estimating the impact of a researcher's work in a particular time instance. A researcher has a trend Hirsch index h^t if h^t of his N_p articles get a score of $S^t(i) >= h^t$ each, and the rest $(N_p - h^t)$ articles get a score of $S^t(i) <= h^t$. For an article i, the score $S^t(i)$ is defined as:

$$S^t(i) = \gamma * \sum_{\forall x \in C(i)} (Year(now) - Year(x) + 1)^{-\delta} \qquad (9)$$

The γ and δ parameters are set to 4 and 1, respectively.

- **Individual Hirsch index** of the author, computed by dividing the value of the standard Hirsch index by the average number of authors in the articles that contribute to the Hirsch index of the author, in order to reduce the effects of frequent co-authorship with influential authors [3].
- The **a-index** of the author/institution, measuring the magnitude of the most influential articles. For an author or institution with an Hirsch index of h that has a total of $N_{c,tot}$ citations toward his papers, we say that he has an a-index of $a = N_{c,tot}/h^2$.
- The **g-index** of the author/institution, also quantifying scientific productivity with basis on the publication record [9]. Given a set of articles associated with the author/institution, ranked in decreasing order of the number of citations that they received, the g-index is the unique largest number g such that the top g articles received on average at least g citations.
- The **e-index** of the author [26] which represents the excess amount of citations of an author. The motivation behind this index is that we can complement the h-index by taking into account these excess amounts of citations which are ignored by the h-index. The e-index is given by the Equation 10, where cit_j are the citations received by the $j_t h$ paper and h is the h-index.

$$e = \sum_{j=1}^{h} \sqrt{cit_j - h^2} \qquad (10)$$

Besides the above features, and following the ideas of Chen et al. [5], we also considered a set of graph features that estimate the influence of individual authors using PageRank, a well-known graph linkage analysis algorithm that was introduced by the Google search engine.

PageRank assigns a numerical weighting to each element of a linked set of objects (e.g., hyperlinked Web documents or articles in a citation network) with the purpose of measuring its relative importance within the set. The PageRank value of a node is defined recursively and depends on the number and PageRank scores of all other nodes that link to it (i.e., the incoming links). A node that is linked to by many nodes with high PageRank receives a high rank itself.

Formally, given a graph with N nodes $i = 1, 2, \cdots, N$, with L directed links that represent references from an initial node to a target node with weights $\alpha = 1, 2, \cdots, L$, the PageRank Pr_i for the ith node is defined by:

$$Pr_i = \frac{0.5}{N} + 0.5 \sum_{j \in inlinks(L,i)} \frac{\alpha_j Pr_j}{outlinks(L,j)} \qquad (11)$$

In the formula, the sum is over the neighboring nodes j in which a link points to node i. The first term represents the random jump in the graph, giving a uniform injection of probability into all nodes in the graph. The second term describes the propagation of probability corresponding to a random walk, in which a value at node j propagates to node i with probability $\frac{\alpha_j Pr_j}{outlinks(L,j)}$.

The features that we considered correspond to the sum and average of the PageRank values associated to the papers of the author that contain the query terms, computed over a directed graph representing citations between papers. Each citation link in the graph is given a score of $1/N$, where N represents the number of authors in the paper. Authors with high PageRank scores are more likely to be considered experts.

5   Experimental Validation

The main hypothesis behind this work is that learning to rank approaches can be effectively used in the context of expert search tasks, in order to combine different estimators of relevance in a principled way, this way improving over the current state-of-the art. To validate this hypothesis, we have built a prototype expert search system, reusing existing implementations of state-of-the-art learning to rank algorithms, namely the SVM*rank*[1] implementation by Thorsten Joachims [12] and the SVM*map*[2] implementation by Yue et al [25].

We implemented the methods responsible for computing the features listed in the previous section, using *Microsoft SQL Server 2008* (e.g., the full-text search capabilities for computing the textual similarity features) and several existing Java software packages (e.g., the LAW[3] package for computing PageRank).

The validation of the prototype required a sufficiently large repository of textual contents describing the expertise of individuals within a specific area. In this work, we used a dataset for evaluating expert search in the Computer Science research domain, corresponding to an enriched version of the DBLP[4] database made available through the Arnetminer project.

DBLP data has been used in several previous experiments regarding citation analysis [19, 20] and expert search [8]. It is a large dataset covering both journal and conference publications for the computer science domain, and where substantial effort has been put into the problem of author identity resolution, i.e., references to the same persons possibly with different names. Table 1 provides a statistical characterization of the DBLP dataset.

To train and validate the different learning to rank methods, we also needed a set of queries with the corresponding author relevance judgments. For the Computer Science domain, we used the relevant judgments provided by Arnetminer[5] which have already been used in other expert finding experiments [24].

---

[1] http://www.cs.cornell.edu/people/tj/svm_light/svm_rank.html
[2] http://projects.yisongyue.com/svmmap/
[3] http://law.dsi.unimi.it/software.php
[4] http://www.arnetminer.org/citation
[5] http://arnetminer.org/lab-datasets/expertfinding/

Table 1. Statistical characterization of the DBLP dataset used in our experiments

Property	Value
Total Authors	1 033 050
Total Publications	1 632 440
Total Publications containing Abstract	653 514
Total Papers Published in Conferences	606 953
Total Papers Published in Journals	436 065
Total Number of Citations Links	2 327 450

The Arnetminer dataset comprises a set of 13 query topics from the Computer Science domain, each associated to a list of expert authors. In order to add negative relevance judgments (i.e., complement the dataset with unimportant authors for each of the query topics), we searched the dataset with the keywords associated to each topic, retrieving the top $n/2$ authors according to the BM25 metric and retrieving $n/2$ authors randomly selected from the dataset, where n corresponds to the number of expert authors associated to each particular topic. This way, we obtained twice the relevant judgments provided by Arnetminer, ending up with 2794 records for all 13 queries. Table 2 shows the distribution for the number of experts associated to each topic, as provided by Arnetminer.

The test collection was used in a leave-one-out cross-validation methodology, in which different experiments used 9 different queries to train a ranking model, which was then evaluated over the remaining queries. The averaged results from the four different cross-validation experiments are finally used as the evaluation result. To measure the quality of the results produced by the different learning to rank algorithms, we used two different performance metrics, namely the Precision@k (P@k) and the Mean Average Precision (MAP).

Precision at rank k is used when a user wishes only to look at the first k retrieved domain experts. The precision is calculated at that rank position through Equation 12.

$$P@k = \frac{r(k)}{k} \qquad (12)$$

In the formula, $r(k)$ is the number of relevant authors retrieved in the top k positions. $P@k$ only considers the top-ranking experts as relevant and computes the fraction of such experts in the top-k elements of the ranked list.

Table 2. Characterization of the Arnetminer dataset of Computer Science experts

Query Topics	Rel. Authors	Query Topics	Rel. Authors
Boosting (B)	46	Natural Language (NL)	41
Computer Vision (CV)	176	Neural Networks (NN)	103
Cryptography (C)	148	Ontology (O)	47
Data Mining (DM)	318	Planning (P)	23
Information Extraction (IE)	20	Semantic Web (SW)	326
Intelligent Agents (IA)	30	Support Vector Machines (SVM)	85
Machine Learning (ML)	34		

Table 3. Results of the SVM*map* and SVM*rank* methods

	P@5	P@10	P@15	P@20	MAP
SVM*rank*	0.9333	**0.9104**	**0.8848**	0.8698	**0.8150**
SVM*map*	**0.9458**	0.8979	0.8778	**0.8721**	0.8131

Table 4. The results obtained with different sets of features and comparison with other approaches

	P@5	P@10	P@15	P@20	MAP
Text Similarity + Profile + Graph	**0.9333**	**0.9104**	**0.8848**	**0.8698**	**0.8150**
Text Similarity + Profile	0.6917	0.6583	0.6861	0.6552	0.6601
Text Similarity + Graph	0.9250	0.8934	0.8167	0.7896	0.7677
Profile + Graph	0.8667	0.8250	0.8273	0.8125	0.7943
Text Similarity	0.7042	0.6646	0.6597	0.6511	0.6569
Profile	0.7500	0.7646	0.7389	0.7313	0.7464
Graph	0.8750	0.8438	0.8181	0.8021	0.7846
h-*b*-Index	0.7385	0.7077	0.6821	0.6700	0.6053
Expert Finding (Yang et al.) [24]	0.5500	0.6000	0.6333	–	0.6356

The Mean of the Average Precision over test queries is defined as the mean over the precision scores for all retrieved relevant experts. For each query r, the Average Precision (AP) is given by:

$$AP[r] = \frac{\sum_{k=1}^{n} P@k[r] \times I\{g_{r_k} = \max(g)\}}{\sum_{k=1}^{n} I\{g_{r_k} = \max(g)\}} \tag{13}$$

As before, n is the number of experts associated with query q and g_{rk} is the relevance grade for author k in relation to the query r. In the case of our datasets, $\max(g) = 1$ (i.e., we have 2 different grades for relevance, 0 or 1).

Table 3 presents the obtained results over the DBLP dataset. The obtained results attest for the adequacy of both learning to rank approaches, showing that SVM*rank* and SVM*map* achieve a similar performance, with SVM*rank* slightly outperforming SVM*map* in our experiments in terms of MAP.

In a separate experiment, we attempted to measure the impact of the different types of ranking features on the quality of the results. Using the best performing learning to rank algorithm, SVM*rank*, we separately measured the results obtained by ranking models that considered (i) only the textual similarity features, (ii) only the profile features, (iii) only the graph features, (iv) only a representative graph feature, namely the h-*b*-index, (v) textual similarity and profile features, (vi) textual similarity and graph features and (vii) profile and graph features. Table 4 shows the obtained results, also presenting the previous results reported by Yang et al. [24] over the same dataset, as well as the results obtained by the h-*b*-index bibliographic index.

As we can see, the set with the combination of all features has the best results. The results also show that, individually, textual similarity features have the poorest results. This means that considering only textual evidence provided by query topics, together with article's titles and abstracts, may not be enough to

Table 5. Top five people returned by the system for four different queries

Best Results			Worst Results
Neural Networks	Machine Learning	SVMs	Boosting
Geoffrey E. Hinton	Robert E. Schapire	Thorsten Joachims	J. Ross Quinlan
Erkki Oja	Vladimir Vapnik	Robert E. Schapire	B. Han
Yann LeCun	Thomas G. Dietterich	Vladimir Vapnik	W. Shireen
Thomas G. Dietterich	Michael I. Jordan	Christopher J. C. Burges	L. Carlos de Freitas
Michael I. Jordan	Manfred K. Warmuth	Tomaso Poggio	Robert E. Schapire

determine if some authors are experts or not, and that indeed the information provided by citation and co-authorship patterns can help in expert retrieval. Finally, the results show that the different combinations of all features proposed in this paper outperform the previously proposed learning to rank approach for expert finding made by Yang et al. [24]

Figure 2 plots the obtained average precision in each of the individual query topics for the best performing approach, namely SVM*rank* with the combination of all features. The figure presents the query topics in the same order as they are given in Table 2. The horizontal dashed line corresponds to the MAP obtained in the same experiment. The results show that there are only slightly variations in performance for the different queries.

Finally, Table 5 shows the top five people which were returned by the system for four different queries, corresponding to the best and worst results in terms of the P@5 metric. The system performed well for the queries Neural Networks, Machine Learning and Support Vector Machines (SVMs). Although these are very related topics, the system managed to distinguish between them and still identify the relevant experts in these areas correctly. However, worse results were returned for the query Boosting. These poor results can be explained by the absence of the query topics in the titles and abstracts of the publications of authors working in the area. We realized that the authors which were judged as relevant, and therefore considered experts, did not have too many query topics present in their publication's titles or abstracts, leading to misclassifications.

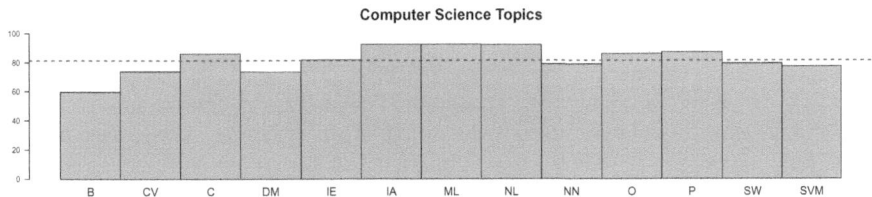

Fig. 2. Average precision over the different query topics

6   Conclusions

This paper explored the usage of learning to rank methods in the context of expert searching within digital libraries of academic publications. We argue that learning to rank provides a sound approach for combining multiple estimators of expertise, derived from the textual contents, from the graph-structure of the community of experts, and from expert profile information. Experiments on datasets of academic publications show very good results in terms of P@5 and MAP, attesting for the adequacy of the proposed approaches.

Despite the interesting results, there are also many ideas for future work. Recent advancements in the area of learning to rank for information retrieval are, for instance, concerned with query-dependent ranking (i.e., using different ranking models according to the type of queries being issued) and it would be interesting to test these techniques in expert searching tasks.

Our approach to the expert finding problem can also be generalized to any type of entity search. The introduction of Entity Ranking Track in INEX 2007, with basis on a *Wikipedia* dataset, provides a good platform for general entity search evaluation [7]. For future work, it would be interesting to experiment with learning to rank methods, similar to the ones proposed in this paper, over the more general entity search problem.

References

1. Balog, K., Azzopardi, L., de Rijke, M.: Formal models for expert finding in enterprise corpora. In: Proceedings of the 29th Annual International ACM SIGIR Conference on Research and Development in Information Retrieval (2006)
2. Banks, M.: An extension of the Hirsch index: Indexing scientific topics and compounds. Scientometrics 69(1) (2006)
3. Batista, P.D., Campiteli, M.G., Kinouchi, O., Martinez, A.S.: Is it possible to compare researchers with different scientific interests? Scientometrics 68(1) (2006)
4. Cao, Y., Liu, J., Bao, S., Li, H.: Research on expert search at enterprise track of TREC 2005. In: Proceedings of the 14th Text Retrieval Conference (2006)
5. Chen, P., Xie, H., Maslov, S., Redner, S.: Finding scientific gems with Google's page rank algorithm. Journal of Informetrics 1(1) (2007)
6. Craswell, N., de Vries, A.P., Soboroff, I.: Overview of the TREC-2005 enterprise track. In: Proceedings of the 14th Text Retrieval Conference (2006)
7. de Vries, A.P., Vercoustre, A.-M., Thom, J.A., Craswell, N., Lalmas, M.: Overview of the INEX 2007 Entity Ranking Track. In: Fuhr, N., Kamps, J., Lalmas, M., Trotman, A. (eds.) INEX 2007. LNCS, vol. 4862, pp. 245–251. Springer, Heidelberg (2008)
8. Deng, H., King, I., Lyu, M.R.: Formal models for expert finding on DBLP bibliography data. In: Proceedings of the 8th IEEE International Conference on Data Mining (2008)
9. Egghe, L.: Theory and practise of the *g*-index. Scientometrics 69(1) (2006)
10. Fang, H., Zhai, C.: Probabilistic models for expert finding. In: Proceedings of the 29th European Conference on Information Retrieval Research (2007)
11. Hirsch, J.E.: An index to quantify an individual's scientific research output. Proceedings of the National Academy of Sciences USA 102(46) (2005)
12. Joachims, T.: Training linear SVMs in linear time. In: Proceedings of the ACM Conference on Knowledge Discovery and Data Mining, KDD (2006)
13. Liu, T.: Learning to rank for information retrieval. Foundations and Trends in Information Retrieval 3(3) (2009)

14. Liu, X., Bollen, J., Nelson, M.L., Van de Sompel, H.: Co-authorship networks in the digital library research community. Information Processing and Management 41(6) (2005)
15. Macdonald, C., Ounis, I.: Voting techniques for expert search. Knowledge and Information Systems 16(3) (2008)
16. Petkova, D., Croft, W.B.: Proximity-based document representation for named entity retrieval. In: Proceedings of the 16th ACM Conference on Information and Knowledge Management (2007)
17. Serdyukov, P.: Search for Expertise: Going Beyond Direct Evidence. PhD thesis, University of Twente (2009)
18. Sidiropoulos, A., Katsaros, D., Manolopoulos, Y.: Generalized h-index for disclosing latent facts in citation networks. Scientometrics (2006)
19. Sidiropoulos, A., Manolopoulos, Y.: A citation-based system to assist prize awarding. ACM SIGMOD Record 34(4) (2005)
20. Sidiropoulos, A., Manolopoulos, Y.: Generalized comparison of graph-based ranking algorithms for publications and authors. Journal for Systems and Software 79(12) (2006)
21. Soboroff, I., de Vries, A.P., Craswell, N.: Overview of the TREC-2006 enterprise track. In: Proceedings of the 15th Text Retrieval Conference (2007)
22. Zhu, J., Song, D., Rüger, S., Huang, X.: Modeling document features for expert finding. In: Proceedings of the 17th ACM Conference on Information and Knowledge Management (2008)
23. Tsochantaridis, I., Joachims, T., Hofmann, T., Altun, Y.: Large margin methods for structured and interdependent output variables. Journal of Machine Learning Research 6 (2005)
24. Yang, Y., Tang, J., Wang, B., Guo, J., Li, J., Chen, S.: Expert2Bole: From expert finding to bole search. In: Theeramunkong, T., Kijsirikul, B., Cercone, N., Ho, T.-B. (eds.) PAKDD 2009. LNCS, vol. 5476. Springer, Heidelberg (2009)
25. Yue, Y., Finley, T., Radlinski, F., Joachims, T.: A support vector method for optimizing average precision. In: Proceedings of the 30th ACM SIGIR international Conference on Research and Development in Information Retrieval (2007)
26. Zhang, C.-T.: The e-index, complementing the h-index for excess citations. Public Library of Science One 4 (2009)
27. Zhu, J., Song, S., Rüger, S., Eisenstadt, M., Motta, E.: The open university at TREC 2006 enterprise track expert search task. In: Proceedings of the 15th Text Retrieval Conference (2007)

Thematic Fuzzy Clusters
with an Additive Spectral Approach

Susana Nascimento[1], Rui Felizardo[1], and Boris Mirkin[2,3]

[1] Department of Computer Science and Centre for Artificial Intelligence (CENTRIA),
Faculdade de Ciências e Tecnologia, Universidade Nova de Lisboa, Caparica, Portugal
[2] Department of Computer Science, Birkbeck University of London, London, UK
[3] School of Applied Mathematics and Informatics, Higher School of Economics,
Moscow, RF

Abstract. This paper introduces an additive fuzzy clustering model for similarity data as oriented towards representation and visualization of activities of research organizations in a hierarchical taxonomy of the field. We propose a one-by-one cluster extracting strategy which leads to a version of spectral clustering approach for similarity data. The derived fuzzy clustering method, FADDIS, is experimentally verified both on the research activity data and in comparison with two state-of-the-art fuzzy clustering methods. Two developed simulated data generators, affinity data of Gaussian clusters and genuine additive similarity data, are described, and comparison of the results over this data are reported.

1 Introduction

Relational data have become popular in several important application areas such as bioinformatics [25,24,34,16], recommendation systems (e.g. [32]), Web mining and text analysis [22,14,27,6]. Our motivation comes from our interest in mapping the activities of a research organization to a taxonomy of the field. The prime objects here are topics of the taxonomy rather than the individual members or teams in the organization, and the information is organized as an index of similarity between the topics rather than the members. In such a setting, it seems rather natural to assume an additive action of the hidden research patterns as the underlying mechanism for the generation of the similarity index. This leads us to develop a novel relational fuzzy clustering method, the Fuzzy Additive Spectral Clustering (FADDIS), by combining a model-based approach of additive clustering and the spectral clustering approach.

In spite of the fact that many relational fuzzy clustering algorithms have been developed already [2,3,5,7,9,10,13,26,33,35], they all involve manually specified parameters such as the number of clusters or threshold of similarity without providing any guidance for choosing them. Our method does provide guidance for choosing the number of clusters. Moreover, it appears, it is quite competitive in comparison to the state of the art fuzzy clustering algorithms.

The method itself is described in a technical report [18] and briefly outlined in [20]. The main goal of this paper is to experimentally compare the

L. Antunes and H.S. Pinto (Eds.): EPIA 2011, LNAI 7026, pp. 446–461, 2011.
© Springer-Verlag Berlin Heidelberg 2011

FADDIS algorithm with two state-of-the-art fuzzy clustering algorithms differently extending fuzzy c-Means to the relational data. One of these fuzzy clustering algorithms combines fuzzy c-means with a recently proposed fast-mapping technique proved superior to many other techniques, the Fast Map Fuzzy c-Means (FMFCM) [5], and the other is an extension of the c-means to dissimilarity data, the Non-Euclidean Relational Fuzzy c-Means (NERFCM) [10].

To be comprehensive in the experimentation, we developed two different cluster structure generators, each involving a controlled extent of noise. The first of them generates Gaussian entity-to-feature clusters with a different extent of intermix. The second produces genuine similarity data according to the additive fuzzy clustering model. Although the FADDIS does outperform the two other algorithms in our experiments, it also shows some unexpected behavior, which is yet to be investigated.

The rest of the paper is organized as follows. Section 2 describes the additive model and FADDIS method. Section 3 describes the experiment and its results over entity-to-feature Gaussian cluster sets. Section 4 describes the experiment and its results over genuine similarity datasets generated according to the additive fuzzy clustering data model. Section 5 illustrates application of FADDIS to the representation of thematic clusters of research activities in a hierarchic taxonomy of the field. Section 6 concludes the paper.

2   Additive Fuzzy Clustering Model and Spectral FADDIS Algorithm

The similarity, or relational, data is a matrix $W = (w_{tt'})$, $t, t' \in T$, of similarity indexes $w_{tt'}$, between objects t, t' from a set of objects T. Specifically, the elements of T can be leaves of a taxonomy tree such as a related hierarchical taxonomy such as Classification of Computer Subjects by ACM (ACM-CCS) [1] (see [18]). Then individual projects or members of a research organization can be represented with fuzzy membership profiles over the subjects (leaves) of the taxonomy. Given a project-to-subject profile matrix F, the similarity matrix can be defined as $W = F^T F$ so that w_{tt} is the inner product of subject columns t and t'. These subject-to-subject similarity values are assumed to be manifested expressions of some hidden patterns represented by fuzzy clusters. To develop an additive model, we formalize a relational fuzzy cluster as represented by: (i) a membership vector $\mathbf{u} = (u_t)$, $t \in T$, such that $0 \leq u_t \leq 1$ for all $t \in T$, and (ii) an intensity $\mu > 0$ that expresses the extent of significance of the pattern corresponding to the cluster. The intensity applies as a scaling factor to $\mathbf{u}$ so that it is the product $\mu \mathbf{u}$ that expresses the hidden pattern rather than its individual co-factors. Given a value of the product μu_t, to separate μ and u_t, a conventional scheme applies: the scale of the membership vector $\mathbf{u}$ is constrained on a constant level by a condition such as $\sum_t u_t = 1$ or $\sum_t u_t^2 = 1$; then the remaining factor defines the value of μ. As will be seen from formula (4), the latter normalization suits our fuzzy clustering model well and thus is accepted further on. Also, to admit a possible pre-processing transformation of the given similarity matrix W, we denote the matrix involved in the process of clustering as $A = (a_{tt'})$.

The additive fuzzy clustering model in (1) follows that of [29,17,28] and involves K fuzzy clusters that reproduce the input similarities $a_{tt'}$ up to additive errors:

$$a_{tt'} = \sum_{k=1}^{K} \mu_k^2 u_{kt} u_{kt'} + e_{tt'}, \tag{1}$$

where $\mathbf{u}_k = (u_{kt})$ is the membership vector of cluster k, μ_k its intensity ($k = 1, 2, ..., K$), and $e_{tt'}$ is the residual similarity not explained by the model.

The item $\mu_k^2 u_{kt} u_{kt'}$ in (1) is the product of $\mu_k u_{kt}$ and $\mu_k u_{kt'}$ expressing the impacts of t and t', respectively, in cluster k. This value adds up to the others to form the similarity $a_{tt'}$ between topics t and t'. The value μ_k^2 summarizes the contribution of the intensity and will be referred to as the cluster's weight.

To fit the model in (1), the least-squares approach is applied, thus minimizing the sum of all $e_{tt'}^2$. Within that, the one-by-one principal component analysis strategy is attended for finding one cluster at a time by minimizing the corresponding one-cluster criterion

$$E = \sum_{t,t' \in T} (b_{tt'} - \xi u_t u_{t'})^2 \tag{2}$$

with respect to the unknown positive ξ weight and fuzzy membership vector $\mathbf{u} = (u_t)$, given similarity matrix $B = (b_{tt'})$.

In the beginning, matrix B is taken to be equal to matrix A. Each found cluster $(\mu, \mathbf{u})$ is subtracted from B, so that the residual similarity matrix applied for obtaining the next cluster is defined as $B - \mu^2 \mathbf{u} \mathbf{u}'$. In this way, A indeed is additively decomposed according to formula (1) and the number of clusters K can be determined in the process.

The optimal value of ξ at a given $\mathbf{u}$ is proven to be

$$\xi = \frac{\mathbf{u}' B \mathbf{u}}{(\mathbf{u}' \mathbf{u})^2}, \tag{3}$$

which is obviously non-negative if B is semi-positive definite.

By putting this ξ in equation (2), one arrives at $E = S(B) - \xi^2 (\mathbf{u}' \mathbf{u})^2$, where $S(B) = \sum_{t,t' \in T} b_{tt'}^2$ is the similarity data scatter.

By denoting the last item as

$$G(\mathbf{u}) = \xi^2 (\mathbf{u}' \mathbf{u})^2 = \left(\frac{\mathbf{u}' B \mathbf{u}}{\mathbf{u}' \mathbf{u}} \right)^2, \tag{4}$$

the similarity data scatter is decomposed as $S(B) = G(\mathbf{u}) + E$ where $G(\mathbf{u})$ is the part of the data scatter that is explained by cluster $(\mu, \mathbf{u})$, and E, the unexplained part. Therefore, an optimal cluster is to maximize the explained part $G(\mathbf{u})$ in (4) or its square root

$$g(\mathbf{u}) = \xi \mathbf{u}'\mathbf{u} = \frac{\mathbf{u}'B\mathbf{u}}{\mathbf{u}'\mathbf{u}}, \tag{5}$$

which is the celebrated Rayleigh quotient: its maximum value is the maximum eigenvalue of matrix B, which is reached at its corresponding eigenvector, in the unconstrained problem.

This shows that the spectral clustering approach can be applied to find a suboptimal maximizer of (5). According to this approach, one should find the maximum eigenvalue λ and corresponding normed eigenvector z for B, $[\lambda, z] = \Lambda(B)$, and take its projection to the set of admissible fuzzy membership vectors.

A number of criteria for halting the process of sequential extraction of fuzzy clusters follow from the above. The process stops if either of the conditions is true:

S1 The optimal value of ξ (3) for the spectral fuzzy cluster becomes negative.
S2 The contribution of a single extracted cluster to the data scatter becomes less than a pre-specified $\tau > 0$ threshold.
S3 The residual data scatter becomes smaller than a pre-specified $\epsilon > 0$ proportion of the original similarity data scatter.

The described one-by-one Fuzzy ADDItive-Spectral cluster extraction method is referred to as FADDIS. It combines three different approaches: additive clustering [29,17,28], spectral clustering [30,23,15,36], and relational fuzzy clustering [9,2,3,7,5]. Since FADDIS extracts clusters one-by-one, in the order of their contribution to the data scatter, the algorithm is supposed to be oriented at cluster structures at which the clusters contribute differently the more different, the better. We refer to this supposed property of the data as the property of different contributions.

To make the cluster structure in the similarity matrix sharper, one may apply the spectral clustering approach to pre-process a raw similarity matrix W into A by using the so-called normalized Laplacian transformation which is related to the popular clustering criterion of normalized cut [15]. The normalized cut criterion can be expressed, in a relaxed form, as the minimum non-zero eigenvalue of the Laplacian matrix. To change this to the criterion of maximum eigenvalue in (5), we further transform this matrix to its pseudo-inverse matrix, which also increases the gaps between eigenvalues.

3   Experimentally Testing FADDIS on Relational Data Derived from the Entity-to-Feature Data

In this section, FADDIS is compared to two most effective methods for fuzzy clustering that are extensions of the popular c-means fuzzy clustering method to relational data: NERFCM [10] and FMFCM [5]. The NERFCM has been derived as an analogue to the classical c-means at the situation in which the Euclidean distance data is derived from the original entity-to-feature data. The FMFCM also starts from the distance data to produce a number of approximating features after which the fuzzy c-means itself applies to the extracted entity-to-feature data. FADDIS applies to the affinity data derived by using the Gaussian kernel:

$$w_{tt'} = exp\left(-\frac{\sum_{v=1}^{V}(y_{tv} - y_{t'v})^2}{2\sigma^2}\right),$$

where $Y = (y_{tv})$ is a data matrix over $t \in T$ and $v = 1, 2, ..., V$, with V the number of features. The diagonal elements are set to be equal to 0: $w_{tt} = 0$ [30,23]. The parameter σ is chosen by empirical tuning [21].

This study has been conducted with generated data based on the data generator used in [5]. Specifically, 4 clusters of data points are generated from a bivariate spherical Gaussian distribution with the standard deviation $\sigma = 950$. The centers of the clusters are defined as $c_1 = (1500, 1500)$, $c_2 = (-1500, 1500)$, $c_3 = (-1500, -1500)$, $c_4 = (1500, -1500)$, so that they are located on bisectors of the quadrants of the Cartesian plane at the same distance from the origin. The clusters have cardinalities of $50, 100, 200, 150$ data points, respectively, 500 entities altogether. In this paper, a scale parameter sn is introduced as a factor to the center of the cluster to be added to all data points, to model stretching the data points to or out of the origin. At $sn < 0$, the clusters stretch in to the origin, whereas they move out from the origin at $sn > 0$. Figure 1 illustrates the type of generated data for different values of the scale parameter sn. A data set generated at $sn = 0$ on the left, and a stretched out dataset generated at $sn = 1$ on the right.

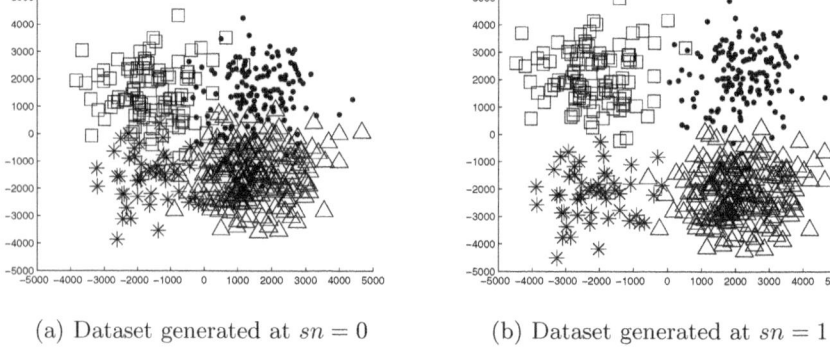

(a) Dataset generated at $sn = 0$ (b) Dataset generated at $sn = 1$

Fig. 1. Dataset with two different scales of noise

The entity-feature generated data sets are pre-processed into dissimilarity data, to be given as input to the NERFCM algorithm, by using the matrix D of Euclidean distances between generated data points. For the FADDIS algorithm, the generated entity-to-feature data is transformed into affinity data using the

Gaussian kernel defined as $w_{ij} = exp(-d^2(y_i, y_j)/p/18)$, where d is Euclidean distance and p is the dimensionality of the data set (in our study $p = 2$). Then the Laplace Pseudo-Inverse transformation applies to transform the affinity data matrix W into the matrix A to which FADDIS algorithm is applied, which does sharpen the cluster structure in this case, as previous studies have shown [18,8].

Ten data sets have been generated for each of the values of the scale parameter sn. The three algorithms have been run and the results have been evaluated according to the Adjusted Rand index (ARI) [12] to score the similarity between generated and computed clusterings. Also, we tested the ability of FADDIS to recover the number of clusters. In the case of FMFCM and NERFCM the number of clusters K must be prespecified; these algorithms have been applied at $K = 3, 4, 5$, after which the results have been evaluated by the extended Xie-Beni validation index [31].

Table 1 shows the means and standard deviations of the ARI index for the 10 data sets generated at each level of the scale parameter. In each row the highest ARI value is marked in boldface and (*). For the FADDIS algorithm the mode of the number of clusters retrieved by the algorithm is also presented.

The results show that FADDIS algorithm always recovers the correct number of clusters with stop condition (S2). Also, FADDIS finds the best ARI values for the data sets generated with the higher levels of cluster intermix ($sn \leq 0$). In these cases the NERFCM and FMFCM found their best partitions for a wrong number of clusters ($K = 3$). In contrast the NERFCM and FMFCM outperform the FADDIS algorithm for lower levels of cluster intermix ($sn > 0$)[1]. Yet, one should notice that the number of clusters is an input to the former algorithms.

Table 1. Bivariate Normal DG with different scale values of cluster intermix – Adjusted Rand Index (ARI) avg/std for FADDIS, NERFCM and FMFCM

	FADDIS			NERFCM			FastMap FCM	
sn	GK+Lapin	K	K = 3	K = 4	K = 5	K = 3	K = 4	K = 5
-5	**0.47/0.048***	4	**0.47/0.05***	0.44/0.05	0.37/0.035	**0.47/0.047***	0.44/0.045	0.37/0.03
0	**0.68/0.029***	4	0.66/0.034	0.64/0.058	0.53/0.032	0.66/0.035	0.61/0.096	0.54/0.013
5	0.83/0.022	4	0.76/0.018	**0.84/0.016***	0.67/0.036	0.76/0.018	**0.84/0.016***	0.67/0.031
10	0.91/0.029	4	0.82/0.015	**0.93/0.021***	0.74/0.025	0.82/0.015	**0.93/0.021***	0.75/0.029
20	0.98/0.022	4	0.86/0.008	**0.99/0.009***	0.85/0.07	0.86/0.008	**0.99/0.009***	0.82/0.067
50	**1/0***	4	0.87/0.007	**1/0***	0.87/0.075	0.87/0.007	**1/0***	0.87/0.07

4   Testing FADDIS with Genuine Similarity Data

4.1   The Fuzzy Cluster Core Data Generator

In this section, we propose a similarity data generator following the additive model (1). As usual in fuzzy clustering, we assume that each entity has one "core" cluster to which it belongs most. Therefore, the data generation process

---

[1] The values of the extended Xie-Beni index are concordant with the ARI values for both NERFCM and FMFCM.

starts with the generation of the "core" clusters. Then we apply the same three algorithms to the generated data.

Given the size N of an entity set I, and the number of clusters K, the proposed Fuzzy Cluster Core Data Generator (FCC DG), generates an $N \times N$ similarity data matrix G according to the underlying (FADDIS) model $W = U\Lambda U^T$, as follows:

$$G = U\Lambda U^T + \alpha E, \tag{6}$$

where:

- $N \times K$ fuzzy membership matrix U is randomly generated using a fuzzy "core" clusters generating procedure.
- Positive real valued $K \times K$ diagonal weight matrix Λ with diagonal positive values λ_k of the cluster weights equal to $\lambda_k = \mu_k^2$ is defined according to model (1). Since the vectors $\mathbf{u}_k$ in (1) are assumed normed, the weights take in the norms of the generated vectors $\mathbf{u}_k$. To test the supposed property of different contributions of the FADDIS, the weights are also made proportional to $(K - k + 1)^\beta$, for $k = 1, 2, \ldots, K$, so that the greater the $\beta > 0$, the greater the difference. Therefore, the weights are defined by $\lambda_k = (K - k + 1)^\beta * \|\mathbf{u}_k\|$.
- Elements of symmetric $N \times N$ error matrix E are independently generated from a Gaussian distribution $N(0, 1)$, and then symmetrized so that $e_{tt'} = (e_{tt'} + e_{t't})/2$.
- The value $\alpha \in [0, 1]$ is the parameter that controls the level of error introduced into the model $W = U\Lambda U^T$.

This generator builds a fuzzy cluster structure by conventionally relaxing a crisp partition. Given a crisp partition R of the entity set I, where $R = R_1, \ldots, R_K$ with non-overlapping clusters R_k, a fuzzy relaxation builds each k-th fuzzy cluster $\mathbf{u}_k$ having the corresponding crisp cluster R_k as its core in such a way that the maximum membership values u_{ik} will be at entities $i \in R_k (k = 1, \ldots, K)$ while the other components of $\mathbf{u}_k$ are close to 0.

Given the number K of core clusters covering the entire data set, I, the data generator builds each core cluster by filling it in with fuzzy membership values, such that: (a) the membership values of k-th fuzzy cluster $\mathbf{u}_k$ are very high at k-th core (e.g. $u_{ik} > 2/3$ for $i \in R_k$); and (b) the fuzzy clusters form a fuzzy partition so that $\sum_k u_{ik} = 1$ at each entity $i \in I$. After all the membership vectors $\mathbf{u}_k$ are generated, the norms of $\mathbf{u}_k$'s are computed and assigned as factors in the clusters' weights, in order to "adjust" them to the additive fuzzy clustering model. Then the final membership matrix has its membership vectors $\mathbf{u}_k$ normalized.

An example of a data set generated from the FCC DG for $K = 3$ clusters, $N = 700$ entities, and $\beta = 0.0$, visualized according to the Visual Assessment of Cluster Tendency (VAT) tool [4], is shown in Figure 2. The 3 clusters are shown in the main diagonal in dark grey, and their relative sizes can be seen. The clusters form a clear-cut structure.

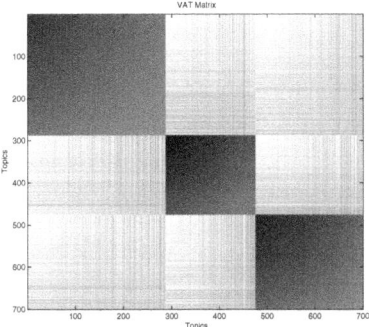

Fig. 2. VAT visualization of the cluster structure for a data set generated by the FCC DG for $K = 3$ and $N = 700$

4.2 Setting of the Experiment and Its Results

The main goal of this experimental study is to compare the FADDIS algorithm with FMFCM and NERFCM in recovering the cluster structures generated by the FCC DG for different levels of generated Gaussian noise.

Particular attention is given to the FADDIS algorithm, whose analysis of the clustering results are made according to the following parameters:

(i) Number of clusters retrieved by the FADDIS algorithm and corresponding stop condition achieved;
(ii) Per generated cluster k and corresponding computed cluster $\widehat{k}$, measure:
  (a) Recovery membership error (RME) of generated cluster k with membership vector $\mathbf{u}_k = [u_{ik}]$, and computed membership, $\widehat{\mathbf{u}}_k = [\widehat{u}_{ik}]$:

$$RME\left(\mathbf{u}_k\right) = \sum_{i=1}^{N} u_{ik}^2 \frac{|u_{ik} - \widehat{u}_{ik}|}{u_{ik}}$$

such that,

$$\sum_{i=1}^{N} u_{ik}^2 = 1$$

Notice that the RME error is an averaged relative difference weighted by u_{ik}^2, in order to normalize the error measure. The maximum value of the error is one.
  (b) Recovery intensity error (RIE) of generated and computed intensities, μ_k, and $\widehat{\mu}_k$,

$$RIE(\mu_k) = \frac{|\mu_k - \widehat{\mu}_k|}{\mu_k}.$$

  (c) Percentage of the matching between generated R_k cores ($k = 1, 2, \ldots, K$) and the crisp cores retrieved from the computed partitions, after defuzzification by maximum membership;

(iii) Similarity between generated and found partitions, made according to the Adjusted Rand index (ARI) (this is to compare all the three algorithms).

The datasets have been generated in three groups corresponding to three different numbers of clusters: $K = 3, 4, 5$. The experiments were cross-combined according to the following settings: (i) Total number of entities of the data set $N = 50, 200, 400, 700$; (ii) α values of the standard deviation of noise, $\alpha = \{0, 0.05, 0.1, 0.15, 0.25, 0.5\}$. (iii) For each value of K, 10 distinct datasets had been generated for each tuple (N, α, β), resulting in a total of 720 datasets for each K value, and so a total of 2160 datasets. In the case of NERFCM, the similarity data matrix G (6) is transformed into a dissimilarity matrix D, such that, $D = max(G) - G$.

In our preliminary experiments, we observed that the ability to recover a cluster structure significantly decreases for the values of $\alpha > 0.1$. Thus, the statistics are presented for $\alpha \in \{0, 0.05, 0.1\}$ only. In the next tables, the best value in each row is marked with (*).

Table 2 shows the means/std and mode values of the recovered number of clusters by the FADDIS algorithm. For $K = 3, 4, 5$ one can see that when the β value increases from $\beta = 0.0$ to $\beta = 1.0$ the percentage of data sets for which the correct number of clusters is recovered also increases. The only exception occurs for $K = 5, N = 200$, where the best values are achieved for $\beta = 0.5$. In all the cases, the most working stop condition of the FADDIS algorithm is condition $S2$.

By analysing the Recovery Membership Error (RME) and the Recovery Intensity Error (RIE) (Table 3), one can see that the minimum values are achieved for $\beta = 1.0$ for the collections of data sets with $k = 3$ and $k = 4$ clusters. For the

Table 2. FCC DG - Summary data of the percentage avg/std of correct extracted clusters and mode of the number of extracted clusters for std of added Gaussian noise=[0, 0.1] for FADDIS in best conditions for $K = \{3, 4, 5\}$

		FADDIS					
		$\beta = 0.0$		$\beta = 0.5$		$\beta = 1.0$	
	N	(%)	Mode	(%)	Mode	(%)	Mode
$K = 3$	50	50.0/0.0	3	62.5/9.6	3	85.0/5.8*	3
	200	60.0/0.0*	3	32.5/20.6	2	60.0/0.0*	3
	400	30.0/21.6	3	62.5/15.0	3	80.0/0.0*	3
	700	17.5/17.1	2	40.0/35.6	2	65.0/19.1*	3
$K = 4$	50	47.5/9.6	4	60.0/8.2	4	70.0/18.3*	4
	200	50.0/35.6	4	50.0/0.0	4	65.0/5.8*	4
	400	27.5/18.9	5	55.0/10.0	4	72.5/5.0*	4
	700	17.5/20.6	1	67.5/5.0	4	77.5/5.0*	4
$K = 5$	50	40.0/21.6	5	60.0/8.2	5	67.5/5.0*	5
	200	37.5/26.3	5	52.5/5.0*	5	40.0/8.2	5
	400	45.0/46.5	5	50.0/0.0	5	65.0/10.0*	5
	700	25.0/23.8	1	35.0/5.8	6	42.5/5.0*	5

data sets with $k = 5$ clusters the minimum values are obtained for parameter $\beta = 0.5$. Indeed, for $\beta = 1.0$ the RME and RIE mean errors are always inferior to 0.2 which is a good value (the only exception is at $K = 3$ and $N = 700$). Also, the errors almost always decrease with the increase of β, which is in accord with the expected property of different contributions of FADDIS.

Table 3. Summary Table of the RME and RIE errors' avg/std for std of added Gaussian noise=[0, 0.1] for FADDIS in best conditions for $K = \{3, 4, 5\}$

		RME			RIE		
	N	$\beta = 0.0$	$\beta = 0.5$	$\beta = 1.0$	$\beta = 0.0$	$\beta = 0.5$	$\beta = 1.0$
$K = 3$	50	0.25/0.08	0.24/0.02	0.14/0.02*	0.14/0.03	0.15/0.01	0.08/0.01*
	200	0.28/0.08	0.54/0.12	0.15/0.01*	0.13/0.02	0.29/0.08	0.07/0.00*
	400	0.45/0.34	0.18/0.11	0.14/0.01*	0.30/0.27	0.09/0.05*	0.09/0.00*
	700	0.56/0.33	0.39/0.18	0.21/0.05*	0.35/0.24	0.25/0.14	0.13/0.05*
$K = 4$	50	0.22/0.07	0.13/0.02*	0.13/0.05*	0.11/0.01	0.07/0.01*	0.08/0.04
	200	0.44/0.35	0.12/0.01*	0.12/0.01*	0.29/0.30	0.06/0.00*	0.06/0.00*
	400	0.41/0.33	0.20/0.01	0.10/0.03*	0.28/0.29	0.10/0.00	0.05/0.01*
	700	0.59/0.36	0.13/0.05*	0.17/0.01	0.43/0.30	0.07/0.02*	0.07/0.00*
$K = 5$	50	0.28/0.12	0.14/0.02*	0.17/0.01	0.15/0.04	0.07/0.01*	0.07/0.00
	200	0.36/0.26	0.14/0.04	0.13/0.02*	0.21/0.18	0.07/0.01	0.06/0.01*
	400	0.40/0.34	0.07/0.01*	0.12/0.01	0.28/0.29	0.04/0.00*	0.05/0.01
	700	0.49/0.37	0.17/0.03*	0.18/0.01	0.35/0.33	0.10/0.01	0.06/0.00*

Table 4 presents the ARI index values for the three algorithms under consideration, FADDIS, FMFCM, and NERFCM. The highest values are marked with (*) and boldface: they always correspond to the FADDIS results. Specifically, the higher ARI values are achieved for data sets generated with $\beta = 1.0$ for the data sets with $K = 3$ and $K = 4$ clusters. For $K = 5$, the best values are achieved at $\beta = 0.5$, in contrast to the expected property of different contributions.

Complementary, and in order to compare the results obtained by the FMFCM and NERFCM algorithms the (*) mark indicates the highest ARI value between the results of these two algorithms. In almost all the cases the NERFCM outperforms FMFCM for the data sets generated with $\beta = 0.0$, which is in contrast to the case of the entity-to-feature data at which FMFCM outperforms NERFCM [5]. This illustrates the idea that the NERFCM is a genuine relational clustering algorithm whereas the FMFCM is not.

Finally, the best values for the percentages of the crisp core matching are concordant with the ARI index (not shown here).

5   Representation of Activities in a Taxonomy of the Field

As has been pointed out above, the motivation in developing the FADDIS method comes from a novel methodology of visualization of the activities of

Table 4. FCC DG - Summary Table for ARI avg/std for std of added Gaussian noise=[0, 0.1] for all algorithms in best conditions for $K = \{3,4,5\}$

	N	FADDIS			FMFCM			NERFCM		
		$\beta = 0.0$	$\beta = 0.5$	$\beta = 1.0$	$\beta = 0.0$	$\beta = 0.5$	$\beta = 1.0$	$\beta = 0.0$	$\beta = 0.5$	$\beta = 1.0$
$K = 3$	50	0.88/0.14	0.84/0.21	**0.90/0.19***	0.72/0.30	0.80/0.29	0.85/0.24*	0.78/0.19	0.56/0.25	0.48/0.19
	200	0.74/0.19	0.70/0.21	**0.81/0.18***	0.45/0.35	0.56/0.31	0.62/0.32	0.69/0.23*	0.58/0.22	0.53/0.22
	400	0.87/0.10	0.87/0.10	**0.91/0.11***	0.46/0.38	0.62/0.35	0.79/0.28	0.80/0.12*	0.72/0.17	0.68/0.17
	700	0.79/0.16	0.70/0.19	**0.80/0.20***	0.36/0.36	0.51/0.38	0.58/0.32	0.70/0.21*	0.64/0.14	0.56/0.18
$K = 4$	50	0.92/0.07	0.91/0.07	**0.93/0.1***	0.65/0.28	0.77/0.18*	0.74/0.19	0.73/0.15	0.55/0.24	0.54/0.22
	200	0.92/0.09	0.91/0.09	**0.94/0.11***	0.49/0.34	0.64/0.24	0.63/0.17	0.68/0.14*	0.44/0.16	0.43/0.17
	400	0.87/0.14	0.91/0.14	**0.93/0.13***	0.42/0.35	0.59/0.28	0.7/0.23	0.72/0.13*	0.56/0.17	0.55/0.21
	700	0.84/0.15	**0.93/0.08***	0.83/0.17	0.36/0.36	0.51/0.3	0.64/0.24	0.71/0.16*	0.53/0.15	0.52/0.15
$K = 5$	50	0.92/0.08	**0.95/0.08***	0.78/0.23	0.66/0.31	0.67/0.18*	0.63/0.16	0.65/0.14	0.52/0.18	0.47/0.16
	200	0.89/0.12	**0.93/0.09***	0.83/0.17	0.48/0.33	0.63/0.22	0.6/0.16	0.64/0.16*	0.38/0.13	0.32/0.1
	400	0.87/0.22	**0.95/0.06***	0.88/0.15	0.41/0.37	0.58/0.27	0.63/0.2	0.67/0.17*	0.47/0.14	0.45/0.2
	700	0.89/0.13	**0.94/0.06***	0.84/0.13	0.36/0.36	0.52/0.29	0.61/0.19	0.66/0.16*	0.44/0.16	0.38/0.14

Table 5. A fuzzy cluster of research activities undertaken in a research centre by FADDIS

Membership value	Code	ACM-CCS Topic
0.69911	I.5.3	Clustering
0.3512	I.5.4	Applications in I.5 PATTERN RECOGNITION
0.27438	J.2	PHYSICAL SCIENCES AND ENGINEERING (Applications in)
0.1992	I.4.9	Applications in I.4 IMAGE PROCESSING AND COMPUTER VISION
0.1992	I.4.6	Segmentation
0.19721	H.5.1	Multimedia Information Systems
0.17478	H.5.2	User Interfaces
0.17478	H.5.3	Group and Organization Interfaces
0.16689	H.1.1	Systems and Information
0.16689	I.5.1	Models in I.5 PATTERN RECOGNITION
0.16513	H.1.2	User/Machine Systems
0.14453	I.5.2	Design Methodology (Classifiers)
0.13646	H.5.0	General in H.5 INFORMATION INTERFACES AND PRESENTATION
0.13646	H.0	GENERAL in H. Information Systems

a research organization such as a University department by mapping them to a related hierarchical taxonomy such as Classification of Computer Subjects by ACM (ACM-CCS) [1].

Our method generalizes the individual member/project profiles in two steps. First step finds fuzzy clusters of the taxonomy subjects according to the working of the organization. Second step maps each of the clusters to higher ranks of the taxonomy in a parsimonious way. An expository outline of this strategy, its motivations and potential benefits, made before the FADDIS has been developed, can be found in [19].

As the FADDIS found clusters are not necessarily consistent with the taxonomy, each is considered as a query set to be interpreted in the taxonomy by lifting each cluster to higher ranks of the taxonomy. The lifting is done by our recursive algorithm for minimizing a penalty function that involves "head subjects" on the higher ranks of the taxonomy together with their "gaps" and "offshoots" [20].

To illustrate the approach, Table 5 presents a fuzzy cluster obtained in our project, on the data from a survey[2] involving 16 respondents and covering 46 ACM-CCS topics, by applying the FADDIS algorithm. This cluster is then mapped to and parsimoniously generalized by the lifting method over the ACM-CCS taxonomy in terms of "head subjects" (i.e *H.-Information Systems* and *I.5-PATTERN RECOGNITION*), their "gaps" (e.g. *H.2-DATABASE MAN-AGEMENT, H.3-INFORMATION STORAGE AND RETRIEVAL*), and "off-shoots" (e.g. *I.4.6- Segmentation, J.2- PHYSICAL SCIENCES AND ENGI-NEERING*). The generalized representation of the cluster resulting from the lifting method is visualized in Figure 3, pointing out its "head subjects", "gaps", and "offshoots".

---

[2] Survey conducted in Centre for Artificial Intelligence (CENTRIA) of Faculdade de Ciências e Tecnologia, Universidade Nova de Lisboa in 2009.

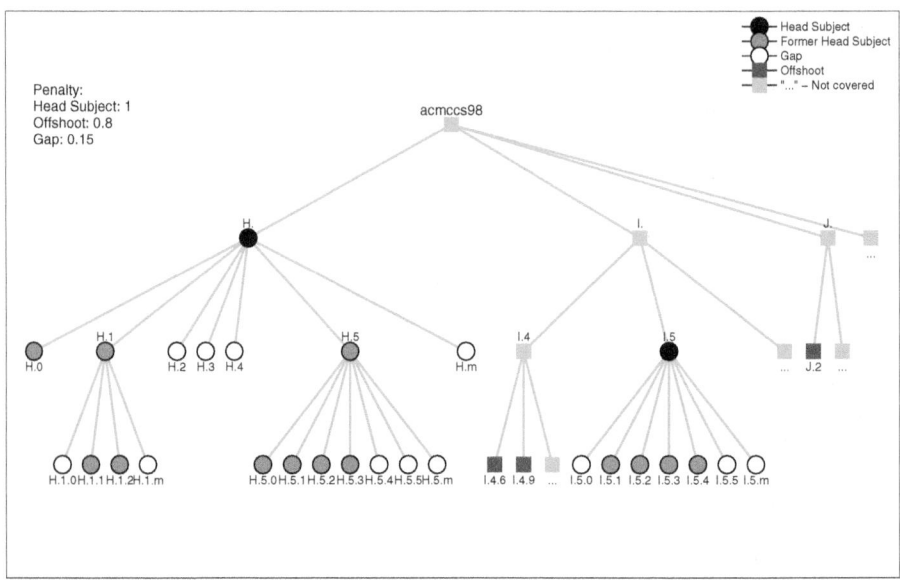

Fig. 3. Visualization of the optimal lift of the cluster in Table 1 in the ACM-CCS tree; the irrelevant tree leaves are not shown for the sake of simplicity

6   Conclusion

The paper introduces and experimentally verifies an unconventional model of fuzzy clusters in which the products of entity membership values contribute towards similarity between the entities. This is motivated by the idea that the similarity between research topics is obtained by adding up the working of different groups on them so that the clusters according to this model can be considered thematic clusters indeed. The model leads to a spectral fuzzy clustering method FADDIS that is accompanied with a set of model-based cluster extracting stop-conditions. This paper demonstrates that FADDIS is competitive on two types of generated cluster structures. Moreover, FADDIS can be used sometimes for recovering the correct number of clusters. Yet, there are some irregularities in its working that deserve to be investigated further. One of the irregularities is the experimentally observed deviations from the property of different contributions. According to the definition of FADDIS, the more different the cluster weights in the data, that is, the greater the β at the genuine similarity data generator, the better should be the correspondence between the generated clusters and those FADDIS-computed. This is true in most cases, but sometimes it is not. We are going to address this in our future work. The other direction of further developments is applying FADDIS for visualization of activities to be captured by the analysis of web posted documents.

Acknowledgments. This work has been supported by project grant PTDC/ EIA/69988/2006 from the Portuguese Foundation for Science & Technology. The partial support of the Laboratory for Analysis and Choice of Decisions in the framework of the Programme of Fundamental Studies of the National Research University Higher School of Economics, Moscow RF, to BM is acknowledged. The authors are indebted to the anonymous reviewers for multiple comments taken into account in the final version.

References

1. ACM Computing Classification System (1998), http://www.acm.org/about/class/1998 (Cited September 9, 2008)
2. Bezdek, J., Hathaway, R., Windham, M.: Numerical comparisons of the RFCM and AP algorithms for clustering relational data. Pattern Recognition 24, 783–791 (1991)
3. Bezdek, J., Keller, J., Krishnapuram, R., Pal, T.: Fuzzy Models and Algorithms for Pattern Recognition and Image Processing. Kluwer Academic Publishers, Dordrecht (1999)
4. Bezdek, J.C., Hathaway, R.J.: VAT: a tool for visual assessment of (cluster) tendency. In: Procs. of the 2002 International Joint Conference on Neural Networks (IJCNN 2002), pp. 2225–2230 (2002)
5. Brouwer, R.: A method of relational fuzzy clustering based on producing feature vectors using FastMap. Information Sciences 179, 3561–3582 (2009)
6. Castellano, G., Torsello, M.A.: How to derive fuzzy user categories for web personalization. In: Castellano, G., Jain, L.C., Fanelli, A.M. (eds.) Web Personalization in Intelligent Environments. SCI, vol. 229, pp. 65–79. Springer, Heidelberg (2009)
7. Davé, R., Sen, S.: Robust fuzzy clustering of relational data. IEEE Transactions on Fuzzy Systems 10, 713–727 (2002)
8. Felizardo, R.: A study on parallel versus sequential relational fuzzy clustering methods, Master thesis, Faculdade de Ciências e Tecnologia, Universidade Nova de Lisboa, p. 212 (2011)
9. Hathaway, R., Davenport, J., Bezdek, J.: Relational duals of the c-means algorithms. Pattern Recognition 22, 205–212 (1989)
10. Hathaway, R.J., Bezdek, J.C.: NERF c-means: Non-Euclidean relational fuzzy clustering. Pattern Recognition 27, 429–437 (1994)
11. Huang, L., Yan, D., Jordan, M.I., Taft, N.: Spectral clustering with perturbed data. In: Koller, D., Schuurmans, D., Bengio, Y., Bottou, L. (eds.) Proceedings of the Twenty-Second Annual Conference on Neural Information Processing Systems. Advances in Neural Information Processing Systems, vol. 21, pp. 705–712. MIT Press, Vancouver (2009)
12. Hubert, L.J., Arabie, P.: Comparing partitions. Journal of Classification 2, 193–218 (1985)
13. Inoue, K., Urahama, K.: Sequential fuzzy cluster extraction by a graph spectral method. Pattern Recognition Letters 20, 699–705 (1999)
14. Krishnapuram, R., Joshi, A., Nasraoui, O., Yi, L.: Low-complexity fuzzy relational clustering algorithms for Web mining. IEEE Transactions on Fuzzy Systems 9(4), 595–607 (2001)

15. von Luxburg, U.: A tutorial on spectral clustering. Statistics and Computing 17, 395–416 (2007)
16. Masullia, F., Mitra, S.: Natural computing methods in bioinformatics: A survey. Information Fusion 10(3), 211–216 (2009)
17. Mirkin, B.: Additive clustering and qualitative factor analysis methods for similarity matrices. Journal of Classification 4(1), 7–31 (1987)
18. Mirkin, B., Nascimento, S.: Analysis of Community Structure, Affinity Data and Research Activities using Additive Fuzzy Spectral Clustering. Technical Report 6, School of Computer Science, Birkbeck University of London (2009)
19. Mirkin, B., Nascimento, S., Pereira, L.M.: Cluster-lift method for mapping research activities over a concept tree. In: Koronacki, J., Raś, Z.W., Wierzchoń, S.T., Kacprzyk, J. (eds.) Advances in Machine Learning II. SCI, vol. 263, pp. 245–257. Springer, Heidelberg (2010)
20. Mirkin, B., Nascimento, S., Fenner, T., Pereira, L.M.: Constructing and Mapping Fuzzy Thematic Clusters to Higher Ranks in a Taxonomy. In: Bi, Y., Williams, M.-A. (eds.) KSEM 2010. LNCS (LNAI), vol. 6291, pp. 329–340. Springer, Heidelberg (2010)
21. Nadler, B., Lafon, S., Coifman, R.R., Kevrekidis, I.G.: Diffusion Maps, Spectral Clustering and Reaction Coordinates of Dynamical Systems. Applied and Computational Harmonic Analysis (21), 113–127 (2006)
22. Nasraoui, O., Frigui, H.: Extracting Web User Profiles Using Relational Competitive Fuzzy Clustering. International Journal on Artificial Intelligence Tools (IJAIT) 9(4), 509–526 (2000)
23. Ng, A., Jordan, M., Weiss, Y.: On spectral clustering: analysis and an algorithm. In: Ditterich, T.G., Becker, S., Ghahramani, Z. (eds.) Advances in Neural Information Processing Systems, vol. 14, pp. 849–856. MIT Press, Cambridge (2002)
24. Pal, N.R., Aguan, K., Sharma, A., Amari, S.: Discovering biomarkers from gene expression data for predicting cancer subgroups using neural networks and relational fuzzy clustering. BMC Bioinformatics, 8(1)(5) (2007)
25. Popescu, M., Keller, J.M., Mitchell, J.A.: Fuzzy Measures on the Gene Ontology for Gene Product Similarity. Journal IEEE/ACM Transactions on Computational Biology and Bioinformatics (TCBB) 3(3), 263–274 (2006)
26. Roubens, M.: Pattern classification problems and fuzzy sets. Fuzzy Sets and Systems 1, 239–253 (1978)
27. Runkler, T.A., Bezdek, J.C.: Web mining with relational clustering. International Journal of Approximate Reasoning, Elsevier Science 32(2-3), 217–236 (2003)
28. Sato, M., Sato, Y., Jain, L.C.: Fuzzy Clustering Models and Applications. Physica, Heidelberg (1997)
29. Shepard, R.N., Arabie, P.: Additive clustering: representation of similarities as combinations of overlapping properties. Psychological Review 86, 87–123 (1979)
30. Shi, J., Malik, J.: Normalized cuts and image segmentation. IEEE Transactions on Pattern Analysis and Machine Intelligence 22(8), 888–905 (2000)
31. Sledge, I.J., Bezdek, J.C., Havens, T.C., Keller, J.M.: Relational Generalizations of Cluster Validity Indices. IEEE Transactions on Fuzzy Systems 18(4), 771–786 (2010)
32. Suryavanshi, B.S., Shiri, N., Mudur, S.P.: An Efficient Technique for Mining Usage Profiles Using Relational Fuzzy Subtractive Clustering. In: Procs. of the International Workshop on Challenges in Web Information Retrieval and Integration (WIRI 2005), pp. 23–29 (2005)

33. Windham, M.P.: Numerical classification of proximity data with assignment measures. Journal of Classification 2, 157–172 (1985)
34. Xu, D., Keller, J.M., Popescu, M., Bondugula, R.: Applications of Fuzzy Logic in Bioinformatics. Imperial College Press, London (2008)
35. Yang, M., Shih, H.: Cluster analysis based on fuzzy relations. Fuzzy Sets and Systems 120, 197–212 (2001)
36. Zhang, S., Wang, R.-S., Zhang, X.-S.: Identification of overlapping community structure in complex networks using fuzzy c-means clustering. Physica A 374, 483–490 (2007)

Automatically Enriching a Thesaurus with Information from Dictionaries

Hugo Gonçalo Oliveira and Paulo Gomes

CISUC, University of Coimbra, Portugal
{hroliv,pgomes}@dei.uc.pt

Abstract. Regarding that information in broad-coverage knowledge bases, such as thesauri, is usually incomplete, merging information from different sources is a good option to amplify coverage. We propose a method for the enrichment of a thesaurus with information acquired automatically from dictionaries: pairs of synonyms are assigned to candidate synsets and, the pairs whose elements are not in the thesaurus are clustered to identify new synsets. This method was used in the enrichment of a Brazilian Portuguese thesaurus with synonyms from a European Portuguese dictionary, and resulted in a larger and broader thesaurus with new words and new concepts. The assignments and the obtained synsets were manually evaluated and yielded correction scores higher than 71% and 85% respectively.

Keywords: thesaurus, synonymy, lexico-semantic knowledge, clustering.

1 Introduction

Lexical knowledge bases are key resources for the development of natural language processing (NLP) tools at the semantic level. In opposition to domain ontologies, this kind of knowledge base is typically broad-coverage and structured on the words of a language and their meanings. The importance of these resources is evidenced for English, where Princeton WordNet [3] is widely used by the NLP and knowledge management communities in the achievement of tasks that are becoming more and more common, including the determination of similarities [1], word sense disambiguation [6], or question answering [20].

WordNet is a handcrafted resource based on synsets, which are groups of synonymous words that can be seen as natural language concepts. Each synset has a gloss, which is similar to a dictionary definition, and several types of lexico-semantic relations (eg. hypernymy, part-of) between synsets are represented. Besides this interesting way of handling words and meanings, the public domain character of WordNet was very important for increasing its popularity and community of users and thus to contribute for a huge development of tools working at the semantic level of English.

The success of WordNet lead to its adaptation to other languages around the world, including the ones in the EuroWordNet [30] project. However, some of the latter resources are not freely available for researchers.

L. Antunes and H.S. Pinto (Eds.): EPIA 2011, LNAI 7026, pp. 462–475, 2011.

For instance, for Portuguese, as far as we know, the freely available resources that are closest to a wordnet are:

- two public domain thesaurus: TeP [14], for Brazilian Portuguese, which is thought to be the synset-base of a future wordnet for Brazilian Portuguese; and OpenThesaurus.PT[1], a collaborative thesaurus approximately four times smaller than TeP, used in the OpenOffice[2] word processor for suggesting synonyms;
- Portuguese Wiktionary[3], a collaborative dictionary by the Wikimedia foundation where, besides definitions, it is possible to add information on semantic relations for each entry. For Portuguese, however, this resource is still small and, besides other problems, most entries do not have information about semantic relations;
- PAPEL [9], a public domain lexical network consisting of instances of several types of semantic relations, extracted automatically from a Portuguese proprietary dictionary. The main differences between PAPEL and a wordnet is that PAPEL was created automatically and it is not structured on synsets, but on simple words.

Furthermore, there are recent reports on efforts towards the automatic construction of a lexical ontology for Portuguese [7]. More information on some of the aforementioned resources and their comparison to other Portuguese wordnet projects can be found in [26]. A comparison between the verbs in TeP, OpenThesaurus.PT, PAPEL and the Portuguese Wiktionary is presented in [27]. Both the later works concluded that, although all of the former resources are broad-coverage, their contents are more complementary than overlapping and it would be fruitful to merge some of them in a unique broader resource. This problem is common to most languages and there have been attempts to merge, or to align, different knowledge bases, or to enrich existing knowledge bases with information extracted from other sources (see examples in section 5).

In this paper, we propose a flexible method to automatically integrate synonymy information, extracted from dictionary text, in an existing thesaurus. More precisely, we use the synonymy pairs (hereafter synpairs) connecting nouns, verbs and adjectives in PAPEL and, when possible, assign them to TeP synsets. For synpairs with both elements new to TeP, synsets are established after discovering clusters. This results in a new and larger thesaurus with augmented TeP synsets and some new synsets. Furthermore, TeP, which was originally made for Brazilian Portuguese, becomes closer to European Portuguese, as most of the information in PAPEL is from the latter variant, making it more suitable to be used in tasks for European Portuguese.

We start by presenting our approach in three stages. Then, before reporting on the results of our work, we analyse the coverage of PAPEL's synpairs by TeP and also the properties of the graphs established by the synpairs to be clustered.

---

[1] http://openthesaurus.caixamagica.pt/
[2] http://www.openoffice.org/
[3] http://pt.wiktionary.org/

Together with the results in numbers, we present the results of the manual evaluation of the assignment and clustering stages, as well as some examples of assignments and discovered synsets. Finally, before concluding, we introduce some related work on the automatic enrichment of lexical knowledge bases.

2   Proposed Approach

In this section we propose an approach for enriching an existing thesaurus with information extracted from electronic dictionaries. This approach consists of three stages: (i) extraction of synpairs from dictionary definitions; (ii) assignment of synpairs to existing synsets of the thesaurus; (iii) discovery of new synsets after clustering the remaining synpairs.

2.1   Extraction of Synonymy from Dictionaries

Besides dictionaries, synpairs can be extracted from other kinds of text. However, since real synonyms do not co-occur frequently in corpora text, it is simpler to acquire them from dictionaries, where synpairs can be extracted from plain definitions consisting of only one word, enumerations or definitions using synonymy textual patterns as the following:

- **mente**, n: *cérebro, cabeça, intelecto*
  [**mind**, n: *brain, head, intellect*]
  - (*cérebro, mente*) (*cabeça, mente*) (*intelecto, mente*)
    [(*brain, mind*) (*head, mind*) (*intellect, mind*)]
- **máquina**, n: *o mesmo que computador*
  [**machine**, n: *the same as computer*]
  - (*computador, máquina*)
    [(*computer, machine*)]

Since there is not a well-defined criteria for the division of meanings into word senses, word senses in different dictionaries do not always match [2] [12] [22]. Therefore, sense numbers are not considered in this extraction procedure.

2.2   Assigning Synpairs to Synsets

This stage looks for candidate synsets, in the thesaurus, suitable for each synpair, extracted previously. It assumes that, if a thesaurus contains a word, it contains all its possible senses. We recall that this might not be true, especially because words do not have a finite number of senses [12]. However, broad coverage lexical knowledge bases typically limit the number of senses of each word. By creating artificial boundaries on meanings, the former knowledge bases become more practical for computational applications.

Assigning a synpair $p = (w_x, w_y)$ to a synset $S_a = (w_1, w_2, ..., w_n)$ means that w_x and w_y are added to S_a, which becomes $S_a = (w_1, w_2, ..., w_n, w_x, w_y)$. A synpair $p = (w_x, w_y)$ is already in the thesaurus T if there is a synset $S_b \in T$

containing both w_x and w_y. In this case, no additional assignment is needed. Otherwise, if the thesaurus does not contain the synpair, we assign the latter to the most similar synset.

Before running the assignment algorithm, a synonymy graph G is established by all the extracted synpairs, such that a synpair $p = (w_x, w_y)$ establishes an edge between nodes w_x and w_y. Synonymy graphs are structures $G = (N, E)$, with $|N|$ nodes and $|E|$ edges, $E \subset N^2$. Each node $w_i \in N$ represents a word and each edge connecting w_i and w_j, $E(w_i, w_j)$, indicates that, in some context, words w_i and w_j may have the same meaning and are thus synonyms. Synonymy graphs can be represented, for instance, as a binary sparse matrix $M(|N| \times |N|)$, where 1s denote adjacencies.

We now summarise the assignment algorithm in seven steps. For each synpair $p = (w_x, w_y)$:

1. If there is a synset $S_i \in T$ containing both elements of p, $w_x \in S_i \wedge w_y \in S_i$, the synpair is already represented in T, so nothing is done.
2. Otherwise, select all the candidate synsets $C_j \in C : C \subset T, C = \{C_1, C_2, ..., C_n\}$ containing at least one of the elements of p, $\forall (C_j \in C) : w_x \in C_j \vee w_y \in C_j$.
3. If $|C| = 1$, assign p to C_1.
4. Otherwise, compute the adjacency vector of p by summing the adjacencies of both of its elements in G, $[p] = [w_x] + [w_y]$. The adjacency vector of a word is a column of the matrix M, $[w_j] = [M_j]$;
5. Compute the adjacency vector of each synset $C_j \in C$ by summing all the adjacencies of its words in the synonymy graph G. $[C_j] = \sum_{k=1}^{|C_j|} [w_k] : w_k \in C_j$;
6. Select the most similar candidate synset $C_{best} : sim(p, C_{best}) = max(sim(p, C_j))$;
7. Assign p to C_{best}.

Any measure for computing the similarity between two vectors can be used. In our experiments, reported in section 3, we have used the cosine similarity based on the adjacencies of the synonymy graph, computed according to expression 1. As this kind of similarity is measured according to the information given by the graph, larger and more complete graphs will provide better results. So, if these methods are used in the assignment of a small sets of synpairs, other sources of information, such as occurrences in a corpus, should be considered as alternatives to compute the similarity.

$$sim(w_a, w_b) = cos([M_a], [M_b]) = \frac{[M_a] \cap [M_b]}{|M_a||M_b|} = \frac{\sum_{i=1}^{|N|} M_{ai} M_{bi}}{\sum_{i=1}^{|N|} \sqrt{M_{ai}^2} \sum_{i=1}^{|N|} \sqrt{M_{bi}^2}} \qquad (1)$$

2.3   Clustering Remaining Words

For the remaining synpairs, the thesaurus does not contain any of its elements, so, there are no assignment candidates. Therefore, we propose to discover new synsets on the graph established by the former synpairs, G'.

While G' might be constituted by several small disconnected subgraphs where all words have a common meaning, some subgraphs might mix more than one meaning, which needs to be handled. However, since synonymy graphs extracted from dictionaries tend to have a clustered structured [5] [16], we propose to run a clustering algorithm on G' to discover clusters which can be seen as synsets.

At this stage, most ambiguous words, which are the ones with more senses and thus with more connections in the synonymy graph, are expected to be already in synsets of the thesaurus. Therefore, a simple clustering procedure, such as the following, is suitable for discovering synsets in G'. For each disconnected subgraph in G':

1. Create a new sparse matrix $M'(|N| \times |N|)$;
2. Fill each cell M'_{ij} with the similarity between the adjacency vectors of the word w_i with the adjacency vectors of w_j, $M'_{ij} = sim(w_i, w_j)$;
3. Extract a cluster S_i from each row M'_i, consisting of the words w_j where $M'_{ij} > \theta$, a selected threshold. A lower θ leads to larger synsets and higher ambiguity, while a larger θ will result on very small synsets or no synsets at all.
4. For each cluster S_i with all elements included in a larger cluster S_j ($S_i \cup S_j = S_j$ and $S_i \cap S_j = S_i$), S_i is discarded.

One important point about this procedure is that clusters might be overlapping, which is desirable as the same word might have different meanings and thus belong to more than one synset.

3   Enriching TeP with Synonymy in PAPEL

In this section, we describe the steps performed towards the enrichment of TeP, with words in a dictionary, encoded in PAPEL as synpairs between nouns, verbs and adjectives. These two resources were chosen because, for Portuguese, they are, as far as we know, the largest public domain broad-coverage resources of their kind. As referred in section 1, TeP is synset-based thesaurus for Brazilian Portuguese, while PAPEL is a lexical network automatically extracted from an European Portuguese dictionary.

We start by reporting on the coverage of the synpairs of PAPEL by TeP. Then, we look at the properties of the graphs established by the remaining synpairs which would be the target for clustering. The effective results of this experimentation and their evaluation are presented in section 4.

3.1   Coverage of the Synpairs by the Thesaurus

There are different kinds of synpairs, according to their coverage by TeP. Some of the synpairs are already represented in TeP, which means that TeP has at least one synset containing both elements of the synpair. For other synpairs, there is only one synset with one of its elements ($|C| = 1$), or several synsets containing one of the synpair elements ($|C| > 1$). Finally, there are synsets without candidate synsets in TeP ($|C| = 0$).

Table 1 summarises the former numbers according to the part-of-speech of the PAPEL synpairs. Furthermore, it presents the average number of candidate synsets for each synpair not represented in TeP ($|\overline{C}|$), which gives an idea of the possible choices involved in this task.

For all parts-of-speech, more than a fourth of the synpais are already represented in TeP. For verbs, the former proportion is the highest and reaches 43%. On the other hand, the number of synpairs without candidate synsets in TeP is very low. Still, it is much higher for nouns (14.98%) than for verbs (1.34%) and adjectives (5.58%). In the next stage, this set of synpairs will be clustered.

Table 1. Synpairs assigned to the thesaurus

POS	Synpairs	In TeP	$|C| = 0$	$|C| = 1$	$|C| > 1$	$|\overline{C}|$
Nouns	37,452	10,255 (27.38%)	5,609 (14.98%)	4,499 (12.01%)	17,089 (45.63%)	3.86
Verbs	21,465	9,232 (43.01%)	287 (1.34%)	868 (4.04%)	11,088 (51.66%)	6.64
Adjectives	19,073	7,172 (37.60%)	1,064 (5.58%)	1,567 (8.22%)	9,270 (48.60%)	4.26

3.2   Properties of the Graphs to Be Clustered

Before clustering the remaining synpairs, whose elements were not in TeP, we analysed some properties of the graphs they established. Table 2 has the latter properties, which include: the number of nodes $|N|$ and edges $|E|$, the number of disconnected subgraphs $|DS|$, the average degree $\overline{deg}(G)$ of the graph, average clustering coefficient $\overline{CC}_{lds}$ and the number of nodes of the largest disconnected subgraph $|N_{lds}|$.

The average degree (expression 2) is the ratio between the number of nodes $|N|$ and the number of edges $|E|$ of the graph. The average clustering coefficient (expression 3) measures the degree to which nodes tend to cluster together as a value in [0-1]. In random graphs, this coefficient is close to 0. The local clustering coefficient $CC(n_i)$ (expression 4) of a node n_i quantifies how connected its neighbours are.

$$\overline{deg}(G) = \frac{1}{|N|} \times \sum_{i=1}^{|N|} deg(n_i) = \frac{1}{|N|} \times \sum_{i=1}^{|N|} |E(n_i, n_j)| : n_i, n_j \in N \tag{2}$$

$$\overline{CC} = \frac{1}{|N|} \times \sum_{i=1}^{|N|} CC(n_i) \tag{3}$$

$$CC(n_i) = \frac{2 \times |E(n_j, n_k)|}{K_i(K_i - 1)} : \forall(n_j \in G), n_k \in neigh(n_i) \wedge K_i = |neigh(n_i)| \tag{4}$$

The numbers in Table 2 confirm that the noun graph is much larger than the others. It is composed by a large disconnected subgraph (*lds*) of 287 nodes, while the others have much smaller *lds*.

Table 2. Data of the synonymy graphs

POS	$|N|$	$|E|$	$|DS|$	$\overline{deg(G)}$	$|N_{lds}|$	$\overline{CC}_{lds}$
Nouns	8,548	5,608	3,111	1.31	287	0.07
Verbs	502	287	216	1.14	6	0.25
Adjectives	1,858	1.15	1,064	803	14	0.10

The lds for nouns and adjectives have lower average clustering coefficients $\overline{CC}$, than those of synonymy graphs automatically extracted from Wiktionaries [16], which are between 0.2 and 0.28, $\overline{CC}_{lds}$. The $\overline{CC}_{lds}$ are as well lower than for lds of a synonymy network extracted from three Portuguese dictionaries [8], which would indicate that these graphs are not suitable for clustering.

However, there cannot be a straight comparison because our graphs are established by the remaining synpairs, and not by all the original synpairs extracted from the dictionary. Therefore, even though the $\overline{CC}_{lds}$ are low, the number of disconnected subgraphs $|DS|$ is the same order of magnitude as the number of the total number of nodes $|N|$, which indicates that these graphs have many small disconnected subgraphs. Consequently, the latter subgraphs can be exploited in the identification of clusters, as some of them establish one cluster.

4   Results and Evaluation

In this section, we report on the results of the experimentation described in section 3, and their evaluation. We start by presenting the evolution of the original thesaurus, in terms of words and synsets, after each stage. Then, we describe and show the results of the evaluation of the assignments and clustering stages, together with examples of assignments and discovered clusters.

4.1   Resulting Thesaurus

After analysing the coverage of the synpairs by TeP and the properties of the graphs established by the synpairs not covered, we ran the assignment algorithm and the clustering procedure. Tables 3 and 4 present the properties of four thesauri, obtained in each of the stages, in terms of words and synsets respectively. They first present the properties of the original thesaurus, TeP, then TeP after the assignments, the thesaurus consisting of the synsets discovered after clustering, and, finally, the resulting thesaurus.

On the words of each thesaurus (table 3) we present the quantity of unique words (Total), the number of words with more than one sense (Ambiguous), the number of average senses per word (Avg(senses)) and the number of senses of the most ambiguous word (Most ambiguous). On the synsets (table 4), we present their quantity (Total), their average size in terms of words (Avg(size)), the number of synsets of size 2 ($size = 2$) and size greater than 25 ($size > 25$), which are less likely to be useful, and, finally, the size of the largest synset ($max(size)$).

Table 3. Thesaurus comparison in terms of words

Thesaurus	POS	Words			
		Total	Ambiguous	Avg(senses)	Most ambig.
TeP 2.0	Nouns	17,158	5,805	1.71	20
	Verbs	10,827	4,905	2.08	41
	Adjectives	14,586	3,735	1.46	19
After assignments	Nouns	23,775	10,418	2.09	37
	Verbs	12,818	7,094	2.64	42
	Adjectives	17,158	6,294	1.83	22
Clusters	Nouns	8,546	701	1.15	8
	Verbs	502	8	1.02	3
	Adjectives	1,858	39	1.03	4
Final thesaurus	Nouns	30,369	12,045	1.96	38
	Verbs	13,090	7,221	2.62	42
	Adjectives	18,525	6,550	1.80	23

Table 4. Thesaurus comparison in terms of synsets

Thesaurus	POS	Synsets				
		Total	Avg(size)	$size = 2$	$size > 25$	$max(size)$
TeP 2.0	Nouns	8,254	3.56	3,079	0	21
	Verbs	3,978	5.67	939	48	53
	Adjectives	6,066	3.50	3,033	19	43
After assignments	Nouns	8,254	6.01	1,930	179	150
	Verbs	3,978	8.50	702	217	148
	Adjectives	6,066	5.17	2,369	120	110
Clusters	Nouns	3,524	2.78	2,247	0	13
	Verbs	220	2.34	174	0	6
	Adjectives	820	2.33	656	0	10
Final thesaurus	Nouns	11,778	5.05	4,177	179	150
	Verbs	4,198	8.18	876	217	148
	Adjectives	6,886	4.84	3,025	120	110

After the assignments, the number of words grows but the number of synsets remains the same. Ambiguity becomes higher, as well as the size of the synsets, as the latter are augmented.

For clustering the remaining synsets, we empirically defined $\theta = 0.4$. The thesauri obtained in this stage are smaller and less ambiguous than the original thesaurus. This happens because the words not covered by TeP tend to be less frequent words, which are typically more specific and thus less ambiguous.

Finally, the synsets of the resulting thesaurus are the sum of the thesaurus after assignments with the synsets discovered after clustering. Its words are just slightly more ambiguous than in TeP and the most ambiguous words have more senses than TeP, especially for nouns. In opposition to synsets, the quantity of unique words of the new thesaurus is not the sum of the words after the assignments with the words in the discovered clusters, because some of the latter could have been added to TeP during the assignments. For instance, considering the sypairs $p_a = (a, b)$ and $p_b = (b, c)$, if TeP contains a, but it does not contain words b and c, p_a is assigned to one TeP synset while p_b is not and will thus be included in G', the graph for clustering.

Furthermore, the resulting thesaurus is much larger than TeP, with almost two times more nouns, 2,263 more verbs and 3,939 more adjectives. The number of synsets is higher too, which means that the new thesaurus is broader also in terms of covered natural language concepts.

4.2 Evaluation of the Assignment Results

For the evaluation of the assignments stage, we asked human judges to classify random assignments either as correct or incorrect. Table 5 shows the results of the evaluation of 100 assignments of each part-of-speech, nouns, verbs and adjectives. Each assignment was classified by two different native Portuguese judges. The agreement, also provided, is given by the proportion of assignments that received the same classification from both judges.

The lowest correction rate is given for verbs (71%). As shown in table 4, this is the category both with larger synsets and more ambiguous words, which makes evaluation more difficult and contributes to the former value. On the other hand, the highest correction is given for nouns (76.5%), which is the category with the largest synonymy graph. This points out the importance of the size of the synonymy graph – more information enables better assignment decisions. The best agreement rate is given for nouns (77%), which means that the judges did not agree in the classification of 24 assignments. For verbs and adjectives, this value is slightly lower (75% and 74%).

Table 5. Assignments evaluation

POS	Sample	Correct	Incorrect	Agreement
Nouns	100 assignments	153 (76.50%)	47 (23.50%)	77.00%
Verbs	100 assignments	142 (71.00%)	58 (29.00%)	74.00%
Adjectives	100 assignments	151 (75.50%)	49 (24.50%)	75.00%

Table 6 contains some examples of assignments and their classifications by the two judges, where 1s stand for correct and 0s for incorrect assignments. The word of the synpair in bold is the word presented to the judges.

4.3 Evaluation of the Clustering Results

Clustering was evaluated after selecting random pairs of words belonging to the same synset. Then, once again, we asked human judges to classify the pairs as correct, if both words could have the same meaning in some context and thus be synonyms, or incorrect, otherwise. Table 7 shows the results of the evaluation of 105 pairs for each part-of-speech. Similarly to the assignments, each pair was classified by two different judges and the agreement is provided. The correction rates in table 7, as well as the agreement rates, are quite high, which shows once again that clustering synpairs not covered by TeP is a suitable way for identifying new synsets and that there was no need for a clustering procedure more complex than the one presented in section 2.3.

Table 6. Sample assignments and their evaluation

POS	Synpair	Synset	Judge 1	Judge 2
	(**escrutínio**,votação)	votação;voto;sufrágio	1	1
Nouns	(decisão,**desempate**)	resolução;objetivação;tenção;intenção	0	1
	(plano,**gizamento**)	planície;chã;chanura;plaino;plano;planura	0	0
	(venerar,**homenagear**)	venerar;cultuar;adorar;idolatrar	1	1
Verbs	(atacar,**combater**)	atacar;inciar	0	1
	(**obter**,rapar)	depilar;despelar;pelar;raspar;rapar;rascar	0	0
	(grandioso,**épico**)	admirável;fabuloso;grandioso	1	1
Adjectives	(delicado,**requintado**)	difcil;complicado;delicado	0	1
	(**falido**,queimado)	queimado;incendiado	0	0

Table 7. Evaluation of clustering

POS	Sample	Correct	Incorrect	Agreement
Nouns	105 × 2 pairs	179 (85.24%)	31 (14.76%)	91.43%
Verbs	105 × 2 pairs	193 (91.90%)	17 (8.10%)	87.62%
Adjectives	105 × 2 pairs	189 (90.00%)	21 (10.00%)	85.71%

Figure 1 shows examples of connected subgraphs for nouns, verbs and adjectives, respectively. Different shades of grey denote different clusters. The first subgraph mixed the concepts of a pre-historic construction and a tree/fruit, which can both be denoted by the word *mamoa*. After clustering, these two concepts are separated and *mamoa* is only considered as a tree/fruit. The second subgraph lead to a single cluster, meaning *to reassure*. Finally, two different but close senses of the adjectives *tónico* and *refectivo*, more precisely invigorating and restorative, are discovered.

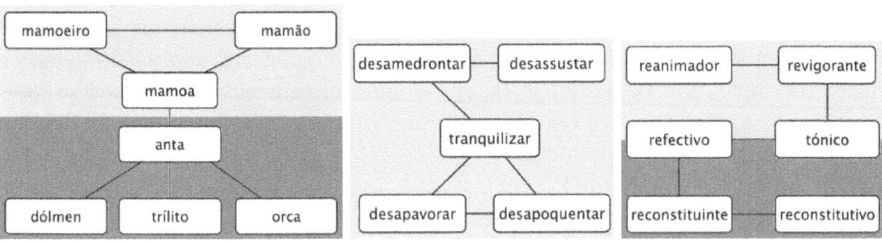

Fig. 1. Examples of connected subgraphs and resulting clusters

5   Related Work

Researchers soon noticed that, although it is a broad-coverage resource, the knowledge encoded in Princeton WordNet was not enough for several purposes, especially domain specific tasks. As a lexical knowledge base, most of the information in WordNet is about the words of a language and their meanings.

In order to overcome the former limitation, several researchers have worked on the enrichment of WordNet with knowledge automatically acquired from other sources. New domain concepts [17] [19] and hypernymy relations [11] have been extracted from corpora. The glosses of the WordNet synsets have been exploited for the extraction of new hypernymy relations [10] and concepts [17]. The WordNet's structure has also been augmented by connecting nominal and verbal senses of the same word, taking advantage of the definitions of these words in a dictionary [15] (eg. connect the noun *ship* with the verb *to ship*).

More recently, the collaborative encyclopedia Wikipedia has become a popular resource for the automatic acquisition of encyclopedic and domain knowledge. There are several works on the enrichment of WordNet with information in Wikipedia, including the alignment of Wikipedia categories with WordNet synsets, in order to provide more information on named entities to WordNet [29] or to improve the taxonomy of Wikipedia categories [23]. Furthermore, there is work on the automatic alignment of Wikipedia articles with WordNet synsets [25], aiming to enrich the semantic relations of WordNet [24] or to refine and augment WordNet's sense inventory [18].

Another alternative to increase the coverage of an existing knowledge base is to link it with other knowledge bases. For instance, WordNet has been linked to the upper ontology SUMO [21], to the descriptive ontology DOLCE [4], and there are researchers working on the mapping between WordNet and other lexical resources such as FrameNet [28].

As for the automatic discovery of synsets from scratch, several authors have shown that the identification of word clusters in text [13], or in synonymy networks extracted from dictionaries [5] [8], is a suitable alternative to identify concepts described by words, useful in the establishment of synsets.

6   Concluding Remarks

We have presented a flexible method for enriching a thesaurus with synonymy information automatically acquired from text. This method was used for enriching a thesaurus for Brazilian Portuguese with information from an European Portuguese dictionary. We believe that our results are interesting and that our approach can thus be applied to integrate other public domain resources in a unique broader resource for Portuguese and, eventually, for other languages. Besides PAPEL and TeP, for Portuguese, there are other candidate resources such as the Portuguese Wiktionary and OpenThesaurus.PT.

The assignment approach is an alternative to the automatic discovery of synsets in the complete synonymy graphs extracted from dictionaries (see [8]).

On the one hand, it has the advantage of using a manually created broad-coverage thesaurus as a starting point, which should guarantee a higher synset precision rate. On the other hand, the approach presented here is very dependent on the coverage of the thesaurus used as a starting point, and does not provide the discovery of new concepts of words already in the latter resource.

Another important contribution of this work is that TeP, originally made for Brazilian Portuguese, is enriched with words from a dictionary whose main entries are in European Portuguese. This way, besides being larger, the new thesaurus has a better coverage of European Portuguese, which will enable a better integration with projects in European Portuguese and will be a suitable alternative for OpenThesaurus.PT as the thesaurus of the OpenOffice word processor.

Acknowledgements. We would like to thank all the reviewers involved in the manual evaluation of this work. Hugo Gonçalo Oliveira is supported by the FCT grant SFRH/BD/44955/2008 co-funded by FSE.

References

1. Agirre, E., Alfonseca, E., Hall, K., Kravalova, J., Paşca, M., Soroa, A.: A study on similarity and relatedness using distributional and WordNet-based approaches. In: Proc. Human Language Technologies: 2009 Annual Conference of the North American Chapter of ACL (NAACL-HLT), pp. 19–27. ACL, Stroudsburg (2009)
2. Dolan, W.B.: Word sense ambiguation: clustering related senses. In: Proc. 15th Conference on Computational Linguistics (COLING), pp. 712–716. ACL, Morristown (1994)
3. Fellbaum, C. (ed.): WordNet: An Electronic Lexical Database (Language, Speech, and Communication). The MIT Press (May 1998)
4. Gangemi, A., Guarino, N., Masolo, C., Oltramari, A.: Interfacing WordNet with DOLCE: towards OntoWordNet. In: Ontology and the Lexicon: A Natural Language Processing Perspective, ch.3. Cambridge University Press (2010)
5. Gfeller, D., Chappelier, J.C., Rios, P.D.L.: Synonym Dictionary Improvement through Markov Clustering and Clustering Stability. In: Proc. International Symposium on Applied Stochastic Models and Data Analysis (ASMDA), pp. 106–113 (2005)
6. Gomes, P., Pereira, F.C., Paiva, P., Seco, N., Carreiro, P., Ferreira, J.L., Bento, C.: Noun sense disambiguation with wordnet for software design retrieval. In: Proc. Advances in Artificial Intelligence, 16th Conference of the Canadian Society for Computational Studies of Intelligence, Halifax, Canada, pp. 537–543 (2003)
7. Gonçalo Oliveira, H., Gomes, P.: Onto.PT: Automatic Construction of a Lexical Ontology for Portuguese. In: Proc. 5th European Starting AI Researcher Symposium (STAIRS 2010). IOS Press (2010)
8. Gonçalo Oliveira, H., Gomes, P.: Automatic discovery of fuzzy synsets from dictionary definitions. In: Proc. 22nd International Joint Conference on Artificial Intelligence (IJCAI), Barcelona, Spain (2011)
9. Gonçalo Oliveira, H., Santos, D., Gomes, P.: Extracção de relações semânticas entre palavras a partir de um dicionário: o PAPEL e sua avaliação. Linguamática 2(1), 77–93 (2010)

10. Harabagiu, S.M., Moldovan, D.I.: Enriching the WordNet taxonomy with contextual knowledge acquired from text. In: Natural Language Processing and Knowledge Representation: Language for Knowledge and Knowledge for Language, pp. 301–333. MIT Press, Cambridge (2000)
11. Hearst, M.: Automated Discovery of WordNet Relations. In: Fellbaum, C. (ed.) WordNet: An Electronic Lexical Database and Some of its Applications, pp. 131–153. MIT Press, Cambridge (1998)
12. Kilgarriff, A.: Word senses are not bona fide objects: implications for cognitive science, formal semantics. In: Proc. 5th International Conference on the Cognitive Science of Natural Language Processing, NLP, pp. 193–200 (1996)
13. Lin, D., Pantel, P.: Concept discovery from text. In: Proc. 19th International Conference on Computational Linguistics (COLING), pp. 577–583 (2002)
14. Maziero, E.G., Pardo, T.A.S., Felippo, A.D., Dias-da-Silva, B.C.: A Base de Dados Lexical e a Interface Web do TeP 2.0 - Thesaurus Eletrônico para o Português do Brasil. In: VI Workshop em Tecnologia da Informação e da Linguagem Humana (TIL), pp. 390–392 (2008)
15. Nastase, V., Szpakowicz, S.: Augmenting WordNet's Structure Using LDOCE. In: Gelbukh, A. (ed.) CICLing 2003. LNCS, vol. 2588, pp. 281–294. Springer, Heidelberg (2003)
16. Navarro, E., Sajous, F., Gaume, B., Prévot, L., Hsieh, S., Kuo, T.Y., Magistry, P., Huang, C.R.: Wiktionary and NLP: Improving synonymy networks. In: Proc. 2009 Workshop on The People's Web Meets NLP: Collaboratively Constructed Semantic Resources, pp. 19–27. ACL, Suntec (2009)
17. Navigli, R., Velardi, P., Cucchiarelli, A., Neri, F.: Extending and enriching WordNet with OntoLearn. In: Proc. 2nd Global WordNet Conference (GWC), pp. 279–284. Masaryk University, Brno (2004)
18. Niemann, E., Gurevych, I.: The people's web meets linguistic knowledge: Automatic sense alignment of wikipedia and WordNet. In: Proc. International Conference on Computational Semantics (IWCS), Oxford, UK, pp. 205–214 (2011)
19. Pantel, P.: Inducing ontological co-occurrence vectors. In: Proc. 43rd Annual Meeting of the Association for Computational Linguistics, pp. 125–132. ACL Press (2005)
20. Pasca, M., Harabagiu, S.M.: The informative role of WordNet in open-domain question answering. In: Proc. NAACL 2001 Workshop on WordNet and Other Lexical Resources: Applications, Extensions and Customizations, Pittsburgh, USA, pp. 138–143 (2001)
21. Pease, A., Fellbaum, C.: Formal ontology as interlingua: the SUMO and WordNet linking project and global WordNet linking project and global WordNet. In: Ontology and the Lexicon: A Natural Language Processing Perspective, ch.2., Cambridge University Press (2010)
22. Peters, W., Peters, I., Vossen, P.: Automatic sense clustering in EuroWordnet. In: Proc. 1st International Conference on Language Resources and Evaluation (LREC), Granada, pp. 409–416 (May 1998)
23. Ponzetto, S.P., Navigli, R.: Large-scale taxonomy mapping for restructuring and integrating Wikipedia. In: Proc. 21st International Joint Conference on Artificial Intelligence (IJCAI), Pasadena, California, pp. 2083–2088 (2009)
24. Ponzetto, S.P., Navigli, R.: Knowledge-rich word sense disambiguation rivaling supervised systems. In: Procs. of 48th Annual Meeting of the Association for Computational Linguistics, pp. 1522–1531. ACL Press, Uppsala (2010)

25. Ruiz-Casado, M., Alfonseca, E., Castells, P.: Automatic Assignment of Wikipedia Encyclopedic Entries to WordNet Synsets. In: Szczepaniak, P.S., Kacprzyk, J., Niewiadomski, A. (eds.) AWIC 2005. LNCS (LNAI), vol. 3528, pp. 380–386. Springer, Heidelberg (2005)
26. Santos, D., Barreiro, A., Costa, L., Freitas, C., Gomes, P., Gonçalo Oliveira, H., Medeiros, J.C., Silva, R.: O papel das relações semânticas em português: Comparando o TeP, o MWN.PT e o PAPEL. In: Actas do XXV Encontro Nacional da Associação Portuguesa de Linguística (forthcomming, 2010)
27. Teixeira, J., Sarmento, L., Oliveira, E.: Comparing Verb Synonym Resources for Portuguese. In: Computational Processing of the Portuguese Language, 9th International Conference Proc. (PROPOR), Porto Alegre, Brasil, pp. 100–109 (2010)
28. Tonelli, S., Pighin, D.: New features for FrameNet: WordNet mapping. In: Proc. 13th Conference on Computational Natural Language Learning (CoNLL), pp. 219–227. ACL, Stroudsburg (2009)
29. Toral, A., Muñoz, R., Monachini, M.: Named Entity Wordnet. In: Proc. International Conference on Language Resources and Evaluation (LREC). ELRA, Marrakech (2008)
30. Vossen, P.: EuroWordNet: a multilingual database for information retrievaleuroWordNet: a multilingual database for information retrieval. In: Proc. DELOS workshop on Cross-Language Information Retrieval, Zurich (1997)

Visualizing the Evolution of Social Networks

Márcia Oliveira and João Gama

Faculty of Economics, University of Porto, Porto, Portugal
and LIAAD - INESC Porto L.A.
Rua de Ceuta 118, 4050-190 Porto, Portugal
marcia@liaad.up.pt,
jgama@fep.up.pt
http://www.liaad.up.pt

Abstract. In recent years we witnessed an impressive advance in the social networks field, which became a "hot" topic and a focus of considerable attention. Also, the development of methods that focus on the analysis and understanding of the evolution of data are gaining momentum. In this paper we present an approach to visualize the evolution of dynamic social networks by using Tucker decomposition and the concept of trajectory. Our visualization strategy is based on trajectories of network's actors in a bidimensional space that preserves its structural properties. Furthermore, this approach can be used to identify similar actors by comparing the shape and position of the trajectories. To illustrate the proposed approach we conduct a case study using a set of temporal friendship networks.

Keywords: Data Evolution, Data Visualization, Social Networks, Trajectories, Tucker3 model.

1 Introduction

A social network is constructed from relational data and can be defined as a set of social entities, such as people, groups and organizations, with some pattern of interactions between them. One of the key features of these social structures is their dynamic nature, which is being increasingly taken into account in the development of visualization methods and tools for the analysis of social networks [1]. This trend arose mainly as a consequence of the growing interest in evolving social networks and the recognition that visualization of time-oriented data gleans insights into the dynamics of the underlying phenomena, allowing for a better understanding of temporal relations. Therefore, in this paper we propose the definition of trajectories of social entities in a low-dimensional space, which accounts for most variation in the original data, based on the output of a Tucker tensor decomposition. The main goal is to propose a natural, compact and simple visual representation of the evolution of multidimensional relational data along a time horizon.

In Section 2 we address the problem of representation of dynamic social networks and introduce our approach to visualize their evolution using trajectories.

L. Antunes and H.S. Pinto (Eds.): EPIA 2011, LNAI 7026, pp. 476–490, 2011.
© Springer-Verlag Berlin Heidelberg 2011

In Section 3 we show and discuss a case study using temporal friendship networks and in Section 4 we briefly present some related work. Both summary and conclusions of the paper are presented in Section 5.

2   Evolution of Social Networks

In this section we present two alternatives to represent social networks and argue which one is the best for representing time-oriented sociometric data. We also introduce our visualization approach, which is based on the development of trajectories of social actors in low-dimensional subspaces. These trajectories are obtained by exploring the compressed information produced by an estimated Tucker3 model.

2.1   Social Networks as Three-Order Tensors

In this work, our endeavor is primarily directed towards the production of an accurate, yet compact, visualization approach to represent the evolution of actors engaged in a given type of social relation. By compact we mean that this representation is not only able to be grasped by human eye, through projections in low dimensional subspaces, but it is also focused in the structural and relevant information contained in raw sociometric data. This is an important aspect that guided our understanding of how a social network should be represented in order to meet requirements such as interpretability of results. In fact, if we choose to represent a social network, comprised of n nodes (or actors), by means of an adjacency matrix $\mathbf{A}_{n \times n}$, and factorize it, we obtain new entities that are difficult to interpret, especially if we are dealing with directed networks, where the rows and the columns represent different concepts. In such cases, where the adjacency matrix is not symmetric, the rows are closely related to the out-neighborhood of social actors and the columns are related to their in-neighborhood. On the other hand, when working with symmetric adjacency matrices, the new entities returned by the decomposition are typically the same for both modes[1] A and B, thus creating redundant results and less valuable information than one would expect from rich network structures. To overcome the barriers posed by the adoption of a standard social network representation, we suggest to compute SNA-specific actor-level statistics (e.g. degree, closeness, betweenness, authority and hub scores), to embed richer structural knowledge into the *snapshot matrices* (these matrices represent snapshots of the state of the world for a specific point in time). In the context of social networks, these statistics are useful in the sense that they provide us higher-level information, regarding network's structure, thus giving us insight about the position and importance of each node in the network, without the need to look to the corresponding graph. Moreover, this strategy helps mitigate the differences between directed and undirected

---

[1] *Mode* is the number of dimensions of a tensor, also referred to as its *order*. These concepts will be later introduced in this paper.

networks and it greatly improves the interpretability of the component matrices yielded by factorial methods.

After using this strategy to build the snapshot matrices, the process of converting them into a tensor (also known as hypermatrices and multidimensional arrays) becomes trivial. To do so, one just needs to introduce an additional mode C, by ordering them by time and putting them together into an unique structure, thus obtaining a three-order tensor $\mathcal{X} \in \mathbb{R}^{I \times J \times K}$, where I ($i = 1, ..., I$) denotes the number of entities of the first mode A (the *row-entities*, which are defined along the horizontal axis), J ($j = 1, ..., J$) refers to the number of entities of the second mode B (the *column-entities*, which are defined along the vertical axis) and K ($k = 1, ..., K$) indicates the number of entities of the third mode C (the *fiber-entities*, which are defined along the depth axis). In the social networks context, the mode A is the dimension of social actors, mode B is the dimension of actor-level network's metrics and mode C is the dimension of time. We adopt tensorial representations of dynamic social networks since they are able to explicitly model the time dimension without collapsing the data and, therefore, without losing the mutual dependencies between all data modes.

2.2   Trajectories in Social Networks

A trajectory can be defined as a set of time-ordered states of an object in a dynamical system. Typically, these trajectories are defined in low-dimensional representative subspaces and are graphically represented by a line that connects the coordinates of an object for different time points. It is common to resort to 2D, instead of 3D subspaces, since they are simpler to analyze and, at the same time, allow for an effective data analysis. Thus, we use two-dimensional projections and encode the third dimension as a trajectory over the plane. In such way, we are able to map a given individual's trajectory along time, by simply using two-dimensional projections, thus producing a compact, clear and informative representation of data evolution.

The appealing feature of trajectories is that they render temporal visualization more appealing to human eye, promoting an efficient dissemination of temporal results. Besides, they help achieve a faster insight into the temporal behavior of an individual and allow for an intuitive detection of structural changes that may occur. When all the trajectories of a group of actors in a network are represented in the same plot, the trajectory is also able to show the relative position of each actor compared to all other actors.

In order to define the trajectories of each actor we decompose the original three-order tensor by estimating a Tucker3 model [2,3,4,5]. The Tucker3 model is a technique devised to decompose three-order tensors into a set of matrices $\mathbf{A} \in \mathbb{R}^{I \times P}$, $\mathbf{B} \in \mathbb{R}^{J \times Q}$ and $\mathbf{C} \in \mathbb{R}^{K \times R}$, known as component or coefficient matrices, and a small core tensor $\mathcal{G} \in \mathbb{R}^{P \times Q \times R}$ (P, Q and R are parameters of the Tucker3 model and represent the number of components/factors in the first, the second and the third mode of the tensor). In Figure 1 we illustrate the idea behind Tucker3 decomposition. Tucker suggested interpreting the core tensor as

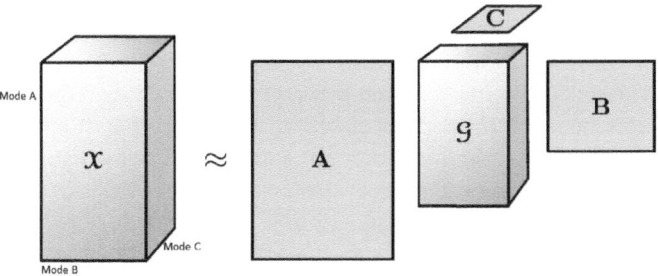

Fig. 1. The Tucker3 decomposition of a tensor [2]

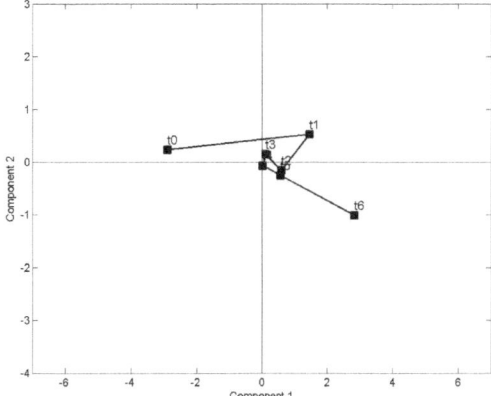

Fig. 2. Example of an actor's trajectory for a time horizon of 7 time points $T = \{t_0, t_1, t_2, t_3, t_4, t_5, t_6\}$ in the space spanned by the two most representative components of matrix **B**

describing the latent structure in data, since it has information about the level of interaction between the different components, and the component matrices as mixing this structure to give the observed data [4]. The core tensor can also be interpreted as a generalization of the eigenvalues, or of the singular values, of the SVD, and it constitutes a further partitioning of the "explained" variation as is indicated by the eigenvalues of the standard PCA. The square of each entry of $\mathcal{G}$ is proportional to the amount of variance that the entry explains and its value indicates how the various components relate to each other. Matrices **A**, **B** and **C** have, in general, less dimensionality than the original corresponding modes, but are able to reconstruct the most important information contained in data, which is given by the sum of squares explained by the model. Also, they are assumed to be columnwise orthogonal. The orthogonality is desirable since it facilitates the analysis and hastens the computation of the decomposition. Usually, this model is estimated using an alternating least squares algorithm.

After decomposing the tensor, we consider the two-dimensional subspace spanned by the two most representative components of matrix $\mathbf{B}$, and define the x and y coordinates for each time point k ($k = 1, ..., K$) of the trajectory. We obtain these coordinates for each actor i ($i = 1, ..., I$), by computing the **dot product** between $x_{i,:,k}$ (horizontal fibers of $\mathcal{X}$) and each column of component matrix $\mathbf{B}$ (the first and the second components are assigned to the x-axis and y-axis, respectively). This vector operation returns the coordinates, or the bidimensional position vector, of the time points, for each actor of the network. For illustration purposes, in Figure 2 we present an example of an actor's trajectory.

The last step of this analysis is to interpret the social micro-evolution[2], through critical observation of its trajectory. The movement, or trajectory, of a given actor can be characterized by a **direction** (upwards, downwards, leftwards or rightwards), that can be more regular or more irregular; and by an **amplitude**, which can be higher, thus covering a larger space area, or lower, by keeping its position in the plane almost unchanged over time. Also, both the **shape** and the **position** of the trajectory can be used to identify actors with similar evolutions. We will take these features into consideration when analyzing the trajectories in the case study.

In short, we can say that our visualization approach maps each snapshot of the social network of a given actor into a point in the Tucker decomposition subspace, and links these points in order to define a trajectory that represents the dynamic behavior of this individual. Using this strategy it is possible to naturally take temporal information into account by adopting tensorial representations of social networks.

Our approach also holds when dealing with large social networks. In such cases, one resorts to sparse representations of tensors and apply algebraic operations appropriate for sparse data structures. Regarding the computational complexity, using the Tucker-ALS algorithm, it has been shown in [7] to be $\mathcal{O}(n^3r + n^2r^2 + nr^3)$, where n is the mode size[3] and r is the number of retained components (assuming $r = P = Q = R$).

3   Social Networks Case Study

In this section we conduct a small case study using a set of friendship social networks collected at seven different moments in time, for the same individuals. The idea is to verify the suitability of the proposed visualization approach for representing the dynamics of social networks and, at the same time, to yield meaningful insights into social micro-evolution (at the actor-level). To do so, we perform the following sequential steps: first, we preprocess the original raw data and compute the actor-level statistics; then, we model the resulting matrices as a three-order tensor, where the third dimension is time; the third step consists

---

[2] In this paper, we use the term *social micro-evolution* to refer to the study of the temporal behavior of a social network's actor.

[3] Note that if the size of, at least, one mode, differs from the size of the others, n will be regarded as $n = \max\{I, J, K\}$.

in applying the Tucker3 decomposition to this tensor; finally, the trajectories are defined and interpreted based on the results of the decomposition. Note that this case study serves to demonstrate the applicability of the introduced approach, not to support specific theoretical claims.

3.1 Data Description

The data we use is comprised of several directed friendship networks, collected by Gerhard Van de Bunt [11] among a group of university freshmen, in order to provide an explanation of some of the important factors behind friendship formation. This information was gathered by questionnaires which were delivered to 49 students in seven different occasions asking them to rate their social relationships in a six point scale. The mentioned occasions are not equally spaced in time, once the first four time points are three weeks apart, and the last three time points are six weeks apart. There were also changes in the number of students, which drops from 49 to 32, due to university "drop-outs" and nonresponse of questionnaires in at least four of the total seven occasions. It is also important to note that, with a few exceptions, the respondents were initially mutual strangers (freshmen), which reflects in a sparse adjacency matrix for the first measurement ($time = t_0$). Originally, the ties linking each pair of students were weighted according to a specific coding scale, representing the rating of the established relationships. For the sake of simplicity, and in order to improve the interpretability of the trajectories, we transform the initial network into an unweighted one, by replacing codes $6 = $ *item non-response* and $9 = $ *actor non-response* by 0 and codes $1 = $ *best friend*, $2 = $ *friend*, $3 = $ *friendly relation*, $4 = $ *neutral* and $5 = $ *troubled relation* by 1. Therefore, the absence of a tie, coded by 0, represents the absence of relationship between a given pair of students, and the existence of a tie, coded by 1, means that there is a relationship or, at least, acquaintance between the corresponding pair of students. Formally, each entry of the adjacency matrix $\mathbf{X}$, generated in time point t_k ($t = 1, ..., K$), can assume the following values: $x_{ij} \in \{0, 1\}$. The established relationships can be positively or negatively connoted, since the original code did take into account not only friendly relationships but also troubled ones. Besides sociometric data, also additional information about students' gender, smoking behavior, residence and education program, is available. These variables can be useful to achieve a further macro understanding of the results.

The main reason behind the choice of this data is the fact that it includes a temporal dimension, which makes it possible to track changes over time in students's social behavior, namely, in their popularity and prestige.

3.2 Definition of Trajectories and Interpretation of the Axes

In order to obtain the trajectories for each freshman student, we first need to transform the initial adjacency matrices into snapshot matrices embedded with structural information about the individuals. To do so, one needs to define which network's metrics should be used to characterize these individuals. In this case study, we resort to the following actor-level metrics and, when needed, to the specific versions of the standard measures for directed networks. Here we provide a brief description of the metrics within the context of the students' friendship networks:

- **In-degree**: indicates the number of students who claim to know actor i ($i = 1, ..., n$), thus being a measure of *support*;
- **Out-degree**: indicates the number of students that actor i states he/she knows, thus being a measure of *influence*;
- **Closeness**: measure of reachability that measures how fast can a given actor reach everyone in the network;
- **Betweenness**: measures the extent to which student i lies between other students in the network. Students with high betweenness occupy critical roles in the network structure, since they usually have a network position that allow them to work as an interface between tightly-knit groups, being "vital" elements in the connection between different regions of the network;
- **Authority score**[4]: authority is a student that receives many inward links or, in other words, is a node with a high in-degree;
- **Hub score**: hub is a student that nominates many other students or, alternatively, a student with high out-degree;
- **Clustering coefficient**: the local version of this metric quantifies the transitivity in the neighborhood of actor i, by indicating the level of cohesion between their neighbors.

These variables were standardized (z-scores) in order to nullify the effect of different magnitudes in the computation of tensor decomposition.

After selecting the metrics and organize the snapshot matrices into a tensorial representation, we need to decompose our three-order tensor $\mathcal{X} \in \mathbb{R}^{32 \times 7 \times 7}$, where the first mode is constituted by the 32 students, the second mode refers to the 7 actor-level network's metrics and the third mode is related to the 7 different time moments when the questionnaires were applied, into a small core tensor $\mathcal{G}$ and a set of component matrices $\mathbf{A}$, $\mathbf{B}$ and $\mathbf{C}$. To obtain them, we estimate a Tucker3 model of order $(4 \times 2 \times 3)$, which explains 63.2% of the total data variation. This order is a parameter of the model and refers to the number of components retained in each mode ($P = 4$, $Q = 2$ and $R = 3$). Its choice was guided by the analysis of a scree plot that indicates the potential ability of a Tucker3 model to explain the original data, for each possible combination of number of components. We have used the N-way toolbox [8] of MATLAB in all the experiments reported here. Learning times are not relevant, given that all the experiments took only few seconds.

The *core tensor* $\mathcal{G}$ contains the weights of all possible triads (combination of components, for the three modes) and these weights reflect the importance of the interaction between components, thus revealing the underlying variation pattern. The results tell us that the interaction of components that explains the

---

[4] Both the *authority* and the *hub* scores were computed by the HITS algorithm.

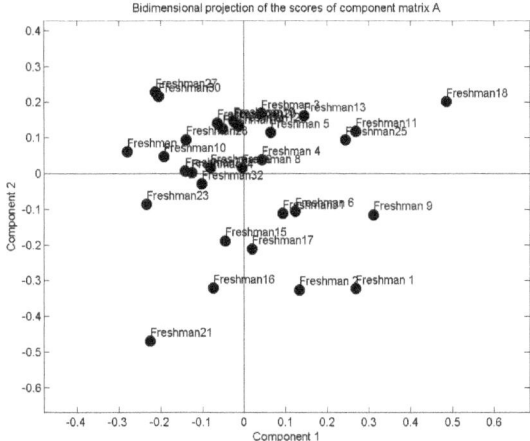

Fig. 3. Projection of the coefficients of matrix **A** in the bidimensional space defined by the two most representative components of mode A, which is associated to the row-entities (i.e. students) of the original tensor $\mathcal{X}$

Table 1. Mean of the in-degree and the authority scores, for all analyzed time points

Metric	In-Degree (-)	Authority Score (-)
All students	13.67	0.0313
Student **7** (-)	19.57	0.0381
Student **18** (+)	1.57	0.0041

higher portion of the sum of squares and, therefore, is the most important for understanding the data structure, is the interaction $(1, 1, 1)$ (explains 61.75% of the initial 63.2% variation). In turn, the entries of the *component matrices* **A**, **B** and **C**, represent the weights (also referred to as *scores* or *coefficients*) of the corresponding entities (actors, metrics and time points, respectively) in a given level of a given mode. Note that these *component matrices* have as many columns, or levels, as the number of components defined in the order of the estimated Tucker3 model.

Before presenting the trajectories, we first need to interpret the meaning of each component of mode B that will define the plane where we represent the social micro-evolution. To help this interpretation, we project the coefficients of each actor-level metric in the space spanned by the two most representative components of the mentioned mode, as shown in Figure 4, and we focus on metrics having extreme scores, since those are the ones with higher contribution to the formation of the axis. We perform the same analysis with component matrix **A**, in order to find the students associated to mode B's components.

Based on the analysis of the scores of the first component of matrix **B**, denoted by $\mathbf{B}_{:,1}$, we deduce that this axis is associated with both the *authority* (*score* = -0.4410) and the *in-degree* statistics (*score* = -0.4407). Conceptually, these

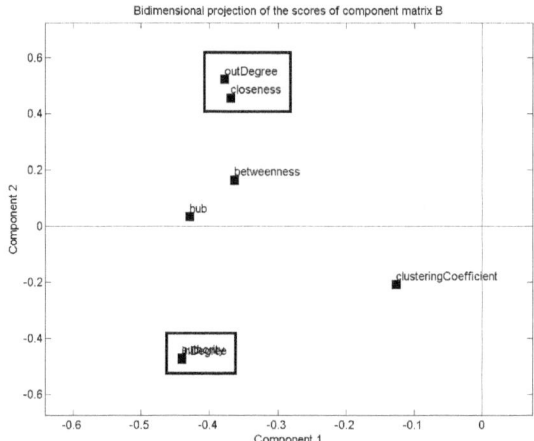

Fig. 4. Projection of the coefficients of matrix **B** in the bidimensional space defined by the two most representative components of mode B, which is associated to the column-entities (i.e. actor-level statistics) of the original tensor $\mathcal{X}$

Table 2. Mean of the out-degree, closeness, in-degree and authority statistics, for all analyzed time points

Metric	Out-degree (+)	Closeness (+)	In-Degree (-)	Authority Score (-)
All students	13.67	0.5657	13.67	0.0313
Student 21 (-)	18.71	0.7527	18.14	0.0574
Student 27 (+)	13.14	0.4740	19.71	0.0383

metrics are very similar, since they are given by the number of inward links of a given actor. Thus, we define this axis as being the *popularity axis*, with large negative values being associated to popular students and, analogously, large positive values being associated to unpopular ones. Examples of such students are *student* 7 ($A_{7,1} = -0.2792$), on the popular side, and *student* 18 ($A_{18,1} = 0.4845$), on the unpopular side. In Table 1 are given the values of the mean of these metrics, considering all time points, for the mentioned students and for all students, in order to provide a basis of comparison. These values corroborate the graphical findings.

Regarding the second component of the same matrix, denoted by $\mathbf{B}_{:,2}$, the same kind of analysis led to the interpretation of this axis as being the *prestige axis*, since it opposes the *out-degree* and *closeness* statistics, strongly associated to influence and social power, to the *in-degree* and *authority* statistics, which can be understood as measures of social support. The first statistics are associated to the positive side of the axis and to *student* 27 ($A_{27,2} = 0.2288$). The second pair of statistics are related to the negative side of the axis and to *student* 21 ($A_{21,2} = -0.4682$). The mean values are summarized in Table 2. From the analysis of the table we can verify that both students have higher in-degrees and authority scores than the corresponding overall mean. This is related to the fact that they are located in the negative side of the first component axis. Other aspect that can be drawn from the table is that both out-degree and closeness increase when one moves from the positive side of the second axis to its negative side. This means that an actor moving upwards loses influence and reachability in the network. Concerning the in-degree and the authority, since these metrics are the ones that define the first axis, their impact in the second axis is not meaningful.

In short, we can conclude that the best position in the component's space is the third quadrant, associated to powerful and prestigious actors (high values in all relevant metrics), and the worst position is the first quadrant, associated to less socially skilled actors (low values in all metrics). Therefore, students whose trajectories take the **direction** of the third quadrant are improving their social status. Otherwise, if moving in the direction of the first quadrant, they are losing their social power. Regarding the **amplitude**, if most of the time points of a given student's trajectory have the same coordinates, then one can assume that his/her social position is stable. An analogous reasoning holds for the opposite scenario. After decomposing the tensor and assign a meaning to the components, we define the trajectories of each student following the procedure described in Section 2.2.

3.3   Analysis of Trajectories

As already stated by Van de Bunt [11], in these networks there is a clear transition between the moment where almost no student knows each other to the moment where there is a significant number of ties between groups of them. This fact is corroborated by the students' trajectories in the bidimensional projection of the scores of **B**, depicted in Figure 5. By projecting all trajectories, we observe that the origin of most of them (indicated by t_0) lies close to the border between the first and the fourth quadrants of the plane, more specifically, in the coordinates marked by the black circles. Such coordinates are indicative of the initial weak social power of the students within the group under study. Nevertheless, there are a few exceptions related to old acquaintances of some students that attended the same former school.

Due to space constraints, in this paper we will only focus on the analysis of two trajectories, namely, the trajectories of freshman 18 and freshman 21. We choose these students because they represent the most extreme social positions that can be found in the analyzed networks. The associated trajectories represent extreme social positions, with student 18 having the worst social status (positive quadrants of the plane), and student 21 having the best social situation (negative quadrants of the plane).

The trajectory of freshman 18, depicted in Figure 6, is somewhat irregular, showing frequent changes in direction from time to time. However, if we consider an overall view of this trajectory we verify that the trend is to move upwards

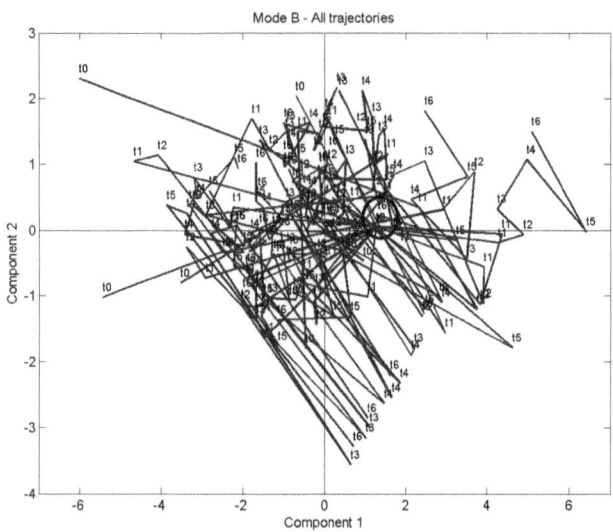

Fig. 5. Trajectories of all students in the space spanned by the two most representative components of mode **B**

and slightly rightwards, with most of the time points being located in the first quadrant of the plane. With concern to amplitude, though the trajectory is not very close, it reveals some stability of the social position of student 18 over the analyzed period, once he tends to be positioned within the space of the first quadrant. Therefore, we can deduce that this student is a bit unsociable, at least within the social circle of the university freshmen, showing a low social status and weak closeness to the remaining members of the network. Other interesting fact is that, along time, the sociability of this student tends to deteriorate, since he moves to the northeast of the plane, contrary to what would be expected. To better understand this outlying behavior we analyzed additional information and we found out that student 18 is a non-smoker and was attending the 2-year program. Since only 6 of the 32 freshmen are enrolled in this program, we infer that he had fewer opportunities to interact than students from other programs and, due to his personality traits derived from the analysis, he may not used them as a way to create bonds and improve his social position.

At the opposite extreme, we have a female smoker student, the so-called *freshman 21*, that was attending the 3-year program. Her trajectory, though having a low amplitude, is quite irregular, oscillating from the second to the third quadrant of the plane, as can be ascertained from Figure 7. The lack of trend in trajectory's direction, especially along the y-axis, can be indicative of some social instability with respect to the influence that exerts in the group and the easiness in reaching others in the network. In t_0 she occupies the best position in the plane, compared with the subsequent time points. This is explained by the

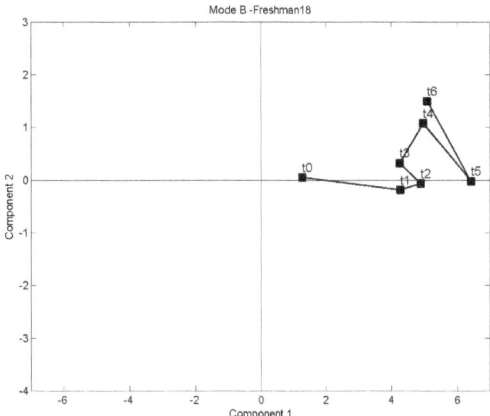

Fig. 6. Trajectories of student 18 in the plane defined by the two components of matrix **B**

fact that she had a few acquaintances from the former school, while most of her colleagues did not. However, three weeks after the first questionnaire, i.e. in t_1, the situation changes since meanwhile people had time to know each other and establish ties. Consequently, the relative social status of student 21 decreases, mostly because of the increase of other students' social behavior. This reflects in a displacement of the student in the plane to a position closer to the origin. In short, and comparing the first time point with the last one (t_6), we observe that freshman 21 slightly decreased her popularity, by moving rightwards, and decreased her influence and reachability, by moving upwards. Nevertheless, she was able to maintain her social power during the time horizon of the study, being the most popular student of the network.

In this paper we also mentioned that the comparison of different trajectories can help us identify actors with similar evolution in dynamic networks. We show an example of such analysis by comparing the trajectories of student 27 and student 30, illustrated in Figure 8. At a first glance, the shape of the trajectories does not look very much alike. Nevertheless, if we take into consideration the position of the time points, we observe that, in both cases, they are positioned in the same quadrants (e.g. t_4/t_5 are located in the second quadrant, t_1/t_2 in the third quadrant, and t_3/t_6 are located in the fourth quadrant). Also, the movement of the time points in the plane is quite identical (e.g. in time point t_3, when student 27 moves to the southeast direction of the space, student 30 also takes the same direction). Based on this kind of information we assume that students 27 and 30 show similar social behavior over time. Note that they also have identical scores in the two first components of matrix **A**, as can be seen from Figure 3.

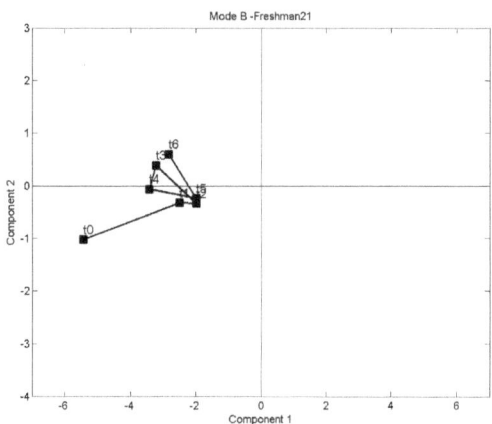

Fig. 7. Trajectories of student 21 in the plane defined by the two components of matrix **B**

4   Related Work

Traditional approaches to represent high-dimensional data include projections of data entities in low-dimensions spaces, that preserve most of the original data variation. Examples of such representations are the ones yielded by PCA, where the entities are usually plotted in a bidimensional subspace spanned by the first and the second principal components. On the other hand, when dealing with tensorial representations of data, the analysis of their decompositions are commonly performed through the analysis of graphical representations of the component matrices of each mode, where the x-axis represent the entities of the corresponding mode and the y-axis represent the entries of the component matrix. However, both these approaches do not meet the requirements we would like to embed in a visualization scheme for evolving data, which are compactness, substance and simplicity.

A good survey on time-oriented visualization approaches appear in [6]. The work of Sun et al.[10] also discusses a topic similar to the one addressed in this paper. In this work [10], they propose a hybrid approach that summarizes and extracts patterns from tensorial representations of large content-based social networks, by first performing a high-order dimensionality reduction using Tucker decomposition and, then, clustering the dimensions of each one of the decomposed modes. The extracted patterns are presented to the user by means of a hierarchical graph visualization, which is build upon the results of the cluster analysis. Nevertheless, they do not define temporal trajectories of entities, or focus on the study of their evolution.

To the best of our knowledge the most related work to ours is the STATIS method, proposed by Lavit [9]. One of the steps of this method consists in projecting the trajectories of individuals, over time, in the so-called "compromise

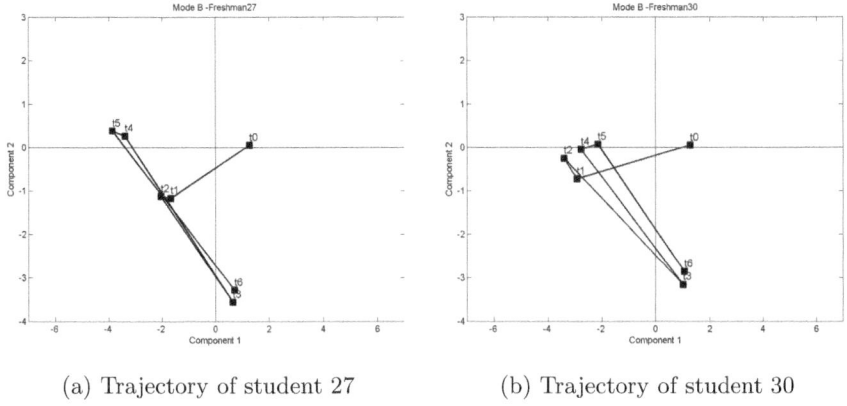

(a) Trajectory of student 27   (b) Trajectory of student 30

Fig. 8. Trajectories of (a)-student 27 and (b)-student 30 in the plane defined by the two components of matrix **B**

space". The main difference between our trajectories and the STATIS trajectories lies in the space where we define them which, in our case, is based on the decomposition yielded by the Tucker3 model. Besides, in STATIS the interpretation of the trajectories is made in relation to the average trajectory of a fictitious individual, while in our approach the interpretation is based on the properties of the trajectories themselves (e.g. direction, shape, amplitude, etc.).

5   Summary and Conclusions

In this paper we discussed a new approach to visualize dynamic social networks, which is based on the extraction of temporal trajectories from tensorial representations of time-evolving structures. Our main contribution in this work is the definition of trajectories, at the actor-level, in a bidimensional space spanned by the components yielded by Tucker tensor decomposition. A case study using a set of self-reported friendship networks among university freshmen revealed that the proposed approach has several desirable features, being concomitantly simple, informative and compact, thus allowing to intuitively understand patterns and structural changes in the evolution of dynamic social network.

Acknowledgments. Thanks to LIAAD - INESC Porto L.A. and to the financial support of the project Knowledge Discovery from Ubiquitous Data Streams (PTDC/EIA-EIA/098355/2008).

References

1. Moody, J., McFarland, D., Bender-deMoll, S.: Dynamic Network Visualization. American Journal of Sociology 110(4), 1206–1241 (2005)
2. Kolda, T.G., Bade, B.W.: Tensor Decompositions and Applications. SIAM Review 51(3), 455–500 (2009)

3. Smilde, A.K.: Three-way Analyses Problems and Prospects. Chemometrics and Intelligent Laboratory Systems 15, 143–157 (1992)
4. Tucker, L.: Some Mathematical Notes on Three-Mode Factor Analysis. Psychometrika 31(3), 279–311 (1966)
5. Kroonenberg, P.M.: Three-mode Principal Component Analysis: Theory and Applications. DSWO Press, Leiden (1983)
6. Aigner, W., Miksch, S., Muller, W., Schumann, H., Tominski, C.: Visualizing Time-Oriented Data - a Systematic View. Computers and Graphics 31, 401–409 (2007)
7. Oseledets, I., Savostyanov, D., Tyrtyshnikov, E.: Linear Algebra for Tensor Problems. Computing 85(3), 169–188 (2009)
8. Bader, B., Kolda, T.: MATLAB Tensor Toolbox Version 2.4 (March 2010), http://csmr.ca.sandia.gov/tgkolda/TensorToolbox/
9. Lavit, C., Escoufier, Y., Sabatier, R., Traissac, P.: The ACT (STATIS method). Computational Statistics and Data Analysis 18, 97–119 (1994)
10. Sun, J., Papadimitriou, S., Lin, C., Cao, N., Liu, S., Qian, W.: Multivis: Content-based Social Network Exploration through Multi-way Visual Analysis. In: Proceedings of the 2009 SIAM International Conference on Data Mining (SDM 2009), pp.1063–1074 (2009)
11. Van de Bunt, G.G., van Duijn, M.A.J., Snijders, T.A.B.: Friendship Networks through Time: An Actor-Oriented Statistical Network Model. Computational and Mathematical Organization Theory 5, 167–192 (1999)
12. Michell, L., Amos, A.: Girls, pecking order and smoking. Social Science and Medicine 44, 1861–1869 (1997)

Using Data Mining Techniques to Predict Deformability Properties of Jet Grouting Laboratory Formulations over Time

Joaquim Tinoco[1], António Gomes Correia[1], and Paulo Cortez[2]

[1] Department of Civil Engineering/C-TAC, University of Minho, Campus de
Azurém, 4800-058 Guimarães, Portugal
{jabtinoco,agc}@civil.uminho.pt
[2] Centro Algoritmi, Departamento de Sistemas de Informação, Universidade do
Minho, Campus de Azurém, 4800-058 Guimarães, Portugal
pcortez@dsi.uminho.pt

Abstract. Jet Grouting (JG) technology is one of the most used soft-soil improvements methods. When compared with other methods, JG is more versatile, since it can be applied to several soil types (ranging from coarse to fine-grained soils) and create elements with different geometric shapes (e.g. columns, panels). In geotechnical works where the serviceability limit state design criteria is required, deformability properties of the improved soil need to be quantified. However, due to the heterogeneity of the soils and the high number of variables involved in the JG process, such design is a very complex and hard task. Thus, in order to achieve a more rational design of JG technology, this paper proposes and compares three data mining techniques in order to estimate the different moduli that can be defined in an unconfined compressed test of JG Laboratory Formulations (JGLF). In particular, we analyze and discuss the predictive capabilities of Artificial Neural Networks, Support Vector Machines or Functional Networks. Furthermore, the key parameters in modulus estimation are identified by performing a 1-D sensitivity analysis procedure. We also analyze the effect of such variables in JGLF behavior.

Keywords: Ground improvement, Jet Grouting, Young Modulus, Regression, Artificial Neural Networks, Support Vector Machines, Functional Networks.

1 Introduction

Given the growth of the human population and the finite resources of the planet Earth, nowadays we are forced to use soft-soils as a soil foundation for construction works. In this context, there is a need to improve the soil foundation physical and mechanical properties. Following this need, important grouting methods for soil improvement have being proposed: compacting grouting, permeation grouting, hydraulic fracture grouting and Jet Grouting (JG) [1]. This paper focuses

L. Antunes and H.S. Pinto (Eds.): EPIA 2011, LNAI 7026, pp. 491–505, 2011.
© Springer-Verlag Berlin Heidelberg 2011

JG, which has been widely applied in the last few years. JG was first introduced by Yamakoda brothers in 1965 and since then this technology was improved [2]. JG is characterized by its great versatility, allowing to improve the mechanical and physical properties of several types of soil (ranging from coarse to fine-grained soils) and to obtain different geometric shapes (e.g. columns, panels).

The fundamental of JG is to produce a soil-cement mixture, often termed as soilcrete, by injecting grout, with or without other fluids (e.g. air or water), at high pressure and velocity into the subsoil. The fluids are injected through small nozzles placed at the end of a rod that, after introduced at the intended depth, is continually rotated and slowly removed up to the surface. The obtained improved mass of soil presents an enhancement in terms of strength, stiffness and permeability. According to the number of fluids injected, three systems are conventionally used: JET 1, JET 2 and JET 3.

This paper focus in the initial JG stage, where a set of laboratory formulations, which are function of the soil type to be treated and the design properties, are used to set the soil-cement mixture that will be used in construction works. In particular, these formulations allow the definition of the grout water/cement ratio, the amount of cement for cubical meter of treated soil and the cement type, needed to satisfy the designing and economical requirements. However, the actual design of JG columns is almost performed based on empirical rules [3,4]. Therefore, and since these empirical rules are often too conservative, the quality and the cost of the treatment can be compromised. Hence, and bearing in mind the high versatility of JG technology and its role in important geotechnical works, there is need to develop reliable and rational methods to estimate the effect of the different variables involved in JG process. Consequently, the number of laboratory formulations can be reduced, the constructive process optimized and economy efficiency improved.

One interesting approach to deal with the high number of variables involved in JG process and the heterogeneity of the soil is to apply Data Mining (DM) techniques, such as Artificial Neural Networks (ANNs), Support Vector Machines (SVMs) and Functional Networks (FNs) [5,6]. Such DM algorithms are powerful tools to automatically learn nonlinear relationships between several inputs and the target variable and have been successfully applied in the Civil Engineering domain [7,8,9,10,11,12,13,14]. Yet, within our knowledge, these techniques have not been applied for JG prediction, namely for the secant deformability modulus at 50% of the maximum applied stress ($E_{sec50\%}$) and the maximum secant deformability modulus (E_{max}). The main criticism of black box DM techniques, such as ANNs and SVMs, is the lack of explanatory power, i.e. the data-driven models are difficult to interpret by humans [7]. However, to overcome this drawback a sensitivity analysis procedure can be applied [15]. Such procedure measures the model's response changes when a given input variable is varied through its domain, allowing a better understanding of how each input affects the target variable.

In geotechnical structures where the Serviceability Limit State (SLS) design criteria is required, deformability properties of the improved soil need to be quantified. In the present paper, the $E_{sec50\%}$ and E_{max} moduli are estimated by ANN, SVM and FN algorithms. The obtained results are exposed and compared with previous studies where the same DM algorithms were trained to estimate initial modulus at very small strains (E_0) [11] and tangent deformability modulus at 50% of the maximum applied stress ($E_{tg50\%}$) [13] of JGLF over time. Moreover, the key variables in stiffness estimation are identified, compared and discussed by applying a 1-D sensitivity analysis procedure. In addition, the effect of the key variables in both $E_{sec50\%}$ and E_{max} estimation is quantified.

2   Data Mining Techniques

There are several regression DM algorithms that can be applied, each one with its own advantages and disadvantages. This paper focuses on three nonlinear models, ANNs, SVMs and FNs [16,17,4], that tend to produce high predictive results [5]. ANNs and SVMs were implemented in the **rminer** library of the **R** tool, which is particularly suited for these two types of models [18]. Before fitting the ANN and SVM models, the data attributes were standardized to a zero mean and one standard deviation and before analyzing the predictions, the outputs post-processed with the inverse transformation [5]. The FNs were formulated and solved in the free version of the GAMS tool [19]. For a baseline comparison, the classic Multiple Regression (MR) method was also tested (using the **R** tool).

ANNs are inspired in some aspects of the human brain, which processes information by means of interactions among several neurons. Here, the ANN consists in a fully connected multilayer perceptron, with one hidden layer with H processing units, bias connections and logistic activation functions $1/(1 + e^{(-x)})$. To find the best value for H, we use a grid search within the range $\{2, 4, ..., 10\}$ under an internal (i.e. applied over training data) 5-fold cross validation [5]. During this grid search, we select the H value that produces the lowest mean absolute error and then the selected ANN is retrained with all training data.

SVM was initially proposed for classification tasks (i.e., to model a discrete labeled output). After the introduction of the ϵ-insensitive loss function, it was possible to apply SVM to regression tasks [17]. SVM has theoretical advantages over ANN, such as the absence of local minima in the learning phase, i.e., the model always converges to the optimal solution. The main idea of the SVM is to transform the input data into a high-dimensional feature space by using a nonlinear mapping that is dependent of a kernel. Then, the SVM finds the best hyperplane within the feature space. We adopted the popular gaussian kernel. To reduce the search space, we adopted the heuristics proposed in [20] to set the complexity penalty parameter, C=3, and the size of the insensitive tube, $\epsilon = \hat{\sigma}/\sqrt{N}$, where $\hat{\sigma} = 1.5/N \cdot \sum_{i=1}^{N} (y_i - \hat{y}_i)^2$, y_i is the measured value, $\hat{y}_i$ is the value predicted by a 3-nearest neighbor algorithm and N the number of examples. The most important SVM parameter, the kernel parameter γ [5], was set using a grid search within $\{2^{-15}, 2^{-13}, ..., 2^3\}$, under the same 5-fold internal cross validation scheme.

FNs are a general framework useful for solving a wide range of problems (e.g. statistics and engineering applications). When compared with ANNs, FNs present the advantage of being possible to use a priori knowledge to design models that are more understandable and closer to the physical domain. In the geotechnical domain, several of the conventional analytical predictive models are of exponential nature (e.g. EC2) [11]. Following this knowledge, we adopted the following generic expression for the FN models:

$$\hat{y} = \beta_0 \cdot \prod_{i=1}^{I} x_i^{\alpha_i}, \quad \text{minimizing } Q = \sum_{k=1}^{N} \delta_k^2 = \sum_{k=1}^{N} \left(y_k - \beta_0 \cdot \prod_{i=1}^{I} x_i^{\alpha_i} \right)^2 \quad (1)$$

where $\{x_1, ..., x_I\}$ denotes the set of input variables and $\{\beta_0, \alpha_1, ..., \alpha_I\}$ are the coefficients to be adjusted.

A regression model aims to induce a model that minimizes an error metric between observed and predicted values considering N examples. For this purpose three common metrics were calculated [12]: Mean Absolute Deviation (MAD), Root Mean Squared Error (RMSE) and Coefficient of Correlation (R^2). The first two metrics should present lower values, while R^2 should be close to the unit value. The main difference between RMSE and MAD is that former is more sensitivity to extreme values. The regression error characteristic (REC) curve [21], which plots the error tolerance on the x-axis versus the percentage of points predicted within the tolerance on the y-axis was also adopted to compare the predictive results.

The overall generalization performance was accessed by using 20 runs under a Leave-One-Out approach [5], where successively one example is used to test the model and the remaining examples are used to fit the model. Under this scheme, all of the data are used for training and testing. Yet, this method requires approximately N (the number of data samples) times more computation, because N models must be fitted. Leave-One-Out validation allows a robust generalization estimate when small datasets are modeled, which is the case of this study (see Section 3). The final model generalization estimate is evaluated by computing the MAD, RMSE and R2 metrics for all test samples (N).

Besides obtaining a high predictive performance, it is also important to extract human understandable knowledge from the data-driven models. For such purpose, we applied a 1-D sensitivity analysis procedure [15]. Such procedure is applied after the training phase and analyzes the model responses when a given input is changed. This procedure can be applied to any supervised DM model and allows to quantify the relative importance of each input parameter and also to measure the average effect of a given input on the target variable. Such quantification is determined by successively holding all inputs at a given baseline (e.g. their average values), except one input attribute (x_a) that is varied through its range with $j \in \{1, ..., L\}$ levels. The obtained responses $\hat{y}_{a,j}$ are stored. Higher response changes indicate a more relevant input. In particular, following the results achieved in [15], we adopt the gradient measure (S_a) to access the input relevance (R_a) of the attribute (x_a):

$$R_a = S_a / \sum_{i=1}^{I} S_i \cdot 100(\%), \quad \text{where } S_a = \sum_{j=2}^{L} |\hat{y}_{a,j} - \hat{y}_{a,j-1}| / (L-1) \quad (2)$$

The higher the gradient, the higher is the input importance. For a more detailed input influence analysis, the use the Variable Effect Characteristic (VEC) curve [15]. For a given input variable, the VEC curve plots the attribute L level values (x-axis) versus the sensitivity analysis responses (y-axis). In this paper, we set L=12.

3   Deformability Modulus Data

All DM models were trained using a database that came from a large labora-tory experimental program, carried out at University of Minho. This experimen-tal program was focused on the analysis of the influence of several parameters in mechanical properties of JG laboratory mixtures. Fig. 1 shows the different moduli that can be defined in a nonlinear stress strain relationship. In previous studies, we have shown how DM tools can be used in the estimation of the ini-tial modulus at very small strains (E_0) [11] and tangent elastic young modulus at 50% of the maximum applied stress ($E_{tg50\%}$) [13] of JGLF over time. In the present paper, we show how such DM algorithms can be used in the prediction of $E_{sec50\%}$ and E_{max} of JGLF over time and compare its performance with E_0 and $E_{tg50\%}$ estimation. All these four different moduli were measured in unconfined compression tests with on sample strain instrumentation [22]. Table 1 summa-rizes the number of records of each dataset used during the experiments, as well as the number of different formulations used in both $E_{sec50\%}$ and E_{max} studies. The dataset is rather small, with only 48 samples. Yet, it should be stressed that the acquisition of each data example requires considerable costs and amount of time, demanding laboratory work and materials (e.g. cement). The input vari-ables were chosen based on the expert knowledge related to soil-cement mixtures [23] and authors' knowledge obtained upon previous works: water/cement ratio (W/C); age of the mixture (t, typically there are two different ages for a given formulation); Relation between the mixture porosity and the volumetric content of cement ($n/(C_{iv})^d$); Cement content of the mixture ($\%C$); Percentage of sand ($\%Sand$); Percentage of silt ($\%Silt$); Percentage of clay ($\%Clay$); and Percentage of organic matter ($\%OM$). The main statistics of the input and target variables used in both $E_{sec50\%}$ and E_{max} datasets are shown in Table 2. The details in Table 1 and 2 for E_0 and $E_{tg50\%}$ can be found in [11,13].

For a physical characterization of the natural, some samples were collected and tested in laboratory. Although all soils were of clayed nature, they have different percentages of sand, silt, clay and organic matter. A detailed description of the distinct soil types can be found in Table 3. In the table, the first column denotes the construction site, while the third column shows the number of records that contain such soil. All laboratory formulations, in both $E_{sec50\%}$ and E_{max} datasets, were prepared with cement type CEM I 42.5R and CEM II 42.5R.

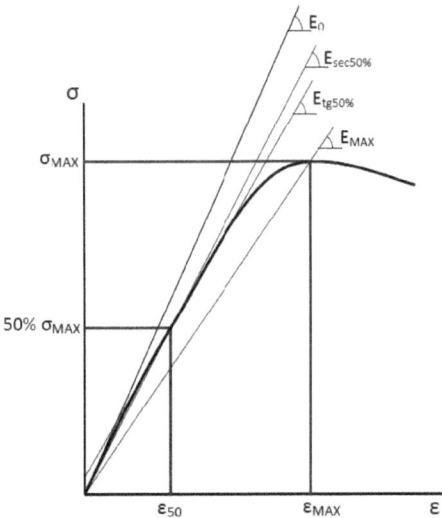

Fig. 1. Illustration of the different deformability properties (i.e. moduli) that can be defined in an unconfined compressed test (x-axis denotes the strain ε and y-axis the stress σ)

Table 1. Number of records and formulations of $E_{sec50\%}$ and E_{max} datasets

	$E_{sec50\%}$	E_{max}
Number of records	48	48
Number of formulations	8	8

4   Results

The predictive performance of ANN, SVM and FN algorithms in $E_{sec50\%}$ and E_{max} estimation is very similar when compared with those reached by these three algorithms for E_0 [11] and $E_{tg50\%}$ [13] prediction. The average hyperparameter and fitting time values (and respective 95% level confidence intervals according to a t-student distribution) are shown in Table 4. The coefficients of Equation 1, as optimized by FN algorithm, can be found in 5. The predictive results are presented in Table 6, which compares all DM models for each of the four properties (E_0, $E_{tg50\%}$, $E_{sec50\%}$ and E_{max}) based on the MAD, RMSE and R^2 metrics computed for the test data (mean value and 95% confidence intervals). The table results show high predictive capabilities, especially for ANN and SVM models. The relation of observed values versus predicted ones for the SVM model for both $E_{sec50\%}$ and E_{max} is shown in Fig. 2 and 3. As we can see, the

Table 2. Summary of the input and output variables in both $E_{sec50\%}$ and E_{max}

Parameter	Minimum	Maximum	Mean	Standard Deviation
W/C	0.69	1.11	0.98	0.12
t (days)	28.00	84.00	64.75	19.29
$n/(C_{iv})^d$	51.21	75.04	64.36	7.82
$\%C$	24.19	64.86	44.55	11.87
$\%Sand$	0.00	39.00	14.40	13.67
$\%Silt$	33.00	57.00	49.90	8.32
$\%Clay$	22.50	45.00	35.52	7.40
$\%OM$	0.40	8.30	3.70	2.45
$E_{sec50\%}$	1.50	5.67	3.17	1.11
E_{max}	1.50	7.00	3.44	1.30

Table 3. Soil types present in the collected data

Site	Soil type	Frequency	$\%Sand$	$\%Silt$	$\%Clay$	$\%MO$
A	Lean clay (CL)	9	39.0	33.0	27.0	8.3
B	Organic lean clay (OL)	6	6.0	57.0	37.0	1.8
C	Fat clay (CH)	22	7.0	53.0	40.0	3.2
D	Silty clay (CL-ML)	6	25.0	52.5	22.5	0.4
E	Lean clay (CL)	5	0.0	55.0	45.0	3.9

SVM predictions are very close to the target values. Fig. 2 and 3 also show the model accuracy in function of the absolute deviation (REC curve, dashed line). We can see that for an absolute deviation around 20%, the SVM model is almost 100% accurate. The relations plotted in Fig. 2 and 3 for E_0 and $E_{tg50\%}$ can be found in [11] and [13], respectively.

Fig. 4 and 5 show the relative importance of each variable in $E_{sec50\%}$ and E_{max} estimation, as measured by the 1-D sensitivity analysis, with the correspondent t-student 95% confidence intervals for all 20 runs performed. According to ANN and SVM models, which show a similar relative importance distributions, $n/(C_{iv})^d$ is the key parameter in both in $E_{sec50\%}$ and E_{max} estimation of JGLF over time. The soil properties, mainly $\%Clay$, are also important in the stiffness behavior. It is also interesting to observe that the age of the mixtures takes the lowest relevance in both in $E_{sec50\%}$ and E_{max}. These results make sense if we take into account the range of t in both datasets. It is known that, when leading with soil-cement mixtures, after 28 days time of cure its stiffness (as well as its uniaxial compress strength) is almost not affect by age. Regarding the FN model, the obtained results for $E_{sec50\%}$ and E_{max} of JGLF are less interesting. According to this model, the soil properties are the only variables that

Table 4. Hyperparameters and computation time for each fitted model

Model		Hyperparameters	time (s)
MR	E_0	-	10.8±0.0
	$E_{tg50\%}$	-	2.6±0.0
	$E_{sec50\%}$	-	2.6±0.0
	E_{max}	-	2.7±0.0
ANN	E_0	$H = 7 \pm 1$	869.9±1.0
	$E_{tg50\%}$	$H = 3 \pm 1$	130.8±1.1
	$E_{sec50\%}$	$H = 5 \pm 1$	134.9±0.3
	E_{max}	$H = 3 \pm 1$	136.3±0.0
SVM	E_0	$\gamma = 0.86 \pm 0.21$, $\epsilon = 0.06 \pm 0.00$	1168.9±1.0
	$E_{tg50\%}$	$\gamma = 0.82 \pm 0.34$, $\epsilon = 0.02 \pm 0.00$	194.5±0.2
	$E_{sec50\%}$	$\gamma = 0.26 \pm 0.05$, $\epsilon = 0.03 \pm 0.00$	202.4±0.2
	E_{max}	$\gamma = 0.38 \pm 0.09$, $\epsilon = 0.02 \pm 0.00$	201.8±1.7
FN	E_0	-	56.6±0.0
	$E_{tg50\%}$	-	17.3±0.0
	$E_{sec50\%}$	-	16.6±0.0
	E_{max}	-	14.0±0.0

Table 5. Fitted coefficients of the FN model

Model		β_0	α_1	α_2	α_3	α_4	α_5	α_6	α_7	α_8
FN	E_0	10.0^{10}	-0.11	-9.80	4.60	-1.99	-1.03	0.23	1.10	-0.73
	$E_{tg50\%}$	633.38	-0.10	-1.09	-1.88	-0.15	0.61	0.23	1.02	0.24
	$E_{sec50\%}$	691.27	-0.09	0.46	-2.65	0.12	-0.73	0.18	0.90	-0.49
	E_{max}	10.0^{10}	-0.03	-8.58	4.92	-1.72	-0.57	0.13	0.94	-1.96

Table 6. Error metrics for all DM models in E_0, $E_{tg50\%}$, $E_{sec50\%}$, E_{max} estimation (test set values, best values in **bold**)

Model		MAD	RMSE	R^2
MR	E_0	0.35±0.00	0.48±0.00	0.87±0.00
	$E_{tg50\%}$	0.31±0.01	0.38±0.01	0.83±0.01
	$E_{sec50\%}$	0.30±0.00	0.39±0.00	0.87±0.00
	E_{max}	0.31±0.01	0.42±0.01	0.90±0.00
ANN	E_0	**0.17±0.01**	**0.24±0.05**	**0.97±0.01**
	$E_{tg50\%}$	0.27±0.05	0.50±0.22	0.67±0.26
	$E_{sec50\%}$	**0.12±0.01**	**0.16±0.01**	**0.98±0.00**
	E_{max}	**0.18±0.01**	**0.26±0.02**	**0.96±0.01**
SVM	E_0	0.18±0.01	0.25±0.02	0.96±0.01
	$E_{tg50\%}$	0.19±0.01	0.28±0.02	0.91±0.01
	$E_{sec50\%}$	0.15±0.01	0.21±0.03	0.96±0.01
	E_{max}	**0.18±0.00**	0.31±0.01	0.94±0.00
FN	E_0	0.22±0.00	0.30±0.00	0.95±0.00
	$E_{tg50\%}$	**0.18±0.00**	**0.24±0.00**	**0.93±0.00**
	$E_{sec50\%}$	0.20±0.00	0.25±0.00	0.95±0.00
	E_{max}	0.20±0.00	0.27±0.00	0.95±0.00

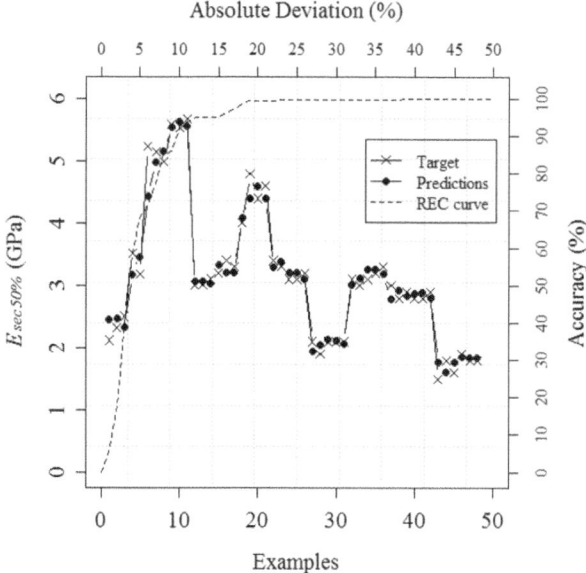

Fig. 2. Relation between $E_{sec50\%}$ observed versus predicted by SVM model (the dashed line represents the REC curve and should be read on the top horizontal and right vertical axis)

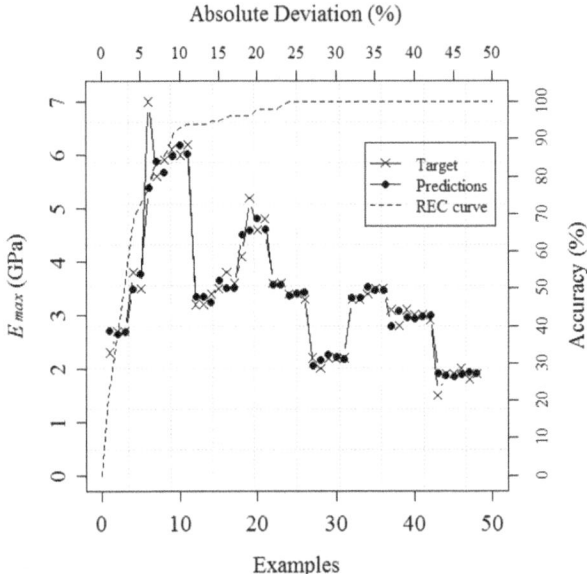

Fig. 3. Relation between E_{max} observed versus predicted by SVM model (the dashed line represents the REC curve and should be read on top horizontal and right vertical axis)

control deformability modulus of JGLF and it is known in within the geotechnical domain that it is not true.

Making a global appreciation of ANN and SVM models, i.e., comparing metrics values and the relative importance of each variable, we can affirm that the latter is more interesting. While achieving a similar predictive performance, the relative input importances of the SVM model are more coherent than in ANN in terms of what is known empirically in the JG domain. This confirms previous results, where the SVM model was also the most interesting when modeling the E_0 [11] and $E_{tg50\%}$ [13] moduli.

Hence, using the SVM model as a reference, we can identify the main variables that control stiffness behavior of JGLF over time. Observing Fig. 6, which compares the relative importance of each variable in stiffness estimation (E_0, $E_{tg50\%}$, $E_{sec50\%}$ and E_{max}) according to SVM model, we can see that the relation $n/(C_{iv})^d$ and the soil properties, mainly its clay and sand content are the key parameters in JGLF stiffness behavior. The W/C ratio also has a strong impact in $E_{tg50\%}$ prediction. The variable t is only strongly relevant (the second more relevant) to predict E_0. Such observation can be justified by t values in E_0 dataset, which range from 3 days to 56 days [11]. Indeed, this is confirmed by the empirical knowledge about soil-cement mixtures, where t plays an important role in stiffness behavior until 28 days time of cure.

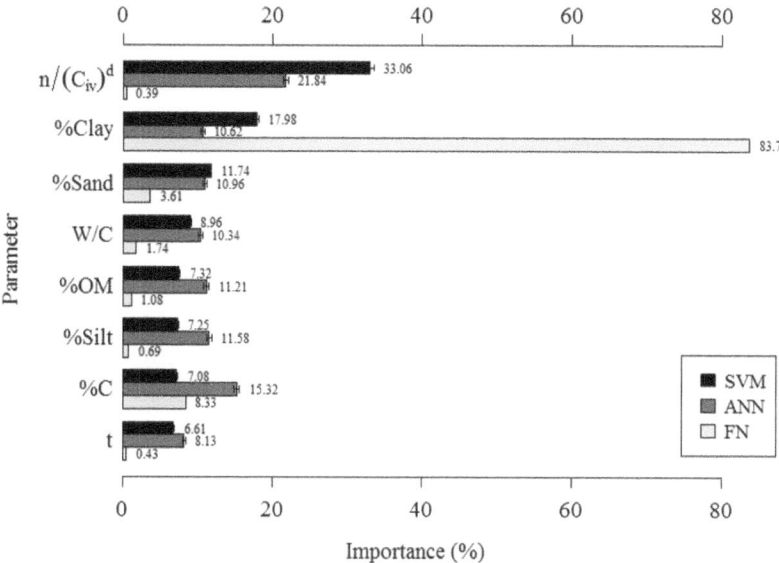

Fig. 4. Relative importance of each variable in $E_{sec50\%}$ estimation according to SVM, ANN and FN models, quantified by a 1-D sensitivity analysis

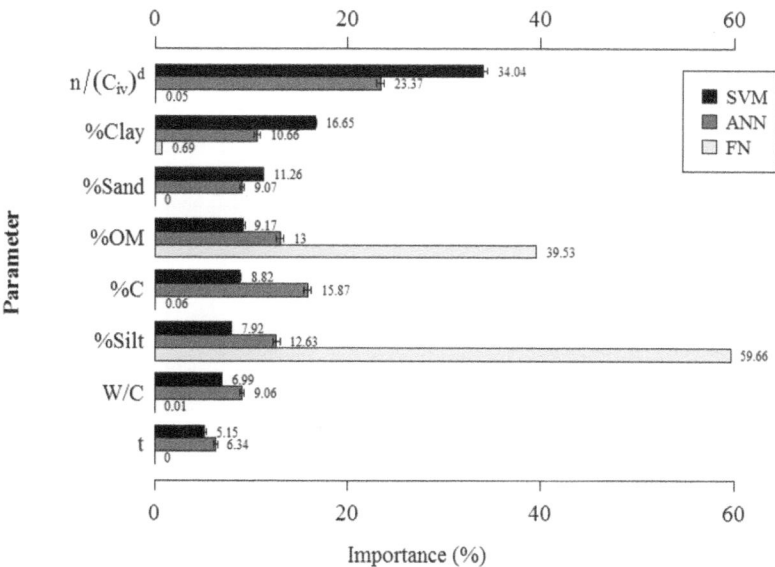

Fig. 5. Relative importance of each variable in E_{max} estimation according to SVM, ANN and FN models, quantified by a 1-D sensitivity analysis

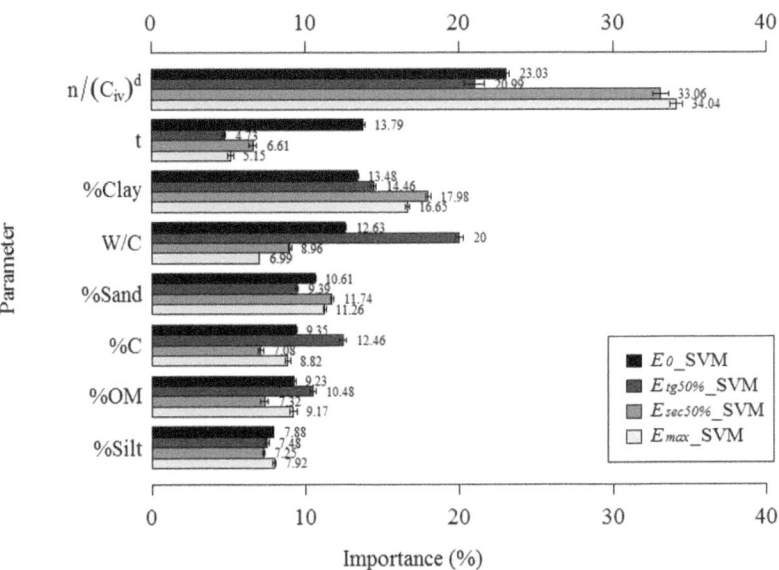

Fig. 6. Comparison of the relative importance of each variable in E_0, $E_{tg50\%}$, $E_{sec50\%}$ and E_{max} estimation according to SVM model, quantified by a 1-D sensitivity analysis

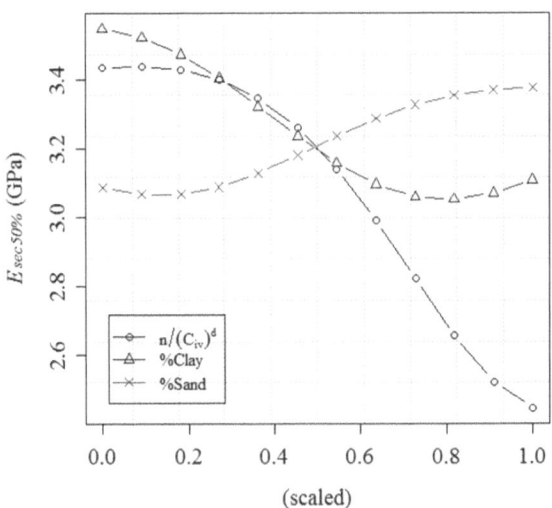

Fig. 7. VEC curves for the most relevant input variables in $E_{sec50\%}$ prediction (y-axis), according to SVM model measured by 1-D sensitivity analysis with L=12 levels (x-axis)

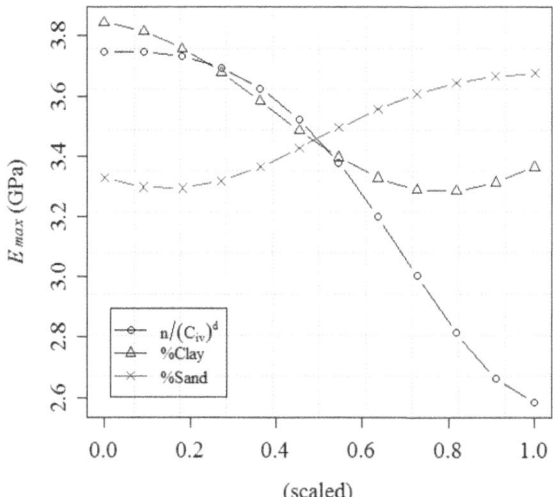

Fig. 8. VEC curves for the most relevant input variables in E_{max} prediction (y-axis), according to SVM model measured by 1-D sensitivity analysis with L=12 levels (x-axis)

Besides the relative importance of each input variable, we also quantify the average key input effects when predicting $E_{sec50\%}$ and E_{max}. Fig. 7 and 8 plot the VEC curves for the three main variables in $E_{sec50\%}$ and E_{max} estimation according to SVM. As expected, both $n/(C_{iv})^d$ and $\%Clay$ have a negative impact in stiffness prediction. In the other hand, when increasing the sand content of the soil, an improvement in the soil-cement mixture stiffness can be observed. In VEC curve of $\%Clay$, it is interesting to observe that for high values of clay percentage, the stiffness increases with clay content. This occurrence can be explained by iteration between the input variables. Indeed, when the clay content of the soil is very high, we are forced to add a high level of cement in JG formulations. Thus, it is anticipated that the modulus increases despite the high amount of clay in the soil. Such phenomenon was also observed in E_0 study [14].

5 Conclusions

The different moduli of Jet Grouting Laboratory Formulations (JGLF) can be successfully predicted by using Data Mining (DM) techniques. In particular, Support Vector machine (SVM) and Artificial Neural Network (ANN) models are able to learn with high accuracy the complex relationships between elastic, tangent and secant deformability modulus and its contributing factors. And such models are capable of predicting material deformability predictions from 28 to 84 days in advance (i.e. the range values of variable t).

By performing a 1-D sensitivity analysis procedure, we have shown that the relation between mixture porosity and the volumetric content of cement $(n/(C_{iv})^d)$ and the soil properties, mainly its clay content of soil ($\%Clay$), play an important role in both $E_{sec50\%}$ and E_{max} prediction, as well as in E_0 and $E_{tg50\%}$ estimation over time. It was also proved that the age of the mixture only produces a high importance in deformability properties of JGLF until 28 days time of cure. Also, the three most relevant input variables in both $E_{sec50\%}$ and E_{max} estimation show a nonlinear effect in such properties. Within these inputs, the sand content is the only one that has a positive impact.

The obtained results are valuable for geotechnical civil engineering, in particular for the project stage. The better understanding of the behavior of JG material based on few parameters (e.g. soil and cement) was achieved. As result of this knowledge, the quality, speed and the cost of the JG treatment can be improved by reducing the number of laboratory formulations that need to be prepared. The next step, which will be addressed in future work, is to apply this DM approach to define predictive models of the final diameter and mechanical properties of real JG columns.

Acknowledgments. The authors wish to thank to Portuguese Foundation for Science and Technology (FCT) the support given through the doctoral grant SFRH/BD/45781/2008. Also, the authors would like to thank the interest and financial support by Tecnasol-FGE.

References

1. Nikbakhtan, B., Ahangari, K., Rahmani, N.: Estimation of jet grouting parameters in Shahriar dam. Iran, Mining Science and Technology 20(3), 472–477 (2010)
2. Xanthakos, P., Abramson, L., Bruce, D.: Ground Control and Improvement. John Wiley & Sons Inc. (1994)
3. Croce, P., Flora, A.: Analysis of single-fluid jet grouting. Géotechnique 51, 905–906 (2001)
4. Kirsh, F., Wolfgang, S.: Ground Improvement and its Numerical Analysis. In: Proceedings of 15th International Conference on Soil Mechanics and Geotechnical Engineering, vol. 3, pp. 1775–1778 (2001)
5. Hastie, T., Tibshirani, R., Friedman, J.: The Elements of Statistical Learning: Data Mining, Inference, and Prediction, 2nd edn. Springer, Heidelberg (2008)
6. Castillo, E., Cobo, A., Gutierrez, J., Pruneda, R.: Functional networks with applications: a neural-based paradigm. Springer, Heidelberg (1999)
7. Lai, S., Serra, M.: Concrete strength prediction by means of neural network. Construction and Building Materials 11(2), 93–98 (1997)
8. Goh, A., Goh, S.: Support vector machines: Their use in geotechnical engineering as illustrated using seismic liquefaction data. Computers and Geotechnics 34(5), 410–421 (2007)
9. Rezania, R., Javadi, A.: A new genetic programming model for predicting settlement of shallow foundations. Canadian Geotechnical Journal 44(12), 1462–1473 (2007)
10. Tinoco, J., Gomes Correia, A., Cortez, P.: A data mining approach for Jet Grouting Uniaxial Compressive Strength prediction. In: Proceedings of World Congress on Nature & Biologically Inspired Computing (NaBIC 2009), pp. 553–558 (2009)

11. Tinoco, J., Gomes Correia, A., Cortez, P.: Application of Data Mining Techniques to Estimate ElasticYoung Modulus Over Time of Jet Grouting Laboratory Formulations. In: Proceedings of 1st International Conference on Information Technology in Geo-Engineering (ICITG - Shanghai 2010), pp. 92–100 (2010)
12. Tinoco, J., Gomes Correia, A., Cortez, P.: Application of data mining techniques in the estimation of the uniaxial compressive strength of jet grouting columns over time. Constructions and Building Material 25, 1257–1262 (2011)
13. Tinoco, J., Gomes Correia, A., Cortez, P.: Application of Data Mining Techniques in the Estimation of Mechanical Properties of Jet Grouting Laboratory Formulations over Time. Advances in Intelligence and Soft Computing 96, 283–292 (2011)
14. Tinoco, J., Gomes Correia, A., Cortez, P.: A Data Mining Approach for Predicting Jet Grouting Geomechanical Parameters. In: Chen, D., Fu, H. (eds.) Proceedings of GeoHunan International Conference II, vol. 223, pp. 97–104. Geotechnical Special Publications (GSP), Hunan (2011)
15. Cortez, P., Embrechts, M.: Opening Black Box Data Mining Models Using Sensitivity Analysis. In: Proceedings of the 2011 IEEE Symposium on Computational Intelligence and Data Mining (CIDM 2011), pp. 341–348 (2011)
16. Kenig, S., Ben-David, A., Omer, M., Sadeh, A.: Control of properties in injection molding by neural networks. Engineering Applications of Artificial Intelligence 14(6), 819–823 (2001)
17. Smola, A.J., Schölkopf, B.: A tutorial on support vector regression. Statistics and Computing 14(3), 199–222 (2004)
18. Cortez, P.: Data mining with neural networks and support vector machines using the r/rminer tool. In: Perner, P. (ed.) ICDM 2010. LNCS, vol. 6171, pp. 572–583. Springer, Heidelberg (2010)
19. GAMS Development Corporation, On-line Documentation, http://www.gams.com/docs/document.htm (accessed January 2011)
20. Cherkassky, V., Ma, Y.: Practical Selection of SVM Parameters and Noise Estimation for SVM Regression. Neural Networks 17(1), 113–126 (2004)
21. Bi, J., Bennett, K.: Regression Error Characteristic Curves. In: Proceedings of the Twentieth International Conference on Machine Learning, CD-ROM (2003)
22. Gomes Correia, A., Valente, T., Tinoco, J., Falcão, J., Barata, J., Cebola, D., Coelho, S.: Evaluation of mechanical properties of jet grouting columns using different test methods. In: Proceedings of 17th International Conference on Soil Mechanics and Geotechnical Engineering, pp. 2179–2171 (2009)
23. Shibazaki, M.: State of practice of jet grouting. In: Proceedings of Grouting and Ground Treatment, pp. 198–217 (2004)

Doubtful Deviations and Farsighted Play[*]

Wojciech Jamroga and Matthijs Melissen

University of Luxembourg

Abstract. Nash equilibrium is based on the idea that a strategy profile is stable if no player can benefit from a unilateral deviation. We observe that some locally rational deviations in a strategic form game may not be profitable anymore if one takes into account the possibility of further deviations by the other players. As a solution, we propose the concept of farsighted pre-equilibrium, which takes into account only deviations that do not lead to a decrease of the player's outcome even if some other deviations follow. While Nash equilibria are taken to include plays that are certainly rational, our pre-equilibrium is supposed to rule out plays that are *certainly irrational*. We prove that positional strategies are sufficient to define the concept, study its computational complexity, and show that pre-equilibria correspond to subgame-perfect Nash equilibria in a meta-game obtained by using the original payoff matrix as arena and the deviations as moves.

1 Introduction

The optimal strategy for an agent depends on his prediction of the other agents' behavior. For example, in security analysis, some predictions of the users' (or even the intruders') behavior can be useful when designing a particular solution. However, if the users (resp. intruders) do not behave in the predicted way, this solution might give rise to new vulnerabilities. To obtain a 'more secure' solution concept, we therefore weaken the assumptions made by agents when playing Nash equilibrium, and introduce a new solution concept based on these weaker assumptions.

Nash equilibrium (NE) defines a play to be stable when, if the players knew what the others are going to do, they would not deviate from their choices unilaterally. Conversely, if some player can beneficially deviate from strategy profile s, then the profile is assumed to describe irrational play. In this paper, we point out that some of these deviations may not be profitable anymore if one takes into account the possibility of further deviations from the opponents. As a solution, we propose the concept of *farsighted pre-equilibrium (FPE)* which takes into account only those deviations of player i that do not lead to decrease of i's outcome, even if some other deviations follow. In consequence, we argue that the notion of irrational play can be meaningfully relaxed.

Rational vs. Irrational Play. We call the new concept *pre-equilibrium* because we do not imply that all FPEs are necessarily stable. Our point is rather that all strategy profiles outside FPEs are certainly *un*stable: a rational player should deviate even if he considers it possible that other players react to his change of strategy. Formally, FPE is strictly

[*] This work was supported by the FNR (National Research Fund) Luxembourg under projects S-GAMES, C08/IS/03 and GMASec, PHD/09/082.

L. Antunes and H.S. Pinto (Eds.): EPIA 2011, LNAI 7026, pp. 506–520, 2011.
© Springer-Verlag Berlin Heidelberg 2011

weaker than NE, with the following intuition: Nash equilibria correspond to play which is certainly rational, strategy profiles that are *not* pre-equilibria are certainly irrational, and the profiles in between can be rational or not, depending on the circumstances.

Farsighted Reasoning about Strategies. The term "farsighted" refers to the type of reasoning about strategic choice that players are supposed to conduct according to FPE. Unlike Nash equilibrium, which assumes "myopic" reasoning (only the immediate consequences of a deviation are taken into account), farsighted pre-equilibrium looks at further consequences of assuming a particular choice. This type of strategic reasoning has been already studied for coalitional games in [2,3,4]. There have been also some attempts at farsighted stability in noncooperative games [5,6], but, as we argue in Section 5, they were based on intuitions from coalitional game theory, incompatible with the setting of noncooperative games.

Assumptions about Opponents' Play. Our assumptions about the way in which players react to another player's deviation are minimal: we only assume that the reactions are locally rational. Our view of local rationality is standard for noncooperative games, i.e., it concerns an *individual* change of play that increases the payoff of the deviating player. In particular, we do not take into account scenarios where a coalition of players makes a sequence of changes that leads to a beneficial state, but leads through nodes where the payoff of some members of the coalition decreases. As we see it, such a scenario can be rational only when the coalition can commit to executing the sequence, which is not possible in noncooperative games.

Farsighted Play vs. Repeated Games. An interpretation of Nash equilibrium is that a player forms an expectation about the other players' behavior based on his past experience of playing the game [1]. Then he chooses his best response strategy to maximize his immediate gain in the next instance of the game, assuming that this move will not influence future plays of the game. In other words, it is assumed that the other players do not best respond to a deviation from the expectation when the game is repeated. In contrast, in repeated games [7], it is assumed that once a player decides to deviate, the deviation will be observed by the opponents, and they will adapt to it accordingly. Then, the player would observe and adapt to their change of behavior, and so on.

In farsighted pre-equilibria, neither of these assumptions are made, as we are looking for a *weak* notion of rationality. This means that a farsighted deviation must succeed against agents that best respond farsightedly (as in the standard setting of repeated games), agents that best respond myopically, and against ones that satisfy only minimal rationality constraints (deviations must be profitable).

Structure of the Paper. We begin by defining the concept of farsighted pre-equilibrium formally and discussing some examples in Section 2. We investigate how the concept behaves on the benchmark case of the n-Player Prisoner's Dilemma, provide an alternative characterization of FPE, and propose a polynomial algorithm for verifying pre-equilibria. In Section 4, we show that FPEs can be seen as subgame-perfect solutions of specific extensive form games ("deviation games"). Finally, we compare our proposal to existing work (Section 5), and conclude in Section 6.

2   Farsighted Pre-equilibria

We begin by presenting the central notions of our proposal.

2.1   Deviation Strategies and Farsighted Stability

Let $G = (N, \Sigma_1, \ldots, \Sigma_n, out_1, \ldots, out_n)$ be a strategic game with $N = \{1, \ldots, n\}$ being a set of players, Σ_i a set of strategies of player i, and $out_i : \Sigma \to \mathbb{R}$ the payoff function for player i where $\Sigma = \Sigma_1 \times \cdots \times \Sigma_n$ is the set of strategy profiles. We use the following notation: s_i is player i's part of strategy profile s, s_{-i} is the part of $N \setminus \{i\}$, and $s \xrightarrow{i} s'$ denotes player i's deviation from strategy profile s to s' (with the obvious constraint that $s'_{-i} = s_{-i}$). Sometimes, we write $(out_1(s), \ldots, out_n(s))$ instead of s.

Definition 1. *Deviation $s \xrightarrow{i} s'$ is locally rational iff $out_i(s') > out_i(s)$. Function $F_i : \Sigma^+ \to \Sigma$ is a deviation strategy for player i iff for every finite sequence of profiles $s^1, \ldots, s^k$ we have that $s^k \xrightarrow{i} F_i(s^1, \ldots, s^k)$ is locally rational or $F_i(s^1, \ldots, s^k) = s^k$. A sequence of locally rational deviations $s^1 \to \ldots \to s^k$ is F_i-compatible iff $s^n \xrightarrow{i} s^{n+1}$ implies $F_i(s^n) = s^{n+1}$ for every $1 \leq n < k$.*

Locally rational deviations turn G into a graph in which the transition relation corresponds to Nash dominance in G. Deviation strategies specify how a player can (rationally) react to rational deviations done by other players.

Definition 2 (Farsighted pre-equilibrium). *Strategy profile s is a farsighted pre-equilibrium (FPE) if and only if there is no player i with a deviation strategy F_i such that: 1) $out_i(F_i(s)) > out_i(s)$, and 2) for every finite F_i-compatible sequence of locally rational deviations $F_i(s) = s^1 \to \ldots \to s^k$ we have $out_i(F_i(s^1, \ldots, s^k)) \geq out_i(s)$.*

This means that a strategy profile s is potentially *unstable* if there is a deviation strategy of some player i such that the first deviation is strictly advantageous, and however the other players react to his deviations so far, i can always recover to a profile where he is not worse off than he was originally in s.

Example 1. Consider the Prisoner's Dilemma game:

	C	D
C	**(7, 7)**	$(0, 8)$
D	$(8, 0)$	**(1, 1)**

The farsighted pre-equilibria are printed in bold font. The locally rational deviations are $(7, 7) \xrightarrow{1} (8, 0)$, $(0, 8) \xrightarrow{1} (1, 1)$, $(7, 7) \xrightarrow{2} (0, 8)$ and $(8, 0) \xrightarrow{2} (1, 1)$. This implies that $(1, 1)$ is an FPE because there is no player i with a deviation strategy F_i such that $out_i(F_i(1, 1)) > out_i(1, 1)$. On the other hand, $(8, 0)$ is not an FPE because $F_2(\ldots, (8, 0)) = (1, 1)$ is a valid deviation strategy. By symmetry, $(0, 8)$ is neither an FPE. Finally we show that $s = (7, 7)$ is an FPE. All deviation strategies F_1 for player 1 with $out_1(F_1(7, 7)) > out_1(7, 7)$ specify $F_1(7, 7) = (8, 0)$. Still, player 1 cannot recover from the F_1-compatible sequence of locally rational deviations $(7, 7) \xrightarrow{1} (8, 0) \xrightarrow{2} (1, 1)$ which makes his payoff drop down to 1. The same holds for deviation strategies of player 2 by symmetry. Therefore, $(7, 7)$ is an FPE.

Theorem 1. *Every Nash equilibrium is an FPE.*

Proof. Assume s is an NE. Then there exists no deviation $s \xrightarrow{i} s'$ to a strategy profile s' such that $out_i(s') > out_i(s)$. Therefore, there exists no player i with a deviation strategy F_i such that $out_i(F_i(s)) > out_i(s)$, so s is an FPE.

Corollary 1. *FPE is strictly weaker than Nash equilibrium.*

2.2 n-Person Prisoner's Dilemma

As we saw in Example 1, the Prisoner's Dilemma has two farsighted pre-equilibria: the NE profile where everybody defects, and the "intuitive" solution where everybody cooperates. This extends to the n-player Prisoner's Dilemma as defined in [5].

Definition 3 (n-Player Prisoner's Dilemma). *Let $N = \{1, 2, \ldots, n\}$ be the set of players. Each player has two strategies: C (cooperate) and D (defect). The payoff function of player i is defined as $out_i(s_1, \ldots, s_n) = f_i(s_i, h)$ where h is the number of players other than i who play C in s, and f_i is a function with the following properties:*

1. *$f_i(D, h) > f_i(C, h)$ for all $h = 0, 1, \ldots, n - 1$;*
2. *$f_i(C, n - 1) > f_i(D, 0)$;*
3. *$f_i(C, h)$ and $f_i(D, h)$ are increasing in h.*

The first requirement says that defecting is always better than cooperating, assuming the other players do not change their strategy. The second requirement specifies that the situation where everyone cooperates is better than the situation where everyone defects. The third requirement says that the payoff increases for a player when a larger number of the other players cooperate.

Theorem 2. *If G is a n-Player Prisoner's Dilemma, the strategy profiles $(C, \ldots, C)$ and $(D, \ldots, D)$ are FPEs in G.*

We leave out the proof because of lack of space.

Example 2. We look at an instance of the 3-player Prisoner's Dilemma. Player 1 selects rows, player 2 columns and player 3 matrices.

C	C	D		D	C	D
C	$(\mathbf{3,4,4})$	$(1,5,2)$		C	$(1,2,5)$	$(0,3,3)$
D	$(\mathbf{5,2,2})$	$(4,3,0)$		D	$(4,0,3)$	$(\mathbf{2,1,1})$

The unique Nash equilibrium is (D, D, D), the strategy profile where everyone defects, so this strategy profile is also an FPE. Furthermore, also the strategy profile where everyone cooperates, i.e., (C, C, C), is an FPE. Finally, (D, C, C) is an FPE, showing that also other FPEs can exist.

We can interpret these results as follows. A population where every player defects might be stable: being the first to cooperate is not necessarily advantageous, as the other players might not follow. A population where all players cooperate might also be stable if the players consider long-term consequences of damaging the opponents' payoffs: if one player starts defecting, the other players might follow. Finally, a strategy profile might also be stable if there are only a couple of defecting agents in the population, and the cooperating players all receive payoffs above some minimal "threshold of fairness" (which is usually the player's payoff in the Nash equilibrium $(D, \ldots, D)$). Hence the asymmetry: (D, C, C) is farsighted stable, but (C, C, D) and (C, D, C) are not, because they provide player 1 with an "unfair" payoff, and player 1 is better off heading for the NE. Another motivation for (D, C, C) to be stable, is that player 1 does not want to cooperate in the hope that players 2 and 3 do not change their strategy (as is assumed by NE), while players 2 and 3 do not want to defect out of fear for follow-ups of the other players (as is assumed in repeated games).

3   Characterizing and Computing Farsighted Pre-equilibria

In general, deviation strategies determine the next strategy profile based on the full history of all preceding deviations. In this section, we show that it suffices for the definition of FPE to consider only *positional* deviation strategies, i.e. strategies that determine the next deviation only based on the current strategy profile, independently of what previously happened.

Definition 4. *A positional deviation strategy for player i is a strategy F_i such that $F_i(s^1, \ldots, s^k) = F_i(t^1, \ldots, t^k)$ whenever $s^k = t^k$. We will sometimes write $F_i(s^k)$ instead of $F_i(s^1, \ldots, s^k)$ for such strategies. A positional FPE is an FPE restricted to positional deviation strategies.*

Theorem 3. *A strategy profile $s \in \Sigma$ is an FPE iff it is a positional FPE.*

Proof. It suffices to prove that there is no player i with a deviation strategy such that conditions 1) and 2) from Definition 2 hold iff there is no player i with a *positional* deviation strategy such that these conditions hold. Every positional deviation strategy is a deviation strategy, so the 'only if' direction is trivial. We prove the 'if' direction by contraposition. Assume there exists a player i with a deviation strategy such that conditions 1) and 2) from Definition 2 hold in s. Now we define a positional deviation strategy F' as follows. For all $s' \in \Sigma$ for which there exist finite F_i-compatible sequences of locally rational deviations $F_i(s) = s^1 \rightarrow \ldots \rightarrow s^k = s'$, let $F_i(s) = t^1 \rightarrow \ldots \rightarrow t^k = s'$ be a shortest F_i-compatible sequence of locally rational deviations. Then we set $F'(s^k) = F(s^0, s^1, \ldots, s^k)$. For all other $s' \in \Sigma$, we set $F'_i(s') = s'$. The function F' is clearly positional and a deviation strategy. Because $F'_i(s)$ is defined based on the shortest sequence which is s (the only sequence of length 1), $F_i(s) = F'_i(s)$, and since we assumed that condition 1) holds for F_i, it also holds for F'_i. Finally we need to check that condition 2) holds. Assume $F'_i(s) = s^1 \rightarrow \ldots \rightarrow s^k$ is a finite F'_i-compatible sequence of locally rational deviations. Then by definition of F'_i, there exists also a finite F_i-compatible sequence of locally rational deviations

$F_i(s) = t^1 \rightarrow \ldots \rightarrow t^{k-1} \rightarrow s^k$ with $F_i(t^1, \ldots, t^{k-1}, s^k) = F'_i(s^k)$. By assumption, $out_i(F_i(t^1, \ldots, t^{k-1}, s^k)) \geq out_i(s)$, so also $out_i(F'_i(s^k)) \geq out_i(s)$.

The following theorem provides an alternative characterization of farsighted play.

Theorem 4. *L is the set of FPEs iff for all $s \in L$, all $i \in N$ and all positional deviation strategies F_i with $F_i(s) \neq s$, there exists a finite F_i-compatible sequence of locally rational deviations $s \xrightarrow{i} s^1 \rightarrow \ldots \rightarrow s^k$ such that $out_i(F_i(s^1, \ldots, s^k)) < out_i(s)$.*

Proof. By Definition 2, a strategy profile s is an FPE iff there is no player i with a deviation strategy F_i such that: 1) $out_i(F_i(s)) > out_i(s)$, and 2) for every finite F_i-compatible sequence of locally rational deviations $F_i(s) = s^1 \rightarrow \ldots \rightarrow s^k$ we have $out_i(F_i(s^1, \ldots, s^k)) \geq out_i(s)$. Because F_i is a deviation strategy, condition 1) is equivalent to $F_i(s) \neq s$ by Definition 1. By using this equivalence and moving the negation inwards, we find that a strategy profile s is an FPE iff for every player i and all deviation strategies F_i such that $F_i(s) \neq s$, there exists a finite F_i-compatible sequence of locally rational deviations $s = s^1 \rightarrow \ldots \rightarrow s^k$ such that $out_i(F_i(s^1, \ldots, s^k)) < out_i(s)$. By Theorem 3, the theorem follows.

Now we will present a procedure that checks if the strategy profile s is a farsighted pre-equilibrium in game G. Procedure $dev(G, i, s)$ returns *yes* if player i has a successful deviation strategy from s in G, and *no* otherwise:

1. **forall** $j \in N$ **do** compute $\prec_j \in \Sigma \times \Sigma$ st. $t \prec_j t'$ iff $\exists t \xrightarrow{j} t'$. $out_j(t) < out_j(t')$;
2. let $\prec_{-i} := \bigcup_{j \neq i} \prec_j$ and let $\ll^*$ be the transitive closure of $\prec_{-i}$;
3. let $Good := \{t \mid out_i(t) \geq out_i(s)\}$;   /profiles at least as good as s/
4. **repeat**
   $Good' := Good$;
   forall $t \in Good$ **do**
      if $\exists t' \gg^* t. (t' \notin Good \land \forall t' \xrightarrow{i} t''. t'' \notin Good)$ **then** remove t from $Good'$;
   until $Good' = Good$;
5. **if** $\exists t \in Good. s \prec_i t$ **then** return *yes* **else** return *no*.

The following is straightforward.

Theorem 5. *Strategy profile s is an FPE in G iff $dev(G, i, s) = no$ for all $i \in N$.*

Note that the procedure implements a standard greatest fixpoint for a monotonic transformer of state sets. As a consequence, we get the following.

Theorem 6. *Checking if s is a farsighted pre-equilibrium in G can be done in polynomial time with respect to the number of players and strategy profiles in G.*

4   Deviations as a Game

Deviations can be seen as moves in a "meta-game" called *deviation game* that uses the original payoff matrix as arena. Transitions in the arena (i.e., players' moves in the meta-game) are given by domination relations of the respective players. In such a setting, *deviation strategies* can be seen as strategies in the deviation game. A successful deviation strategy for player i is one that gets i a higher payoff immediately (like in the case of NE) but also guarantees that i's payoff will not drop below the original level after possible counteractions of the opponents. A node in the original game is an FPE exactly when no player has a winning strategy in the deviation game.

4.1   Deviation Games

A deviation game D is constructed from a strategic game G and a strategy profile s in G, and consists of two phases. In the first phase, each player can either start deviating from s or pass the turn to the next player. If no player deviates, all players get the "neutral" payoff 0 in D. If a player i deviates, the game proceeds to the second phase in which i tries to ensure that his deviation is successful, while all other players try to prevent it. This phase is strictly competitive: if i succeeds, he gets the payoff of 1 and all the other players get -1; if i fails, he gets -1 and the other players get 1 each.

Formally, given a strategic game G and a strategy profile s, the deviation game is an extensive form game $T(G, s) = (N, H, P, out'_1, \ldots, out'_n)$, where N is the set of players as in G, H is the set of histories in the deviation game, P is a function assigning a player to every non-terminal history, and for every $i \in N$, out'_i is a function assigning the payoff for player i to every terminal history. A history in H is a sequence of nodes of the form (i, t, j), with the intended meaning that $i \in N \cup \{-\}$ is the player whose deviation strategy is currently tested (where "$-$" means that no deviation has been made yet), $t \in \Sigma$ is the current strategy profile under consideration, and $j \in N \cup \{\bot\}$ is the player currently going to play (i.e., $P(\ldots, (i, t, j)) = j$), where $\bot$ indicates that the game has terminated. The initial state is $(-, s, 1)$. For every player j, we define out'_j as follows:

- $out'_j(\ldots, (-, t, \bot)) = 0$;
- if $out_i(t) \geq out_i(s)$, then $out'_j(\ldots, (i, t, \bot)) = 1$ when $j = i$, otherwise $out'_j(\ldots, (i, t, \bot)) = -1$;
- if $out_i(t) < out_i(s)$, then $out'_j(\ldots, (i, t, \bot)) = -1$ when $j = i$, otherwise $out'_j(\ldots, (i, t, \bot)) = 1$.

Now we recursively define the set of histories H, where $\underline{i}$ is defined as $\min(N \setminus \{i\})$.

1. $(-, s, 1) \in H$.
2. If $h = \ldots, (-, s, i) \in H$ and $i + 1 \in N$ then $h, (-, s, i + 1) \in H$.
3. If $h = \ldots, (-, s, i) \in H$ and $i = \max(N)$, then $h, (-, s, \bot) \in H$.
4. If $h = \ldots, (-, s, i) \in H$, $s \xrightarrow{i} s'$ is a locally rational deviation and $i' \in N \setminus \{i\}$ then $h, (i, s', \underline{i}) \in H$.
5. If $h = \ldots, (i, s', i) \in H$ and $s' \xrightarrow{i} s''$ is a locally rational deviation, then $h, (i, s'', \underline{i}) \in H$.
6. If $h = \ldots, (i, s', i) \in H$, then $h, (i, s', \bot) \in H$.
7. If $h = \ldots, (i, s', i') \in H$, $i' \in N \setminus \{i\}$ and either $s' \xrightarrow{i} s''$ is a locally rational deviation or $s' = s''$, then $h, (i, s'', i') \in H$ whenever both $h, (i, s'', i') \notin H$ and $i' = i$ implies $h, (-, s'', i') \notin H$.

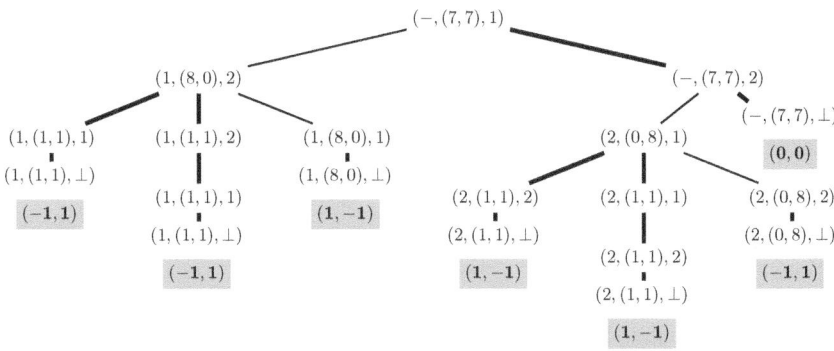

Fig. 1. Deviation game for strategy profile $(7, 7)$ in Prisoner's Dilemma (Example 1)

Statement 1 specifies the initial history. Statements 2–4 say that if nobody has deviated so far, player i can embark on a deviation strategy or refrain from deviating and pass the token further. If no player deviates, the game ends. If player i initiates deviations, the strategy profile changes, and the token goes to the first opponent. Statement 5 says that the latter also applies during execution of the deviation strategy. Furthermore, 6 indicates that player i can stop the game if it is his turn (note that this can only be the case if the opponents do not want to deviate anymore). Finally, 7 states that an opponent player can make a locally rational deviation or do nothing if it is his turn, and pass the turn to another player i' (as long as the player has not had the turn in the new strategy profile before, to guarantee finite trees).

Now we can see an *opponent strategy* against player i as a set of strategies for players $N \backslash \{i\}$ such that every deviation is locally rational, and in every strategy profile, not more than one player deviates. Formally, an opponent strategy against player i is a function $F_{-i} : N \backslash \{i\} \times \Sigma^* \to \Sigma$ such that for every player $j \in N \backslash \{i\}$, $s \xrightarrow{j} F_{-i}(j, (\ldots, s))$ is a locally rational deviation or $F_{-i}(j, (\ldots, s)) = s$ and such that $F_{-i}(j, (\ldots, s)) \neq s$ for some j implies $F_{-i}(j', (\ldots, s)) = s$ for all $j' \neq j$.

Example 3. Figure 1 depicts the deviation game $T(G, s)$ where G is the Prisoner's Dilemma and s is $(7, 7)$. The moves selected by the minimax algorithm are printed as thick lines. The minimax algorithm selects outcome $(0, 0)$, so no player has a strategy yielding more than 0, which indicates that $(7, 7)$ is an FPE.

4.2   Correspondence to FPE

Now we will prove that a strategy profile in the original game is an FPE exactly when no player has a strategy that guarantees the payoff of 1 in the deviation game. We say that a sequence of strategy profiles $s^1, \ldots, s^k$ is *(F_i, F_{-i})-compatible* if for all $k' < k$ either $F_{-i}(j, (s^1, \ldots, s^{k'})) = s^{k'+1}$ for some $j \in N \backslash \{i\}$ or both $F_i(s^1, \ldots, s^{k'}) = s^{k'+1}$ and $F_{-i}(j, (s^1, \ldots, s^{k'})) = s^{k'}$ for all $j \in N \backslash \{i\}$. Furthermore, a sequence of strategy profiles $s^1, \ldots, s^k$ is *loop-free* if $s^n \neq s^{n'}$ for $1 \leq n \leq n' \leq k$.

Let G be a strategic form game, s be a strategy profile and $i \in N$ be a player. Now we say that a deviation strategy F_i is successful against an opponent strategy F_{-i}, if 1) $out_i(F_i(s)) > out_i(s)$, and 2) for every loop-free (F_i, F_{-i})-compatible sequence of strategy profiles $F_i(s) = s^1, \ldots, s^k$, it holds that $out_i(F_i(s^1, \ldots, s^k)) \geq out_i(s)$. The following lemma shows that it is indeed sufficient to look at loop-free (F_i, F_{-i})-compatible sequences.

Lemma 1. *Strategy profile s is an FPE in game G iff there does not exist a player i with a deviation strategy F_i that is successful against all opponent strategies F_{-i}.*

Proof. First we prove the 'only if' direction by contraposition. Assume there exists a player i with a deviation strategy F_i that is successful against all opponent strategies F_{-i}, i.e., 1) $out_i(F_i(s)) > out_i(s)$, and 2) for every loop-free (F_i, F_{-i})-compatible sequence $F_i(s) = s^1, \ldots, s^k$, it holds that $out_i(F_i(s^1, \ldots, s^k)) \geq out_i(s)$.

Let $F_i(s) = s^1 \to \ldots \to s^k$ be a loop-free F_i-compatible sequence of locally rational deviations. We define opponent strategy F_{-i} such that $F_{-i}(j, s^k) = s^{k+1}$ whenever $s^k \xrightarrow{j} s^{k+1}$ for $j \in N\setminus\{i\}$. Then $s^1, \ldots, s^k$ is (F_i, F_{-i})-compatible, so we have $out_i(F_i(s^1, \ldots, s^k)) \geq out_i(s)$ by assumption. Therefore, for every loop-free F_i-compatible sequence of locally rational deviations $F_i(s) = s^1 \to \ldots \to s^k$, it holds that $out_i(F_i(s^1, \ldots, s^k)) \geq out_i(s)$ $(*)$.

Now let $F_i(s) = s^1 \to \ldots \to s^k$ be a finite F_i-compatible sequence of locally rational deviations. Then we can construct a loop-free sequence $t^1, \ldots, t^k$ with $t_1 = s_1$ and $t_k = s_k$. Now $out_i(F_i(t^1, \ldots, t^k)) = out_i(F_i(s^k))$ because F_i is positional, and $out_i(F_i(t^1, \ldots, T^k)) \geq out_i(s)$ by $(*)$. Therefore, for every finite F_i-compatible sequence of locally rational deviations $F_i(s) = s^1 \to \ldots \to s^k$, it holds that $out_i(F_i(s^1, \ldots, s^k)) \geq out_i(s)$. This shows that there exists a player i with a deviation strategy F_i such that 1) $out_i(F_i(s)) > out_i(s)$, and 2) for every finite F_i-compatible sequence of locally rational deviations $F_i(s) = s^1 \to \ldots \to s^k$, it holds that $out_i(F_i(s^1, \ldots, s^k)) \geq out_i(s)$, i.e., s is not an FPE.

The 'if' direction follows from the fact that every loop-free sequence of strategy profiles is finite, and the fact that when a sequence of strategy profiles is (F_i, F_{-i})-compatible, it is also an F_i-compatible sequence of deviations.

We proceed by defining a bijection ϕ between strategy F_i in G and strategy Φ_i in $T(G, s)$ as follows.

If $F_i(s) = s$ then $\Phi_i(-, s, i) = (-, s, i + 1)$ where $i + 1 \in N$;
If $F_i(s) = s$ then $\Phi_i(-, s, i) = (-, s, \perp)$ where $i = \max(N)$;
If $F_i(s) = s'$ then $\Phi_i(-, s, i) = (i, s', \underline{i})$ where $s \neq s'$;
If $F_i(s') = s'$ then $\Phi_i(i, s', i) = (i, s', \perp)$; where $s \neq s'$;
If $F_i(s') = s''$ then $\Phi_i(i, s', i) = (i, s'', \underline{i})$ where $s \neq s' \neq s''$.

We call a set of strategies Φ_{-i} for players $N\setminus\{i\}$ *non-initially-deviating* whenever $\Phi_{i'}(-, s, i') = (-, s, i'')$ where $i \neq i'$. Then an *opponent strategy* Φ_{-i} in the deviation game is a set of non-initially-deviating strategies Φ_j for players $j \in N\setminus\{i\}$ such that in every strategy profile, not more than one player in $N\setminus\{i\}$ deviates and the other players always give the turn to the deviating player, i.e., $\Phi_j(i, s', j) = (i, s'', j')$ with $s' \neq s''$ for some $j, j' \in N\setminus\{i\}$ implies $\Phi_i(i, s', j'') = (i, s', j)$ for all $j'' \neq j$. Now we define a bijection ψ between an opponent strategy F_{-i} in T and an opponent strategy Φ_{-i} in $T(G, s)$. Let $\psi(F_{-i}) = \Phi_{-i}$, where Φ_{-i} is defined as follows:

- If $F_{-i}(i', s') \neq s'$ for some $i' \in N\backslash\{i\}$, then $\Phi_{i'}(i, s', i') = (i, s'', i')$ and $\Phi_{i''}(i, s', i'') = (i, s', i')$ for $i'' \neq i'$.
- If $F_{-i}(i', s') = s'$ for all $i' \in N\backslash\{i\}$, then $\Phi_{i'}(i, s'', i') = (i, s', i)$.

It can easily be checked that ϕ and ψ are indeed bijections.

Let $out_i(\Phi_i, \Phi_{-i})$ be the outcome of the game for player i when player i plays strategy Φ_i and players $N\backslash\{i\}$ play strategy Φ_{-i}. When $out_i(\Phi_i, \Phi_{-i}) = u_i$ and $out_j(\Phi_i, \Phi_{-i}) = u_{-i}$ for $j \in N\backslash\{i\}$, we sometimes write $out_{i,-i}(\Phi_i, \Phi_{-i}) = (u_i, u_{-i})$.

Lemma 2. *If $i \in N$ is a player with a deviation strategy F_i and F_{-i} is an opponent strategy, then F_i is successful against F_{-i} in game G and strategy profile s if and only if $out_{i,-i}(\phi(F_i), \psi(F_{-i})) = (1, -1)$ in $T(G, s)$.*

Proof. By construction of ψ and ϕ, we have run $s = s^1 \xrightarrow{i^1} s^2 \xrightarrow{i^2} \ldots \xrightarrow{i^{k-1}} s^k$ in G iff $(-, s, 1), \ldots, (-, s, i), (i, s', \underline{i}), \ldots, (i, s^k, i), (i, F_i(s^k), \underline{i}), (i, F_i(s^k), i), (i, F_i(s^k), \perp)$ is a run in $T(G, s)$. Therefore a run ends in s^k in G with $out_i(F_i(s^1, \ldots, s^k)) \geq out_i(s)$ iff a run ends in $(i, F_i(s^k), \perp)$ in $T(G, s)$ with $out_i(F_i(s^k)) \geq out_i(s)$. Therefore, F_i is successful against F_{-i} if and only if $out_i(\phi(F_i), \psi(F_{-i})) = 1$.

Theorem 7. *Strategy profile $s \in \Sigma$ is an FPE in game G if and only if all subgame-perfect Nash equilibria in $T(G, s)$ yield $(0, \ldots, 0)$.*

Proof. To prove the 'only if' direction, assume strategy profile s is an FPE in game G. By Lemma 1, there does not exist a player i with a deviation strategy F_i that is successful against all opponent strategies F_{-i}. Because f and g are bijections, by Lemma 2 there does not exist a player i with a strategy Φ_i such that for every opponent strategy Φ_{-i} it holds that $out_{i,-i}(\Phi_i, \Phi_{-i}) = (1, -1)$. This means that for every player i with a strategy Φ_i, there exists an opponent strategy Φ_{-i} such that $out_{i,-i}(\Phi_i, \Phi_{-i}) \neq (1, -1)$ (∗). Now we prove that every subgame-perfect Nash equilibrium (SPNE) $(\Phi_1, \ldots, \Phi_n)$ in the subgame starting at $(-, s, i)$ yields $(0, \ldots, 0)$ by backwards induction on $i \in (1, \ldots, n, \perp)$. The base case, where $i = \perp$, follows from the definition of out. Now assume that the claim holds for $i + 1$ (where $n + 1 = \perp$). To show that $\Phi_i(-, s, i) = (i, s, i + 1)$, we assume that $\Phi_i(-, s, i) = (i, s', \underline{i})$ for some s' and derive a contradiction. Let Φ_{-i} be an opponent strategy. Now $out_{i,-i}(\Phi_i, \Phi_{-i})$ is either $(1, -1)$ or $(-1, 1)$. If $out_{i,-i}(\Phi_i, \Phi_{-i}) = (1, -1)$, by (∗), there exists an opponent strategy Φ'_{-i} such that $out_{i,-i}(\Phi_i, \Phi'_{-i}) \neq (1, -1)$ and thus $out_{i,-i}(\Phi_i, \Phi'_{-i}) = (-1, 1)$. Now $out_{-i}(\Phi_i, \Phi'_{-i}) > out_{-i}(\Phi_i, \Phi_{-i})$, which contradicts the assumption that (Φ_i, Φ_{-i}) is an NE. If $out_{i,-i}(\Phi_i, \Phi_{-i}) = (-1, 1)$, let Φ'_i be a strategy such that $\Phi'_i(-, s, n) = (-, s, n + 1)$ and Φ'_i is a SPNE strategy in the subgame starting at $(-, s, n + 1)$. Then $out_i(\Phi'_i, \Phi'_{-i}) \geq 0$ for all opponent strategies Φ'_{-i} by i.h.. This implies that $out_i(\Phi'_i, \Phi_{-i}) > out_i(\Phi_i, \Phi_{-i})$, contradicting the assumption that (Φ_i, Φ_{-i}) is an NE. This implies that the assumption $\Phi_i(-, s, i) = (i, s', \underline{i})$ is false, so $\Phi_i(-, s, i) = (i, s, i + 1)$ or $\Phi_i(-, s, i) = (i, s, \perp)$. By i.h., all SPNE in the subgame starting at $(-, s, i+1)$ yield $(0, \ldots, 0)$. Therefore, all SPNE in $T(G, s)$ yield $(0, \ldots, 0)$.

We prove the 'if' direction by contraposition. Assume strategy profile $s \in \Sigma$ is not an FPE in game G. By Lemma 1, there exists a player i with a strategy F_i that is successful against all opponent strategies F_{-i}. Because ϕ and ψ are bijections, by Lemma 2 there exists a player i with a strategy Φ_i such that for every opponent strategy Φ_{-i} it holds that $out_{i,-i}(\Phi_i, \Phi_{-i}) = (1, -1)$ (†). Now let (Φ'_i, Φ'_{-i}) be a strategy profile such that $out_i(\Phi'_i, \Phi'_{-i}) = 0$. Then there exists a strategy Φ_i such that $out_i(\Phi_i, \Phi'_{-i}) = 1$ by (†), so $out_i(\Phi_i, \Phi'_{-i}) > out_i(\Phi'_i, \Phi'_{-i})$, and therefore (Φ'_i, Φ'_{-i}) is not an SPNE. An extensive game always has an SPNE and (Φ'_i, Φ'_{-i}) is the only strategy profile yielding $(0, \ldots, 0)$, so there exist SPNEs not yielding $(0, \ldots, 0)$, which implies that not all SPNEs yield $(0, \ldots, 0)$.

Note that Theorem 7 provides an alternative way of checking pre-equilibria: s is an FPE in G iff the minimaxing algorithm [1] on $T(G, s)$ returns 0 for every player. However, the deviation game for G can be exponentially larger than G itself, so the algorithm proposed in Section 3 is more efficient.

5   Comparing Farsighted Solution Concepts

There has been a number of solution concepts with similar agenda to FPE. In this section, we discuss how they compare to our new proposal.

5.1   Related Work

The discussion on myopic versus farsighted play dates back to the *von Neumann-Morgenstern stable set* (VNM) in coalitional games [2], and Harsanyi's *indirect dominance* of coalition structures, leading to the *strictly stable set* (SSS) [3]. More recent proposals are the *noncooperative farsighted stable set* (NFSS) [6] and the *largest consistent set* (LCS) [4]. Other similar solution concepts include [8,5,9]. Also Halpern and Rong's *cooperative equilibrium* [10] can be seen as a farsighted solution concept.

Definitions. In order to define the concepts, we introduce three different dominance relations between strategy profiles. *Direct dominance* of x over y means that player i can increase his own payoff by deviating from strategy profile x to strategy profile y. *Indirect dominance* of x over y says that a coalition of players can deviate from strategy profile x to strategy profile y, possibly via a number of intermediate strategy profiles, such that every coalition member's final payoff is better than his payoff before his move. Finally, *indirect dominance in Harsanyi's sense* is indirect dominance with the additional requirement that each individual deviation is locally rational. Formally:

- We say that y directly dominates x through player i ($x \prec_i y$) if there is a locally rational deviation $x \xrightarrow{i} y$. We also write $x \prec y$ if $x \prec_i y$ for some $i \in N$.
- We say that y indirectly dominates x ($x \ll y$) if there exists a sequence of (not necessarily locally rational) deviations $x = x^0 \xrightarrow{i_1} x^1 \ldots \xrightarrow{i_p} x^p = y$ such that $out_{i_r}(x^{r-1}) < out_{i_r}(y)$ for all $r = 1, 2, \ldots, p$.
- We say that x indirectly dominates y in Harsanyi's sense ($x \ll_H y$) if there exists a sequence of locally rational deviations $x = x^0 \xrightarrow{i_1} x^1 \ldots \xrightarrow{i_p} x^p = y$ such that $out_{i_r}(x^{r-1}) < out_{i_r}(y)$ for all $r = 1, 2, \ldots, p$.

It can easily be seen that $x \prec y$ implies $x \ll_{\mathrm{H}} y$, and $x \ll_{\mathrm{H}} y$ implies $x \ll y$.

Example 4. In the Prisoner's Dilemma (Example 1), we have $(7,7) \prec_1 (8,0)$, $(0,8) \prec_1 (1,1)$, $(7,7) \prec_2 (0,8)$ and $(8,0) \prec_2 (1,1)$ In addition, $(7,7)$ indirectly dominates $(1,1)$ in Harsanyi's sense, i.e., $(1,1) \ll_{\mathrm{H}} (7,7)$.

With these definitions, we can introduce four main farsighted solution concepts.

- A subset K of Σ is a *von Neumann-Morgenstern stable set* (VNM) if it satisfies the following two conditions: (a) for all $x, y \in K$, neither $x \prec y$ nor $y \prec x$; (b) for all $x \in \Sigma \setminus K$, there exists $x \in K$ such that $x \prec y$ [2]. In fact, a VNM corresponds to stable extensions in the argumentation theory $(\Sigma, \prec')$, where $\prec'$ is the converse of $\prec$, in Dung's argumentation framework [11].
- If we replace in VNM the direct dominance relation $\prec$ by the indirect dominance relation $\ll$, we obtain the *noncooperative farsighted stable set* (NFSS) [6].
- Furthermore, a subset S of Σ is a *strictly stable set* (SSS) if it is a VNM such that for all $x, y \in S$, neither $x \ll_{\mathrm{H}} y$, nor $y \ll_{\mathrm{H}} x$ [3].
- Finally, a subset L of Σ is consistent in Chwe's sense if $(x \in L$ iff for all deviations $x \xrightarrow{i} y$ there exists $z \in L$ such that $[y = z$ or $y \ll z]$ and $out_i(x) \geq out_i(z))$. Now the *largest consistent set* (LCS) is the union of all the consistent sets in Σ [4].

Another solution concept that has been recently proposed is *perfect cooperative equilibrium* (PCE) [10]. Like FPE, PCE aims at explaining situations where cooperation is observed in practice. A player's payoff in a PCE is at least as high as in any Nash equilibrium. However, a PCE does not always exist. Every game has a Pareto optimal maximum PCE (M-PCE), as defined below. We only give the definition for 2-player games; the definition for n-player games can be found in [10].

Given a game G, a strategy s_i for player i in G is a best response to a strategy s_{-i} for the players in $N \setminus \{i\}$ if $U_i(s_i, s_{-i})) = \sup_{s_i' \in \Sigma_i} U_i(s_i', s_{-i})$. Let $BR_i(s_{-i})$ be the set of best responses to s_{-i}. Given a 2-player game, let BU_i denote the best payoff that player i can obtain if the other player j best responds, that is $BU_i = \sup_{s_i \in \Sigma_i, s_j \in BR(s_i)} U_i(s)$. A strategy profile is a PCE if for $i \in \{1, 2\}$ we have $U_i(s) \geq BU_i$. A strategy profile is an α-PCE if $U_i(s) \geq \alpha + BU_i$ for all $i \in N$. The strategy profile s is an M-PCE if s is an α-PCE and for all $\alpha' > \alpha$, there is no α'-PCE.

Example 5. In the Prisoner's Dilemma (Example 1), there is one VNM ($\{(1,1), (7,7)\}$) and one NFSS ($\{(7,7)\}$). There is no SSS, and the LCS is $\{(1,1), (7,7)\}$.

Regarding PCE, we have $BR_1(C) = \{D\}$, $BR_1(D) = \{D\}$, $BR_2(C) = \{D\}$, and $BR_2(D) = \{D\}$. This implies that $BU_1 = D$ and $BU_2 = D$. Thus, the set of PCE outcomes is $\{(7,7), (1,1)\}$, and $(7,7)$ is the unique M-PCE (with $\alpha = 6$).

5.2   FPE vs. Other Farsighted Concepts

The main idea of all introduced farsighted solution concepts (except PCE) is very similar. One can test whether a given strategy profile is stable by checking whether a player or group of players can deviate from the strategy profile in a profitable way, given a possible follow-up from the other players. However, there are also many differences

F	L	R
T	$(3,1,2)$	$(\mathbf{0},\mathbf{2},\mathbf{2})$
B	$(0,0,0)$	$(0,0,0)$

	L	R
T	$(3,1)$	$(\mathbf{0},\mathbf{2})$
B	$(\mathbf{1},\mathbf{1})$	$(0,0)$

(a)

S	L	R
T	$(3,1,1)$	$(0,0,0)$
B	$(\mathbf{1},\mathbf{1},\mathbf{1})$	$(0,0,0)$

(b)

	L	R
T	$(3,1)$	$(\mathbf{2},\mathbf{2})$
B	$(1,1)$	$(0,0)$

(c)

	L	C	R
T	$(0,0)$	$(3,1)$	$(\mathbf{0},\mathbf{2})$
B	$(\mathbf{1},\mathbf{3})$	$(\mathbf{2},\mathbf{4})$	$(0,0)$

(d)

Fig. 2. Example games (FPEs are printed in bold)

between the concepts. In this section, we will compare FPE with other farsighted solution concepts in various aspects.

Scope of Farsightedness. In farsighted reasoning about strategies, players consider further consequences of their deviations, as opposed to reasoning in myopic solution concepts like Nash equilibrium. Consider for example the game in Fig. 2a. Strategy profile $(1,1)$ is not an NE because $(1,1) \overset{1}{\rightarrow} (3,1)$ is locally rational. However, this deviation is not necessarily *globally* rational, as it might trigger player 2 to follow up with the deviation $(3,1) \overset{2}{\rightarrow} (0,2)$. Unlike Nash equilibrium, which only considers (0,2) stable, the set $\{(1,1),(0,2)\}$ is considered stable in all presented farsighted solution concepts (VNM, NFSS, SSS, LCS and FPE).

The degree of farsightedness is different across the concepts. The least farsighted concept is VNM. Here, players only look at whether they can recover from a *single* deviation of the opponents, as the game in Fig. 2b illustrates. The deviation $(1,1,1) \overset{1}{\rightarrow} (3,1,1)$ is locally rational but might intuitively be wrong because deviations $(3,1,1) \overset{3}{\rightarrow} (3,1,2) \overset{2}{\rightarrow} (0,2,2)$ can spoil its effect. However, since VNM does not take sequences of deviations into account, it does not consider $(1,1,1)$ stable ($\{(0,2,2),(3,1,1)\}$ being the only stable set). The concepts NFSS, LCS and FPE have a "more farsighted" view, and consider sequences of follow-up deviations. In consequence, they all deem the profile $(1,1,1)$ stable.

Furthermore, the solution concepts evaluate follow-ups differently. In VNM and NFSS, a follow-up deviation from the opponents is always considered undesirable, even if it gives a higher payoff for the first deviating player. In LCS and FPE, beneficial follow-ups only strengthen the success of the original deviation. Consider the game in Fig. 2c. After $(1,1) \overset{1}{\rightarrow} (3,1)$, the follow-up $(3,1) \overset{2}{\rightarrow} (2,2)$ still leaves player 1 with a payoff higher that his initial one. Thus, both LCS and FPE deem $(1,1)$ unstable, which matches intuition, while VNM and NFSS consider $(1,1)$ stable.

Type of Solution Concept. The concepts also yield objects of different types. LCS and FPE both return a set of strategy profiles, thus ascribing rationality to *individual profiles*. On the other hand, VNM, NFSS and SSS return a set of sets of profiles each, hence ascribing rationality to *sets* of strategy profiles. In the latter case a rational set of profiles can be understood as a set of collective decisions to whom the grand coalition of players can consistently stick. Clearly, this makes sense in coalitional games, but is less suitable for noncooperative games where the players' control over collective choice is limited.

Deviation Strategy. VNM, SSS and FPE are built on a pessimistic view of the follow-up to the first deviation, as they make no assumptions about the other players' rationality. In particular, it is not assumed that opponents will help to increase the initiator's outcome, even if it is also to their advantage. In consequence, these solution concepts assume that the deviations of the initiator must always be locally rational. In contrast, NFSS assumes that a player can make deviations which are not locally rational if he hopes that other players will further increase his outcome. The game in Fig. 2a illustrates this. The set $\{(1,1),(0,2)\}$ is a VNM, SSS, LCS and collects all FPEs. On the other hand, $\{(0,2)\}$ is the only NFSS. PCE and M-PCE may also require players to deviate in a locally irrational way because they do not take into account the domination relation explicitly. For example, $(0,2)$ in Fig. 2d is neither PCE nor M-PCE, although it is a Nash equilibrium and hence no player has a locally rational deviation in it. All the other solution concepts considered here deem $(0,2)$ stable.

Expected Behavior of Opponents. Different solution concepts imply different opponent models. We have already mentioned that the initiator of deviations can either be optimistic or pessimistic about the follow-up by the opponents. Another distinction is whether the deviator expects the opponents to be farsighted as well, or whether they might be regular best-response players. Consider the game in Fig. 2d. Intuitively, a farsighted player 1 would not deviate $(2,4) \overset{1}{\to} (3,1)$, because the follow-up deviation $(3,1) \overset{2}{\to} (0,2)$ can damage his payoff. Therefore player 2 can safely play $(1,3) \overset{2}{\to} (2,4)$ if he is sure that player 1 is farsighted. However, if player 1 plays best response, the deviation $(1,3) \overset{2}{\to} (2,4)$ might harm player 2, because player 1 will deviate $(2,4) \overset{1}{\to} (3,1)$ afterwards. Therefore, if player 2 has no information about the kind of behavior of player 1, it might be better to stick to strategy profile $(1,3)$. FPE is the only solution concept that captures this intuition by considering $(1,3)$, $(2,4)$ and $(0,2)$ to be (potentially) stable; the other formalisms (VNM, SSS, NFSS, LCS) all result in the stable set $\{(1,4),(0,2)\}$.

Summary. The main difference between our farsighted pre-equilibrium and the other solution concepts discussed in this section lies in the perspective. It can be argued that the type of rationality defined in [2,3,4,5,6] is predominantly coalitional. This is because those proposals ascribe stability to *sets* of strategy profiles, which does not have a natural interpretation in the noncooperative setting. Moreover, some of the concepts are based on coalitional rather than individual deviations. On the other hand, the concept of cooperative equilibrium [10] is *not* based on reasoning about possible deviations. In this sense, FPE is the first truly noncooperative solution concept for farsighted play that we are aware of.

6   Conclusions

We have proposed a new solution concept that we call *farsighted pre-equilibrium*. The idea is to "broaden" Nash equilibrium in a way that does not discriminate solutions that look intuitively appealing but are ruled out by NE. Then, Nash equilibrium may be interpreted as a specification of play which is certainly rational, and strategy profiles that are *not* farsighted pre-equilibria can be considered certainly *irrational*. The area in between is the gray zone where solutions are either rational or not, depending on the detailed circumstances.

Our main motivation is predictive: we argue that a solution concept that makes too strong assumptions open up ways of possible vulnerability if the other agents do not behave in the predicted way. Nash equilibrium seems too restrictive in many games (Prisoner's Dilemma being a prime example). We show that FPE does select non-NE strategy profiles that seem sensible, like the "all cooperate" strategy profile in the standard as well as the generalized version of Prisoner's Dilemma. Moreover, we observe that FPE favors solutions with balanced distributions of payoffs, i.e., ones in which no player has significantly higher incentive to deviate than the others.

A natural way of interpreting deviations in strategy profiles is to view the deviations as moves in a "deviation game" played on the metalevel. We show that farsighted pre-equilibria in the original game correspond to subgame-perfect Nash equilibria in the meta-game. This is a strong indication that the concept that we propose is well rooted in game-theoretic tradition of reasoning about strategic choice.

Farsighted play has been investigated in multiple settings, starting from von Neumann and Morgenstern almost 70 years ago. Our proposal is (to our knowledge) the first truly noncooperative solution concept for farsighted play. In particular, it is obtained by reasoning about *individual* (meta-)strategies of *individually* rational players, rather than by reconstruction of the notion of *stable set* from coalitional game theory.

References

1. Osborne, M., Rubinstein, A.: A Course in Game Theory. MIT Press (1994)
2. von Neumann, J., Morgenstern, O.: Theory of Games and Economic Behaviour. Princeton University Press, Princeton (1944)
3. Harsanyi, J.: Interpretation of stable sets and a proposed alternative definition. Management Science 20, 1472–1495 (1974)
4. Chwe, M.: Farsighted coalitional stability. Journal of Economic Theory 63, 299–325 (1994)
5. Suzuki, A., Muto, S.: Farsighted stability in an n-person prisoners dilemma. International Journal of Game Theory 33, 431–445 (2005)
6. Nakanishi, N.: Purely noncooperative farsighted stable set in an n-player Prisoners Dilemma. Technical Report 707, Kobe University (2007)
7. Mailath, G., Samuelson, L.: Repeated Games and Reputations: Long-Run Relationships. Oxford University Press (2006)
8. Greenberg, J.: The theory of social situations: an alternative game-theoretic approach. Cambridge University Press (1990)
9. Diamantoudi, E., Xue, L.: Farsighted stability in hedonic games. Social Choice and Welfare 21, 39–61 (2003)
10. Halpern, J.Y., Rong, N.: Cooperative equilibrium. In: International Conference on Autonomous Agents and Multi Agent Systems, pp. 1465–1466 (2010)
11. Dung, P.M.: On the acceptability of arguments and its fundamental role in nonmonotonic reasoning, logic programming and n-person games. Artificial Intelligence 77, 321–357 (1995)

Uncertainty and Novelty-Based Selective Attention in the Collaborative Exploration of Unknown Environments

Luis Macedo, Miguel Tavares, Pedro Gaspar, and Amílcar Cardoso

University of Coimbra,
Centre for Informatics and Systems of the University of Coimbra, Department of
Informatics Engineering, Polo II, 3030 Coimbra, Portugal
{macedo,amilcar}@dei.uc.pt, {mgt,pgaspar}@student.dei.uc.pt

Abstract. We propose a multi-agent approach to the problem of exploring unknown environments that relies on providing the agents with a measure of interest for the viewpoints of the surrounding environment. Such measure of interest takes into account the expected decrease in uncertainty provided by acquiring the information of objects seen from a viewpoint and the novelty of the potential class label of those objects. This allows the agents to visit selectively the objects that populate the environment. This single agent exploration strategy is combined with a multi-agent exploration strategy relying on a brokering system that allows the coordination of the agent team according to the agents's personal interest and their distance to the viewpoints. The advantages of these forms of selective attention, together with those of the collaborative multi-agent exploration strategy, are tested in several scenarios, comparing our approach against classical ones.

Keywords: Exploration of Unknown Environments, Interest, Curiosity, Active Learning, Selective Attention, Classification, Coordination, Collaboration.

1 Introduction

The exploration of unknown environments is a specific kind of active learning [23,21]. It can be defined as the process of selecting and executing actions in such a way that a maximum of knowledge of a given domain is acquired (e.g., [20]). In the case of physical exploration, the result is the acquisition of a model of the physical environment. Because exploring unknown environments requires resources such as time and energy, there is always a trade-off between the amount of knowledge that can be acquired and the costs of acquiring it. Therefore, exploration strategies that minimize costs and maximize knowledge acquisition have been proposed for artificial agents. These strategies have been grouped into two main categories: undirected and directed exploration [20]. Strategies belonging to the former group (e.g.: random walk exploration, Boltzman distributed exploration) use no exploration-specific knowledge and ensure

L. Antunes and H.S. Pinto (Eds.): EPIA 2011, LNAI 7026, pp. 521–535, 2011.

exploration by merging randomness into action selection. On contrary, strategies belonging to the latter group rely heavily on exploration specific-knowledge for guiding the learning process. Several techniques have been proposed and tested either in simulated and real, indoor and outdoor environments, using single or multiple agents (e.g., [1,2,5,9,10,18,26,23,22,21,20,27,28,30,31]). The exploration domains include planetary exploration (e.g., Mars, Titan or lunar exploration) (e.g., [19,3,29]), the search for meteorites in Antarctica (e.g., [15]), underwater mapping, volcano exploration, map-building of interiors (e.g., [24,26,28]), etc. The main advantage of using artificial agents in those domains instead of humans is that most of them are extreme environments making their exploration a dangerous task for human agents. However, there is still much to be done especially in dynamic environments such as those mentioned above.

Real exploration environments contain objects. For example, office environments possess chairs, doors, garbage cans, etc., cities are comprised of many different kinds of buildings (houses, offices, hospitals, churches, etc.), as well as other objects such as cars. Many of these objects are non-stationary, that is, their locations may change over time. This observation motivates research on a new generation of mapping algorithms, which represent environments as collections of objects [7,8]. At a minimum, such object models would enable a robot to track changes in the environment. For example, a cleaning robot entering an office at night might realize that a garbage can has moved from one location to another. It might do so without the need to learn a model of this garbage can from scratch, as would be necessary with existing robot mapping techniques. Object representations offer a second, important advantage, which is due to the fact that many environments possess large collections of objects of the same type. For example, most office chairs are examples of the same generic chair and therefore look alike, as do most doors, garbage cans, and so on. As these examples suggest, attributes of objects are shared by entire classes of objects, and understanding the nature of object classes is of significant interest to mobile robotics. In particular, algorithms that learn properties of object classes would be able to transfer learned parameters (e.g., appearance, motion parameters) from one object to another in the same class. This would have a profound impact on the accuracy of object models, and the speed at which such models can be acquired. If, for example, a cleaning robot enters a room it has never visited before, it might realize that a specific object in the room possesses the same visual appearance of other objects seen in other rooms (e.g., chairs). The robot would then be able to acquire a map of this object much faster. It would also enable the robot to predict properties of this newly seen object, such as the fact that a chair is non-stationary, without ever seeing this specific object move.

To our knowledge, the classification methods used to achieve such object models are mostly non-memory based [11,26,25,28]. However, given that most of the environments in which exploration occurs lack a domain theory and are characterized by unpredictability or uncertainty, memory-based classification methods are suitable to classify objects of those environments [6]. Previously, Macedo and Cardoso [12,14] addressed this issue, but they used a single agent approach. On contrary,

[4,18,12,13] provided evidence that multi-agent approaches are better in comparison with single agent ones, reducing the time required to explore the environment.

However, the exploration strategies used by these multi-agent teams cannot be the same as those used by a single agent. Having other agents mapping the same environment, these agents must take into account the behavior of its fellow "explorers" and the locations that have already been mapped by them when deciding the next step to make, in order to reduce the redundancy in the mapping and reduce the time taken in the process. Therefore, some sort of coordination is needed in order achieve a truly collaborative behavior between the agents. On this aspect, regarding the coordination of the agents, [4,18,12,13] achieved significant results by reducing the redundancy in the mapping and directing each agent to the location most favorable to be explored by it. This is achieved by evaluating the "frontiers" of the currently mapped environment and having each agent bid on the next location to explore until a consensus is reached between all agents. In addition, no location is picked by more than one agent and each agent gets the most favorable location to explore, given the existence of the other agents. The work of [12,13] proposes the use of motivational agents with various "feelings" (surprise, curiosity and hunger) in the decision-making process that defines the exploration behavior of the agents. In all these approaches, the goal is always to map an entire environment, with the (mostly time) costs associated with such an exhaustive method. Also, these simulation works never take into account the cost of fully identifying an object. In a real situation, in which object identification must be done, usually it is costly to have a clear classification of an object based on its characteristics as it is necessary to query large amounts of data or to prompt for input from a human being, both of which are extremely time-consuming [16].

In this paper we focus on the selective attention and coordination aspects of an exploration strategy used by the agents to decide the next viewpoints. The coordination aspects of the problem are dealt with a brokering system. Regarding the selective attention, our approach consists of providing the exploring agents with reasoning capabilities that allow them to rate unknown objects in the environment according to an interest level determined by the explorer. By making the agents "ignore" objects with a low interest value and making a predictive identification of such an objects at distance instead of approaching them to fully identify them, we aim to reduce mapping times of full environments, while keeping a non significant misclassification level.

The next section presents a multi-agent approach for the exploration of unknown environments. Section 3 describes the functionalities of each type of agent of the multi-agent system. Section 4 presents the experimental tests. Finally, section 5 discusses the results and presents conclusions.

2 Overview of the Multi-agent System for the Collaborative Exploration of Unknown Environments

Our approach to the exploration problem relies on ensuring that locations that are highly unlikely to possess new and useful information for the mapping, such as empty space or locations with a known geometry, won't be given much attention by the exploring agents, resulting in a partial mapping of the environment.

In order to efficiently map the environment, we chose a master-slave architecture. Mapping and exploration are coordinated by two separate agents with no physical presence: the *mapper* and the *broker*. The slave agents are called *explorers*. The mapper is in charge of merging everyone's maps and sending the global map back to each explorer. The broker assigns next moves to every explorer, based on the interesting locations they spotted. The explorers analyze the environment they inhabit, send their local map to the mapper, pick points of interest based on their current knowledge of the area, send them to the broker, and finally move to the location assigned by the broker.

In our simulated environment, there are a set of properties explorers have access to, including its position, and the list of objects it senses. Much like a physical agent requires a set of sensors in order to gather information about the surrounding world, our simulated explorer has access through simulated sensors to some properties of the objects it can see. The simulated agent can see objects in all directions within a certain distance.

The environment used is considered as a discrete, two dimensional grid, in which each cell can be populated with one object, no objects or an object and an exploring agent at once. This is done because in the classification of the environment, each cell is marked as identified or not, while in the former case the class of the object identified is placed in an auxiliary grid.

Each object that populates the environment has a set of core attributes. In this case, size and color were chosen for the sake of simplicity and in order to keep the experiments simple and understandable. Apart from this, each object has a class, that identifies it as a member of a kind of object family. For example, there could be classes for trees, deer, bushes, grass and rabbits, all these would have a size and color that can distinguish them from other classes or not (bushes and grass have similar colors). These two core attributes can be sensed by the agents at distance and determine the probability distribution assigned by an agent to an unknown object.

3   Agents' Description

In this section, we describe the various agents implemented and their in-depth functionalities.

3.1   Mapper Agent

The mapper agent, as stated before, has the task of collecting all the data from the environment that the explorer agents supply and aggregating that data into a map of the environment. To that purpose, the agent has in its memory a Sparse Object Grid, where it stores the objects that all the agents have seen so far and its location (by the index in the grid) and an auxiliary array that contains information whether a given object at a given location has been identified or if its type is still unknown. This last array is meant to keep the explorers from extracting interest from objects that have already been identified, allowing them to skip through fully identified locations into more interesting areas.

The mapper also stores a list of *prototypes*, i.e., an abstract representation of the different objects witnessed by the explorers so far, with information about the average characteristics of an object (a kind of *idea* of an object, something that we can compare to an actual object and obtain a high correlation). These prototypes are to be used by the explorer agents when they wish to assign a probability distribution to an unknown object. This is explored in depth in the section about the explorer agent.

3.2 Broker Agent

The broker agent acts, as the name indicates, as a broker for all the explorer's requests for new targets to explore. As such, it must maintain a list of interesting locations provided by the explorers, so that it can select from that list when an explorer agent requests a new target. This approach is highly influenced by the work of Simmons et al. [18], being an adaptation of their brokering system. To implement this behavior, the broker has methods to receive information from the explorers, namely the location of an interest point and its corresponding interest, and to remove points from its memory – for example, when an explorer arrives at the point and identifies the object there, no longer that point is interesting.

The brokering itself takes place when an explorer agent requests a new target from the broker. Upon this request, the broker will determine the relevance of each interest point in its memory to the given agent. This relevance is a Benefit minus Cost function presented as follows that takes into account the interest of the point ($Interest$), the distance of the agent to the point ($Dist$), and the maximum distance that an agent is willing to travel before arriving at its target (Max_{dist}) (Equation 1). This formula assures that an agent will pursue the most informative viewpoints that are closer to it.

$$Utility = Interest - \frac{Dist \times 100}{Max_{dist}} \tag{1}$$

The interesting points are sorted by decreasing relevance, and the most relevant one is passed to the requesting explorer as its next target. This point is then removed from the list, so we can prevent situations where the broker assigns the same target to multiple agents, effectively disrupting the entire agent coordination.

3.3 Explorer Agent

The explorer agent is the core of this multi-agent system. It performs the heavy-duty work of travelling around the environment, collecting information about the objects it detects and trying to extract knowledge from it. To do so, it has movement and sensing capabilities, represented by a speed and a view range, inside which the agent is capable of detecting general characteristics of an object. In each simulation step, the agent moves towards a given target (a location on the environment) sensing its surroundings. Whenever and object in the environment comes inside its view range, the agent starts a process to attain its characteristics, assign a probability distribution to that object and, based on that probability, calculate the interest of that object to the agent, which will determine what to do next: either the agent identifies it probabilistically, or it marks it as an interesting spot to be visited later.

Note that these are different ways to identify an object: if the agent arrives at its target and there is an object there, it spends some time on it, identifying it (emulating the behavior of a real robot sending information to a central processing unit where either a large database is queried or human input is required to identify an object). In this case, the agent adds the object description to the prototype of the identified class, because it has an exact knowledge that the current object is of that class. However, the agent also possesses capabilities to identify an object at distance: if, while analyzing it, the explorer determines that an object is not interesting enough to be visited, it rather chooses to identify it at distance, by picking the most likely class from the probability distribution of classes computed for the object. This approach does not add the object attributes to the class prototype, because the agent cannot be sure that it is making a correct classification. Thus, this prevents the agent from reinforcing a bad classification, leading to disastrous results.

To attain the object's characteristics, the agent simply queries the environment as to what are the basic attributes of the object (e.g., size and color). This emulates a real-world scenario where an agent has a limited sight and is only capable of identifying some basic attributes of objects in the environment at distance.

Classifying Objects. Afterwards, the explorer agent has the task of classifying the unknown objects, by assigning it a probability distribution based on the knowledge of the worlds it has so far, namely the prototypes that it has developed. These are abstract representations of an object (more precisely, a class), which contain the average characteristics of the witnessed objects of that class. A good example is the idea of "tree": it's tall and it's greenish. Not all trees are really tall or green, but, on average, this is an accurate representation of trees.

The explorer agent browses through a list of prototypes, either its own list or a collaboratively filled list that all the agents share in the mapper agent, and correlates the attributes of the unknown object to each of the prototypes. This correlation is done by doing a weighed average of the Euclidean distance between each attribute of the prototype and the object. This gives us a good approximation of the probability for the object to be an instance of each of the prototypes.

However, this cannot be the only measure we use. If it were, by witnessing one tree, the agent would assume to possess all knowledge of trees that existed in the world, and this is not the case. This representation must take in the fact that only after several observations of the same object, some understanding of it can be extrapolated. As such, a saturation factor (Equation 2) is applied to the correlation to take into account the number of occurrences (n_{occurs}) of that object that have been witnessed.

$$Saturation_{corr} = \tanh(\frac{\frac{(n_{occurs}-5)}{2} + 1}{2})$$ (2)

This formula was designed to saturate at near 1 when the number of occurrences of that given object approaches 10. So, we are assuming that the agent needs to see at least 10 instances of a given object to have any real understanding of the object as a class. This can be changed, of course, to represent different levels of learning rates, by increasing or decreasing the denominator of the fraction inside the hyperbolic tangent.

All these calculations, however, don't account for the possibility that the agent might not have witnessed any instance of the class of the object it is now trying to identify as it might be a new kind of object that needs its own prototype. This is addressed by the agent by calculating an unknown correlation value that determines the possibility that the current object is not yet known. This value is automatically calculated to 1 when no prototypes are present (at the beginning of the simulation) and using the following formula (Equation 3) when there is already n prototypes, i.e., there is already some knowledge of the world:

$$sim(O_{new}, O_{unknown}) = 1 - \max_{i} sim(O_{new}, O_i)$$ (3)

where $sim(O_{new}, O_i)$ represents the correlation or similarity of the object O_{new} with feature vector x_{new} to the ith prototype O_i with feature vector x_i and class label y_i and is computed as the Euclidean distance between O_{new} and O_i, and more precisely between x_{new} and x_i.

This gives us a fair measure of the chances that this new object O_{new} is something new, given the knowledge the agent has. This value is maximum when all the correlations are 0, and minimum when at least one of the correlations is 1.

Finally, the explorer agent translates these correlations into a probability distribution, using Equation 4.

$$p(y_{new} = y_j | x_{new}) = \frac{sim(O_{new}, O_j)}{\sum_{i=1}^{n+1} sim(O_{new}, O_i)}$$ (4)

The explorer agent now has a probability distribution, which sums to 1, that tells the agent, for each known class and for the unknown class, the probability that the current object O_{new} is an instance of that class y_j.

Determining Interest. Given a probability distribution, the explorer agent assigns an interest to such a probability distribution. This is done in two steps, that represent two different interesting situations: one where the agent sees there is a high probability of finding a new object, and one where the agent detects that the current object has similar characteristics to several known classes.

The first case, the interest for the unknown, is mapped using only the probability that the object is none of the known classes. The interest is given by Equation 5.

$$I_{unknown}(O_{new}) = \tanh(2 \times P(y_{new} = y_{unknown})) \tag{5}$$

where O_{new} is an unlabeled new object with feature vector x_{new}, and $P(y_{new} = y_{unknown}|x_{new})$ is the probability that the new object is from a new, unknown class.

This formula is used instead of simply mapping probability to informativeness because we assume that a $x\%$ probability of being something new has more than an informative value of x in a scale of [0-100]. This formula increases the rate of climb of the informative value, saturating when approaching 1, the maximum probability, representing the maximum informativeness, here valued at 1.

However, this is not the only situation where the agent discerns something interesting. It also needs to identify an interest for the chaotic, i.e., an object that matches several of the known classes well, or at least some of them. In this case, we have a simple solution: the explorer agent calculates the entropy of the probability distribution using Shannon's approach [17]. Higher entropy values represent more chaotic situations, where all the interest of the agent should be focused. As such, we calculate the entropy (Equation 6) of the probability distribution for that object, and map that value proportionally to an interest value.

$$I_{uncertainty}(O_{new}) = -\sum_{j=1}^{n+1} P(y_{new} = y_j) \times \log(P(y_{new} = y_j)) \tag{6}$$

After calculating these two interest values, $I_{uncertainty}$ and $I_{unknown}$, the explorer agent picks the one that represents the maximum interest and assigns it as the interest value for the object under scrutiny. Note however that other approaches may be considered such as the mean of those values or even their sum (the comparison between these approaches is a future work).

Agent Reasoning. After determining the interest for the current object, the explorer agents makes a decision: either determines that the object has an high enough interest and sends its location to the broker agent so it can be identified as an interesting position; or it determines that the object is not interesting enough to be visited and classifies it at distance. This last option is the result we expect from our agent after exploring a bit of the map, becoming aware that this option is less time consuming and it does not require the agent to be in the same exact location of the object. In order to make this decision, each explorer agent has an interest threshold under which he identifies the object at distance and over which it sends the location to the broker, along with its interest.

The experimental tests that follow allows to see whether this approach is favorable to more classical approaches, and how much error this behavior introduces in the mapping of the environment.

4 Experiments

4.1 Experimental Setup

To test our approach, we run a team of explorer agents in several different scenarios (maps with 400 units of width by 300 units of height) with different configurations to test out several aspects of our system. The testing environments were populated with 5 different kinds of objects: Water, Trees, Bushes, Houses, and Walls. These objects were randomly distributed by the environment. Figure 1 presents an example of these testing environment. Objects of different classes are represented with different colors. What needs to be known about these objects is that each of these classes has a central value for its attributes, and varies slightly in each instance of the object. However, some of them vary a lot in one attribute and a little in others (Water, Houses) and some vary averagely in all attributes (Tree, Bushes). The classes Tree and Bush are very similar to one another, with the Bushes being smaller and darker than the Trees, allowing us to test the system in an extreme situation where closely related objects aren't supposed to be classified as the same. In order to test different configurations, we varied the number of agents in the exploring team, the threshold at which an agent finds an object informative, the mode of sharing knowledge (either sharing their knowledge – global knowledge – or not sharing it – local knowledge), and the starting positions of the agents.

The runs of the simulation have been limited to 5000, and we specified that in order to classify an object in place, an agent must spend 10 steps in that position (representing the time it takes to identify it). In addition to that, the agents have a 40-unit view range radius, which will be constant throughout the experimentation.

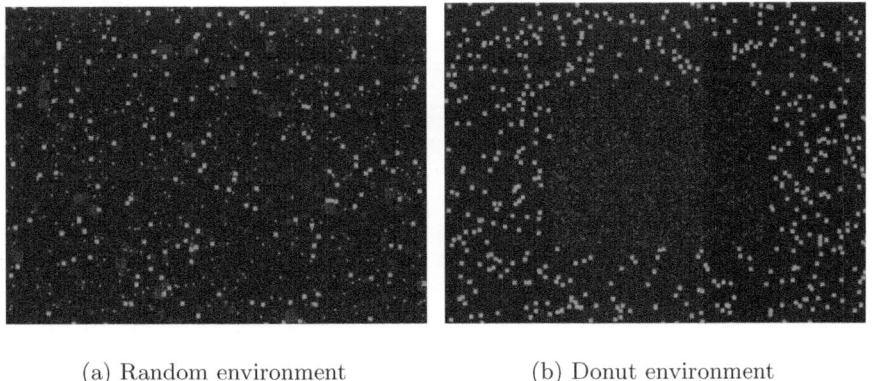

(a) Random environment   (b) Donut environment

Fig. 1. Two kinds of environment

4.2 Experimental Results

We present the results of the exploratory study about the influence of the informativeness threshold and number of agents on the percentage of objects classified or misclassified.

Figure 2 illustrates the results of varying the threshold in the set {0, 50, 75, 90}, using teams of two agents. The agents share their knowledge, so every agent has the same knowledge of the environment. We can clearly see that our system works as hypothesized: in a classical approach where every object must be visited (threshold = 0), the system doesn't have the time to classify half of the objects, but the identification is always correct. However, by increasing the threshold, we achieve faster exploration (more objects are identified), but also with an also climbing error rate. We can see that with a threshold of 90 over 20% of the objects are misclassified. So, the threshold must be balanced to achieve good temporal performances while still maintaining a good level of certainty on the classification. It can be seen that with a threshold of 50 a good compromise is reached between exploration time and error rate.

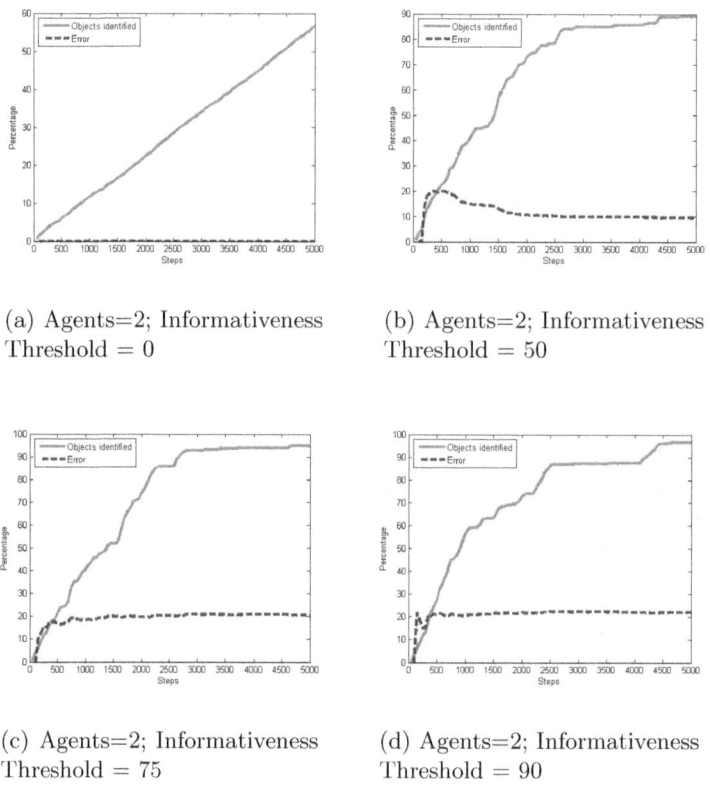

(a) Agents=2; Informativeness Threshold = 0

(b) Agents=2; Informativeness Threshold = 50

(c) Agents=2; Informativeness Threshold = 75

(d) Agents=2; Informativeness Threshold = 90

Fig. 2. Results for 2 agents and different informativeness threshold

Figure 3 shows the results of varying the starting positions of the agents and the strategy of sharing or not sharing their knowledge. This is done with 2 explorer agents and a threshold of 30. It also shows a comparison between using 2 and 8 agents, while keeping constant the starting positions of the agents and the sharing knowledge mode.

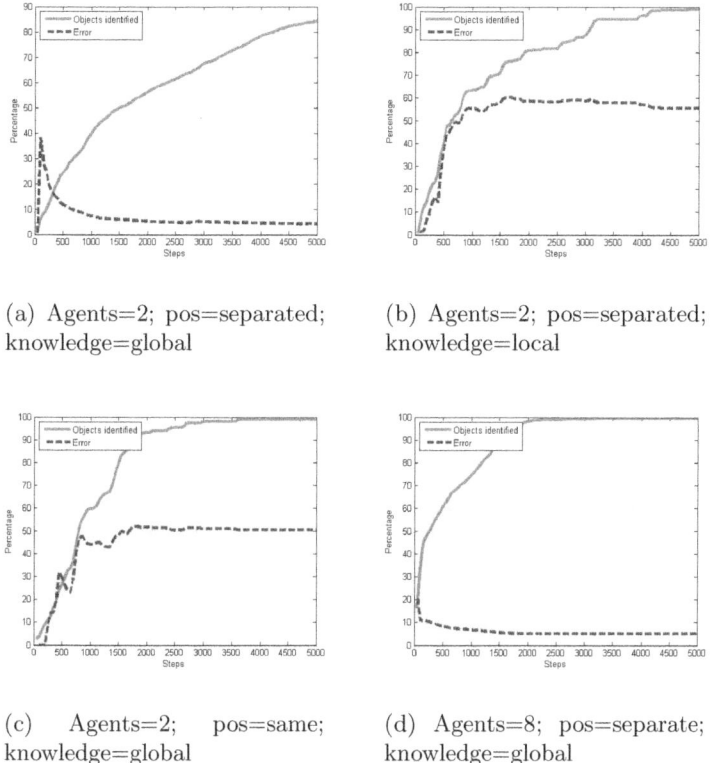

(a) Agents=2; pos=separated; knowledge=global

(b) Agents=2; pos=separated; knowledge=local

(c)   Agents=2;   pos=same; knowledge=global

(d) Agents=8; pos=separate; knowledge=global

Fig. 3. Results in the donut environment for different sharing modes of knowledge, different starting positions for the agents, and different number of agents

As it can be seen, when both agents start in similar positions, the results are worse, with 50% error indicating that one of the classes is misclassified every time. This is because the classes are very similar and clearly one of them saturates its prototype much faster and its correlation is rapidly high with the other class. By separating the agents at start, we can see that the results aren't much brighter using local knowledge: each agent basically has knowledge of only one of the classes and identifies each object it sees as that class. It is, however, an improvement we cannot see in these results, if we take as a fact that the error is distributed between the classes, with half of the instances of each class being misclassified, as opposed to a class as a whole not being identified correctly.

Using global knowledge this is mitigated, as it can seen in the graphs: by sharing their knowledge and starting in separate clusters, the agents gain knowledge of both of the classes at the same time and are able to correctly identify most of the objects efficiently.

As it can be seen, increasing the number of agents with global knowledge, the results are better time and error-wise, because the necessary knowledge for a correct classification is much quickly attained by a large number of agents than by a single one. The starting positions of the 8 agents are: 4 of the agents are in the center cluster and the remaining 4 start in the environment corners.

The results are much better than those attained with the previous approaches, in a rather difficult scenario for our multi-agent system. The results show evidence for the importance of agent placement and number of agents, as well as for the benefit of using global knowledge. The performance of the system increases with the number of agents introduced. Also, the error of the exploration decreases with the number of increasing explorers. This is partially because of the global knowledge implementation we are using: with many agents starting in various locations, a better knowledge of the world is quickly gained by the team as a whole, as their shared experiences provide a more accurate knowledge of their surroundings than the knowledge from a single agent.

5   Discussion and Conclusions

We proposed a multi-agent system for the classification of objects that populate unknown environments, in which each explorer agent is equipped with a classifier that selects for labeling those unlabeled objects that are more informative. The informativeness of an unlabeled object is measured in terms of the decrease in uncertainty by labeling it and also in terms of the novelty of its potential class label.

There is evidence indicating that the approach proposed in this paper is superior to a classic approach, in which identifying an object is a time-consuming process in the exploration paradigm. However, this does not come without a cost: this approach, using a predictive, at distance identification, introduces some error derived from classification mistakes. The approach is also clearly flawed in the aspect that, in the absence of informative objects to be explored, the agents roam through the environment, when some approach could be taken for the agents to visit unvisited areas.

Still, this approach has its merit and, the core of it, the application of an informativeness threshold to know when to approach or not an object, reveals itself as a more than capable way to explore efficiently and quickly and unknown environment. The experimental results obtained with different sizes for the team of agents each one with that selective classification of objects agree with the previous results obtained in previous studies.

Note that this entire work is based on the assumption that no previous knowledge of the environment is given, nor of the possible objects that the agents may encounter. If some information is given, more suitable approaches that rely on more robust classification systems could be used. The approach here presented promises to be effective when used in real unknown environments where communicating with large knowledge centers is a costly process and where the agents are resource bounded by not being able to store a large amount of information about their surroundings. In these circumstances, we believe this approach is a robust one, which is able to deliver interesting results.

Some extra work should be done in the future in order to enhance this approach, especially towards reducing the error obtained and enhancing the brokering algorithm. For instance, some work could be developed towards a self-policing behavior of the agents, checking periodically for classification errors and adapting the system accordingly, for instance by resetting the knowledge base or having its weight decrease through time, with this aging effect preventing consecutive misclassifications.

Further experimental tests are required to study the influence of the kind of the environment in the performance of the agents. What are the results obtained in structured environments? What happens with environments of different complexity? Moreover, more statistical techniques such as ANOVA should be carried out to assess the significance of the results. In this case, we may find conclusions about the influence and interaction of the independent variables (the number of agents, informativeness threshold, starting positions of the agents, and knowledge sharing mode) on the dependent variable (classification correctness). Preliminary results obtained with a structured environment indicate evidence of the same behaviour achieved in unstructured environments. Furthermore, more information metrics may be used, and their influence on the performance of the classifier may be studied using factorial experiments.

References

1. Amat, J., Mantaras, R., Sierra, C.: Cooperative autonomous low-cost robots for exploring unknown environments. In: Khatib, O., Salisbury, K. (eds.) Proceedings of the 4th International Symposium on Experimental Robotics IV. LNCIS, pp. 40–49. Springer, Stanford (1997)
2. Anguelov, D., Biswas, R., Koller, D., Limketkai, B., Sanner, S., Thrun, S.: Learning hierarchical object maps of non-stationary environments with mobile robots. In: Darwiche, A., Friedman, N. (eds.) Proceedings of the 17th Annual Conference on Uncertainty in Artificial Intelligence, pp. 10–17. Morgan Kaufmann Publishers, Inc., Alberta (2002)
3. Bresina, J., Dorais, G., Golden, K., Smith, D., Washington, R.: Autonomous rovers for human exploration of mars. In: Proceedings of the Mars Society Founding Convention. Boulder, Colorado (1999)
4. Burgard, W., Fox, D., Moors, M., Simmons, R., Thrun, S.: Collaborative multi-robot exploration. In: Proceedings of the IEEE International Conference on Robotics and Automation, pp. 476–481. IEEE Computer Society, San Francisco (2000)
5. Burgard, W., Moors, M., Schneider, F.: Collaborative exploration of unknown environments with teams of mobile robots. In: Beetz, M., Hertzberg, J., Ghallab, M., Pollack, M.E. (eds.) Dagstuhl Seminar 2001. LNCS (LNAI), vol. 2466, pp. 52–70. Springer, Heidelberg (2002)
6. Gupta, K.M., Aha, D.W., Moore, P.: Case-based collective inference for maritime object classification. In: McGinty, L., Wilson, D.C. (eds.) ICCBR 2009. LNCS, vol. 5650, pp. 434–449. Springer, Heidelberg (2009)
7. Hähnel, D., Burgard, W., Thrun, S.: Learning compact 3d models of indoor and outdoor environments with a mobile robot. Robotics and Autonomous Systems 44(1), 15–27 (2001)
8. Hähnel, D., Triebel, R., Burgard, W., Thrun, S.: Map building with mobile robots in dynamic environments. In: Proceedings of the International Conference on Robotics and Automation, pp. 1557–1563. IEEE Computer Society, Taipei (2003)

9. Lee, D.: The map-building and exploration strategies of a simple, sonar-equipped mobile robot; an experimental, quantitative evaluation. Phd, University College of London (1996)
10. Lee, D., Recce, M.: Quantitative evaluation of the exploration strategies of a mobile robot. International Journal of Robotics Research 16(4), 413–447 (1994)
11. Liu, Y., Emery, R., Chakrabarti, D., Burgard, W., Thrun, S.: Using em to learn 3d models of indoor environments with mobile robots. In: Proceedings of the Eighteenth International Conference on Machine Learning, pp. 329–336. Morgan Kaufmann Publishers, Inc., Williams College (2001)
12. Macedo, L.: The Exploration of Unknown Environments by Affective Agents. Phd thesis, University of Coimbra (2007)
13. Macedo, L., Cardoso, A.: Exploration of unknown environments with motivational agents. In: Jennings, N., Tambe, M. (eds.) Proceedings of the Third International Joint Conference on Autonomous Agents and Multiagent Systems, pp. 328–335. IEEE Computer Society, New York (2004)
14. Macedo, L., Cardoso, A.: Using CBR in the exploration of unknown environments with an autonomous agent. In: Funk, P., González Calero, P.A. (eds.) ECCBR 2004. LNCS (LNAI), vol. 3155, pp. 272–286. Springer, Heidelberg (2004)
15. Moorehead, S., Simmons, R., Apostolopoulos, D., Whittaker, W.: Autonomous navigation field results of a planetary analog robot in antarctica. In: International Symposium on Artificial Intelligence, Robotics and Automation in Space. Noordwijk, Holland (1999)
16. Settles, B.: Active learning literature survey. Computer Sciences Technical Report 1648, University of Wisconsin–Madison (2009)
17. Shannon, C.: A mathematical theory of communication. Bell System Technical Journal 27, 379–423, 623–656 (1948)
18. Simmons, R., Apfelbaum, D., Burgard, W., Fox, D., Moors, M., Thrun, S., Younes, H.: Coordination for multi-robot exploration and mapping. In: Proceedings of the Seventeenth National Conference on Artificial Intelligence, Austin, Texas, USA, pp. 852–858 (2000)
19. Simmons, R., Krotkov, E., Chrisman, L., Cozman, F., Goodwin, R., Hebert, M., Katragadda, L., Koenig, S., Krishnaswamy, G., Shinoda, Y., Whittaker, W.: Experience with rover navigation for lunar-like terrains. In: IROS (1995)
20. Thrun, S.: Efficient exploration in reinforcement learning. Tech. Rep. CMU-CS-92-102, Carnegie Mellon University, Computer Science Department (1992)
21. Thrun, S.: The role of exploration in learning control. In: White, D., Sofge, D. (eds.) Handbook of Intelligent Control: Neural, Fuzzy and Adaptive Approaches, pp. 527–559. Van Nostrand Reinhold, New York (1992)
22. Thrun, S.: Exploration and model building in mobile robot domains. In: Proceedings of the International Conference on Neural Networks, San Francisco, CA, pp. 175–180 (1993)
23. Thrun, S.: Exploration in active learning. In: Arbib, M. (ed.) Handbook of Brain Science and Neural Networks, pp. 381–384. MIT Press, Cambridge (1995)
24. Thrun, S.: Learning maps for indoor mobile robot navigation. Artificial Intelligence (1997)
25. Thrun, S.: Probabilistic algorithms in robotics. Tech. Rep. CMU-CS-00-126, School of Computer Science, Carnegie Mellon University (2000)
26. Thrun, S.: Robotic mapping: a survey. In: Lakemeyer, G., Nebel, B. (eds.) Exploring Artificial Intelligence in the New Millenium, pp. 1–35. Morgan Kaufmann Publishers, Inc., San Francisco (2002)

27. Thrun, S., Burgard, W., Fox, D.: A real-time algorithm for mobile robot mapping with applications to multi-robot and 3d mapping. In: Proceedings of the 2000 IEEE International Conference on Robotics and Automation, San Francisco, CA, USA, pp. 321–328 (2000)

28. Thrun, S., Thayer, S., Whittaker, W., Baker, C., Burgard, W., Ferguson, D., Hhnel, D., Montemerlo, M., Morris, A., Omohundro, Z., Reverte, C.: Autonomous exploration and mapping of abandoned mines. IEEE Robotics and Automation Magazine 11(4), 79–91 (2005)

29. Washington, R., Bresina, J., Smith, D., Anderson, C., Smith, T.: Autonomous rovers for mars exploration. In: Proceedings of the 1999 IEEE Aerospace Conference, Aspen, CO, USA (1999)

30. Yamauchi, B.: Frontier-based exploration using multiple robots. In: Proceedings of the Second International Conference on Autonomous Agents, pp. 47–53. Minneapolis, MN (1998)

31. Yamauchi, B., Schultz, A., Adams, W.: Integrating exploration and localization for mobile robots. Adaptive Systems 7(2), 217–230 (1999)

A Dynamic Agents' Behavior Model for Computational Trust

Joana Urbano, Ana Paula Rocha, and Eugénio Oliveira

LIACC / Departamento de Engenharia Informática
Faculdade de Engenharia, Universidade do Porto
Rua Dr. Roberto Frias, 4200-465 Porto, Portugal
{joana.urbano,arocha,eco}@fe.up.pt

Abstract. The development of computational trust models is growing in attention in the community of multi-agent systems and these models are currently seen as of extreme importance in social networks, electronic business and grid computing, among others. However, one of the biggest limitations in validating the existing computational trust models is the absence of realistic models of the behavior of agents. In fact, most of the work done in this area assumes that agents behave following simple and static probabilistic models. In this paper, we present a formal model of behavior of business agents that entail in inter-organizational exchanges, taking as basis diverse literature on socio-economic theories. With this model, we empirically show that some of the computational trust approaches which are more cited in the literature are not able to capture the temporal dynamics in the behavior of the business agents. Based on the results obtained from this study, we enumerate different properties that must be present in computational trust models in order to couple with realistic agents' behavior.

Keywords: Computational Trust, Behavior Models.

1   Introduction

Trust is a complex issue that is being studied in several areas of research, from sociology and psychology to economics and computer science. In computer science, computational models of trust and reputation are being proposed in order to support several processes associated with the decision making in social networks, electronic business and distributed resources. Namely, processes that imply the contact with strangers, where uncertainty and vulnerability, as well as the associated risk, are high.

In the concrete case of business exchanges, uncertainty is hard to reduce through personal relations. In the same way, vulnerability is also hard to reduce due to the presence of power relations [2] and to the existence of information asymmetry. Uncertainty and vulnerability leads to opportunism, which, in the words of Wathne and Heide (2000), can be defined as "some form of cheating or undersupply relative to an implicit or explicit contract" [14].

L. Antunes and H.S. Pinto (Eds.): EPIA 2011, LNAI 7026, pp. 536–550, 2011.

The opportunism problem in interfirm relationships is widely studied in economics literature and different governance mechanisms are proposed to manage opportunism, from specific forms of control and monitoring to selection mechanisms, that may include certification and reputation.

Therefore, the behavior of buyers and suppliers engaged in interfirm exchanges is built from a difficult balance between several complex factors that include trust, risk, opportunism and power. In order to estimate the trustworthiness of a potential partner to an exchange, computational trust models that infer future behaviors from past evidence must take these factors into account. However, many of these models tend to aggregate the past evidence using simple statistical-based techniques. Besides, as they are usually empirically evaluated using simple statistical models of agents' behavior, it is difficult to extrapolate their performance to real situations.

In this paper, we give further insight on this issue, by proposing a model of behavior for business agents participating in interfirm purchasing activities. Although simple, this model is based on important concepts associated with trust in interfirm relationships, and incorporates behavioral dynamics that are not present in statistical-based models used in most of the computational trust research.

Then, we evaluate three representative computational trust approaches using our model of behavior, and show that these approaches fail to capture the dynamics of trust building and maintenance. As a results of this study, we pinpoint characteristics that we think must be present in computational trust models in order for them to be applied in real scenarios.

This paper is structured as follows: Section 2 presents related work on computational trust models and on the models of behavior used to validate these computational approaches. Section 3 presents the model of behavior that we have developed taking into consideration diverse literature on trust in interfirm relationships. In Section 4, we perform an empirical analysis of the limitations of current computational trust models when addressing models of dynamic behavior of agents that engage in interfirm relationships. Finally, Section 5 concludes the paper.

2   Related Work

There are several computational models of trust. Some of the more cited in the literature are able to estimate the trustworthiness of the agents in evaluation by aggregating past evidence on these agents using simple statistical techniques. Paradigmatic examples of these models are REGRET [9] and FIRE [3], which compute the *direct trust* on an agent weighing all the existent evidence on the agent according to their recency.

The evaluation of these models is done using relative simple models of the behavior of agents. In [3], the testbed built to evaluate the FIRE approach is a multi-agent system consisting of agents that provide services and agents that consume these services. It considers three different types of providers, whose performance level vary within a range of five possible values with a given standard deviation. Furthermore, there is a direct mapping between these values of performance and the utility gained by the consumers. Although the model allows for some dynamics (e.g. at every round, the suppliers can change the average level of performance by a given amount or even switch to a new profile), it is however a pure statistical model and does not reflect known theories of behavior in consumer-provider scenarios.

TRAVOS [7] and The Beta Reputation System [5] are other well known models that use Beta distributions to aggregate trust evidences. In TRAVOS, the behavior of an agent acting as a provider is given by a probability that it will participate in a successful interaction (trustworthy behavior) and a probability that it will perform an unsuccessful interaction (untrustworthy behavior). This behavior governs the tendency of an agent to fulfill or default on its obligations to the other party in the interaction.

More recently, a new trend of situation-aware computational trust models has appeared. These models also rely on past evidence of the agents in evaluation to compute their trustworthiness, however here every piece of evidence is assigned a different level of importance taking into account the similarity of its context with the current situation under evaluation. One of these models is the Context Space and Reference Contexts model (CSRC), presented in [8]. This approach defines, for each one of the agents in evaluation, a multi-dimensional context space representing the possible situations in evaluation. When aggregating the past evidence, it calculates the similarity of the context of each evidence with the context of the situation in evaluation, and uses the level of similarity to weight the relevance of the piece of evidence being aggregated. This model is evaluated in an interesting scenario where humanitarian aid organizations require transportation services from local transporters after a major disaster. The selection of the transporters is based on their bid prices and on their trustworthiness. The transporters are modeled with bid prices based on transportations costs and profit margins and on their competence in specific scenarios. However, this represents a static behavior that does not allow the agents in evaluation to evolve with time.

Finally, the model presented in [13] is also situational, however it works in a rather different way than the CSRC model. It is constituted by a heuristic-based aggregator (Sinalpha) and a situation-aware tuner (Contextual Fitness). The model is evaluated in an electronic commerce scenario where supplier agents have fixed handicaps in performing specific context-based tasks. Once again, this model of behavior is probabilistic and does not evolve over time.

3   The Model

In this section, we present a model of behavior for agents that assume the roles of buyers and suppliers in buyer-supplier relationships. This model is based essentially on literature on trust in interfirm relationships.

Due to the complexity of the thematic, and also to the absence of consensual definitions and approaches among the different branches of research that study this issue, we opted to perform a significant amount of simplifications and hard assumptions. Nevertheless, the model is sufficiently expressive to allow for realist dynamics in the agents' behavior and to permit behaviors that evolve over time. Therefore, its utilization is adequate to the purpose of this paper, which consist in testing our suspicion that current computational trust are inadequate in capturing the dynamics of behavior of the agents in evaluation.

3.1 Background

The scenario underlying our model is an Electronic Institution (EI) where buyer-supplier relationships are developed between firm agents registered in the institution. In this scenario, buyer agents announce their business needs at a regular pace and supplier agents answer by sending a proposal. The buyers then select the best proposals by weighing up the utility associated to each proposal with the trustworthiness of the proponent agent. Finally, a simple purchasing contract – stipulating the good to deliver and the associated quantity, price, and delivery time conditions – is automatically drafted by the EI to regiment the exchange interaction.

The EI has limited monitoring capabilities. More specifically, for each contract established, the institution controls if the delivery time is respected by the supplier. When the transaction is over, the EI builds a *contractual evidence*, which includes the outcome o of the transaction ($o = f$ when the good is provided on time, $o = fd$ when the delivery is delayed, and $o = v$ if the good is not delivered at all). Other costly monitoring activities (e.g. about the quality of delivered products) are not performed by the EI or by any one of the corporate agents registered in the EI.

3.2 Agents

There are two types of agents: the ones that play the role of buyers, and the ones that play the role of suppliers (Equation 1). In this section, we start by describing the latter, and then proceed to the description of buyers.

$$Agents = Buyers \cup Suppliers \qquad (1)$$

Supplier Agents. These agents are characterized by the following properties: *dimension, reciprocity, subcontracting power* and *willingness to promote goodwill*.

In the scope of this paper, the *dimension* property measures the recent contractual activity of the supplier, i.e. the ratio of the number of contracts won by the supplier in the recent past to the number of proposals it had made in the same period of time. The contractual activity of a supplier may vary through the agent's life, due to several factors (e.g. its current trustworthiness, the utility of the proposals it makes, and whether or not the agent is engaged in goodwill relationships). Hence, the dimension of supplier agents may evolve over time, switching from small to big and vice-versa. Equation 2 shows the properties that identify the dimension of a supplier. In the Equation, $CAThreshold$ is a threshold parameter that can be configured in the experiments.

$$\begin{aligned} ContractualActivity(x) < CAThreshold \Rightarrow Small(x) \\ ContractualActivity(x) \geqslant CAThreshold \Rightarrow Big(x) \end{aligned} \tag{2}$$

The *reciprocity* property indicates the ability of the supplier agent to reciprocate to the buyer in current contract when the latter shows goodwill towards the former. The *subcontracting power* property indicates the ability/willingness of the supplier to subcontract when the current order made by the buyer is bigger than the supplier's current capacity to fulfill it, i.e., when the supplier is small. Finally, the *willingness to promote goodwill* property is related with the previous property. It indicates the willingness of the supplier agent to subcontract when the current order made by the buyer is bigger than the supplier's current capacity to fulfill it, even when the buyer has not shown goodwill towards it.

According to the aforementioned properties, we define three distinct categories of suppliers by providing the necessary and sufficient conditions for membership through predicates' formulas, as shown in Equation 3.

$$\begin{aligned} Suppliers &= Endeavored \cup Reciprocal \cup Opportunistic \\ Endeavored(x) &\Leftrightarrow Reciprocate(x) \wedge CanSubcontract(x) \wedge PromoteGoodwill(x) \\ Reciprocal(x) &\Leftrightarrow Reciprocate(x) \wedge CanSubcontract(x) \wedge \neg PromoteGoodwill(x) \quad (3) \\ Opportunistic(x) &\Leftrightarrow \neg Reciprocate(x) \wedge \neg CanSubcontract(x) \\ &\quad \wedge \neg PromoteGoodwill(x) \end{aligned}$$

We are now in conditions to model the behavior of supplier agents. This behavior is defined in terms of the ability and willingness of the supplier to provide the good to the buyer within the deadline stipulated in the contract.

The decision of whether to delay or not the delivery as stated in current contract takes into consideration the category of the supplier (as defined in Equation 3), its current dimension (cf. Equation 2), the size of the order, and the agent's belief about the goodwill motivation of the buyer towards it.

Equation 4 illustrates the behavior of suppliers of type *Endeavor* concerning the property of delaying the delivery of goods. These agents normally do not delay contracts. They tend to subcontract every time they get a contract that is bigger than their current capacity, even if they loose some utility in the short term due to subcontracting.[1] In other words, they show an attitude of goodwill towards the buyer seeking a longterm relationship with it. However, as these suppliers grow in dimension and are eventually considered big, they delay the delivery of small contracts if the corresponding buyers have not showed any goodwill towards them.

---

[1] In this model, we do not model the loss of utility of suppliers due to subcontracting.

$$Endeavored(x) \land Big(x) \land Contract(z) \land Size(z, Small) \land Buyer(y) \land$$
$$\neg ShowGoodwill(y) \Rightarrow Delay(x)$$
$$Endeavored(x) \land Small(x) \lor$$
$$Endeavored(x) \land Contract(z) \land Size(z, Medium) \lor \tag{4}$$
$$Endeavored(x) \land Big(x) \land Contract(z) \land Size(z, Small) \land Buyer(y) \land$$
$$ShowGoodwill(y) \Rightarrow \neg Delay(x)$$

Equation 5 shows the behavior of suppliers of type *Reciprocal*. There are two situations where the suppliers may delay the delivery to the buyer: when the size of the order is bigger than their current supplying capacity, and when they are considered big suppliers and the current order is of small magnitude. In both situations, these suppliers opt to not delay the delivery (even if it implies that some subcontracting must be done, in the first case) if the corresponding buyers have already shown goodwill attitudes towards them. Otherwise, they do delay these deliveries.

$$Reciprocal(x) \land Small(x) \land Contract(z) \land Size(z, Big) \land Buyer(y) \land$$
$$\neg ShowGoodwill(y) \lor$$
$$Reciprocal(x) \land Big(x) \land Contract(z) \land Size(z, Small) \land Buyer(y) \land$$
$$\neg ShowGoodwill(y) \Rightarrow Delay(x)$$
$$Reciprocal(x) \land Small(x) \land Contract(z) \land Size(z, Big) \land Buyer(y) \land \tag{5}$$
$$ShowGoodwill(y) \lor$$
$$Reciprocal(x) \land Big(x) \land Contract(z) \land Size(z, Small) \land Buyer(y) \land$$
$$ShowGoodwill(y) \lor$$
$$Reciprocal(x) \land Contract(z) \land Size(z, Medium) \Rightarrow \neg Delay(x)$$

Finally, Equation 6 illustrates the behavior of the last category of suppliers, *Opportunistic*. This behavior reflects the fact that these suppliers do not have subcontracting power ability.

$$Opportunistic(x) \land Small(x) \land Contract(z) \land Size(z, Big) \lor$$
$$Opportunistic(x) \land Big(x) \land Contract(z) \land Size(z, Small) \Rightarrow Delay(x)$$
$$Opportunistic(x) \land Small(x) \land Contract(z) \land Size(z, Small) \lor \tag{6}$$
$$Opportunistic(x) \land Big(x) \land Contract(z) \land Size(z, Big) \lor$$
$$Opportunistic(x) \land Contract(z) \land Size(z, Medium) \Rightarrow \neg Delay(x)$$

Buyer Agents. In the previous section, we showed that supplier agents base their delivery decisions taking into account their own idiosyncrasies and contextual information that they sense in the environment. In this section, we describe the behavior of buyer agents taking into account how they react to a delay in delivery. Buyer agents can be classified as *Benevolent* or *NonBenevolent*, taking into consideration two different properties: *reciprocity* and *ability to promote goodwill* relationships (cf. Equation 7).

$$Buyers = Benevolent \cup NonBenevolent$$
$$Benevolent(x) \Leftrightarrow Reciprocate(x) \wedge PromoteGoodwill(x) \tag{7}$$
$$NonBenevolent(x) \Leftrightarrow \neg Reciprocate(x) \wedge \neg PromoteGoodwill(x)$$

The *reciprocity* property indicates whether or not the buyer will reciprocate to existent manifestations of goodwill from suppliers. The *ability to promote goodwill* property indicates the propensity of the buyer to *initiate* relationships based on reciprocity that eventually will lead to goodwill.

Given the categories of the buyers, we are now in conditions to describe their behavior. This latter translates in decisions about whether the buyers will denounce, or not, a contract whose delivery is delayed (Equation 8).[2] As can be observed in Equation 8, in the case of buyers of type *Benevolent*, the decision to denounce the delay depends on whether or not the buyer has already established a goodwill relationship with the supplier.

$$Benevolent(x) \wedge Supplier(y) \wedge Delayed(y) \wedge HasGoodwill(x) \Rightarrow \neg Denounce(x)$$
$$Benevolent(x) \wedge Supplier(y) \wedge Delayed(y) \wedge \neg HasGoodwill(x) \vee \tag{8}$$
$$NonBenevolent(x) \wedge Supplier(y) \wedge Delayed(y) \Rightarrow Denounce(x)$$

Although the behavior of buyers is defined in simpler terms than the behavior of suppliers, it also allows for important dynamics in the behavior of agents.

3.3 Theoretical Motivation of the Model

In the previous section we gave an intuitive description of the agents' behavior model. We are now ready to present the main theoretical socio-economic notions underlying this model.

Electronic Institutions, Contract Drafting and Monitoring. In our work we use the paradigm of electronic institutions (EI). These institutions, which can be seen as an umbrella for different types of virtual organizations, strategic alliances and organizational networks, are indeed a promising trend for inter-organizational relationships [1].

The contracts established in the EI lack the formal detail. This is common within the textile industry, where contracts can even be relational [11]. Moreover, designing detailed contracts involves substantial drafting costs [14] if it is not considered impossible [15]. Also, the EI has limited monitoring functionalities. Mechanisms for control and coordination are costly and we use instead trust as a complementary control mechanism, as suggested in [4].

---

[2] When a supplier does not deliver within the deadline, the EI registers a delay. However, it is up to the buyer to denounce this delay (resulting in outcome $o = v$) or to be benevolent with it ($o = fd$).

Finally, we assume that buyer agents are able to access information about all contracts established in the electronic institution. There is not enough empirical knowledge supporting this assumption. However, we verify similar concepts in highly, though open, regulated markets, such as the stock market exchange, where the information about firms is made public. Moreover, an agent that is evaluating a business party may acquire information about contracts established between the party and other agents through reputation, opinions and recommendations. The way that this information is acquired and its reliability does not affect the aggregation models, *per se*, and is out of the scope of this paper.

Behavior of Agents. The behavior of suppliers is based on the study of opportunism. Following Wathne and Heide (2000), opportunism can be passive or active and applies under existent or new conditions (either present in an informal agreement or in a legal contract) [14]. In existent conditions, it takes the form of evasion of obligations (passive) and violation (active), and under new conditions it takes the form of refusal to adapt to new circumstances (passive) and forced renegotiation (active).

In our work, we adopt the notion of passive opportunism under existent conditions, where obligations can be evaded. This can happen in two distinct situations. In the first situation, small suppliers bid to provide big quantities, even when they know that they are not able to fulfill what was stipulated in the contract. This is the case for suppliers of type *Opportunistic* and for suppliers of type *Reciprocal* in specific conditions. A common problem in interfirm relationships that is reflected in the proposed categorization is adverse selection, where suppliers hide their true attributes from the buyer. This happened in the famous Ford vs. Lear case, where Lear committed to supply the seats for all Ford Taurus versions, withholding the information about its lack of adequate resources. As a result, "Lear missed deadlines, failed to meet weight and price objectives, and furnished parts that did not work (Walton 1997)" and Ford incurred in substantial transaction costs [14].

Reciprocation and Goodwill Trust. As mentioned by D. Ireland and Webb (2007), when unanticipated contingencies surface, business partners may opt to show some goodwill towards the target agent, or to select tougher forms of action, depending on the magnitude of the contingency and the level of trust existing between the partners [4]. As described by Sako (2002), goodwill trust is a form of trust that develops within long-term relationships through repeated exchanges [10]. It can be built upon reciprocation of initial favors that allow the establishment of the norms and shared values that characterize relational behavior and continued interactions between the partners (Jones (2001), referenced in [4]).

Dimension of Buyers and Suppliers. The dimension of buyers and suppliers is oversimplified in our model. Other properties may be added in the future, such as those derived from the complex concept of power in business relations.

4   Empirical Analysis

This section describes the empirical study we have performed on the adequacy of existent computational trust models in capturing the dynamics of behavior that exist in interfirm relationships, as modeled in the previous section. In this study, we analyzed three distinct computational trust models, as described in the next section.

4.1   Computational Trust Models

The first model we analyzed was FIRE, more concretely, its component of aggregation of both direct experiences and reputation [3]. We chose this model because it is widely cited in the literature and also because it includes in its formula a temporal dimension given by weighing by recency. Therefore, it could be a good choice for dealing with populations whose behavior evolves over time. The formula used in the experiments to estimate the trustworthiness of agents ag is depicted in Equation 9. In this formula, w_i is the weight of current evidence, v_i is the value of the outcome of the evidence and Δ_i is the time elapsed between the occurrence of the evidence and current time.

$$trustworthiness(ag) = \frac{\sum_{i=0}^{N-1} w_i.v_i}{\sum_{i=0}^{N-1} v_i}, w_i = e^{\frac{-\Delta t_i}{\lambda}}, \lambda = \frac{-0.5}{\ln(0.5)}. \tag{9}$$

The second model we analyzed was the situation-aware CSRC model [8] referred to in the related work. As FIRE, it also aggregates evidences using weighting means. However, this model weights the evidences by the similarity between the context of the evidence being aggregated and the context of the current situation in assessment (cf. Equations 10, 11 and 12).

$$trustworthiness(ag, s) = \frac{\sum_{i=0}^{N-1} w_i.v_i}{\sum_{i=0}^{N-1} v_i}, w_i = e^{-d(c_1,c_2)}. \tag{10}$$

$$d^{fabric}(c_1, c_2) = \begin{cases} 0, \text{ if } fabric_1 = fabric_2, \\ 1, \text{ if } fabric_1 \neq fabric_2. \end{cases} \tag{11}$$

$$d^{attr}(c_1, c_2) = |ln(attr_1) - ln(attr_2)|. \tag{12}$$

Finally, we chose as the third model another situation-aware approach (which we name here SACF), which includes the heuristic-based aggregator Sinalpha [12] and the situation-aware tuner, Contextual Fitness [13]. We chose to evaluate this model because it extracts tendencies of failure in an incremental way and therefore, as happened with the first two approaches, it could be a good choice to tackle the problem of dynamic populations of agents. Equations 13 and 14 show the main formulas of this model. The algorithm underlying $CF(ag, s)$, as well as a detailed explanation of Sinalpha's parameters, are documented in [13].

$$trustworthiness(ag, s) = sinalpha(ag) * CF(ag, s), \tag{13}$$

$$sinalpha(\alpha, ag) = \delta \cdot sin\alpha + \delta, \alpha = \alpha + \lambda \times \omega. \tag{14}$$

4.2   Testbed

All experiments described in this paper were performed using the Repast simulation tool [6]. We ran three different experiments, one for each computational trust model considered.

At every round, a fixed number of buyer agents selected the best suppliers of textile fabric using a simple one round, multi-attribute negotiation protocol. Every buyer had a business need randomly assigned at setup, consisting of a fabric and associated values of quantity, price and delivery time.

The set of possible fabrics is given by $\{cotton, chiffon, voile\}$. The values of quantity, price and delivery time are assigned randomly from sets $\{q \in \mathbb{N} : q \in [v_{quant,min}, v_{quant,max}]\}$, $\{p \in \mathbb{N} : p \in [v_{price,min}, v_{price,max}]\}$ and $\{d \in \mathbb{N} : d \in [v_{dtime,min}, v_{dtime,max}]\}$, respectively. The values $v_{i,min}$ and $v_{i,max}$ define the minimum and maximum values allowed in the simulation for each attribute i, respectively.

All suppliers registered in the EI were able to provide any type of fabric. When a buyer announced its need (in the form a call for proposals – cfp) to a defined set of suppliers, each one of these suppliers generated a proposal with its own values for the quantity, price and delivery time attributes. These values were generated randomly following a uniform distribution in the range $[v_{i,p,min}, v_{i,p,max}]$, where $v_{i,p,min}$ and $v_{i,p,max}$ are defined in Equation 15.

$$v_{i,p,min} = \max \left((1 - \zeta) \times v_{i,cfp}, v_{i,min} \right),$$
$$v_{i,p,max} = \min \left((1 + \zeta) \times v_{i,cfp}, v_{i,max} \right). \tag{15}$$

In Equation 15, $v_{i,cfp}$ is the value defined in the cfp for attribute i (quantity, price or delivery time), and ζ is a *dispersion* parameter that allows to define how distant the generated proposal is from the preferences of the buyer, as stated in the cfp.

Selection Strategy. After receiving the proposals from the suppliers, the buyer calculated the utility of each one of them. The utility of a proposal, $\mu_p \in [0, 1]$, was given by the complement of the *deviation* between the client preferences specified in the cfp, for all the negotiable items price, quantity and delivery time, and what is offered in the received proposal (cf. Equation 16).

$$\mu_p = 1 - \frac{1}{k} \times \left(\sum_i^k \frac{|v_{i,cfp} - v_{i,p}|}{v_{i,max} - v_{i,min}} \right). \tag{16}$$

In Equation 16, $v_{i,p}$ is the value of the negotiation attribute i of the current proposal in evaluation and k is the number of negotiation attributes considered in this paper.

After calculating the utilities of all received proposals, the buyer selected the *best* proposal, by sorting the proposals by the weighted sum of their utility and the trustworthiness of the corresponding proponents, and by choosing the one that presented the highest value for this weighted sum. This assumes that, previous to the evaluation phase, the buyer estimated the trustworthiness of all suppliers that presented a proposal using one of the computational trust algorithms described above. Equation 17 illustrates the weighted sum, where τ stands for the estimated trustworthiness value and the weighting parameter $\omega_\tau \in [0,1]$ allows to configure the importance assigned to the trustworthiness component in the selection.

$$\omega_\tau \times \tau + (1 - \omega_\tau) \times \mu_p. \tag{17}$$

Finally, after the selection of the best proposal, the buyer establishes a contract with the selected supplier, stipulating that the latter must provide the fabric at the conditions of quantity, price and delivery time described in its proposal.[3]

Configuration Parameters. In all experiments, we used 10 buyer agents and 20 supplier agents. The behavior of these agents was extensively described in Section 3.2. Each experiment was composed of 30 *episodes*, and at every episode each buyer started a new negotiation cycle by issuing a new *cfp*. At the first episode of each experiment, the repository of trust evidences was cleaned, which means that the trustworthiness of all suppliers was set to zero. Finally, we ran every experiment 10 times.

Additional configuration parameters can be seen in Table 1.

4.3   Evaluation Metrics

In these experiments, we considered five different performance metrics. The first metric was the *utility of the transaction*, $\mu_t \in [0,1]$, given by Equation 18, which was further averaged over all buyers and all episodes.

$$\mu_t = \begin{cases} \mu_p, \text{ if } o = f \ , \\ 0, \text{ if } o = v \ . \end{cases} \tag{18}$$

The second metric was the *number of positive outcomes* ($o^+ \in [0,1]$) obtained by a buyer agent in an episode, averaged over all buyers and all episodes. The third metric was the *number of different suppliers* ($\Delta_{sup} \in [0,1]$) selected by all buyers in one episode, averaged over all episodes. Finally, the fourth and the fifth metrics measured the *trustworthiness of the supplier* and the *utility of the proposal* selected by a buyer in one episode ($\tau_s \in [0,1]$ and $\mu_s \in [0,1]$, respectively), averaged over all buyers and all episodes.

---

[3] The negotiation mechanism we present in this paper is deliberately simple, as it does not constitute the focus of this work. We assume that the conclusions derived from our study using this mechanism are still valid in the presence of others, more complex negotiation protocols.

Table 1. Configuration parameters

parameter	value	appears in
ζ	1.0	Eq. 15
vi	0.0 $(o = f)$, 0.5 $(o = fd)$, 1.0 $(o = v)$	Eq. 9
δ	0.5	Eq. 14
λ	1.0 $(o = f)$, 0.5 $(o = fd)$, -1.5 $(o = v)$	Eq. 9
ω	$\pi/2$	Eq. 9
ω_τ	0.5	Eq. 17

4.4 Results

Table 2 presents the results obtained in the experiments for the metrics described in Section 4.3.

Table 2. Results obtained in the experiments

#	Trust Model	μ_t	o^+	Δ_{sup}	τ_s	μ_s
1	SACF	0.64	0.80	0.37	0.75	0.79
	(stdev)	(0.06)	(0.07)	(0.03)	(0.06)	(0.01)
2	FIRE	0.55	0.67	0.37	0.76	0.83
	(stdev)	(0.06)	(0.05)	(0.06)	(0.06)	(0.02)
3	CSRC	0.59	0.70	0.41	0.76	0.84
	(stdev)	(0.05)	(0.05)	(0.05)	(0.04)	(0.01)

From the results presented in Table 2, we verified that the FIRE approach got the worse results, both in terms of utility of transaction ($\mu_t = 0.55$) and of positive outcomes ($o^+ = 0.67$), even with the obtained values for the number of distinct suppliers and trustworthiness of the selected supplier being approximately the same as those of the model that performed best (SACF).

Also, we verified that, between the situation-aware approaches, the SACF got better results than the CSRC approach, both in terms of utility of transaction ($\mu_t = 0.64$ vs. $\mu_t = 0.59^4$) and of positive outcomes ($o^+ = 0.80$ vs. $o^+ = 0.70$). This happened even with the latter allowing for a broader selection of partners than the former (CSRC: $\Delta_{sup} = 0.41$ vs. SACF: $Delta_{sup} = 0.37$), which in turn increased the utility of the proposals received by the buyers (CSRC: $\mu_s = 0.84$ vs. SACF:$\mu_s = 0.79$).

4.5 Interpretation of the Results

From the results depicted in the previous section, we verified that recency is not enough to capture evolving behavior of agents, as the FIRE model got the worse

---

[4] We used the two-sample t-test to infer about the difference in μ_t between SACF and FIRE and between SACF and CSRC. The test results were statistically significant at the $\alpha = 0.05$ level of significance.

results. We also verified that the dynamic extraction of failure tendencies of the
SACF approach seemed to be more adequate to evolving behaviors than the use
of reference contexts of the CSRC approach.

However, looking at the traces of the experiments, we observed that even
the best approach (SACF) had a relatively poor performance in detecting the
changes on the behavior of buyer and suppliers agents, based on reciprocation
and goodwill relations that formed, and also by the evolving dimension of suppliers.

In order to clarify this question, we present next a simple trace extracted from
the experiments with the SACF model, representing an evaluation performed by
agent $C0$ (of type *Benevolent*) to a supplier agent of type *Recriprocate*. At the
evaluation time, agent $C0$ had access to the following contractual evidences of
the supplier in evaluation:

episode	client	cfp	outcome
1	C2	chiffon low medium	f
2	C2	chiffon low low	f
2	C3	voile medium big	f
3	C1	chiffon low medium	f
3	C2	chiffon medium low	f
3	C3	voile high medium	v
3	C9	chiffon low low	f
4	C0	voile low big	f
4	C2	chiffon medium big	f
4	C6	cotton low low	v
4	C7	voile low low	v
4	C9	chiffon medium medium	f

Using the SACF algorithm, the model was able to detect the following tendencies of failure: $(cotton, *, *)$, $(voile, *, low)$ and $(voile, *, medium)$. Also, the
algorithm inferred that the supplier had the following tendency to *not* delaying
contracts: $(chiffon, *, *)$ and $(voile, *, big)$

In this extreme case, when the number of evidences available is reduced, the
model was not able to understand the shift that occurred in the dimension of
the suppliers. In other situations (which traces we do not reproduce here), this
model was not also able to detect goodwill relationships between the interfirm
partners.

5   Conclusions

In this paper, we raised the suspicion that current computational trust models
were not able to detect the dynamics of agents that engage in trust relationships. In order to confirm this suspicion, we presented a conceptual model for
the behavior of trading agents that was based on existent literature on interfirm relationships. Although the presented model is simple, it is able to model

evolving behaviors that are due to the establishment of reciprocation and goodwill relationships between business partners and to changes in the dimension of supplier agents.

Next, we evaluated three computational trust models, including the representative FIRE model that embeds the recency of the trust evidences on its algorithm and two situation-aware approaches. The results have shown that these latter models performed better than FIRE in evolving populations like the ones modeled in this paper. Also, we have observed that even the model that performed best was not able to detect the complex web of relations that happens between interfirm business partners.

From this study, we concluded that computational trust models must interpret the past evidence in light of the *context* where these evidences occurred. Also, the models must take into account the (inferred) relation between the partners to the exchange, as the outcome of the evidences may vary substantially depending on the evaluator and its relationship with the evaluated. Finally, computational trust models must process the past evidence through a temporal lens, and ideally they must infer the dynamics of behavior that evolve over time.

Acknowledgments. This research is supported by Fundação para a Ciência e a Tecnologia (FCT), under project PTDC/EIA-EIA/104420/2008. The first author is supported by FCT under grant SFRH/BD/39070/2007.

References

1. Bachmann, R.: Trust, power and control in trans-organizational relations. Organization Studies 22(2), 337–365 (2001)
2. Heimer, C.: Solving the problem of trust. In: Cook, K.S. (ed.) Trust in Society. Russell Sage Foundation Series on Trust, pp. 40–88 (2001)
3. Huynh, T.D., Jennings, N.R., Shadbolt, N.R.: An integrated trust and reputation model for open multi-agent systems. Autonomous Agents and Multi-Agent Systems 13, 119–154 (2006)
4. Ireland, R.D., Webb, J.W.: A multi-theoretic perspective on trust and power in strategic supply chains. Journal of Operations Management 25(2), 482–497 (2007); special Issue Evolution of the Field of Operations Management SI/ Special Issue Organisation Theory and Supply Chain Management
5. Ismail, R., Josang, A.: The beta reputation system. In: Proceedings of the 15th Bled Conference on Electronic Commerce (2002); paper 41
6. North, M.J., Howe, T.R., Collie, N.T., Vos, J.R.: A declarative model assembly infrastructure for verification and validation. In: Takahashi, S., Sallach, D.L., Rouchier, J. (eds.) Advancing Social Simulation: The First World Congress, Springer, Heidelberg (2007)
7. Patel, J.: A Trust and Reputation Model for Agent-Based Virtual Organisations. Ph.D. thesis, University of Southampton (2006)
8. Rehák, M., Pěchouček, M.: Trust modeling with context representation and generalized identities. In: Klusch, M., Hindriks, K.V., Papazoglou, M.P., Sterling, L. (eds.) CIA 2007. LNCS (LNAI), vol. 4676, pp. 298–312. Springer, Heidelberg (2007)

9. Sabater, J., Sierra, C.: Regret: reputation in gregarious societies. In: Proceedings of the Fifth International Conference on Autonomous Agents, AGENTS 2001, pp. 194–195. ACM, New York (2001)
10. Sako, M.: Does trust improve business performance? (2002),
    http://hdl.handle.net/1721.1/1462
11. Tokatli, N.: Global sourcing: insights from the global clothing industry - the case of zara, a fast fashion retailer. Journal of Economic Geography (2007)
12. Urbano, J., Rocha, A.P., Oliveira, E.: Computing confidence values: Does trust dynamics matter? In: Lopes, L.S., Lau, N., Mariano, P., Rocha, L.M. (eds.) EPIA 2009. LNCS, vol. 5816, pp. 520–531. Springer, Heidelberg (2009)
13. Urbano, J., Rocha, A.P., Oliveira, E.: Trustworthiness tendency incremental extraction using information gain. In: Proceedings of the 2010 IEEE/WIC/ACM International Conference on Web Intelligence and Intelligent Agent Technology, WI-IAT 2010, vol. 02, pp. 411–414. IEEE Computer Society, Washington, DC (2010)
14. Wathne, K.H., Heide, J.B.: Opportunism in interfirm relationships: Forms, outcomes, and solutions. The Journal of Marketing 64(4), 36–51 (2000)
15. Williamson, O.E.: Transaction-cost economics: The governance of contractual relations. Journal of Law and Economics 22, 233–261 (1979)

The BMC Method for the Existential Part of RTCTLK and Interleaved Interpreted Systems

Bożena Woźna-Szcześniak, Agnieszka Zbrzezny, and Andrzej Zbrzezny

IMCS, Jan Długosz University. Al. Armii Krajowej 13/15, 42-200 Częstochowa, Poland
{b.wozna,a.zbrzezny,agnieszka.zbrzezny}@ajd.czest.pl

Abstract. In the paper, we focus on the formal verification of multi-agent systems – modelled by interleaved interpreted systems – by means of the bounded model checking (BMC) method, where specifications are expressed in the existential fragment of the Real-Time Computation Tree Logic augmented to include standard epistemic operators (RTECTLK). In particular, we define an improved SAT-based BMC for RTECTLK, and present performance evaluation of our newly developed BMC method by means of the well known train controller and generic pipeline systems.

1 Introduction

The problem of model checking [4] is to check automatically whether a structure M defines a model for a modal (temporal, epistemic, etc.) formula p. Bounded model checking (BMC) is a verification technique designed for finding counterexamples, and whose main idea is to consider a model curtailed to a specific depth. BMC via SAT was first proposed for linear-time temporal logic (LTL, LTL+Past) [1,2], and then it was extended, among others, to the universal fragment of CTL [18,22], and to the branching time epistemic logic, called ACTLK [13,17].

Multi-agent systems (MASs) are composed of many intelligent agents that interact with each other. The agents can share a common goal or they can pursue their own interests. Also, the agents may have deadlines or other timing constraints to achieve intended targets. As it was shown in [8], knowledge is a useful concept for analyzing the information state and the behaviour of agents in multi-agent systems. In particular, it is useful to reason about and to verify the evolution over time of epistemic states [10]. Thus reasoning about knowledge of agents and multi-agent systems has always been a core issue in artificial intelligence. Therefore, many logical formalisms and verification techniques, especially the one based on model checking, have been developed and refined over the years, among others, [8,9,11,13,14,19,20].

An existential fragment of the soft real-time CTL (RTECTL) [7] is a propositional branching-time temporal logic with bounded operators, which was introduced to permit specification and reasoning about time-critical correctness properties. More specifically, existential CTL formulae allow for the verification of properties such as *"there is a computation such that φ will eventually occur"*, or *"there is a computation such that φ will never be asserted"*. However, it is not possible to directly express bounded properties like for example *"there is a computation such that φ will occur in less than 30 unit*

L. Antunes and H.S. Pinto (Eds.): EPIA 2011, LNAI 7026, pp. 551–565, 2011.

time", or "*there is a computation such that φ will always be asserted between 10 and 30 unit time*". While it is true that properties like the above can be expressed using nested applications of the next state operators, the resulting CTL formula can be very complex and cumbersome to work with. RTECTL defeats this restriction by allowing bounds on all temporal operators to be specified, and provides a much more compact and convenient way of expressing time-bounded properties. The RTECTLK language is an epistemic soft real-time computation tree logic that is the fusion [3] of the two underlying languages: RTECTL and a multi-modal logic $S5_n$ for knowledge that satisfies the following properties: the truth axiom, the distribution axiom, the necessitation axiom, the positive introspection axiom, and the negative introspection axiom.

A version of the BMC method for specifications expressed in RTECTLK, and MASs modelled by interpreted systems [8], in which agents have time-limits or other explicit timing constraints to accomplish intended goals, has been published in pre-proceedings of CEE-SET'2009 ([21]). However, this method not only does not take into account the development related to the BMC algorithm for ECTL [22], but also it is based on unnecessarily complex bounded semantics. Moreover, it has not been implemented and experimentally evaluated.

The main idea of the [21] translation is the following. Given are a RTECTLK formula φ and a bound $k \in \mathbb{N}$. First the number of paths of length k (k-paths) that are sufficient for checking the formula φ is computed; this is done by means of the function f_k. Next, the set of k-paths of size $f_k(\varphi)$ is created and used to translate every subformula of the formula φ. This means that this translation uses all the $f_k(\varphi)$ k-paths both for the formula φ and for each of its proper subformulae. In the paper we propose an improvement of this translation, which is based on the BMC method for ECTL [22]. Its main idea consists in translating every subformula ψ of the formula φ using only $f_k(\psi)$ paths of length k. So, our new BMC algorithm uses a reduced number of paths, what results in significantly smaller and less complicated propositional formulae that encode the RTECTLK properties.

Specifically, in the paper we make three contributions: first, we present the BMC method (which is based on the improved ECTL translation [22]) for specifications expressed in RTECTLK, and multi-agent systems modelled by interleaved interpreted systems (IIS) introduced in [15] (a subclass of standard interpreted systems [8]); second, we implement under the same semantics, i.e., interleaved interpreted systems, both BMC translations the one presented in [21], and our new one; third, we present performance evaluation of the two implemented BMC algorithms for the verification of several properties expressed in RTECTLK.

The [21] BMC translation could be also improved by adopting the SAT-based BMC for ECTLK [13], which according to the authors significantly improves the [22] BMC encoding. However, for this paper we have decided not to make use of the method presented in [13], but we plan to research this possibility in the future.

The structure of the paper is as follows. In Section 2 we shortly introduce interleaved interpreted systems and the RTECTLK logic. In Section 3 we define an improved SAT-based BMC for RTECTLK, and we prove its correctness. In Section 4 we present performance evaluation of our newly developed SAT-based BMC algorithm for RTECTLK. In Section 5 we conclude the paper.

2 Preliminaries

In this section we first define *interleaved interpreted systems (IIS)*, and next we introduce syntax and semantics of RTECTLK. The formalism of IIS was introduced in [15] to model multi-agent system (MASs) that are composed of multiple agents, each of which is an independently operating entity, and to reason about the agents' epistemic and temporal properties. The interleaved interpreted system is a special subclass of standard interpreted systems introduced in [8], which allow for modelling asynchronous MASs. In this formalism, each agent is modelled using a set of local states, a set of actions, a local protocol, and a local (interleaved) evolution function. In IIS only one action at a time is performed in a global transition. Specifically, if several agents act in the global action, then they have to perform the same action, thereby synchronising at that particular time step. Thus, the local protocols are defined in such a way that if an agent has the action being performed in its set of actions, then it must be able to perform the action in order to allow the global evolution of the whole system.

Interleaved Interpreted Systems. We assume that a MAS consists of n agents[1], and by $Ag = \{1, \ldots, n\}$ we denote the non-empty set of agents. Further, we assume that each agent $c \in Ag$ is in some particular local state at a given point in time, and that a set L_c of local states for agent $c \in Ag$ is non-empty and finite (this is required by the model checking algorithms). An agent's local state encapsulates all the information the agent has access to, and which are required to completely characterise its state. Further, with each agent $c \in Ag$ we associate a finite set of *possible actions* Act_c such that a special action ϵ_c, called "null", belongs to Act_c; as it will be clear below the local state of agent c remains the same if the null action is performed. We do not assume that the sets Act_c (for all $c \in Ag$) are disjoint. We define the following sets $Act = \bigcup_{c \in Ag} Act_c$ and $Agent(a) = \{c \in Ag \mid a \in Act_c\}$. Next, for each agent $c \in Ag$ we associate a protocol that defines rules, according to which actions may be performed in each local state. The protocol for agent $c \in Ag$ is a function $P_c : L_c \to 2^{Act_c}$ such that $\epsilon_c \in P_c(l)$ for any $l \in L_c$, i.e., we insist on the null action to be enabled at every local state. Note that the above definition of the protocol enables more than one action to be performed for a given local state. This means that an agent selects non-deterministically which action to perform. Finally, for each agent c, there is defined a (partial) evolution function $t_c : L_c \times Act_c \to L_c$ such that for each $l \in L_c$ and for each $a \in P_c(l)$ there exists $l' \in L_c$ such that $t_c(l, a) = l'$; moreover, for each $l \in L_c, t_c(l, \epsilon_c) = l$. Note that the local evolution function considered here differs from the standard one (see [8]) by having the local action instead of the join action as the parameter.

A *global state* $g = (l_1, \ldots, l_n)$ is a tuple of n local states, one per each agent, corresponding to an instantaneous snapshot of the MAS at a given time. By $l_c(g)$ we denote the local component of agent $c \in Ag$ in a global state $g = (l_1, \ldots, l_n)$, and by G we denote a set of global states. It is assumed that, in every state, agents evolve simultaneously. Thus the *global interleaved evolution function* $t : G \times Act_1 \times \cdots \times Act_n \to G$ is defined as follows: $t(g, a_1, \ldots, a_n) = g'$ iff there exists an action $a \in$

---

[1] Note in the present study we do not consider the environment component. This may be added with no technical difficulty at the price of heavier notation.

$Act \setminus \{\epsilon_1, \ldots, \epsilon_n\}$ such that for all $c \in Agent(a)$, $a_c = a$ and $t_c(l_c(g), a) = l_c(g')$, and for all $c \in Ag \setminus Agent(a)$, $a_c = \epsilon_c$ and $t_c(l_c(g), a_c) = l_c(g)$. In brief we write the above as $g \xrightarrow{a} g'$. Now, for a given set of agents Ag and a set of propositional variables $\mathcal{PV}$, which can be either true or false, an *interleaved interpreted system* is a tuple: $IIS = (\iota, < L_c, Act_c, P_c, t_c >_{c \in Ag}, \mathcal{V})$, where $\iota \in G$ is an initial global state, and $\mathcal{V} : G \rightarrow 2^{\mathcal{PV}}$ is a valuation function. With such an interleaved interpreted system IIS it is possible to associate a *Kripke model* $M = (\iota, S, T, \{\sim_c\}_{c \in Ag}, \mathcal{V})$, where ι is the initial global state; $S \subseteq G$ is a set of reachable global states that is generated from ι by using the global interleaved evolution functions t; $T \subseteq S \times S$ is a global transition (temporal) relation on S defined by: sTs' iff there exists an action $a \in Act \setminus \{\epsilon_1, \ldots, \epsilon_n\}$ such that $s \xrightarrow{a} s'$. We assume that the relation is total, i.e., for any $s \in S$ there exists an $a \in Act \setminus \{\epsilon_1, \ldots, \epsilon_n\}$ such that $s \xrightarrow{a} s'$ for some $s' \in S$; $\sim_c \subseteq S \times S$ is an indistinguishability relation for agent c defined by: $s \sim_c s'$ iff $l_c(s') = l_c(s)$; and $\mathcal{V} : S \rightarrow 2^{\mathcal{PV}}$ is the valuation function of IIS subtracted to the set S. $\mathcal{V}$ assigns to each state a set of propositional variables that are assumed to be true at that state. For more details we refer to [8].

Syntax of RTECTLK. Let $p \in \mathcal{PV}$, $c \in Ag$, $\Gamma \subseteq Ag$, and I be an interval in $\mathbb{N} = \{0, 1, 2, \ldots\}$ of the form: $[a, b)$ and $[a, \infty)$, for $a, b \in \mathbb{N}$ [2]. Hereafter by $left(I)$ we denote the left end of the interval I, i.e., $left(I) = a$, and by $right(I)$ the right end of the interval I, i.e., $right([a, b)) = b - 1$ and $right([a, \infty)) = \infty$. The language RTECTLK is defined by the following grammar:

$$\varphi := \textbf{true} \mid \textbf{false} \mid p \mid \neg p \mid \varphi \wedge \varphi \mid \varphi \vee \varphi \mid \text{EX}\varphi \mid \text{E}(\varphi \text{U}_I \varphi) \mid \text{EG}_I \varphi \mid$$
$$\overline{\text{K}}_c \varphi \mid \overline{\text{D}}_\Gamma \varphi \mid \overline{\text{E}}_\Gamma \varphi \mid \overline{\text{C}}_\Gamma \varphi$$

U_I and G_I are the operators, resp., for bounded "Until" and "Always". The formula $\text{E}(\alpha \text{U}_I \beta)$ is read as "there exists a computation such that β holds in the interval I at least in one state and always earlier α holds", the formula $\text{EG}_I \alpha$ is read as "there exists a computation such that α always holds in the interval I". The remaining bounded temporal operators are introduced in the standard way: $\text{E}(\alpha \text{R}_I \beta) \stackrel{def}{=} \text{E}(\beta \text{U}_I (\alpha \wedge \beta)) \vee \text{EG}_I \beta$, $\text{EF}_I \alpha \stackrel{def}{=} \text{E}(\textbf{true}\text{U}_I \alpha)$. $\overline{\text{K}}_c$ is the operator dual for the standard epistemic modality K_c ("agent c knows"), so $\overline{\text{K}}_c \alpha$ is read as "agent c does not know whether or not α holds". Similarly, the modalities $\overline{\text{D}}_\Gamma, \overline{\text{E}}_\Gamma, \overline{\text{C}}_\Gamma$ are the dual operators for $\text{D}_\Gamma, \text{E}_\Gamma, \text{C}_\Gamma$ representing distributed knowledge in the group Γ, "everyone in Γ knows", and common knowledge among agents in Γ.

Semantics of RTECTLK. Let $M = (\iota, S, T, \{\sim_c\}_{c \in Ag}, \mathcal{V})$ be a model. Then, a *path* in M is an infinite sequence $\pi = (s_0, s_1, \ldots)$ of states such that $(s_j, s_{j+1}) \in T$ for each $j \in \mathbb{N}$. For a path $\pi = (s_0, s_1, \ldots)$, we take $\pi(j) = s_j$. By $\Pi(s)$ we denote the set of all the paths starting at $s \in S$. For the group epistemic modalities we also define the following. If $\Gamma \subseteq Ag$, then $\sim_\Gamma^E \stackrel{def}{=} \bigcup_{c \in \Gamma} \sim_c$, $\sim_\Gamma^C \stackrel{def}{=} (\sim_\Gamma^E)^+$ (the transitive closure of $\sim_\Gamma^E$), and $\sim_\Gamma^D \stackrel{def}{=} \bigcap_{c \in \Gamma} \sim_c$. Given the above, the formal semantics of RTECTLK is defined recursively as follows:

---

[2] Note that the remaining forms of intervals can be defined by means of $[a, b)$ and $[a, \infty)$.

- $M, s \models \textbf{true}$, • $M, s \not\models \textbf{false}$, • $M, s \models p$ iff $p \in \mathcal{V}(s)$, • $M, s \models \neg p$ iff $p \notin \mathcal{V}(s)$,
- $M, s \models \alpha \wedge \beta$ iff $M, s \models \alpha$ and $M, s \models \beta$,
- $M, s \models \alpha \vee \beta$ iff $M, s \models \alpha$ or $M, s \models \beta$,
- $M, s \models \text{EX}\alpha$ iff $(\exists \pi \in \Pi(s))(M, \pi(1) \models \alpha)$,
- $M, s \models \text{E}(\alpha \text{U}_I \beta)$ iff $(\exists \pi \in \Pi(s))(\exists m \in I)[M, \pi(m) \models \beta$ and $(\forall j < m) M, \pi(j) \models \alpha]$,
- $M, s \models \text{EG}_I \alpha$ iff $(\exists \pi \in \Pi(s))$ such that $(\forall m \in I)[M, \pi(m) \models \alpha]$,
- $M, s \models \overline{\text{K}}_c \alpha$ iff $(\exists s' \in S)(s \sim_c s'$ and $M, s' \models \alpha)$,
- $M, s \models \overline{Y}\alpha$ iff $(\exists s' \in S)(s \sim s'$ and $M, s' \models \alpha)$, where $\overline{Y} \in \{\overline{\text{D}}_\Gamma, \overline{\text{E}}_\Gamma, \overline{\text{C}}_\Gamma\}$, and $\sim \in \{\sim_\Gamma^D, \sim_\Gamma^E, \sim_\Gamma^C\}$.

We end the section by defining the notions of validity and the model checking problem. Namely, a RTECTLK formula φ is *valid* in M (denoted $M \models \varphi$) iff $M, \iota \models \varphi$, i.e., φ is true at the initial state of the model M. The *model checking problem* asks whether $M \models \varphi$.

3   SAT-Based BMC for RTECTLK

As it was already mentioned, the BMC method for RTECTLK presented in [21] is based on unnecessarily complicated bounded semantics, and it does not take into account the BMC encoding of [22]. In this section, we present a new bounded semantics for RTECTLK, show its equivalence to the unbounded one, and define an improved SAT-based BMC method for RTECTLK, which is based on the BMC encoding presented in [22]. As usual, we start by defining k-paths and (k, l)-*loops*. Next we define a bounded semantics, which is later used for translation to SAT.

Bounded Semantics. Let $M = (\iota, S, T, \{\sim_c\}_{c \in Ag}, \mathcal{V})$ be a model and $k \geq 0$. A k-*path* π_k in M is a finite sequence of states $(s_0, \ldots, s_k)$ such that $(s_j, s_{j+1}) \in T$ for each $0 \leq j < k$. By $\Pi_k(s)$ we denote the set of all the k-paths starting at s in M. A k-path π_k is a (k, l)-*loop* iff $\pi_k(l) = \pi_k(k)$ for some $0 \leq l < k$; note that (k, l)-loop π generates the infinite path of the following form: $u \cdot v^\omega$ with $u = (\pi(0), \ldots, \pi(l-1))$ and $v = (\pi(l), \ldots, \pi(k-1))$. Since in the bounded semantics we consider finite prefixes of paths only, the satisfiability of all the temporal operators depends on whether a considered k-path is a loop. Thus, as customary, we introduce a function $loop : \bigcup_{s \in S} \Pi_k(s) \to 2^{\mathbb{N}}$, which identifies these k-paths that are loops. The function is defined as: $loop(\pi_k) = \{l \mid 0 \leq l < k$ and $\pi_k(l) = \pi_k(k)\}$.

Definition 1. *Given are a bound* $k \in \mathbb{N}$, *a model* M, *and* RTECTLK *formulae* α, β. $M, s \models_k \alpha$ *denotes that* α *is* $k-true$ *at the state* s *of* M. *The relation* $\models_k$ *is defined inductively as follows:*

- $M, s \models_k \textbf{true}$,   • $M, s \not\models_k \textbf{false}$,
- $M, s \models_k p$ iff $p \in \mathcal{V}(s)$,   • $M, s \models_k \neg p$ iff $p \notin \mathcal{V}(s)$,
- $M, s \models_k \alpha \vee \beta$ iff $M, s \models_k \alpha$ or $M, s \models_k \beta$,
- $M, s \models_k \alpha \wedge \beta$ iff $M, s \models_k \alpha$ and $M, s \models_k \beta$,
- $M, s \models_k \text{EX}\alpha$ iff $k > 0$ and $(\exists \pi \in \Pi_k(s)) M, \pi(1) \models_k \alpha$,
- $M, s \models_k \text{E}(\alpha \text{U}_I \beta)$ iff $(\exists \pi \in \Pi_k(s))(\exists 0 \leq m \leq k)(m \in I$ and $M, \pi(m) \models_k \beta$ and $(\forall 0 \leq j < m) M, \pi(j) \models_k \alpha)$,

- $M, s \models_k \mathrm{EG}_I \alpha$ iff $(\exists \pi \in \Pi_k(s))((k \geq right(I)$ and $(\forall j \in I)\ M, \pi(j) \models_k \alpha)$ or $(k < right(I)$ and $(\exists l \in loop(\pi))(\forall min(left(I), l) \leq j < k)\ M, \pi(j) \models_k \alpha))$,
- $M, s \models_k \overline{\mathrm{K}}_c \alpha$ iff $(\exists \pi \in \Pi_k(\iota))(\exists 0 \leq j \leq k)(M, \pi(j) \models_k \alpha$ and $s \sim_c \pi(j))$,
- $M, s \models_k \overline{\mathrm{Y}} \alpha$ iff $(\exists \pi \in \Pi_k(\iota))(\exists 0 \leq j \leq k)(M, \pi(j) \models_k \alpha$ and $s \sim \pi(j))$, where $\overline{\mathrm{Y}} \in \{\overline{\mathrm{D}}_\Gamma, \overline{\mathrm{E}}_\Gamma, \overline{\mathrm{C}}_\Gamma\}$ and $\sim \in \{\sim_\Gamma^D, \sim_\Gamma^E, \sim_\Gamma^C\}$.

A RTECTLK formula φ is *valid in model M with bound k* (denoted $M \models_k \varphi$) iff $M, \iota \models_k \varphi$, i.e., φ is $k-$true at the initial state of the model M. The *bounded model checking problem* asks whether there exists $k \in \mathbb{N}$ such that $M \models_k \varphi$.

By straightforward induction on the length of a RTECTLK formula φ we can show that the following two lemmas hold.

Lemma 1. *Given are a model M, a RTECTLK formula φ, and a bound $k \geq 0$. Then, for each s in M, $M, s \models_k \varphi$ implies $M, s \models_{k+1} \varphi$.*

Lemma 2. *Given are a model M, a RTECTLK formula φ, and a bound $k \geq 0$. Then, for each s in M, $M, s \models_k \varphi$ implies $M, s \models \varphi$.*

Lemma 3. *Given are a model M, and a RTECTLK formula φ. Then, for each s in M, $M, s \models \varphi$ implies that there exists $k \geq 0$ such that $M, s \models_k \varphi$.*

Proof (By induction on the length of φ). The lemma follows directly for the propositional variables and their negations. Next, assume that the hypothesis holds for all the proper sub-formulae of φ. If φ is equal to either $\alpha \wedge \beta$, $\alpha \vee \beta$, or $\mathrm{EX}\alpha$, then it is easy to check that the lemma holds. For the epistemic operators, i.e., $\varphi = \overline{\mathrm{K}}_c \alpha, \overline{\mathrm{E}}_\Gamma \alpha, \overline{\mathrm{D}}_\Gamma \alpha$, $\overline{\mathrm{C}}_\Gamma \alpha$, the proof is like in [17] (see Lemma 2). So, consider φ to be of the following forms:

- Let $\varphi = \mathrm{EG}_I \alpha$ and $M, s \models \varphi$. By the definition of the unbounded semantics we have that there exists path $\pi \in \Pi(s)$ such that $(\forall j \in I)(M, \pi(j) \models \alpha)$. We have to consider two cases for the form of the interval I.

(a) $right(I) < \infty$. By the inductive assumption we have that for each $j \in I$ there exists k_j such that $M, \pi(j) \models_{k_j} \alpha$. Let $k = max\{right(I), max\{k_j \mid j \in I\}\}$. It follows by Lemma 1 that for each $j \in I$, $M, \pi(j) \models_k \alpha$. Now, consider the prefix π_k of length k of π. We have that $\pi_k \in \Pi_k(s)$, so by Definition 1 we conclude that $M, s \models_k \mathrm{EG}_I \alpha$.

(b) $right(I) = \infty$. Since number of states of M is finite, there exists $k' \geq left(I)$ such that for some $l < k'$, $\pi(k') = \pi(l)$. Now, let $u = (\pi(0), \ldots, \pi(l-1))$, $v = (\pi(l), \ldots, \pi(k'-1))$, and $\pi' = u \cdot v^\omega$. It is clear that for each $j \in I$, $M, \pi'(j) \models \alpha$. Therefore, taking into consideration the form of π', we get that for each j such that $l \leq j \leq k'$, $M, \pi'(j) \models \alpha$. Now, by the inductive assumption we get that for each $j \in I$ there exists k_j such that $M, \pi'(j) \models_{k_j} \alpha$. Then, let k be the least natural number such that for each $j \in \{l, \ldots, k'\}$, $k \geq k_j$, and moreover $k = l + m(k' - l)$, for some natural number $m > 0$. It follows by Lemma 1 that for each $j \in \{l, \ldots, k\}$, $M, \pi(j) \models_k \alpha$. Now, consider the prefix π'_k of length k of π'. Obviously, $l \in loop(\pi'_k)$ and $M, \pi'_k(j) \models_k \alpha$, for each j such that $min\{left(I), l\} \leq j < k$. We have that $\pi_k \in \Pi_k(s)$, so by Definition 1 we conclude that $M, s \models_k \mathrm{EG}_I \alpha$.

– Let $\varphi = \mathrm{E}(\alpha \mathrm{U}_I \beta)$ and $M, s \models \varphi$. By the definition of the unbounded semantics we have that there exists path $\pi \in \Pi(s)$ and $m \in I$ such that $M, \pi(m) \models \beta$ and $(\forall j < m) M, \pi(j) \models \alpha$. Thus, by the inductive assumption we have that there exists k' such that $M, \pi(m) \models_{k'} \beta$ and for each $0 \leq j < m$ there exists k_j such that $M, \pi(j) \models_{k_j} \alpha$. It follows by Lemma 1 that $M, \pi(m) \models_k \beta$ and $M, \pi(j) \models_k \alpha$, where $k = max\{k', k_0, \ldots, k_m\}$. Now, consider the prefix π_k of length k of π. We have that $\pi_k \in \Pi_k(s)$. Since $m \in I$, we conclude that $M, s \models_k \mathrm{E}(\alpha \mathrm{U}_I \beta)$.

The following theorem states that for a given model and formula there exists a bound k such that the model checking problem ($M \models \varphi$) can be reduced to the bounded model checking problem ($M \models_k \varphi$). Its proof follows from Lemmas 2 and 3.

Theorem 1. *Let M be a model and φ a RTCTLK formula. Then, the following equivalence holds: $M \models \varphi$ iff there exists $k \geq 0$ such that $M \models_k \varphi$.*

Now we show how to reduce BMC for RTECTLK to the propositional satisfiability problem. This reduction allows us to use efficient SAT solvers to perform model checking. We begin by introducing a function f_k that gives a bound on the number of k-paths of M, which are sufficient to validate a given RTECTLK formula. Namely, the function $f_k : \text{RTECTLK} \rightarrow \mathbb{N}$ is defined as follows:

- $f_k(\mathbf{true}) = f_k(\mathbf{false}) = f_k(p) = f_k(\neg p) = 0$, where $p \in \mathcal{PV}$,
- $f_k(\alpha \wedge \beta) = f_k(\alpha) + f_k(\beta)$,
- $f_k(\alpha \vee \beta) = max\{f_k(\alpha), f_k(\beta)\}$,
- $f_k(Y\alpha) = f_k(\alpha) + 1$, for $Y \in \{\mathrm{X}, \overline{\mathrm{K}}_c, \overline{\mathrm{D}}_\Gamma, \overline{\mathrm{E}}_\Gamma\}$,
- $f_k(\mathrm{E}(\alpha \mathrm{U}_I \beta)) = k \cdot f_k(\alpha) + f_k(\beta) + 1$,
- $f_k(\mathrm{EG}_I \alpha) = (k+1) \cdot f_k(\alpha) + 1$,
- $f_k(\overline{\mathrm{C}}_\Gamma \alpha) = f_k(\alpha) + k$.

By straightforward induction on the length of a RTECTLK formula φ we can show that φ is k−true in M if and only if φ is k−true in M with a number of k−paths reduced to $f_k(\varphi)$.

The New Translation of RTECTLK to Propositional Formulae. Now we present our translation of a RTECTLK formula into a propositional formula. Given are a model $M = (\iota, S, T, \{\sim_c\}_{c \in Ag}, \mathcal{V})$, a RTECTLK formula φ, and a bound $k \geq 0$. It is well known that the main idea of the BMC method consists in translating the bounded model checking problem, i.e., $M \models_k \varphi$, to the problem of checking the satisfiability of the following propositional formula:

$$[M, \varphi]_k := [M^{\varphi, \iota}]_k \wedge [\varphi]_{M,k} \tag{1}$$

The formula $[M^{\varphi, \iota}]_k$ constrains the $f_k(\varphi)$ symbolic k-paths to be valid k-paths of M, while the formula $[\varphi]_{M,k}$ encodes a number of constraints that must be satisfied on these sets of k-paths for φ to be satisfied. Once this translation is defined, checking satisfiability of a RTECTLK formula can be done by means of a SAT-solver.

In order to define the formula $[M, \varphi]_k$ we proceed as follows. We assume that each state s of M is encoded by a bit-vector whose length, say r, depends on the number of agents' local states. Thus, each state s of M we can represent by a vector $w = (u_1, \ldots, u_r)$ of propositional variables (usually called *state variables*), to which we refer to as *symbolic states*. A finite sequence $(w_0, \ldots, w_k)$ of symbolic states is called a *symbolic k-path*. Since, in general, we may need to consider more than one symbolic k-path, we introduce a notion of the j-th symbolic k-path, which is denoted by $(w_{0,j}, \ldots, w_{k,j})$, where $w_{i,j}$ are symbolic states for $0 \leq j < f_k(\varphi)$ and $0 \leq i \leq k$. Note that the exact number of necessary symbolic k-paths depends on the checked formula φ, and it can be calculated by means of the function f_k.

For a given infinite set SV of state variables and a *valuation of state variables* $\sigma :$ $SV \rightarrow \{0,1\}$ (a *valuation* for short), its extension to vectors of states variables $\boldsymbol{\sigma} :$ $SV^r \rightarrow \{0,1\}^r$ is defined in the following way:

$$\boldsymbol{\sigma}((u_{j_1}, \ldots, u_{j_r})) = (\sigma(u_{j_1}), \ldots, \sigma(u_{j_r})).$$

In what follows for a symbolic state w, by $SV(w)$ we denote the set of all the state variables occurring in w.

Now, let w and w' be symbolic states such that $SV(w) \cap SV(w') = \emptyset$. We define the following propositional formulae:

- $I_s(w)$ is a formula over $SV(w)$ that is true for a valuation σ iff $\boldsymbol{\sigma}(w) = s$.
- $p(w)$ is a formula over w that is true for a valuation σ
  iff $p \in \mathcal{V}(\boldsymbol{\sigma}(w))$ (encodes a set of states of M in which $p \in \mathcal{PV}$ holds).
- $H(w, w')$ is a formula over $SV(w) \cup SV(w')$ that is true for a valuation σ
  iff $\boldsymbol{\sigma}(w) = \boldsymbol{\sigma}(w')$ (encodes equivalence of two global states).
- $H_c(w, w')$ is a formula over $SV(w) \cup SV(w')$ that is true for a valuation σ
  iff $l_c(\boldsymbol{\sigma}(w)) = l_c(\boldsymbol{\sigma}(w))$ (encodes equivalence of local states of agent c).
- $\mathcal{R}(w, w')$ is a formula over $SV(w) \cup SV(w')$ that is true for a valuation σ
  iff $(\boldsymbol{\sigma}(w), \boldsymbol{\sigma}(w')) \in T$ (encodes the transition relation of M).
- Let $j \in \mathbb{N}$, and I be an interval. Then $In(j, I) := \begin{cases} \textbf{true}, & \text{if } j \in I \\ \textbf{false}, & \text{if } j \notin I \end{cases}$.

The propositional formula $[M^{\varphi, \iota}]_k$ is defined over state variables in the set $\bigcup \{SV(w_{i,j}) \mid 0 \leq i \leq k \text{ and } 0 \leq j < f_k(\varphi)\}$, in the following way:

$$[M^{\varphi, \iota}]_k := I_\iota(w_{0,0}) \wedge \bigwedge_{j=0}^{f_k(\varphi)-1} \bigwedge_{i=0}^{k-1} \mathcal{R}(w_{i,j}, w_{i+1,j}) \tag{2}$$

The next step of the translation is the transformation of a RTECTLK formula φ into a propositional formula $[\varphi]_{M,k} := [\varphi]_k^{[0,0,F_k(\varphi)]}$, where $F_k(\varphi) = \{j \in \mathbb{N} \mid 0 \leq j < f_k(\varphi)\}$, and $[\varphi]_k^{[m,n,A]}$ denotes the translation of φ at the symbolic state $w_{m,n}$ using k-paths from the set A.

Following [22], to translate an RTECTLK formula with an operator Q (where Q $\in$ $\{\text{EX}, \text{EU}_I, \text{EG}_I, \overline{\text{K}}_1, \ldots, \overline{\text{K}}_n, \overline{\text{D}}_\Gamma, \overline{\text{E}}_\Gamma\}$), we want exactly one path to be chosen for translating the operator Q, and the remaining k-paths to be used to translate arguments of Q. To accomplish this goal we need some auxiliary functions. However, before we define them, we first recall a definition of a relation $\prec$ that is defined on the power set of $\mathbb{N}$ as follows: $A \prec B$ iff for all natural numbers x and y, if $x \in A$ and $y \in B$, then $x < y$. Notice that from the definition of $\prec$ it follows that $A \prec B$ iff either $A = \emptyset$ or $B = \emptyset$ or $A \neq \emptyset$, $B \neq \emptyset$, $A \cap B = \emptyset$ and $max(A) < min(B)$. Now, let $A \subset \mathbb{N}$ be a finite nonempty set, $k, p \in \mathbb{N}$, and $m \in \mathbb{N}$ such that $m \leq |A|$. Then,

- $g_l(A, m)$ denotes the subset B of A such that $|B| = m$ and $B \prec A \setminus B$.
- $g_r(A, m)$ denotes the subset C of A such that $|C| = m$ and $A \setminus C \prec C$.
- $g_s(A)$ denotes the set $A \setminus \{min(A)\}$.
- If $k+1$ divides $|A|-1$, then $h_{\mathrm{G}}(A, k)$ denotes the sequence $(B_0, \ldots, B_k)$ of subsets of $A \setminus \{min(A)\}$ such that $\bigcup_{j=0}^{k} B_j = A \setminus \{min(A)\}$, $|B_0| = \ldots = |B_k|$, and $B_i \prec B_j$ for every $0 \leq i < j \leq k$. If $h_{\mathrm{G}}(A, k) = (B_0, \ldots, B_k)$, then $h_{\mathrm{G}}(A, k)(j)$ denotes the set B_j, for every $0 \leq j \leq k$.
- If k divides $|A| - 1 - p$, then $h_{\mathrm{U}}(A, k, p)$ denotes the sequence $(B_0, \ldots, B_k)$ of subsets of $A \setminus \{min(A)\}$ such that $\bigcup_{j=0}^{k} B_j = A \setminus \{min(A)\}$, $|B_0| = \ldots = |B_{k-1}|$, $|B_k| = p$, and $B_i \prec B_j$ for every $0 \leq i < j \leq k$. If $h_{\mathrm{U}}(A, k, p) = (B_0, \ldots, B_k)$, then $h_{\mathrm{U}}(A, k, p)(j)$ denotes the set B_j, for every $0 \leq j \leq k$.

In order to explain the purpose of the auxiliary functions defined above let us recall that each RTECTLK formula φ will be translated by using a set of exactly $f_k(\varphi)$ symbolic k-paths. In what follows we will say that a set A of positive natural numbers is used to translate a formula φ instead of saying that the set of symbolic k-paths with indices in A is used to translate the formula φ. Moreover, saying that a set A is used to translate a formula φ assumes that $|A| = f_k(\varphi)$.

The function g_l is used in the translation of the formulae with the main connective being a disjunction: for a given RTECTLK formula $\alpha \vee \beta$, if a set A is to be used to translate this formula, then the set $g_l(A, f_k(\alpha))$ is used to translate the subformula α and the set $g_l(A, f_k(\beta))$ is used to translate the subformula β.

The functions g_l and g_r are used in the translation of the formulae with the main connective being a conjunction: for a given RTECTLK formula $\alpha \wedge \beta$, if a set A is to be used to translate this formula, then the set $g_l(A, f_k(\alpha))$ is used to translate the subformula α and the set $g_r(A, f_k(\beta))$ is used to translate the subformula β.

The function g_s is used in the translation of the formulae with the main connective $\mathrm{Q} \in \{\mathrm{EX}, \overline{\mathrm{K}}_1, \ldots, \overline{\mathrm{K}}_n, \overline{\mathrm{D}}_\Gamma, \overline{\mathrm{E}}_\Gamma\}$: for a given RTECTLK formula $\mathrm{Q}\alpha$, if a set A is to be used to translate this formula, then the path of the number $min(A)$ is used to translate the operator Q and the set $g_s(A)$ is used to translate the subformula α.

The function h_{G} is used in the translation of the formulae with the main connective EG_I: for a given RTECTLK formula $\mathrm{EG}_I\alpha$, if a set A is to be used to translate this formula, then the path of the number $min(A)$ is used to translate the operator EG_I and the set $h_{\mathrm{G}}(A, k)(j)$, for every $0 \leq j \leq k$, is used to translate the formula α at the j-th symbolic state of the symbolic k-path of the number $min(A)$. Notice that if $k + 1$ does not divide $|A| - 1$, then $h_{\mathrm{G}}(A, k)$ is undefined. However, for every set A such that $|A| = f_k(\mathrm{EG}_I\alpha)$, it follows from the definition of f_k that $k + 1$ divides $|A| - 1$.

The function h_{U} is used in the translation of the formulae with the main connective EU_I: for a given RTECTLK formula $\mathrm{E}(\alpha \mathrm{U}_I \beta)$ if a set A is to be used to translate this formula, then the path of the number $min(A)$ is used to translate the operator EU_I, the set $h_{\mathrm{U}}(A, k, f_k(\beta))(j)$, for every $0 \leq j \leq k$, is used to translate the formula β at the symbolic state $w_{j,min(A)}$ and the set $h_{\mathrm{U}}(A, k, f_k(\beta))(i)$, for every $0 \leq i < j$, is used to translate the formula α at the symbolic state $w_{i,min(A)}$. Notice that if k does not divide $|A| - 1 - p$, then $h_{\mathrm{U}}(A, k, p)$ is undefined. However, for every set A such that $|A| = f_k(\mathrm{E}(\alpha \mathrm{U}_I \beta))$, it follows from the definition of f_k that k divides $|A| - 1 - f_k(\beta)$.

Definition 2 (Translation of RTECTLK formulae). *Let φ be a RTECTLK formula, and $k \geq 0$ a bound. We define inductively the translation of φ over path number $n \in F_k(\varphi)$ starting at symbolic state $w_{m,n}$ as shown below.*

- $[\mathbf{true}]_k^{[m,n,A]} := \mathbf{true}$,   • $[\mathbf{false}]_k^{[m,n,A]} := \mathbf{false}$,
- $[p]_k^{[m,n,A]} := p(w_{m,n})$,   • $[\neg p]_k^{[m,n,A]} := \neg p(w_{m,n})$,
- $[\alpha \wedge \beta]_k^{[m,n,A]} := [\alpha]_k^{[m,n,g_l(A,f_k(\alpha))]} \wedge [\beta]_k^{[m,n,g_r(A,f_k(\beta))]}$,
- $[\alpha \vee \beta]_k^{[m,n,A]} := [\alpha]_k^{[m,n,g_l(A,f_k(\alpha))]} \vee [\beta]_k^{[m,n,g_l(A,f_k(\beta))]}$,
- $[\mathrm{EX}\alpha]_k^{[m,n,A]} :=$
    - (1) $H(w_{m,n}, w_{0,min(A)}) \wedge [\alpha]_k^{[1,min(A),g_s(A)]}$, if $k > 0$
    - (2) $\mathbf{false}$,   *otherwise*
- $[\mathrm{E}(\alpha\mathrm{U}_I\beta)]_k^{[m,n,A]} := H(w_{m,n}, w_{0,min(A)}) \wedge \bigvee_{i=0}^{k}([\beta]_k^{[i,min(A),h_{\mathrm{U}}(A,k,f_k(\beta))(k)]}$
    $\wedge\ In(i,I) \wedge \bigwedge_{j=0}^{i-1}[\alpha]_k^{[j,\,min(A),h_{\mathrm{U}}(A,k,f_k(\beta))(j)]})$,
- $[\mathrm{EG}_I\alpha]_k^{[m,n,A]} := H(w_{m,n}, w_{0,min(A)})\wedge$
    - (1) $\bigwedge_{j=left(I)}^{right(I)}[\alpha]_k^{[j,min(A),h_{\mathrm{G}}(A,k)(j)]}$, if $right(I) \leq k$
    - (2) $\bigvee_{l=0}^{k-1}(H(w_{k,min(A)}, w_{l,min(A)}) \wedge \bigwedge_{j=min(left(I),l)}^{k-1}[\alpha]_k^{[j,min(A),h_{\mathrm{G}}(A,k)(j)]})$,
      otherwise.
- $[\overline{\mathrm{K}}_c\alpha]_k^{[m,n,A]} := I_\iota(w_{0,min(A)}) \wedge \bigvee_{j=0}^{k}([\alpha]_k^{[j,min(A),g_s(A)]} \wedge H_c(w_{m,n}, w_{j,min(A)}))$,
- $[\overline{\mathrm{D}}_\varGamma\alpha]_k^{[m,n,A]} := I_\iota(w_{0,min(A)}) \wedge \bigvee_{j=0}^{k}([\alpha]_k^{[j,min(A),g_s(A)]}$
    $\wedge \bigwedge_{c\in\varGamma} H_c(w_{m,n}, w_{j,min(A)}))$,
- $[\overline{\mathrm{E}}_\varGamma\alpha]_k^{[m,n,A]} := I_\iota(w_{0,min(A)}) \wedge \bigvee_{j=0}^{k}([\alpha]_k^{[j,min(A),g_s(A)]}$
    $\wedge \bigvee_{c\in\varGamma} H_c(w_{m,n}, w_{j,min(A)}))$,
- $[\overline{\mathrm{C}}_\varGamma\alpha]_k^{[m,n,A]} := [\bigvee_{j=1}^{k}(\overline{\mathrm{E}}_\varGamma)^j\alpha]_k^{[m,n,A]}$.

The following two lemmas state the correctness and the completeness of the new translation respectively. Lemma 4 claims that if some valuation V satisfies the translation of a RTECTLK formula for some k, then this formula is k-true at the state corresponding to the valuation V. On the other hand, Lemma 5 claims that if a RTECTLK formula is, for some k, k-true at some state of the model, then its translation is a satisfiable propositional formula.

Let $[M]_k^{F_k(\varphi)} = \bigwedge_{j=0}^{f_k(\varphi)-1} \bigwedge_{i=0}^{k-1} \mathcal{R}(w_{i,j}, w_{i+1,j})$. From now on, for every RTECTLK formula φ and every subformula α of φ, we denote by $[\alpha]_k^{[\varphi,m,n,A]}$ the propositional formula $[M]_k^{F_k(\varphi)} \wedge [\alpha]_k^{[m,n,A]}$. We shall write $\sigma \Vdash \xi$ if the valuation σ satisfies the propositional formula ξ. Moreover, we shall write $s \models_k \alpha$ instead of $M, s \models_k \alpha$ and $\sigma_{i,j}$ instead of $\sigma(w_{i,j})$.

Lemma 4 (Correctness of the translation). *Let M be a model, φ be a RTECTLK formula and $k \in \mathbb{N}$. Then for every subformula α of the formula φ, every $A \subseteq F_k(\varphi)$ such that $|A| = f_k(\alpha)$, every $(m,n) \in \{(0,0)\} \cup \{0, \dots, k\} \times \mathbb{N}_+$ and every valuation σ such that $\sigma_{m,n}$ is a state of M the following condition holds: if $\sigma \Vdash [\alpha]_k^{[\varphi,m,n,A]}$, then $M, \sigma_{m,n} \models_k \alpha$.*

Proof. The proof is analogous to the proof of Lemma 3.1 from [22] and use induction on the complexity of α.

Lemma 5 (Completeness of the translation). *Let M be a model, φ be a* RTECTLK *formula and $k \in \mathbb{N}$. Then for every subformula α of the formula φ, every $A \subseteq F_k(\varphi)$ such that $|A| = f_k(\alpha)$, every $(m,n) \in \{(0,0)\} \cup \{0,\ldots,k\} \times (\mathbb{N}_+ \setminus A)$ and every state s of M the following condition holds: if $M, s \models_k \alpha$, then there exists a valuation σ such that $\boldsymbol{\sigma}_{m,n} = s$ and $\sigma \Vdash [\alpha]_k^{[\varphi,m,n,A]}$.*

Proof. The proof is analogous to the proof of Lemma 3.2 from [22] and use induction on the complexity of α.

The next theorem easily follows from Lemmas 4 and 5.

Theorem 2. *Let M be a model, and φ a* RTECTLK *formula. Then for every $k \in \mathbb{N}$, $M \models_k \varphi$ if, and only if, the propositional formula $[M, \varphi]_k$ is satisfiable.*

Proof. Recall that $[M,\varphi]_k = I(w_{0,0}) \wedge [M]_k^{F_k(\varphi)} \wedge [\varphi]_k^{[0,0,F_k(\varphi)]} = I(w_{0,0}) \wedge [\varphi]_k^{[\varphi,0,0,F_k(\varphi)]}$. If $M \models_k \varphi$, then $M, \iota \models_k \varphi$. By Lemma 5, there exists a valuation σ such that $\boldsymbol{\sigma}_{0,0} = \iota$ and $\sigma \Vdash [\varphi]_k^{[\varphi,0,0,F_k(\varphi)]}$. Hence, the propositional formula $[M, \varphi]_k$ is satisfiable.

If the propositional formula $[M, \varphi]_k$ is satisfiable, then there exists a valuation σ such that $\sigma \Vdash I(w_{0,0}) \wedge [\varphi]_k^{[\varphi,0,0,F_k(\varphi)]}$. Thus, $\boldsymbol{\sigma}_{0,0} = \iota$, and by Lemma 4 it follows that $M, \boldsymbol{\sigma}_{0,0} \models_k \varphi$. Hence, $M \models_k \varphi$.

Now, from Theorems 1 and 2 we get the following

Corollary 1. *Let M be a model, and φ a* RTECTLK *formula. Then, $M \models \varphi$ if, and only if, there exists $k \in \mathbb{N}$ such that the propositional formula $[M, \varphi]_k$ is satisfiable.*

4 Experimental Results

In this section we consider two scalable multi-agent systems, and present performance evaluation of both SAT-based BMC algorithms for RTECTLK, the new and the old one. Unfortunately, we cannot compare our experimental results to others, simply because, to the best of our knowledge, there is no tool implementing the BMC method for RTECTLK. In order to appraise the behaviour of our new algorithm, we have tested it on several RTECTLK properties. An evaluation of both BMC algorithms, which have been implemented in C++ under the same semantics (i.e., interleaved interpreted systems), is given by means of the running time, the memory used, the number of generated variables and clauses. All the benchmarks together with an instruction how to reproduce our results can be found at the webpage http://ajd.czest.pl/~modelchecking/.

For the tests we have used a computer equipped with AMD phenom(tm) 9550 Quad-Core 2200 MHz processor and 4 GB of RAM, running Ubuntu Linux with kernel version 2.6.35-28-generic-pae, and we have set the timeout to 15000 seconds, and memory limit to 3072 MB. We have used the state of the art SAT-solver MiniSat 2 [6,5].

Generic Pipeline Paradigm. The first benchmark we consider is a generic pipeline paradigm (GPP) [16], which consists of three parts: Producer producing data, Consumer receiving data, and a chain of n intermediate Nodes that transmit data produced by Producer to Consumer. The local states for each agent (Producer, Consumer, and intermediate Nodes), and their protocols are shown on Fig. 1. The evaluation of both BMC algorithms for RTECTLK with respect to the GPP system has been done by means of the following RTECTLK specifications:

$$\varphi_1 = \mathrm{EF}_{[0,\infty]}\,(ProdSend \wedge \mathrm{EG}_{[a,\infty]}\overline{\mathrm{K}}_C\overline{\mathrm{K}}_P(Received)),\ \text{where}\ a = 2n+1,\ n \geq 1;$$
$$\varphi_2 = \mathrm{EF}_{[0,\infty]}\overline{\mathrm{K}}_P(ProdSend \wedge \mathrm{EF}_{[0,3]}(Received));$$
$$\varphi_3 = \mathrm{EF}_{[0,\infty]}\overline{\mathrm{K}}_P(ProdSend \wedge \mathrm{EF}_{[n,n+3]}(Received)),\ \text{where}\ n \geqslant 1.$$

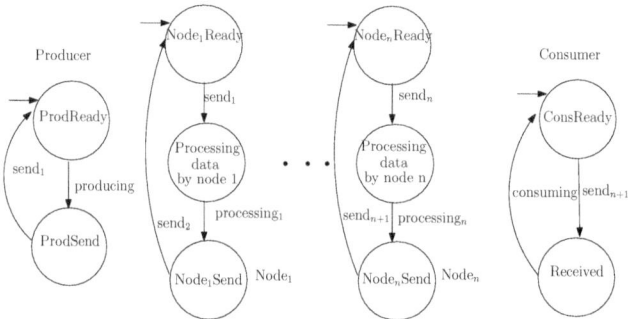

Fig. 1. The GPP system

The formula φ_1 states that it is not true that if Producer generates a product, then ultimately at time later or equal a Consumer knows that Producer does not know that Consumer has the product. The formula φ_2 expresses that it is not true that Producer knows that if he produces a product, then always within the first three time units later Consumer does not have the product. The formula φ_3 represents that it is not true that Producer knows that if he generates a product, then always within interval $[n, n+3]$ Consumer does not have the product. The above formulae are true in the model for GPP.

A train controller system (TC). The second benchmark we consider is a standard train controller (TC) system [12], which consists of n trains (for $n \geq 2$) and a controller. In the system it is assumed that each train uses its own circular track for travelling in one direction. At one point, all trains have to pass through a tunnel, but because there is only one track in the tunnel, trains arriving from each direction cannot use it simultaneously. There are traffic lights on both sides of the tunnel, which can be either red or green. All trains notify the controller when they request entry to the tunnel or when they leave the tunnel. The controller controls the colour of the traffic lights.

The local states for each agent (trains and the controller), and their protocols are shown on Fig. 2. The evaluation of both BMC algorithms for RTECTLK with respect to the TC system has been done by means of the following RTECTLK specifications:

$$\varphi_1 = \mathrm{EF}_{[0,\infty]}(\overline{\mathrm{K}}_{Train_1}(InTunnel_1 \wedge \mathrm{EG}_{[1,\infty]}(\neg InTunnel_1))),$$
$$\varphi_2 = \mathrm{EF}_{[0,\infty]}(InTunnel_1 \wedge \overline{\mathrm{K}}_{Train_1}(\mathrm{EG}_{[1,n+1]}(\textstyle\bigwedge_{i=1}^{n}(\neg InTunnel_i)))),$$

where n is the number of considered trains.

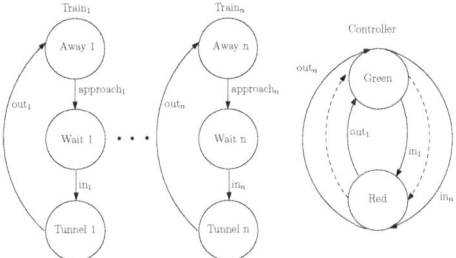

Fig. 2. Agents for train controller system

The formula φ_1 states that it is not true that it is always the case that agent Train 1 knows that whenever he is in the tunnel, it will be in the tunnel once again within a bounded period of time, i.e., within n time units for $n \geq 1$. The formula φ_2 represents that it is not true that it is always the case that if Train 1 is in the tunnel, then he knows that either he or other train will be in the tunnel during the next $n + 1$ time units. All above formulae are true in the model for TC.

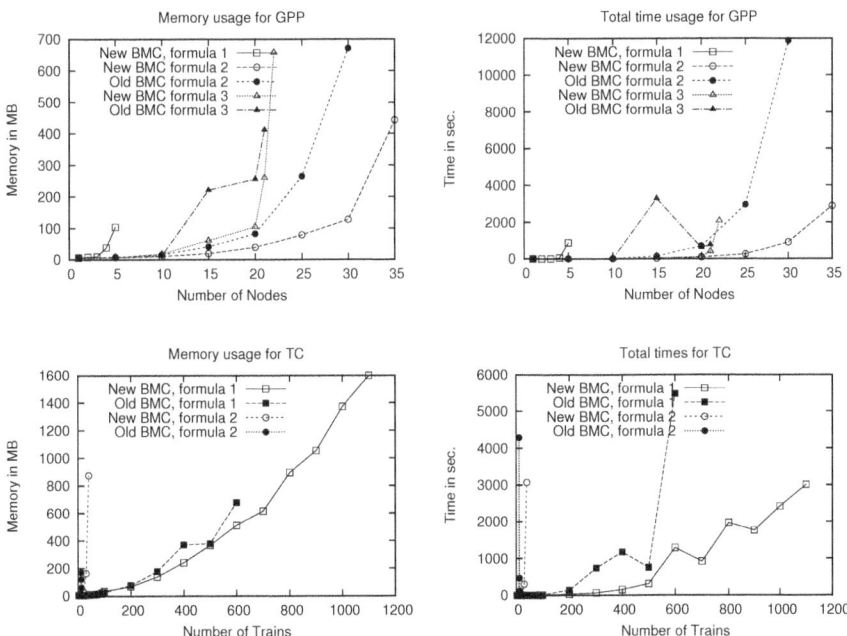

Performance Evaluation. The experimental results show that the improved BMC method for RTECTLK outperforms the old BMC method for RTECTLK in both the memory consumption and the execution time (as shown below in the line plots). This is so, because the new method produces a significantly smaller set of clauses (as shown in Table 1), and the SAT solver is given this smaller set. The reason for this is that the new translation does not use all the $f_k(\varphi)$ k-paths both for the formula φ and for each of its proper subformulea. Moreover, the produced set of clauses is not only smaller, but also easier for a SAT solver, which further boosts the performance of the improved BMC method. Therefore using the old translation only smaller systems can be model-checked. Notice, that for the GPP system it was even impossible to generate the translation of the formula φ_1 by using the old method.

As is well known, the SAT-based BMC method is divided into two steps: in the first step the set of clauses is generated that describes both the checked model and the formula under consideration, and in the second step the SAT solver checks the satisfiability of the generated set of clauses. Both methods, the new and the old, consumed comparable memory in the first step, but the old method took significantly more time in the second step. Also, the memory consumed by the SAT solver was significantly larger for the set of clauses generated by the old method, which suggests that the propositional formula obtained by the old method is bigger and more complicated in comparison with the one generated by the new method.

Table 1. Results for selected witnesses generated by the new and old BMC translations

Checked formula	Which translation	(Max) number of components	Length of the witnesses	Number of paths	Number of variables	Number of clauses
$gpp - \varphi_1$	new	1	5	14	3210	9153
$gpp - \varphi_1$	new	5	13	30	61926	178381
$gpp - \varphi_2$	old	30	63	3	443621	1359604
$gpp - \varphi_2$	new	30	63	3	301286	887929
$gpp - \varphi_2$	new	35	73	3	444741	1312844
$gpp - \varphi_3$	old	21	24	3	95015	290305
$gpp - \varphi_3$	new	21	24	3	65957	192865
$gpp - \varphi_3$	new	22	25	3	73532	215155
$tc - \varphi_1$	old	500	4	3	1716818	5159566
$tc - \varphi_1$	new	500	4	3	1647670	4918036
$tc - \varphi_1$	new	1100	4	3	7584670	22699036
$tc - \varphi_2$	old	13	14	3	29123	89905
$tc - \varphi_2$	new	13	14	3	17140	49765
$tc - \varphi_2$	new	40	41	3	217372	638392

5   Conclusions

In the paper we have presented a compact and elegant bounded semantics for RTECTLK, an improved SAT-based BMC algorithm for RTECTLK properties of interleaved interpreted systems, and we have shown the correctness of both, the new bounded semantics for RTECTLK and the new BMC encoding. Further, we have implemented, tested, and compared with each other on two standard benchmarks both BMC translations for RTECTLK, the new and the old one. Our experiments have shown that the new translation is clearly superior; it is much faster, and consumes less memory.

Our future work will involve an implementation of the method also for other models of multi-agent systems, for example for standard interpreted systems. Moreover, we are going to define a BDD-based BMC algorithm for RTECTLK, and compare it with the method presented in this paper.

References

1. Biere, A., Cimatti, A., Clarke, E., Fujita, M., Zhu, Y.: Symbolic model checking using SAT procedures instead of BDDs. In: Proc. of DAC 1999, pp. 317–320 (1999)
2. Biere, A., Heljanko, K., Junttila, T., Latvala, T., Schuppan, V.: Linear Encodings of Bounded LTL Model Checking. Logical Methods in Computer Science 2(5:5), 1–64 (2006)
3. Blackburn, P., de Rijke, M., Venema, Y.: Modal Logic. Cmbridge Tracts in Theoretical Computer Science, vol. 53. Cambridge University Press (2001)
4. Clarke, E.M., Grumberg, O., Peled, D.A.: Model Checking. The MIT Press, Cambridge (1999)
5. Eén, N., Sörensson, N.: MiniSat, http://minisat.se/MiniSat.html
6. Eén, N., Sörensson, N.: MiniSat - A SAT Solver with Conflict-Clause Minimization. In: Bacchus, F., Walsh, T. (eds.) SAT 2005. LNCS, vol. 3569. Springer, Heidelberg (2005)
7. Emerson, E.A., Sistla, A.P., Mok, A.K., Srinivasan, J.: Quantitative temporal reasoning. Real-Time Systems 4(4), 331–352 (1992)
8. Fagin, R., Halpern, J.Y., Moses, Y., Vardi, M.Y.: Reasoning about Knowledge. MIT Press, Cambridge (1995)
9. Fagin, R., Halpern, J.Y., Vardi, M.Y.: What can machines know? On the properties of knowledge in distributed systems. Journal of the ACM 39(2), 328–376 (1992)
10. Halpern, J.Y., Vardi, M.Y.: The complexity of reasoning about knowledge and time 1: lower bounds. Journal of Computer and System Sciences 38(1), 195–237 (1989)
11. van der Hoek, W., Wooldridge, M.J.: Model checking knowledge and time. In: Bošnački, D., Leue, S. (eds.) SPIN 2002. LNCS, vol. 2318, pp. 95–111. Springer, Heidelberg (2002)
12. van der Hoek, W., Wooldridge, M.: Cooperation, knowledge, and time: Alternating-time temporal epistemic logic and its applications. Studia Logica 75(1), 125–157 (2003)
13. Huang, X., Luo, C., van der Meyden, R.: Improved Bounded Model Checking for a Fair Branching-Time Temporal Epistemic Logic. In: van der Meyden, R., Smaus, J.-G. (eds.) MoChArt 2010. LNCS (LNAI), vol. 6572, pp. 95–111. Springer, Heidelberg (2011)
14. Kacprzak, M., Lomuscio, A., Niewiadomski, A., Penczek, W., Raimondi, F., Szreter, M.: Comparing BDD and SAT based techniques for model checking Chaum's dining cryptographers protocol. Fundamenta Informaticae 63(2,3), 221–240 (2006)
15. Lomuscio, A., Penczek, W., Qu, H.: Partial order reduction for model checking interleaved multi-agent systems. In: AAMAS, pp. 659–666. IFAAMAS Press (2010)
16. Peled, D.: All from one, one for all: On model checking using representatives. In: Courcoubetis, C. (ed.) CAV 1993. LNCS, vol. 697, pp. 409–423. Springer, Heidelberg (1993)
17. Penczek, W., Lomuscio, A.: Verifying epistemic properties of multi-agent systems via bounded model checking. Fundamenta Informaticae 55(2), 167–185 (2003)
18. Penczek, W., Woźna, B., Zbrzezny, A.: Bounded model checking for the universal fragment of CTL. Fundamenta Informaticae 51(1-2), 135–156 (2002)
19. Raimondi, F., Lomuscio, A.: Automatic verification of multi-agent systems by model checking via OBDDs. Journal of Applied Logic 5(2), 235–251 (2005); Special issue on Logic-based agent verification
20. van der Meyden, R., Su, K.: Symbolic model checking the knowledge of the dining cryptographers. In: Proc. of CSFW 2004, pp. 280–291. IEEE Computer Society, Los Alamitos (2004)
21. Woźna-Szcześniak, B.: Bounded model checking for the existential part of Real-Time CTL and knowledge. In: Pre-Proc. of CEE-SET 2009, pp. 178–191. AGH Krakow, Poland (2009)
22. Zbrzezny, A.: Improving the translation from ECTL to SAT. Fundamenta Informaticae 85(1-4), 513–531 (2008)

Building Spatiotemporal Emotional Maps
for Social Systems

Pedro Catré, Luis Cardoso, Luis Macedo, and Amílcar Cardoso

University of Coimbra,
Centre for Informatics and Systems of the University of Coimbra, Department of Informatics
Engineering, Polo II, 3030 Coimbra, Portugal
{catre,lfac}@student.dei.uc.pt, {macedo,amilcar}@dei.uc.pt

Abstract. In this paper we present a system that gathers from a society of
agents their emotional arousal and self-rated motivations as well as their
location in order to plot a map of a city or geographical region with information
about the motivational and emotional state of the agents that inhabit it. This tool
provides the visualization of the agent's reactions to the external world as well
as the ultimate reasons behind them, namely the goals and beliefs, making it
possible to know for instance where a society feels stressed and excited and
which desires are predominant in specific parts of a geographical region. With a
careful analysis of this information, further conclusions may be reached about
particular aspects of the agent society.

Keywords: Social systems visualization tool, emotion and motivation map.

1 Introduction

Understanding what happens in the interplay between specified individual behaviours
and the interactions among the agents of the simulated population and between the
agents and the environment is of great importance for social simulation. In fact, those
interactions make emerge a collective output. Thus, the emotions resulting from those
interactions are a valuable aspect to explore in order to improve methods and tools
for visualizing agent-based social simulation. The advances in psychological and
philosophical theories of emotion, and more recently in affective computing [1]
provide ways to better understand those emotions in a computational manner, namely
their causes. Therefore, not only the emotional information and its analysis can be
visualized, but also their explanation, which may provide a more complete
information of a social system. According to the cognitive theory of emotion [2], the
dominant theory of emotion, emotions result from the confrontation between agents'
desires/goals and the new beliefs [3], [4]. Computational models such as the OCC [5]
or CBDTE [4] provide information on the nature of those confrontations for each
specific emotion, and therefore, if taken into account, provide more detailed
information behind the emotion and thereby a more complete information about the
interactions among the agents and between the agents and the environment of the
social system.

L. Antunes and H.S. Pinto (Eds.): EPIA 2011, LNAI 7026, pp. 566–580, 2011.
© Springer-Verlag Berlin Heidelberg 2011

Bio Mapping/Emotional Mapping is a methodology and tool for visualizing people's reactions to the external world [6]. Typically this consists of some sort of Bio Sensor capable of calculating the wearer's Galvanic Skin Response, which is a simple indicator of emotional arousal, in conjunction with their geographical location. That sensor is attached to each individual in a group in order to gather data that enables the creation of a map of emotion arousal. What this implies is that this tool has immense potential and can be used in as many contexts, recreational or otherwise, such as art, community development, science research, architectural planning and large scale political consultations [6]. Even though the most recent maps created by Christian Nold present data with greater detail and even context, no information is presented on the actual emotions, goals and beliefs of the individuals at each moment in time. Moreover, such a methodology involves later analysis of the data that was gathered. To our knowledge, there are no tools that can process data retrieved from emotional agents and generate a spatiotemporal map visualizing both the spatiotemporal position and the mental state of the agents.

In this paper we present a system that gathers from several agents their emotional arousal and self-rated motivations in conjunction with the location of the agents in order to plot a two dimensional map of the average level of arousal, kinds of emotions and motivations on a city or region. By gathering this data, such a system can construct maps for visualizing agent's reactions to the external world as well as their causes (agents' goals and beliefs). This way it is possible to know, for instance, where a community feels stressed and excited and which desires are predominant in specific parts of a city. We also describe a simulation module created to generate data for testing the spatiotemporal mapper. This simulation module plays a simple test scenario and it is composed by emotional agents (agents that generate emotions) that receive information about the world from a master agent that represents the environment and send the generated data to this agent. The emotional component of the agents implements the computational model of the Belief-Desire Theory of Emotions (BDTE) described in recent literature [3], [4]. Several reasons justify our choice of BDTE, namely that it provides plausible answers to central explanatory challenges posed by emotional experience, including: the phenomenal quality, intensity and object-directedness of emotional experience, the function of emotional experience and its relation to cognition and motivation, and the relation between emotional experience and emotion, while at the same time it manages to avoid most objections that have been raised against cognitive theories of emotion.

The rest of the paper is structured as follows. Section 2 describes the emotional mapper and its implementation. Section 3 presents an example of a possible simulation module to be used with the system. The experiments and results are presented in Section 4, and Section 5 discusses those results. In Section 6 we summarize the most important ideas.

2   The Emotional Mapper

The Emotional Mapper is a data visualization tool that provides a wide variety of emotion data visualizations using data sets with information about agents' goals, beliefs, geographic location, and new beliefs acquired by the agents when interacting with the environment. These data should be produced by a social simulator or captured from a real social system. This means the Emotional Mapper was created to be an independent module that can be reused freely so any application or user can produce a data file and provide it for the creation of a map. The map is only plotted after all data is gathered, which means the visualization is not synchronized with the data generation. The main reason for this was the importance of making the Emotional Mapper independent from any other applications.

To construct the map, the module need to receive, for each time unit (frame), the position, the identifier (*id*) and intensity of the emotions, goals and beliefs of each agent. This information will be passed in an XML file. Information about the emotions is required at all times, while the other (goals and beliefs) is optional. Also, every agent that appears in the first frame must be present in all subsequent frames.

The map generated shows the complete set of agents in a determined instant of time. It also allows the user to navigate the 2D map (zoom in/out, panning) as well as the time dimension (play, pause, stop, seek, change frame rate). An image file can be defined as background. Also, it is possible to export images of the maps created.

By clicking an agent, the user can view its position, emotions, goals and beliefs as well as graphics of the evolution of this data. All those graphics can be combined with each other for comparison (even from different agents). Each agent is identified by unique identifier (*id*). In the map, an agent is represented by a circle (the size can be defined) and the color is that of the most intense emotion. The lightness of the color is adapted to the intensity of the emotion. There is a default emotion color mapping and this mapping can be updated and expanded. There is also the possibility of viewing a global graph that shows the average of each emotion (for all agents) for every time unit. In all the graphs we can zoom in/out and export to an image file. Also, for better analysis of the data, the plots are constructed in synchrony with the map (there is an option to show the entire plot).

The main components of the Emotional Mapper and their dependencies are shown in Fig. 1. As can be seen, the XML interpreter only depends of the input file and thus can be used in other applications. Moreover, the well-defined dependencies help to maintain and use the modules in other applications.

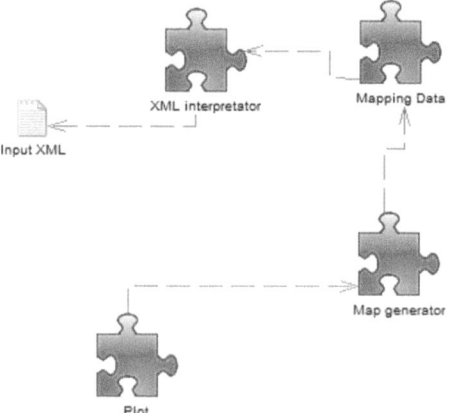

Fig. 1. Emotional mapper's main components

The structure of the XML input file has its basis on raw video codecs: it is divided in frames, each one representing a time unit and containing all the information necessary to render that particular frame. This aproach simplifies both the implementation of the application itself and the creation of input files. For the map generation, the Qt Graphics View Framework was used which provides a highly scalable 2D drawing surface and various operations (e.g. scaling and rotating). To synchronize the "playback" we used a timer: at each tick we advance one frame. This approach allowed us to easily implement play, pause and stop as well as enabling us to change the frame rate on the fly. The agent's circle color is calculated by changing the lightness (Hsl color model) of the original color using the Formula 1.

$$newLightness = originalLightness + \frac{(1 - (emotionIntensity \times 0{,}7 + 0{,}3))}{2} \qquad (1)$$

We can see the resulting effect in Fig. 2 where the agent with the grey circumference is happy with intensity 1 and the other agent is happy with intensity 0.2.

Fig. 2. Agents showing the emotion happy with different intensities

For each agent we can plot its emotions, goals and beliefs separately. The emotions are plotted as dots and the goals and beliefs as continuous lines. Various series of different types (emotions, goals, beliefs) and from different agents can be combined.

It is worth mentioning that graphic visualization of data is animated to be constructed in relation to the animation of the simulation. However, it is possible to skip this animation to the complete construction of the graphic by clicking in "Show All". Also, as already mentioned, any data can be plotted on the same graphic by checking the Add box before clicking the Plot button (this is not limited to the data from a single agent, you can plot together different data from several agents).

More details of the Emotional Mapper, including a video and illustrative figures can be found on http://student.dei.uc.pt/~lfac/stem/.

3   An Example of a Simulation Module

To illustrate the use of the Emotional Mapper, we developed an agent-based simulation system with the Jadex Framework [7] that produces the data required by the Emotional Mapper. The application is composed by two types of agents: master and emotional (worker). The master agent collects information to be stored in a data sink and acts like a sensory input for all the emotional agents in the simulation being responsible for sending that information and providing a numerical value with the percentage of certainty or probability that the information provided is true. This makes the master agent, among other things, the environment component of this simulation. The emotional (worker) agents' job is to generate data – the current emotions, beliefs, goals, and position of the agent. They do not interact with each other, except with the master.

The implementation of the emotional component of the emotional agents follows the outline of the computational explication of the BDTE [3]. In this emotion theory, beliefs and desires are the causes of emotions. Our belief-desire system is a propositional representation system hardwired with mechanisms that monitor, and, if necessary, update the system in response to newly acquired information (new beliefs). The monitoring mechanisms compare the newly acquired beliefs to existing beliefs and desires. The outputs of these information-processing mechanisms are "non-propositional signals" [3], [8] that carry information about the degree of a match or mismatch of a newly acquired belief with existing beliefs and desires.

In our approach, propositions are represented by the following components: a sentence in mental language, a strength value representing the degree of belief in the proposition (in the interval (0, 1)); a desire value representing the degree of desire of the proposition (in the interval (-100, 100)), and a flag informing if the sentence is affirmative or negative. Typically, we have this flag set to true since it simplifies the simulation without removing any of its potential to represent a given proposition. To illustrate this proposition representation, consider the proposition <the door is closed> <0.6><70><false>, which means one believes the door is not closed (because the flag is set to false corresponding to a negation of the sentence) with a strength of belief of 0.6 (more than 0.5, 50%) and a desire of 70. Such proposition would carry the same information as the proposition <the door is closed><0.4><-70><true>, which means one believes the door is closed with a strength of 0.4 which is less than 0.5 (50%) corresponding to a disbelief in the sentence (so I actually do not believe the door is closed), and a desire of -70 that the door is closed which is the same as a desire of 70 that the door is opened (not closed). The values for the intensity of an emotion are normalized to the interval between 0 and 1.

To generate emotions we implemented the Belief-Desire Comparator (BDC) [3], a mechanism that compares a newly acquired belief to the preexisting desires looking for a match or mismatch (a match means that a desire has been fulfilled whereas a mismatch means that a desire has been frustrated), and the Belief-Belief Comparator (BBC) [3], a mechanism that compares the newly acquired belief to the preexisting beliefs for a match or mismatch (here a match means that a preexisting belief is confirmed by the new information, whereas a mismatch means that a preexisting belief is disconfirmed).

Belief and desire update occurs considering the output of the BDC-BBC system in response to the newly acquired belief. It consists in adding new beliefs and desires to the system, as well as to abandon existing beliefs if they turn out to be false, and existing desires if they are fulfilled. For instance, when a proposition reaches belief strength 1 (one is 100% sure it is true) then its desire is fulfilled, and thereby the value of the strength of this desire is updated to 0. On the other hand, abandoning existing beliefs when they turn out to be false means that we update their belief strength value to 0. However, we do not remove those beliefs from the agents belief store because in our proposition representation they carry relevant information. For example, if the strength of a belief is 0 we know that it is false and that is the information we do want to keep.

As previously mentioned, beliefs and desires are modeled as quantitative variables and used to calculate the intensity of emotions. The BDC and the BBC make use of these values to calculate the degrees of belief-desire and belief-belief match. The intensity of an emotion is given by these degrees. There is a threshold value that is a parameter of the simulation for deciding if an emotion is intense enough to be considered relevant. Table 1 shows the qualitative formulation of the Belief-Desire Theory of Emotion, in which is described the mapping between the variables "Belief at time t", "Desire at time t", and "Belief at time t-1" and the generated emotion (for more details about the qualitative and quantitative formulation of the BDTE see [3]).

Table 1. Belief-Desire Theory of Emotion, Qualitative Formulation. Table adapted from [3].

Emotion	Belief at t	Desire at t	Belief at t-1
happiness(p, t)	Certain(p, t)	Des(p, t)	
unhappiness(p, t)	Certain(p, t)	Des(¬p, t)	
hope(p, t)	Uncertain(p, t)	Des(p, t)	
fear(p, t)	Uncertain(p, t)	Des(¬p, t)	
surprise(p, t)	Certain(p, t)	irrelevant	Bel(¬p, t-1)

Notation: Bel(p, t) → believes p at time t; Certain(p, t) → firmly believes (strength=1) p at t
Uncertain(p, t) = Bel(p, t) & ¬Certain(p, t) & ¬Certain(¬p, t)
Des(p, t) → desires p at t; Des(¬p, t) → desire not-p at t (≈is aversive against p at t)

The BDC and BBC mechanisms are implemented as modified pattern matchers [3], [9]. New beliefs are compared in parallel with all preexisting beliefs and desires in working memory for identity versus opposition of content. In this case this means a new proposition is compared to all propositions in working memory with respect to identity or contrariness of the element sentence of each. If a match or a mismatch between two sentences is detected, then the belief strength of the new proposition is integrated with the belief and desire values by means of a hardwired procedure (in our case conditional statements) that generates one or more nonpropositonal output signals. These signals are a representation of the degree of "expectedness versus unexpectedness" of the new proposition (computed as a function of the belief strength of both the belief in the working memory and the new belief) and the degree of "desire versus undesire" of the new proposition. The output signals can also be a combination of these two comparisons like in the case of disappointment [3].

When an agent is integrating a new belief in the working memory, and that belief already exists, the belief strength is recomputed as the average of the strengths of the two beliefs. Here we use the calculation of the relative frequency of a belief to update the belief strength, in other words the more frequent an event is the more likely it is that it occurs. For example, if 9 out of 10 times one sees the door open one is more inclined to believe (without looking) that the door is opened right now. If the proposition is new in the working memory we simply add it to the system.

We also made considerations concerning other issues that were suggested in [3], [4]. One of the most important decisions is concerned with the emotion revision that occurs all the time the BDC-BBC system produces an output. The new intensity values of each emotion have to be updated considering the previous values. In addition there is the problem of how to integrate opposite emotions. To solve this problem, we assume that it would not make sense for an agent to have opposite emotions at the same time (e.g. an agent cannot be happy and unhappy simultaneously) or for an agent to have the same emotion repeated in is current set of emotions. Therefore we created a mechanism for overriding emotions that works as follows:

- The intensity of an emotion is reinforced positively when it occurs again according to Equation 2. Simply adding the values of both intensities would not be coherent since typically these would cause unrealistic situations with emotions having exaggerated intensities. Note that, since we normalized the values, there is no emotion with a negative intensity. We also limit the result of the *New Intensity* to the maximum value of 1. A more sophisticated approach which is not in the scope of this paper could better adapt this calculation to the characteristics of each agent. Our concern here was simply to eliminate any inconsistency through a coherent method.
- If opposite emotions are detected the most intense emotion remains but its intensity is subtracted by the value of the intensity of the discarded emotion.

$$\text{New Intensity} = \text{Highest Intensity} + \frac{\text{Lowest Intensity}}{10} \tag{2}$$

We considered that hope, fear, and surprise could coexist and are not opposite to any other emotions our system could generate. On the other hand happy/unhappy, disappointed/relieved seem unable to coexist. These assumptions can easily be changed if further analysis reveals that there are better alternatives.

Finally, we deal with the fact that the intensity of an emotion decreases with time. We followed the approach for response decay described in [1], [10], which explains how emotions felt by the agent decay, in absence of any new specific stimulus. When a new emotion is created its initial intensity and time of creation is stored. The emotional agent updates the intensity of the emotion using a negative exponential model as follows:

$$\text{True Intensity} = \text{original intensity} \times e^{\left(-\frac{\text{Current Time}}{100}\right)} \tag{3}$$

4 Experiment

4.1 Setup

We carried out an experiment comprising the two applications: the emotional mapper application and the simulation application. These two modules are completely independent. The simulation module executes a simulation creating a data file that can be used on the mapper module for analysis of the data. Fig. 3 illustrates the system's topology:

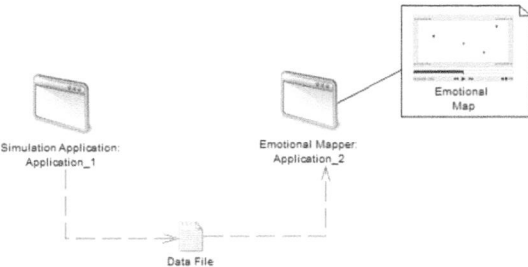

Fig. 3. Experimental setup

As the diagram shows, the two modules are completely decoupled and any other application or user can produce a data file and provide it to the Emotional Mapper for the creation of a map. As it was already mentioned in Section 2, the map is only plotted after all data is gathered.

We defined a simple scenario in the simulation module in which an emergency occurs forcing the agents involved to run to the exit. Using the simulation application we can define new beliefs that are going to be communicated from the master, i.e., the environment, to the emotional agents. The structure of the messages is <moment><ID>request_addNewBelief<sentence><strength><desire><affirmation> in which:

- *Moment* defines the iteration when the information is going to be transmitted (a simulation lasts a pre-specified number of iterations);
- *ID* is the agent's unique identifier;
- *request_addNewBelief* is the identification of the request;
- *Sentence* is the information to transmit in mental language;
- *Strength* is the value of certainty of the belief (between 0 and 1);
- *Desire* is the value of desire between -100 and 100;
- *Affirmation* is a flag informing if the sentence is affirmative or negative.

In a similar way it is possible to inject emotions to a specific agent in a defined iteration with a message <moment><ID> injectEmotion <type><intensity>. If the emotion already exists it is updated with the new intensity.

The parameters of the simulations are:

- The exit to where all agents converge is the point x = 0.0 and y = 0.0 (corresponding to the lower left corner);
- There are 40 agents in the simulation.

Additionally we defined a list of sensory input to be fed to pre-specified agents (with identifiers 1, 2, 3, and 4) in the simulation and also a request for injecting an emotion in one (agent 4) of the 40 agents, as follows:

- 2 1 request_addNewBelief the_exit_door_is_open 0.05 80 true
- 10 1 request_addNewBelief the_exit_door_is_open 1 80 true
- 20 1 request_addNewBelief the_fire_is_spreading 0.6 -60 true
- 30 1 request_addNewBelief the_fire_is_spreading 1 -60 true

- 3 2 request_addNewBelief the_building_is_collapsing 0.7 -75 true
- 10 2 request_addNewBelief the_building_is_collapsing 0 -75 true
- 4 3 request_addNewBelief the_emergency_response_unit_arrived 0.6 60 true
- 15 3 request_addNewBelief the_emergency_response_unit_arrived 0 60 true
- 3 4 injectEmotion happy 0.98 0 true

Finally, for the purpose of this experiment, all agents begin with a current emotion of type fear, with intensity 0.8 and a belief in working memory with the following contents:

- Sentence: get.to.the.door
- Strength: 0.2
- Desire: 100
- Affirmation: true

This belief is actually more of a goal with corresponding strength and desire, but we represent it this way as a simplification.

At the moment an agent reaches the exit point, it gets a new belief:

- Sentence: get.to.the.door
- Strength: 1.0
- Desire: ---
- Affirmation: true

4.2   Results

Fig. 4 shows an image of the map generated.

Fig. 4. Map of the simulation in a given time frame

We plotted the emotions and beliefs (a combination of this data in a multiple plot) for agents 1, 2, 3 and 4 (Fig. 5-8):

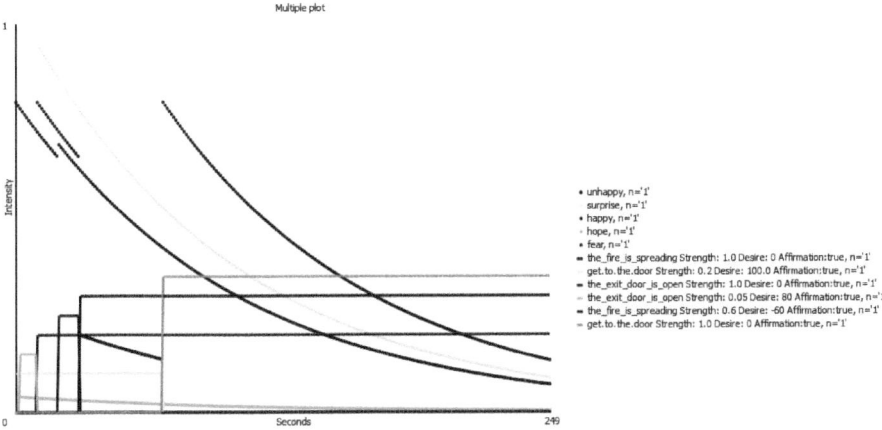

Fig. 5. Plot of the emotions and beliefs for agent 1

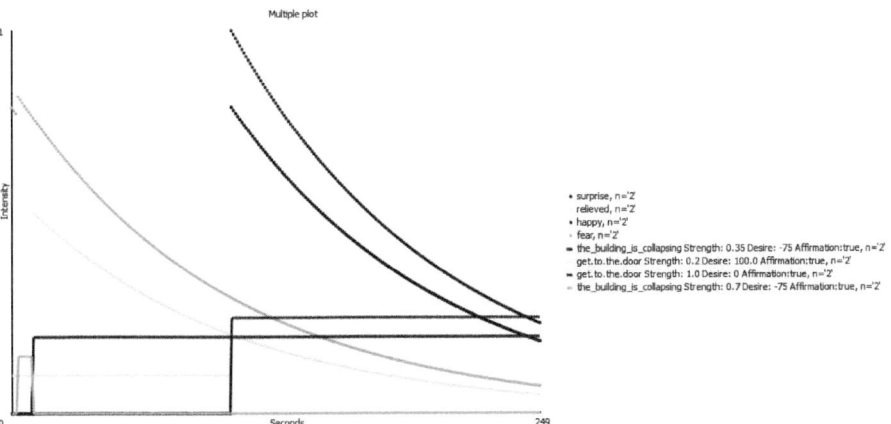

Fig. 6. Plot of the emotions and beliefs for agent 2

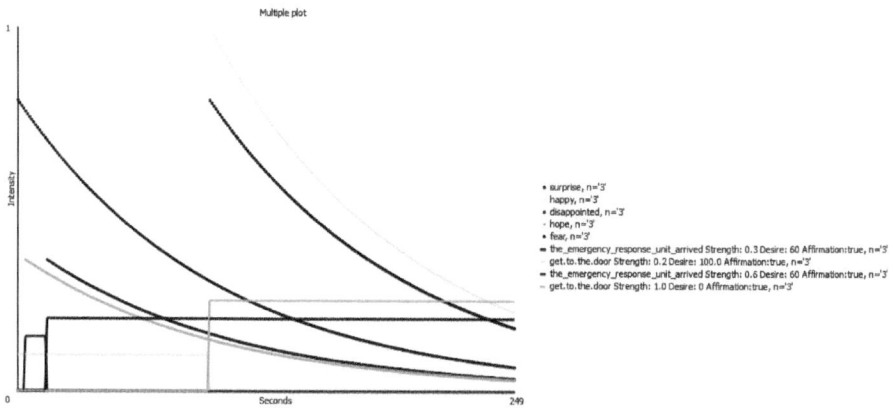

Fig. 7. Plot of the emotions and beliefs for agent 3

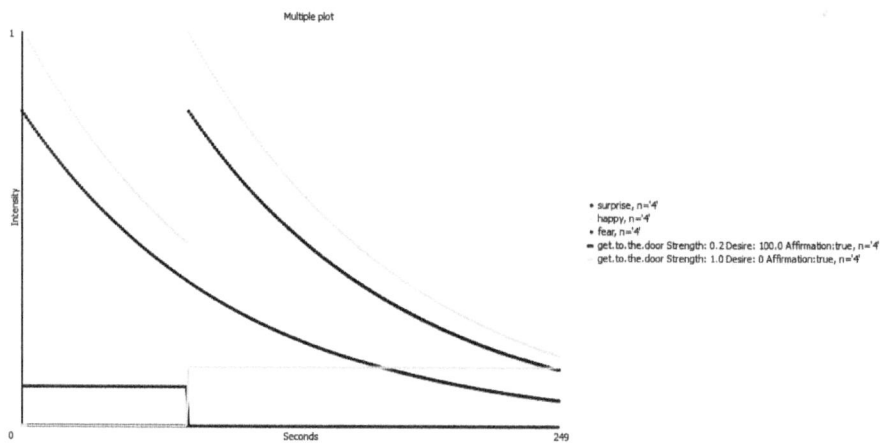

Fig. 8. Plot of the emotions and beliefs for agent 4

Also, we plotted the graph of agent 14 (Fig. 9), who did not get any sensory input from the master (this was an agent chosen randomly), we plotted the emotions graph of agent 1 and 4 together (Fig. 10), and also the graphs for the average of all emotions of all agents in the simulation (Fig. 11):

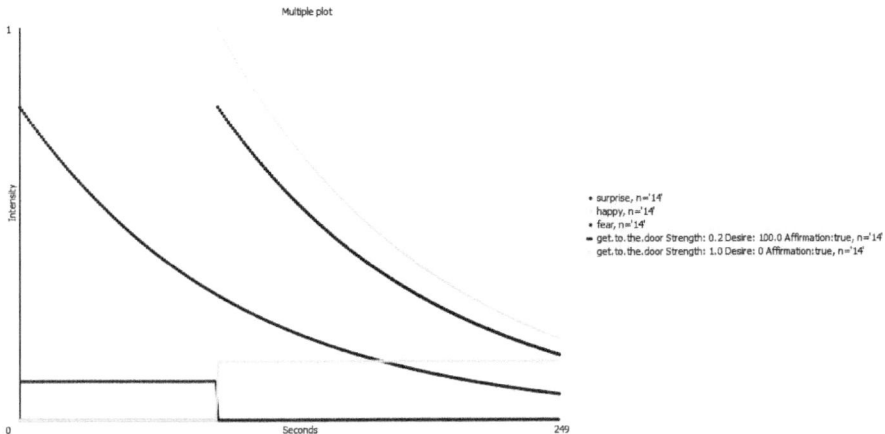

Fig. 9. Plot of the emotions and beliefs for agent 14

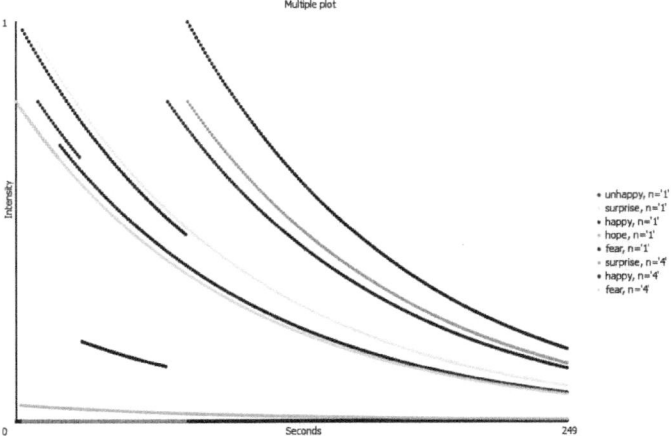

Fig. 10. Plot of the combined emotions for agents 1 and 4

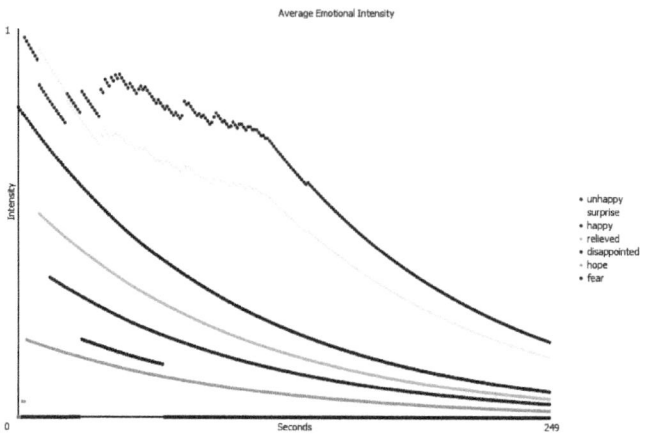

Fig. 11. Plot of the average intensity of the emotions of all agents

Analyzing Fig. 5 we can see that when agent 1 receives the second sensory input (message 2 in the list) it gets happy and surprised. In fact it did not expect the exit door to be open (initial strength 0.05, message 1) and it intensely desired this (desire of 80). Note also that the desire of this belief is revised to 0 because it was fulfilled. This is shown in the caption of the graph on Fig. 5.

When agent 1 first learns the fire might be spreading (message 3) it is slightly more afraid (it was already afraid to begin with). And when it learns the fire is definitely spreading it is unhappy about this, which overrides the felling of happiness it previously had. From the same figure you can also notice the unhappiness is overridden by a feeling of happiness when the agent gets to the exit.

In Fig. 6, upon receiving message 5, the agent learns the building might be collapsing (notice the fear increases), but when it receives message 6, it is relieved to find that this is not the case.

In Fig. 7 the agents believes the emergency response unit has arrived when it receives sensory input 7. However message 8 informs it this is not the case so it is disappointed.

Fig. 8 shows agent 4 was injected with an emotion of type happiness with intensity 0.98. Observe also that, although the intensity of the emotion is decaying with time, it goes back to a very high intensity when the agent gets to the exit.

Fig. 9 is only meant to represent all the other agents in the simulation who did not receive sensory input. They begin the simulation afraid and eventually get to the exit which is surprising and also generates happiness.

Fig. 10 intends to show the ability of the mapper to combine plots from different agents. We can see how such an analysis would be useful for determining, for example, that agent 4 had higher intensity of happiness most of the time, however it was never hopeful or unhappy and it did not experience so much fear during the simulation.

In all graphs it is possible to realize the decay of emotions. The graph of Fig. 11 also shows this for the overall simulation. Note how the unhappy emotion only exists for a limited period of time. In fact, agent 1 is the only one who feels unhappy and it only generates that emotion when it receives message 4 (the fire is definitely spreading) and that emotion is overridden when it gets to the exit point. Another interesting thing to notice is the erratic behavior of surprise and happy curves in the first half of the simulation and also how these curves show some similarities. This was expected since these emotions are felt by all the agents in the simulation, as all of them get to the exit door and are surprised and happy about it, and they do so, mostly, in the first half of the simulation. In the second half most of the agents already got to the exit so the average is much more balanced and makes a smoother curve.

6 Discussion and Conclusions

In summary this paper presents a system which, when given the result of a simulation or real-world data, is able to plot a map of the average level of emotion intensity, kinds of emotions and motivations on a given geographic area. The results provided by our system is intuitive and makes it easy to explore the data available to conduct studies with practical purposes. Additionally, the user has access to the coordinates, goals, beliefs, and emotions of any agent at any moment in time. The modules composing our system are created to be easily integrated with other systems and also easily extended for implementation of novel features.

The mapper has the potential to be used in a number of contexts from real-world data analysis to investigation and testing of emotional models. The experiments made using a simulator implementation based on the BDTE model provided an immediate example of a valid and realistic use for our system, namely its worth as a tool for supporting the development, testing, experimenting and observation of an emotional model. We believe the rich set of analysis features offered by our tool as well as its ease of use enables a meaningful user experience.

There are several interesting aspects that can be extended in our system, such as that it would be helpful adding support for other time units. It would also be interesting to allow geographic coordinates and even to automatically download satellite images of the area the agents are in. The plots could also use some improvements in order to provide additional features for more powerful analysis and greater usability.

Acknowledgments. The authors gratefully acknowledge the support and advice provided by Rainer Reisenzein.

References

1. Picard, R.: Affective Computing. The MIT Press, Cambridge (2000)
2. Scherer, K.R., Schorr, A., Johnstone, T.: Appraisal processes in emotion: Theory, methods, research. Oxford University Press, Oxford (2001)
3. Reisenzein, R.: Emotions as Metarepresentational States of Mind: Naturalizing the Belief-Desire Theory of Emotion. Cognitive Systems Research 10(1), 6–20 (2009)

4. Reisenzein, R.: Emotional Experience in the Computational Belief-Desire Theory of Emotion. Emotion Review 1(3), 214–222 (2009)
5. Ortony, A., Clore, G.L., Collins, A.: The cognitive structure of emotions. Cambridge University Press, Cambridge (1988)
6. Bio Mapping / Emotion Mapping, http://www.biomapping.net/
7. Jadex BDI Agent System, http://jadex-agents.informatik.uni-hamburg.de/xwiki/bin/view/BDI+Tutorial/01+Introduction
8. Oatley, K., Johnson-Laird, P.N.: Towards a cognitive theory of emotions. Cognition and Emotion 1, 29–50 (1987)
9. Anderson, J.R., Lebiere, C.: The atomic components of thought, 1st edn. Psychology Press (1998)
10. Carofiglio, V., de Fiorella, R., Roberto, G.: Mixed Emotion Modeling. Department of Informatics, University of Bari (2000)

An Exploratory Study on the Impact of Temporal Features on the Classification and Clustering of Future-Related Web Documents

Ricardo Campos[1,2,4], Gaël Dias[1,3], and Alípio Jorge[4]

[1] HULTIG, University of Beira Interior, Covilhã, Portugal
[2] Polytechnic Institute of Tomar, Tomar, Portugal
[3] DLU/GREYC, Univeristy of Caen Basse-Normandie, Caen, France
[4] LIAAD – INESC Porto LA, University of Porto, Porto, Portugal
`ricardo.campos@ipt.pt, ddg@di.ubi.pt, amjorge@fc.up.pt`

Abstract. In the last few years, a huge amount of temporal written information has become widely available on the Internet with the advent of forums, blogs and social networks. This gave rise to a new challenging problem called future retrieval, which consists of extracting future temporal information, that is known in advance, from web sources in order to answer queries that combine text of a future temporal nature. This paper aims to confirm whether web snippets can be used to form an intelligent web that can detect future expected events when their dates are already known. Moreover, the objective is to identify the nature of future texts and understand how these temporal features affect the classification and clustering of the different types of future-related texts: informative texts, scheduled texts and rumor texts. We have conducted a set of comprehensive experiments and the results show that web documents are a valuable source of future data that can be particularly useful in identifying and understanding the future temporal nature of a given implicit temporal query.

Keywords: Temporal Information Retrieval, Prospective Search, Temporal Web Mining, Temporal Classification, Temporal Clustering.

1 Introduction

With the advent of the World Wide Web, a huge amount of temporal data became available on the Internet ready to be exploited. This gave rise to the emergence of a new research area called Temporal Information Retrieval (T-IR). The purpose of this research area is to detect temporal data in documents and rank Internet search results based on temporal information. Alongside this, a new topic called future retrieval (FR) was introduced by Baeza-Yates [2] in 2005, with the specific goal of searching for future temporal references within web documents in order to answer queries that combine text and time. Such a system should include three components: (1) an information extraction module that recognizes temporal expressions, (2) an information retrieval system that indexes articles together with time segments and (3) a text mining system that given a time query, finds the most important topics associated with that time segment. An example of this type of system would be a

L. Antunes and H.S. Pinto (Eds.): EPIA 2011, LNAI 7026, pp. 581–596, 2011.
© Springer-Verlag Berlin Heidelberg 2011

system that would return information, such as *Dacia Coming in 2012* or *Dacia plans 8 new models by 2015,* for the query *Dacia*. This would benefit a number of users who are looking for future-related contents. Despite the relevance of this topic, little research has been conducted on using temporal information features for future search purposes, and the only known temporal analytics engine is Recorded Future [13].

1.1 Motivation

Although we cannot know the future, a lot can be deduced about it by mining huge collections of texts such as weblogs and microblogs (e.g. *Twitter, Facebook*). It is possible to look for events that are planned in advance, based on existing information. The following sentences show examples of three types of texts: informative texts, texts about scheduled events and rumors.

1. *Sony Ericsson Yendo Release Postponed for February 2011 Due to Software Issues.* (Informative – not predictable)
2. *The 2022 FIFA World Cup will be the 22nd FIFA World Cup, an international football tournament that is scheduled to take place in 2022 in Qatar.* (Schedule - predictable)
3. *Avatar 2? Arriving in 2013? James Cameron intends to complete his next film, another 3D epic, within three to four years.* (Rumor)

Based on this information we could, for example, decide whether or not to buy a property given the presumed tax increase, or to re-direct business negotiations based on economic predictions. Understanding the future temporal intent of web documents is therefore of the utmost importance. This is a particularly difficult task that has been mostly supported by a reliable collection of web news articles, annotated with a timestamp and that mainly consists of informative and scheduled texts. However this can also be based on several other types of sources such as web documents. In contrast to web news articles, web documents, especially those from social networks, suffer however from the problem of containing a large number of comments, predictions or plans, all expressed by means of rumors. This has led some authors, such as Adam Jatowt et al. [9], to question its credibility. However, what apparently seems to be a drawback, can actually constitute a great opportunity to infer the user's interests. For example, James Cameron may discover that people are interested in another 3D Avatar movie; mobile companies may redirect their core business to the development of mobile applications due to the growth of this industry that is expected to reach an impressive $35 billion by 2014; environmentalists may be interested to know that the easyJet airline plans to cut CO_2 emissions by 50% by 2015.

Despite the relevance of this information, none of the proposed works to date have focused on this type of future-related documents, either of an informative nature, or rumors or scheduled texts. Therefore, developing an effective model to classify web documents with regard to their future intent, based on the temporal features incorporated in the text, is extremely important. Consequently, two challenging issues need to be considered: (1) Do web documents have enough temporal information for future analysis? (2) Can text classification and clustering be improved based on the existing future-related information in web documents? The aim is to answer these questions in this paper. We are particularly interested in considering a specific type of query, the so-called implicit temporal query, which as stated by Campos et. al. [5], constitutes approximately 35% of all queries. This work takes place within the context of ephemeral clustering. The goal is to develop a language independent solution. As a consequence, this research focuses on the identification and extraction of numerical dates with years. Moreover, the study is based on the analysis of web content, rather than on a metadata-based approach. As noted by Klaus et al. [3], this is an interesting direction for future research, for which there is not yet a clear solution. Our study is based on web snippets which can be a powerful data source for future prediction as they reflect the views of society, as this paper will demonstrate.

1.2 Overview

The motivation behind this research is that: (1) to the best of our knowledge, this is the first work based on comprehensive future data analysis with web documents as a data source and implicit temporal queries (2) it is also the first work that aims to understand the impact of temporal features on both classification and clustering based on three genres of future-related texts (informative texts, scheduled texts and rumors).

Extensive experiments have been conducted to perform both types of analysis. In particular, the distribution of year dates present in snippets, in titles and their respective URLs was studied. Five different classification algorithms (Naive Bayes, Multinomial Naive Bayes, K-NN, Weighted K-NN, Multi-Class SVM) and one clustering algorithm (K-means) were then used to explore the main ideas. It must be noted that the main objective is not to attain a higher level of accuracy in the results, but instead to understand the impact of temporal features on different learning paradigms. The results of this paper show that web snippets are a valuable source of future data that can be particularly useful in identifying and understanding the future temporal nature of a given implicit temporal query. This exploratory study also shows that to some extent, the use of temporal features has an impact on the classification and clustering of future-related texts.

2   Related Work

Little research has been conducted so far in this area. However, there are some studies that do focus on this domain. Kira Radinsky et al. [12] use patterns in web search queries to predict whether an event will appear in tomorrow's news. Gilad Mishne et al. [11] predict movie sales through blogger sentiment analysis. Yang Liu et al. [10], focus on the same line of research and attempt to predict sales performance.

Indeed, it seems that only Baeza-Yates [2] and Jatowt et al. [8] [9] are concerned with Future Retrieval in a more general way. For example, Baeza-Yates [2] was the first to define this problem and to introduce a basic method for searching, indexing and ranking documents according to their future features. Each document is represented by a tuple consisting of a time segment and a confidence probability that measures whether the event will actually happen or not in this time segment. On the other hand, Jatowt et al. [9] propose the generation of visual summaries of future-related information on user queries using two methods. The first method is based on calculating the probability of the next instances of periodical events appearing in the future, through the analysis of past data, such as the statistics on document creation over time from the Google News Archive search engine. The second method relies on the analysis of explicit future-related information contained in documents. Future events are clustered using a partitioning clustering algorithm in order to answer queries on named entities, such as the names of people or places. For that purpose, they propose the linear interpolation between two documents d_i and d_j as a new time-related similarity measure. This is illustrated in Equation 1.

$$Dist\left(d_i, d_j\right) = (1 - \beta).TermDist\left(d_i, d_j\right) + \beta.TimeDist(d_i, d_j) \tag{1}$$

The best results in terms of precision occur for $\beta = 0.2$. Consequently, it is clear that the impact of future-related features is relatively reduced. Finally, Adam Jatowt et al. [8] conducted an exploratory analysis of future-related information on the Internet. For that purpose, they gathered a set of queries in English, composed of temporal expressions. The queries (873.054) with a year reference ranging from 2010 to 2050 belong to the yearly dataset and the queries (39.312) with a month reference and a year ranging from 2010 to 2011 belong to the monthly dataset. Each query is then executed on Bing which is set to retrieve up to 1000 results, resulting in two sets of 1.044.224 and 770.715 unique Internet search results. Their analysis relies on the average number of hits obtained from the search engine for all of the queries and they show that (1) future-related information clearly decreases after a few years, with some occasional peaks, and that (2) most of the near future-related contents are related to expected international events. However, distant years are mostly linked to predictions and expectations that relate to issues such as the environment and climate change.

3   Measuring the Future Temporal Nature of Web Documents

This section assesses whether web snippets are a valuable source of data that can help deduce the future temporal intent of queries that do not specify a year. Unlike Jatowt et al. [8], the analysis is not based on the execution of future temporal explicit queries (queries with temporal expressions), but it is based on implicit queries. Subsequently, restrictions have not been placed on the language, type and topic of the query. Furthermore, this analysis is not based on the number of hits reported by the search engine, but on the detection and manual analysis of future dates that occur within the set of results retrieved. Moreover, in accordance with the work produced by Jatowt et al. [9], the impact of introducing future features on the process of classifying and clustering future-related web contents will be studied. However, unlike this work, more than 20 queries are used, and each text is classified according to three possible genres: informative web snippets, scheduled texts and rumors. This framework consists of four steps. Fig. 1. outlines the overall evaluation framework.

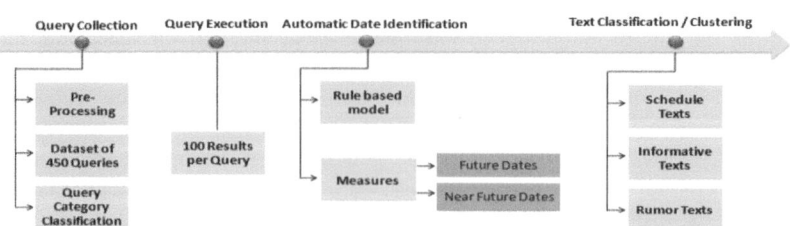

Fig. 1. Overview of the framework

3.1 Query Collection and Query Execution

Our dataset was collected by crawling the web in response to a set of diverse queries (see Table 1) selected from Google Insights for Search [7], which registers the most common searches performed worldwide. These queries are from the period January-October 2010. Twenty queries were manually selected for each of the 27 categories, which resulted in a total number of 540 queries. Since this research is looking at how web snippets temporally behave towards implicit temporal queries, explicit queries have not been included. Therefore, the final query collection consists of 450 queries (without duplicates) mostly belonging to the categories of the Internet.

Table 1. Example of queries

dacia duster	toyota recall	hairstyles	unemployment
avatar	lady gaga	tour de france	bank of america

To build the dataset, Yahoo! and Bing APIs were used and defined to retrieve a total of 200 web results per query resulting in a unique set of 62.842 web snippets. Each web snippet includes a title, a text (also known as a snippet), and a link, known as the URL.

3.2 Automatic Date Identification

In order to extract the temporal information from these data, a pattern matching methodology was performed, as proposed in [5]. Since the aim is to detect dates in the form of numerical years, to make the search efficient, and to keep the system language-independent, a custom built rule-based model was developed, which achieves results of almost 96% accuracy in the detection of dates within documents, particularly within titles and web snippets [5]. From this labeled data set, 5.777 web snippets, titles and URLs containing year dates were extracted. Each text was then manually labeled as indicative of a *near* or *distant future* purpose depending on the dates found in the text. If the date identified was from 2011, the text was labeled as a *near future* intention. If the date was later than 2011, or if the text had both a near and a distant future date, the text was labeled as having a *distant future* nature. The function *NearFuture*, is computed to all the queries, as the ratio between the number of dates retrieved labeled with the year 2011, divided by the total number of dates retrieved (see Equation 3).

$$NearFuture(q) = \frac{\#\ 2011\ Dates\ Retrieved}{\#\ Future\ Dates\ Retrieved} \tag{3}$$

In order to understand the future temporal value of each item more clearly and determine its value more easily, a basic measure represented by the function *FutureDates(q)* was defined, which is computed as the ratio between the number of dates retrieved with a future nature, for a given query *q*, divided by the number of dates retrieved for the same query (see Equation 2).

$$FutureDates(q) = \frac{\#\ Future\ Dates\ Retrieved}{\#\ Dates\ Retrieved} \qquad (2)$$

A date in a document is considered of a future nature, if, regardless of the document timestamp, the focus time (the time of the content) is superior to the reading time (the time of the query). In this paper, the reading time is December 2010. As such, all years later than 2010 are considered future dates in the web snippets (e.g. *in 2014 the World Cup...*), with regard to the execution of an implicit query (e.g., *World Cup*). The final dataset consists of 508 web snippets, 419 titles and 195 URLs containing future dates. This data set will make it possible to perform classification and clustering tasks in order to understand the impact of temporal features.

3.3 Text Classification

Each of these texts (508 web snippets, 419 titles, 195 URLs) was then manually classified by three annotators who were asked to place them in three future temporal classes: informative texts, schedule texts or rumor texts. Fleiss' Kappa statistic [6] was used in order to measure the consistency between the different annotators. Results show Kappa was found to be 0.93, meaning an almost perfect agreement between the raters. Most of the differences occur in the classification of rumor and scheduled texts.

3.4 Data Analysis and Discussion of Results

This section outlines a number of issues on future temporal web mining analysis. The issues discussed include for example, the temporal value of future dates with regard to a given future year, the frequency of occurrence in a near future temporal window, related categories and text genres. Unlike conventional T-IR systems, where the amount of temporal information available is relatively significant, in a future retrieval system, values are naturally lower. That is perfectly clear in Table 2, where from a total number of 62.842 web snippets retrieved, 5.777 have temporal features and only 508 are of a future nature. This means that 9.2% of the web snippets contain year dates, but only 0.81% contain future dates. This is due to the fact that people talk about the past more than the future, which hampers the extraction of information in large quantities from a future retrieval system. Overall, these results would clearly be higher if the execution of explicit temporal queries had also been included (e.g. *hairstyles 2010* and subsequently *hairstyles 2011*). A recent work [5] has shown that 3.49% of the queries in a web query log collection have a future temporal intent, and that this value is inherently linked to the occurrence of higher results.

Nevertheless, it must be noted that the nature of a search in a conventional system is naturally different from a search in a future retrieval system, in which a person does not need much information to meet their objectives. Subsequently, it is important to note that albeit in a reduced scale, 149 queries, from the total number of 450 queries issued, retrieved at least one future date within the web snippet item (see Table 3), of which 32 had more than one future date. This means that of the 33.1% queries that retrieved a future date in a web snippet, 21.4% had more than one future date.

Table 2. Future dates analysis. *A* represents Absolute values, *R* represents relative values.

Item	Dates	Future Dates			Near Future Dates			Categories	
		#	A	R	#	A	R		
Web Snippet	5777	9.2%	508	0.8%	8.7%	419	0.6%	82.4%	Automotive	33.1%
									Society	15.5%
									Finance	11.0%
Title	2058	3.2%	419	0.6%	20.3%	373	0.5%	88.7%	Automotive	49.3%
									Beauty	28.5%
									Finance	23.8%
URL	3512	5.5%	195	0.3%	5.5%	167	0.2%	85.6%	Automotive	23.6%
									Computer	9.2%
									Sports	8.0%

Table 3. Classification of queries in terms of the occurance of future dates

Item	One Future Date		> One Future Date	
Web Snippet	149	33.11%	32	21.47%
Title	113	25.11%	14	12.38%
URL	75	16.67%	10	13.33%

Two of these cases are illustrated in the two following sentences: (1) *Japan plans to establish a robot moon base by 2020 with a landing by 2015* and (2) *FIFA denied that the process for the 2018-2022 World Cup was corrupt*. A closer study also shows that most of the future dates with relative (R) values occur in titles. Indeed, from a total number of 2.058 items tagged with dates, 20.3% (see Table 2) have a future temporal intent. This constitutes a rich set of information that could be considered when trying to infer the temporal nature of implicit queries. Regardless of a continuous shortage of future dates as we move forward in the calendar, a great number of references to far distant years are still found. The occurrence of dates is largely predominant in 2011, but consistent until 2013. Thereafter, there are some quite small peaks in 2014 and 2022 that mostly relate to the Football World Cup, which coincides with the results of [8]. Overall, the occurrence of future dates is very common in items retrieved in response to queries belonging to the categories of Automotive (e.g., *dacia duster*), Finance & Insurance (e.g., *Bank of America*), Beauty & Personal Care (e.g., *Hairstyles*), Sports (e.g., *football*) and Computer & Electronics (e.g., *hp*). A more detailed analysis of each of the three items: titles, snippets and URLs will now be presented.

Titles. On average, more than 90% of the future dates are related to the near future. This information is mostly related to economic forecasts, such as the expected growth of India, or the prediction that 2011 will be a good year to buy property. Some other examples are related to IT companies, for example the release date for electronic devices, or sport events, as illustrated in these titles: (1) *2011 will be best year to buy a home, says BSA*, (2) *Experts bet on India growth story in 2011*, (3) *Tour de France organizers unveil climb-heavy 2011 route* and (4) *Nokia to launch tablet in Q3 2011*. As we move forward in the calendar, reference years become more scarce such as with scheduled events, including the Football World Cup or rumors relating to environmental issues or company previews: (1) *Mobile App Revenue Estimated at $35*

Billion by 2014, (2) *Octopus Paul joins England's 2018 World Cup bid* and (3) *Qatar Plans 'Island Stadium' For 2022 World Cup*.

Snippets. Unlike in the titles, the occurrence of future dates in web snippets is not very common. Still, they occur in 8.79% of cases and they mostly include a short temporal window, such as 2011. Once again, we can note that most of the texts are related to economic forecasts concerning the worldwide crisis we are currently facing. References to upcoming events can also be seen, such as the Detroit Auto Show and an interesting political text on a visa agreement between Turkey and Azerbaijan: (1) *Honda is planning a major jump in hybrid sales in Japan in 2011*, (2) *Next-generation Ford 2012 Escape unveiled at the 2011 Detroit Auto Show* and (3) *Visa agreement expected to be signed between Turkey and Azerbaijan in 2011*. As with titles, business plans prevail in far distant years. References to PayPal accounts can be seen as well as sales of mobile applications or Adidas plans. Even those related to scheduled events have an economic nature, such as the Qatar Football World Cup reference. In addition, there are other quite interesting examples, one related to the translation of the Bible, another to the environment and another with the calendar of holidays until 2070. Some examples include: (1) *Avatar 2? in 2013? Cameron intends to complete his next film in 3 to 4 years*, (2) *IDC predicts sales of mobile apps will be a $35 billion industry by 2014*, (3) *Wycliffe's mission is to see a Bible translation in every language by 2025* and (4) *Calendar of all legal Public and Bank Holidays worldwide, until 2070*.

URLs. As expected, the occurrence of future dates in URLs is scarce when compared to web snippets or even titles. Indeed, only 5.6% of the links have a future temporal nature. Regardless of the fact that future dates are very uncommon in URLs, they can still be very useful in some specific cases. A careful observation of the list below leads to the conclusion that future dates in URLs are as descriptive as in titles or even in web snippets. Predictions are mostly related to IT companies, economic forecasts, and automotives, as this example shows (1) *http://www.grist.org/article/2010-11-15-fords-first-electric-car-to-be-sold-in-20-cities-in-2011*. Finally, references to far distant dates also appear in URLs such as (1) *http://www.london2012.com/* and (2) *http://msn.foxsports.com/foxsoccer/worldcup/story/world-cup-bid-usa-loses-2022-world-cup-bid-to-qatar*.

Web Snippets Genre. The distribution of items was also analyzed according to the three categories: informative texts, scheduled texts and rumor texts. On average (see Table 4), almost 77% of the texts have either an informative nature or concern a scheduled event which has a very high probability of taking place. The remaining 23% relate to rumor texts, which lack confirmation in the future. Some examples are listed here: (1) *WebOS tablet will arrive in March 2011. Details are not officially* (Rumor), (2) *Tickets for Lady Gaga 2011 Tour* (Scheduled Event) and (3) *Latest Hairstyles 2011* (Informative).

Table 4. Classification of texts according to genre

Item	#Items	Scheduled Events		Informative		Rumor	
web snippet	508	136	26.77%	255	50.20%	117	23.03%
Title	419	85	20.29%	248	59.19%	86	20.53%
URL	195	38	19.49%	101	51.79%	56	28.72%

Table 5. Classification of texts according to genre for near and distant future dates

	Near Future			Distant Future		
Item	Schedule	Informative	Rumor	Schedule	Informative	Rumor
Web Snippet	25.7%	55.8%	18.3%	31.4%	23.6%	44.9%
Title	15.0%	65.4%	19.5%	63.0%	8.7%	28.2%
URL	13.7%	56.8%	29.3%	53.5%	21.4%	25.0%

While informative texts mostly occur with near future dates, schedule events and rumor texts occur more frequently with far distant years (see Table 5). Words such as *latest, new, review, information, schedule, announce, official* and *early* are usually used to describe the near future in informative texts, such as information on product releases (e.g., *Dacia Duster, Audi, Toyota, Ford, Honda, Nissan, Nokia, Microsoft*) and upcoming scheduled events (e.g. *Auto Show*).

As we move forward in the calendar, it is more common for texts to be related to events planned in advance and to also be of a rumor nature. These are associated with events that require confirmation in the future, as shown in Table 5. Long term schedule events such as the Olympic Games in London or the FIFA Football World Cup in Brazil and also in Qatar, and rumor words such as *planning, report, preview, coming, expecting, rumor, scenarios, reveal* and *around* often replace words with a near future nature, such as *early* or *new*. Another interesting issue to note is the fact that future dates are mostly year related and fewer are related to months or days. This becomes more apparent further into the future. Exceptions only occur with scheduled events. The following sentence is an illustrative example: *Tour de France: from Saturday July 2nd to Sunday July 24th 2011, the 98th*.

4   Classification and Clustering of Future-Related Texts

This paper aims to understand whether data features influence the classification and clustering of future-related texts according to their nature: informative, scheduled or rumor. It is important to note that the goal is not to achieve high accuracy results but to understand if these three genres can be discovered by simply using specific linguistic features, thus avoiding the importance of time for these tasks, or if instead, the inclusion of temporal features plays an important role.

4.1   Classification of Future-Related Texts

This study includes cross-domain experiments by selecting and issuing queries for the set of 27 categories available. The Aue et al. [1] and Boey et al. [4] model that suggests training a classifier on a domain-mixed set, in order to tackle cross-domain learning, was used. Experiments are based on two collections[1]: one consisting of 508 snippets and another consisting of 419 text titles, both tagged with future dates. URL texts were not included in this experiment. A selection of 117 text snippets of Informative nature, 117 of Scheduled intent and 117 of Rumor purpose were collected, together with 86 text titles of Informative nature, 86 of Scheduled intent and 86 of Rumor purpose. The result is a set of 351 balanced texts snippets and 258

---

[1] Available at http://www.ccc.ipt.pt/~ricardo/software [17th June, 2011].

Table 6. Datasets structure

Dataset	Web Snippet		Near/Distant Future	Class
	Unigram	Year Dates		
D1	x	x		x
D2	x			x
D3	x	x	x	x
D4	x		x	x

Table 7. Web Snippet classification results for the boolean and tf.idf cases

Algorithm	Case	Dataset	Accuracy	Scheduled		Informative		Rumor	
				Precis.	F-Mea	Precis.	F-Mea	Precis.	F-Mea
Naïve Bayes	Boolean	D1	78.1%	84.2%	75.4%	77.8%	77.8%	74.1%	80.5%
	Boolean	D2	77.2%	80.8%	74.1%	78.6%	78.6%	73.3%	78.6%
K-NN	Boolean	D1	58.1%	52.0%	58.1%	56.7%	51.4%	67.9%	64.6%
	Boolean	D2	57.0%	48.2%	60.3%	67.3%	43.0%	68.3%	63.3%
Multi-Class SVM	Boolean	D1	79.2%	87,3%	81,3%	75,2%	77,7%	76,6%	78,8%
	Boolean	D2	79.8%	87,0%	80,2%	75,6%	78,7%	78,2%	80,5%
Multi-Class SVM	TF.IDF	D1	75.2%	83,0%	76,5%	69,5%	72,7%	74,8%	76,7%
	TF.IDF	D2	74.4%	85,6%	77,6%	66,9%	71,2%	73,6%	74,8%
M. Naïve Bayes	TF.IDF	D1	76.4%	78.6%	78.6%	79.4%	72.0%	72.3%	78.0%
	TF.IDF	D2	75.8%	76.0%	77.3%	79.6%	72.6%	72.7%	77.1%
Weighted K-NN	TF.IDF	D1	59.3%	87.5%	61.9%	65.3%	49.7%	48.8%	63.3%
	TF.IDF	D2	51.0%	51.5%	55.0%	66.7%	35.2%	46.9%	56.2%
Naïve Bayes	Boolean	D3	78.6%	84.4%	76.1%	77.8%	77.8%	73.9%	80.0%
	Boolean	D4	78.1%	83.5%	75.7%	79.1%	78.4%	73.4%	79.7%
K-NN	Boolean	D3	62.7%	59.1%	62.7%	57.1%	57.6%	74.0%	68.2%
	Boolean	D4	57.6%	50.0%	59.5%	60.7%	50.0%	59.7%	57.2%
Multi-Class SVM	Boolean	D3	78.6%	86,3%	80,4%	73,8%	76,5%	77,2%	79,2%
	Boolean	D4	79.2%	87,1%	80,7%	74,2%	77,6%	77,9%	79,5%
Multi-Class SVM	TF.IDF	D3	74.9%	83.7%	76.3%	67.7%	72.0%	75.8%	76.8%
	TF.IDF	D4	79.2%	87.1%	80.7%	74.2%	77.6%	77.9%	79.5%
M. Naïve Bayes	TF.IDF	D3	75.5%	78.3%	77.6%	78.4%	71.0%	71.2%	77.3%
	TF.IDF	D4	76.5%	75.2%	75.2%	82.8%	73.3%	77.0%	76.1%
Weighted K-NN	TF.IDF	D3	56.4%	86.8%	54.1%	66.7%	49.5%	46.3%	61.3%
	TF.IDF	D4	57.5%	50.0%	59.5%	60.7%	50.7%	68.4%	61.3%

balanced text titles, from which four datasets D1, D2, D3 and D4 (see Table 6) were built, respectively. Each dataset is labeled with the respective text genre/class. In particular, (D1) consists of texts with their year dates, (D2) consists of texts withdrawing their year dates, (D3) consists of texts with their year dates plus the mention of their belonging to a near or distant future and (D4) consists of texts without their year dates plus the mention of their belonging to a near/distant future.

Experiments are run on the basis of a 5-fold cross-validation for boolean and tf.idf unigram features for five different classifiers: the Naive Bayes algorithm (boolean), the K-NN (k = 10, boolean), the Multinomial Naive Bayes algorithm (tf.idf), the Weighted K-NN (K = 10 and weight=1/distance, tf.idf) and the Multi-Class SVM (boolean and tf.idf). Results are presented in Table 7 and show that the importance of temporal features in the classification task is heterogeneous, as it depends on the learning algorithm and on text representation.

In general, all of the algorithms (see Fig. 2), with the exception of SVM (boolean) show improved results in terms of accuracy with the simple use of explicit year dates. The greatest difference is in the Weighted K-NN algorithm. However, both Naïve Bayes and SVM (boolean) largely outperform the Weighted K-NN in terms of accuracy. In contrast, the dates do not have a great impact if combined with near/distant future knowledge. Indeed, Multi-Class SVM (boolean and tf.idf), Multinomial Naïve Bayes and Weighted K-NN provide better results for D4 than D3. Equally, in the comparison between D1 and D2, the greatest difference in accuracy occurs with the K-NN algorithm. Once again, the Naïve Bayes and SVM (boolean) achieve the best results.

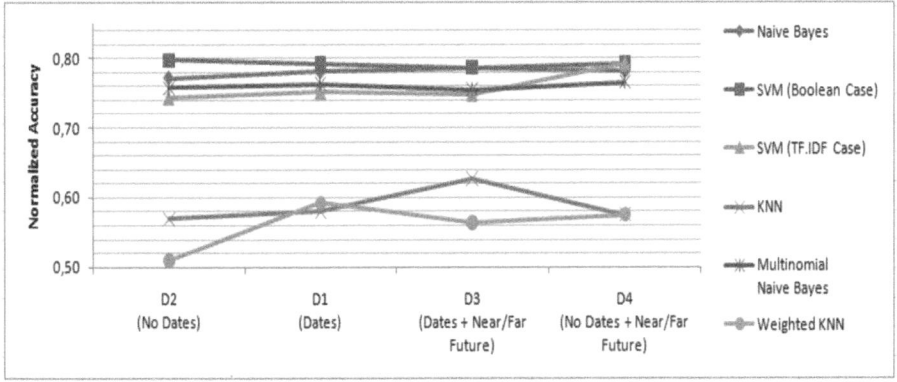

Fig. 2. Overall analysis of global accuracy for Web Snippets texts

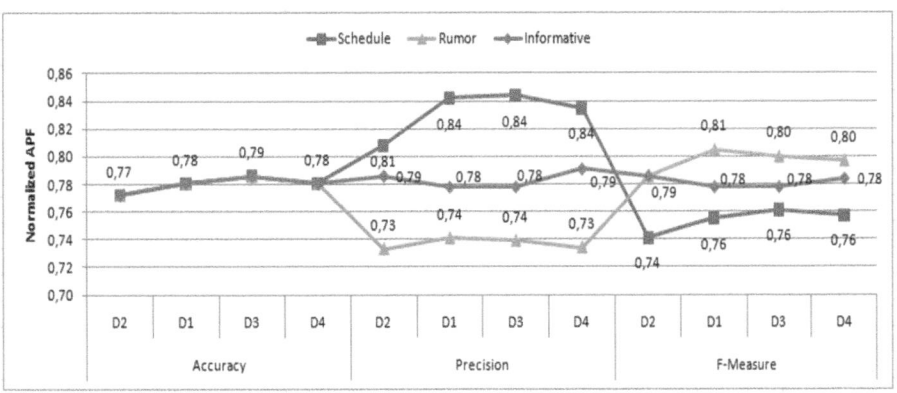

Fig. 3. Text genre analysis for Naïve Bayes (D1,D2) and (D3,D4) comparison

An individual analysis of each text genre (informative, scheduled, rumor) also led to the conclusion that the introduction of temporal features has an overall positive impact on precision in the classification of scheduled texts. In contrast, the classification of informative texts is more accurate without dates and this is uncertain in the case of rumor texts. Overall these conclusions are confirmed by F-Measure for scheduled and informative texts, but interestingly, not for rumor texts, which show an overall positive impact with F-Measure with the introduction of time features. The best results, however, occur for the SVM algorithm (boolean) without the use of any temporal features. Fig. 3 shows the results for the specific case of Naïve Bayes.

The same experiments performed on the web snippets were then performed on the set of 258 balanced text titles. The results are shown in Table 8.

Table 8. Title classification results for the boolean and tf.idf cases

Algorithm	Case	Dataset	Accuracy	Scheduled		Informative		Rumor	
				Precis.	F-Mea	Precis.	F-Mea	Precis.	F-Mea
Naïve Bayes	Boolean	D1	78.1%	83.5%	75.7%	79.1%	78.4%	73.4%	79.7%
	Boolean	D2	79.9%	77.8%	83.2%	74.1%	80.0%	96.4%	75.2%
K-NN	Boolean	D1	54.3%	71.6%	62.7%	44.7%	60.4%	100%	24.5%
	Boolean	D2	55.4%	56.9%	67.0%	51.2%	61.0%	100%	17,0%
Multi-Class SVM	Boolean	D1	74,4%	75,0%	75,9%	66,7%	72,3%	85,3%	75,3%
	Boolean	D2	76,4%	74,7%	78,5%	70,5%	74,0%	86,8%	76,6%
Multi-Class SVM	TF.IDF	D1	72.9%	71.4%	73.4%	66.7%	70.3%	83.1%	75.2%
	TF.IDF	D2	76.4%	73.5%	78.3%	71.3%	74.4%	87.9%	76.3%
M. Naïve Bayes	TF.IDF	D1	77.9%	78.9%	80.7%	70.4%	78.4%	90.0%	74.0%
	TF.IDF	D2	76.4%	76.5%	81.5%	69.3%	74.9%	88.1%	71.7%
Weighted K-NN	TF.IDF	D1	53.1%	70.0%	62.8%	43.8%	59.1%	100%	20,8%
	TF.IDF	D2	53.1%	53.5%	64.2%	50.8%	60.0%	100%	11,0%
Naïve Bayes	Boolean	D3	72.9%	71,8%	71,3%	63,6%	74,4%	96,2%	72,5%
	Boolean	D4	77.9%	75,3%	79,8%	71,0%	78,8%	96,3%	74,3%
K-NN	Boolean	D3	53.9%	71,9%	61,3%	44,5%	60,4%	100%	24,5%
	Boolean	D4	52.7%	70,4%	63,7%	43,6%	58,9%	100%	17,0%
Multi-Class SVM	Boolean	D3	75,2%	75,9%	76,3%	67,3%	73,7%	86,6%	75,8%
	Boolean	D4	75,6%	76,4%	77,7%	66,7%	73,3%	89,1%	76,0%
Multi-Class SVM	TF.IDF	D3	73,6%	73.0%	74.3%	67.0%	73.0%	84.8%	73.7%
	TF.IDF	D4	74.4%	75.0%	75.9%	65.7%	73.2%	88.7%	74.3%
M. Naïve Bayes	TF.IDF	D3	77.1%	77.8%	79.5%	70.0%	78.6%	89.7%	72.2%
	TF.IDF	D4	77.1%	76.5%	81.5%	71.3%	77.0%	88.1%	71.1%
Weighted K-NN	TF.IDF	D3	52.3%	69.7%	60,5%	43,4%	59,0%	100%	46,8%
	TF.IDF	D4	51.1%	62.7%	61,5%	44,1%	58,6%	100%	43,7%

Overall, it is clear that most of the algorithms (see Fig. 4) perform worst in terms of accuracy with the introduction of temporal features, meaning that time characteristics do not have a great impact on the classification task. This does not happen with the Multinomial Naïve Bayes, which has one of the best overall results, only supplanted by the Naïve Bayes algorithm.

This is confirmed by a detailed analysis of all three types of text genres, where the Multinomial Naïve Bayes algorithm shows successful results. Overall, for almost all of the algorithms, scheduled texts benefit with the introduction of temporal features, which is not as clear in the case of informative texts. Another interesting result is that

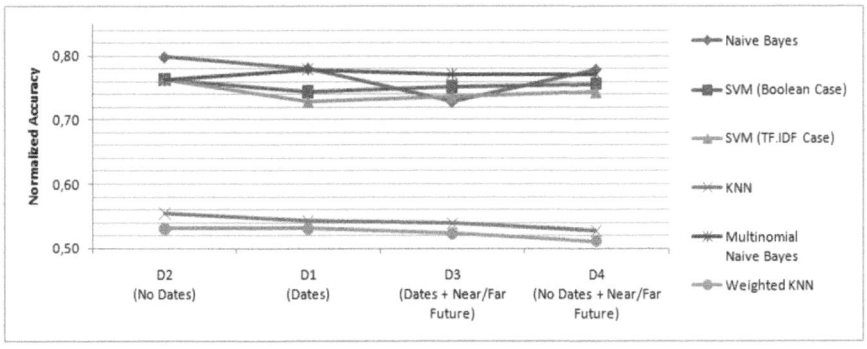

Fig. 4. Overall analysis of global accuracy for Titles texts

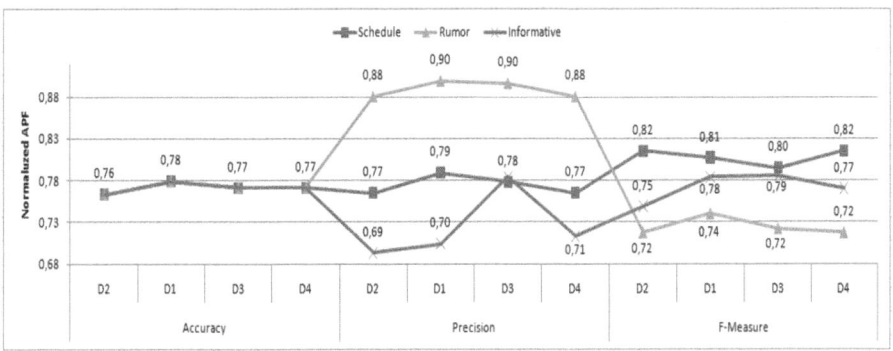

Fig. 5. Text genre analysis for Multinomial Naïve Bayes (D1,D2) and (D3,D4) comparison

precision in rumor texts is very high. However, with the exception of the Multinomial Naïve Bayes algorithm, time features do not have an overall impact on the classification task. The following figure (see Fig. 5) shows these results for the specific case of the Multinomial Naïve Bayes algorithm.

4.2 Clustering of Future-Related Texts

Finally, a set of experiments based on the well known K-means clustering algorithm was proposed in order to understand the impact of temporal features within this process. The idea is to automatically retrieve three different clusters (informative, scheduled and rumors) based on the same representations of web snippets, the D1, D2, D3 and D4. As in the classification case, experiments for the boolean and tf.idf cases, and for snippets and text titles are shown.

Results for text snippets are presented in Table 9 and show that they are more sensitive to the near/distant future feature, as the best results, for the Boolean case, are obtained for D3. However, the best overall results are obtained by using the K-means over D4, which only takes into account a coarse-grained temporal feature. It must also be noted that scheduled texts have a very high precision rate of almost 85% with a positive impact on the use of temporal features.

Table 9. Web snippet clustering results for the K-means in the boolean and tf.idf cases

Algorithm	Case	Dataset	Correctly Clustered	Scheduled Precision	Informative Precision	Rumor Precision
K-Means	Boolean	D1	43.59%	34.7%	59.5%	41.1%
		D2	43.59%	34.7%	59.5%	41.1%
		D3	45.02%	36.0%	55.8%	50.0%
		D4	41.88%	33.9%	46.6%	43.6%
	tf.idf	D1	39.04%	84.6%	35.6%	20.0%
		D2	35.90%	83.3%	34.4%	29.4%
		D3	40.74%	25.0%	38.0%	50.6%
		D4	51.00%	43.4%	50.5%	58.4%

This is a clear contrast to text titles clustering, as the best results occur for D3 in the tf.idf representation, with nearly a 13% impact when compared to D4 (Table 10). Moreover, the use of temporal features, either alone or combined with near/distant future knowledge, show a positive impact in the clustering task, but for rumor texts they reach an impressive value of almost 85% in terms of precision. The results obtained were not conclusive for D1 and D3 (Boolean case), in that more than two clusters were not found. A more detailed analysis led to the conclusion that this is mostly because the system appears to have some difficulties in splitting schedule texts from those of a rumor nature.

Table 10. Title clustering results for the K-means in the boolean and tf.idf cases

Algorithm	Case	Dataset	Correctly Clustered	Scheduled Precision	Informative Precision	Rumor Precision
K-Means	Boolean	D1	39,54%			
		D2	42,25%	34.9%	47.5%	84.5%
		D3	39,54%			
		D4	42,25%	34.9%	47.5%	84.5%
	tf.idf	D1	41,87%	34.7%	37.6%	82.4%
		D2	41,87%	37.1%	37.0%	79.3%
		D3	53,49%	68.0%	45.0%	82.8%
		D4	41,87%	37.5%	35.8%	79.3%

5   Conclusion

In this paper, we conducted an exploratory analysis of future information on the Internet. Results show that titles, particularly in the near future, contain a broad range of temporal information, which is still significant in the case of text snippets and URLs. In addition, we conclude that texts are more often of a scheduled and rumor nature as we move forward in the calendar, contrary to what happens with informative texts, which are unlikely to appear. The high precision of these results and the work presented by Adam Jatowt et al. [9], who has shown that temporal features can help cluster future-related web snippets, led to our final experiments. We performed a set of exhaustive classification and clustering tests based on the three different future-related text genres (informative, scheduled and rumors). The results obtained from our analysis are subject to discussion. Indeed, depending on the representation of the text and on the algorithm family, the temporal issue may or may not have any influence.

For the classification task, the SVM and the Naïve Bayes provide the best overall results for text snippets and text titles respectively. However, none of these results was obtained using temporal features. Moreover, the probabilistic learning and the lazy learning families always show the best results for the classification of text snippets when any time feature is used, with the exception of the Multinomial Naive Bayes and the Weighted K-NN for D3. This is the opposite of what happens with the classification of text titles, where most of the algorithms perform better without temporal features. Furthermore, we can also conclude that in general, the introduction of temporal features has an overall positive impact on the classification of scheduled texts, both in snippets as well as in text titles. Interestingly we can also note that the detection of rumor texts benefits from the introduction of temporal features, particularly in the probabilistic algorithms. For the clustering task, and in particular for the K-means algorithm, the impact of temporal features is more apparent in D1 for snippets and in D3 for text titles. Moreover, the identification of schedule texts is particularly easy in text snippets, while rumor texts are easily identified in text titles.

We believe that this information will serve to improve temporal knowledge in terms of the aims of the user's query, and is a step towards the formation of a future search engine, where the returned documents relate to future periods of time. As such, time features must definitely be treated in a special way and further experiments must be carried out with different representations of time-related features in the learning process, to reach final conclusions and to assess new exhaustive results in the clustering process.

Acknowledgments. This research was part-funded by the PhD grant with reference SFRH/BD/63646/2009 from the Portuguese Foundation for Science and Technology (FCT). This work was also supported by the VIPACCESS project with reference PTDC/PLP/72142/2006 funded by the FCT.

References

1. Aue, A., Gamon, M.: Customizing Sentiment Classiers to New Domains: a Case Study. In: RANLP 2005, Borovets, Bulgaria, September 21-23 (2005)
2. Baeza-Yates, R.: Searching the Future. In: MFIR 2005 associated to SIGIR 2005, Salvador, Brazil, August 15-19 (2005)
3. Berberich, K., Bedathur, S., Alonso, O., Weikum, G.: A language modeling approach for temporal information needs. In: Gurrin, C., He, Y., Kazai, G., Kruschwitz, U., Little, S., Roelleke, T., Rüger, S., van Rijsbergen, K. (eds.) ECIR 2010. LNCS, vol. 5993, pp. 13–25. Springer, Heidelberg (2010)
4. Boey, E., Hens, K., Deschacht, K., Moens, M.-F.: Automatic Sentiment Analysis of On-Line Text. In: ELPUB 2007, Vienna, Austria, June 13-15 (2007)
5. Campos, R., Dias, G., Jorge, A.M.: What is the Temporal Value of Web Snippets? In: TWAW2011 Associated to WWW 2011, Hyderabad, India, March 28 (2011)

6. Fleiss, J.L.: Measuring Nominal Scale Agreement Among many Raters. Psychological Bulletin 76(5), 378–382 (1971)
7. Google Insights for Search, http://www.google.com/insights/search
8. Jatowt, A., Kawai, H., Kanazawa, K., Tanaka, K., Kunieda, K.: Analyzing Collective View of Future, Time-referenced Events on the Web. In: WWW 2010, Raleigh, USA, April 26 - 30, pp. 1123–1124 (2010)
9. Jatowt, A., Kawai, H., Kanazawa, K., Tanaka, K., Kunieda, K.: Supporting Analysis of Future-Related Information in News Archives and the Web. In: JCDL 2009, Austin, USA, June 15-19, pp. 115–124 (2009)
10. Liu, Y., Huang, X., An, A., Yu, X.: ARSA: A Sentiment-Aware Model for Predicting Sales Performance Using Blogs. In: SIGIR 2007, Amsterdam, Netherlands, pp. 607–614 (July 2007)
11. Mishne, G., Glance, N.: Predicting Movie Sales from Blogger Sentiment. In: CAAW 2006 Associated to AAAI 2006, Boston, USA, July 16-20 (2006)
12. Radinsky, K., Davidovich, S., Markovitch, S.: Predicting the News of Tomorrow Using Patterns inWeb Search Queries. In: WIC 2008, Sydney, Australia, pp. 363–367 (2008)
13. Recorded Future, http://www.recordedfuture.com/

Using the Web to Validate
Lexico-Semantic Relations

Hernani Pereira Costa, Hugo Gonçalo Oliveira, and Paulo Gomes

Cognitive and Media Systems Group, CISUC
University of Coimbra, Portugal
{hpcosta,hroliv,pgomes}@dei.uc.pt

Abstract. The evaluation of semantic relations acquired automatically from text is a challenging task, which generally ends up being done by humans. Despite less prone to errors, manual evaluation is hardly repeatable, time-consuming and sometimes subjective. In this paper, we evaluate relational triples automatically, exploiting popular similarity measures on the Web. After using these measures to quantify triples according to the co-occurrence of their arguments and textual patterns denoting their relation, some scores revealed to be highly correlated with the correction rate of the triples. The measures were also used to select correct triples in a set, with best F_1 scores around 96%.

1 Introduction

During the last decades, there have been several attempts to discover knowledge automatically from text (see [16] [6] [12] [20] [9] [22]). Regardless the kind of knowledge, information extraction (IE) systems generally acquire entities (e.g. e_1, e_2) and relations between them, represented as triples (e_1, r, e_2), where r identifies the type of relation. For instance, considering named entities such as people, places or organisations, born-in or headquarters-of are possible kinds of relations. As for lexico-semantic knowledge, hyponymy and part-of are typical types of relations held by word meanings.

Knowledge discovered automatically is useful to create or to enrich existing ontologies, such as WordNet [13] or CyC [17], however its evaluation is a challenging task, especially when dealing with broad-coverage open-domain knowledge. Despite the existence of, at least, four distinct approaches for evaluating (domain) ontologies (see [5]): manual, gold standard, comparison with a source of data, and indirect – most evaluations end up needing an excess of human intervention. The comparison with a collection of documents evaluates only the coverage of one or several domains and hardly applies to broad-coverage knowledge, while indirect evaluation does not depend exclusively on the quality of the knowledge but also on the way it is used. Also, most of the times there are no available gold standards for the needed evaluation, thus requiring the manual creation of one, which is an expensive task. Furthermore, once again, when it comes to broad-coverage knowledge it is more difficult to create gold standards due to the huge quantity of knowledge these should contain, as well as

L. Antunes and H.S. Pinto (Eds.): EPIA 2011, LNAI 7026, pp. 597–609, 2011.

other structural issues. So, even though time-consuming, hardly repeatable and sometimes subjective, manual evaluation of a representative set of the extracted knowledge is the most common choice (as in [6] [20] [22]).

An automatic alternative to the aforementioned evaluation approaches is to search, in large collections of text, for support on the knowledge to be evaluated. Since a semantic relation can be denoted by several textual patterns (e.g. *"is a"* in *"car is a vehicle"*, for hyponymy, or *"of the"* in *"the wheel of the car"*, for part-of) the quality of a relational triple can be extrapolated based on the frequency of its entities connected by one or more patterns denoting its relation. Still, since some entities are more frequent than others and because natural language is ambiguous, we should not rely only on the latter frequency and more data should be combined to validate relational triples.

The goal of this paper is to ascertain how well-suited similarity measures based on the distribution of words on the Web are for evaluating relational triples, in a completely automatic fashion. After collecting a small set of patterns denoting the relations to validate, similarity measures are adapted – instead of looking for occurrences of the entities alone, the measures are used to search for these entities followed by or after the patterns. Our assumption is that the scores given by the measures can be used to filter incorrect or less probable triples.

Therefore, in order to verify how similarity measures could be exploited, we have used WordNet to collect sets of correct and incorrect hyponymy and part-of triples to calculate how their correction correlates with the scores given by the measures. We then used the latter sets as a gold standard and selected the correct triples based on the same scores. Not only some correlation coefficients were very high, but some measures were capable of selecting the correct triples with F_1 scores around 96%. This is very promising and sets these measures as a new automatic approach for evaluating semantic relations.

In the rest of the paper, distributional similarity measures are introduced, our experiments are described, and their results presented. Before concluding, we refer some related work.

2 Web-Based Similarity Measures

Some of the most popular methods for computing the semantic similarity of words involve mathematical models based on the distribution of words in large corpora, or in the World Wide Web. The latter infrastructure is very attractive because it is a huge and heterogeneous source of knowledge, probably the largest available. Furthermore, search engines are efficient interfaces to interact with the contents of the Web. This provides an easy access to information on the frequency and distribution of words, which can thus be used to infer similarities.

In this section, we present five common Web-based similarity measures. Some of them are simple adaptations of popular co-occurrence measures. In their expressions, we use q for denoting a query and $P(q)$ to denote the number of pages returned by a search engine (hereafter page counts) for q. So, $P(e_1 \cap e_2)$ represents the page counts for the query consisting of the entities e_1 and e_2, more precisely "e_1 AND e_2".

The WebJaccard measure, in expression 1, is an adaptation of the Jaccard coefficient, given by the number of documents in which e_1 and e_2 co-occur, divided by the number of documents where each one occurs. The WebOverlap (expression 2) and WebDice (expression 3) measures are two variations of WebJaccard, respectively for measuring the overlap and the mean overlap of two sets. More precisely, the Overlap minimises the effect of comparing two objects of different sizes, so the number of co-occurrences is divided by the lowest number of page counts, $\min(P(e_i), P(e_j))$.

$$WebJaccard(e_1, e_2) = \frac{P(e_1 \cap e_2)}{P(e_1) + P(e_2) - P(e_1 \cap e_2)} \tag{1}$$

$$WebOverlap(e_1, e_2) = \frac{P(e_1 \cap e_2)}{\min\left(P(e_1), P(e_2)\right)} \tag{2}$$

$$WebDice(e_1, e_2) = \frac{2 * P(e_1 \cap e_2)}{P(e_1) + P(e_2)} \tag{3}$$

The WebPMI measure (expression 4) stands for Pointwise Mutual Information (PMI) and quantifies the statistical dependence between two entities [21]. In its expression, N is the total number of pages indexed by the search engine which, for Google search engine, can be roughly estimated to 10^{10} [4] [3]. If entities e_1 and e_2 are statistically independent, the probability that they co-occur is given by $P(e_1) * P(e_2)$. On the other hand, if they tend to co-occur, $P(e_1 \cap e_2)$ will be higher than $P(e_1) * P(e_2)$, and the PMI will thus be greater.

$$WebPMI(e_i, e_j) = \log_2 \left(\frac{P(e_i \cap e_j)}{P(e_i) * P(e_j)} * N \right) \tag{4}$$

The Normalised Web Distance (NWD, expression 5) [7] is an approximation of the Normalised Information Distance [1] and measures the distance of two entities, based on their co-occurrences on the Web. Therefore, if the entities always co-occur, this means they are very similar and NWD is 0. On the other hand, although NWD most of the times ranges from 0 to 1, if the entities never co-occur, NWD is $+\infty$.

If we invert NWD and bound it to the [0-1] range [14], we can measure the similarity of two entities, in a measure that we will, from now on, call Normalised Web Similarity (NWS, expression 6).

$$NWD(e_1, e_2) = \frac{\max\left(\log P(e_1), \log P(e_2)\right) - \log P(e_1 \cap e_2)}{\log N - \min\left(\log P(e_1), \log P(e_2)\right)} \tag{5}$$

$$NWS(e_1, e_2) = e^{-2*NWD(e_1,e_2)} \tag{6}$$

3 Experimentation

Our experimentation was performed to analyse how well Web distributional measures, presented in section 2, suit the task of validating semantic relations. Despite quantifying the semantic similarity between two entities alone, the measures were adapted to quantify the similarity of the entities attached to textual patterns denoting semantic relations. In the first experiment, the correlation between the measures and the correction rate of the triples is calculated. The second is an information retrieval task, where the measures are used to identify correct triples from a set. Looking at the results, we believe that some of the measures can be used in future evaluations of semantic relations and, eventually, replace manual evaluation.

3.1 Set-up

All the measures presented in section 2 were implemented. However, having in mind the validation of semantic relations, we followed previous hints [19] and adapted the measures to quantify the similarity between entities connected by patterns expressing semantic relations. Therefore, for validating the triple $t = (e_1, r, e_2)$, a pattern π_{ri}, indicative of relation r, is selected. The expressions of the measures are changed according to the following:

- $P(e_1)$ = page counts for query: "$e_1 \pi_{ri}$";
- $P(e_2)$ = page counts for query: "$\pi_{ri} e_2$";
- $P(e_1 \cap e_2)$ = pages counts for query: "$e_1 \pi_{ri} e_2$".

For instance, if $e_1=\{planet\}$, $e_2=\{Mars\}$ and $\pi_{ri}=\{such\ as\}$, we would have $P(e_1)=\{planet\ such\ as\}$, $P(e_2)=\{such\ as\ Mars\}$, $P(e_1 \cap e_2)=\{planet\ such\ as\ Mars\}$. To this end, a set of indicative patterns for the relations we were validating, hyponymy and part-of, was created (table 2 and 3). To increase the coverage of the patterns, we used not only those conveying the direct relation (e.g. e_1 *and other* e_2, for hyponymy), but also the indirect (e.g. e_2 *such as* e_1, for hypernymy).

As semantic relations can be expressed by several different textual patterns, we could not select one best pattern. So, we decided to use two sets, Π_h and Π_p, consisting of the most frequent hyponymy and part-of patterns. Furthermore, as the measures accept only one pattern at once, we computed the final scores by four distinct methods. Considering that Π_r has all the patterns for relation r, we sort a list, $S_m : |S_m| = |\Pi_r|$, containing the scores given by a measure m, with each pattern $\pi_{ri} \in \Pi_r$, such that the best score is in S_{m1}. The final score is then given by:

- a baseline consisting of simple co-occurrence, without including the patterns in the expressions (NP);
- the score of the best pattern (B), S_{m1};
- the average of the scores given by the two best patterns (2B), $\frac{S_{m1}+S_{m2}}{2}$;
- the average of the scores given by all patterns (Av), $\frac{\sum_{i=1}^{|S_m|} S_{mi}}{|S_m|}$.

3.2   Datasets

We have used WordNet 2.0[1] for collecting sets of hyponymy and part-of triples. In order to reduce noise due to ambiguities, we took advantage of the organisation of WordNet, which has the synsets ordered by the most frequent senses of the words and created the sets in the following way:

1. We selected all the relation instances between synsets which denote the first sense of their most frequent word.
2. For each of the latter, we defined relational triples held by the first word in the connected synsets. For example, the instance {*corporation.1, corp.1*} hyponym-of {*firm.1, house.2, business firm.1*} originates the triple {*corporation*, hyponym-of, *firm*}.
3. To create the final sets, we ranked the triples according to the frequency of their arguments in Google web search engine[2] and selected the first 1,100 hyponymy triples and 1,100 part-of triples, respectively H and P.

Sets H and P contain only correct triples, but regarding the need for incorrect triples, we created a third set, I, with 1,010 random pairs of words, which we made sure to be not related by hyponymy nor by part-of. In table 1, we present examples of triples in the datasets and their classification.

Table 1. Examples of triples in the datasets

Classification	Examples
Correct (C)	*fight* hyponym-of *conflict* *hour* part-of *day*
Incorrect (I)	*towel* hyponym-of *engineer* *ibuprofen* part-of *light*
Wrong Relation (WR)	*eye* hyponym-of *face* *hometown* part-of *town*

3.3   Preliminary Analysis

Before applying the measures, we analysed the page counts for the entities connected by the patterns. Both sets of patterns, Π_h and Π_p, for hyponymy and part-of respectively, were used with the sets of triples H, P and I.

Regarding that IE systems may extract triples held by related entities but fail on identifying the relation, our correct triples were searched with patterns for other relations, making it possible to compare the page counts of this kind of triples with completely correct and completely incorrect triples. Table 2 presents the average (Av) and the standard deviation (SD) of page counts of the patterns in Π_h connecting the entities in H (Correct), P (WrongRel) and I (Incorrect). Similarly, table 3 presents the same statistical measures, this time for page counts of the patterns in Π_p connecting the entities in P (Correct), H (WrongRel) and I (Incorrect).

In both tables (2 and 3), page counts tend to be higher for correct triples, a little lower for the ones with a wrong relation and even lower or 0 for incorrect

---

[1] Available through `http://wordnet.princeton.edu`
[2] $t = (e_1, r, e_2)$, $score(t) = log(P(e_1) + P(e_2))$.

Table 2. Page counts for hyponymy patterns

R	Textual Pattern (π_h)	Correct Av	Correct SD	WrongRel Av	WrongRel SD	Incorrect Av	Incorrect SD
Y hypo X	is a|an|one|the kind of	0.46	73.38	0.01	5.46	$9.90E^{-4}$	0.99
	is a|an|one|the	274.7	44560.8	8.09	1007.66	0.53	175.74
	is a|an|one|the variety of	0.01	5.28	0.0	0.0	0.0	0.0
	is a|an|one|the type of	0.77	191.2	0.07	23.87	0.0	0.0
	is a|an|one|the form of	0.99	510.96	$4.5E^{-3}$	3.31	0.0	0.0
	and|or other	66.9	15512.9	15.15	2748.7	0.36	23.34
X hyper Y	such as	27.48	6832.4	18.15	2620.2	0.16	6.3
	like	42.60	6486.2	14.29	3264.8	0.02	11.41
	including	26.47	7307.9	81.63	10414.3	$9.90E^{-4}$	8.74
	especially	2.79	570.8	21.1	4147.7	0.03	11.41

Table 3. Page counts for part-of patterns

R	Textual Pattern (π_p)	Correct Av	Correct SD	WrongRel Av	WrongRel SD	Incorrect Av	Incorrect SD
Y part X	of	208.28	49053.4	313.58	65352.7	26.13	15971.7
	of a|an|one|the	546.28	119150.7	319.91	73777.7	3.99	1390.4
	from a|an|one|the	43.07	18873.5	71.42	43370.6	0.21	100.90
	in	646.38	43269.8	152.86	45098.7	9.12	6785.4
	is part of	2.08	418.33	0.11	47.64	$2.97E^{-3}$	2.23
	is member of	$2.72E^{-3}$	1.73	$2.73E^{-3}$	2.99	0.0	0.0
	part of a|an|one|the	1.45	251.7	1.72	161.95	0.13	120.98
	member of a|an|one|the	0.27	122.73	0.89	17.12	$4.95E^{-3}$	2.64
	is a|one|the part of	0.99	290.33	0.06	19.94	$1.98E^{-3}$	1.99
	is a|an|one|the member of	1.33	1439.3	0.01	7.19	0.0	0.0
	is a|an|one|the part of a|one|the	0.91	218.01	0.08	13.92	$1.98E^{-3}$	1.99
	is a|an|one|the member of a|one|the	0.42	301.84	0.12	64.59	0.0	0.0
X has Y	's	550.9	188279.1	243.92	159845.1	5.23	3233.0
	has a|an|one|the	9.17	1809.0	9.48	2577.7	0.25	177.89
	contains a|an|one|the	1.12	309.75	2.67	871.14	$9.90E^{-4}$	4.12
	consists of	0.61	111.26	2.36	1228.2	$6.93E^{-3}$	0.99
	is made of	0.02	8.74	0.11	68.13	0.0	0.0

triples. There are however exceptions, especially for ambiguous patterns. For instance, the patterns *"including"* and *"especially"* were used as hypernymy indicators, but they can sometimes denote the part-of relation. Furthermore, for hyponymy, 26 correct triples, 135 triples with the wrong relation and 947 incorrect triples do not have page counts with any pattern, as well as 17 correct, 241 with the wrong relation, and 859 incorrect part-of triples. Correct triples with no page counts are acceptable because we only selected a set with the most frequent patterns, such as the ones proposed by Hearst [16] for hyponymy. Yet, these relations can be expressed by many other ways.

3.4   Correlation Analysis

The parallelism between the correctness of the triples and the scores given by the similarity measures is quantified by Spearman's coefficient, $\rho : -1 \leq \rho \leq 1$ (expression 7) between these two variables.

Table 4. Correlations between the correctness of the triples and the similarity measures

Relation		nHits	Jaccard	Overlap	Dice	PMI	NWS	hasHits
Hyponymy $(C + I)$	NP	0.11	0.14	0.15	0.14	0.13	0.63	**0.77**
	B	0.16	0.17	0.34	0.18	**0.93**	0.87	0.92
	2B	0.18	0.19	0.36	0.20	**0.92**	0.86	0.86
	Av	0.18	0.20	0.35	0.22	**0.78**	0.72	-
Hyponymy $(C + I + WR)$	NP	$-5.3E^{-3}$	-0.11	-0.11	-0.11	-0.17	-0.14	**0.39**
	B	0.04	0.16	0.29	0.17	**0.76**	0.74	**0.76**
	2B	0.04	0.17	0.32	0.19	0.73	**0.74**	0.69
	Av	0.07	0.20	0.34	0.21	**0.69**	0.67	-
Part-of $(C + I)$	NP	0.19	0.22	0.35	0.23	0.29	0.71	**0.76**
	B	0.16	0.18	0.21	0.23	**0.89**	0.85	0.85
	2B	0.17	0.19	0.34	0.23	**0.90**	0.86	0.88
	Av	0.18	0.21	0.26	0.25	**0.78**	0.72	-
Part-of $(C + I + WR)$	NP	0.13	0.24	0.33	0.25	0.33	**0.65**	0.39
	B	0.16	0.17	0.21	0.20	**0.82**	0.69	0.72
	2B	0.17	0.17	0.24	0.20	**0.82**	0.68	0.72
	Av	0.18	0.15	0.25	0.16	**0.57**	0.42	-

$$\rho(m_i, x_i) = \frac{\sum_i (m_i - \overline{m})(x_i - \overline{x})}{\sqrt{\sum_i (m_i - \overline{m})(x_i - \overline{x})}} \qquad (7)$$

Our references, x, consisted of arrays with $1, 0.5$ and 0, respectively for correct triples, triples with a wrong relation and incorrect triples. Table 4 shows the correlations between x and the scores given by the measures, m, calculated by the four methods described in section 3.1, namely, a baseline using no patterns (NP), using only the best pattern (B), the average of the two best patterns (2B), and the average of all patterns (Av). Besides the similarity measures, we have used two simpler measures – one considers the number of page counts (nHits); the other marks the triple as correct if there is at least one page count for $P(e_1 \cap e_2)$, using no pattern (NP), one pattern (B), or two patterns (2B) (hasHits). The results are shown for both studied relations, hyponymy and part-of, and, for each of them, we have calculated the coefficient using just the correct and incorrect triples $(C + I)$, and adding the triples with a wrong relation $(C + I + WR)$.

All the measures, except for the baseline method with hyponymy, are positively correlated with the correctness of the triples. Still, just some are highly correlated. For both relations, WebPMI is the measure most correlated with the triples correctness. Other highly correlated measures are NWS and hasHits. It should be remarked that a measure as simple as hasHits outperforms all the measures except WebPMI. Also worth noticing is that, for the highest correlated measures, using the average of the scores with all the patterns leads generally to lower correlations. Furthermore, while for hyponymy it seems to be enough to use only the best patterns, for part-of the correlation is higher when using the two best patterns.

As expected, all correlations drop when the triples with wrong relations (WR) are included. Still, WebPMI and hasHits correlation coefficients are noteworthy.

3.5   Identification of Correct Triples

In the second experiment, we performed an information retrieval task, where the measures were used to filter incorrect triples automatically from our dataset. According to the score of each measure, we tested several cut points (θ), used to select the triples with a score higher than θ. Then, we computed the precision, recall and F_1 in the following manner, for measuring the quality and the quantity of triples selected:

$$Precision = \frac{Selected\_correct\_triples}{Selected\_triples} \qquad Recall = \frac{Selected\_correct\_triples}{Total\_correct\_triples}$$

$$F_1 = \frac{2*Precision*Recall}{Precision+Recall}$$

Table 5. Best F_1 measures and respective θ

R		Jaccard		Overlap		Dice		PMI		NWS		hasHits
		F_1	θ	F_1	θ	F_1	θ	F_1	θ	F_1	θ	F_1
Hypo	NP	70.5	$1E^{-4}$	68.5	$1E^{-4}$	77.5	$2E^{-4}$	68.5	0	68.5	0	**89.0**
	B	94.9	$2E^{-4}$	68.5	0	95.4	$2E^{-4}$	**96.2**	26	96.1	0.15	96.0
	2B	94.1	$2E^{-4}$	68.5	0	95.1	$2E^{-4}$	**96.3**	16	96.0	0.05	92.6
	Av	86.3	$2E^{-4}$	68.5	0	90.9	$2E^{-4}$	**96.2**	3	91.5	0.05	-
Part	NP	91.5	$2E^{-4}$	68.5	$1E^{-4}$	**93.7**	$2E^{-4}$	68.5	0	91.6	0.05	88.9
	B	93.9	$2E^{-4}$	80.7	$2E^{-4}$	94.0	$2E^{-4}$	94.3	32	**94.7**	0.25	92.8
	2B	93.8	$2E^{-4}$	75.6	0.05	93.8	$2E^{-4}$	94.7	33	**94.9**	0.2	94.1
	Av	86.7	$2E^{-4}$	68.5	0	90.3	$2E^{-4}$	**94.5**	4	87.1	0.05	-
Including triples with wrong relation (WR)												
Hypo	NP	51.0	$1E^{-4}$	51.0	$1E^{-4}$	51.0	$1E^{-4}$	51.0	0	51.0	0	**61.8**
	B	75.3	$2E^{-4}$	54.7	$2E^{-4}$	**75.3**	$3E^{-4}$	69.9	28	75.2	0.25	69.5
	2B	74.8	$2E^{-4}$	51.1	0	**74.9**	$5E^{-4}$	69.9	16	74.1	0.25	68.2
	Av	72.7	$2E^{-4}$	51.1	0	**75.1**	$2E^{-4}$	71.9	4	74.8	0.05	-
Part	NP	70.6	$2E^{-4}$	51.0	$1E^{-4}$	70.7	$4E^{-4}$	62.1	1	**72.5**	0.05	61.5
	B	65.5	$2E^{-4}$	62.1	$2E^{-4}$	65.2	$2E^{-4}$	**86.9**	42	62.3	0.2	65.7
	2B	65.4	$2E^{-4}$	59.4	0.05	65.6	$2E^{-4}$	**85.5**	41	67.5	0.2	68.8
	Av	59.7	$2E^{-4}$	51.0	0	62.3	$2E^{-4}$	**68.6**	4	60.7	0.05	-

For each measure, table 5 has the best F_1 scores and the respective θ. These values are presented, first, using only the correct and incorrect triples ($C + I$), and then adding the triples with the wrong relations ($C + I + WR$), which, for this task, were considered to be incorrect. This way, we compare how the similarity measures behave ideally or in a more realistic scenario, where, sometimes, extracted triples only fail on identifying the type of the relation.

Even though some of the measures had low correlations in the previous experiment, most of them achieve high F_1, and significantly outperform the baseline. Yet, the best F_1 measure without the triples with the wrong relation is achieved by the WebPMI using the two best patterns, with $\theta=16$ for hyponymy and $\theta=33$ for part-of. When triples with the wrong relation are added, WebPMI is still the best for part-of triples. Using only the best pattern with $\theta=42$, it achieves 86.9% F_1. WebPMI is, however, worse for hyponymy, where WebJaccard, WebDice and NWS outperform it, in this order.

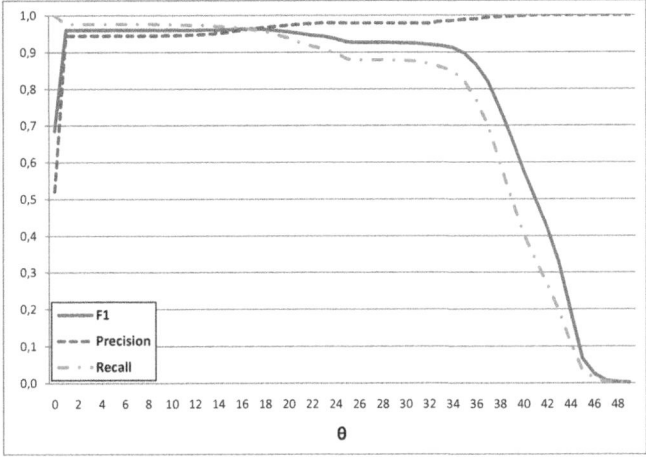

Fig. 1. Identification of the correct hyponymy triples from a set including correct and incorrect triples, with WebPMI using the two best patterns

In the latter experimentation scenario, F_1 measures are lower, around 75%. This is still good, considering that, sometimes, if we change the type of a relation from hyponymy to part-of, we still get acceptable relations, as the following situations:

- (*economy* hyponym-of *system*) ◇ (*economy* part-of *system*);
- (*computer* hyponym-of *machine*) ◇ (*computer* part-of *machine*).

Figures 1 and 2 are examples of how precision, recall and F_1 change with θ, in two situations where WebPMI is used in the identification of correct triples.

4   Related Work

Classical IE systems rely on textual patters that frequently denote semantic relations (see [16]). However, some trade-off is often needed in the selection of patterns because, on the one hand, some of them occur rarely and, on the other hand, the most frequent are usually ambiguous. In order to increase the recall and to minimise the effort needed to encode the patterns, state of the art IE systems typically have a (weakly) supervised pattern learning component (e.g. [20] [22]), which, nevertheless, is prone to extract more noise.

Therefore, some mathematical models have been proposed to filter incorrect triples [6] or to estimate the reliability of learned patterns [20]. This lead directly to higher precision and, eventually by using more ambiguous patterns, higher recall. Having in mind the distributional hypothesis [15], these filters are generally based on distributional similarity measures, which quantify the similarity of words according to their distribution in large corpora.

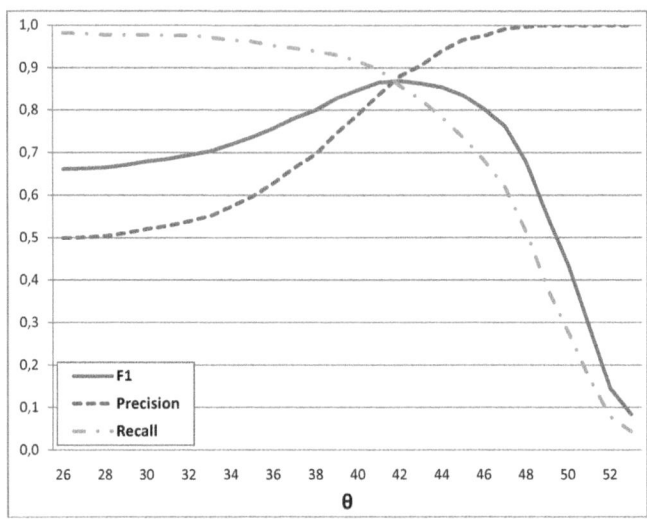

Fig. 2. Identification of the correct part-of triples from a set including correct, incorrect and triples with a wrong relation, with WebPMI using the best pattern

In the last decade, the Web became an attractive target for the extraction of huge quantities of knowledge (e.g. in [12] [9] [2] [22]). It started as well to be seen as an interesting infrastructure for quantifying and validating knowledge extracted automatically, not only because of its size, variety of subjects and redundancy, but also because web search engines provide an efficient interface.

Distributional similarity measures were adapted for the Web and were used, for instance, to rank semantic relations [9] [11]. They have also been exploited as features and combined with lexical patterns indicating synonymy in a robust metric which was claimed [4] to outperform all web-based similarity metrics.

PMI-IR [21] is a popular measure for searching the Web for pairs of similar words. Variations of the PMI have been used to reduce the noise in information extracted from the Web [12] [2]. However, in the latter works, the similarity measures were adapted and, instead of searching for the entities alone, they were used to measure the similarity between the entities and indicative patterns. The obtained scores can thus be exploited to assess the likelihood of the triples or the quality of the patterns. For instance, Etzioni et al. [12] compute the correction likelihood of hyponymy triples, held between classes and named entities. They rely on the ratio between the hits of the named entity in a hyponymy pattern connecting it to the class (e.g. *Liege is a city*), and the hits of the entity alone.

A very interesting work on this topic [11] defines a probabilistic model for evaluating the impact of redundancy, sample size and different extraction rules on the correctness of extracted information. This model is claimed to outperform models based on PMI, but was evaluated in four relations (Corporations, Countries, CEO of a company, and Capital of a Country) which are simpler and less ambiguous than lexico-semantic relations. For instance, all the former relations either have a static argument (X is-a *country*, Y is-a *corporation*) or are a one-to-one correspondence (a country has only one capital and a company one CEO). On the other hand, lexico-semantic relations such as hypernymy have always variable arguments (e.g. most lexical entities have hyponyms or hypernyms and a lexical entity may have several hyponyms). Furthermore, hypernymy can be held between entities on different levels of a hierarchy (e.g. *animal* hypernym-of *mammal*, *animal* hypernym-of *dog*, *animal* hypernym-of *boxer*, *mammal* hypernym-of *dog*, *dog* hypernym-of *boxer*, etc.).

More than assigning probabilities, the similarity measures can take advantage of the redundancy of the Web to validate knowledge, including not only semantic triples, but also question-answer pairs [18]. In order to verify how well automatic web-based validation performs, it has been put side-by-side with the manual evaluation of triples obtained after analysing the sequence of search engine queries in the same session [10].

Besides validation, web-based similarity measures have been used for other tasks, such as suggesting hyponymy relations between named entities and the concepts of an ontology [8] or to identify aliases of named entities [3].

5   Concluding Remarks

We have conducted several experiments to confirm if several web-based distributional similarity measures were well suited to validate lexico-semantic relations. These measures were applied to sets of correct and incorrect triples from WordNet. First, we confirmed that the scores given by some measures are highly correlated to the correction of triples. Then, we performed an information retrieval task consisting of the identification of correct triples, based on the scores of the measures. All the measures had high F_1 scores, some of them higher than 96% for hyponymy and higher than 94% for part-of.

These results are promising and we believe that the best performing measures can be used as an alternative to manual evaluation of relational triples, extracted automatically from textual resources. Even though our experiments were performed for English hyponymy and part-of triples, we intend to make our framework available for the validation of other kinds of relations (e.g. causation-of, purpose-of, headquarters-of, founded-by), eventually, in other languages.

Acknowledgements. Hernani Pereira Costa is supported by the FCT scholarship grant BII/FCTUC/C2008/CISUC/2^{nd}Phase. Hugo Gonçalo Oliveira is supported by the FCT scholarship grant SFRH/BD/44955/2008, co-funded by FSE.

References

1. Bennett, C.H., Gacs, P., Gcs, P., Member, S., Li, M., Vitanyi, P.M.B., Zurek, W.H.: Information Distance. IEEE Transactions on Information Theory 44, 1407–1423 (1998)
2. Blohm, S., Cimiano, P., Stemle, E.: Harvesting relations from the web: quantifiying the impact of filtering functions. In: Proc. 22nd National Conf. on Artificial Intelligence, pp. 1316–1321. AAAI (2007)
3. Bollegala, D., Honma, T., Matsuo, Y., Ishizuka, M.: Mining for personal name aliases on the web. In: Proc. 17th International Conf. on the World Wide Web, pp. 1107–1108. ACM (2008)
4. Bollegala, D., Matsuo, Y., Ishizuka, M.: Measuring semantic similarity between words using web search engines. In: Proc. 16th International Conf. on the World Wide Web, pp. 757–766. ACM, New York (2007)
5. Brank, J., Grobelnik, M., Mladenić, D.: A survey of ontology evaluation techniques. In: Proc. Conf. on Data Mining and Data Warehouses, SIKDD (2005)
6. Cederberg, S., Widdows, D.: Using LSA and Noun Coordination Information to Improve the Precision and Recall of Automatic Hyponymy Extraction. In: Proc. Conf. on Computational Natural Language Learning, pp. 111–118 (2003)
7. Cilibrasi, R., Vitanyi, P.M.B.: Normalized Web Distance and Word Similarity. Computing Research Repository, ArXiv e-prints (2009)
8. Cimiano, P., Staab, S.: Learning by googling. SIGKDD Explorations Newsletter 6(2), 24–33 (2004)
9. Cimiano, P., Wenderoth, J.: Automatic Acquisition of Ranked Qualia Structures from the Web. In: Proc. 45th Annual Meeting of the Association of Computational Linguistics, pp. 888–895. ACL, Prague (2007)
10. Costa, R.P., Seco, N.: Hyponymy extraction and web search behavior analysis based on query reformulation. In: Geffner, H., Prada, R., Machado Alexandre, I., David, N. (eds.) IBERAMIA 2008. LNCS (LNAI), vol. 5290, pp. 332–341. Springer, Heidelberg (2008)
11. Downey, D., Etzioni, O., Soderland, S.: A probabilistic model of redundancy in information extraction. In: Proc. 19th International Joint Conf. on Artificial Intelligence, pp. 1034–1041. Morgan Kaufmann Publishers Inc., San Francisco (2005)
12. Etzioni, O., Cafarella, M., Downey, D., Popescu, A.M., Shaked, T., Soderland, S., Weld, D.S., Yates, A.: Unsupervised named-entity extraction from the web: an experimental study. Artificial Intelligence 165(1), 91–134 (2005)
13. Fellbaum, C. (ed.): WordNet: An Electronic Lexical Database (Language, Speech, and Communication). MIT (May 1998)
14. Gracia, J.L., Mena, E.: Web-Based Measure of Semantic Relatedness. In: Bailey, J., Maier, D., Schewe, K.-D., Thalheim, B., Wang, X.S. (eds.) WISE 2008. LNCS, vol. 5175, pp. 136–150. Springer, Heidelberg (2008)
15. Harris, Z.: Distributional structure. In: Papers in Structural and Transformational Linguistics, pp. 775–794. D. Reidel Publishing Comp., Dordrecht (1970)
16. Hearst, M.A.: Automatic acquisition of hyponyms from large text corpora. In: Proc. 14th Conf. on Computational Linguistics, pp. 539–545. ACL, Morristown (1992)
17. Lenat, D.: CYC: A Large-Scale Investment in Knowledge Infrastructure. Communications of the ACM 38, 33–38 (1995)
18. Magnini, B., Negri, M., Prevete, R., Tanev, H.: Is It the Right Answer? Exploiting Web Redundancy for Answer Validation. In: Proc. 40th Annual Meeting of the Association for Computational Linguistics, pp. 425–432 (2002)

19. Oliveira, P.C.: Probabilistic Reasoning in the Semantic Web using Markov Logic, pp. 67–73. University of Coimbra, Faculty of Sciences and Technology, Department of Informatics Engineering (July 2009)
20. Pantel, P., Pennacchiotti, M.: Espresso: Leveraging Generic Patterns for Automatically Harvesting Semantic Relations. In: Proc. 21st International Conf. on Computational Linguistics and 44th Annual Meeting of the Association for Computational Linguistics (COLING-ACL), pp. 113–120. ACL, Sydney (2006)
21. Turney, P.D.: Mining the Web for Synonyms: PMI-IR versus LSA on TOEFL. In: Flach, P.A., De Raedt, L. (eds.) ECML 2001. LNCS (LNAI), vol. 2167, pp. 491–502. Springer, Heidelberg (2001)
22. Wu, F., Weld, D.S.: Open Information Extraction Using Wikipedia. In: Proc. 48th Annual Meeting of the Association for Computational Linguistics, pp. 118–127. ACL, Uppsala (2010)

A Resource-Based Method for Named Entity Extraction and Classification*

Pablo Gamallo and Marcos Garcia

Centro de Investigação em Tecnologias da Informação (CITIUS),
Universidade de Santiago de Compostela, Galiza, Spain
{pablo.gamallo,marcos.garcia.gonzalez}@usc.es

Abstract. We propose a resource-based Named Entity Classification
(NEC) system, which combines named entity extraction with simple
language-independent heuristics. Large lists (gazetteers) of named
entities are automatically extracted making use of semi-structured information from the Wikipedia, namely infoboxes and category trees.
Language-independent heuristics are used to disambiguate and classify
entities that have been already identified (or recognized) in text. We
compare the performance of our resource-based system with that of a
supervised NEC module implemented for the FreeLing suite, which was
the winner system in CoNLL-2002 competition. Experiments were performed over Portuguese text corpora taking into account several domains
and genres.

1 Introduction

Named Entity Recognition and Classification (NERC) is the process of identify-
ing and classifying proper names of people, organizations, locations, and other
Named Entities (NEs) within text. NERC is a crucial task in several natural
language applications, namely Question Answering and Information Extraction.
This paper will be focused on the second step of the task, i.e., on Named En-
tity Classification (NEC). Note that we use here the term NER (Named Entity
Recognition) in a narrow sense: it is defined as the process of just identifying
NEs.

Most approaches for NEC adopt machine learning techniques as a way to
automatically induce statistic classifiers starting from a collection of training
examples. The main drawback of these supervised techniques is the requirement
of a large amount of annotated corpora. The unavailability of such corpora and
the high cost required to build them lead to search for alternative resource-based
methods.

In this paper, we propose a NEC system requiring no human intervention
such as manually labeling training data (supervised learning) or manually creating gazetters (i.e., repositories of named entities). This system combines named

---

* This work has been supported by Ministerio de Educació y Ciencia (Spain), within
the project Ontopedia, with reference : FFI2010-14986.

L. Antunes and H.S. Pinto (Eds.): EPIA 2011, LNAI 7026, pp. 610–623, 2011.

entity extraction with simple language-independent heuristics for named entity disambiguation. In order to extract named entities, we use semi-structured information from the Wikipedia, namely infoboxes and category trees. This technique allows us to create large gazetteers of entities, such as lists of persons, organizations, and locations. The second step uses language-independent rules to classify entities in the context of a given text (i.e., entity disambiguation). Only disambiguation among different homonyms is considered (e.g., "Austin" as town or person). Polysemy and metonymy of proper names are not taken into account. We compare the performance of our resource-based system with that of a supervised NEC module implemented for the FreeLing suite [7,9], which was the winner system in CoNLL-2002 competition. Experiments were performed over Portuguese text corpora considering several domains and genres.

More precisely, the major contributions of this paper are the following:

– adding a new Portuguese NEC module to the FreeLing package,
– comparing a supervised NEC method with a resource-based strategy,
– analyzing the portability of both strategies to new domains and textual genres.

The article is organized as follows. The following section (2) introduces some related work. Then, Section 3 describes and justifies the classification criteria employed by our NEC systems. Next, Section 4 describes in detail the resource-based method and, in Section 5, we report the experiments performed on several Portuguese corpora. Finally, some conclusions are put forward in 6.

2 Related Work

The current dominant methods to named entity identification and classification are based on supervised learning. This is evidenced by the fact that most of the 28 systems presented at both CoNLL-2002 and CoNLL-2003 rely on supervised strategies. These learning strategies consist in creating disambiguation rules (classifier) based on discriminative features found in an annotated corpus (training corpus). The variants of this general strategy include different learning algorithms: Boosting [9], Support Vector Machines (SVM) [3], or Conditional Random Fields (CRF) [17].

Alternative systems are based on resource-based techniques that can classify named entities without prior training. These techniques require external resources such as WordNet [1] or gazetteers [18]. The latter work is very close to our proposal. As our system, [18] describes a method consisting of two modules. The first one automatically creates gazetteers of entities, and the second one uses language and domain independent heuristics for entity classification. However, there is a significant difference between their system and our proposal: their extractor of gazetteers needs some manual supervision. In particular, their extraction strategy requires some lists of seed named entities to generate queries and retrieve Web pages containing occurrences of the seed entities. By contrast, we do not need to manually define any prior seeds since our extraction strategy takes advantage of the semi-structured information found in Wikipedia.

In this sense, we must mention there exists recent interesting work using Wikipedia as gold standard corpora to train supervised NEC classifiers [19].

Finally, there also exist several NEC systems for Portuguese language. Most are rule-based, language dependent approaches [5,2,23,10], few supervised systems [11], and even one hybrid (stochastic and rule-based) method [12]. None of them is based on a resource-based strategy.

3   Classification Criteria

Since the MUC-6 competition [13], the main three semantic classes of proper names used for the NEC task are "persons", "locations", and "organizations". These classes were known as "enamex". Then, in CoNLL 2002 [24] the type "miscellaneous" was included to encode proper names falling outside the three "enamex" categories. Temporal expressions and some numerical expressions such as amounts of money and other types of units are also accepted as NEs in many NEC tasks.

The criteria given to the annotators to tag proper names in context, using a predefined set of classes, may change considerably according to the guidelines of the competition. One of the main problems arising in annotation is metonymy/polysemy. For instance, it is common to use names of countries, cities, or other locations to make reference to some kind of organization (*"Germany* signed the treaty", "In the morning *Tokyo* lost 3.7%"), a group of people (*"Spain* is against the Iraq war"), or even abstract entities such as economic systems (*"Portugal* grew 0.2% in the second quarter"), complex structures and cultural entities ("I miss *Portugal*"), etc. Different classification criteria have been used in previous competitions: in MUC-6 metonymy/polysemy was not taken into account, only homonyms were considered. In CoNLL, only some metonymy types were distinguished. For instance, countries referring to their governments are annotated as organizations: in *"Germany* signed the treaty" the proper name "Germany" is classified as an organization and not as a location. However, metonymies dealing with other types of organizations were not considered: in "Tokyo lost 3.7%", "Tokyo" is taken as the name of a city and then a location. Finally, in HAREM[22] were considered many types of metonymy and polysemy. For example, the proper name "Portugal" in "I miss *Portugal*" is annotated as an abstract entity. In this competition, countries are perceived as very ambiguous words and, in consequence, may be annotated not only as locations, but also as organizations, groups of people, or even abstract entities.

In our experiments, we decided to annotate training and test corpora considering only homonymy, and disregarding metonymic interpretations of NEs. Such a decision was motivated by several reasons, which are described in the following subsections.

3.1   Criticism in Lexical Semantics

NEC is perceived as a specific type of Word Sense Disambiguation (WSD). One of the main drawbacks of most WSD tasks is that they are based on the naive enumerative model of word meaning [21]. According to this model, the meaning of an ambiguous word is a set of senses, and one of them is selected in the context of the utterance. Differences between unrelated and related senses, that is, between homonymy and metonymy/polysemy, are not taken into account by the disambiguation procedure. By contrast, other lexical models try to represent the range of possible word senses in a more compact way than by enumeration. In particular, [20] distinguishes two representations:

- underspecified representations for a polysemous word,
- a list of alternative (or disjoint) readings for homonyms.

Polysemous words are specified (or *precisified*) under certain conditions, while homonyms are enumerated and disambiguated as in the well-known WSD models. Similar assumptions can be found in [15]. So, following these approaches, we will use classic WSD techniques, not to identify metonymy interpretations or very specific senses of NEs, but just to disambiguate their homonyms. Metonymy resolution and sense *precisification* are much more complex tasks requiring more complex and sophisticated techniques which are beyond the objective of our work.

3.2   Polysemy and Homonymy in Psycholinguistics

Psycholinguistic experiments seem to prove that polysemy and homonymy are different phenomena. In [4], the experiments performed offer neurophysiological support for modelling homonymy by means of different mental entries, while polysemy is compacted in single entries. Similarly, the experiments described in [16] revealed different cognitive processing strategies depending on the type of lexical ambiguity (homonymy or polysemy). So, the use of the same WSD technique for dealing with both lexical ambiguities seems to be not very appropriate.

3.3   Identity Criteria in Formal Ontology

There are some work on Formal Ontology [6,14] distinguishing between lexical (metonymy/polysemy) and ontological relations (IS-A, which is used to classify entities). The Formal Ontology framework criticizes the abuse of IS-A roles and multiple inheritance to deal with lexical polysemy. This results in overloading ontologies. To simplify ontologies and classification, they define the IS-A relation by means of the notion of *identity criteria*. Individuals with different identity criteria are in different classes, even if they can be semantically related by different kinds of dependencies, colocalizations, etc. For instance, a university is a social organization, and not a place or building, according to their functional criteria of identity. If its functional identity is destroyed, then the organization ceases to exist, even if the place or building is not destroyed. From this viewpoint, a university should be always classified as a functional organization, even if its *dependent* entities (location or group of people) can be highlighted in some linguistic contexts.

3.4   Encyclopedic Organization

As in the case of traditional dictionaries, in encyclopedias only homonyms are separated in different entries. For instance, if the same name is used for two different individuals, the encyclopedia defines two separated entries. By contrast, the location, population, and government of a country are not separated in different entries. All this information is organized within a single one. But there exist some borderline cases where the difference between homonymy and polysemy is not very clear, for instance the case of national football teams. Should they be a component of countries? or should be considered as separated entries? In Wikipedia, a national football team is assigned a single entry, so it is not perceived as a part of the country. We will follow the same convention.

3.5   Commercial NEC Systems

Finally, we must point out that many commercial NEC systems make classification without polysemy: Alchemy[1], Extractiv[2], and Daedalus[3].

4   Resource-Based NEC System

The NEC system we propose classifies 4 types of named entities: persons, locations, organizations, and other entities ("miscellaneous"). It can be considered as a resource-based strategy which consists of two tasks. First, three large lists of NEs (persons, locations, and organizations) are automatically generated with the aid of semi-structured information from Wikipedia. Second, some disambiguation rules are applied on previously identified NEs, in order to solve homonyms and unknown NEs. Even if our experiments will be focused on Portuguese, the disambiguation rules we propose can be considered as (almost) totally independent on a specific language and knowledge domain.

4.1   Automatic Generation of Gazetteers and Trigger Words

The main objective here is to generate three lists of NEs, one for each semantic class, by exploiting both the category trees and infoboxes of Wikipedia. In particular, category trees and infoboxes will allow us to identify common nouns referring to different subclasses of persons, locations, and organizations. This way, the extraction task consists of two steps: first, we select common nouns (trigger words) denoting subclasses of the three target classes and, then, by means of these subclasses we extract the lists of NEs (gazetteers).

---

[1] http://www.alchemyapi.com/api/entity/
[2] http://extractiv.com
[3] http://www.daedalus.es/productos/stilus/stilus-ner.html

Table 1. Portuguese subclasses of Persons, Locations, and Organizations

Persons	Locations	Organizations
Misses, Políticos, Designers, Professores, Personagens, Criminosos, Chefes, Escritores, Artistas, Treinadores	Terra, Planeta, Mundos, Locais, Ilhas, Cidades, Subdivisões, Rios, Países, Monumentos, Hoteis	Instituições, Partidos, Federações, Associações, Sindicatos, Clubes, Entidades, Empresas, Cooperativas

Subclasses. In the first step, the goal is to search into the category tree of Wikipedia a set of categories which are subclasses of persons, locations, and organizations. The strategy is the following: we identify the categories containing in the *head* position the words "People" "Places", and "Organizations" as well as their synonyms, and then, we extract the *head* of their hyponyms. Let us see how, in Portuguese, we extract `Partidos` (Parties) as a subclass of `Organizações` (Organizations). First, we select the generic category `Organizações políticas` (political organizations), since it contains in the *head* position the target category `Organizações`. Then, we search its hyponyms and identify, among others, the category `Partidos políticos` (political parties). Finally, we extract the head of its expression, namely `Partidos` (parties), and put it in the list of subclasses of organizations. Table 1 shows a sample of the Portuguese subclasses selected for each target class.

These lists of nouns will be used, on the one hand, as *trigger* words (after lemmatization) in the disambiguation process (see section 4.2) and, on the other, as seeds to generate the gazetteers.

Gazetteers. The second step consists in extracting those NEs considered as instances of the selected subclasses. Two strategies were implemented.

The first one verifies whether the set of categories of each Wikipedia article contains one or more of the selected subclasses. If a subclass is contained in the set of categories, then the title of the article, which is a named entity, is classified as an instance of this subclass and, then, of the corresponding generic class. For instance, let us suppose that the article with the title `Rui Zink` is assigned the category `Escritores de Portugal` (writers of Portugal). As this category contains `Escritores` (writers), which is a subclass of *Persons*, then we add `Rui Zink` to the list of persons.

The second strategy follows the same procedure but, instead of checking the set of categories of an article, we search within the "attribute-value" structure of infoboxes. If one of the subclasses is contained in the value of an infobox, then we add the title of the article to the list of the corresponding generic class. To filter out noise, it is possible to restrict the search by using only those values tagged with some specific infobox attributes, e.g., "type", "occupation", etc.

This process let us generate three gazetteers of NEs: a list of people, a list of locations, and a list of organizations. In case of homonymy, a NE can be in more than one list. Finally, the list of subclasses are lemmatized and used as *trigger words* in the process of disambiguation.

4.2   Disambiguation and Classification

The input of our NEC system is PoS tagged text containing single and composite proper names already identified as NEs. In addition, two external resources are required: both the gazetteers and trigger words automatically generated from Wikipedia.

Given an identified NE, the algorithm we use to select a semantic class can be informally described as follows:

list lookup strategy: if the NE matches an entry appearing in only one gazetteer, then it can be considered as an unambiguous NE and can be assigned the class of the gazetteer.

contextual checking: if the NE appears in various gazetteers (homonymy) or it is an unknown NE (missing in gazetteers), then we search within its linguistic context for relevant trigger words. In particular, we check if the words appearing to the left and to the right of the target NE (in our experiments, the window size is 3) match the lists of trigger words. For instance, the NE "Austin" will be classified as a location in the context "Austin, a town in ..." because the common noun "town" is a trigger word in the list of locations. If there are several trigger words of different classes in the context of the target NE, we give preference to the closest one. If there are two triggers at the same distance, the preference is given to the left position. Finally, if there is a preposition between the trigger and the NE, then the trigger is not considered. For instance, in "the king of Spain", the trigger "king" is a *person* but the NE "Spain" is a *location*. This last heuristic is motivated by the fact that prepositions tend to be used to syntactically relate nouns and NEs of different semantic classes.

class ranking: if the NE is ambiguous and cannot be disambiguated by contextual checking (previous step), then we select a single class by taking into account our ranking of classes: *person > location > organization*. That is, if the NE appears in the gazetteers of persons and locations, we select the *person* reading. If it is in gazetteers of locations and organizations, the preference is given to the *location* class. This ranking was not set *ad hoc*. It was defined by taking into account the distribution of classes within the gazetteers extracted from the Wikipedia.

internal checking: if the NE is unknown and cannot be assigned a class by contextual checking, then we check some of its constituent expressions. In particular, we check whether the first expression of a NE matches the first expression of a NE in a gazetteer or a common noun in a trigger list. For instance, since the first expression of the NE "University of Alberta" is a trigger word ("university") for organizations, the target NE is classified as an organization. In case of several options, we give preference to gazetteers over trigger words and follow the class ranking defined above.

default rule: if no rule is applied, the NE is classified as "miscellaneous".

Note that these rules are language and domain independent. We do not make use of specific cues such as organizational designators (e.g., Corp.) or personal suffixes (e.g., Jr.).

5 Experiments

The resource-based method described in this paper was compared with a supervised learning system, namely the NEC module of FreeLing [7]. FreeLing is an open source suite of modules for natural language processing: lemmatization, PoS tagging, named entity recognition/identification (NER), named entity classification (NEC), chunking, etc. The NEC module was ranked among the top performing systems in the CoNLL-2002 competition. It was based on a boosting algorithm (AdaBoost) which consists in combining many base classifiers. It may make use of external resources such as gazetteers and trigger words to define specific features. In the experiments reported in this paper, both our resource-based NEC method and the supervised FreeLing system take as input the basic NER module of FreeLing, an heuristic rule based strategy, which takes into account capitalization patterns, functional words and dictionary lookup. This module achieves 90% precision [8].

The overall organization of our experiments is the following: first, we generate the different lists of gazetteers and triggers that will be used in the experiments. Then, we train the NEC module of FreeLing for Portuguese language. Next, five different test corpora, belonging to different domains and genres, are annotated. Finally, the two systems are applied to the five test corpora, and the results obtained are compared.

5.1 Gazetteers and Triggers

We follow the two strategies defined above in section 4, with the aim of generating two versions of gazetteers. First, three lists with $115,650$ named entities were built by checking the categories of the Wikipedia articles (first strategy). Second, the attribute-value structure of the infoboxes was used to select three lists with $37,445$ NEs (second strategy). In addition, a third version was also considered for our experiments, namely the gazetteers freely available in the Spanish NEC module of FreeLing.

To compare the impact of gazetteers in terms of size in a NEC task, we will make use of three sets of gazetteers (see Table 2): *es* stands for the gazetteers taken from the Spanish version of FreeLing, *es+infobox* corresponds to the union of *es* with the gazetteers extracted from infoboxes, and finally *es+infobox+cat* consists of the previous set and the NEs extracted using the article's categories. Table 2 shows the number of persons (PER), locations (LOC), and organizations (ORG) found in each gazetteer version.

Table 2. Size of the three gazetteers used in the experiments

	es	es+infobox	es+infobox+cat
PER	2,598	17,600	64,735
LOC	7,312	23,732	58,305
ORG	2,263	4,586	13,599
TOTAL	12,173	45,918	136,639

Concerning trigger words, only 419 nouns (57 types of organizations, 82 types of locations, and 320 types of persons) will be used in the experiments.

5.2   Training the NEC Module of FreeLing

We trained both the PoS tagged and NEC modules of FreeLing on an manually annotated European Portuguese corpus. In particular, we selected 87,000 tokens from Bosque 8.0[4], containing about $5,000$ proper names which we have manually classified according to the classification criteria defined in Section 3, that is, only homonymy was considered. The training corpus consists of news of a Portuguese newspaper (*Público*). The NEC module for Portuguese will be freely available in the next version of FreeLing.

5.3   Test Corpora

In order to perform experiments in different domains and textual genres, 5 different test corpora were elaborated:

bosque: 50,000 test tokens from Bosque 8.0 (part of CETEMPúblico), which is constituted by news of *Público* (journalistic genre, open domain)
wiki: 30,000 tokens from Portuguese Wikipedia (first paragraph per article) (encyclopedic genre, open domain)
europarl: 30,000 tokens from the Portuguese version of the parallel corpus Europarl[5] (formal genre, political domain)
br: 24,000 tokens from the Brazilian Portuguese part of European Corpus Initiative Multilingual Corpus I (ECI/MCI)[6] (technical genre, economical domain)
harem: 70,000 tokens from HAREM competition [22] (open genre, open domain)

The proper names contained in *bosque, wiki, europarl,* and *br* were manually annotated by us following the simple criterion of homonymy disambiguation. No metonymy was considered. By contrast, *harem,* which is the corpus used as reference in the Portuguese NEC competition, was annotated by other linguists according to more complex criteria, since many types of metonymy were taken into account. We had to adapted the set of categories used in HAREM competition to the four main categories of our experiments.

The heterogeneity of these test corpora will allow us to verify whether the two compared systems may be ported to new domains or textual genres without losing their performance.

5.4   Results

The two NEC systems were applied on the five test corpora provided with the same sets of gazetteers. In terms of computational efficiency, the resource-based system turned out to be about 40% speeder than the supervised one.

---

[4] http://www.linguateca.pt/floresta/corpus.html
[5] http://www.statmt.org/europarl/
[6] http://www.elsnet.org/eci.html

Table 3. F-score$_{\beta=1}$ values of the two compared systems (and a baseline), provided with four different gazetteers. Experiments performed on five test corpora

	null	es	es+infobox	es+infobox+cat
baseline (*bosque*)	.33	.33	.33	.33
resource (*bosque*)	.33	.54	.69	.74
superv (*bosque*)	.74	.75	.77	**.78**
baseline (*wiki*)	.57	.57	.57	.57
resource (*wiki*)	.32	.48	.75	**.92**
superv (*wiki*)	.79	.80	.83	.88
baseline (*br*)	.35	.35	.35	.35
resource (*br*)	.61	.73	.74	**.75**
superv (*br*)	.50	.53	.62	.63
baseline (*europarl*)	.45	.45	.45	.45
resource (*europarl*)	.46	.48	.48	.74
superv (*europarl*)	.77	**.78**	.76	.76
baseline (*harem*)	.41	.41	.41	.41
resource (*harem*)	.26	.39	.53	.58
superv (*harem*)	.56	.56	.59	**.60**
baseline (**average**)	.42	.42	.42	.42
resource (**average**)	.40	.52	.64	**.75**
superv (**average**)	.67	.68	.71	.73

The performance (f-score$_{\beta=1}$) of the two NEC systems (and a baseline) are presented in Table 3. The resource-based system is noted *resource* and the supervised is *superv*. As a baseline, we include the results obtained by using the strategy based on the most frequent category. In each column, we show the results obtained with different sets of gazetteers. In column *null* no gazetteer is used. Column *es* shows the results with the gazetteers taken from the Spanish NEC. Column *es+infobox* shows the results by combining *es* with the infobox extraction technique, and finally in column *es+infobox+cat*, the gazetteers also include the NEs learnt with the article's categories. Precision, recall, and f-score$_{\beta=1}$ were computed by means of the evaluation script (*conlleval*) used in CoNLL competition, considering only for evaluation those NEs correctly identified by the NER.

We can observe in Table 3 that no system is clearly better than other. On the one hand, the supervised strategy performs better on three test corpora: *bosque*, *europarl*, and *harem*. The good results obtained from *bosque* were expected, since this test is constituted by the same type of documents as in the training corpus. On the one hand, the resourceised system performs better on *wiki* and *br*. The high score obtained from *wiki* (92%) is not a surprise because most gazetteers were extracted from that corpus. In average, the resourceised technique achieves a slightly better f-score with the largest set of gazetteers: 75% against 73%. Let us note the low scores of *harem* are due to the fact that this test corpus was annotated according to different criteria as those used to elaborate both the training corpus of *superv* and the disambiguation rules of *resource*.

The main difference between the two approaches concerns the degree of dependence on knowledge-rich gazetteers. While the performance of *superv* remains quite stable regardless the size of gazetteers, *resource* requires very rich gazetteers to reach acceptable performance. Figure 1 shows how the performance of the two systems improves as the size of gazetteers increases, but the improvement curve is clearly more marked in the case of *resource*. It means that, as it was expected, the resource-based strategy is much more dependent on external gazetteers. By contrast, the supervised one relies on many different features, being only those defined from external resources a small part of the decision model. This is in accordance with the experiments reported in [8], where the same supervised NEC system merely improved 2 or 3 points its performance when it was trained with external resources such as gazetteers.

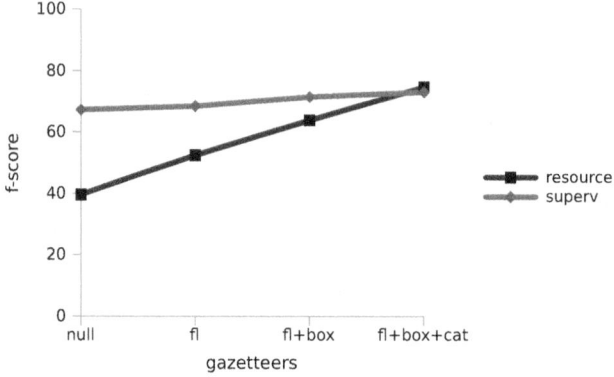

Fig. 1. Improvement curve of the two systems in function of the size of gazetteers (average f-score)

We observe a similar tendency in the case of trigger words (see Table 4). The resource-based strategy is clearly more dependent on the contextual information provided by trigger words than the supervised one. In brackets, we add the difference (in percentage points) between these results and those obtained by the same system but provided with trigger words. In average, *resource* decreases 5 percentage points while *superv* only 2.

Table 4. F-score$_{\beta=1}$ of the two systems without triggers (and with gazetteers *es+infobox+cat*)

	bosque	wiki	br	europarl	harem
resource (*no triggers*)	.69 (-5)	.92 (=)	.67 (-8)	.65 (-9)	.56 (-2)
superv (*no triggers*)	.76 (-2)	.88 (=)	.59 (-4)	.74 (-2)	.59 (-1)

Table 5. Results of the two systems for the four semantic categories. They were obtained from the *bosque* text corpus, with the largest gazetteers (*es+infobox+cat*).

resource (*bosque*)	precision	recall	f-score$_{\beta=1}$
PER	.86	.89	.88
LOC	.76	.54	.63
ORG	.79	.68	.73
MISC	.33	.51	.40
overall	74.48%	74.19%	74.33%

superv (*bosque*)	precision	recall	f-score$_{\beta=1}$
PER	.88	.94	.91
LOC	.85	.57	.68
ORG	.81	.70	.75
MISC	.35	.52	.42
overall	77.77%	77.54%	77.65%

Finally, in order to analyze in more detail the behaviour of the two systems, Table 5 breaks down the results into four categories: PER, LOC, ORG, and MISC. In particular, this table shows the results obtained with the largest gazetteers (*es+infobox+cat*), and the *bosque* text corpus. We can observe that the best performance is reached with the PER category, while the MISC category turns out to be the most hard to predict. The low values of MISC could be caused by either the difficulty of identifying a so general and heterogeneous category, or by the fact that we have not created any specific gazetteers with *miscellaneous* named entities.

5.5 Comparing with Related Work

It is difficult to establish a fair comparison between our systems and those described in other works, due to the specificities of each evaluation setup. However we should note that the performance of *superv* on *bosque* (about 78% f-score) is similar to that achieved by the same NEC system trained for Dutch [9]. It should also be noted that our training data (87, 000) is three times smaller than that available in CoNLL-2002 for Dutch and Spanish. This could be the reason of the slightly lower score obtained by our supervised method in comparison to the Spanish NEC system. On the other hand, a full comparison of our results with those obtained by the participants in the HAREM competition is not possible because the semantic classification criteria do not coincide. Besides, some errors could be produced when converting the original set of categories of HAREM to the basic set used in our experiments. However, we observed that our resource-based method achieves the same f-score (58%) as the best NEC system in that competition.

6   Conclusions

We have presented a named entity recognition system that avoids the need for supervision by making use of some language independent rules on automatically extracted external resources, namely gazetteers and trigger words. When comparing with a supervised system, we made the two following observations: First, the supervised strategy performs better when both the test and training corpora are similar (same genre and same domain). Second, our resource-based strategy is not worse than the supervised system when they are applied on a great variety of texts, especially if the domains and genres of these texts are not found in the training corpus. So, we conclude that if we need to work on domain-specific texts, it is worth manually annotating a corpus to tune a supervised system. Nevertheless, if we require a more generic NEC with acceptable performance on any type of text, our resource-based system could be a reasonable solution.

References

1. Alfonseca, E., Manandhar, S.: An unsupervised method for general named entity recognition and automated concept discovery. In: International Conference on General WordNet (2002)
2. Aranha, C.: O cortex e a sua participação no HAREM. In: Santos, D., Cardoso, N. (eds.) Reconhecimento de Entidades Mencionadas Em Português: Documentação e actas do HAREM, pp. 113–122 (2007)
3. Asahara, M., Matsumoto, Y.: Japanese named entity extraction with redundant morphological analysis. In: Human Language Technology Conference - North American Chapter of the Association for Computational Linguistics (2003)
4. Beretta, A., Fiorentino, R., Poeppel, D.: The effets of homonymy and polysemy on lexical access: an MEG study. Cognitive Brain Research 24, 57–65 (2005)
5. Bick, E.: Functional Aspects in Portuguese NER. In: Vieira, R., Quaresma, P., Nunes, M.d.G.V., Mamede, N.J., Oliveira, C., Dias, M.C. (eds.) PROPOR 2006. LNCS (LNAI), vol. 3960, pp. 80–89. Springer, Heidelberg (2006)
6. Borgo, S., Guarino, N., Masolo, C.: Stratified ontologies: the case of physical objects. In: Workshop on Ontological Engineering, ECAI 1996 (1996)
7. Carreras, X., Chao, I., Padró, L., Padró, M.: An Open-Source Suite of Language Analyzers. In: 4th International Conference on Language Resources and Evaluation (LREC 2004), Lisbon, Portugal (2004)
8. Carreras, X., Marquez, L., Padró, L.: Wide coverage spanish named entity extraction. In: Garijo, F.J., Riquelme, J.-C., Toro, M. (eds.) IBERAMIA 2002. LNCS (LNAI), vol. 2527, pp. 674–683. Springer, Heidelberg (2002)
9. Carreras, X., Marquez, L., Padró, L., Padró, M.: Named entity extraction using adaboost. In: COLING 2002 Proceedings of the 6th Conference on Natural Language Learning (2002)
10. Mota, C., Silberztein, M.: Em busca da máxima precição sem almanaques. o stencil/nooj no HAREM. In: Santos, D., Cardoso, N. (eds.) Reconhecimento de Entidades Mencionadas em Português: Documentação e actas do HAREM, pp. 191–208 (2007)
11. Ferrández, O., Kozareva, Z., Toral, A., Mu noz, R., Montoyo, A.: Tackling HAREM's portuguese named entity recognition task with spanish resources. In: Santos, D., Cardoso, N. (eds.) Reconhecimento De Entidades Mencionadas Em Português: Documentação e Actas Do HAREM, pp. 137–144 (2007)
12. Ferreira, E., Balsa, J., Branco, A.: Combining rule-based and statistical methods for named entity recognition in portuguese. In: V Workshop em Tecnologia da Informação e da Linguagem Humana, pp. 1615–1624 (2007)

13. Grishman, R., Sundheim, B.: Message understanding conference -6: A brief history. In: International Conference on Computational Linguistics (1996)
14. Guarino, N.: The role of identity conditions in ontology design. In: IJCAI 1999 Workshop on Ontologies and Problem-Solving Methods, Stockholm, Sweden (1998)
15. Kleiber, G.: Problèmes de Sémantique. Presses Universitaires Septentrion, Lille (1999)
16. Klepousniotou, E.: The processing of lexical ambiguity: Homonymy and polysemy in the mental lexicon. Brain and Language 81, 205–223 (2002)
17. McCallum, A., Li, W.: Early results for named entity recognition with conditional random fields, features induction and web-enhanced lexicons. In: Conference on Computational Natural Language Learning (CoNLL 2003), pp. 117–126 (2003)
18. Nadeau, D., Turney, P., Matwin, S.: Unsupervised named entity recognition: Generating gazetteers and resolving ambiguity. In: Canadian Conference on Artificial Intelligence (2006)
19. Nothman, J., Murphy, T., Curran, J.R.: Analysing wikipedia and gold-standard corpora for ner training. In: 12th Conference of the European Chapter of the ACL, Athens, Greece, pp. 612–620 (2009)
20. Pinkal, M.: Vagueness, ambiguity, and underspecification. In: Conference on Semantics and Linguistic Theory, pp. 181–201 (1996)
21. Pustejovsky, J.: The Generative Lexicon. MIT Press, Cambridge (1995)
22. Santos, D., Seco, N., Cardoso, N., Vilel, R.: HAREM: An advanced NER evaluation contest for portuguese. In: 5th International Conference on Language Resources and Evaluation - LREC 2006, Genova, Italy, pp. 1986–1981 (2006)
23. Sarmento, L.: O SIEMÊS e a sua participação no HAREM e no mini-HAREM. In: Santos, D., Cardoso, N. (eds.) Reconhecimento de Entidades Mencionadas em Português: Documentação e actas do HAREM, pp. 173–189 (2007)
24. Tjong, K.S., Erik, F.: Introduction ot the CoNLL-2002 shared task: Language independent named entity recognition. In: Conference on Natural Language Learning (2002)

Measuring Spelling Similarity
for Cognate Identification

Luís Gomes and José Gabriel Pereira Lopes

Centro de Informática e Tecnologias da Informação (CITI)
Universidade Nova de Lisboa, 2829-516 Caparica, Portugal
luismsgomes@gmail.com, gpl@fct.unl.pt

Abstract. The most commonly used measures of string similarity, such
as the Longest Common Subsequence Ratio (LCSR) and those based
on Edit Distance, only take into account the number of matched and
mismatched characters. However, we observe that cognates belonging to
a pair of languages exhibit recurrent spelling differences such as "ph" and
"f" in English-Portuguese cognates "phase" and "fase". Those differences
are attributable to the evolution of the spelling rules of each language
over time, and thus they should not be penalized in the same way as
arbitrary differences found in non-cognate words, if we are using word
similarity as an indicator of cognaticity.

This paper describes SpSim, a new spelling similarity measure for
cognate identification that is tolerant towards characteristic spelling differences that are automatically extracted from a set of cognates known
apriori. Compared to LCSR and EdSim (Edit Distance -based similarity),
SpSim yields an F-measure 10% higher when used for cognate identification on five different language pairs.

Keywords: String Similarity, Cognate Identification, Translation
Extraction.

1 Introduction

The most commonly used measures of spelling similarity for the task of cognate identification, such as the Longest Common Subsequence Ratio (LCSR)
and Edit-Distance-based measures, treat all spelling differences in the same way,
disregarding the fact that cognates exhibit spelling differences that are characteristic of each pair of languages. For example, "ph" and "f" consistently appear
in corresponding positions in English and Portuguese words such as "phase" and
"fase". Such characteristic spelling differences are often due to the evolution of
spelling rules of each language, and thus they should not be penalized (for the
purpose of identifying cognates) in the same way as arbitrary differences.

Measures that account for characteristic spelling differences have been proposed earlier [8,1,4], but they are complex, computationally expensive, and there
are no readily available implementations.

This paper describes SpSim, a new spelling similarity measure that is tolerant towards characteristic spelling differences that are automatically extracted

L. Antunes and H.S. Pinto (Eds.): EPIA 2011, LNAI 7026, pp. 624–633, 2011.

from a set of cognates known apriori. SpSim is conceptually simple, computationally light, and as shown in the paper, it performs better than LCSR and EdSim (a similarity measure based on Edit Distance). Our implementation is open source[1].

Cognates are words with a common etymological origin. Given the fact that many cognates belonging to different languages have similar spellings and the same meaning, the spelling similarity has been used as an indicator of cognaticity, and thus translational equivalence, between words of different languages. Cognate identification in parallel texts (texts that are translation of each other) is usually accomplished by combining a measure of spelling similarity with a statistical association measure like Dice, that take into account the co-occurrence frequency of both words. Some authors [8,4] have tried to identify cognate translations solely by looking at words' spelling, but it is easy to find examples of translations and non-translations that cannot be distinguished by their spelling similarity. For example ("phase", "fase") are a translation pair, while ("phrase", "frase") are *false friends* (words of different languages with similar spellings but different meanings).

In this paper we focus on improving the spelling similarity analysis between putative cognates. We present a similarity measure that recognizes substitution patterns that are characteristic of each pair of languages. As an example of such substitutions let's consider the English word 'pharmacy' and its Portuguese cognate 'farmácia'. We can transform 'pharmacy' into 'farmácia' by replacing 'ph' with 'f' and adding an acute accent to the second 'a'. Each of those substitutions appears in many other cognates, like for example in 'photographic' and 'fotográfica', 'phase' and 'fase', 'abacus' and 'ábaco', etc. These substitutions are characteristic of English-Portuguese cognates, and in general, each pair of languages has their own set of characteristic substitutions. We present a very simple and effective method to extract these substitution patterns from known cognates.

The main contribution in this paper is a new spelling similarity measure that is tolerant towards the substitutions patterns that it already knows, thus improving substantially the recall of cognate extraction, compared to other measures that have been commonly used throughout related literature.

2   Previous Work

Many heuristics have been presented in literature to assess the spelling similarity between words. Those heuristics range from being very basic, like the 4 character prefix rule introduced by [7] that takes as possible cognates those words that share a common prefix of at least 4 characters; to more sophisticated ones, like the longest common subsequence ratio (LCSR) introduced by [5], that considers as cognates those words that have a LCSR greater than some empirically fixed threshold. The longest common subsequence ratio is determined by the equation

---

[1] available from `http://hlt.di.fct.unl.pt/luis/spsim/`

$$LCSR(w_1, w_2) = 1 - \frac{|LCS(w_1, w_2)|}{max(|w_1|, |w_2|)} \tag{1}$$

where $LCS(w_1, w_2)$ is the longest (not necessarily contiguous) subsequence that is common to both words.

A similar measure was suggested by [3], based on the edit distance between words. Since the authors didn't give a name to that measure we call it EdSim (Edit-Distance-based Similarity measure), as we will use is as a baseline measure in our experiment. It is given by the equation

$$EdSim(w_1, w_2) = 1 - \frac{ED(w_1, w_2)}{max(|w_1|, |w_2|)} \tag{2}$$

where w_1 and w_2 are the words under consideration, and $ED(w_1, w_2)$ is the edit distance between them.

2.1   Dynamic Approaches

The technique for cognate identification presented by [6] is very sophisticated and unique, as it tries to learn spelling features that are characteristic of the pair of languages under consideration. Their approach is divided in two steps. First, they use a tool called SENTA to extract, from a bilingual corpus, sequences of characters (which may contain gaps) that are recurrent and cohesive in both languages. Afterwards, they take as possible cognates pairs of words that match together one or more characteristic sequences.

We classify this technique as being *dynamic* because it tries to identify characteristic spelling features and use them to automatically adapt itself to the pair of languages at hand. Conversely, the measures presented earlier are *static* as they lack the ability to adapt themselves.

It seems that the dynamic behaviour is closer to the way that we (people) identify cognates, as our brain is able to ignore some spelling differences that are recurrent in cognates of a pair of languages. This allows us to immediatly recognize cognates even in cases that get low scores from the static measures presented earlier. For example, LCSR and EdSim give both a score of just 0.5 to the English-Portuguese cognates 'accepted' and 'aceitou', and 'abruptly' and 'abruptamente'. However, any Portuguese speaker can promptly identify these words as being cognates with high confidence in that assessment. On the other hand, we tried to identify cognates between languages that are completely outside our sphere of knowledge and we performed worse than the static measures because all words look (just the same) alien to us. This leads us to speculate that our brain is not well suited to find similarities between things that are very distant from anything we know. We will close this philosophic discourse, which hopefully has given some food for thought, with the remark that in the remainder of the paper we describe a dynamic measure that *identifies and tolerantes dissimilarities that are recurrent in cognates* of the pair of languages under consideration.

3   SpSim

SpSim is based on standard string alignment using dynamic programming.

The spelling similarity is measured by the total length of misaligned segments that are not characteristic.

The next subsection formalizes the concept of substitution pattern, and shows how we extract them from cognates. Then, we present the SpSim equation to compute the spelling similarity between words, and we show how it (closely) relates to EdSim.

3.1   Substitution Patterns

SpSim relies on standard string alignment using dynamic programming (a good textbook is [2]). We use the standard weigthing table for computing the best global alignment for the two strings.

To aid the explanation we will use two English-Portuguese cognate word pairs as an example: 'photographic' and 'fotográfica', and 'achromatic' and 'acromático'. The alignment of these cognates, is shown below:

```
photographic        achromatic
::||||||:::||:      ||:|||:|||:
fotográf ica        ac romático
```

In this representation the pipe ('|') character marks aligned matches, while the colon (':') character marks aligned mismatches. If we eliminate all matched characters, we are left with the mismatched segments that contain some characteristic substitutions like ('ph', 'f'):

```
ph      aph         h      a
::      :::   :      :      :    :
f       áf    a             á    o
```

However, because we removed the context around the mismatched segments, some of these patterns are too generic: the insertion of 'a', the deletion of 'h' and the insertion of 'o', are patterns that are likely to appear even if the words are totally unrelated.

To improve the specificity of the patterns we include a fixed radius context of characters around each pattern. Empirically, we setled on a radius of one character, as it provides a good balance between specificity and generality. Furthermore, we have introduced two special characters, the caret (^) and the dollar sign ($) at the beginning and end of the aligned strings, respectively. These two characters are not allowed to appear in the words, and they allow us to distinguish patterns that appear as prefixes, infixes or suffixes. The patterns extracted with context of radius one are presented below:

```
^pho    raphi c $        chr mat c $
|::|    |:::| |:|        |:| |:| |:|
^ fo    rá fi ca$        c r mát co$
```

Now we have patterns much more specific, such as: the insertion of an 'a' at the end of a word whose last character is a 'c'. The complete list of patterns extracted from these two words is: ('^pho', '^fo'), ('raphi', 'ráfi'), ('c$', 'ca$'), ('chr', 'cr'), ('mat', 'mát'), and ('c$', 'co$').

3.2   Generalization of Substitution Patterns

The context characters of these patterns may be dropped if we find the same substitution in a different context, making the pattern more general. For example, applying the method given in the previous subsection to 'phase' and 'fase' we extract the pattern ('^pha', '^fa'), containing the substitution ('ph', 'f') that is also found in the pattern ('^pho', '^fo'), extracted in the previous example.

The generalization is done by dropping the context characters that are different between patterns containing the same substitution. In this case, we drop the right context character from both patterns and we obtain the generalized pattern is ('^ph', '^f').

3.3   SpSim Equation

The equation for computing SpSim is:

$$SpSim(w_1, w_2) = 1.0 - \frac{D(w_1, w_2)}{max(|w_1|, |w_2|)} \qquad (3)$$

where $D(w_1, w_2)$ is the length of all mismatched segments containing substitutions that are not known, $|w_1|$ is the length of w_1, and $|w_2|$ is the length of w_2.

$$D(w_1, w_2) = ED(w_1, w_2) - \sum d_i, \forall d_i \in M(w_1, w_2) \qquad (4)$$

To understand this distance function, lets start by considering the situation where $M(w_1, w_2)$ is an empty set. That happens whenever the mismatched segments between the two words are not a substitution pattern that we know about. In this situation, the value of $D(w_1, w_2)$ is exactly the same as $ED(w_1, w_2)$.

Conversely, whenever $M(w_1, w_2)$ is non empty, it means that mismatched segments contain one or more substitutions that we known. Each d_i is the length of the mismatched segment within the pattern, ie, the length of the pattern excluding the context characters. It is easy to see that the inequation

$$D(w_1, w_2) \leq ED(w_1, w_2) \qquad (5)$$

is always true, which means that SpSim always assigns similarity values higher or equal than EdSim. This is inline with the experimental results presented later in this paper showing that SpSim has greater recall than EdSim.

3.4   Examples

Next, let us consider that we have a dictionary containing the patterns from the examples given earlier in Subsection. 3.1.

To compute the SpSim of the words 'phonetic' and 'fonético' we align them and we extract the substitution patterns as described earlier:

```
^pho net c $
|::| |:| |:|
^ fo nét co$
```

Then, we look up each pattern in the dictionary and we sum the d values. In this case, the dictionary contains the fist and last patterns, which have d values of 2 and 1, respectively.

The edit distance of the two words is easily obtained by counting the number of mismatched characters in the aligment, which is 4. The two words have length 8. Thus, substituting all values in equation 3 we obtain a SpSim measure of 0.875. By contrast, the EdSim measure for these two words is only 0.5.

The next table presents some cognates (except last two rows) and the their EdSim and SpSim scores (obtained using a larger dictionary).

English	Portuguese	EdSim	SpSim
recommended	recomendada	0.72	1.00
decembers	dezembros	0.66	0.88
efficiency	eficiência	0.60	1.00
billion	bilião	0.57	1.00
masters	mestres	0.57	0.85
memory	memória	0.57	1.00
used	usados	0.50	1.00
transborder	transfronteiriços	0.47	0.56
both	ambas	0.00	0.00

From this table we observe that SpSim assigns a score of 1.0 very often. That happens whenever all the substitution patterns in a pair of words are already known.

Nevertheless, as we can see in the last rows of the table it does not give high scores for all pairs of words.

3.5   Operational Remarks

There are two fundamental operations: (1) extending the dictionary with new substitution patterns from a list of known cognates, and (2) measuring the spelling similarity between two given words.

Both operations start by aligning the two strings and extracting the substitution patterns around the mismatched segments, using the same predefined context radius.

To compute the similarity between two words we look up in the dictionary each substitution pattern that was extracted from the word pair, and we sum the d values of the patterns that were not found in the dictionary.

The time cost of the whole method is dominated by the cost of the dynamic programming string alignment, and therefore the overall cost is the same as for computing the Edit-Distance or LCSR. Unlike the static measures, this method requires some memory for the hashtable containing the substitution patterns. The prototype implementation used less than 1MB in the experiment reported in the next Section, which is a low requirement for today's hardware.

4 Evaluation

This section presents the results of an experiment carried over 5 language pairs, to compare the performance of SpSim, EdSim and LCSR in the task of cognate identification. The task is specified as follows: given a list of word pairs that may be translations of each other, compute the spelling similarity of each pair and accept as translations the pairs with a score higher than a given threshold. The list of putative translations was extracted from parallel texts (texts that are translation of each other) by thresholding a statistical association measure (Dice).

The Dice association statistic for a pair of words (w_1, w_2) is given by the equation

$$Dice(w_1, w_2) = \frac{2 * F(w_1, w_2)}{F(w_1) + F(w_2)} \tag{6}$$

where $F(w_1)$ is the frequency of w_1 in the text of the first language and $F(w_2)$ is the frequency of w_2 in the text of the second language; $F(w_1, w_2)$ is the *joint frequency* of the words in parallel segments, ie, the number of times that both words occur in corresponding segments (like sentences or paragraphs) of the parallel texts.

The text of the European Constitution was carefully translated to many languages, making it a good resource for multilingual experiments that require parallel texts. For this experiment, the English, German, Spanish, French, Portuguese, and Italian versions of the document were manually aligned at paragraph level (882 paragraphs). The number of tokens in the text of each language is given below.

English	German	Spanish	French	Portuguese	Italian
17415	16704	18750	19908	18726	18142

The extraction was performed for 5 language pairs, English-Spanish (EN-ES), English-French (EN-FR), English-Portuguese (EN-PT), German-English

(DE-EN), and French-Italian (FR-IT), by computing the Dice measure for all words of each language pair and accepting as possible translation those pairs with Dice ≥ 0.6.

Then, each putative translation pair was manually classified as correct or incorrect. Note that the classification was performed with regard to the translational equivalence of words and not their cognaticity, because in general it is harder to tell if two words are cognates than if they are translations.

Table 1 presents the results of the extraction and the classification. Translations that where classified as correct and having EdSim ≥ 0.6 were selected as examples to train SpSim for each language pair.

Table 1. Results of the extraction by thresholding Dice above 0.6 for each language. The second column gives the total number of putative translations extracted for each language pair. The third and fourth columns give the percentage of putative translations that were classified as correct and incorrect respectively. The last column gives the percentage of translations that were used to train SpSim for each language pair.

Language Pair	Extracted	Correct	Incorrect	Examples
English-Spanish	1153	35.3%	64.6%	28.9%
English-French	1206	31.9%	68.0%	25.2%
English-Portuguese	1207	34.5%	65.4%	28.9%
German-English	1148	23.9%	76.0%	13.0%
French-Italian	1609	39.9%	60.0%	36.5%

Note that the purpose of this experiment is not to evaluate the performance of the method being used for translation extraction, which is deliberatedly simple to make it easier to understand the contribution of spelling similarity measure to the final results. Instead, this experiment is intended to give insight on the relative performance gains by using SpSim as a replacement for EdSim or LCSR in the more or less sophisticated methods that employ those measures for cognate identification.

Table 2 presents the precision, recall, and F-measure (the harmonic mean of precision and recall) for various thresholds of spelling similarity used to decide if the words are cognates. The table includes the results for EdSim, LCSR, and SpSim (the three major columns) for each language pair (the five major rows). The highest precision, recall and F-measure for each similarity measure and language pair appear in **boldface**.

As expected, increasing the threshold tends to increase the precision of all measures, but the recall decays. However, we see that recall of EdSim and LCSR decays much more than the recall of SpSim.

SpSim yields the highest F-measure for all language pairs by an absolute margin close to 10%. Furthermore, the highest F-measure is attained by SpSim at the same thresholds that it obtains the highest precision. This means that, not only the precision is the highest, but also that the precision and recall are balanced at that point.

Table 2. Precision (column P), Recall (R) and F-measure (F) for thresholds (T) ranging from 0.1 to 0.9 for each of the 5 language pairs, for the various similarity measures. The highest values of each column for each language pair appear in **boldface**.

L1-L2	T	EdSim P	EdSim R	EdSim F	LCSR P	LCSR R	LCSR F	SpSim P	SpSim R	SpSim F
EN-ES	0.1	42.7	**91.9**	58.3	37.5	**96.6**	54.0	42.6	**92.2**	58.3
	0.2	59.1	81.9	68.7	46.0	91.2	61.2	57.4	83.8	68.1
	0.3	81.5	71.1	**75.9**	63.5	80.1	70.9	74.5	83.1	78.6
	0.4	91.5	63.2	74.8	81.1	70.6	**75.5**	86.8	82.1	84.4
	0.5	95.7	60.5	74.2	91.3	64.2	75.4	92.3	82.1	86.9
	0.6	97.2	50.2	66.2	96.0	53.4	68.7	96.5	81.9	88.6
	0.7	97.7	41.7	58.4	97.8	44.1	60.8	97.4	81.9	88.9
	0.8	**100.0**	25.5	40.6	**100.0**	28.4	44.3	98.8	81.9	89.5
	0.9	100.0	4.4	8.5	100.0	5.4	10.2	**99.7**	81.9	**89.9**
EN-FR	0.1	38.0	**90.4**	53.5	33.6	**95.6**	49.8	37.7	**90.4**	53.3
	0.2	54.3	79.0	64.3	41.1	90.1	56.4	51.4	81.3	63.0
	0.3	75.4	68.3	**71.7**	60.1	80.5	68.8	69.6	80.3	74.5
	0.4	85.6	61.6	71.6	76.1	69.4	**72.6**	82.0	79.5	80.7
	0.5	91.0	57.9	70.8	86.7	61.0	71.6	88.7	79.2	83.7
	0.6	92.0	47.8	62.9	90.4	51.2	65.3	93.0	79.0	85.4
	0.7	**93.2**	39.0	54.9	93.1	42.1	58.0	95.0	79.0	86.2
	0.8	93.1	24.4	38.7	**93.6**	26.8	41.6	96.5	79.0	86.9
	0.9	88.6	8.1	14.8	89.2	8.6	15.6	**97.1**	79.0	**87.1**
EN-PT	0.1	42.7	**92.6**	58.5	37.1	**98.1**	53.9	42.7	**93.3**	58.6
	0.2	60.3	83.7	70.1	45.7	91.1	60.9	57.9	85.6	69.1
	0.3	82.9	72.2	**77.2**	64.3	79.1	71.0	78.5	84.9	81.6
	0.4	89.1	64.5	74.8	81.9	70.5	**75.8**	86.8	83.7	85.2
	0.5	92.2	59.5	72.3	89.8	63.3	74.3	90.9	83.7	87.1
	0.6	94.4	48.7	64.2	92.7	51.8	66.5	95.4	83.7	89.1
	0.7	95.9	33.8	50.0	95.0	36.7	52.9	96.9	83.7	89.8
	0.8	95.1	13.9	24.3	95.9	16.8	28.6	98.3	83.7	90.4
	0.9	**100.0**	3.8	7.4	**100.0**	5.0	9.6	**99.1**	83.7	**90.8**
DE-EN	0.1	28.0	**81.1**	41.6	26.8	**97.1**	42.0	28.1	**82.9**	42.0
	0.2	37.8	54.5	44.6	32.4	83.6	46.7	37.7	64.4	47.5
	0.3	66.0	38.9	**49.0**	41.3	53.5	46.6	60.6	58.2	59.4
	0.4	87.4	32.7	47.6	63.8	37.8	**47.5**	80.8	56.7	66.7
	0.5	92.6	27.3	42.1	81.3	31.6	45.5	91.0	55.3	68.8
	0.6	95.3	22.2	36.0	94.1	23.3	37.3	95.5	54.5	69.4
	0.7	95.8	16.7	28.5	96.2	18.5	31.1	98.0	54.5	70.1
	0.8	**100.0**	9.1	16.7	**100.0**	9.8	17.9	**99.3**	54.5	**70.4**
	0.9	100.0	1.5	2.9	100.0	1.5	2.9	99.3	54.5	70.4
FR-IT	0.1	49.1	**98.0**	65.5	42.5	**99.2**	59.5	48.8	**98.1**	65.2
	0.2	66.6	91.4	77.1	49.9	96.4	65.8	63.0	93.6	75.3
	0.3	85.4	84.6	85.0	68.0	90.0	77.5	76.9	92.4	84.0
	0.4	93.4	79.0	**85.6**	86.2	83.7	**84.9**	87.4	92.1	89.7
	0.5	94.4	70.8	80.9	92.1	75.9	83.2	91.6	92.1	91.9
	0.6	95.3	56.9	71.3	95.3	60.5	74.0	95.5	91.8	93.6
	0.7	95.5	39.7	56.0	95.6	44.3	60.6	97.0	91.8	94.3
	0.8	94.1	19.8	32.6	94.6	21.8	35.4	97.5	91.4	94.4
	0.9	**100.0**	2.2	4.3	**100.0**	2.3	4.6	**98.2**	91.4	**94.7**

5   Conclusions

In this paper we presented a new measure for spelling similarity that dynamically adapts to recognize and tolerate substitution patterns that are characteristic of the pair of languages under consideration. SpSim is conceptually simpler than other dynamic measures (but this assessment is arguable, of course).

The comparative evaluation of SpSim against two baseline measures, EdSim and LCSR, shows that SpSim performs significantly better than both for all 5 language pairs (English-Spanish, English-French, English-Portuguese, German-English, and French-Italian). The empirical results show that SpSim improves the signal-to-noise ratio of cognate extraction, by improving substantially the recall while maintaining a precision close to the baseline measures.

SpSim is language independent as the patterns are obtained automatically from a list of cognates, without requiring any linguistic knowledge.

Acknowledgements. This work was supported by the Portuguese Foundation for Science and Technology (FCT/MCTES) through individual PhD grant SFRH/BD/65059/2009 (LG), and funded research projects VIP-ACCESS (ref. PTDC/PLP/71142/2006) and ISTRION (ref. PTDC/EIA-EIA/114521/2009).

References

1. Bergsma, S., Kondrak, G.: Alignment-based discriminative string similarity. In: Annual Meeting – Association for Computational Linguistics, vol. 45, page 656 (2007)
2. Gusfield, D.: Algorithms on Strings, Trees and Sequences – Computer Science and Computational Biology. Cambridge University Press (1997)
3. Ildefonso, T., Pereira Lopes, J.G.: Longest Sorted Sequence Algorithm for Parallel Text Alignment. In: Moreno Díaz, R., Pichler, F., Quesada Arencibia, A. (eds.) EUROCAST 2005. LNCS, vol. 3643, pp. 81–90. Springer, Heidelberg (2005)
4. Kondrak, G.: Identification of Cognates and Recurrent Sound Correspondences in Word Lists. Traitement Automatique des Langues 50(2), 201–235 (2009)
5. Dan Melamed, I.: Bitext maps and alignment via pattern recognition. Comput. Linguist. 25(1), 107–130 (1999)
6. Ribeiro, A., Dias, G., Lopes, G.P., Mexia, J.T.: Cognates alignment. In: Maegaard, B. (ed.) Proceedings of the Machine Translation Summit VIII (MT Summit VIII), Santiago de Compostela, Spain, September 18-22, pp. 287–292. European Association of Machine Translation (2001)
7. Simard, M., Foster, G., Isabelle, P.: Using cognates to align sentences in parallel corpora. In: Proceedings of the Fourth International Conference on Theoretical and Methodological Issues in Machine Translation, pp. 67–81 (1992)
8. Tiedemann, J.: Automatic construction of weighted string similarity measures. In: Proceedings of the Joint SIGDAT Conference on Empirical Methods in Natural Language Processing and Very Large Corpora, pp. 213–219 (1999)

Identifying Automatic Posting Systems in Microblogs

Gustavo Laboreiro[1], Luís Sarmento[1,2], and Eugénio Oliveira[1]

[1] Faculdade de Engenharia da Universidade do Porto - DEI - LIACC, Portugal
[2] SAPO Labs Porto, Portugal
{gustavo.laboreiro,las,eco}@fe.up.pt

Abstract. In this paper we study the problem of identifying systems that automatically inject non-personal messages in micro-blogging message streams, thus potentially biasing results of certain information extraction procedures, such as opinion-mining and trend analysis. We also study several classes of features, namely features based on the time of posting, the client used to post, the presence of links, the user interaction and the writing style. This last class of features, that we introduce here for the first time, is proved to be a top performer, achieving accuracy near the 90%, on par with the best features previously used for this task.

Keywords: User-Generated Content, UGC, Microblogging, Twitter, Spam, Bot, Automatic Identification, User Classification, Stylistic Analysis, Noisy Text.

1   Introduction

Microblogging systems — of which Twitter is probably the best known example — have become a new and relevant medium for sharing spontaneous and personal information. Many studies and applications consider microblogs as a source of data, precisely because these characteristics can confer authenticity to results. For example, *trend detection* ([8]), *opinion-mining* ([6]) or *recommendation* ([4]).

Because of its popularity, Twitter is also part of the on-line communication strategy of many organizations, which use a Twitter account for providing updates on news, initiatives, commercial information (e.g. promotions, advertisements and spam) and various other types of information people may find interesting (like weather, traffic, TV programming guides or events).

Messages conveyed by these automatized accounts – which we will now refer to as *robot accounts* or, simply, *bots* – can easily become part of the stream of messages processed by information extraction applications. Since bots provide content aimed at being consumed by the masses instead of the personal messages that information extraction systems consider meaningful (for example, for trend detection), automatic messages may bias the results that some information extraction systems try to generate. For this reason, from the point of view of these systems, messages sent by bots can be considered noise.

L. Antunes and H.S. Pinto (Eds.): EPIA 2011, LNAI 7026, pp. 634–648, 2011.

The number of such robot accounts is extremely large and is constantly growing. Therefore, it is practically impossible to manually create and maintain a list of such accounts.

Even considering that the number of messages typically produced by a bot each day is not significantly larger than the number of messages written by an active user in the same period, we must remember that bots are capable of sustaining their publication frequency for longer periods than most humans (that can stop using the service temporarely or permanently after some time). Thus, in the long run, bots are capable of producing a larger set of messages than an active person.

In this work we propose a system that can identify these robot accounts using a classification approach based on a number of observable features related to activity patterns and message style. We evaluate its performance, and compare it with some of the more common approaches used for this task, such as the client used to post the messages and the regularity of new content.

1.1   Types of Users

Based on the work of Chu et al.[5], we start by distinguishing between three types of users. The term *human* is used to refer to users that author all or nearly all their messages. They usually interact with other users, post links on some of the messages, use abbreviations, emoticons and ocasionally misspell words. Many employ irregular writing styles. The subject of their messages can be different, but they tend to express personal opinions. Below we have examples of two human users:

- Who's idea was it to take shots of tequila? You are in so much trouble.
- I forgot to mention that I dropped said TV on my finger. ouchie.
- Heard that broseph. RT @ReggaeOCD: So bored with nothing to do. #IHateNotHavingFriends
- Just being a bum today. http://twitpic.com/4y5ftu

- aw, grantly:'destroys only happy moment in fat kids life' when talking about food :@
- @ryrae HAHAHAHAHHAHA :')
- JOSH IS IN SEASON 5 OF WATERLOO ROAD! WHEN DID THIS HAPPEN?

Bots, on the other hand, are in place to automate the propagation of information. The content is generally written by a person, although in some cases the entire process is automated (e.g. sensor readings).

We should note that what we are distinguishing here is more a matter of content than a matter of process or form. It is possible that an account where a person writes the message directly at the Twitter website be labeled as a bot, if the messages are written in the cold objective way seen in the examples below, from three different accounts.

- Social Security and Medicare to run short sooner than expected. http://on.cnn.com/lauSNv
- For Louisiana town, a collective gasp as it braces for floodwaters. http://on.cnn.com/mb481c
- Jindal: Morganza Spillway could open within the next 24 hours. http://on.cnn.com/j2jIBs

- 96kg-Bosak takes 7th place at the University Nationals
- 84kg-Lewnes takes 2nd to Wright of PSU and qualifies for the world team trials in Oklahoma Ciry
- Bosak loses his consolation match 0-1, 1-3 to Zac Thomusseit of Pitt.

- #Senate McConnell: Debt limit a 'great opportunity' http://bit.ly/kanlc9 #Politics
- #Senate Wisconsin Sen. Kohl to retire http://bit.ly/kQpAsS #Politics
- #Senate Ensign may face more legal problems http://bit.ly/mN7XsB #Politics

Many accounts are not run entirely by a person nor are they completely controlled by a machine. We label these mixed accounts as *cyborgs*, the term used by Chu et al. [5], that describes them as a "bot-assisted human or human-assisted bot". For example, an enterprise can have an automatic posting service, and periodically a person provides the user interaction to maintain a warmer relation with the followers, and foster a sense of community. Another possibility occurs when a person uses links to websites that post a message on the account of that person (for example, "share this" links). If these pre-written content are noticeable among the original messages, the user is labeled as a cyborg. If barely no original content is present in the user's timeline, they will be considered a bot. Below we show examples of a cyborg account.

- Explore The Space Shuttle Era http://go.nasa.gov/gzxst5 and immerse yourself in the Space Shuttle Experience http://go.nasa.gov/iHVfGN
- Track the space shuttle during launch and landing in Google Earth using real-time data from Mission Control http://go.nasa.gov/mwO9Ur
- RT @Rep_Giffords: Gabrielle landed safely @NASAKennedy. For more details go to www.fb.me/GabrielleGiffords. #NASATweetUp
- Space shuttle Endeavour's preferred launch time moved two seconds later! Now 8:56:28 a.m. EDT Monday.
- @Angel_head NASA frequently tweaks the shuttle launch time by seconds based on the latest space station tracking data to use the least fuel.

As we will explain next, we used these guidelines to construct a Ground Truth that will be used in our experiments. The details of this task are given in Section 4.1.

To address the problem of automatic posting, we study different sets of features that allow us to classify Twitter users into the three user categories described. These features explore characteristics exhibited by the users, such as their posting times, the microblog client application they use, their interaction with other users, the content of their messages, and their writing style. The main goal of the work presented in this paper is to access the usefulness and robustness of the different types of features proposed. We discuss the features in Section 3.

We describe our experiment and its parameters in Section 4 and evaluate our results in Section 5. In Section 6 we present our conclusions and future work.

2   Related Work

Most literature addressing the identification of automated systems in microblogs is related with the detection of spam. While there is some overlapping between spam and automated posting systems (spammers often employ automation to help them in their work), we feel that the problem we are addressing is much more general.

Wang [10] presents an effort to detect spamming bots in Twitter using a classification based approach. The author explored two sets of features: (i) information about the number of followers, friends and the follower per friend ratio for the social network aspect of Twitter; and (ii) information about duplicate content and number of links present in the last 20 messages of a given user account. The author used a manually annotated corpus to train a Naive-Bayes classifier. The classifier achieved slightly over 90% Precision, Recall and F-measure in a 10-fold cross validation experiment. However, since the training corpus was biased towards non-spam users (97% of the examples), any classifier that only reported "non-spam" would be almost always correct, so results are not really significant.

Grier et al. [7] analyze several features that indicate spamming on Twitter. They looked for automated behavior by inspecting the precision of timing events (minute-of-the-hour and second-of-the-minute), and the repetition of text in the messages across a user's history. They also studied the Twitter client application used to write the messages, since some allow to pre-schedule tweets at specific intervals.

Zhang and Paxson [11] present a study where they try to identify bots by looking only at the minute-of-the-hour and the second-of-the-minute. If the posting times are either too uniform or not uniform enough, there is the possibility of the account being automated. This analysis is similar to the one present in Grier et al. [7], a work where Zhang and Paxson participated.

The authors present no validation of their results (since it is not possible to determine for sure if the account is automated or not). However, they claim that "11% of accounts that appear to publish exclusively through the browser are in fact automated accounts that spoof the source of the updates".

Chu et al. [5] propose to distinguish between humans, bots or cyborgs, but much of the effort was put into spam detection. They claim that more sophisticated bots unfollow users that do not follow back, in an effort to keep their friends to followers ratio close to 1, thus reducing the effectiveness of features based on the social network of Twitter. However, by discarding the uncertain and ambiguous examples from their Ground Truth set, the authors seriously reduced the usefulness of their results.

Contrary to previous work, we focus on a problem that is much more generic than spam-detection, since a very large number of bots belong to newspapers and other organization, which are voluntarily followed by users. Our goal is to separate potentially opinionated and highly personal content from content injected in the Twittosphere by media organizations (mostly informational or promotional). One other point where we distinguish ourselves from the mentioned works is in our attempt to expand the set of features used in the detection, now including a vast array of stylistic markers. In our opinion, this opens a new field for exploration and study.

3   Methodology

Most of our discussion is centered around distinguishing human users and bots. We propose five sets of features, described below, that are intended to help to discriminate between these two poles. A cyborg user, by definition, exhibits characteristics typical of both classes of users.

3.1   Chronological Features

One of the characteristics of automatic message posting systems is that they can be left running indefinitely. Therefore, we can expect to see different chronological patterns between human users and bots. To address these points, we defined a number of features, divided into the four following sub-classes.

Resting and Active Periods. Constant activity throughout the day is an indication that the posting process is automated or that more than one person is using the same account — something we expect to be unusual for individual users. Figure 1 was drawn using information from our manually classified users (described in Section 4.1). It shows how human and cyborg activity is reduced between 1 and 9 AM. Bot activity is also reduced between 11 PM and 7 AM, but it never approaches zero.

At the same time, other things can keep people from blogging. This can lead to a certain hour of preferred activity, such as evenings, as suggested in Figure 1. Bots appear to have a more evenly spaced distribution across the day, with smaller fluctuations in the level of activity. This can be a conscious choice, to allow more time for their followers to read each post.

Since we tried to limit our crawling efforts to Portugal, most of the observations are expected to fall within the same time zone (with the exception of the Azores islands, which accounts for 2% of the population). We believe that the problem of users in different time zones cannot be avoided completely. For example, we do not expect users to correct their Twitter profile when traveling.

To detect the times at which the user is more or less active, we define 24 features that measure the fraction of messages they posted at each hour. These values should reflect the distributions represented at Figure 1. We also analyze

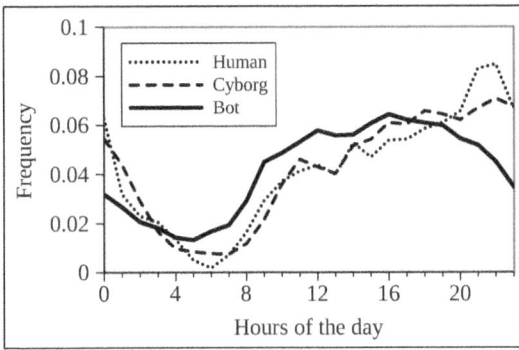

Fig. 1. Comparison of Twitter activity between bots, cyborgs and humans as a function of the hour

the average and standard deviation of these values. We expect that the standard deviation of a bot is lower than that of a human.

Finally, we register the 10 hours with the highest and lowest activity, and the average number of messages that the user posts per day (as a floating-point number).

Long term Activity. Days are not all equal. This is true for both humans and bots. For example, as shown in Figure 2, most activity happens at Thursdays and Fridays. This trend matches the result published by Hubspot [1].

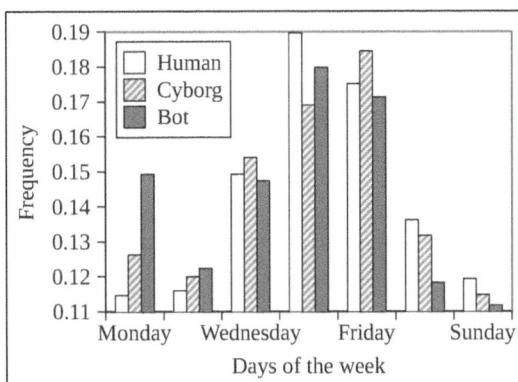

Fig. 2. Comparison of Twitter activity between bots, cyborgs and humans as a function of the day of the week

We can see that bots are less active during the weekends, while they dominate on Mondays and lead on Tuesdays.

We define seven features related to the frequency of the messages posted across each day of the week, that should reflect the proportions in Figure 2. We also calculate the workday and weekend posting frequency, and rank the days of the week by the frequency of posting.

Inactivity Periods. From direct observation of Twitter messages, we can see that bots tend to be more regular on their updates than humans. It is known that irregular accounts can lose popularity quickly. At the same time, normal people need to rest, get occupied with other matters, and can lose interest in blogging for some time.

To make use of this information, we measure the periods of inactivity in minutes, and record the length of the 10 longest intervals, in decreasing order. We also calculate the average and standard deviation of all these values. From our observations, we expect that bots will have lower variation in inactivity periods (lower standard deviation) and a higher average.

For example, considering a user that only blogged at 1 PM, 2 PM, 3:30 PM and 7 PM on the same day, we would have the following features:

Feature name	Value (minutes)
top_inactive_period_1	210.00
top_inactive_period_2	90.00
top_inactive_period_3	60.00
average_top_inactive_period	120.00
standard_deviation_top_ inactive_period	79.37

Humans are unable to match the speed at which bots can create new messages. For this reason, we also calculate the analogous features for the minimum inactivity periods (i.e. the 10 shortest inactive periods).

Posting Precision. Since some automatic posting systems work based on a fixed periodicity (e.g. TV programming guides), we decided to calculate the frequency of messages that are created at each minute (60 features) and second (another 60 features). This approach is a simpler version of other works [7,11].

For a human, we expect their posts to be evenly spread across both these measurements. Some bots, on the other hand, are expected to concentrate their activity around the 0 seconds mark. They may also do the same around some particular minutes (e.g. 0 or 30).

As before, we also calculate the average and standard deviation of these measurements, where humans should result in a lower average and higher standard deviation compared to bots.

For both minutes and seconds we take note of the 10 most frequent values — that is, when most activity occurs.

3.2   The Client Application Used

It makes sense that the Twitter client used to post the messages be a relevant aspect in identifying automated processes. There are many clients and methods of accessing the microblogging system (e.g. web interface, several applications, etc.). From the point of view of automation, some of these methods are easier or more convenient than others. Also, most microblogging systems have an open API, meaning that it is possible to interact with them directly. In Twitter, unregistered clients are identified as "API", while those that were registered are identified by their name.

We track the number of different clients used to post the messages, and the proportion of messages posted with each client. Cyborgs are expected to have the largest variety of clients used (as they usually post automatically from several sources). Some humans can use more than one client, for example, a mobile client and the website. Bots, on the other hand, can adhere to a single, exclusive client that is tied with their on-line presence; or may use a general client that imports messages from an RSS feed, for example.

3.3   The Presence of URLs

We can make a distinction between two types of bots: information bots, which only intend to make their readers aware of something (e.g. weather forecast, TV scheduling and traffic information), and link bots, whose main purpose is to generate traffic towards their website (e.g. news, advertisements and spam). Information bots usually don't have URLs in their messages, while some link bots are capable of truncating the text of the message to make room for the URL. Most URLs shared by a bot usually have the same domain, i.e. they were all created by the same URL shortening system, or point to the same website.

Humans are also capable of introducing many URLs, but our observation reveals that cyborgs are more likely to do so; and to vary the domains of said URLs. We can observe both types of linking behaviour represented in the bot and human examples presented in the introduction, in Section 1.1.

We defined a feature that represents the ratio of URLs shared per total of messages written, and also keep track of the proportion of the domains associated to the URLs.

3.4   User Interaction

Bots usually have one main objective that is to spread information regularly. While they may be programmed to do more complex actions (such as follow other users), automating user interaction can have undesired repercussions for the reputation of the account holder. Thus, *reblogs*[1] (to post a copy of another user's message) and *replies* (directing a message at a user) are shuned by most bots. The main exception are some spamming bots, that send several messages directed at users [7]. To include the name of other users in the message (*mentions*) also seems to be uncommon in automated accounts.

However, a number of users also avoid some types of interaction, such as the ones previously mentioned. Therefore, while we expect this information to help identify humans, they may be less helpful in identifying bots.

---

[1] Called "retweets" on Twitter, often shortened to "RT".

With our features we keep track of the proportion of reblogs, replies and mentions, as well as the average number of users mentioned per message written.

3.5   Writing Style

Stylistic information has been successfully used to distinguish the writing style of different people on Twitter [9]. Thus, we believe it to be helpful in distinguishing between automated and non automated messages since, as observed in the examples in Section 1.1, these users adopt different postures. The austere writing style may help with the readability, and also credibility, associated to the account, while many humans do not seem too concerned about that.

We identify the frequency (per message) of a large number of tokens, as listed below.

Emotion Tokens. Bot operators wish to maintain a serious and credible image, and for this reason avoid writing in a style too informal (or even informal). We collect information on the use of various popular variations of smileys and "LOLs".

We also try to identify interjections. While this part of speech is culturally dependent, we try to identify word tokens that have few different letters compared with the word length — if the word is longer than 4 characters, and the number of different letters is less than half the word length, we consider it an interjection.

These three stylistic features were the most relevant features mentioned by Sousa-Silva et al. [9]. Below, we can see example messages containing many emotion tokens:

-   we talked before......... on twitter. **HAHAHAHAHA** RT @Farrahri: @Marcology **LOL** she smiled at me! **Hehehe**, jealuzzz not?
-   **riiiiiight**.... im **offffffffffffffffff**!!!!! bye bye
-   RT **yesssssssss**! That is my **soooong**!!!! @nomsed: You got the **looove** that I **waaaant** RT @LissaSoul: U got that BUTTA **LOOOOoooooVVVEEE**!

Emotion can also be expressed through punctuation, but we include those features in the punctuation feature group, below.

Punctuation Tokens. Humans vary widely in regards to their use of punctuation. Many are not consistent across their publications. This is in opposition to bots, that can be very consistent in this regard.

Punctuation can often be used as a separator between the "topic" and "content" of the message, as can be seen on the first example on Table 1 . Different sources of information may structure their messages differently. Therefore one bot may use more than one separator.

We also notice that some bots publish only the headline of the article they are linking to. These articles are usually blog posts or news at a news website. Since headlines usually do not include a full stop, this feature receives a very low frequency (as seen in example 1).

Table 1. Examples of bot Twitter messages making use of punctuation for structural purposes

Example Text	
1	Tours: Brian Wilson should retire next year http://dstk.me/Oi6
2	Gilberto Jordan at Sustain Worldwide Conference 2011: Gilberto Jordan, CEO of Grupo André Jordan, is the only spea...http://bit.ly/mOOxt1

Question or exclamation marks are usually infrequent in news bots, or bots looking for credibility [2]. Some bots truncate the message to make room for the URL, signaling the location of the cut with ellipsis (some using only two dots). We can see an example on the second example on Table 1.

We measure the frequency of occurrences of:

- Exclamation marks (single and multiple);
- Question marks (single and multiple);
- Mixed exclamation and question marks (e.g. "!?!?!?!?!!?!?!!");
- Ellipsis (normal [i.e. "..."], or not normal [i.e. ".." or "...."]);
- Other punctuation signs (e.g. full stop, comma, colon, ...);
- Quotation marks;
- Parenthesis and brackets (opening and closing);
- Symbols (tokens without letters and digits); and
- Punctuation at the end of the message (both including and excluding URLs).

Word Tokens. This group of tokens is kept small for the sake of language independence. We begin by tracking the average length of the words used by the author. We also define features that track the frequency of words made only of consonants (that we assume to be abbreviations most of the time), and complex words. We consider complex words as those having more than 5 letters and with few repeated characters (more than half). Thus "current" (7 characters in length, 6 different characters) is a complex word, while "lololol" (7 characters in length, 2 different characters) or "Mississippi (11 characters in length, 4 different characters) do not fit the definition.

Word Casing. Bots are usually careful in the casing they use. Careful writing aids with the image that is passed through. We measure the frequencies with which the following is used:

- Upper and lower cased words;
- Short (≤ 3 letters), medium (4–5 letters) and long (≥ 6 letters) upper cased words;
- Capitalized words; and
- Messages that start with an upper cased letter.

Quantification Tokens. We track the use of some numeric tokens. Dates and times are comonly used to mention events. Percentages can be more common on news or advertisements.

- Date (e.g. "2010-12-31" or "22/04/98");
- Time (e.g. "04:23");
- Numbers;
- Percentages; and
- Monetary quantities (e.g. "23,50€", "$10" or "£5.00").

Beginning and Ending of Messages. Some accounts post many messages (some times all or near all their messages) using one or a small number of similar formats. This behavior is specific of bots, that automate their posting procedure. Below we can see two examples.

- **#football** Kenny Dalglish says Liverpool will continue conducting their transfer business in the appropriate manner. http://bit.ly/laZYsO
- **#football** Borja Valero has left West Brom and joined Villarreal on a permanent basis for an undisclosed fee. http://bit.ly/iyQcXs

- **New post:** Google in talks to buy Hulu: report http://zd.net/kzcXFt
- **New post:** Federal, state wiretap requests up 34% http://zd.net/jFzKHz

To determine the pattern associated with the posts, we calculate the frequency with which messages begin with the same sequence of tokens (excluding URLs and user references, that frequently change between messages). We define tokens as words, numbers, punctuation signs, emoticons and other groups of symbols that have a specific meaning.

We group all the messages by their first token. For each group with two or more messages, we store their relative proportion in a feature related to the token. We also register the 10 highest proportions found, in descending order. This entire process is then repeated, looking at the first two tokens, then the first three, and so on.

Once complete, we take note of the largest number of tokens seen, and repeat the entire process, looking at the endings of the messages.

This procedure results in a number of features that are very specific. In the case of humans we collect a relatively small number of features, as their messages can be varied. Some bots will reveal a pattern that is used for *all* their messages (e.g. see the last bot examples in Section 1.1). In the case of cyborgs, it is very useful to detect a number of patterns such as "I liked a @YouTube video *[URL]*" or "New Blog Post …".

Below we can see examples of messages where this approach is useful. The first two messages are from a bot account, while the last two were taken from a cyborg account.

4   Experimental Set-up

Our aim is to compare the level of performance provided by the five sets of features described in Section 3. First we create a Ground Truth by classifying a number of Twitter users manually. This data is then used to both train and test our classification system.

4.1   Creation of a Ground Truth

In late April 2011 we started a Twitter crawl for users in Portugal. We considered only users who would specifically state that they were in Portugal, or, not mentioning a known location, that we detect to be writing in European Portuguese. This collection started with 2,000 manually verified seeds, and grew mostly by following users that are referred in the messages. In this way our collection moved towards the more active users in the country. However, there is no guarantee that we have been collecting all the messages from any of the users.

At the moment we have over 72 thousand users and more than 3 million messages. From this set, we selected 538 accounts that had posted at least 100 messages, and 7 people were asked to classify each user as either a human, a bot or a cyborg, in accordance with our guidelines, as described in Section 1.1.

The annotators were presented with a series of user accounts, displaying the handle, a link to the Twitter timeline, and a sample of messages.

Since the users presented to the annotators were randomly selected, not every annotator saw the users in the same order, and the sample of messages for each user was also different. For each user, we considered the classification that the annotators most often attributed them. In the case of a tie, we asked the annotator to solve them before ending the voting process.

To finalize the voting process, we asked one eighth annotator to solve the ties between annotators. In the end we were left with 2,721 votes, 95% of which from the 4 main annotators, that we used to calculate the agreement. Looking only at the 197 users that were classified by all 4 main annotators, we get a substantial 0.670 Fleiss' kappa value, showing adequate reliability in the classification.

Figure 3 shows the distribution of the users across all three categories. We can see that humans dominate our collection of Twitter (448), while cyborgs were the least numerous (22). In total we identified 68 bots.

4.2   The Classification Experiment

We randomly selected our example users from our manually classified examples. To have a balanced set, we limited ourselves to only 22 users of each type, randomly selected before the experiment. To handle the automatic classification, we opted for an SVM due to its ability to handle a large number of features. We opted for the libSVM [3] implementation.

For each user we selected up to 200 messages to analyze and create the features. Due to the chronological features (Section 3.1), we selected only sequential messages in our collection.

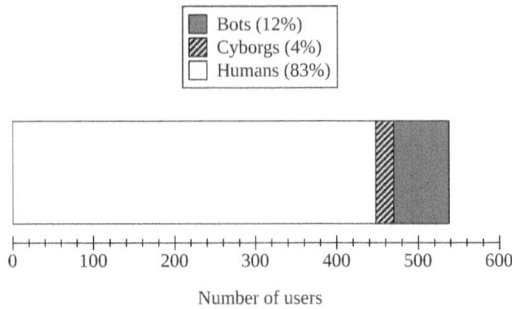

Fig. 3. Distribution of the 538 users in the three classes

We used the radial basis function kernel from libSVM, allowing it to look for the parameters that best fit the data, and normalized the values of the features, allowing for more accurate results. We measured the results using the accuracy, i.e. the ratio between correct classifications and total classifications.

We opted for a 2-fold cross validation system, where we select 11 users of each type to be used in the training set, and the other 11 were part of the testing set. This allows enough testing messages to provide adequate granularity in the performance measurement, and a more reasonable number of messages to train the SVMs. We repeat each experiment 50 times (drawing different combinations in the training and testing set).

Given that we are using a balanced set of examples, we expect that a random classifier would be correct 1/3 of the times. We will be considering this as the baseline in our analysis.

5   Results and Analysis

Our results are represented in Figure 4, representing the minimum, lower quartile, median, upper quartile and maximum accuracy across the 50 runs.

We separate the results in two groups: the first group, that never reaches 100% accuracy, and the second group, that does.

In the first group, the user interaction features outperformed the chronological features, that had two poor runs. However, none of them shows performance similar to the other feature sets.

In the second group, the stylistic features presents the best results, with median accuracy 97%. The feature that identifies the client application also performs adequately, but twice failed 7 or 8 of the 33 examples. The URL features showed more stable results than the client information feature, but generally failed in more cases. Finally, using all the features combined yielded very good results, with 97% median (and 97% upper quartile, hence overlapping in Figure 4), failing once in 5 of the examples.

Over 26,000 features were generated during the experiments, most of them encode stylistic information. While in a large group they can be quite powerful

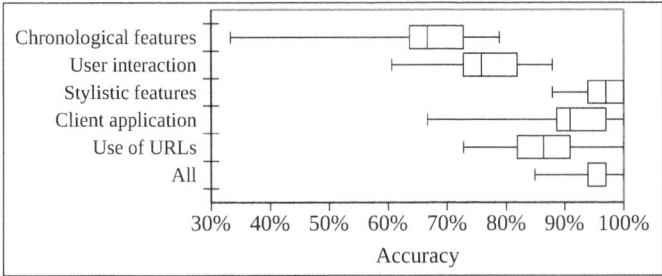

Fig. 4. Box plot showing the results for the classification of users, using 50 2-fold cross validation runs. The limits of the boxes indicate the lower and higher quartile. The line inside the box indicates the median. The extremities of the lines represent the minimum and maximum values obtained.

(as shown), each of these features carries little information. This is in contrast with the URL, user interaction and client application features, where a small number of features can contain very meaningful information.

Most features related with the client application, work almost like a database of microblog applications. That is, except for the number of different clients used, we are only recalling the identification strings present in the training messages. In the presence of an unknown client, the classifier has little information to work with. Hence the cases with low accuracy.

The features related with the URLs and with user interaction obtain information from the presence or absence of certain elements. However, in our implementation we could not encode enough information to address all the relevant cases, especially in the case of user interaction.

It is unfortunate that we are unable to compare our results with other approaches, mentioned in Section 2. There are three reasons for this: (i) their work has a different goal (i.e. spam detection); or (ii) the authors do not provide a quantification that we could use for comparison; or (iii) we consider that their experiment is biased (e.g. excluding some messages because they are more difficult to classify).

6   Conclusion and Future Work

We have shown that automatic user classification into either human, cyborg or bot — as we have defined them — is possible using standard classification techniques. In particular, as we have supposed, stylistic features can be a reliable indicator in this type of classification. In fact, they achieved results as good or better than other, more frequently used, indicators of automatic activity.

Basing the user classification in the client application used raises two problems: first, some applications can have mixed using (as Chu et al. [5] and Grier et al. [7] point out); and second, dealing with the large amount of different clients is difficult. For example, we counted 2,330 different clients in our 73,848 users database (around 1 different client for every 32 users). Thus, while fast and simple, this approach does not appear to hold on its own, and should be combined with other approach.

In the future, we would like to improve our chronological features by adopting the same method Zhang and Paxson used [11,7], as our minute-of-the-hour and second-of-the-minute approach was, perhaps, too simplistic. We would also like to study the scalability of the stylistic approach, as they generate a large number of new features.

References

1. Burnes, R.: When do most people tweet? at the end of the week (January 2010), http://blog.hubspot.com/blog/tabid/6307/bid/5500/When-Do-Most-People-Tweet-At-the-End-of-the-Week.aspx
2. Castillo, C., Mendoza, M., Poblete, B.: Information credibility on twitter. In: Proceedings of the 20th International Conference on World Wide Web, WWW 2011, pp. 675–684. ACM, New York (2011)
3. Chang, C.C., Lin, C.J.: LIBSVM: a library for support vector machines (2001) software, http://www.csie.ntu.edu.tw/~cjlin/libsvm
4. Chen, J., Nairn, R., Nelson, L., Bernstein, M., Chi, E.: Short and tweet: Experiments on recommending content from information streams. In: ACM Conference on Human Factors in Computing. Association for Computing Machinery, Atlanta, GA (04/10/2010)
5. Chu, Z., Gianvecchio, S., Wang, H., Jajodia, S.: Who is tweeting on twitter: human, bot, or cyborg? In: Gates, C., Franz, M., McDermott, J.P. (eds.) Proceedings of the 26th Annual Computer Security Applications Conference, ACSAC 2010, pp. 21–30. ACM, New York (2010)
6. Dey, L., Haque, S.M.: Opinion mining from noisy text data. International Journal on Document Analysis and Recognition 12, 205–226 (2009)
7. Grier, C., Thomas, K., Paxson, V., Zhang, M.: @spam: the underground on 140 characters or less. In: Proceedings of the 17th ACM Conference on Computer and Communications Security, CCS 2010, pp. 27–37. ACM, New York (2010)
8. Mathioudakis, M., Koudas, N.: Twittermonitor: trend detection over the twitter stream. In: Proceedings of the 2010 International Conference on Management of Data, SIGMOD 2010, pp. 1155–1158. ACM, New York (2010)
9. Sousa-Silva, R., Laboreiro, G., Sarmento, L., Grant, T., Oliveira, E., Maia, B.: Twazn me!!!; automatic authorship analysis of micro-blogging messages. In: Procedings of the 16th International Conference on Applications of Natural Language to Information Systems (June 2011)
10. Wang, A.: Detecting spam bots in online social networking sites: A machine learning approach. In: Foresti, S., Jajodia, S. (eds.) Data and Applications Security and Privacy XXIV. LNCS, vol. 6166, pp. 335–342. Springer, Heidelberg (2010)
11. Zhang, C.M., Paxson, V.: Detecting and Analyzing Automated Activity on Twitter. In: Spring, N., Riley, G.F. (eds.) PAM 2011. LNCS, vol. 6579, pp. 102–111. Springer, Heidelberg (2011)

Determining the Polarity of Words through a Common Online Dictionary

António Paulo-Santos[1], Carlos Ramos[1], and Nuno C. Marques[2]

[1] GECAD – Knowledge Engineering and Decision Support Group Institute of Engineering, Polytechnic of Porto, Portugal
pgsa@isep.ipp.pt, csr@dei.isep.ipp.pt
[2] Departamento de Informática, Faculdade de Ciências, Universidade Nova de Lisboa, Monte da Caparica, Portugal
nmm@di.fct.unl.pt

Abstract. Considerable attention has been given to polarity of words and the creation of large polarity lexicons. Most of the approaches rely on advanced tools like part-of-speech taggers and rich lexical resources such as WordNet. In this paper we show and examine the viability to create a moderate-sized polarity lexicon using only a common online dictionary, five positive and five negative words, a set of highly accurate extraction rules, and a simple yet effective polarity propagation algorithm. The algorithm evaluation results show an accuracy of 84.86% for a lexicon of 3034 words.

Keywords: lexicon generation, polarity lexicon, polarity of words.

1 Introduction

Entries on a polarity lexicon are tagged either as positive, negative, or neutral in some works. For instance, *good, beautiful, happiness* can be tagged as positive words, whereas words such as *bad, ugly, sadness* can be tagged as negative words. A polarity lexicon is a resource that can be used for instance to identify, classify and extract sentiment or opinions from sentences or larger units of text.

Most of the approaches to date on polarity lexicons generation have required advanced tools like part-of-speech taggers and rich lexical resources such as WordNet [16]. Most of the research has focused on English, as evidenced by the availability of lexical resources such as Harvard Inquirer [19], SentiWordNet [7], [2], Q-WordNet [1], WordNet-Affect [20].

In works such as [9], [14], [4], and [17] polarity is propagated based on Word-Net [16] and graph propagation algorithms. Our approach is different because we use a common online dictionary. Esuli and Sebastiani [6] use WordNet [16] and Merriam-Webster online dictionary[1] to determine the polarity of terms through gloss classification (i.e. textual definitions classification). It is also worth mentioning the work by Rao and Ravichandran [17] which uses a common synonym

---

[1] http://www.merriam-webster.com

L. Antunes and H.S. Pinto (Eds.): EPIA 2011, LNAI 7026, pp. 649–663, 2011.

dictionary (the OpenOffice thesaurus[2]) to build a graph. We focus on a common online dictionary to retrieve not only synonyms but also antonyms, the latter retrieved by a set of high accurate extraction rules (e.g. if *ugly* appears on dictionary as *opposite of beautiful*, we extract *beautiful* as an antonyms of *ugly*).

This paper presents first results of an empirical study (section 4) by using a common online dictionary to build a moderate-sized polarity lexicon with scarce resources. Beginning with a small seed set of words labeled as positive or negative and a common online dictionary we propagate the positive and negative sense to unlabeled words applying a simple and intuitive graph propagation algorithm (section 2.3). The study focuses on the polarity of Portuguese words but could be applied to other languages. The contribution of this work is to empirically show that with scarce resources and a simple yet effective semi-supervised approach it is possible to build a polarity lexicon.

2   Approach

In this section we present a graph propagation algorithm to propagate the positive and negative sense of a seed set of words labeled as positive or negative to unlabeled words. In this approach, a dictionary is viewed as a graph and the positive or negative sense of labeled words is propagated to their neighboring words.

2.1   Dictionary as a Graph

In a dictionary meanings are given through synonyms, antonyms, or richer semantic relations such as in WordNet [16]. We can think of a dictionary as a graph in which nodes correspond to words and edges correspond to synonyms, antonyms or other semantic relations between words. The representation of a dictionary as a directed graph is done for instance in [4], [10] and as an undirected graph for instance in [17], in both cases using the WordNet [16]. In this study we look to a common dictionary as a directed graph as shown on figure 1.

In figure 1, on the left side we have the meanings or definitions of the word *bad*, *good*, and *pleasant* and on the right side their representation as a directed graph.

2.2   Intuitive and Simple Polarity Propagation - Key Ideas

The first key idea is that viewing a dictionary as a directed graph and starting with the seed set of words manually classified as positive and negative we can propagate their polarity to unlabeled words applying a graph breadth-first traversal. E.g. on figure 1 assuming that good is a positive seed word we can propagate its positive sense to *attractive*, *suitable*, *agreeable*, and *pleasant*. We can then proceed the same way firstly for all seed words (e.g. *bad* and *pleasant*)

---

[2] http://www.openoffice.org

Bad

    Evil, unpleasant, disagreeable,
    not enjoyable, (…)

Good

    Attractive, suitable, agreeable,
    pleasant, (...)

Pleasant

    Good, enjoyable, (...)

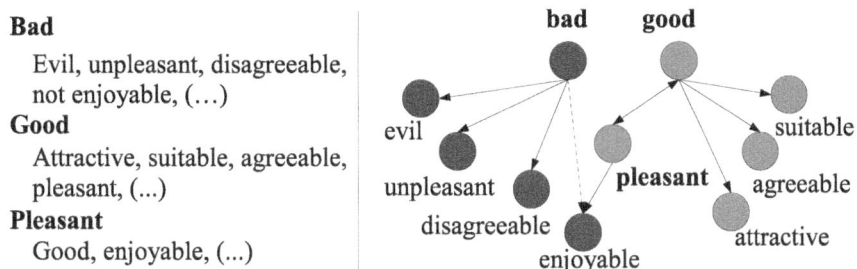

Fig. 1. Representation of the word *bad, good,* and *pleasant* as a directed graph (*dashed arrow = antonym, solid arrow = synonym*)

and then for all other remaining words until we have reached all possible nodes. The positive and negative polarity is propagated based on the assumptions that:

1. Two synonymous or related words have the same or a close sense, so they should have the same polarity. E.g. if *good* is positive the synonymous or related words *attractive, suitable, agreeable,* and *pleasant* should also be.
2. Two antonyms words have opposite sense, so they should have opposite polarity. E.g. if the word *bad* is negative their antonym *enjoyable* should be positive.

The second key idea is that the closer a word is to a seed word, the higher the probability of the propagation being right (section 4.2, table 8). Knowing this, the described sense propagation approach can be extended assigning different weights to words according to their distance to the closest seed word. For instance, supposing that word A and Z (fig. 2) are seed words. A is manually labeled as positive and Z as negative. The word B and Y are those which we want to know whether they are positive or negative (fig. 2 at left).

Fig. 2. Polarity propagation

Following the simple unweighted approach we should: 1.) propagate the positive sense of seed word A to B which is the only neighbor of A; 2.) propagate the negative sense of Z to Y which is only neighbor of Z; and finally 3.) propagate the positive sense of B to Y. While in this approach, word Y should be considered both positive and negative because it has a positive and a negative neighbor (B and Z respectively). In the weighted propagation approach, Y should be considered more negative than positive (or even only negative), because their neighbor Z should be considered more important than B since Z is closer than B from a seed word.

2.3   Intuitive and Simple Polarity Propagation - Algorithm

In this section we present the algorithm applied to propagate the positive and negative sense of a seed set of words and at the same time to compute the short distance between each word and the closest seed word.

Assuming that we have a dictionary represented as a direct graph (fig. 3 at left), we get the graph on the right (the polarity lexicon) by applying the following algorithm:

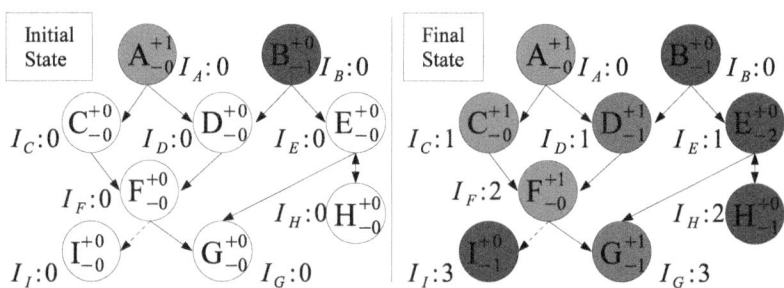

Fig. 3. Polarity propagation *(dashed arrow = antonym, solid arrow = synonym, Ix = Iteration or shortest distance from the closest seed word to word X)*

1. The first step is to label a set of words $W = \{w_1, \ldots, w_n\}$ as positive or negative, and for each word initialize an iteration counter I to 0. Finally add the words in a queue Q.

   Taking as example the graph on the left on the figure 3, suppose that word A is labeled as positive by adding 1 unit to its positive counter and 0 to its negative counter (A_{-0}^{+1}). In a similar way word B is labeled as negative (B_{-1}^{+0}). The iteration counter of each word is initialized to 0 $(I_A : 0$ and $I_B : 0)$, meaning they are seed words. Finally we add both words to a queue $Q = \{A_{-0}^{+1}, B_{-0}^{+1}\}$.

2. Retrieve the first word w_1 from the queue Q and we get all their neighbors $Nb_{w1} = \{nb_1, \ldots, nb_m\}$. For each neighbor nb_i visited for the first time, set the iteration counter to *value of iteration counter of w_1 + 1*.

   Continuing the previous example, word A_{-0}^{+1} is retrieved from Q and we get all their neighbors $Nb_A = \{C_{-0}^{+0}, D_{-0}^{+0}\}$. Since iteration counter of word A_{-0}^{+1} is set to 0 $(I_A : 0)$ and we are visiting their neighbors C_{-0}^{+0} and D_{-0}^{+0} for the first time, the iteration counter of each one is set to $I_A + 1$ leaving us with $I_C : 1$ and $I_D : 1$. The value 1 means that the shortest distance from the closest seed word to word C_{-0}^{+0} and D_{-0}^{+0} is 1.

3. Propagate the positive or negative sense of the word w_1 to all their neighbors Nb_{w1} according to their semantic relation (e.g. synonymous or antonymous). The propagation is done increasing the positive (pos.) or negative (neg.) counter of each neighbor according to the following rules:

$$\text{If } w_1 > 0 \bigwedge w_1 \longrightarrow nb_j \quad \text{Then} \quad \text{pos.} \leftarrow \text{pos.} + 1 \times \text{Weight}$$
$$\text{If } w_1 < 0 \bigwedge w_1 \longrightarrow nb_j \quad \text{Then} \quad \text{neg.} \leftarrow \text{neg.} + 1 \times \text{Weight}$$
$$\text{If } w_1 > 0 \bigwedge w_1 \dashrightarrow nb_j \quad \text{Then} \quad \text{neg.} \leftarrow \text{neg.} + 1 \times \text{Weight}$$
$$\text{If } w_1 < 0 \bigwedge w_1 \dashrightarrow nb_j \quad \text{Then} \quad \text{pos.} \leftarrow \text{pos.} + 1 \times \text{Weight}$$

The *Weight* variable should not be used or set to 1 if we want to apply an unweighted propagation approach. As discussed on section 2.2, the *Weight* should decrease as iteration increases.

Continuing the previous example, since word A_{-0}^{+1} is positive $(w_1 > 0)$ and has two synonymous neighbors C_{-0}^{+0} and D_{-0}^{+0} $(w_1 \longrightarrow nb_j)$, the first rule is applied to both (assuming Weight $= 1$*). This means that the positive counter of C_{-0}^{+0} and D_{-0}^{+0} is increased 1 unit leaving us with C_{-0}^{+1} and D_{-0}^{+1}.*

4. Mark word w_1, as visited, by adding him to a list of visited words V and for each neighbor nb_i of w_1:
   (a) If nb_i already exists on queue Q, update him (replace him);
   (b) If nb_i does not exists on queue Q nor list V, add him to the end of Q.
   Continuing the previous example, $V = \{A_{-0}^{+1}\}$, $Q = \{B_{-1}^{+0}, C_{-0}^{+1}, D_{-0}^{+1}\}$.
5. If the Q is not empty go to step 2 or else the algorithm ends.

To prevent infinite loops a breadth-first traversal is used and each node is visited just once. However the polarity of a word can also be propagated back (namely for words that already have a polarity). A weighted propagation approach may use the distance to the seed word to improve polarity propagation (see step 5 on above algorithm and section 2.2 for a weighted propagation approach). Also, on a real dictionary (graph) a word has several neighbors and the balance of all direct neighbors gives its final polarity. E.g., table 5 shows the word *destruir* (to destroy) as strongly negative because it has 22 negative vs. 5 positive direct neighbors.

At the end of the algorithm we get a polarity lexicon (Fig. 3 at right) where each word has a positive, a negative and an iteration counter. We can use the positive and negative counters to compute not only the polarity but also their relative strength. For instance, word A_{-0}^{+1} is positive with strength 1 *(1/(1 + 0))*. Word E_{-2}^{+0} is negative with strength 1 *(2/(2 + 0))*. Word D_{-1}^{+1} is positive with strength 0.5 *(1/(1 + 1))* and negative with same strength.

3   Polarity Lexicon - Pilot Experiment

In section 2.3 we presented the algorithm to propagate the positive and negative sense of the seed words assuming that we already have a dictionary (lexicon) represented as a graph. In this section we present an approach to propagate the positive and negative sense of each seed word while building the graph at a same time. The short distance between the closest seed word and each word is also computed. The algorithm is:

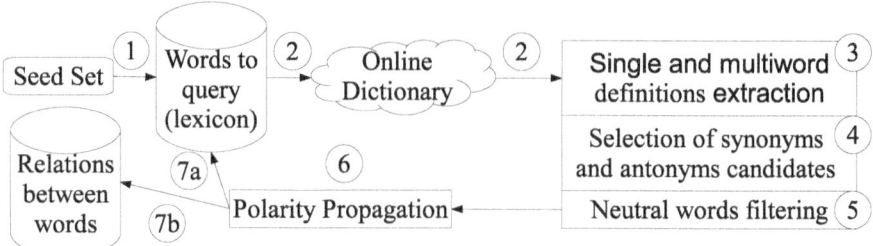

Fig. 4. Lexicon construction from an online dictionary

Step 1. Proceed as described on section 2.3 step 1. In our empirical study we start with a seed set of 5 positive and 5 negative Portuguese words. We tried to choose pairs of opposite words and with a strong positive and negative sense, hoping to minimize the propagation errors on a later stage. We initialized the positive, negative and iteration counter of each seed word as described on 2.3 step 1, and we stored them in the queue "Words to Query".

Table 1. Queue "words to query" (*"+"* = *number of positive senses, "-" = number of negative senses, It. = iteration, V. = visited*).

Word	+	-	It.	V.	Word	+	-	It.	V.
Bom (good)	1	0	0	no	Mau (bad)	0	1	0	no
Positivo (positive)	1	0	0	no	Negativo (negative)	0	1	0	no
Certo (right)	1	0	0	no	Errado (wrong)	0	1	0	no
Alegria (joy)	1	0	0	no	Tristeza (sadness)	0	1	0	no
Justo (fair)	1	0	0	no	Injusto (unfair)	0	1	0	no

Step 2. Query an online or offline dictionary and get the list of definitions for each word. In our study we query a Portuguese public online dictionary[3] and we get a web page for each queried word containing their meaning. For instance, for the queried seed word *injusto* (unfair) we get the list of meanings illustrated by Fig. 5.

Step 3. Extraction of definitions from the html document (web page). In Fig. 5 the definitions to be extracted for the queried word *injusto* (unfair) are rounded by a square. We have 5 single word expressions and 3 multiword expressions.

Step 4 a. Get all single words either by taking all single words, e.g. *arbritrio* (arbitary), *illegal* (illegal), or by extracting them from multi word expressions, e.g. *justiça* (justice) can be extracted from *oposto à justiça* (opposite of justice).

---

[3] http://www.infopedia.pt/lingua-portuguesa/

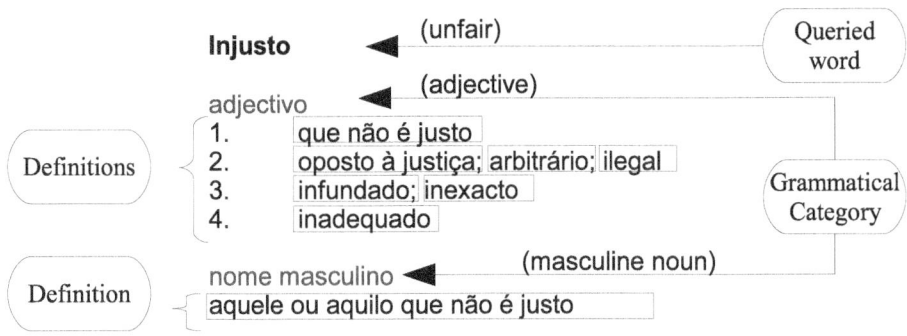

Fig. 5. Part of html page showing the definition (or definitions) of word *injusto* (unfair) *Definitions meaning: 1) opposite of justice; 2) that is not fair; arbitrary; illegal; 3) baseless; inaccurate; 4) inadequated; 5) who or what is not fair.*

Step 4 b. Determination of the semantic relation (e.g. *synonym, antonym*) among the queried word and each extracted single word. Note that single words as opposed to multiword expressions can be used later to obtain new words from the online dictionary.

In our study, step 4 a) and 4 b) are performed by extraction and classification rules as those on table 2.

Table 2. Most frequent rules for single word extraction and semantic relation classification

Extraction and classification rule	Semantic Relation	E.g. based on Fig. 5	Freq.
<single word>	Synonym	ilegal (illegal)	20605
fazer <single word> *(to do <single word>)*	Synonym	—	222
que tem <single word> *(which has <single word>)*	Synonym	—	112
acto de <single word> *(act of <single word>)*	Synonym	—	95
não <single word> *(not <single word>)*	Antonym	—	225
que não é <single word> *(that is not <single word>)*	Antonym	que não é <justo> *(that is not <fair>)*	35

Each extraction and classification rule extracts a <single word> and classifies it as synonym or antonym. For instance, for the word *injusto* (unfair) (Fig. 5), the rule *<single word>*, among others, extracts *ilegal* (illegal) and classifies it as synonym. The rule *que no <single word>* (that is not <single word>) extracts the single word *justo* (fair) and classify it as antonym.

The freq. column indicates the number of matches of a rule. Note that each rule needs to do an exact match to ensure that the extracted single word has meaning. A non exact match could increase the number of noisy single words, for instance, the extraction rule *que tem <single word>* (which has <single word>) applied to *que tem o poder de dicidir* (which has the power to decide) would extract the noisy word *o* (the).

Step 5. Filtration of neutral words (words which are neither positive nor negative). A word is not taken into consideration on next step if it is filtered.

In our study we used a small list of neutral words obtained by observing the dictionary. On example of figure 5, none of the words is filtered in this step.

Step 6. Propagation of the positive or negative polarity of queried word (qw) to all their directed neighbors according the rules presented on section 2.3, step 3. Determining the distance between the closest seed word to each single words sw_i as described on section 2.3, step 2.

Continuing the previous example, the negative polarity of *injusto* (unfair) is propagated according to the rules above, for instance to *ilegal* (illegal), *justo* (fair), and all others single words extracted on step 4 and not filtered on step 5. Since the word *justo* (fair) already exists on queue "words to query", its iteration counter remains unchanged (value 0). For all other single words the iteration counter is set to 1 (0 + 1, where 0 is the value of iteration counter of the word *injusto* (unfair)).

Step 7 a. Mark the queried word as visited and save all its meanings to be queried in future iterations.

Continuing the previous example, the word *injusto* (unfair) is marked as visited and its meanings saved on queue "Words to Query". The result of applying this step is shown in table 3.

Table 3. State of queue "Words to Query" after this step (*It.* = *Iteration, V.* = *Visited,* ... = *Other Seed Words*)

Word	+	-	It.	V.	Word	+	-	It.	V.
...	..	..	..	..	Arbitrário (arbitrary)	0	1	1	No
Injusto (unfair)	0	1	0	Yes	Ilegal (illegal)	0	1	1	No
Justo (fair)	2	0	0	No	...	..	..	..	..
Justiça (justice)	1	0	1	No					

Table 3 maintains a list of words with their positive and negative sense and it also maintains the short distance from the closest seed word to each word. This list forms our polarity lexicon.

Step 7 b. Save the queried word and its synonyms and antonyms. In our study we saved this information in the form of [*queried word, meaning word, semantic relation*].

Table 4. State of "relations between words" list after this step

Queried Word	Semantic Relation	Meaning Word
injusto (unfair)	Synonym	ilegal (illegal)
injusto (unfair)	Antonym	justo (fair)
...	...	...

Table 4 forms an adjacency list (a data structure for representing the dictionary as a graph).

Step 8: Finally, we check if there are more words to query. If there are, we retrieve the first word not yet visited from the queue "words to query" and return to step 3, if not the process ends. In the end, after applying the algorithm, the output is a polarity lexicon (table 5) and an adjacency list (table 4).

Table 5. Sample output of polarity words (*+% = positive strength, -% = negative strength*)

Word	PoS	+	-	It.	+%	-%
fora (strength)	noun	39	1	4	97.5	2.5
conveniente (convenient)	adjective	17	0	2	100	0
inconstante (fickle)	adjective	0	18	2	0	100
perturbar (to disturb)	verb	4	21	4	16	84
destruir (to destroy)	verb	6	22	4	21.43	78.57
confundir (to confuse)	verb	0	15	4	0	100

4   Evaluation

To evaluate the ability of the propagation algorithm to classify a word as positive or negative (algorithm presented on section 3), we first asked two humans to classify a sample of 524 words and to measure the agreement between them (section 4.1). Afterwards, we measured the agreement between them and the algorithm (4.2).

4.1   Inter-human Agreement

The following steps here performed for measure the inter-human agreement:

1. We randomly selected 524 words from the generated lexicon of 9107 words.
2. Initially we asked two native speakers of Portuguese to do a domain and context independent classification of each word to one of four categories: *positive* (+), *negative* (-), *neutral* (0), and *ambiguous* (A) for words positive and negative at same time. Soon there was the need to create the *unknown* category to words which the annotators do not know its meaning.

Table 6. Contingency table for inter-human classification agreement (diagonal elements show the agreement between Humans, off-diagonal elements show the disagreement)

		H1				Total				**H1**		Total
		+	-	0	A					+	-	
	+	188	5	1	4	198		**H2**	+	270	18	288
H2	-	13	157	2	12	184			-	15	157	172
	0	75	13	6	22	116		**Total**		285	175	460
	A	6	4	0	1	11						
Total		282	179	9	39	509						

Of the 524 words, 15 were *unknown* to one or both annotators, therefore they were immediately discarded. For the remaining 509 words belonging to the remaining four categories, the inter-human agreement was 69.16% (352 words) as shown on table 6 on the left (sum of diagonal elements). This is a similar agreement reported by Kim and Hovy [14] for English, and similar to Jijkoun and Hofmann [10] for Dutch, in both cases considering the categories: positive, negative and neutral. Kim and Hovy [14] reported an agreement of 76.19% for adjectives and 62.35% for verbs. Jijkoun and Hofmann [10] reported an agreement of 69%.

For the purpose of this study, we were only interested in evaluating the positive and negative sense assigned by the algorithm. Therefore, we discarded all words marked as ambiguous (like [10]) by one or both humans, and as [14] we merged the positive and neutral categories. In this case the inter-human agreement for the positive and negative categories was 92.83% (*Cohen's k = 0.85*) (table 6 at right). In the same way as [14] and as theoretically expected, the merging of positive and neutral categories increased the inter-human agreement. In our study increased from 69.16% to 92.83%.

Since 49 ambiguous words were removed, we were left with 460 words classified as positive or negative for human-machine agreement evaluation.

4.2   Human-Machine Agreement

In this section we report the agreement between each human (Human1 and Human2) and the algorithm presented on section 3. The results are shown on table 7. On run *H1:M.10seeds* and *H2:M.10seeds* it was used the 10 seed words pointed out on section 3, table 1. Table 7 shows that the agreement between Human1 and Machine is 75.43%, and between Human2 and Machine is 74.78%.

Table 8 shows the accuracy of run *H1:M.10seeds* and *H2:M.10seeds* by iteration. As it is shown, the accuracy decreases at each iteration. Given these results, we concluded that we should stop to propagate the positive and negative sense at earlier iterations.

Considering all the 460 words and all 6 iterations, the total agreement between Human1 and the Machine was 75.43% and between Human2 and Machine was

Table 7. Lexicon evaluation results *(A = Accuracy or Agreement, P = Precision, R = Recall and F1 = F-Measure, for positive and negative classes)*

		Positive Class			Negative Class		
Run	A (%)	P (%)	R (%)	F1 (%)	P (%)	R (%)	F1 (%)
H1:H2	92.83	n/a	n/a	n/a	n/a	n/a	n/a
H1:M.10seeds	75.43	75.22	90.63	82.20	76.11	50.00	60.35
H2:M.10seeds	74.78	74.35	90.63	81.65	76.11	49.14	59.72

Table 8. Human-Machine agreement by iteration for the 460 words *(It. = Iteration, T. = Total)*

	Human1:M.10seeds				Human2:M.10seeds		
It.	Agreement	Disagreement	T.	It.	Agreement	Disagreement	T.
0	10 (100.00%)	0 (0.00%)	10	0	10 (100.00%)	0 (0.00%)	10
1	52 (92.86%)	4 (7.14%)	56	1	51 (91.07%)	5 (8.93%)	56
2	48 (82.76%)	10 (17.24%)	58	2	50 (86.21%)	8 (13.79%)	58
3	54 (81.82%)	12 (18.18%)	66	3	54 (81.82%)	12 (18.18%)	66
4	77 (81.91%)	17 (18.09%)	94	4	70 (74.47%)	24 (25.53%)	94
5	54 (60.67%)	35 (39.33%)	89	5	56 (62.92%)	33 (37.08%)	89
6	52 (59.77%)	35 (40.23%)	87	6	53 (60.92%)	34 (39.08%)	87
T.	347 (75.43%)	113 (24.57%)	460	T.	344 (74.78%)	116 (25.22%)	460

74.78% for a lexicon of 9017 words. However, if we stop at iteration 4 we get an accuracy of 84.86% (241/284) for Human1 and 82.75% (235/284) for Human2 (stopping at iteration 4 reduces the lexicon from 9017 to 3034 words). These accuracies are similar to those obtained by [6] using the Merriam-Webster online dictionary and tested on three different test sets. These last authors have obtained accuracies of 83.71% (for 1,336 positive and negative adjectives), 79.78% (for 3,596 positive and negative terms), and 85.44% (for 663 adjectives).

5   Related Work

Several are the techniques for automatically building general wordnets. For example, Barbu et al [3] which used similar techniques for building a general wordnet for Hungarian. However we further study the polarity of words. In this section we will restrict our discussion to direct related work (polarity lexicons techniques).

There are a number of previous works which focus on building polarity lexicons. There are basically two main approaches to build it:

1. Approaches based on dictionaries. These approaches explore synonyms, antonyms, hypernyms, and hyponyms, among other relations, e.g. [12], [14], [17], or explore glosses (i.e. textual definitions) classification [6], [5]. Most of these approaches are based on WordNet [16]. Our work differs in that

we use a common online dictionary. There are also approaches that derive a polarity lexicon for a new language using an existing lexicon on another language and bilingual dictionaries [13] [23].

2. Approaches based on corpus. These approaches explore the co-occurrence of words, e.g. [8], [22], [21], [11].

5.1 Approaches Based on Dictionaries

Kamps et al. [12] determine the positive or negative semantic orientation of adjectives. They rely on the structure of the WordNet [16] to build a graph on the adjectives based on the WordNet *synonymous* relation. To determine the orientation of an adjective they measure the minimum distance between that adjective and two seed opposite adjectives *good* and *bad*. With this approach, only adjectives connected to any of the two chosen seed adjectives by some path in the synonymy relation graph can be evaluated.

Kim and Hovy [14] determine the strength of the positive and negative orientation of words. They label a small amount of seed words by hand and then use the *synonymous* and *antonymous* relation of WordNet [16] to expand them.

Esuli et al. [6], [5] determine the orientation of subjective terms. The orientation of a term is obtained based on the classification of glosses (i.e. textual definitions) that terms have in an online or offline glossary or dictionary. The method starts with a seed set of positive and negative words. That seed set is then expanded with WordNet [16] exploring the *synonym, direct antonym, indirect antonym, hypernym,* and *hyponym* relations. Then the glosses are extracted from WordNet [16] and from the online version of the Merriam-Webster[4] dictionary, and represented as vectors. Finally, it is applied a binary classifiers (naïve Bayes and Support Vector Machines).

Rao and Ravichandran [17] determine the positive and negative polarity of words. The polarity detection is treated as a semi-supervised label propagation problem in a graph. They try several graph-based semi-supervised learning methods like Mincuts, Randomized Mincuts, and Label Propagation. The study is done using WordNet [16] and OpenOffice thesaurus.

Silva et al. [18] determine the positive, negative and neutral polarity of human adjectives (adjectives that co-occur with a human subject). The authors build a graph from a seed set of adjectives manually classified and adjective synonyms from the union of multiple open thesauri. They compute the distance in the graph of each adjective whose polarity is unknown to the seed set adjectives already classified. The computed distances are then used to generate input features for an automatic polarity classifier.

5.2 Approaches Based on Corpus

Hatzivassiloglou and McKeown [8] determine the positive or negative semantic orientation of adjectives. They rely on the idea that conjunctions (e.g. *and, or,*

---

[4] http://www.merriam-webster.com/

but, *either-or*, and *neither-nor*) between adjectives provide indirect information about orientation. For instance, while *fair and legitimate* and *corrupt and brutal* have the same orientation and occurs in their corpus, the pairs *fair and brutal* and *corrupt and legitimate* would be semantically anomalous. These last two pairs of adjectives are reversed for *but*, which usually connects two adjectives of different orientations.

Turney and Littman [22] determine the positive or negative semantic orientation of two word phrases containing adjectives or adverbs. They rely on the idea that a phrase has a positive orientation when it has good associations and negative semantic orientations when it has bad associations. For determining these good and bad associations for each phrase, the authors query the Altavista[5] search engine using the phrase, the operator NEAR, and the opposite words *excellent* and *poor* (e.g. "phrase$_n$ NEAR excellent", "phrase$_n$ NEAR poor"). Based on the number of documents returned for each phrase, it is used the PMI-IR (Pointwise Mutual Information and Information Retrieval) to calculate the semantic orientation (for more details about PMI-IR, please refer to the original paper).

Takamura et al. [21] determine the positive or negative semantic orientation of words. They rely on a method that they call "spin model". The authors construct a lexical network by connecting two words and if one word appears in the gloss of the other word, they apply the spin model. The intuition behind this is that if a word has a semantic orientation, then the words in its gloss tend to have the same polarity.

6   Conclusion and Future Work

This paper shows how to build a moderate-sized polarity lexicon using only a common online dictionary; a small seed set of words, a set of high accurate extraction rules, and a simple yet effective graph polarity propagation algorithm. We also show how to infer the antonymy relation using extraction rules. The acquired results are similar to those of related work, but we use less resources and a more direct method. Evidence is given regarding the interest of using more specific, but highly accurate extraction rules of single words. The applied method is language-independent and can easily be applied to other languages for which a common online or offline dictionary exists. The method tries to capture the sense that a word has in most contexts (e.g. usually *good, agreeable* is positive, and *bad, unpleasant* is negative).

Since this pilot experiment was encouraging, we plan, for instance, to: a) increase the polarity lexicon; b) study a way to extend the positive and negative lexicon to include neutral words; c) adapt this domain independent lexicon into a domain-specific lexicon; d) use the polarity lexicon to identify polarity senses of larger units of text such as sentences; e) capture the different meanings of a word, taking into account the context of the sentence using models similar to those used in part-of-speech tagging [15].

---

[5] http://www.altavista.com/

Once our approach is applicable to English, as future work we may compare this approach, for instance, with SentiWordNet [7], [2], Q-WordNet [1].

Acknowledgments. This work was funded by the *FCT - Fundao para a Cincia e Tecnologia* (Portuguese research funding agency) through fellowship SFRH/BD/47551/2008.

References

1. Agerri, R., Garca-Serrano, A.: Q-WordNet: Extracting polarity from WordNet senses. In: Seventh Conference on International Language Resources and Evaluation, Malta (retrieved May 2010)
2. Baccianella, S., et al.: SentiWordNet 3.0: An enhanced lexical resource for sentiment analysis and opinion mining. In: Proceedings of the 7th Conference on Language Resources and Evaluation (LREC 2010), Valletta, MT, pp. 2200–2204 (2010)
3. Barbu, E., Mititelu, V.B.: Automatic building of Wordnets. In: Nicolov, N., Bontcheva, K., Angelova, G., R.M. (eds.) Recent Advances in Natural Language Processing IV (RANLP 2005), pp. 217–226. J. Benjamins Pub. Co., Amsterdam (2005)
4. Blair-Goldensohn, S., et al.: Building a Sentiment Summarizer for Local Service Reviews. Electrical Engineering (2008)
5. Esuli, A., Sebastiani, F.: Determining term subjectivity and term orientation for opinion mining. In: Proceedings the 11th Meeting of the European Chapter of the Association for Computational Linguistics (EACL 2006), pp. 193–200 (2006)
6. Esuli, A., Sebastiani, F.: Determining the semantic orientation of terms through gloss classification. In: Proceedings of the 14th ACM International Conference on Information and Knowledge Management, pp. 617–624. ACM, Bremen (2005)
7. Esuli, A., Sebastiani, F.: SentiWordNet: A publicly available lexical resource for opinion mining. In: Proceedings of the 5th Conference on Language Resources and Evaluation (LREC 2006), Citeseer, Genova, IT, pp. 417–422 (2006)
8. Hatzivassiloglou, V., McKeown, K.R.: Predicting the semantic orientation of adjectives. In: Proceedings of the 35th Annual Meeting of the Association for Computational Linguistics and Eighth Conference of the European Chapter of the Association for Computational Linguistics, pp. 174–181. Association for Computational Linguistics (1997)
9. Hu, M., Liu, B.: Mining and summarizing customer reviews. In: Proceedings of the Tenth ACM SIGKDD International Conference on Knowledge Discovery and Data Mining, pp. 168–177. ACM, New York (2004)
10. Jijkoun, V., Hofmann, K.: Generating a Non-English Subjectivity Lexicon: Relations That Matter. Computational Linguistics 398–405 (April 2009)
11. Kaji, N., Kitsuregawa, M.: Building lexicon for sentiment analysis from massive collection of HTML documents. In: Proceedings of the Joint Conference on Empirical Methods in Natural Language Processing and Computational Natural Language Learning (EMNLP-CoNLL), pp. 1075–1083 (2007)
12. Kamps, J., et al.: Using WordNet to measure semantic orientation of adjectives. In: Proceedings of LREC, pp. 1115–1118 (2004)
13. Kim, J., et al.: Conveying Subjectivity of a Lexicon of One Language into Another Using a Bilingual Dictionary and a Link Analysis Algorithm. International Journal of Computer Processing Of Languages 22, 02 & 03 205 (2009)

14. Kim, S.M., Hovy, E.: Determining the sentiment of opinions. In: Proceedings of the 20th International Conference on Computational Linguistics, p. 1367. Association for Computational Linguistics (2004)

15. Marques, N.C., Pereira Lopes, J.G.: Tagging with Small Training Corpora. In: Hoffmann, F., Adams, N., Fisher, D., Guimarães, G., Hand, D.J. (eds.) IDA 2001. LNCS, vol. 2189, pp. 63–72. Springer, Heidelberg (2001)

16. Miller, G.: WordNet: A lexical database for English. Communications of the ACM 11, 39–41 (1995)

17. Rao, D., Ravichandran, D.: Semi-supervised polarity lexicon induction. In: Proceedings of the 12th Conference of the European Chapter of the Association for Computational Linguistics on EACL 2009, pp. 675–682 (April 2009)

18. Silva, M.J., et al.: Automatic Expansion of a Social Judgment Lexicon for Sentiment Analysis. Technical Report. TR 10-08. University of Lisbon, Faculty of Sciences, LASIGE (2010)

19. Stone, P.J.: The General Inquirer: A Computer Approach to Content Analysis, 1st edn., January 1. M.I.T. Press (1966)

20. Strapparava, C., Valitutti, A.: WordNet-Affect: an affective extension of WordNet. In: Proceedings of LREC, Citeseer, pp. 1083–1086 (2004)

21. Takamura, H., et al.: Extracting Emotional Polarity of Words using Spin Model. In: Proceedings of the Joint Workshop of Vietnamese Society of AI, SIGKBS-JSAI, ICS-IPSJ and IEICE-SIGAI on Active Mining (AM 2004), Hanoi, Vietnam (2004)

22. Turney, P.D.: Thumbs up or thumbs down?: semantic orientation applied to unsupervised classification of reviews. In: Proceedings of the 40th Annual Meeting on Association for Computational Linguistics, pp. 417–424. Association for Computational Linguistics, Morristown (2002)

23. Waltinger, U.: German Polarity Clues: A Lexical Resource for German Sentiment Analysis. In: Proceedings of the Seventh International Conference on Language Resources and Evaluation (LREC), pp. 1638–1642 (2010)

A Bootstrapping Approach for Training a NER with Conditional Random Fields

Jorge Teixeira, Luís Sarmento, and Eugénio Oliveira

LIACC - FEUP/DEI & Labs Sapo UP
Porto, Portugal
{jft,las,eco}@fe.up.pt

Abstract. In this paper we present a bootstrapping approach for training a Named Entity Recognition (NER) system. Our method starts by annotating persons' names on a dataset of 50,000 news items. This is performed using a simple dictionary-based approach. Using such training set we build a classification model based on Conditional Random Fields (CRF). We then use the inferred classification model to perform additional annotations of the initial seed corpus, which is then used for training a new classification model. This cycle is repeated until the NER model stabilizes. We evaluate each of the bootstrapping iterations by calculating: (i) the precision and recall of the NER model in annotating a small gold-standard collection (HAREM); (ii) the precision and recall of the CRF bootstrapping annotation method over a small sample of news; and (iii) the correctness and the number of new names identified. Additionally, we compare the NER model with a dictionary-based approach, our baseline method. Results show that our bootstrapping approach stabilizes after 7 iterations, achieving high values of precision (83%) and recall (68%).

Keywords: Named Entity Recognition, Machine Learning, Conditional Random Fields, Natural Language Processing.

1 Introduction

There are currently many popular machine learning approaches for inferring Named Entity Recognition (NER) systems. Most of these techniques require a relatively large amount of text where entities have been annotated in context. However, annotating such corpora is difficult and expensive, and these factors usually limit both the *size* and the *recency* of such corpora. As a consequence, most available NER-annotated corpora are usually small and are composed of annotations made in text with several years old. From a practical point of view, this raises two problems. First, a small corpus may not be enough to allow inferring robust NER models, since only a relatively small number of contexts are present. Second, models inferred from old data may not be suitable to classify new data [8][7]. As we will show later, by training a classification model based on part of HAREM [10], a relatively small and old (from 1997) annotated NER

L. Antunes and H.S. Pinto (Eds.): EPIA 2011, LNAI 7026, pp. 664–678, 2011.

corpus, and testing the other part on the learned model, we show that modest precision values can be attained. Additionally, by testing the learned model on a dataset of recent news (from May 2011), we obtained even lower accuracy, meaning that the training corpus did not have enough new information to build a reliable model. We will present this data and results in Section 6.

The solution for both these problems would consist in constantly updating the annotated corpus with more recent examples (possibly substituting older annotations). The resulting corpus would become larger, and would contain recent text. But the amount of human effort involved in such task is simply too much for this strategy to become sustainable. Thus, we propose a bootstrapping approach to perform the annotation of entities in a large corpus, while simultaneously inferring a NER model.

We start with a large set of (non-annotated) news items and a dictionary of names that are very frequently found in news. We only consider names that have two or more words (e.g. "name surname"), which we assume to be unambiguously mentioned. Next, we annotate names in the set of news items by considering matches with entries in the dictionary. We then select the subset of sentences in which all the capitalized tokens are part of an annotated name, which can thus be considered *completely* annotated. This set of sentences will serve as the *seed* corpus.

In the second stage, we use the seed corpus to infer a conditional random field (CRF) model for performing name annotation. Such model is then run over the initial seed corpus to increase the number of (completely) annotated sentences. The resulting larger corpus is used to infer a new CRF model. This cycle is repeated until the model stabilizes. In the end, we expect to have a very large corpus of news annotated with high accuracy.

In each iteration, we evaluate three parameters. First, we evaluate the precision and recall of the inferred model in annotating a small gold-standard collection (HAREM) [10]. This allow us to check how robust our classification model is becoming, taking into account a standard (although relatively small and old) reference corpus. Second, we manually evaluate the precision and recall of the annotation over a small sample of news corpus from which we generated the news corpus. This allow us to estimate the accuracy of the annotation that we are producing for the entire news corpus. Finally, we manually check the correctness and the number of new names identified using the inferred model (i.e. not found in the initial dictionary) for assessing the speed at which the system converges to a stable NER model.

The remaining of the paper is organized as follows. In Section 2 we discuss some related work. In Section 3 we describe our Method and in Section 4 the Classification Model and Features Description. The Experimental Set-up will be presented in section 5, the Results obtained are described in Section 6 and its Analysis and Discussion are presented in Section 7. Finally, Conclusions and Future Work are presented in Section 8.

2   Related Work

The difficulty in obtaining manually annotated data for training NER systems has motivated researchers to look for alternative ways of generating annotated data, or for making the best possible use of unlabeled data.

For example, Collins et al. [1] use seven very simple rules to perform the annotation of a seed news corpus. The rules are: "New York", "California" and "U.S." are locations; any name containing Mr. is a person; any name containing Incorporated is an organization; and I.B.M. and Microsoft are organizations. This is the only supervised information used. The approach proposed by the authors is to find a weighted combination of simple (weak) classifiers. The two classifiers are built iteratively: each iteration involves minimizing a continuously differential function which bounds the number of unlabeled examples (around 90,000) on which the two classifiers disagree. The authors used a dataset of approximately 1 million sentences extracted from New York Times and manually evaluated a sub-set of 1,000 examples, assigning one of the four available categories: location, person, organization or noise. The authors report that their system classified names with over 91% accuracy, which was obtained with almost no manual effort involved.

Valchos et al. [13] demonstrated that bootstrapping an entity recognizer for genes from automatically annotated text can be more effective than by using a fully supervised approach based on manually annotated biomedical text. Their system was based on an improvement of a bootstrapping method previously presented by Morgan et al. [6]. The authors started by creating a test set for evaluating the quality of the NER gene recognizer proposed. The test set contained 82 biomedical articles manually annotated, following some pre-determined guidelines and taking special attention for the context around the words to be annotated. The authors then used the previously annotated texts to automatically annotate abstracts based on pattern matching. The resulting corpus, which, contained approximately 117,000 annotated names (17,000 of them unique) was used to train an Hidden Markov Model (HMM) for performing gene NER. Evaluation on the test set achieved an F-score of 81%. The authors also presented three different approaches for improving the results achieved. The first one consists in using state-of-the-art gene dictionary to increase the number of names annotated in the articles. After reapplying their HMM system, they achieved lower F-score (78%), which lead the authors to stress the importance of using naturally occurring data as training material. For improving the results previously obtained, the authors remove all sentences from the training set that did not contain any entities. After retraining the models, the resulting F-score obtained decreased slightly (80%), mainly because the precision decreased considerably, since this strategy deprived the classifier from contexts that could help the resolution of erroneous cases. Lastly, the authors tried to filter the contexts used for substitution and the sentences that were excluded using the confidence values of the HMM system. Results obtained improved slightly (83%) indicating that this was the best approach proposed.

Our work is similar to the one presented by Valchos et al. [13], since we also start with a dictionary of names to perform the seed annotation. However, tackling name recognition in news is a more dynamic problem, since new persons' names may "appear" everyday in news streams, including foreign ones for which no dictionary information may be (even partially) available. Also, in contrast with other works, namely Collins et al. [1] , we iteratively re-annotate our initial corpus using the models that we infer. The bootstrapping cycle has no pre-defined number of iterations, and runs until it reaches stability. This strategy allows our system to deal with an open set of names.

Regarding the impact of using relatively old data to train NER system, the study of Mota and Grishman [7] is one of the most relevant ones. The authors tested the performance of their NER system on a news corpus that spans for 8 years. Their NER tagger was trained and tested on distinct time segments of the news corpus. The main result was that the performance of the tagger clearly decreased as the the time gap between the training data and the test data became larger.

As far as we know, there has not been much work in trying to automatically rebalance a reference corpus with more up to date material. In this work, we also try to tackle this dimension of the problem.

3   Method

3.1   Initial Data

Our initial data is a corpus of news items, $\mathcal{C}^{news}$, and a list of names, $\mathcal{N}^{initial}$. The $\mathcal{C}^{news}$ corpus is composed of 50,000 news items extracted from Portuguese online newswires between the end of April 2011 and the middle of May 2011. Each news item contains a title and a body, and both parts are subject to identification of named entities. On total, this dataset contains approximately 400,000 sentences. The dictionary of names, $\mathcal{N}^{initial}$, is a list of 2,450 persons' names that are frequently found on news, and includes both Portuguese and international names. This list was compiled by scanning a collection of approximately 500,000 news and extracting all sequences of capitalized words that could be found in a context that is very correlated with names of people. The context used was "[Capitalized Word Sequence], [ergonym], ", where ergonym is a word normally included in a job description. Such pattern is frequently used on news to introduce people relevant to the news piece (e.g. "[Nicholas Sarkozy], [president]..."). We only considered capitalized word sequences that were identified more than 3 times on the entire collection, so only 2,450 persons' names where obtained. Although this is a relatively small number, past studies ([5] and [13]) have proven that a small but yet well-known and naturally occurring list of names is more advantageous than large gazetteers of low-frequency names.

3.2   Bootstrapping Cycle

The bootstrapping cycle is summarized in Figure 1. In the first run of the bootstrapping cycle (identified in Figure 1 by Iteration 0), we automatically annotate $\mathcal{C}^{news}$ following a simple dictionary-based approach, using the 2,450 entries stored in $\mathcal{N}^{initial}$. This annotation is performed using the following rules:

1. Exact matches starting by the longest name string from $\mathcal{N}^{initial}$ towards the shortest;
2. Soft matches between $n_i \in \mathcal{N}^{initial}$ on $\mathcal{C}^{news}$, which will allow us to include parts of names in common to both the $n_i \in \mathcal{N}^{initial}$ and $\mathcal{C}^{news}$ (e.g. we consider "Obama" as a soft match of "Barack Obama");

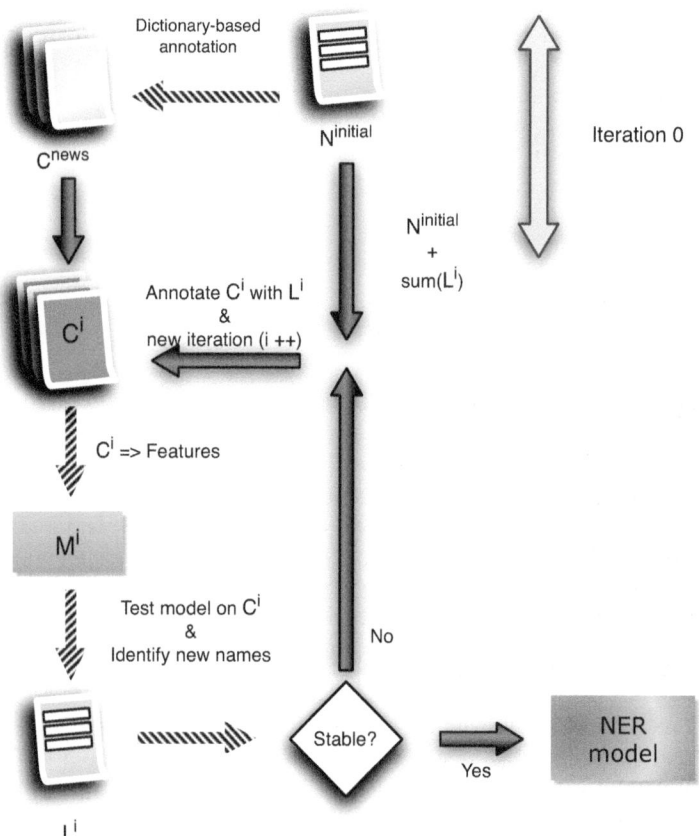

Fig. 1. Bootstrapping method

By following these rules, we were able to automatically annotate $\mathcal{C}^{news}$ and end-up with an annotated news corpus $\mathcal{C}^0$ with 57,642 persons' names, from which 50,514 were annotated in the body b_i of the news and 7,128 from the title t_i. We then used $\mathcal{C}^0$ to learn a classification model based on CRFs. We start by describing each example in the annotated corpus using a rich set of features $\mathcal{F}$, explained in section 4.2. Then, we infer a model $\mathcal{M}^0$. This model will then be applied on our previously used corpus $\mathcal{C}^0$ and we will create a list of the newly identified names, $\mathcal{L}^0$. With this list, together with the initial list of names $\mathcal{N}^{initial}$, we will be able to re-annotate the news corpus $\mathcal{C}^0$ and obtain a new annotated corpus, $\mathcal{C}^1$. The re-annotation process is based on the annotation rules described above.

At this point we will start a new iteration i of the bootstrapping process. This process will finish as soon as the system achieves a stable state.

4   Classification Model and Feature Description

4.1   Conditional Random Fields Models

Although our bootstrapping strategy does not directly depend on the classification algorithms used, we opted for Conditional Random Fields. CRFs are undirected statistical graphic models, and McCallum et al. [4] have shown that are well suited for sequence analysis, particularly on named entity recognition on newswire data.

According to Lafferty et al. [3] and McCallum et al. [4], let $o = \{o_1, o_2, ..., o_n\}$ be a sequence of words from a text with length s. Let $\mathcal{S}$ be a set of states in a finite state machine, each of which is associated with a label $l \in \mathcal{L}$ (e.g.: name, job, etc.). Let $s = \{s_1, s_2, ..., s_n\}$ be a sequence of states that corresponds to the labels assigned to words in the input sequence o. Linear chain CRFs define the conditional probability of a state sequence given an input sequence to be:

$$P(s|o) = \frac{1}{Z_o} exp \left(\sum_{i=1}^{n} \sum_{j=1}^{m} \lambda_j f_j(s_{i-1}, s_i, o, i) \right) \tag{1}$$

where Z_o is a normalization factor of all state sequences, $f_j(s_{i-1}, s_i, o, i)$ is one of the m functions that describes a feature, and λ_j is a learned weight for each such feature function. For this work we only use binary feature functions, a first order Markov independence assumption. A feature function may be defined, for example, to have value 0 in most cases, and have value 1 if and only if state s_{i-1} is state #1 (this state may has, for example, label *verb*) and state s_i is state #2 (for example a state that have label *article*). Intuitively, the learned feature weight λ_j, for each feature f_j, should be positive for features correlated with the target label, negative for features anti-correlated with the label, and near zero for relatively uninformative features, as described by [12]. CRFs are described in more detail by [3].

We used CRF++ (version .054)[1], a customizable implementation of CRFs for segmentation/labeling of sequential data, and we set to 50 the maximum number of iterations of the algorithm. On one hand, the convergence becomes extremely slow for large sets of data, such as the one we are using; and on the other hand 50 iterations are enough for the algorithm to converge in our scenario. We also specify a template that will be used by the CRF++ algorithm

---

[1] Available at: http://crfpp.sourceforge.net/

to learn the model. We opted for using a simple and straightforward template that only describes each of the tokens (usually a word, but may also include punctuation), their positions and their features within a sliding window of size 5. However, templates allow us to make different combinations of each token, its position and its features along with the other tokens from the sliding window. After several tests we conclude that the gains achieved by changing the templates description were very low, thus we used the simplest template approach.

4.2   Features Description

The quality and robustness of the NER model obtained greatly depends on the set of features used to describe the examples [9]. In our case, we decided to use *word-level* features and a window of 2 tokens to the left and to the right of the focus word. Table 1 presents groups of features used:

Table 1. Set of features used for the annotation of $\mathcal{C}^{news}$

	Features	Examples
$\mathcal{F}_{cap}$	Capitalized word	*John* or *Sophie*
$\mathcal{F}_{acr}$	Acronym	*NATO* or *USA*
$\mathcal{F}_{lng}$	Word Length	"musician" - *8*
$\mathcal{F}_{end}$	End of sentence	
$\mathcal{F}_{syn}$	Syntactic Cat.	"said" - *verb*
$\mathcal{F}_{sem}$	Semantic Cat.	"journalist" - *job*
$\mathcal{F}_{names}$	Names of people	*Barack Obama*

For the first group of features from Table 1, "Capitalized Word", "Acronym" and "Word Length", we developed simple and straightforward methods that fit these features. Regarding the "End of sentence" features, we used a tokenizer, developed by Laboreiro et al. [2]. This tool is based on a classification approach and is focused on the Portuguese language. After tokenizing the text, we apply a set of regular expression in order to split sentences and correctly identify the end of the sentences. For the "Syntactic Category" and "Semantic Category" features, we used LSP (Léxico Semântico do Português). LSP is a lexicon developed for the Portuguese Language, able to perform syntactic (and for some words a semantic) analysis of words. This allows us to add, for example, the semantic category "[nationality]" to the word "american" or even the semantic category "[communication verb]" to the word "say". The last set of features, $\mathcal{F}_{names}$ represent a list of names extracted from a Portuguese gazetteer developed by Sarmento et al. [11]. REPENTINO is a gazetteer for the Portuguese language that stores names under nearly 100 categories and subcategories. For this work, we are only interested in names of people, which are identified by the category *HUM* (human), subcategory *EN_SER* (human being entity). The task of extracting names from REPENTINO is thus straightforward and consists simply on building a list of all entities tagged on REPENTINO with the previous described category and subcategory.

Preliminary studies that we have conducted led to the conclusion that the best performance obtained by the trained models for NER tasks is by using all the 7 features together. Thus, we will describe training examples with all the features described in Table 1.

5 Experimental Set-Up

We are interested in: (i) proving that the age of the corpus has an important effect on the performance of NER systems; and (ii) evaluating our bootstrapping method in two different perspectives: by measuring the quality of the CRFs models created at each iteration and by evaluating the performance of our method in annotating a news corpus.

5.1 Measuring the Effect of Age in the Training NER Models

Mota and Grishman [7] had shown that there is a significant effect of the age of an annotated corpus on a NER tagger, and we are interested on evaluating this effect. For that, we will start by using 80% of HAREM annotated corpus, $\mathcal{C}_{train}^{HAREM}$, as our training corpus and the remaining 20% (with the annotations removed) as the test corpus, $\mathcal{C}_{test}^{HAREM}$. Then, for our baseline NER method, we create a dictionary of names from the training corpus, and annotate the test corpus by simply performing string matching operations. The quality of the annotated test set will allow us to calculate a performance measure for our baseline.

For the CRF method, we will train a CRF model with the $\mathcal{C}_{train}^{HAREM}$ and then test this model on $\mathcal{C}_{test}^{HAREM}$. Similar to the previous case, we will measure the performance of the CRF method based on the results of the annotation of the test set. By applying these evaluation methods, we want to prove that HAREM corpus is small and thus insufficient to be used as a model for NER. Then, we will use the same training set - $\mathcal{C}_{train}^{HAREM}$ - but this time the test set will be a small set of 1,000 recent news items, $\mathcal{C}_{test}^{news}$, extracted from the web in May 2011. We apply both the baseline NER model and the CRF NER model on this test set and evaluate the annotations automatically produced. Following the idea of Mota et al [7], with this test we intend to show that the performance of NER systems trained with a corpus that is chronologically distant (14 years) from the test corpus is seriously affected by the age factor. This should reinforce our motivation for proposing the bootstrapping technique we described before. Tests performed over the gold-standard corpus (HAREM) are totally automatic, as we have access to the complete annotated dataset. On the other hand, evaluation tests performed on the test set of recent news are manual, and consist of manually annotating a random sample of 50 different news items extracted from $\mathcal{C}_{test}^{news}$.

5.2 Evaluating the Bootstrapping Process

To measure the performance of our bootstrapping method and its evolution in each iteration, we will calculate the precision and recall of the inferred bootstrapping CRF model in annotating a small gold-standard collection (HAREM), in order to test the robustness of our NER model taking into account a gold-standard corpus (HAREM). Also, we will manually evaluate the precision and recall of the annotation process over a random sample of 20 news items extracted from the automatically annotated set of news. This will allow us to estimate the accuracy of the NER system on annotating a news corpus.

Our experiments will be performed considering the following empirically set conditions:

– The CRF threshold was empirically set to 0.6, so that the system will only assign a new name to the list of new names if its precision value obtained from the CRF bootstrapping model is higher than 0.6.
– The system will only assign a new name to the list of new names if it occurs at least 4 times on the entire test set, thus avoiding incorrect rare names that may introduce noise to the bootstrapping system.
– Persons' names with only one word (this means that the context words were not identified by the NER model as persons' names, or do not exist) will only be considered as valid new names, and thus added to the list of new names, if the precision value obtained by the CRF model is greater than 0.9. (e.g.: "Obama" or "Sócrates").

6   Results

6.1   Results on Evaluating NER by Training with HAREM Dataset

Results obtained for both the baseline NER model and the CRF NER model are presented in Table 2. Both methods were trained with $\mathcal{C}_{train}^{HAREM}$. This allows us to directly compare results obtained by each of them.

Regarding the dictionary-based method (see Table 2 - Dictionary Training Method), one can see that the precision is 1 for both test sets, as the annotation method consists only of string matches. Also, F1-measures obtained are relatively low (54% when tested with $\mathcal{C}_{test}^{HAREM}$ and 21% when testing with $\mathcal{C}_{test}^{news}$) and decrease when we test the model with recent news items. For the results obtained using the CRF NER model (see CRF Training Method on Table 2), the F1-measure values are considerably higher when compared to the dictionary-based method. Additionally, the F1-measure also decreases when the model is tested with the subset of 1,000 recent news, $\mathcal{C}_{test}^{news}$.

Table 2. Results for baseline NER model and the CRF NER model

Training Method	Testset	Precision	Recall	F1-measure
Dictionary	$\mathcal{C}_{test}^{HAREM}$	1.00	0.37	**0.54**
Dictionary	$\mathcal{C}_{test}^{news}$	1.00	0.12	**0.21**
CRFs	$\mathcal{C}_{test}^{HAREM}$	0.93	0.82	**0.87**
CRFs	$\mathcal{C}_{test}^{news}$	0.94	0.40	**0.55**

6.2   Results for the Bootstrapping Method

As far as the bootstrapping method is concerned, we performed two different evaluations, as described in section 5.2. Both evaluations were performed on the bootstrapping CRF model. This model was built from the news corpus C^{news} (composed by 50,000 news) and the initial set of names $N^{initial}$ (containing 2,450 names frequently found on news). Results for the automatic evaluation, performed on the gold-standard corpus (HAREM), are presented in Table 3 (precision P, recall R and F1-measure $F1$).

Table 3. Automatic Evaluation of the performance of the bootstrapping method on HAREM (gold-standard corpus)

Iteration	1	2	3	4	5	6	7	8	9	10	11	12
P	0.89	0.88	0.90	0.88	0.91	0.90	0.90	0.86	0.86	0.89	0.90	0.88
R	0.32	0.36	0.45	0.36	0.41	0.44	0.47	0.49	0.48	0.48	0.56	0.45
$F1$	0.47	0.51	0.60	0.51	0.56	0.59	0.62	0.62	0.62	0.62	0.69	0.60

From these results one can see that the bootstrapping system consistently increases the F1- measure of the NER system along the iterations. Also, after 7 iterations the NER system stabilizes, as the F1-measure obtained for subsequent iterations is mostly constant (62%).

Results obtained for the manual evaluation of the bootstrapping method, performed on a small random subset of recent news, are presented in Table 4.

Table 4. Manual evaluation of the performance of the CRF models trained using a bootstrapping approach

Iteration	1	2	3	4	5	6	7	8	9	10	11	12
P	0.78	0.78	0.74	0.88	0.82	0.78	0.83	0.81	0.77	0.77	0.76	0.78
R	0.42	0.61	0.50	0.53	0.53	0.61	0.68	0.66	0.65	0.66	0.64	0.68
$F1$	0.55	0.68	0.60	0.66	0.64	0.68	0.75	0.73	0.71	0.71	0.70	0.73

From Table 4, one can see that the F1-measure after each bootstrapping iteration grows sustainedly until iteration 7 supported by a near constant growth of recall, despite small fluctuations in precision. From iteration 8 onwards both recall and precision start oscillating resulting in a set of F1 values that oscillate between 0.70 and 0.73. However, the maximum value of F1 is reached at iteration 7.

Additionally, we evaluate the new names identified on each iteration of the bootstrapping method (built from C^{news}). Results are presented in Table 5 and include both the number of new names identified as well as its correctness, measured by the precision measure, of the new names identified.

Table 5. Manual evaluation of the new names identified

Iteration	1	2	3	4	5	6	7	8	9	10	11	12
$\mathcal{P}$	0.90	0.90	1.00	0.95	0.85	1.00	0.95	1.00	1.00	0.80	0.85	0.95
#new names	1,165	500	159	374	28	40	52	101	203	94	52	29

From results presented in Table 5 one can see that the precision values are equal or higher than 85% for the majority of the bootstrapping iterations. Also, one can see that the number of new names identified on each iteration is decreasing from iteration 1 to iteration 7. After iteration 7 we observe small variations of the number of new names. However, the global tendency is a decrease of the number of new names identified.

7   Analysis and Discussion

Table 2 shows that HAREM is not adequate to be used on an up to date NER system, when considering its age. Let us compare results obtained by using $\mathcal{C}_{test}^{HAREM}$ as test set, against $\mathcal{C}_{test}^{news}$. Both tests use the same training set, $\mathcal{C}_{train}^{HAREM}$. For the first case, this represents a chronologically similar test set, when compared to the training set. On the other hand, the second test set, represents a chronologically distant dataset (about 14 years old of difference). In the first case, we obtained a F1-measure of 54%. However, on the second case, F1-measure decreases to 21%. This means that using an old corpus to build a NER model is less efficient when it is applied to new, and chronologically distant, data.

Still observing results from Table 2, it is interesting to compare results obtained by using the baseline method, a straightforward dictionary-based approach, against the ones obtained by using CRF model. As one can see, for both test sets, the F1-measure obtained when using CRFs method is always significantly higher than using the dictionary based approach. From these results we may conclude that: (i) both NER and NER CRF models suffer from the effect of the training set age; (ii) CRFs seems to be more robust to the age effect when compared with the NER model, which is based on a dictionary of names that quickly gets out of date.

We used two different strategies for evaluating the bootstrapping approach we propose. Table 3 shows the results obtained from the automatic evaluation of the performance of the bootstrapping method on HAREM, the gold-standard corpus. From this results one can see that the bootstrapping system stabilizes after 7 iterations, as the F1-measure obtained for subsequent iterations is always constant. These results allows us to say that our bootstrapping approach is robust for the NER task proposed.

Table 4 presents results obtained for the manual evaluation of our bootstrapping method on a set of recent news. These results are coherent (similarly behave) with those achieved for the automatic evaluation of our method with HAREM. One can see that for this evaluation scenario the bootstrapping method also stabilizes after 7 iterations. Interestingly, the F1-measure obtained for iteration 7

(75%) is considerably higher that the one obtained from the automatic evaluation on HAREM (62%). As described in subsection 3.1, the news dataset $\mathcal{C}^{news}$ used for training is recent (from April to May 2011). The chronological distance between the training and testing datasets is very small for this scenario. On the contrary, this distance is considerably higher (14 years old) when comparing it to HAREM test set, used on the automatic evaluation. This clearly shows the effect age of the training set on NER systems.

Also, we manually evaluate the number and correctness of the new names identified by our bootstrapping methods. Results in Table 5 show that this number tends to decrease from iteration to iteration. This is an expected behavior, and ideally this number would tend to zero, meaning that the model would not be able to identify more names. However, from a practical point of view, the gain of new names identified for iterations 8 and more compared with the global accuracy of the system becomes insignificant. Performing a simple error analysis on the new names identified, we found two different types of errors, presented in Table 6.

Table 6. Error analysis on the new names identified on each bootstrapping iteration

Error type	Error description	Example
$\mathcal{E}_{ne}$	Wrong type of named entity	"*General Motors* comment on the crisis"
$\mathcal{E}_{conj}$	Missed name conjunction	"Jorge Nuno Pinto *da* Costa"

The first type of errors, $\mathcal{E}_{ne}$, happens when the named entity (on the example from Table 6, an organization) occurs on a context that is misleading. Considering the example phrase "Barack Obama comment on the crisis", in this case the context around the named entity (Barack Obama) is exactly the same as in the erroneous one. However, while the incorrectly annotated name is a name of an organization, the name of the example is a person's name. This type of error may be reduced if we use additional information as lists of names of organizations. Additionally, when the bootstrapping CRF model is not able to identify the conjunctions in the middle of a persons' name, we are in the presence of errors of type $\mathcal{E}_{conj}$. This error is only common in long names (four or more words) because the context is too broad and the sliding window may not be sufficiently large to capture all the relevant context around the name, and thus correctly identify the boundaries of the name.

In Figure 2 we present a comparative study of the performance of the baseline and the bootstrapping method, measured in terms of the F1-measure. We compare four different methods:

- *Baseline dictionary on news*: We built a model based on a dictionary of names from the training set $\mathcal{C}_{train}^{HAREM}$ and test this model $\mathcal{C}_{test}^{news}$ on a set of 1,000 recent news.
- *Baseline CRF on news*: We built a model based on the same dictionary of names from the previous case and apply it on $\mathcal{C}_{train}^{HAREM}$. This model was then tested on $\mathcal{C}_{test}^{news}$.

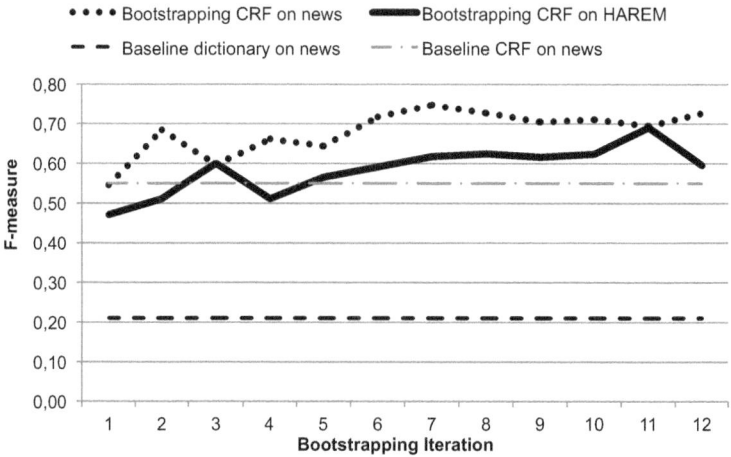

Fig. 2. Comparison of the annotation methods

- *Bootstrapping CRF on HAREM*: We built a bootstrapping CRF model based on the news dataset $\mathcal{C}^{news}$ (50,000 news) and a dictionary of names of people frequently mentioned on news. We automatically tested this model on the gold-standard corpus HAREM.
- *Bootstrapping CRF on news*: We use the same training corpus, but this time tested it on $\mathcal{C}^{news}_{test}$.

From the results obtained for the bootstrapping method (bootstrapping CRF on HAREM *versus* bootstrapping CRF on news) with both test sets, one can see they are comparable. Both results show an evolution of the F1-measure from the first to the seventh bootstrapping iteration, where the method stabilizes. Also, one can see that when training this model with recent news, testing it with an old dataset (HAREM) (see bootstrapping CRF on HAREM) does not decrease its performance (F1-measure) more than 10% when comparing it to $\mathcal{C}^{news}_{test}$ (see bootstrapping CRF on 1,000 recent news). Additionally, by comparing our baseline method based on a dictionary approach and the CRF bootstrapping method, both tested on $\mathcal{C}^{news}_{test}$, one can see that the performance achieved by the CRF bootstrapping method is much higher (73%) than that obtained with the dictionary-based method (21%). This proves, as expected, that the CRF bootstrapping method largely outperforms the dictionary-based one. Finally, considering the results obtained for the bootstrapping CRF on news and the baseline CRF on news, we are directly comparing methods trained with datasets of different sizes and age, but tested with the same data set, $\mathcal{C}^{news}_{test}$. In the first case, we used a training set with 50,000 recent news items, while in the second one, the training test was build from a gold-standard corpus, HAREM, with 14 years old. From these results one may see that the performance achieved by the baseline method is considerably lower (55%) than that obtained for the bootstrapping method, with a F1-measure of 73%. This result let us conclude that the CRF bootstrapping method, without any human effort, clearly outperforms the CRF baseline one, obtained using an - unfortunately old - human annotated corpus.

8   Conclusions and Future Work

We presented a bootstrapping approach for training a Named Entity Recognition (NER) system. We start by automatically annotating a news corpus of 50,000 news with a list of names of persons, with a dictionary-based approach. Then we built a CRF model that was tested on the previously annotated dataset, and we identified new names. These new names, together with the initial list of names, were used to re-annotate the news corpus and train a new model. This process was repeated until the system stabilized.

We were able to prove that typical gold-standard NER corpus (as HAREM) are not suitable for training NER systems for tagging recent texts, since they might not be sufficiently large and up to date. Also, we proved that our bootstrapping approach achieved a higher performance than when using CRFs trained with a limited dataset. Results have shown that our system stabilized after 7 iterations, which we consider a fast convergence, and with relatively high values of precision (83%) and recall (68%), corresponding to a F1-measure of 75%. Finally, using the CRF bootstrapping method we created a large annotated corpus of 50,000 news without any human effort and with a performance that clearly outperformed both the dictionary-based and CRF model approaches.

For future work, we may consider using sliding windows with different sizes, as this may help reducing errors found on the new names identified. Additionally, using lists of semantic categories (lists of jobs, list of organizations, etc.) could be helpful for the NER system to identify other named entities based on the context. Finally, one can think of experimenting and comparing different classification algorithms for this bootstrapping approach.

Acknowledgments. This work was partially supported by Labs Sapo UP from Portugal Telecom.

References

1. Collins, M., Singer, Y.: Unsupervised models for named entity classification. In: Proceedings of the Joint SIGDAT Conference on Empirical Methods in Natural Language Processing and Very Large Corpora, pp. 189–196 (1999)
2. Laboreiro, G., Sarmento, L., Teixeira, J., Oliveira, E.: Tokenizing Micro-Blogging Messages using a Text Classification Approach, AND 2010 - ACM, pp. 81–87 (2010)
3. Lafferty, J., McCallum, A., Pereira, F.: Conditional random fields: Probabilistic models for segmenting and labeling sequence data. In: Machine Learning - International Workshop, Citeseer, pp. 282–289 (2001)
4. McCallum, A., Li, W.: Early results for named entity recognition with conditional random fields, feature induction and web-enhanced lexicons. In: Proceedings of the Seventh Conference on Natural Language Learning at HLT-NAACL 2003, vol. 4, pp. 188–191. Association for Computational Linguistics (2003)

5. Mikheev, A., Moens, M., Grover, C.: Named Entity Recognition without Gazetteers. In: Proceedings of the Ninth Conference on European Chapter of the Association for Computational Linguistics, pp. 1–8. Association for Computational Linguistics (1999)
6. Morgan, A.a., Hirschman, L., Colosimo, M., Yeh, A.S., Colombe, J.B.: Gene name identification and normalization using a model organism database. Journal of Biomedical Informatics 37(6), 396–410 (2004)
7. Mota, C., Grishman, R.: Is this NE tagger getting old? In: Proceedings of the Sixth International Language Resources and Evaluation (LREC 2008), pp. 1196–1202 (2008)
8. Mota, C., Grishman, R.: Updating a name tagger using contemporary unlabeled data. In: Proceedings of the ACL-IJCNLP 2009 Conference Short Papers on - ACL-IJCNLP 2009, 353 (August 2009)
9. Nadeau, D., Sekine, S.: A survey of named entity recognition and classification. Linguisticae Investigationes 30(1), 3–26 (2007)
10. Santos, D., Seco, N., Cardoso, N., Vilela, R.: Harem: An advanced ner evaluation contest for portuguese. In: Odjik, Tapias, D. (eds.) Proceedings of LREC 2006, Genoa, pp. 22–28 (2006)
11. Sarmento, L., Pinto, A., Cabral, L.: REPENTINO A Wide-Scope Gazetteer for Entity Recognition in Portuguese. In: Computational Processing of the Portuguese Language pp. 31–40 (2006)
12. Settles, B.: Biomedical named entity recognition using conditional random fields and rich feature sets. In: Proceedings of the International Joint Workshop on Natural Language Processing in Biomedicine and its Applications - JNLPBA 2004, p. 104 (2004)
13. Vlachos, A., Gasperin, C.: Bootstrapping and evaluating named entity recognition in the biomedical domain. In: Proceedings of the HLT-NAACL BioNLP Workshop on Linking Natural Language and Biology, pp. 138–145. Association for Computational Linguistics, Morristown (2006)

Domain-Splitting Generalized Nogoods from Restarts

Luís Baptista[1,2] and Francisco Azevedo[1]

[1] CENTRIA, Faculdade de Ciências e Tecnologia, Universidade Nova de Lisboa, Portugal
[2] Instituto Politécnico de Portalegre, Portugal
lmtbaptista@gmail.com, fa@di.fct.unl.pt

Abstract. The use of restarts techniques associated with learning nogoods in solving Constraint Satisfaction Problems (CSPs) is starting to be considered of major importance for backtrack search algorithms. Recent developments show how to learn nogoods from restarts and that those nogoods are essential when using restarts. Using a backtracking search algorithm, with 2-way branching, generalized nogoods are learned from the last branch of the search tree, immediately before the restart occurs. In this paper we further generalized the learned nogoods but now using domain-splitting branching and set branching. We believe that the use of restarts and learning of domain-splitting generalized nogoods will improve backtrack search algorithms for certain classes of problems.

Keywords: constraint, restarts, learning, nogoods, domain-splitting, branching.

1 Introduction

Constraint Satisfaction Problems (CSPs) are a well-known case of NP-complete problems [1]. They have extensive application in areas such as scheduling, configuration, timetabling, resources allocation, combinatorial mathematics, games and puzzles, and many other fields of computer science and engineering.

The impressive progress in propositional satisfiability problems (SAT) has been achieved using restarts and nogoods recording. SAT and CSP share many solving techniques [2]. As noted in [3] the interest of the CSP community in restarts and nogood recording is growing.

In this paper we extend the work of Lecoutre on nogood recording from restarts [4], that use a 2-way branching scheme. We present a new form of nogoods recorded from restarts, when a domain-splitting branching scheme is used. We call them domain-splitting generalized nogoods. This is a theoretical improvement on nogoods from restarts. We show that our proposed nogoods have potentially more pruning power. We also show that our nogoods can be used when a set branching scheme is used. This is an important issue since recently set branching has been shown to be an important technique.

The rest of the paper is organized as follows. Section 2 gives background about constraint satisfaction problems, the search algorithm used and about restarts and learning. Section 3 presents the related work. In section 4 we present the state of the art on recording nogoods from restarts and in section 4 we present our contribution. Finally in section 6 we present conclusions and future work.

L. Antunes and H.S. Pinto (Eds.): EPIA 2011, LNAI 7026, pp. 679–689, 2011.

2 Background

2.1 Constraint Satisfaction Problem

A Constraint Satisfaction Problem (CSP) consists of a set of variables, each with a domain of values, and a set of constraints on a subset of these variables.

Based on [1, 5], we define more formally a CSP. Consider a set of variables $X = \{x_1,\ldots,x_n\}$ with respective domains $D = \{d_1,\ldots,d_n\}$ associated with them. Each variable x_i ranges over the domain d_i, not empty, of possible values. Now consider a set of constraints $C = \{c_1,\ldots,c_m\}$ over the variables X. Each constraint c_i involves a subset X' of X, stating the possible values combinations of the variables in X'. If the cardinality of X' is 1 we say that the constraint is unary, and if the cardinality is 2 we say that the constraint is binary. Hence, a CSP is a set X of variables with respective domains D, together with a set C of constraints.

Now we must define for the CSP what a solution is. A problem state is defined by an assignment of value to some (or all) variables. As an example, consider $\{x_i = v_i, x_j = v_j\}$, where v_k is one value of the domain d_k assigned to variable x_k, for $1 \leq k \leq n$. An assignment is said to be complete if every variable of the problem has a value (are instantiated). An assignment that satisfies every constraint (does not violate constraints) is said to be consistent, otherwise it is said to be inconsistent. So, a complete and consistent assignment is a solution to the CSP. A problem is satisfiable if at least one solution exists. More formally, if there exists at least one element from the set $d_1 \times \ldots \times d_n$ which is a consistent assignment. A problem is unsatisfiable if it does not have solution. Formally, in this case, all elements from the set $d_1 \times \ldots \times d_n$ are inconsistent assignments.

In this paper we will consider a CSP with finite domains (CP(FD)).

A propositional satisfiability problem (SAT) is a particular case of a CSP where the variables are Boolean, and the constraints are defined by propositional logic expressed in conjunctive normal form.

2.2 Search Algorithm

Search algorithms for solving a CSP can be complete or incomplete. Complete algorithms will find a solution, if one exists. If a CSP does not have a solution complete algorithms can be used to prove it. Backtrack search is an example of a complete algorithm. Incomplete algorithms may not be able to prove that a CSP does not have a solution, but may be effective at finding a solution if one exists. Local search is an example of an incomplete algorithm. In this paper we will use a complete backtrack search algorithm.

A backtrack search algorithm performs a depth-first search. At each node a uninstantiated variable is selected based on a variable selection heuristic. The branches out of the node correspond to instantiating the variable with a possible value (or a set of values) from the domain, based on a value selection heuristics. The constraints are used to check whether the assignments are consistent.

At each node of the search tree, an important technique, known as constraint propagation, is used to improve efficiency by maintaining local consistency. This technique can remove, during the search, inconsistent values from the domains of the variables and therefore prune the search tree. Also notice that the usually very important heuristics for variable ordering may depend on the outcomes of the constraint propagation mechanism. The variable selection heuristic based on the fail-first principle is an example. At each node of the search tree this heuristic chooses the variable with the smallest domain size. This is a dynamic heuristic, since the constraint propagation mechanism removes inconsistent values from the domain, which will influence the next variable selection.

At each node of the search different branching schemes could be used. Two traditional and widely used branching schemes are the d-way and 2-way. In the first one, at each node, branches are created, one branch for each of the possible values of the domain of the variable associated with the node. Branches correspond to assignments of values to variables. In the 2-way branching scheme two branches are created out of each node. In this scheme a value v_k, is selected from the domain d_k of a variable x_k. The left branch corresponds to the assignment of the value to the variable, and the right branch is the refutation of that value. This can be viewed as adding the constraint $x_k = v_k$ to the problem, the left branch; or, if this fails, adding the constraint $x_k \neq v_k$ to the problem, the right branch. An important difference in these two schemes is that in d-way branching the algorithm has to branch again on the same variables until the values of the domain are exhausted. In 2-way branching, when a value assignment fails, the algorithm can choose to branch on any other unassigned variable.

Another branching scheme is domain splitting [6]. This scheme splits the domain of the variable into two sets, typically based on the lexicographic order of the values. Two branches are crated out of each node, one for each set. In each branch the other set of values is removed from the domain of the variable. The algorithm evolves by reducing the domains of the variables. An assignment occurs when the domain size of a variable is reduced to one. Note that this scheme results in a much deeper search tree. This is useful for optimization problems and when the domains sizes of the variables are very large.

Set Branching refers to any branching scheme that split the domain values in different sets, based on some similarity criterion [7]. The algorithm then branches on those sets. Note that 2-way and domains splitting branching schemes can be viewed as a particular case of set branching.

2.3 Restarts and Learning

A complete backtrack search algorithm is randomized by introducing a fixed amount of randomness in the branching heuristic [8]. Randomization is a key aspect of restart strategies. The utilization of randomization results in different sub-trees being searched each time the search algorithm is restarted.

For many combinatorial problems, different executions of a randomized backtrack search algorithm, on the same instance, can result in extremely different runtimes. This large variability in the runtime of the complete search procedures can be explained by the phenomena of heavy-tail distribution [8-10]. The heavy-tail distribution is characterized by long tails, as we can see in figure 1, for the 8-queens problem. The curve gives the cumulative fraction of successful runs as a function of the number of backtracks [10].

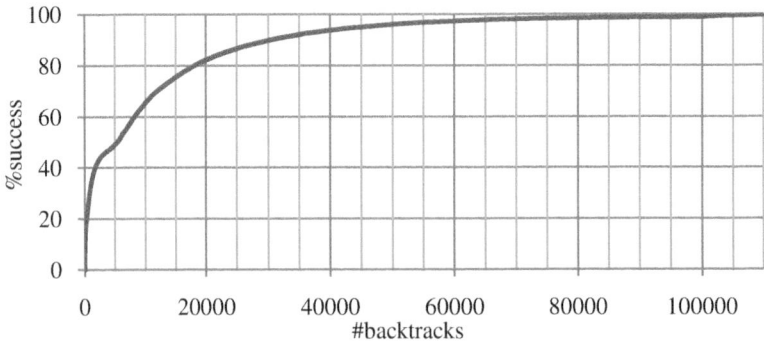

Fig. 1. Heavy-tail distribution example (8-queens)

A randomized complete search algorithm is repeatedly run (restart), each time limiting the maximum number of backtracks to a cutoff value. In practice a good cutoff values eliminates the heavy-tail phenomena, but unfortunately such a value has to be found empirically [8]. If restarts are used with a fixed cutoff value, the resulting algorithm is not complete. A solution to this problem is to implement a policy for increasing the cutoff value [11]. A simple policy is to increment by a constant the cutoff value after each restart. The resulting algorithm is complete, and thus able to prove unsatisfiability [12].

However, the incremental cutoff policy still exhibits a key drawback, because paths in the search tree can be visited more than once. This was addressed for SAT problems in [13] and for CSP in [4, 14], where nogoods are recorded, from the last branch of the search tree before the restart. Those recorded nogoods guarantee that the already visited search space will not be searched again.

3   Related Work

Learning in the context of SAT algorithms is known as conflict clause recording and in the context of CSP algorithms is known as nogood recording.

Nogood recording was introduced in [15], where a nogood is recorded when a conflict occurs during a backtrack search algorithm. Those recorded nogoods were used to avoid exploration of useless parts of the search tree.

Contrary to CSP, leaning is an important feature of SAT solver algorithms. Important progress in SAT solvers was due to the use of restarts, conflict clause recording [12, 16] and the use of very efficient data structures [16].

Standard nogoods correspond to variable assignments, but recently, a generalization of standard nogoods, that also uses value refutations, has been proposed by [17, 18]. They show that this generalized nogood allows learning more useful nogoods from global constraints. This is an important point since state of the art CSP solvers rely on heavy propagators for global constraints. The use of generalized nogoods significantly improves the runtime of the CSP algorithms. It is also important to notice that this generalized nogoods is much like clause recording in SAT solvers, and they show how to adapt other SAT techniques.

Recently the use of standard nogood and restarts in the context of CSP algorithms was studied [4, 14]. They record a set of nogoods after each restart (at the end of each run). Those nogoods are computed from the last branch of the search tree. So, the already visited tree is guaranteed not to be visited again. This approach is similar to one used for SAT, where clauses are recorded, from the last branch of the search tree before the restart (search signature) [13]. Recorded nogoods are considered as a unique global constraint with an efficient propagator. This propagator uses the 2-literal watching technique introduced for SAT. Experimental results show the effectiveness of this approach.

A hybrid CP(FD) solver that combines modeling and search of CP(FD) with learning and restarts of SAT solvers is proposed in [19]. The resulting solver is able to tackle problems that are beyond the scope of CP(FD) and SAT. They conclude that the combination of CP(FD) search with nogoods can be extremely powerful.

In SAT solvers the interplay of learning and restarts has proven to be extremely important for the success of the solvers. In the context of CP(FD) algorithms we are starting to study and understand this relation. But promising results show that learning is starting to be an interesting research area. As noted in [3] the impressive progress in SAT, unlike CSP, has been achieved using restarts and nogood recording (plus efficient lazy data structures). And this is starting to stimulate the interest of the CSP community in restarts and nogood recording.

4 Nogoods in 2-Way Branching

In this section we explain the nogoods presented in [4]. Consider a search tree built by a backtracking search algorithm with a 2-way branching scheme. In figure 2 we can see an example of such a tree just before the restart occurs. The left branch is called the positive decision and corresponds to an assignment. The right branch is called the negative decision and corresponds to a value refutation. A path in the search tree can be seen as a sequence of positive and negative decision.

Given a sequence of decision (positive and negative), an nld-subsequence (negative last decision subsequence) is a subsequence ending with a negative decision. As an example consider the sequence of decisions before the restart (the last branch of the search tree),

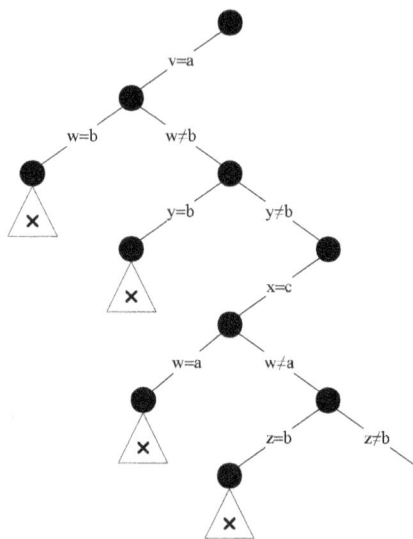

Fig. 2. Partial search tree before the restart, with 2-way branching

$$\langle v{=}a,\ w{\neq}b,\ y{\neq}b,\ x{=}c,\ w{\neq}a,\ z{\neq}b \rangle \tag{1}$$

The nld-subsequences that can be extracted from (1), are

$$\langle v{=}a,\ w{\neq}b \rangle \tag{2}$$

$$\langle v{=}a,\ w{\neq}b,\ y{\neq}b \rangle \tag{3}$$

$$\langle v{=}a,\ w{\neq}b,\ y{\neq}b,\ x{=}c,\ w{\neq}a \rangle \tag{4}$$

$$\langle v{=}a,\ w{\neq}b,\ y{\neq}b,\ x{=}c,\ w{\neq}a,\ z{\neq}b \rangle \tag{5}$$

A set of decisions is a nogood if they make the problem unsatisfiable. So, for any branch of the search tree a nogood can be extracted from each negative decision (nld-subsequence). Consider an nld-subsequence $\langle d_1, d_2, \ldots, d_i \rangle$, the set $\{d_1, d_2, \ldots, \neg d_i\}$ is a nogood (nld-nogood). The nld-nogoods that can be extracted from (1), one for each nld-subsequence (2-5) are,

$$\{v{=}a,\ w{=}b\} \tag{6}$$

$$\{v{=}a,\ w{\neq}b,\ y{=}b\} \tag{7}$$

$$\{v{=}a,\ w{\neq}b,\ y{\neq}b,\ x{=}c,\ w{=}a\} \tag{8}$$

$$\{v{=}a,\ w{\neq}b,\ y{\neq}b,\ x{=}c,\ w{\neq}a,\ z{=}b\} \tag{9}$$

As noted in [4] the nld-nogood corresponds to the definition of generalized nogoods [18], because it contains both positive and negative decisions. They also show that nld-nogoods can be reduced in size, considering only positive decisions. Consider an nld-subsequence $\langle d_1, d_2, ..., d_i \rangle$ and $Pos(\langle d_1, d_2, ..., d_i \rangle)$ denoting the set of positive decisions of the nld-subsequence, then the set $Pos(\langle d_1, d_2, ..., d_i \rangle) \cup \{\neg d_i\}$ is a nogood (reduced nld-nogood).

$$\{v=a, w=b\} \tag{10}$$

$$\{v=a, y=b\} \tag{11}$$

$$\{v=a, x=c, w=a\} \tag{12}$$

$$\{v=a, x=c, z=b\} \tag{13}$$

The advantages of using a reduced nld-nogood are more pruning power and reduction in space complexity. Also notice that the set of reduced nld-nogoods is equivalent to its original set of nld-nogoods. It consists of a more compact and efficient representation of nogoods.

5 Domain-Splitting Generalized Nogoods

In this section we give a generalization of the work presented in [4] about nogood recording from restarts. Adapting the same concepts we define a different kind of nogoods from restarts that use domain splitting instead of assignment. This work is a theoretical contribution for learning nogoods from restarts, in the context of backtrack search algorithms with a domain splitting branching scheme. Recall that we use finite domain.

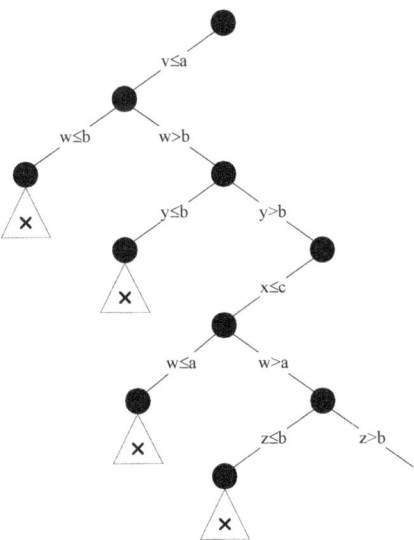

Fig. 3. Partial search tree before the restart, with domain-splitting branching

Consider a search tree built by a backtracking search algorithm with a domain splitting branching scheme. As for the 2-way branching scheme this is also a binary tree. But now the domain is split lexicographically in one of the values. In figure 3 we can see an example of such a tree before the restart occurs (a, b, and c are integers). The left branch is called the positive decision and corresponds to constraining the variable to the left part of the domain (the less than or equal). The right branch is called the negative decision and corresponds to constraining the variable to the right part of the domain (the greater than). The right branch can be seen as negating the decision made in the left branch. A path in the search tree can be seen as a sequence of positive and negative decision.

Consider the sequence of decisions before the restart (the last branch of the search tree),

$$\langle v{\leq}a, w{>}b, y{>}b, x{\leq}c, w{>}a, z{>}b \rangle \tag{14}$$

In a similar way we can extract nld-subsequences from (14),

$$\langle v{\leq}a, w{>}b \rangle \tag{15}$$

$$\langle v{\leq}a, w{>}b, y{>}b \rangle \tag{16}$$

$$\langle v{\leq}a, w{>}b, y{>}b, x{\leq}c, w{>}a \rangle \tag{17}$$

$$\langle v{\leq}a, w{>}b, y{>}b, x{\leq}c, w{>}a, z{>}b \rangle \tag{18}$$

And the equivalent to nld-nogoods, that we call domain-splitting nogoods (ds-nogoods),

$$\{ v{\leq}a, w{\leq}b \} \tag{19}$$

$$\{ v{\leq}a, w{>}b, y{\leq}b \} \tag{20}$$

$$\{ v{\leq}a, w{>}b, y{>}b, x{\leq}c, w{\leq}a \} \tag{21}$$

$$\{ v{\leq}a, w{>}b, y{>}b, x{\leq}c, w{>}a, z{\leq}b \} \tag{22}$$

In the context of ds-nogoods a decision node splits the domain in two sets. In the context of nld-nogoods a decision node can be seen also as splitting the domain in two sets: in the positive decision branch the set has only one value, because of the assignment; and in the negative branch the set has the other values, because of the refutation of the value. In this sense we can say that ds-nogoods are more powerful than nld-nogoods, because they use a more compact representation, since one positive decision can represent more than one value.

Similarly to reduced nld-nogoods, we can also have reduced ds-nogoods, considering only positive decision,

$$\{ v{\leq}a, w{\leq}b \} \tag{23}$$

$$\{ v{\leq}a, y{\leq}b \} \tag{24}$$

$$\{v \leq a, \ x \leq c, \ w \leq a\} \tag{25}$$

$$\{v \leq a, \ x \leq c, \ z \leq b\} \tag{26}$$

Again we can say that reduced ds-nogoods have potentially more pruning power than reduced nld-nogoods. Because they use a more compact representation, since one positive decision can represent more than one value.

5.1 Simplifying ds-nogoods

By construction, a CSP nogood does not contain two opposite decisions, e.g., $x \leq a$ and $x > a$. But a ds-nogood can have more than one decision on the same variable. As an example consider the decision over variable w in the ds-nogood (22), $w > b$ and $w > a$. It is easy to see, from the search tree in figure 3, that the decision $w > a$ subsumes $w > b$; because decision $w > a$ is made after $w > b$ we know that $a > b$. So for each ds-nogood (including the reduced version) a subsume procedure must be applied to remove unimportant decisions and thus simplify the nogood. For each variable, a ds-nogood only has to keep its last negative decision (if any) and its last positive decision (if any). In the case of reduced ds-nogoods, only the last (positive) decision has to be kept (for each of its variables). Thus, a great compaction can be obtained with such ds-nogoods.

5.2 Generalizing to dsg-nogoods

But ds-nogoods suffer from a possible key drawback, domain-splitting branching uses lexicographic order, which limits the expressive power of the search and consequently the learned nogoods. Namely, values are split in two sets lexicographically. It would be better for the sake of the flexibility of the search if the values could be split in any order.

We now assume a broad definition of domain-splitting branching scheme, which splits the domain D in two disjoint sets, s1 and s2 (i.e. $D = s1 \cup s2$), not necessarily in lexicographic order. The positive decision considers set s1 as the domain, in the left branch. If this fails the negative decision considers set s2 as the domain, in the right branch. More formally, for a variable x, the left branch corresponds to adding the constraint $x \in s1$, and the right branch to adding the constraint $x \in s2$ (the negative decision is $x \notin s1$ which is the same as $x \in s2$). The construction of the nogoods applies trivially to this more generic case, and we call these dsg-nogoods (domain-splitting generalized nogood) and reduced dsg-nogoods.

Note that (reduced) nld-nogoods are a particular case of (reduced) dsg-nogoods, since we can simulate 2-way branching with this broad definition of domain-splitting. The assignment branch corresponds to a set with the value of the assignment. The refutation branch corresponds to a set with the remaining values of the domain.

Recent developments have shown the importance of backtracking search algorithms using set branching schemes [7, 20]. The use of restarts and dsg-nogoods (and the reduced version) in those algorithms is direct if we only have two sets. But even if we have more than two sets, we can use a 2-way style set branching [7], where the sets are tried in a series of binary choices. The positive decision considers one of the sets as the domain, if this fails the negative decision considers the removal of that set from the domain. The use of dsg-nogoods is thus direct.

6 Conclusions and Future Work

The utilization of restarts with nogoods recording in backtrack search algorithms for solving constraint satisfaction problems is starting to be considered of great importance. We present an extension of Lecoutre's work on recording nogoods from restarts, the so called nld-nogoods. This paper is a theoretical contribution. We generalized the nld-nogoods to work in the context of backtracking search algorithms with domain-splitting and set branching schemes. Additionally, we gave evidences that our proposed nogoods have potentially more pruning power. We call these new nogoods domain-splitting generalized nogoods (dsg-nogoods).

We believe that the use of restarts and nogoods in a backtracking search algorithm with domain-splitting branching scheme will boost the performance of the algorithm. The problems tackled by domain-splitting are mostly optimization problems and especially with bigger domain sizes. Recent successful applications of set branching to optimization problems [20] makes us believe that adding restarts and nogoods will improve the algorithm performance.

In the near future we expect to empirically evaluate the use of dsg-nogoods in the context of restarts. But this work is included in a wider research project, whose aim is to study the use of restarts in constraint programming with finite domains. We will evaluate the interplay of different techniques associated with restarts, namely, nogoods, search restart strategies, randomization and heuristics.

Acknowledgments. This paper was submitted to SDIA 2011 - 3rd Doctoral Symposium on Artificial Intelligence. This work is included in a PhD program, supervised by Professor Francisco Azevedo, which began in September 2009 and is expected to be concluded in 2013. This work is supported by the Portuguese "Fundação para a Ciência e a Tecnologia" (SFRH/PROTEC/49859/2009).

References

1. Apt, K.R.: Principles of constraint programming. Cambridge University Press (2003)
2. Bordeaux, L., Hamadi, Y., Zhang, L.: Propositional Satisfiability and Constraint Programming: A comparative survey. ACM Comput. Surv. 38, 12 (2006)
3. Lecoutre, C.: Constraint Networks: Techniques and Algorithms. Wiley-ISTE (2009)
4. Lecoutre, C., Sais, L., Tabary, S., Vidal, V.: Nogood recording from restarts. In: Proceedings of the 20th International Joint Conference on Artifical Intelligence, pp. 131–136. Morgan Kaufmann Publishers Inc., Hyderabad (2007)
5. Russell, S., Norvig, P.: Artificial Intelligence: A Modern Approach. Prentice Hall (2002)
6. Dincbas, M., Hentenryck, P.V., Simonis, H., Aggoun, A., Graf, T., Berthier, F.: The Constraint Logic Programming Language CHIP. In: FGCS 1988, pp. 693–702 (1988)
7. Balafoutis, T., Paparrizou, A., Stergiou, K.: Experimental Evaluation of Branching Schemes for the CSP. 1009.0407 (2010)

8. Gomes, C.P., Selman, B., Kautz, H.: Boosting combinatorial search through randomization. In: Proceedings of the Fifteenth National Conference on Artificial Intelligence, pp. 431–437. American Association for Artificial Intelligence, Madison (1998)

9. Gomes, C., Selman, B., Crato, N.: Heavy-Tailed Distributions in Combinatorial Search. Principles and Practices of Constraint Programming, 121–135 (1997)

10. Gomes, C.P., Selman, B., Crato, N., Kautz, H.: Heavy-Tailed Phenomena in Satisfiability and Constraint Satisfaction Problems. Journal of Automated Reasoning 24, 67–100 (2000)

11. Walsh, T.: Search in a Small World. In: Proceedings of the Sixteenth International Joint Conference on Artificial Intelligence, pp. 1172–1177. Morgan Kaufmann Publishers Inc. (1999)

12. Baptista, L., Marques-Silva, J.: Using Randomization and Learning to Solve Hard Real-World Instances of Satisfiability. In: Dechter, R. (ed.) CP 2000. LNCS, vol. 1894, pp. 489–494. Springer, Heidelberg (2000)

13. Baptista, L., Lynce, I., Marques-Silva, J.: Complete Search Restart Strategies for Satisfiability. In: IJCAI Workshop on Stochastic Search Algorithms, IJCAI-SSA (2001)

14. Lecoutre, C., Saïs, L., Tabary, S., Vidal, V.: Recording and Minimizing Nogoods from Restarts. Journal on Satisfiability, Boolean Modeling and Computation 1, 147–167 (2007)

15. Dechter, R.: Enhancement Schemes for Constraint Processing: Backjumping, Learning, and Cutset Decomposition. Artif. Intell. 41, 273–312 (1990)

16. Moskewicz, M.W., Madigan, C.F., Zhao, Y., Zhang, L., Malik, S.: Chaff: engineering an efficient SAT solver. In: Proceedings of the 38th Annual Design Automation Conference, pp. 530–535. ACM, Las Vegas (2001)

17. Katsirelos, G., Bacchus, F.: Unrestricted Nogood Recording in CSP Search. CP. pp. 873–877 (2003)

18. Katsirelos, G., Bacchus, F.: Generalized NoGoods in CSPs, pp. 390–396. AAAI (2005)

19. Feydy, T., Stuckey, P.J.: Lazy clause generation reengineered. In: Gent, I.P. (ed.) CP 2009. LNCS, vol. 5732, pp. 352–366. Springer, Heidelberg (2009)

20. Kitching, M., Bacchus, F.: Set Branching in Constraint Optimization. In: Proceedings of the 21st International Joint Conference on Artificial Intelligence (IJCAI 2009), pp. 532–537 (2009)

A Proposal for Transactions
in the Semantic Web

Ana Sofia Gomes and José Júlio Alferes

CENTRIA and Departamento de Informática
Faculdade de Ciências e Tecnologia
Universidade Nova de Lisboa
2829-516 Caparica, Portugal

Abstract. The success of the Semantic Web project has triggered the emergence of new challenges for the research community. Among them, relies the ability of evolving the web by means of actions and updates in accordance with some standard proposals as RIF or SPARQL-Update. However, from the moment that actions and updates are possible, the need to ensure properties regarding the outcome of performing such actions emerges. Moreover, this need also leaves open the specification of such properties and requirements that an intended solution should comply to.

In this paper we motivate the need for employing transactional properties in this new Web and delineate a proposal for the requirements that such solution should provide. Afterwards, we develop a logic, based on the well-known Transaction Logic, that partially achieves such requirements, as a first step of an ongoing work.

1   Introduction

The World Wide Web has dramatically changed the way we communicate and share knowledge by aiding all kinds of users to access documents and participate with their own content. This, along with the simplicity inherent of the Web (which is in fact the most crucial factor for its popularity), enabled the explosion of both range and quantity of Web contents.

However, this sheer growth has triggered the appearance of more efficient ways to locate and search resources, and as a result, in the past decade, a large trend has been developed to evolve from a static and informative web to a linked web of data. Traditionally, data published on the Web was designed mainly for humans and made available in strict formats such as CSV or XML, or marked up as HTML tables, sacrificing much of its structure and semantics. As a consequence, the exploration of web content by automated processes became a synonym of a complex and painful task.

A community effort started with the advent of the Semantic Web, aims to disrupt this tendency by explicitly defining the semantics of contents and links for machine consumption. Underpinning this evolution is a demand to provide data as "raw" and thereby enabling the construction of powerful linked data mashups

L. Antunes and H.S. Pinto (Eds.): EPIA 2011, LNAI 7026, pp. 690–703, 2011.

across heterogeneous data source collections, without further programming effort. This project known as Linked Data[1] envisages to link arbitrary things in the web, relying mainly on documents containing data in RDF [20] (Resource Description Framework) format. This movement has gained such popularity that today the amount of links between datasets as well as the quality of these links has largely increased, paving the way to a *Web of Data* with the ultimate goal to use the web like a single global database. Nevertheless, in order to achieve such realization there are still several research quests that need to be addressed, and before using the web as a huge database it is essential to ensure (at least some) properties that one is used to see in standard databases [?]. Among these features is the ability to perform transactions which is the subject of this paper.

2   Transactions on a Web of Data

2.1   The Evolving Web

The growth of popularity of the semantic web has triggered the appearance of query languages capable of extracting knowledge from a complete distributed and heterogeneous database. One example is the SPARQL language [32] endorsed by W3C and able to perform SQL-like queries in RDF. This kind of languages has the power to relate and extract information from completely different applications, sites and/or services such as RSS feeds, government data or individual Friend of a Friend files.

Nonetheless, the (semantic) web, as envisioned, should be able to perform more activities than just querying. Communication platforms such as wikis (where several users can modify the same document), or online market places, are examples of existing web applications that require updates according to client requests or actions. However, if today's web evolution means evolution of individual web sites that are updated locally, lots of effort has been made to embrace the idea of cooperative evolution [2].

Crucial to the success of the semantic web is a cooperative behavior to publish web content in standard format (OWL [24] and RDF) in order to make intelligent access possible. Likewise, cooperative evolution relates to a joint effort to provide the web with the tools it needs in order to automatically evolve. Consequently, along with OWL and RDF standards, others have been proposed to break up with the conventional idea that the web is for read-only operations. In fact, the same way it is intended to give intelligent agents tools to consult the web, it is also the goal to allow these intelligent agents to perform automatic updates according to some rules. One example of this is the RIF-PRD [14] specification that intends to provide semantics for executing actions on the web in accordance with some given production rules (**If** *Condition* **then** *Action*).

In a similar way, other proposals have been made to give the web not only the capability to act but also to react automatically to changes and events. In this context, Event-Condition-Action languages, which represent an intuitive

---

[1] http://linkeddata.org/

and powerful paradigm for programming reactive system, have been largely used
to provide such semantics to the web [4,30,3].

As a result, from the original proposal of the Semantic Web by Tim Berners-
Lee in 1998 [6], a big effort has been made from the research community to shift
from a static web to a network of autonomous data sources capable of reacting
to changes and self-updating [1,22,11,21].

2.2 Motivating Transactions on the Web

With the adoption of the web of data as a new paradigm for the web, several
problems arise. Particularly, from the moment that actions and updates are
possible, it appears the need to ensure some properties regarding the outcome of
performing such actions. As an illustration, imagine some RIF-PRD production
rule stating that: **If** *a customer reaches $5000 of cumulative purchases during
the current year* **then** *its status becomes* Gold *and a golden customer card will be
printed and sent to him within one week.* In RIF-PRD syntax this is translated
into:

```
Prefix(ex <http://example.com/2008/prd1#>)
Forall ?customer ?purchasesYTD (
  If    And( ?customer#ex:Customer
             ?customer[ex:purchasesYTD->?purchasesYTD]
             External(pred:numeric-greater-than(?purchasesYTD 5000)))
  Then Do( And( Modify(?customer[ex:status->"Gold"])
               Execute(act:printCard(?customer,"Gold")))))
```

One obvious requirement of applying such rule is atomicity, that is, if the action
could not be performed completely, then it should not be performed at all. In
this particular example, a customer should not become a gold customer without
the emission of the corresponding card, neither a card should be delivered to a
customer whose status is not *gold.*

This kind of problems is generally solved in databases with the use of transactions. Transactions ensure atomicity, consistency, isolation and durability of a
special set of actions. These properties, known as ACID, play a fundamental role
in providing reliability to standard databases. Atomicity requires a transaction
to follow an all-or-nothing rule – all operations should be performed as a unit
which means that if one part of the transaction fails then the entire transaction
must fail as well. The standard way to handle a failure of an ACID transaction
is by a rollback, i.e. by restoring the state of the system before the execution
of the failed transaction. Consistency ensures that a transaction either achieves
a state where consistency of data is preserved, or returns to the original (consistent) state without changing data. Isolation guarantees that, even though a
set of transactions can be executed concurrently, the outcome is equivalent to
execute each transaction one-by-one in a given sequence. Durability states that
once a transaction has been committed, its changes will not be lost, even in the
event of some system failure[2].

---

[2] Durability is usually achieved by low-level software management and is hence outside
the scope of this study.

2.3   Why the ACID Model is not Enough

Albeit the advantages of the ACID model, it imposes severe demands [15] that are not always suitable for some systems, e.g. Web Sources, as argued in [25,29].

Particularly, atomicity requires all the steps of the transactions to rollback when the transaction fails. However, when these steps include external actions like sending an email or printing some document, this property is no longer possible to be guaranteed, since there is no way to revert these operations. In addition, there are situations where a transaction may involve iterated information exchange between different actors such as web services, human agents, databases, etc., potentially lasting for hours or days. These transactions are denoted as *long running transactions* or *sagas* [15]. Since it is impossible to perform a rollback when a transaction with such characteristics fails, other mechanisms are required. The usual approach is to define *compensation* operations for each operation to be performed. The idea is that these compensations lead the database into a state that is considered equivalent to the initial one, thus achieving some weaker form of atomicity. It is worth noting that the traditional rollback of an ACID transaction can be seen also as a form of compensation.

Since the traditional ACID model as found in standard databases is not suitable for the new context of the Semantic Web, we argue that transaction's properties must be redefined taken into consideration the characteristics of this peculiar context. Nevertheless, it is unquestionable that ensuring at least a relaxed model of such properties will provide crucial reliability to this emergent web of data.

2.4   Requirements for a New Web

After motivating the need for implementing transactional properties on the web, and discussing the limitations of the standard ACID model, we now present some argument on the necessary requirements for a new transactional model aimed for the new Web.

Transparency and collaboration are key features of anything that is related to the Web. One example of this concern is the W3C Recommendation RIF [18] which was designed to provide a standard for rule interchange in the web. In such context, declarative languages can play a fundamental role by providing ways to define programs which are clear and quickly understandable by their users. In fact, declarative languages have the advantage of being substantially more concise and self-explanatory, as they state what is to be computed rather than how it is to be computed. As a result, they are inherently high-level where programs can be viewed as theories and the details of the computation are left to the abstract machine.

However, as motivated by the proposal of several dialects for RIF, a language for the web needs not only to provide means to represent and reason about knowledge, but also to allow the execution of rules and actions. In this sense we are aiming for a declarative language that allow us to represent programs, but also to execute them.

Moreover, in order to abandon the concept of a read-only web, it is necessary to provide it the ability to react and respond to changes. In fact, the web is mainly dynamic, in the sense that its resources may change their content over time, and thus, a strong motivation exists in defining languages to specify updates and to detect them immediately. Furthermore, it is important to not only enable the detection of atomic events, but also of complex ones, in order for the solution to be useful in real scenarios.

Additionally, when defining a language which is intended to specify transactions it is obvious that this language must provide the ability to guarantee ACID properties, and to this end, it is necessary to have some notion of state change embedded in the theory. However, the web as a huge database consists of an agglomerate of different sources accessed and updated by an unpredictable large number of users, and thus it is impossible to control all the knowledge as well as to guarantee its consistency. As a result, if by one hand it is important to ensure ACID properties for local usage where one has total (or at least a high) control of the data and its accesses; on the other hand, and continuing the argumentation provided in Section 2.3, we believe that these properties are too strong in cases where it is not intended to restrict who can update the knowledge base, and/or what is updated. Note that this coincides with the reasons for combining open-world assumption with closed-world assumption for the semantic web context [13]. This way, our argumentation is that the two possibilities must coexist together, and the intended semantics must allow one to switch between pure database transactions (which ensure all ACID properties), and a weaker model of transaction designed for the web and loosening some properties like Isolation and Atomicity.

Finally, it is necessary to take into account that the Web is accessed by an unpredictable large number of users. This way, anything conceived for this kind of context, needs to be scalable and provide concurrency features. Nevertheless, it is worth noting that we are not interested here on "low-level" algorithms or efficient implementations for the problem of integrating transactions in the Semantic Web context. Particularly, the main goal of this proposal is concerned with knowledge representation and what are the requirements that a language and semantics must have in order to express transactions for this context. Therefore, when we refer to properties such as scalability and concurrency, we are arguing that these properties must be part of the intended semantics in a similar sense as Concurrent Transaction Logic ($\mathcal{CTR}$) [9] or the Calculus of Communicating Systems (CCS) [26].

In summary, we believe that a semantics which defines transactional properties needs to provide the following properties in order to be considered suitable for the context of a Web of Data: (1) *Declarativity*; (2) *Reactivity*; (3) *Transactional Properties* (ACID model); (4) *Weaker model of Transactional Properties*; (5) *Concurrency*.

3   A Logic for Transactions with External Actions

As a first step to achieve the requirements proposed we propose a novel logic that allows for the combination of standard ACID transactions with external actions.

Such logic has two main components, an "internal" component following the standard ACID model that interacts and executes actions with an "external" component. However, since it is impossible to rollback operations in a system that is *external*, the logic ensures a relaxed form of atomicity in the external domain by means of compensation operations.

As a starting point of the logic, we use Transaction Logic, a unique logic for specifying transactions in a very flexible way.

3.1   Transaction Logic

Transaction Logic ($\mathcal{TR}$) is an extension of predicate logic originally proposed in [8] which provides a logical foundation to reason about state changes in arbitrary logical theories (such as databases, logic programs or other knowledge bases), and particularly, to deal with ACID transactions. Contrary to most logics that reason about state change, $\mathcal{TR}$ does not use a separate procedural language to specify programs, as programs are specified in the logic itself. $\mathcal{TR}$ is thus a single representation language that can be used in two ways: to reason about programs (and the properties that they need to satisfy), and to execute them. When used for reasoning, one can, for instance, infer that a particular program preserves the integrity constraints of a knowledge base; or that under certain conditions, a transaction program is guaranteed not to abort. $\mathcal{TR}$ thus comes with a natural model theory (to perform reasoning) and a sound and complete proof theory (to specify and execute programs).

Moreover, reasoning in $\mathcal{TR}$ is *flexible* in the sense that it does not commit to any particular logical theory. To achieve this flexibility, $\mathcal{TR}$ is parameterized by a pair of oracles that encapsulate elementary knowledge base operations of querying and updating, thus allowing $\mathcal{TR}$ to reason about states and updates while accommodating a wide variety of semantics [8]. As a result, there is no distinction in $\mathcal{TR}$ between formulas that query the knowledge base and formulas that update it. As in classical logic, every formula has a truth value, but it also may have a side effect by changing the state of the knowledge base. It is thus the oracles' responsibility to decide if the formula can be executed and its corresponding effects.

Example 1 (Financial Transactions). As illustration of $\mathcal{TR}$, consider a knowledge base of a bank [8] where the balance of an account is given by the relation *balance(Acnt, Amt)*. To modify it we have a pair of elementary update operations: *balance(Acnt, Amt).ins* and *balance(Acnt, Amt).del* (denoting the insertion, resp. deletion, of a tuple of the relation). With these elementary updates, one may define several transactions, e.g. for making deposits in an account, make transfers from one account to another, etc. In $\mathcal{TR}$ one may define such transactions by the rules below where, e.g the first one means that one possible way to succeed the transfer of *Amt* from *Acnt* to *Acnt'* is by first withdrawing *Amt* from *Acnt*, followed by (denoted by $\otimes$) depositing *Amt* in *Acnt'*.

$transfer(Amt, Acnt, Acnt') \leftarrow withdraw(Amt, Acnt) \otimes deposit(Amt, Acnt')$
$withdraw(Amt, Acnt) \leftarrow balance(Acnt, B) \otimes changeBalance(Acnt, B, B - Amt)$
$deposit(Amt, Acnt) \leftarrow balance(Acnt, B) \otimes changeBalance(Acnt, B, B + Amt)$
$changeBalance(Acnt, B, B') \leftarrow balance(Acnt, B).del \otimes balance(Acnt, B').ins$

State change and evolution in $\mathcal{TR}$ is caused by executing ACID transactions, i.e. by posing logical formulas into the system in a Prolog-like style as e.g. $? - transfer(10, a_1, a_2)$. Since every formula is assumed as a transaction, by posing $transfer(10, a_1, a_2)$ we know that either $transfer(10, a_1, a_2)$ can be executed respecting all ACID properties evolving the knowledge base from an initial state D_0 into a state D_n (passing through an arbitrary number of states n); or $transfer(10, a_1, a_2)$ cannot be executed under these conditions and so the knowledge base does not evolve and remains in the state D_0. Also, in $\mathcal{TR}$ it is possible to have several rules (or rule instances) for defining one transaction, thereby allowing for the specification non-deterministic transactions.

3.2   External Transaction Logic

The characteristics shown make $\mathcal{TR}$ a unique logic for specifying transactions as it provides a unifying framework combining declarative knowledge *and* execution, whilst achieving a high flexible semantics. As argued, these are some of the features desired for transactional languages for the web.

Unfortunately $\mathcal{TR}$ is not suitable to model situations that require relaxing the standard ACID model. This limitation makes it impossible to express in $\mathcal{TR}$ *external actions*, that is, actions that are executed in an external entity. Furthermore, since $\mathcal{TR}$ only considers internal knowledge, i.e. transactions can only be executed in an internal knowledge base where one has a complete control and there is no way to interact with external entities, it becomes impossible for $\mathcal{TR}$ to *react* to external changes. In fact, as it is, $\mathcal{TR}$ is passive in the sense that transactions are only executed upon requests and internally. This conventional pattern has already been considered insufficient by the database community [31], and today most DBMS provide reactive features.

With the goal to provide a solution for the aforementioned limitations, and as a first step in achieving the requirements proposed in Section 2.4 we propose External Transaction Logic ($\mathcal{ETR}$). $\mathcal{ETR}$ is an extension of $\mathcal{TR}$ to accommodate interactions with an external domain. This interaction requires relaxing the traditional ACID model, since one has no control over the *external* domain in which actions are executed, external actions cannot be rollbacked, and thus, it is no longer possible to ensure the standard ACID model. To address this, $\mathcal{ETR}$ follows the proposal of [15]. The idea is to define a compensation for each operation to be performed. If the transaction fails and these compensations are performed in backward order, then they lead the database into a state that is considered equivalent to the initial one, thus ensuring a weaker form of atomicity.

3.3   Syntax and Oracles

$\mathcal{ETR}$ operates over a knowledge base which includes both an internal knowledge base, and an external domain, on which actions may be performed. For that, formally $\mathcal{ETR}$ works over two propositional languages: $\mathcal{L}_P$ (states language), and $\mathcal{L}_a$ (action language). Propositions in $\mathcal{L}_a$ denote actions that can be executed in the external domain. Propositions in $\mathcal{L}_P$ represent fluents that are true (or false) in the internal knowledge base, as well as in the external domain.

To build complex logical formulas, $\mathcal{ETR}$ uses the usual classical logic connectives (of conjunction, disjunction, etc) plus a special connective $\otimes$ to denoted *serial conjunction*. Informally, the formula $\phi \otimes \psi$ represents an action composed of an execution of ϕ followed by an execution of ψ. To allow for the specification of external actions $\mathcal{ETR}$ uses a special kind of formula $\mathbf{ext}(a, a^{-1})$ known as *external*. In $\mathbf{ext}(a, a^{-1})$, a is an external action formula, and a^{-1} its corresponding compensation which can be internal and/or external actions as to make possible for more flexible compensations.

Example 2. Consider the system of the web shop where clients submit orders. In the end of each order, a final confirmation is asked to the client that may or not confirm the transaction. If the client accepts it, the order ends successfully. Otherwise, the transaction fails and consistency must be preserved. In this case, it means that we need to rollback the update of the stock, and to compensate for the executed payment. The obvious compensation here is to simply ask the bank to refund the charged money. However, note that the transaction may fail sooner. E.g. the transaction may fail if the bank cannot charge the given amount in the credit card, or if the product is out-of-stock. This situation can be modeled in $\mathcal{ETR}$ by the rules:

$$buy(Prdt, Card, Amt) \leftarrow \mathbf{ext}(chargeCard(Card, Amt), refundCard(Card, Amt))$$
$$\otimes\ updateStock(Prdt) \otimes \mathbf{ext}(confirmTransaction(Product, Card, Amt), ())$$

$$updateStock(Prdt) \quad \leftarrow product(Prdt, N, WHouse) \otimes\ N > 0$$
$$\otimes\ product(Prdt, N, WHouse).del \otimes\ product(Prdt, N-1, WHouse).ins$$

To reason about elementary updates, $\mathcal{ETR}$ is parameterized by a triple of oracles $\mathcal{O}^d$, $\mathcal{O}^t$ and $\mathcal{O}^e$ respectively denoted the data, the transition oracle and the external oracle. These oracles encapsulate the elementary knowledge base operations, allowing the separation of elementary operations from the logic of combining them. As a result of this separation, $\mathcal{ETR}$ does not commit to any particular theory of elementary updates. An $\mathcal{ETR}$ program is then defined as follows.

Definition 1 ($\mathcal{ETR}$ actions, atoms, formulas and programs). *Given propositional languages $\mathcal{L}_P$ and $\mathcal{L}_a$, an $\mathcal{ETR}$ action is either a proposition in $\mathcal{L}_a$, or $\mathbf{ext}(a, b)$ where a is a proposition in $\mathcal{L}_a$ and b is either a proposition in $\mathcal{L}_a$ or a proposition in $\mathcal{L}_P$. An $\mathcal{ETR}$ atom is either a proposition in $\mathcal{L}_P$ or an $\mathcal{ETR}$ action. $\mathcal{ETR}$ formulas are inductively defined as follows:*

- *an $\mathcal{ETR}$ atom is an $\mathcal{ETR}$ formula;*
- *if ϕ and ψ are $\mathcal{ETR}$ formulas, then $\neg\phi$, $\phi \wedge \psi$, $\phi \vee \psi$, $\phi \leftarrow \psi$ and $\phi \otimes \psi$ are $\mathcal{ETR}$ formulas; nothing else is an $\mathcal{ETR}$ formula.*

A set of of $\mathcal{ETR}$ formulas is called an $\mathcal{ETR}$ program.

3.4   Model Theory

An important concept in $\mathcal{ETR}$'s model theory is the notion of *compensation*. A compensation occurs when the executed transaction ϕ contains external actions and fails. Since, in such a case, it is not possible to simply rollback to the initial state before executing ϕ, a series of compensating actions are executed to restore the consistency of the external knowledge base.

A transaction is, as usual, a sequence of actions that need to be performed (among other things) in a all-or-nothing way, making the internal knowledge base evolve from an initial state D_1 into a state D_n. In $\mathcal{ETR}$, since an external domain is also considered, a transaction may also make this external domain evolve from an initial state E_1 into a state E_m. During the execution of such a transaction both the internal and the external domain pass through an arbitrary number of intermediate states $D_1, D_2, \ldots, D_{n-1}, D_n$ and $E_1, \ldots, E_m$. This notion of sequence of states, denoted as *path*, is central to $\mathcal{ETR}$'s model theory as it represents the basic structure on which formulas are evaluated.

Given an $\mathcal{ETR}$ theory, formulas are evaluated on paths of internal states (as also in the original $\mathcal{TR}$), together with paths of external states, and with sequences of (external) actions (as we shall see, needed to perform compensations). For that, an *interpretation* is defined as a mapping from such a pair of paths and a sequence of actions into a set of $\mathcal{ETR}$ formulas (those true under that interpretation). As in $\mathcal{TR}$, interpretations are restricted such that formulas classically true in a state are true in (internal) paths just with that state (evaluated by the state data oracle $\mathcal{O}^d$), and such that, if a formula φ forces the internal state to evolve from D_1 into D_2 (evaluated by the state transition oracle $\mathcal{O}^t$), then it is true in the path D_1, D_2. Moreover, to account for the behavior of the external domain, we further restrict interpretations to obey to an external oracle $\mathcal{O}^e$ that detects external changes, and in which the external actions are modeled.

Definition 2 (Interpretations). *An interpretation is a mapping M that given a path of internal states, a path of external states and a sequence of actions, returns a set of transaction formulas (or $\top$)[3]. This mapping is subject to the following restrictions:*

1. $\varphi \in M(\langle D \rangle, \langle E \rangle, \emptyset)$, *for every* φ *such that* $\mathcal{O}^d(D) \models \varphi$
2. $\varphi \in M(\langle D_1, D_2 \rangle, \langle E \rangle, \emptyset)$ *if* $\mathcal{O}^t(D_1, D_2) \models \varphi$
3. $A \in M(\langle D \rangle, \langle E_1, \ldots, E_p \rangle, \langle A \rangle)$
   if $\mathcal{O}^e(E_1, \ldots, E_p) \models A$ *and* $p > 1$

The definition of satisfaction of $\mathcal{ETR}$ formulas, over general paths, requires the prior definition of operations on paths. These take into account how sequences of action are satisfied, and how to construct the correct compensation.

Definition 3 (Paths and Splits). *A path of length k, or a k-path, is any finite sequence of states (where the Ss are all either internal or external states),*

---

[3] Similar to $\mathcal{TR}$, for not having to consider partial mappings, besides formulas, interpretation can also return the special symbol $\top$. The interested reader is referred to [10] for details.

$\pi = \langle S_1, \ldots, S_k \rangle$, where $k \geq 1$. A split of π is any pair of subpaths, π_1 and π_2, such that $\pi_1 = \langle S_1, \ldots, S_i \rangle$ and $\pi_2 = \langle S_i, \ldots, S_k \rangle$ for some i $(1 \leq i \leq k)$. In this case, we write $\pi = \pi_1 \circ \pi_2$.

Definition 4 (External action split). *A split of a sequence of external actions* $\alpha = \langle A_1, \ldots, A_j \rangle$ $(j \geq 0)$ *is any pair of subsequences,* α_1 *and* α_2, *such that* $\alpha_1 = \langle A_1, \ldots, A_i \rangle$ *and* $\alpha_2 = \langle A_{i+1}, \ldots, A_j \rangle$ *for some* i $(0 \leq i \leq k)$. *In this case, we write* $\alpha = \alpha_1 \alpha_2$.

Note that there is a significant difference between Definitions 3 and 4. In fact, splits for sequences of external actions can be empty, and thus, it is possible to define splits of empty sequences, whereas a split of a path requires a sequence with at least length 1.

Besides the general definition of paths and splits of states and actions, we also define special operations over internal paths and external actions to handle compensations. The idea is that, if a transaction formula that contains external actions fails, then the compensations of each external action performed need to be executed in the backward order and the internal path rollbacked, i.e. the initial internal state is restored as the current state. These notions are made precise as follows.

Definition 5 (Rollback split). *A rollback split of* $\pi = \langle D_1, \ldots, D_k \rangle$ *is any pair of finite subpaths,* π_1 *and* π_2, *such that* $\pi_1 = \langle D_1, \ldots, D_i, D_1 \rangle$ *and* $\pi_2 = \langle D_1, D_{i+1}, \ldots, D_k \rangle$.

Definition 6 (Inversion). *An external action inversion of a sequence* α *where* $\alpha = (\mathbf{ext}(a_1, a_1^{-1}), \ldots, \mathbf{ext}(a_n, a_n^{-1}))$, *denoted* α^{-1}, *is the corresponding sequence of compensating external actions performed in the inverse way as* $(a_n^{-1}, \ldots, a_1^{-1})$.

Note that inversion is only defined for sequences where all action have are of the form $\mathbf{ext}(a, a^{-1})$. In fact, if for one action in the sequence no compensation is defined, then it is impossible to compensate the whole sequence. Building on these definitions, we formalize what (complex) formulas are true on what paths.

Definition 7 (Satisfaction). *Let* M *be an interpretation,* π *be an internal path,* ϵ *be an external path and* α *be a sequence of external actions. If* $M(\pi, \epsilon, \alpha) = \top$ *then* $M, \pi, \epsilon, \alpha \models \phi$ *for every* $\mathcal{ETR}$ *formula* ϕ; *otherwise:*

1. **Base Case:** $M, \pi, \epsilon, \alpha \models p$ *if* $p \in M(\pi, \epsilon, \alpha)$ *for any* $\mathcal{ETR}$ *atom* p
2. **Negation:** $M, \pi, \epsilon, \alpha \models \neg\phi$ *if it is not the case that* $M, \pi, \epsilon, \alpha \models \phi$
3. **"Classical" Conjunction:** $M, \pi, \epsilon, \alpha \models \phi \wedge \psi$ *if* $M, \pi, \epsilon, \alpha \models \phi$ *and* $M, \pi, \epsilon, \alpha \models \psi$.
4. **Serial Conjunction:** $M, \pi, \epsilon, \alpha \models \phi \otimes \psi$ *if* $M, \pi_1, \epsilon_1, \alpha_1 \models \phi$ *and* $M, \pi_2, \epsilon_2, \alpha_2 \models \psi$ *for some split* $\pi_1 \circ \pi_2$ *of path* π, *some split* $\epsilon_1 \circ \epsilon_2$ *of path* ϵ, *and some external action split* $\alpha_1 \circ \alpha_2$ *of external actions* α.
5. **Compensating Case:** $M, \pi, \epsilon, \alpha \models \phi$ *if* $M, \pi_1, \epsilon_1, \alpha_1 \alpha_1^{-1} \rightsquigarrow \phi$ *and* $M, \pi_2, \epsilon_2, \alpha_2 \models \phi$ *for some split* $\pi_1 \circ \pi_2$ *of* π, *some split* $\epsilon_1 \circ \epsilon_2$ *of path* ϵ, *and some external action split* $\alpha_1 \alpha_1^{-1}$, α_2 *of* α.
6. *For no other* $M, \pi, \epsilon, \alpha, \phi$ *it holds that* $M, \pi, \epsilon, \alpha \models \phi$.

In the sequel we also mention the satisfaction of disjunctions and implications, where as usual $\phi \vee \psi$ means $\neg(\neg\phi \wedge \neg\psi)$, and $\phi \leftarrow \psi$ means $\phi \vee \neg\psi$.

Note that Definition 7 requires the definition of *Consistency Preserving Path*. Intuitively, $M, \pi, \epsilon, \alpha\alpha^{-1} \rightsquigarrow \phi$ (defined below) means that, in the *failed* attempt to execute ϕ, a sequence α of external actions were performed. Although the internal state before the execution of the transaction is restored (by rollback split), since it is impossible to perform external rollbacks, consistency is ensured by performing a sequence of compensating actions α^{-1} in backward order (in case such compensating actions were defined, i.e. if there is an inversion of α). A formula ϕ is said to succeed over a compensating case if, although the execution failed, it is possible to construct a consistency preserving path (cf. Definition 8) and further *succeed* on an alternative execution.

Definition 8 (Consistency Preserving Path). *Let M be an interpretation, π be an internal path, ϵ an external path, and α be a non-empty sequence of external actions such that α^{-1} is defined. Let π_1 and π_2 be a rollback split of π. The path π_1' is obtained from $\pi_1 = \langle D_1, \ldots, D_n \rangle$ by removing the state D_n from the sequence; α^{-1} is a non-empty sequence of external actions obtained from α by inversion; $\epsilon_1 \circ \epsilon_2$ is a split of ϵ. We say that $M, \pi, \epsilon, \alpha\alpha^{-1} \rightsquigarrow \phi$ iff $\exists b_1 \otimes \ldots \otimes b_i \otimes \ldots \otimes b_n$ such that:*

$$M, \pi_1', \epsilon_1, \alpha \models \phi \leftarrow (b_1 \otimes \ldots \otimes b_i \otimes \ldots \otimes b_n)$$
$$M, \pi_1', \epsilon_1, \alpha \models b_1 \otimes \ldots \otimes b_i \otimes \neg \ b_{i+1}$$
$$M, \pi_2, \epsilon_2, \alpha^{-1} \models \bigotimes \alpha^{-1}$$

where $\bigotimes$ represents the operation of combining a sequence of actions using $\otimes$.

The notion of satisfaction allows us to define models of $\mathcal{ETR}$ programs, and entailment of $\mathcal{ETR}$ formulas, in a natural way. Intuitively, a formula ϕ entails another formula ψ if, independently of the sequence of actions and internal and external paths, whenever ϕ is true ψ is also true; similarly for entailment of formulas by programs (i.e. sets of formulas). Formally:

Definition 9 (Models). *An interpretation M is a model of a transaction formula ϕ if $M, \pi, \epsilon, \alpha \models \phi$ for every internal path π, every external path ϵ, and every action sequence α. In this case, we write $M \models \phi$. An interpretation is a model of a set of formulas if it is a model of every formula in the set.*

Definition 10 (Logical Entailment). *Let P be an $\mathcal{ETR}$ program and ϕ be an $\mathcal{ETR}$ formula. Then P entails ϕ if every model of P is also a model of ϕ. In this case we write $P \models \phi$*

4   Comparisons and Related Work

$\mathcal{ETR}$ can be compared to many logics that reason about state change or about the related phenomena of time and action. These include action languages, the situation calculus [23], the event calculus [19], process logic [16] and many others. An extensive comparison of these formalisms with $\mathcal{TR}$ can be found in [8].

However, this kind of logics was not designed to reason about database programs but rather intended to describe changes in dynamic systems where one has little or no control such as external domains. As a result, although these formalisms provide powerful tools to specify changes and reason about their causalities in a very general and abstract way, they are simply inappropriate to model database transactions [8]. Moreover, the flexibility achieved by having an external oracle as a parameter allows for the combination of $\mathcal{ETR}$ with several different languages and semantics for describing the effects of actions in an external knowledge base. As a result, it is our opinion that rather than an alternative to $\mathcal{ETR}$, these solutions, as action languages or situation calculus, should be seen as a possible built-in component, orthogonal to $\mathcal{ETR}$ theory, as they can be used to define the semantics of the external oracle.

On the other hand, one can also compare $\mathcal{ETR}$ to formalisms that involve the notion of long-running transactions or sagas. Generally such formalisms are based on process algebras, a family of algebraic systems for modeling concurrent communicating processes, as Milner's Calculus of Communicating Systems (CCS) and Hoare's Communicating Sequential Processes (CSP), among others. One clear difference between $\mathcal{ETR}$ and such systems is that $\mathcal{ETR}$ does not support concurrency and synchronization. However, extending $\mathcal{ETR}$ to provide such features represents a next obvious step and is in line with what has been done in Concurrent Transaction Logic [9].

Notwithstanding the major difference between $\mathcal{ETR}$ and other proposals based on process algebras as [17,27] is mainly conceptual. In fact, the semantics of these latter systems are mostly focused on the correct execution and synchronization of processes whilst $\mathcal{ETR}$ semantics emphasizes knowledge base states. As a result, process algebras solutions are interested in modeling the correctness evolution of each transaction, thereby possessing a powerful operational semantics, but they are normally not interested in knowing what is true in each state of the knowledge base. In this sense, such solutions enclose powerful operators that in some cases even allow the system to construct the correct compensation for each action "on-the-fly" as in [33]. However, since these solutions, based on process algebras, are designed to define programs and behaviors, they are not suitable to be used as a knowledge representation formalism. Consequently, it is not possible to model what is true at each step of the execution of these processes nor to specify constraints on their execution based on this knowledge.

5   Discussion and Future Work

This work represents a first step towards a more generic goal, that we intend to pursue in our future work. Particularly we are interesting in developing a generic solution able to ensure the requirements as presented in Section 2.4. With this goal we already started to define top-down procedures for the serial-Horn subset of the logic.

Moreover, as it is, $\mathcal{ETR}$ does not support concurrency and synchronization. However, extending $\mathcal{ETR}$ to provide such features represents a next obvious step and is in line with what has been done in Concurrent Transaction Logic [9].

Another important step is to address reactivity. Reactivity denotes the ability to monitor changes and act accordingly to them. Such issue has always been an important concern and several active database language and systems have been proposed so far - a very incomplete list include [35,34,7,28,12,5]. Normally, reactive system are based on an Event-Condition-Action (ECA) paradigm. ECA-rules have the general syntax: **on** *event* **if** *condition* **do** *action*. Intuitively, events are received as an input stream from the external environment. It is then the system's responsibility to interpret which events have occurred and identify which rules should be triggered. The condition is a query to check if the system is in a specific state, where the rule can be applied. Finally, the action part specifies the actions that should be performed after the event occurs and the condition is true. $\mathcal{TR}$ as defined is not suitable as the action component of such reactive language as it does not account the possibility to interact with the external world, and thus it becomes impossible to "receive" events as well as execute actions externally. Contrarily, by having the possibility to interact with an external oracle, $\mathcal{ETR}$ is able to perceive the effects of the external actions performed, but also to model other arbitrary changes that have independently occurred in that domain. These characteristics make $\mathcal{ETR}$ an ideal candidate for modeling the semantics of the action component of an ECA language, providing the possibility of combining internal ACID transactions with a relaxed model of transactions for the accommodation of external actions.

Acknowledgements. This paper was submitted to SDIA 2011 - 3rd Doctoral Symposium on Artificial Intelligence and is part of an ongoing PhD supervised by José Júlio Alferes started on February 2010 and foreseen to be concluded on February 2014. Ana Sofia Gomes is supported by the FCT grant SFRH / BD / 64038 / 2009.

References

1. Alferes, J.J., Eckert, M., May, W.: Evolution and reactivity in the semantic web. In: REWERSE, pp. 161–200 (2009)
2. Antoniou, G., van Harmelen, F.: A Semantic Web Primer. MIT Press, Cambridge (2004)
3. Bailey, J., Bry, F., Eckert, M., Patranjan, P.-L.: Flavours of xchange, a rule-based reactive language for the (semantic) web. In: RuleML, pp. 187–192 (2005)
4. Behrends, E., Fritzen, O., May, W., Schenk, F.: Combining eca rules with process algebras for the semantic web. In: RuleML, pp. 29–38 (2006)
5. Behrends, E., Fritzen, O., May, W., Schenk, F.: Embedding event algebras and process for eca rules for the semantic web. Fundam. Inform. 82(3), 237–263 (2008)
6. Berners-Lee, T.: Semantic web road map (1998),
   http://www.w3.org/DesignIssues/Semantic.html
7. Bertossi, L.E., Pinto, J., Valdivia, R.: Specifying active databases in the situation calculus. In: SCCC, pp. 32–39 (1998)
8. Bonner, A.J., Kifer, M.: Transaction logic programming. Technical Report CSRI-323, Computer Systems Research Institute, University of Toronto (1995)
9. Bonner, A.J., Kifer, M.: Concurrency and communication in transaction logic. In: JICSLP, pp. 142–156 (1996)
10. Bonner, A.J., Kifer, M.: Results on reasoning about updates in transaction logic. Transactions and Change in Logic Databases, 166–196 (1998)

11. Bry, F., Eckert, M.: Twelve theses on reactive rules for the web. In: Event Processing (2007)

12. Bry, F., Patranjan, P.-L.: Reactivity on the web: paradigms and applications of the language xchange. In: Preneel, B., Tavares, S. (eds.) SAC 2005. LNCS, vol. 3897, pp. 1645–1649. Springer, Heidelberg (2006)

13. Viegas Damásio, C., Analyti, A., Antoniou, G., Wagner, G.: Supporting open and closed world reasoning on the web. In: Alferes, J.J., Bailey, J., May, W., Schwertel, U. (eds.) PPSWR 2006. LNCS, vol. 4187, pp. 149–163. Springer, Heidelberg (2006)

14. de Sainte Marie, C., Hallmark, G., Paschke, A.: RIF Production Rule Dialect (June 2010), W3C Recommendation, http://www.w3.org/TR/rif-prd/

15. Garcia-Molina, H., Salem, K.: Sagas. SIGMOD 16, 249–259 (1987)

16. Harel, D., Kozen, D., Parikh, R.: Process logic: Expressiveness, decidability, completeness. In: FOCS, pp. 129–142 (1980)

17. Hoare, C.A.R.: Communicating Sequential Processes. Prentice-Hall (1985)

18. Kifer, M.: Rule Interchange Format: The Framework. In: Calvanese, D., Lausen, G. (eds.) RR 2008. LNCS, vol. 5341, pp. 1–11. Springer, Heidelberg (2008)

19. Kowalski, R.A., Sergot, M.J.: A logic-based calculus of events. New Generation Comp. 4(1), 67–95 (1986)

20. Manola, F., Miller, E.: RDF Resource Description Framework. W3C Recommendation (February 2004), http://www.w3.org/RDF/

21. May, W., Alferes, J.J., Amador, R.: Active rules in the semantic web: Dealing with language heterogeneity. In: RuleML, pp. 30–44 (2005)

22. May, W., Alferes, J.J., Bry, F.: Towards generic query, update, and event languages for the semantic web. In: Ohlbach, H.J., Schaffert, S. (eds.) PPSWR 2004. LNCS, vol. 3208, pp. 19–33. Springer, Heidelberg (2004)

23. McCarthy, J.: Situations, actions, and causal laws. Technical report, Stanford University, Reprinted in MIT Press, Cambridge, Mass, pp. 410–417 (1968)

24. Mcguinness, D.L., van Harmelen, F.: OWL web ontology language overview. W3C Recommendation (February 2004), http://www.w3.org/TR/owl-features/

25. Mikalsen, T., Tai, S., Rouvellou, I.: Transactional attitudes: reliable composition of autonomous web services. In: WDMS (2002)

26. Milner, R.: A Calculus of Communication Systems. LNCS, vol. 92, Springer, Heidelberg (1980)

27. Milner, R.: Calculi for synchrony and asynchrony. Theor. Comput. Sci. 25, 267–310 (1983)

28. Nakamura, M., Baral, C.: Invariance, maintenance, and other declarative objectives of triggers - a formal characterization of active databases. In: Computational Logic, pp. 1210–1224 (2000)

29. Pan, Y.: Will reliability kill the web service composition? Technical report, Department of Computer Science, Rutgers University USA (2009)

30. Papamarkos, G., Poulovassilis, A., Wood, P.T.: Event-condition-action rule languages for the semantic web. In: SWDB, pp. 309–327 (2003)

31. Paton, N.W., Díaz, O.: Active database systems. ACM Comput. Surv. 31, 63–103 (1999)

32. Prud'hommeaux, E., Seaborne, A.: SPARQL Query Language for RDF. W3C Recommendation (June 2006), http://www.w3.org/TR/rdf-sparql-query/

33. Vaz, C., Ferreira, C.: Towards compensation correctness in interactive systems. In: WS-FM, pp. 161–177 (2009)

34. Widom, J.: The starburst active database rule system. IEEE Transactions on Knowledge and Data Engineering 8, 583–595 (1996)

35. Zaniolo, C.: A unified semantics for active and deductive databases. In: Rules in Database Systems, pp. 271–287 (1993)

Author Index

Abdolmaleki, Abbas   340
Aguiar, Bruno   83
Alferes, José Júlio   690
Almeida, Ana   15
Amor Pinilla, Mercedes   29
Anacleto, Ricardo   15
Augusto, Douglas A.   110
Ayala, Inmaculada   29
Azevedo, Francisco   679

Bajo, Javier   59
Banzhaf, Wolfgang   208
Baptista, Luís   679
Baptista, Tiago   125
Barbosa, Helio J.C.   110
Barreira, Luís   268
Barreto, André M.S.   110
Bernardino, Heder S.   110
Bianchi, Reinaldo A.C.   365
Boaventura Cunha, José   196

Calado, Pável   431
Campos, Ricardo   581
Cardoso, Amílcar   521, 566
Cardoso, Luis   566
Carneiro, Davide   44
Castelli, Mauro   138
Castro, António J.M.   83
Catré, Pedro   566
Cavaco, Sofia   268
Chen, Ning   407
Choobdar, Sarvenaz   418
Christensen, Anders Lyhne   153
Collet, Pierre   208
Corchado, Juan M.   59
Correia, Luís   168
Cortez, Paulo   491
Costa, Ernesto   125, 182
Costa, Hernani Pereira   597
Costa, Paulo G.   377
Costa, Vítor Santos   282

da Silva, Joaquim Ferreira   268
Daza-Gonzalez, María-Teresa   297
de la Prieta, Fernando   59

de Moura Oliveira, Paulo B.   196
Dias, Gaël   581
Domingues, Edgar   352
Duarte, Miguel   153

Felizardo, Rui   446
Ferreira, Carlos Abreu   282
Figueiredo, Lino   15
Fuentes, Lidia   29

Gama, João   282, 476
Gamallo, Pablo   610
Garcia, Marcos   610
Gaspar, Pedro   521
Gomes, Ana Sofia   690
Gomes, Luís   624
Gomes, Marco   44
Gomes, Paulo   462, 597
Gomes Correia, António   491
Gonçalo Oliveira, Hugo   462, 597
Gouyon, Fabien   392
Grilo, Carlos   168
Guil-Reyes, Francisco   297
Gurzoni Jr., José Angelo   365

Han, The Anh   254
Hernandez-Morales, Cindy G.   311

Insa, David   224

Jamroga, Wojciech   506
Jorge, Alípio   581
Jose-Garcia, Adan   311

Laboreiro, Gustavo   634
Lau, Nuno   340, 352
Lopes, Rui L.   182
Loureiro, Daniel   1

Macedo, Luis   521, 566
Manzoni, Luca   138
Marques, Nuno C.   649
Marreiros, Goreti   1
Martinho, Carlos   71
Martins, Bruno   431
Melissen, Matthijs   506
Mirkin, Boris   446

Moniz Pereira, Luís   254
Moreira, António Paulo   377
Moreira, Catarina   431
Movahedi, Mostafa   340

Nascimento, Susana   446
Neves, António J.R.   352
Neves, José   1, 44
Novais, Paulo   15, 44

Oliveira, Eugénio   536, 634, 664
Oliveira, João L.   392
Oliveira, Márcia   476
Oliveira, Sancho   153

Paulo-Santos, António   649
Pedrosa, Dulce   98
Pereira Lopes, José Gabriel   624
Pimentel, Bruno   352
Pinto, Andry Maykol   377
Pronobis, Andrzej   326

Raimundo, João   239
Ramos, Carlos   649
Ramos, Jorge Alpedrinha   98
Reis, Luís Paulo   98, 326, 340, 352, 392
Ribeiro, Bernardete   407
Ribeiro, Pedro   418
Rivera-Islas, Ivan   311
Rocha, Ana Paula   536
Rocha, Luís F.   377
Rocha, Ricardo   239
Rodriguez-Cristerna, Arturo   311

Rodríguez González, Sara   59
Romero-Monsivais, Hillel   311

Salehi, Sajjad   340
Santos, Marco   71
Sarmento, Luís   634, 664
Shafii, Nima   352
Silva, Fernando   418
Silva, Josep   224
Solteiro Pires, Eduardo J.   196
Sousa, Paulo   392
Susano Pinto, André   326

Tapia, Dante I.   59
Tavares, Miguel   521
Teixeira, Jorge   664
Tinoco, Joaquim   491
Tonidandel, Flavio   365
Torres, José   83
Torres-Jimenez, Jose   311

Urbano, Joana   536

Vanneschi, Leonardo   138
Vieira, Armando S.   407

Woźna-Szcześniak, Bożena   551

Yamamoto, Lidia   208

Zbrzezny, Agnieszka   551
Zbrzezny, Andrzej   551